GCSE Mathematics
Foundation

Michael Haese
Mark Humphries
Chris Sangwin
Ngoc Vo

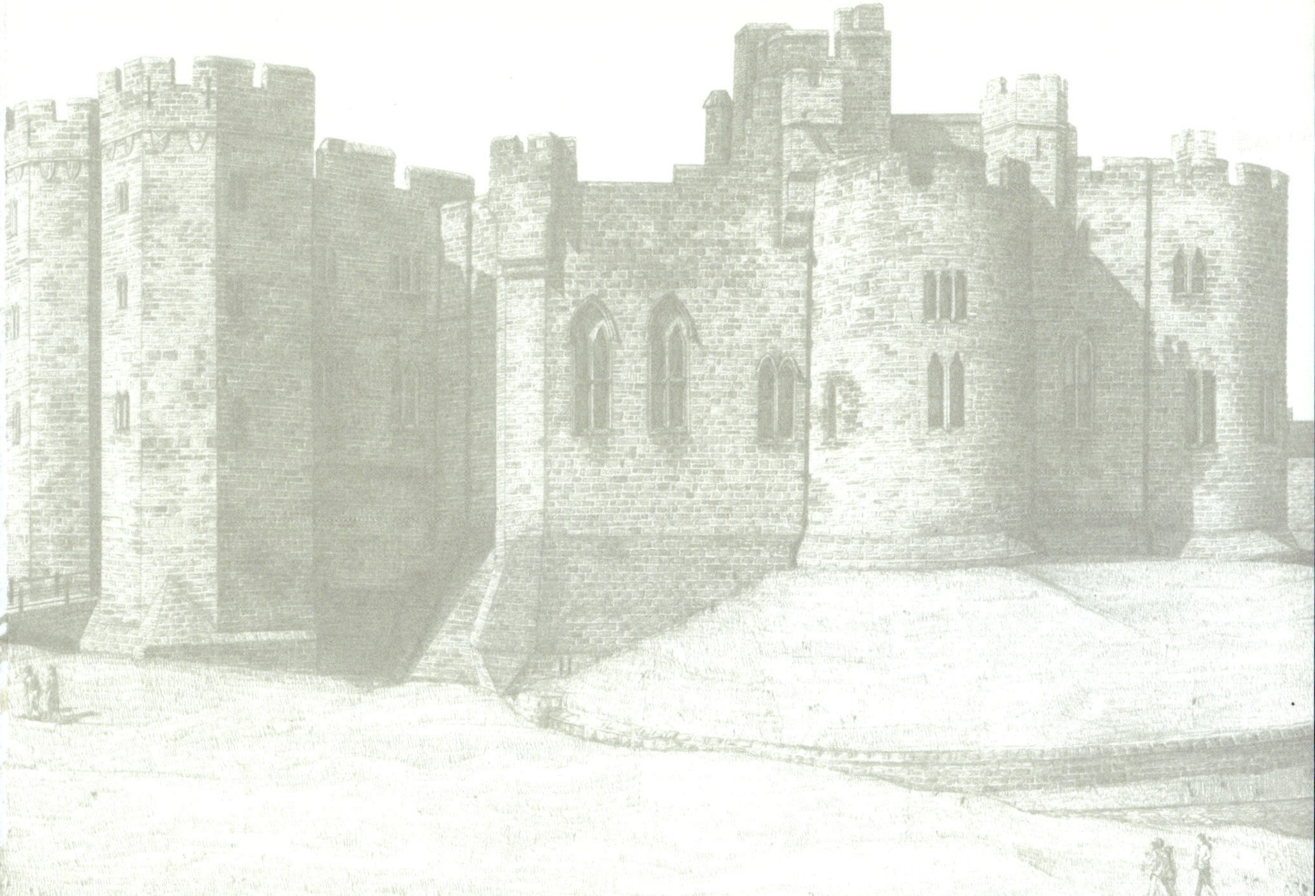

GCSE MATHEMATICS FOUNDATION

Michael Haese   B.Sc.(Hons.), Ph.D.
Mark Humphries   B.Sc.(Hons.)
Chris Sangwin   M.A., M.Sc., Ph.D.
Ngoc Vo   B.Ma.Sc.

Published by Haese Mathematics
152 Richmond Road, Marleston, SA 5033, AUSTRALIA
Telephone: +61 8 8210 4666, Fax: +61 8 8354 1238
Email: info@haesemathematics.com.au
Web: www.haesemathematics.com.au

National Library of Australia Card Number & ISBN 978-1-925489-61-3

© Haese & Harris Publications 2018

     First Edition   2018

Editorial review by Sarah Pearson.

Cartoon artwork by John Martin and Rebecca Huang.

Cover art by Jennifer Court.

Artwork by Brian Houston, Bronson Mathews, and Charlotte Frost.

Computer software by Huda Kharrufa, Brett Laishley, Bronson Mathews, Linden May, Joshua Douglass-Molloy, Jonathan Petrinolis, and Nicole Szymanczyk.

Production work by Sandra Haese, Bradley Steventon, Nicholas Kellett-Southby, and Cashmere Collins-McBride.

Typeset in Australia by Deanne Gallasch and Charlotte Frost. Typeset in Times Roman 10.

Printed in China by Prolong Press Limited.

This book is copyright. Except as permitted by the Copyright Act (any fair dealing for the purposes of private study, research, criticism or review), no part of this publication may be reproduced, stored in a retrieval system, or transmitted in any form or by any means, electronic, mechanical, photocopying, recording or otherwise, without the prior permission of the publisher. Enquiries to be made to Haese Mathematics.

Copying for educational purposes: Where copies of part or the whole of the book are made under Part VB of the Copyright Act, the law requires that the educational institution or the body that administers it has given a remuneration notice to Copyright Agency Limited (CAL). For information, contact the Copyright Agency Limited.

Acknowledgements: While every attempt has been made to trace and acknowledge copyright, the authors and publishers apologise for any accidental infringement where copyright has proved untraceable. They would be pleased to come to a suitable agreement with the rightful owner.

Disclaimer: All the internet addresses (URLs) given in this book were valid at the time of printing. While the authors and publisher regret any inconvenience that changes of address may cause readers, no responsibility for any such changes can be accepted by either the authors or the publisher.

FOREWORD

This book is written for the Foundation tier of the GCSE Mathematics specifications for first assessment in 2017. The book is suitable for any of the awarding organisations.

To reflect the principles on which the course is based, we have attempted to produce a book that embraces understanding and problem solving in order to give students different learning experiences.

The textbook and interactive online features provide an engaging and structured package, allowing students to explore and develop their confidence in mathematics. The material is presented in a clear, easy-to-follow style, free from unnecessary distractions, while effort has been made to contextualise questions so that students can relate concepts to everyday use.

Each chapter begins with an Opening Problem, offering an insight into the application of the mathematics that will be studied in the chapter. Important information and key notes are highlighted, while worked examples provide step-by-step instructions with concise and relevant explanations. Discussions, Activities, Investigations, Puzzles, and Research exercises are used throughout the chapters to develop understanding, problem solving, and reasoning, within an interactive environment.

The interactive online features include our SELF TUTOR software (see p. 4), links to graphing software, statistics software, demonstrations, calculator instructions, and a range of printable worksheets, tables, and diagrams, allowing teachers to demonstrate concepts and students to experiment for themselves.

The authors and publishers would like to thank Sarah Pearson for reviewing the book.

We welcome your feedback.   Email: info@haesemathematics.com.au
                               Web: www.haesemathematics.com.au

PMH, MAH, CS, NV

ONLINE FEATURES

With the purchase of a new textbook, you will gain 27 months subscription to our online product. This subscription can be renewed for a small fee.

Access is granted through **SNOWFLAKE**, our book viewing software that can be used in your web browser or may be installed to your tablet or computer.

Students can revisit concepts taught in class and undertake their own revision and practice online.

COMPATIBILITY

For iPads, tablets, and other mobile devices, some of the interactive features may not work. However, the digital version of the textbook can be viewed online using any of these devices.

REGISTERING

You will need to register to access the online features of this textbook.
Visit www.haesemathematics.com.au/register and follow the instructions. Once registered, you can:
- activate your digital textbook
- use your account to make additional purchases.

To activate your digital textbook, contact Haese Mathematics. On providing proof of purchase, your digital textbook will be activated. **It is important that you keep your receipt as proof of purchase.**

For general queries about registering and subscriptions:
- Visit our **SNOWFLAKE** help page: http://snowflake.haesemathematics.com.au/help
- Contact Haese Mathematics: info@haesemathematics.com.au

ONLINE VERSION OF THE TEXTBOOK

The entire text of the book can be viewed online, allowing you to leave your textbook at school.

SELF TUTOR

Self tutor is an exciting feature of this book.
The ◄⁾ **Self Tutor** icon on each worked example denotes an active online link.

> Simply 'click' on the ◄⁾ **Self Tutor** (or anywhere in the example box) to access the worked example, with a teacher's voice explaining each step necessary to reach the answer.
>
> Play any line as often as you like. See how the basic processes come alive using movement and colour on the screen.

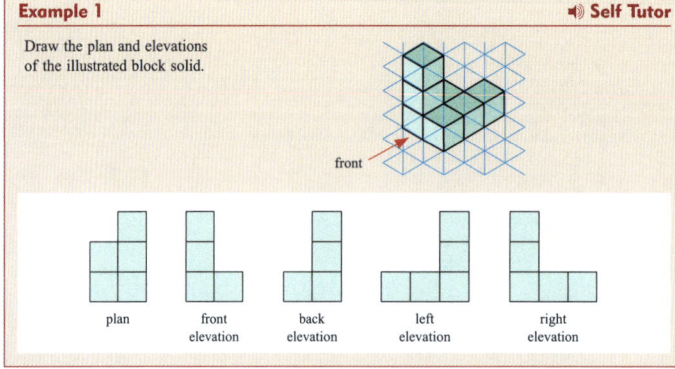

See **Chapter 12, Solids**, p. 240

INTERACTIVE LINKS

Throughout your digital textbook, you will find interactive links to:
- Graphing software
- Statistics software
- Games
- Demonstrations
- Printable pages

ICON

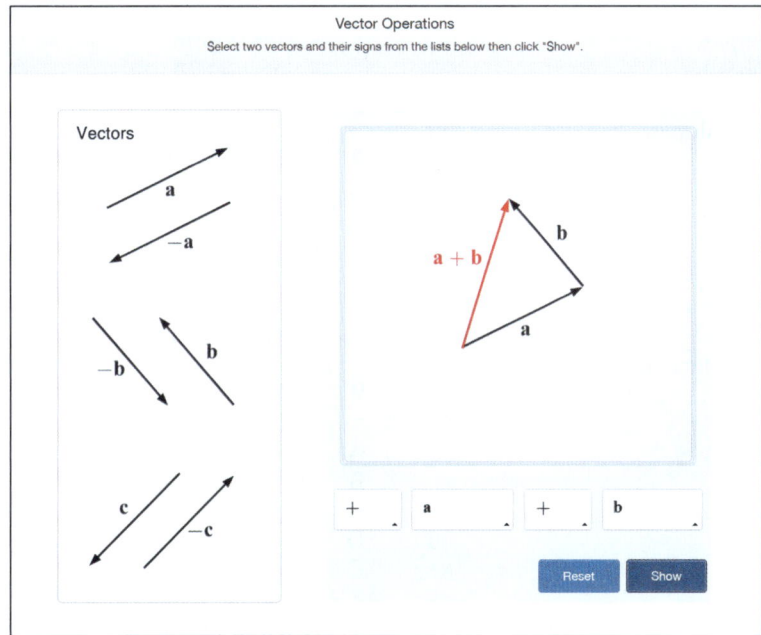

See **Chapter 32**, **Vectors**, p. 633

CALCULATOR INSTRUCTIONS

Printable calculator instruction booklets are available for the **Casio fx-991EX**, **Casio fx-CG20**, and the **TI-84 Plus CE**. Click on the relevant icon below.

CASIO fx-991EX

CASIO fx-CG20

TI-84 Plus CE

When additional calculator help may be needed, specific instructions can be printed from icons within the text.

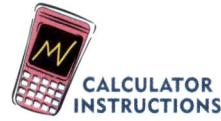

TABLE OF CONTENTS

1	**NUMBER**	**11**
A	The number line	12
B	Adding and subtracting negatives	15
C	Multiplying negative numbers	18
D	Dividing negative numbers	19
E	Index notation	20
F	Square numbers and square roots	23
G	Order of operations	24
H	Factors and multiples	27
I	Primes and composites	28
J	Highest common factor	29
K	Lowest common multiple	30
	Review set 1A	32
	Review set 1B	33
2	**REAL NUMBERS**	**35**
A	Fractions	36
B	Operations with fractions	40
C	Decimal numbers	44
D	Operations with decimal numbers	46
E	Percentage	49
	Review set 2A	54
	Review set 2B	55
3	**UNITS**	**57**
A	Units of length	60
B	Units of area	63
C	Units of volume	65
D	Units of capacity	65
E	Units of mass	67
F	Time	67
	Review set 3A	72
	Review set 3B	73
4	**APPROXIMATION AND ESTIMATION**	**75**
A	Rounding numbers	76
B	Errors from rounding	80
C	Estimation	82
	Review set 4A	84
	Review set 4B	85
5	**ALGEBRA**	**87**
A	Algebraic notation	88
B	The language of mathematics	90
C	Collecting like terms	92
D	Writing expressions	94
E	Generalising arithmetic	96
	Review set 5A	97
	Review set 5B	98
6	**RATIO**	**99**
A	Ratio	100
B	Writing ratios as fractions	102
C	Equal ratios	103
D	Proportions	105
E	Using ratios to divide quantities	107
F	Scale diagrams	108
	Review set 6A	113
	Review set 6B	114
7	**INDICES**	**117**
A	Index laws	118
B	Expansion laws	121
C	The zero index law	123
D	The negative index law	124
E	Standard form	127
	Review set 7A	132
	Review set 7B	133
8	**ALGEBRA: EXPANSION**	**135**
A	The distributive law	136
B	The product $(a+b)(c+d)$	140
C	Perfect square expansion	142
D	Difference between two squares	144
	Review set 8A	145
	Review set 8B	146
9	**LINEAR EQUATIONS AND INEQUALITIES**	**147**
A	Linear equations	148
B	Maintaining balance	149
C	Inverse operations	151
D	Algebraic flowcharts	152
E	Solving linear equations	154
F	Equations with a repeated unknown	157
G	Rational equations	160
H	Problem solving	163
I	Linear inequalities	165
J	Solving linear inequalities	167
	Review set 9A	171
	Review set 9B	172
10	**GEOMETRY**	**175**
A	Points and lines	176
B	Measuring and classifying angles	179
C	Angle properties	182
D	Angle pairs	185
E	Parallel lines	188
F	Bearings	192
G	Geometric construction	194
	Review set 10A	201
	Review set 10B	203
11	**SHAPE**	**205**
A	Polygons	206
B	Triangles	208

C	Isosceles triangles	213
D	Quadrilaterals	216
E	Angles of an n-sided polygon	221
F	Circles	223
	Review set 11A	226
	Review set 11B	228

12 SOLIDS — 231
A	Solids	232
B	Nets of solids	234
C	Drawing rectangular solids	236
D	Views of solids	238
	Review set 12A	244
	Review set 12B	245

13 ALGEBRAIC FACTORISATION — 247
A	Common factors	248
B	Factorising with common factors	250
C	Difference between two squares	253
D	Perfect square factorisation	255
E	Factorising quadratic trinomials	257
F	Miscellaneous factorisation	260
	Review set 13A	261
	Review set 13B	261

14 TABLES, GRAPHS, AND CHARTS — 263
A	Interpreting tables	265
B	Interpreting graphs	269
C	Graphs of categorical data	273
D	Comparing categorical data	277
E	Time series data	281
	Review set 14A	284
	Review set 14B	287

15 PYTHAGORAS' THEOREM — 291
A	Solving $x^2 = k$	292
B	Pythagoras' theorem	294
C	The converse of Pythagoras' theorem	299
D	Problem solving	301
	Review set 15A	304
	Review set 15B	305

16 MEASUREMENT — 307
A	Perimeter	308
B	Area	312
C	Surface area	319
D	Volume	324
E	Capacity	331
	Review set 16A	333
	Review set 16B	335

17 FORMULAE AND FUNCTIONS — 337
A	Number machines	338
B	Formula construction	341
C	Substituting into formulae	345
D	Rearranging formulae	347
E	Rearrangement and substitution	350
F	Predicting formulae	353
	Review set 17A	354
	Review set 17B	356

18 APPLICATIONS OF PERCENTAGE — 357
A	Percentage increase and decrease	358
B	Business calculations	362
C	Chain percentage problems	366
D	Appreciation and depreciation	367
E	Simple interest	368
F	Compound interest	370
	Review set 18A	372
	Review set 18B	373

19 RATES — 375
A	Rates	376
B	Speed	379
C	Density	383
D	Pressure	386
E	Unit cost	387
F	Exchange rates	389
G	Converting rates	390
	Review set 19A	392
	Review set 19B	393

20 COORDINATE GEOMETRY — 395
A	The Cartesian plane	397
B	Linear relationships	399
C	Gradient	402
D	Parallel lines	407
E	Axes intercepts	409
F	The equation of a line	410
G	Graphing lines in the form $y = mx + c$	412
H	Graphing lines in the form $Ax + By = C$	416
I	Vertical and horizontal lines	417
J	Finding the equation of a line	418
	Review set 20A	421
	Review set 20B	423

21 SIMULTANEOUS EQUATIONS — 425
A	Trial and error solution	427
B	Graphical solution	428
C	Solution by equating values of y	429
D	Solution by substitution	430
E	Solution by elimination	431
F	Problem solving with simultaneous equations	434
	Review set 21A	437
	Review set 21B	438

22 TRANSFORMATIONS — 439
A	Translations	441
B	Reflections and line symmetry	445
C	Rotations and rotational symmetry	450
D	Enlargements and reductions	455

	Review set 22A	458
	Review set 22B	460

23 SIMILARITY AND CONGRUENCE 463
A	Similar figures	464
B	Similar triangles	467
C	Problem solving	471
D	Congruent figures	473
E	Congruent triangles	475
F	Proof using congruence	480
	Review set 23A	481
	Review set 23B	483

24 TRIGONOMETRY 485
A	Scale diagrams in geometry	486
B	Labelling right angled triangles	487
C	The trigonometric ratios	488
D	Finding side lengths	491
E	Finding angles	493
F	Problem solving with trigonometry	495
G	The first quadrant of the unit circle	498
	Review set 24A	501
	Review set 24B	503

25 QUADRATIC EQUATIONS AND FUNCTIONS 505
A	Quadratic equations	506
B	The Null Factor law	507
C	Solving quadratic equations	508
D	Problem solving with quadratic equations	512
E	Quadratic functions	514
F	Graphs of quadratic functions	515
G	Axes intercepts	516
H	Axis of symmetry	520
I	Vertex	521
	Review set 25A	523
	Review set 25B	524

26 PROPORTION 525
A	Direct proportion	526
B	Inverse proportion	530
	Review set 26A	534
	Review set 26B	535

27 FURTHER FUNCTIONS 537
A	Reciprocal functions	538
B	Cubic functions	540
C	Other functions	543
	Review set 27A	545
	Review set 27B	546

28 NUMBER SEQUENCES 547
A	Number sequences	548
B	Arithmetic sequences	551
C	Geometric sequences	554
D	Fibonacci-type sequences	555
	Review set 28A	557
	Review set 28B	557

29 PROBABILITY 559
A	Probability	561
B	Experimental probability	563
C	Sample space	568
D	Theoretical probability	569
E	Venn diagrams	571
F	The addition law of probability	575
G	Tables and grids	577
H	Independent events	579
I	Dependent events	581
J	Tree diagrams	583
K	Expectation	586
	Review set 29A	587
	Review set 29B	589

30 STATISTICS 591
A	Populations and samples	593
B	Discrete data	594
C	Continuous data	600
D	Measuring the centre	602
E	Measuring the spread	610
	Review set 30A	612
	Review set 30B	613

31 BIVARIATE STATISTICS 615
A	Scatter graphs	616
B	Correlation	618
C	Line of best fit	622
	Review set 31A	625
	Review set 31B	627

32 VECTORS 629
A	Directed line segment representation	630
B	Vector equality	631
C	Vector addition	633
D	Vector subtraction	635
E	Vectors in component form	636
F	The vector between two points	641
G	Scalar multiplication	642
	Review set 32A	644
	Review set 32B	645

ANSWERS 647

INDEX 719

ABOUT THE AUTHORS

Michael Haese completed a BSc at the University of Adelaide, majoring in Infection and Immunity, and Applied Mathematics. He completed Honours in Applied Mathematics, and a PhD in high speed fluid flows. Michael has a keen interest in education and a desire to see mathematics come alive in the classroom through its history and relationship with other subject areas. He is passionate about girls' education and ensuring they have the same access and opportunities that boys do. His other interests are wide-ranging, including show jumping, cycling, and agriculture. He has been the principal editor for Haese Mathematics since 2008.

Mark Humphries completed a degree in Mathematical and Computer Science, and an Economics degree at the University of Adelaide. He then completed an Honours degree in Pure Mathematics. His mathematical interests include public key cryptography, elliptic curves, and number theory. Mark enjoys the challenge of piquing students' curiosity in mathematics, and encouraging students to think about mathematics in different ways. He has been working at Haese Mathematics since 2006, and is currently the writing manager.

Chris Sangwin completed a BA in Mathematics at the University of Oxford, and an MSc and PhD in Mathematics at the University of Bath. He spent thirteen years in the Mathematics Department at the University of Birmingham, and from 2000 - 2011 was seconded half time to the UK Higher Education Academy "Maths Stats and OR Network" to promote learning and teaching of university mathematics. He was awarded a National Teaching Fellowship in 2006, and is now Professor of Technology Enhanced Science Education at the University of Edinburgh.

His research interests focus on technology and mathematics education and include automatic assessment of mathematics using computer algebra, and problem solving using the Moore method and similar student-centred approaches.

Ngoc Vo completed a BMaSc at the University of Adelaide, majoring in Statistics and Applied Mathematics. Her Mathematical interests include regression analysis, Bayesian statistics, and statistical computing. Ngoc has been working at Haese Mathematics as a proof reader and writer since 2016.

# Number

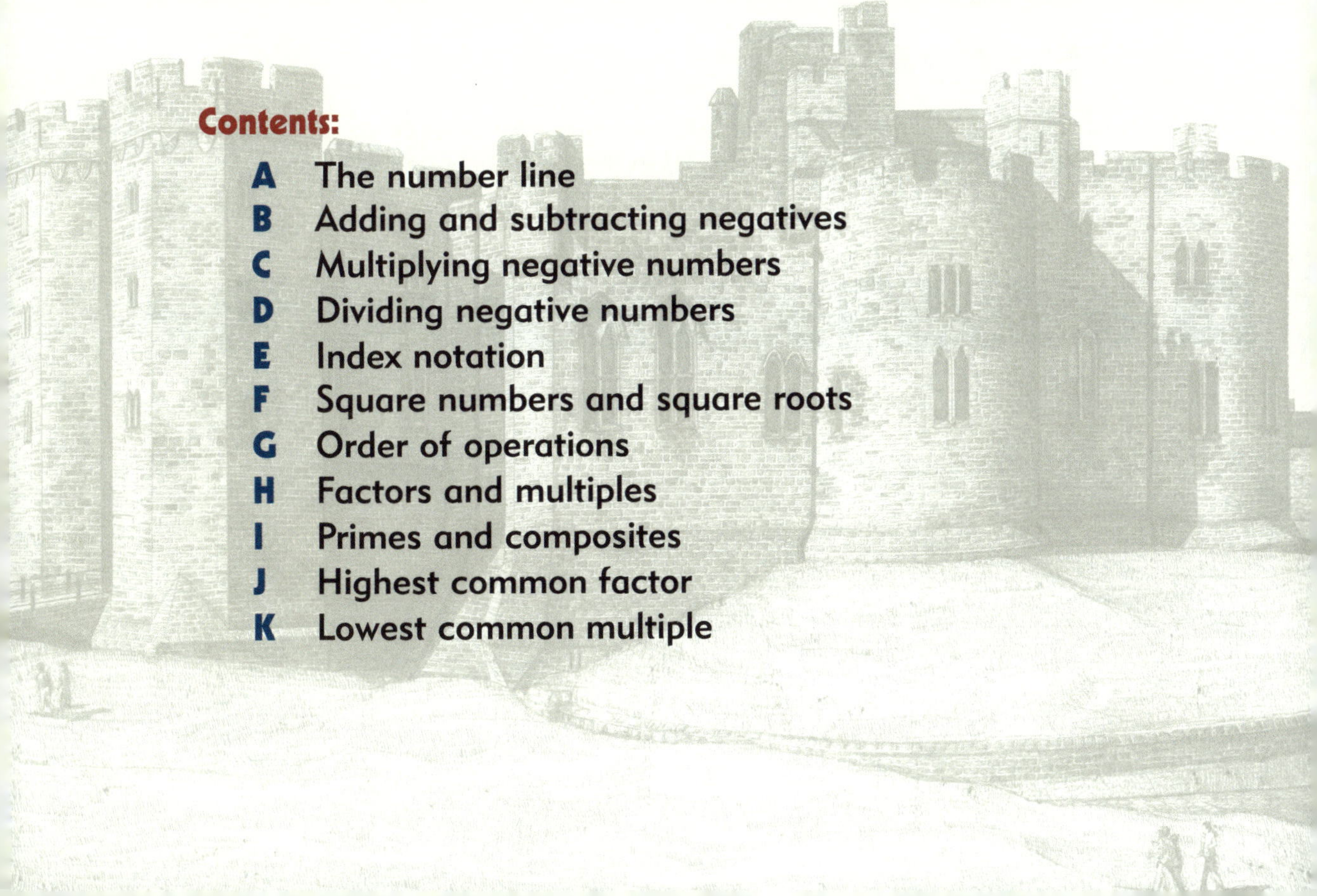

Contents:

- **A** The number line
- **B** Adding and subtracting negatives
- **C** Multiplying negative numbers
- **D** Dividing negative numbers
- **E** Index notation
- **F** Square numbers and square roots
- **G** Order of operations
- **H** Factors and multiples
- **I** Primes and composites
- **J** Highest common factor
- **K** Lowest common multiple

Opening problem

A wildlife park has three shows: 'Brilliant Birds' is 60 minutes long, 'Meet a Monkey' is 90 minutes long, and 'Cuddle a Cat' is 45 minutes long. The shows run continuously throughout the day.

Things to think about:

a What are the *multiples* of 45, 60, and 90?

b If all three shows start together, how long will it be before all three shows again start together?

People have used numbers since prehistoric times. We know this from ancient writings and drawings.

Today, we live in numbered streets, have telephone numbers, registration numbers, bank account numbers, and tax file numbers. We use numbers to describe the value of things, to measure the universe, and plot courses through time and space. We are "tagged" with a number when we are born, and often after we die. Numbers are thus an essential part of our lives.

In this Chapter we will study whole numbers, their properties, and how we can work with them.

We focus on two important number sets:

- The **natural numbers** are the counting numbers 0, 1, 2, 3, 4, 5, 6, 7, 8, 9, ...

- The negative whole numbers, zero, and the positive whole numbers form the set of all **integers**.
  ...., −5, −4, −3, −2, −1, 0, 1, 2, 3, 4, 5, ...

Both of these sets of numbers are endless. We say that they are **infinite sets**.

A THE NUMBER LINE

We can represent all whole numbers on a **number line**. This line extends forever in both directions.

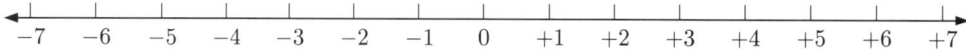

The numbers to the **right of zero** are the **positive** numbers.

The numbers to the **left of zero** are the **negative** numbers.

The negative whole numbers, zero, and the positive whole numbers are together known as **integers**.

Pairs of numbers like −7 and 7 are exactly the same distance from 0 but on opposite sides of zero. They are therefore called **opposites**.

Zero is neither positive nor negative.

Discussion

The **Bakhshali manuscript** is an ancient Indian text consisting of mathematics and Sanskrit text written on 70 leaves of birch bark. It was written in the 3rd or 4th century, and is the oldest known reference to the number we call zero.

What things make zero special?

You may wish to consider:
- where zero is placed on the number line
- the effect of adding or subtracting zero
- the effect of multiplying or dividing by zero.

Example 1

What is the opposite of: **a** $+4$ **b** -9?

a The opposite of $+4$ is -4.
b The opposite of -9 is $+9$.

Numbers which are **opposites** are the same distance from zero on the number line, but on different sides.

We can use a number line to compare the sizes of different numbers and arrange them in order.

- As you move along the number line from *left* to *right*, the numbers increase.

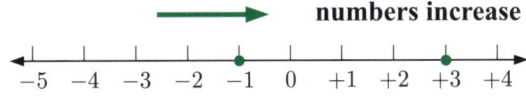

$+3$ is *greater* than -1 because it is further to the right on the number line.
We write: $+3 > -1$

$>$ means "**is greater than**"
$<$ means "**is less than**"

- As you move along the number line from *right* to *left*, the numbers decrease.

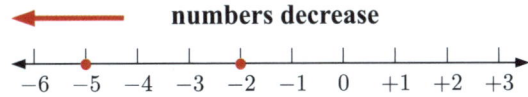

-5 is *less* than -2 because it is further to the left on the number line.
We write: $-5 < -2$

Example 2

a Show $+3$ and -2 on a number line and write a sentence comparing their size.
b Write the statement $-7 > -4$ in words, then state whether it is true or false.

a

Since $+3$ is further to the right, $+3$ is greater than -2.
We could also say -2 is less than $+3$.

b The statement reads "negative 7 is greater than negative 4".
This is false because -7 is to the **left** of -4, and so it is less than -4.

14 Number (Chapter 1)

EXERCISE 1A

1 Write the opposite of each number:
   a -3 **b** 15 **c** -10 **d** -9 **e** 38 **f** -6 **g** $+7$ **h** 0

2 Show -1 and -6 on a number line and write a sentence comparing their size.

3 Write the statement $2 > -5$ in words, then state whether it is true or false.

4 Write *true* or *false* for each of the following statements:
   a $6 > 2$ **b** $-4 < 15$ **c** $17 > 18$
   d $-2 < 19$ **e** $-13 > 5$ **f** $-20 < -12$

5 Insert either $<$ or $>$ in place of □ to make each statement true:
   a $8 \square 6$ **b** $18 \square 7$ **c** $-9 \square -4$
   d $-3 \square 15$ **e** $20 \square -15$ **f** $-6 \square -2$

Example 3   ◀) Self Tutor

Locate the values of -1, 3, 0, 5, and -4 on a number line. Hence find the greatest and least of the numbers.

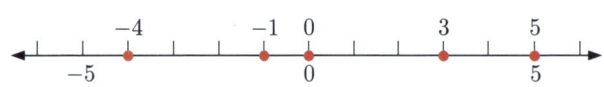

The greatest number is 5. The least number is -4.

> The greatest number is furthest to the right. The least number is furthest to the left.

6 Draw a number line to show each set of numbers. Hence find the greatest and least number in each set.
   a $-9, 2, 5$ **b** $2, 6, 8, -3, -4$
   c $9, -4, -9, 1, 6, -6$ **d** $-3, 2, 5, -5, 0, -1$

7 Display the numbers $4, -2, 1, -1$ on a number line.
Hence arrange the numbers from least to greatest.

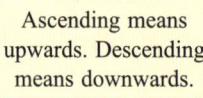

> Ascending means upwards. Descending means downwards.

8 Display the numbers $5, -3, 0, 2, -4, 6, -1$ on a number line.
Hence arrange the numbers in ascending order.

9 Arrange in descending order: $0, -5, 8, -7, -2, 6$.

10 Arrange in ascending order: $0, -10, 8, 7, -7, -2$.

11 The temperatures of six cities were:
Ulaanbaatar 3°C, Singapore 33°C, Melbourne 19°C, Oslo -4°C, Moscow -6°C, Tokyo 1°C.
Place them in order from coldest to hottest.

Example 4

Use a number line to: **a** increase -2 by 5 **b** decrease -1 by 4.

a To *increase* -2 by 5, we move along the number line 5 units to the *right*.
The result is $+3$.

b To *decrease* -1 by 4, we move along the number line 4 units to the *left*.
The result is -5.

12 Copy and complete:

a
$2 + 3 = \ldots$
$1 + 3 = \ldots$
$0 + 3 = \ldots$
$-1 + 3 = \ldots$
$-2 + 3 = \ldots$
$-3 + 3 = \ldots$

b
$4 - 2 = \ldots$
$3 - 2 = \ldots$
$2 - 2 = \ldots$
$1 - 2 = \ldots$
$0 - 2 = \ldots$
$-1 - 2 = \ldots$

c
$2 + 2 = \ldots$
$2 + 1 = \ldots$
$2 + 0 = \ldots$
$2 - 1 = \ldots$
$2 - 2 = \ldots$
$2 - 3 = \ldots$

13 Use a number line to:

a increase -5 by 1   **b** decrease 3 by 4   **c** decrease -5 by 6
d increase -4 by 3   **e** increase -5 by 5   **f** decrease -1 by 6

14 Use a number line to find:

a $3 + 7$ **b** $-3 + 5$ **c** $5 - 9$ **d** $4 - 8$
e $0 - 6$ **f** $-11 + 7$ **g** $-2 - 6$ **h** $-3 - 9$

B   ADDING AND SUBTRACTING NEGATIVES

ADDING A NEGATIVE NUMBER

We know that $4 + 3 = 7$, but what is the value of $4 + -3$?

Consider the following statements:
$4 + 3 = 7$
$4 + 2 = 6$
$4 + 1 = 5$
$4 + 0 = 4$

As the number being added to 4 decreases by 1, the final answer also decreases by 1.

Continuing this pattern gives:
$4 + -1 = 3$ which we compare with: $4 - 1 = 3$
$4 + -2 = 2$ $4 - 2 = 2$
$4 + -3 = 1$ $4 - 3 = 1$
$4 + -4 = 0$ $4 - 4 = 0$

Adding a **negative** number is equivalent to **subtracting** its opposite.

SUBTRACTING A NEGATIVE NUMBER

We know that $4 - 3 = 1$, but what is the value of $4 - -3$?

Consider the following statements:
$$4 - 3 = 1$$
$$4 - 2 = 2$$
$$4 - 1 = 3$$
$$4 - 0 = 4$$

As the number being subtracted decreases by 1, the answer increases by 1.

Continuing this pattern gives:
$4 - -1 = 5$ which we compare with: $4 + 1 = 5$
$4 - -2 = 6$ $4 + 2 = 6$
$4 - -3 = 7$ $4 + 3 = 7$
$4 - -4 = 8$ $4 + 4 = 8$

Subtracting a negative number is equivalent to adding its opposite.

Example 5

Find the value of:
a $4 + -9$ **b** $4 - -9$ **c** $-3 + -5$ **d** $-3 - -5$

a $\quad 4 + -9$
$\quad = 4 - 9$
$\quad = -5$

b $\quad 4 - -9$
$\quad = 4 + 9$
$\quad = 13$

c $\quad -3 + -5$
$\quad = -3 - 5$
$\quad = -8$

d $\quad -3 - -5$
$\quad = -3 + 5$
$\quad = 2$

EXERCISE 1B

1 Simplify if possible, and then evaluate:
- **a** $4 - 3$
- **b** $4 + 3$
- **c** $-4 - 3$
- **d** $-4 + 3$
- **e** $4 + -3$
- **f** $4 - -3$
- **g** $-4 + -3$
- **h** $-4 - -3$

2 Simplify if possible, and then evaluate:
- **a** $2 - 6$
- **b** $2 + 6$
- **c** $-2 - 6$
- **d** $-2 + 6$
- **e** $2 + -6$
- **f** $2 - -6$
- **g** $-2 + -6$
- **h** $-2 - -6$

3 Evaluate:
- **a** $1 + -2$
- **b** $-2 + -6$
- **c** $6 + -8$
- **d** $-3 - -7$
- **e** $3 + -13$
- **f** $4 - -5$
- **g** $-7 - -10$
- **h** $-9 + -8$

4 A maintenance man in an office block starts his working day on the ground floor. To fulfil his duties he goes up 12 floors, down 3 floors, up 5 floors, down 7 floors, up 1 floor, and then down 6 floors. What floor does he end up on?

5 Evaluate:
- **a** $-2 - 7$
- **b** $-7 + 5$
- **c** $-6 + -3$
- **d** $-13 - 8$
- **e** $-7 + -5$
- **f** $-6 - -9$
- **g** $2 + -12$
- **h** $-4 - -3$
- **i** $-11 - -17$

6 Simplify and hence evaluate:

- **a** $4 + 7 + -10$
- **b** $8 - -2 + -4$
- **c** $-4 - -6 - -1$
- **d** $-12 - -9 + 5$
- **e** $-4 - -5 + -8$
- **f** $-3 - -10 + 10$
- **g** $-3 - 7 + -7$
- **h** $-10 + -8 + -9$
- **i** $-1 + 10 - -7$

7 Find the difference between:

- **a** -5 and 4
- **b** -2 and -5
- **c** 8 and -1
- **d** -4 and -2
- **e** 8 and -8
- **f** -7 and 8

The difference between two numbers is the greater number minus the lesser number.

8 York recorded the following maximum temperatures for a week:
Mon 5°C, Tues −3°C, Wed −7°C, Thurs 2°C, Fri 1°C, Sat −4°C, Sun −1°C.
What was the *average* daily maximum temperature for the week?

To find the *average*, add the 7 temperatures and then divide this sum by 7.

Activity 1 — Magic squares

4	3	8
9	5	1
2	7	6

A **magic square** is a square grid filled with **consecutive whole numbers** so that each row, column, and diagonal has the same sum.

For example, this magic square has the **magic sum** of 15 along every row, column, and diagonal.

PRINTABLE WORKSHEET

What to do:

1 Copy and complete the following magic squares:

a
4		8
	7	
		10

b
16	2	3	13
		11	
9	7		12
4		15	1

c
		7	12
15		9	6
	5		
8	11	2	

2 Magic squares may also contain negative numbers.

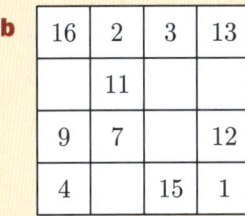

2	−5	0
−3	−1	1
−2	3	−4

- **a** Is the square alongside a magic square? If so, what is the magic sum?
- **b** Make a new magic square by adding 2 to each number in the magic square given. State the new magic sum.
- **c** Make a new magic square by subtracting 3 from each number in the magic square given. State the new magic sum.

3 If 3 was added to each number in a 3×3 magic square, what would happen to the magic sum? Use your answers in **2** to help you.

4 Copy and complete the following magic squares:

a
-4		0
	-1	
		2

b
3			-9
-8			
-7	-4	5	
6	-1	-6	

c
4	11		-5	2
			-6	1
-9	-7	0	7	
-3			8	
-2		-11	-4	

C MULTIPLYING NEGATIVE NUMBERS

Investigation — Multiplying negative numbers

In this Investigation we are aiming to find rules for multiplication with negative numbers.

What to do:

1 Copy and complete each pattern of numbers:

a
$3 \times 3 = 9$
$3 \times 2 = 6$
$3 \times 1 = 3$
$3 \times 0 = \ldots$
$3 \times -1 = \ldots$
$3 \times -2 = \ldots$
$3 \times -3 = \ldots$

b
$3 \times 2 = 6$
$2 \times 2 = 4$
$1 \times 2 = \ldots$
$0 \times 2 = \ldots$
$-1 \times 2 = \ldots$
$-2 \times 2 = \ldots$
$-3 \times 2 = \ldots$

2 Hence complete:
  a A positive times a positive is ......
  b A positive times a negative is ......
  c A negative times a positive is ......

3 What do you think a negative times a negative will be?

4 Use your answer from **1** to help you start this pattern:
$3 \times -3 = \ldots$
$2 \times -3 = \ldots$
$1 \times -3 = \ldots$
$0 \times -3 = \ldots$
$-1 \times -3 = \ldots$
$-2 \times -3 = \ldots$
$-3 \times -3 = \ldots$

5 Hence decide whether a negative times a negative is positive or negative.

From the **Investigation**, you should have discovered that:

- (positive) × (positive) = (positive)
- (positive) × (negative) = (negative)
- (negative) × (positive) = (negative)
- (negative) × (negative) = (positive)

When the signs are the **same**, the answer is **positive**.
When the signs are **different**, the answer is **negative**.

Number (Chapter 1)

Example 6

Find the value of:

a 3×4 **b** 3×-4 **c** -3×4 **d** -3×-4

a $3 \times 4 = 12$ **b** $3 \times -4 = -12$ **c** $-3 \times 4 = -12$ **d** $-3 \times -4 = 12$

EXERCISE 1C

1 Evaluate:
- **a** 6×4
- **b** 6×-4
- **c** -6×4
- **d** -6×-4
- **e** 4×-6
- **f** -4×6
- **g** 4×6
- **h** -4×-6

2 Find the value of:
- **a** 4×9
- **b** 4×-9
- **c** -4×9
- **d** -4×-9
- **e** 3×11
- **f** 3×-11
- **g** -3×11
- **h** -3×-11
- **i** $4 \times 2 \times 7$
- **j** $4 \times -2 \times 7$
- **k** $4 \times -2 \times -7$
- **l** $-4 \times -2 \times -7$

3 Evaluate:
- **a** 3×-2
- **b** -10×3
- **c** -2×-7
- **d** 5×-10
- **e** -6×8
- **f** 5×-9
- **g** -8×11
- **h** 3×-11
- **i** 9×-9
- **j** -12×-2
- **k** 11×-5
- **l** -6×-7

4 Determine the missing number in each of the following:
- **a** $-2 \times \square = -2$
- **b** $\square \times 5 = -10$
- **c** $\square \times 1 = -11$
- **d** $8 \times \square = -32$
- **e** $-3 \times \square = 18$
- **f** $8 \times \square = -16$
- **g** $-2 \times \square = -8$
- **h** $9 \times \square = -9$
- **i** $\square \times -7 = -42$
- **j** $\square \times -10 = 30$
- **k** $4 \times \square = -12$
- **l** $\square \times 12 = -120$

5 Solve the following problems:
- **a** A skydiver falls 70 metres per second for 4 seconds. How many metres does he fall?
- **b** When Tania bought a new bicycle for £540, she borrowed the money from her parents. She repays them £70 per week for 6 weeks. How much does Tania still owe her parents?

6 Evaluate:
- **a** $-5 \times 8 \times 5$
- **b** $-7 \times 3 \times -3$
- **c** $-2 \times 5 \times -2$
- **d** $-8 \times 5 \times -5$
- **e** $-2 \times 9 \times -5$
- **f** $-8 \times 2 \times -3$

D DIVIDING NEGATIVE NUMBERS

The rules for division are identical to those for multiplication. This occurs because multiplication and division are **inverse operations**, which means that one *undoes* the other.

- (positive) ÷ (positive) = (positive)
- (positive) ÷ (negative) = (negative)
- (negative) ÷ (positive) = (negative)
- (negative) ÷ (negative) = (positive)

Dividing numbers with the **same** signs gives a **positive**.
Dividing numbers with **different** signs gives a **negative**.

Example 7

Find the value of:

a $14 \div 2$
b $14 \div -2$
c $-14 \div 2$
d $-14 \div -2$

a $14 \div 2 = 7$
b $14 \div -2 = -7$
c $-14 \div 2 = -7$
d $-14 \div -2 = 7$

EXERCISE 1D

1 Evaluate:

a $15 \div 3$
b $15 \div -3$
c $-15 \div 3$
d $-15 \div -3$
e $45 \div 9$
f $-45 \div -9$
g $-45 \div 9$
h $45 \div -9$
i $6 \div 6$
j $6 \div -6$
k $-6 \div 6$
l $-6 \div -6$
m $44 \div 4$
n $-44 \div 4$
o $-44 \div -4$
p $44 \div -4$

2 Determine the missing number:

a $12 \div \square = -3$
b $\square \div -2 = 3$
c $-4 \div \square = 1$
d $\square \div 5 = -5$
e $-18 \div \square = 2$
f $\square \div 4 = -3$
g $\square \div -2 = 4$
h $30 \div \square = -6$
i $\square \div -8 = 5$
j $36 \div \square = -4$
k $-15 \div \square = -5$
l $\square \div -4 = 7$
m $72 \div \square = -9$
n $\square \div 10 = -12$
o $\square \div -12 = -12$
p $-96 \div \square = -8$

3 Solve the following problems:

a A company owned equally by seven people has a debt of £350 000. What is each person's share of the debt?

b One night in the Gobi Desert, the temperature drops from 33°C to −12°C in five hours. What is the average temperature change per hour?

> The *average* temperature change is the total temperature change divided by the number of hours.

4 Determine the missing number:

a $7 \times \square = -70$
b $\square - 3 = -4$
c $15 + \square = -1$
d $25 \div \square = -5$
e $\square \times -8 = 40$
f $-8 + \square = 2$
g $\square \div 2 = -20$
h $4 - \square = 6$
i $-18 + \square = 0$
j $\square \div -6 = -12$
k $\square \times 10 = 100$
l $9 - \square = 18$

E INDEX NOTATION

Rather than writing $2 \times 2 \times 2$, we can write such a product as 2^3. We call this **index notation**. We say that 2 is the **base** and that 3 is the **index**, **power**, or **exponent**.

2^3 reads "two cubed" or
"the third power of two" or
"two to the power three".

2^3 ← index, power or exponent
2 ← base

> If n is a positive integer, then a^n is the product of n factors of a.
> $$a^n = \underbrace{a \times a \times a \times a \times a \times \ldots \times a}_{n \text{ factors}}$$

The following table demonstrates correct language when talking about index notation:

Natural number	Product form	Index form	Spoken form
3	3	3^1	three
9	3×3	3^2	three squared
27	$3 \times 3 \times 3$	3^3	three cubed
81	$3 \times 3 \times 3 \times 3$	3^4	three to the fourth
243	$3 \times 3 \times 3 \times 3 \times 3$	3^5	three to the fifth

Example 8 ◀) Self Tutor

Write in index form:

$2 \times 2 \times 2 \times 2 \times 3 \times 3 \times 3$

$\underbrace{2 \times 2 \times 2 \times 2}_{2^4} \times \underbrace{3 \times 3 \times 3}_{3^3}$

$= 2^4 \times 3^3$

EXERCISE 1E

1 Match the following numbers in index form with the correct product:

- **a** 5^4
- **b** 7^5
- **c** 5^7
- **d** 7^1
- **e** 7^3

- **A** $5 \times 5 \times 5 \times 5 \times 5 \times 5 \times 5$
- **B** $7 \times 7 \times 7$
- **C** $7 \times 7 \times 7 \times 7 \times 7$
- **D** $5 \times 5 \times 5 \times 5$
- **E** 7

2 Write in index form:

- **a** $2 \times 3 \times 3$
- **b** $2 \times 2 \times 3 \times 5$
- **c** $2 \times 5 \times 5 \times 5$
- **d** $3 \times 3 \times 5 \times 5 \times 5$
- **e** $2 \times 2 \times 2 \times 5 \times 7$
- **f** $3 \times 3 \times 3 \times 7 \times 7$
- **g** $3 \times 3 \times 3 \times 3 \times 5 \times 5$
- **h** $7 \times 7 \times 7 \times 7 \times 7 \times 11 \times 11 \times 11$

Example 9 ◀) Self Tutor

Find the integer equal to:

a 3^4 **b** $2^4 \times 3^2 \times 7$

a 3^4
$= 3 \times 3 \times 3 \times 3$
$= 81$

b $2^4 \times 3^2 \times 7$
$= 2 \times 2 \times 2 \times 2 \times 3 \times 3 \times 7$
$= 1008$

3 Find the integer equal to:

- **a** 2^3
- **b** 2^4
- **c** 3^3
- **d** 2×5^2
- **e** $2^2 \times 5^2$
- **f** $3^3 \times 5$
- **g** $3^2 \times 5 \times 7$
- **h** $2^2 \times 3^3 \times 11$

4 Use your calculator to write each product as a natural number:

- **a** $2^4 \times 3^6$
- **b** $2^2 \times 5^4 \times 7^5$
- **c** $2^5 \times 3^3 \times 11^2$
- **d** $2^3 \times 3^4 \times 5^2 \times 11$
- **e** $3^4 \times 7^2 \times 11^3$
- **f** $2^3 \times 5^5 \times 13^3$

CALCULATOR INSTRUCTIONS

Example 10

Simplify:

a -4^2 b $(-4)^2$ c -2^3 d $(-2)^3$

a -4^2
$= -4 \times 4$
$= -16$

b $(-4)^2$
$= -4 \times -4$
$= 16$

c -2^3
$= -2 \times 2 \times 2$
$= -8$

d $(-2)^3$
$= -2 \times -2 \times -2$
$= -8$

5 Simplify:
  a -3^2
  b $(-3)^2$
  c -1^3
  d $(-1)^3$
  e $(-5)^2$
  f -5^2
  g $(-2)^4$
  h -2^4

6 Evaluate:
  a $(-1)^2$
  b $(-1)^4$
  c $(-1)^5$
  d $(-1)^7$
  e $(-1)^{10}$

What do you notice?

7 Find, using your calculator:
  a 5^6
  b 9^3
  c 11^5
  d $(-3)^8$
  e $(-4)^7$
  f $(-7)^4$
  g $(-2)^6$
  h 23^2

8 Write in index form with 2 as a base:
  a 2
  b 4
  c 16
  d 64

9 Write in index form with 3 as a base:
  a 3
  b 27
  c 81
  d 729

10 Write in index form with 10 as a base:
  a 100
  b 1000
  c 100 000
  d 1 000 000

11 Write in index form with a single digit base:
  a 8
  b 25
  c 36
  d 125

Activity 2 — Index crossword

Click on the icon to obtain a printable version of this crossword.

Across
1 19^2
3 2^4
4 4^4
5 5^2
6 3^4
7 3^3
8 9^2
9 22^2
11 4^3
12 13^2

Down
1 6^2
2 5^3
3 41^2
5 14^3
8 29^2
10 7^2

CROSSWORD

F SQUARE NUMBERS AND SQUARE ROOTS

SQUARE NUMBERS

When a number is multiplied by itself, we say that the number is **squared**.

For example, since $4 \times 4 = 16$, we say "four squared is equal to sixteen". Using index notation, we write $4^2 = 16$.

When a whole number is squared, the result is a **square number** or **perfect square**.

The first five square numbers are:
$1^2 = 1 \times 1 = 1$
$2^2 = 2 \times 2 = 4$
$3^2 = 3 \times 3 = 9$
$4^2 = 4 \times 4 = 16$
$5^2 = 5 \times 5 = 25$

SQUARE ROOTS

Finding the **square root** of a number is the opposite operation to squaring a number.

> The **square root** of the number a is the positive number which, when squared, gives a.
> We write the square root of a as $\sqrt{a}$.
> $$\sqrt{a} \times \sqrt{a} = a$$

For example, since $4^2 = 16$, $\sqrt{16} = 4$.

EXERCISE 1F

1 Without using a calculator, find:
   a 6^2 **b** 9^2 **c** 11^2 **d** 12^2

2 Use your calculator to find the largest 3 digit square number.

3 Copy and complete:
   a "The square of an even number is ......"
   b "The square of an odd number is ......"

4 Without using your calculator, find:
   a $\sqrt{49}$ **b** $\sqrt{64}$ **c** $\sqrt{121}$ **d** $\sqrt{144}$

5 Use your calculator to find:
   a $\sqrt{289}$ **b** $\sqrt{576}$ **c** $\sqrt{1521}$ **d** $\sqrt{2304}$

6 Consider the pattern:
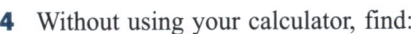
$1 \times 3 + 1 = ...$
$2 \times 4 + 1 = ...$
$3 \times 5 + 1 = ...$
$4 \times 6 + 1 = ...$

   a Copy and complete the pattern, then add three more rows.
   b Use the pattern to find:
      i $19 \times 21 + 1$ **ii** 29×31

24   Number   (Chapter 1)

Discussion

- We have seen that $\sqrt{16} = 4$. What happens if you try to find $\sqrt{-16}$ on your calculator? *Why* does this happen?
- Between which two consecutive integers do you think $\sqrt{70}$ will be? *Why* do you think this?

G ORDER OF OPERATIONS

Some expressions contain more than one operation. To evaluate these expressions correctly, we use a set of rules.

Discussion

- If you were asked to evaluate $3 + 4 \times 2$, what answer do you get if you perform:
  ▸ the addition first   ▸ the multiplication first?
  What rules does your calculator use?
- Why do we need a set of rules for the order of operations?

The rules we use are:

- Perform the operations within **B**rackets first.
- Calculate any part involving **E**xponents.
- Starting from the left, perform all **D**ivisions and **M**ultiplications as you come to them.
- Restart from the left, performing all **A**dditions and **S**ubtractions as you come to them.

The word **BEDMAS** may help you remember this order.

Brackets are **grouping symbols** which are used to indicate a part of an expression which should be evaluated first.

- If an expression contains *one set* of brackets, evaluate that part first.
- If an expression contains *two or more sets* of brackets one inside the other, evaluate the *innermost set* first.
- In a fraction, we assume that the numerator and denominator each have brackets around them. We must evaluate them first before doing the division.

Example 11   ◀) Self Tutor

Simplify:

a $3 + 7 - 5$ **b** $6 \times 3 \div 2$

a $\quad 3 + 7 - 5$ {only $+$ and $-$ involved}   **b** $\quad 6 \times 3 \div 2$ {only $\times$ and $\div$ involved}
$\quad = 10 - 5$ $\quad = 18 \div 2$
$\quad = 5$ $\quad = 9$

Example 12

Simplify:

a $23 - 10 \div 2$

b $3 \times 8 - 6 \times 5$

a $\quad 23 - 10 \div 2 \quad \{\div \text{ before } -\}$
$= 23 - 5$
$= 18$

b $\quad 3 \times 8 - 6 \times 5 \quad \{\times \text{ before } -\}$
$= 24 - 30$
$= -6$

EXERCISE 1G

1 Simplify:

a $4 + 9 - 5$
b $4 - 9 + 5$
c $4 - 9 - 5$
d $3 \times 12 \div 6$
e $12 \div 6 \times 3$
f $6 \times 12 \div 3$

2 Simplify:

a $8 + 9 \times 3$
b $6 \times 3 + 7$
c $14 - 2 \times 4$
d $21 - 6 \times 1$
e $3 \times 2 - 2$
f $30 - 2 \times 2 \times 3$
g $30 \div 2 + 6$
h $4 + 3 + 2 \times 6$
i $12 - 6 \times 3 + 2$
j $26 - 9 \div 3$
k $5 + 12 \div 2$
l $15 \div 3 + 16 \div 2$

Example 13

Use the correct order of operations rules to calculate:

a $5 + -8 \times 3$

b $-5 - 15 \div -5$

a $\quad 5 + -8 \times 3$
$= 5 + -24 \quad$ {multiplication first}
$= 5 - 24 \quad$ {simplify}
$= -19$

b $\quad -5 - 15 \div -5$
$= -5 - -3 \quad$ {division first}
$= -5 + 3 \quad$ {simplify}
$= -2$

3 Find, using the order of operations rules:

a $3 + 4 \div -2$
b $-1 + -3 \times 2$
c $8 \div -2 + 5$
d $-3 \times -2 - 4$
e $2 - 6 \div -3$
f $-2 \times 4 + -7$
g $7 - 3 \times -3$
h $-4 \times -5 - 12$
i $3 - 6 \div -6$

4 Do -3^2 and $(-3)^2$ have the same value? Explain your answer.

5 Min's company makes a £100 000 profit per month for eight months, and then an £80 000 loss for each of the next four months. Find her company's total profit or loss.

6 In indoor cricket, the person batting is penalised 5 runs for each wicket lost.
Josh lost 6 wickets, and scored 17 runs. What was his final score?

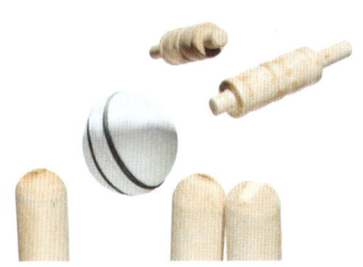

7 A computer store has the following sales record over a six-week period:

Week 1: £388 profit   Week 2: £1373 loss
Week 3: £179 loss   Week 4: £3013 profit
Week 5: £832 profit   Week 6: £1763 loss.

> The *average* weekly profit or loss is the overall profit or loss divided by the number of weeks.

a What was the store's overall profit or loss for this period?
b What was the store's *average* weekly profit or loss during this period?

Example 14   ◆) **Self Tutor**

Simplify: $3 + (11 - 7) \times 2$

$3 + (11 - 7) \times 2$
$= 3 + 4 \times 2$ {evaluate the brackets first}
$= 3 + 8$ {× before +}
$= 11$

---

8 Simplify:

a $15 + (9 - 3)$
b $(15 + 9) - 3$
c $(12 \div 6) - 2$
d $12 \div (6 - 2)$
e $(11 - 8) - 3$
f $11 - (8 - 3)$
g $36 - (9 \div 3)$
h $(36 - 9) \div 3$
i $24 - (6 + 10) - 3$
j $(24 - 6) + (10 - 3)$
k $(20 \div 10) \div 2$
l $20 \div (10 \div 2)$
m $16 - (4 \times 3) - 2$
n $30 \times 6 \div (5 - 2)$
o $28 - (3 \times 8) \div 6$
p $(-3 + 4) \times -7$
q $15 \div (4 - 7)$
r $-3 \times (-2 + 5)$

9 Evaluate using your calculator:

a $16 + 25 \times 9$
b $(16 + 25) \times 9$
c $112 \div 7 + 7$
d $112 \div (7 + 7)$
e 43×-6
f $-256 \div -32$

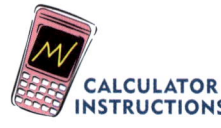
CALCULATOR INSTRUCTIONS

Example 15   ◆) **Self Tutor**

Simplify: $[12 + (9 \div 3)] - 4^2$

$[12 + (9 \div 3)] - 4^2$
$= [12 + 3] - 4^2$ {evaluate the inner brackets first}
$= 15 - 4^2$ {evaluate the outer brackets}
$= 15 - 16$ {evaluate the exponents}
$= -1$

10 Simplify:
  a $9 - [(4-3) + 2 \times 5]$
  b $[21 - (4+5)] \times 2$
  c $12 - [(7+4) + 3]$
  d $[18 - (15 \div 5)] + 6$
  e $108 \div [2 \times (18 \div 3)]$
  f $[(4 \times 2) \div (4 \div 2)] \times 3$

11 Simplify:
  a $5 - 4^2$
  b $(5-4)^2$
  c $2 \times (4-7)^2$
  d $2 \times 4 - 7^2$
  e $(3-8) \times (5+2)^2$
  f $3^2 - 8 \times 5 + 2^3$

12 Using $\times$, $\div$, $+$, or $-$ only, replace each $\square$ so that correct equations result. Remember that the operations must be evaluated in the correct order.
  a $7 \square 3 \square 4 = 6$
  b $4 \square 6 \square 3 = 21$
  c $12 \square 4 \square 3 = 9$

13 Insert grouping symbols where necessary to make the following true:
  a $9 - 7 \times 4 = 8$
  b $80 \div 8 \times 2 = 5$
  c $80 \div 8 \times 2 = 20$
  d $4 \times 8 - 7 - 1 = 26$
  e $4 \times 8 - 7 - 1 = 3$
  f $4 \times 8 - 7 - 1 = 0$
  g $5 + 2 \times 6 - 3 = 39$
  h $5 + 2 \times 6 - 3 = 11$
  i $5 + 2 \times 6 - 3 = 21$

Activity 3

Click on the icon to run the BEDMAS Challenge.

How fast can you go?

BEDMAS CHALLENGE

H FACTORS AND MULTIPLES

FACTORS

The **factors** of a natural number are all the natural numbers which divide exactly into it, leaving no remainder.

For example, the factors of 10 are 1, 2, 5, and 10.

A number may have many factors. When a number is written as a **product** of factors, we say it is **factorised**.

For example, the number 20 has factors 1, 2, 4, 5, 10, and 20. It can be factorised into pairs as:
1×20, 2×10, or 4×5.

20 may also be factorised as the product of 3 factors, for example $20 = 2 \times 2 \times 5$.

MULTIPLES

A **multiple** of any natural number is obtained by multiplying it by another natural number.

For example, the multiples of 3 are 3, 6, 9, 12, 15, 18, ... These are obtained by multiplying 3 by each of the natural numbers in turn: $3 \times 1 = 3$, $3 \times 2 = 6$, $3 \times 3 = 9$, $3 \times 4 = 12$, ...

28 Number (Chapter 1)

EXERCISE 1H

1 **a** Is 5 a factor of 40? **b** Is 4 a factor of 50? **c** Is 7 a factor of 26?
 d Is 8 a factor of 56? **e** Is 6 a factor of 82? **f** Is 3 a factor of 87?

2 List all the factors of:
 a 6 **b** 8 **c** 10 **d** 15 **e** 12
 f 13 **g** 18 **h** 20 **i** 24 **j** 25
 k 42 **l** 54 **m** 63 **n** 23 **o** 72

3 Write all the factor pairs of:
 a 6 **b** 8 **c** 9 **d** 12 **e** 14
 f 28 **g** 45 **h** 48 **i** 72 **j** 100

4 List the numbers which are factors of both 60 and 96.

5 List the first five multiples of:
 a 3 **b** 5 **c** 7 **d** 9 **e** 12

6 Find the:
 a lowest multiple of 7 which is greater than 60
 b highest multiple of 6 which is less than 80.

7 How many factors does each number have?
 a 3 **b** 8 **c** 11 **d** 16 **e** 32

I PRIMES AND COMPOSITES

A **prime number** is a natural number which has **exactly** two distinct factors, 1 and itself.
A **composite number** is a natural number which has more than two factors.

For example:
- 17 is a prime number since it has only 2 factors, 1 and 17.
- 26 is a composite number since it has more than two factors. These are 1, 2, 13, and 26.

Notice that **one** (1) is neither prime nor composite.

THE FUNDAMENTAL THEOREM OF ARITHMETIC

Apart from order, every composite number can be written as a **product of prime factors** in **one and only one way**.

For example, $72 = 2 \times 2 \times 2 \times 3 \times 3$ is the only way of writing 72 as the product of prime factors. Using index notation, we write $72 = 2^3 \times 3^2$. We call this a **prime factorisation** and say the number is written in **prime factored form**.

There are two methods we can use for writing a number in prime factored form:
- In **repeated division**, we continue to divide the number by primes until we are left with 1.
- In a **factor tree**, we find a factor pair for the number, and these factors become branches of the tree. We continue finding factor pairs for each branch until we are left only with prime numbers.

Example 16

Write 600 as the product of prime factors in index form.

Repeated division

2	600
2	300
2	150
3	75
5	25
5	5
	1

$\therefore\ 600 = 2 \times 2 \times 2 \times 3 \times 5 \times 5$
$= 2^3 \times 3 \times 5^2$

or

Factor tree

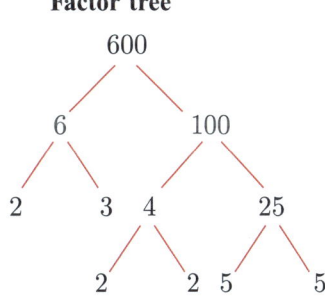

$\therefore\ 600 = 2 \times 3 \times 2 \times 2 \times 5 \times 5$
$= 2^3 \times 3 \times 5^2$

EXERCISE 1I

1. List the set of all primes less than 30.

2. Find two consecutive odd numbers between 60 and 80 which are both prime.

3. A number between 20 and 30 has 2 prime factors and 3 composite factors. Find the number.

4. Write the following as powers of a prime:
   - **a** 8
   - **b** 27
   - **c** 125
   - **d** 128
   - **e** 343
   - **f** 729
   - **g** 361
   - **h** 1331

5. Express the following as the product of prime factors in index form:
   - **a** 54
   - **b** 108
   - **c** 360
   - **d** 228
   - **e** 196
   - **f** 756
   - **g** 936
   - **h** 1225
   - **i** 588
   - **j** 945
   - **k** 910
   - **l** 1274

Game

Click on the icon to play a game which involves writing numbers as the product of prime factors. See if you can get the highest score in your class.

J HIGHEST COMMON FACTOR

A number which is a factor of two or more other numbers is called a **common factor** of these numbers.

For example, 7 is a common factor of 28 and 35.

The **highest common factor (HCF)** of two numbers is the largest factor of *both* of these numbers.

For example: The factors of 18 are 1, 2, 3, 6, 9, and 18.
The factors of 45 are 1, 3, 5, 9, 15, and 45.
The *common* factors of 18 and 45 are 1, 3, and 9, so the highest common factor is 9.

The highest common factor of two numbers can be found by expressing each number as a product of prime factors. We multiply all the factors which are common to both numbers, to find the highest common factor.

Example 17 — Self Tutor

Find the highest common factor (HCF) of 18 and 24.

2	18
3	9
3	3
	1

2	24
2	12
2	6
3	3
	1

$\therefore$ $18 = 2 \times 3 \times 3$
and $24 = 2 \times 2 \times 2 \times 3$

2×3 is common to the factorisations of both 18 and 24.

So, the highest common factor of 18 and 24 is $2 \times 3 = 6$.

EXERCISE 1J

1 Find the highest common factor of:

- **a** 12 and 16
- **b** 9 and 15
- **c** 14 and 56
- **d** 16 and 40
- **e** 24 and 60
- **f** 35 and 50
- **g** 55 and 121
- **h** 24 and 42
- **i** 28 and 70
- **j** 80 and 96
- **k** 64 and 288
- **l** 169 and 208
- **m** 90 and 189
- **n** 252 and 490
- **o** 280 and 308

2 Farmer Giles has a field 45 m × 60 m. He wishes to divide it into square yards of equal size. What is the biggest size the yards could be?

3 Two pieces of copper pipe 315 cm and 225 cm long will be cut into sections. All sections must have equal length. What is the greatest possible length of each section?

4 A florist has 30 lilies, 42 gerberas, and 36 roses to make bouquets with. Each bouquet must be identical. What is the largest number of bouquets that can be made using all of the flowers?

K LOWEST COMMON MULTIPLE

A number which is a multiple of two or more other numbers is called a **common multiple** of these numbers.

For example, 60 is a common multiple of 5 and 6.

The **lowest common multiple** or **LCM** of a set of natural numbers is the smallest multiple which is common to all of them.

Example 18

Find the lowest common multiple of 9 and 12.

The multiples of 9 are: 9, 18, 27, **36**, 45, 54, 63, **72**, 81, ...

The multiples of 12 are: 12, 24, **36**, 48, 60, **72**, 84, ...

∴ the common multiples are 36, 72, ... and 36 is the smallest of these

∴ the LCM is 36.

One way to find the lowest common multiple starts by writing each number as the product of its prime factors. Write one product above the other, matching up any common prime factors. The LCM is the product of the fewest prime factors needed to create a multiple of both numbers.

Example 19

Find the LCM of: **a** 9 and 12   **b** 15, 20, and 24.

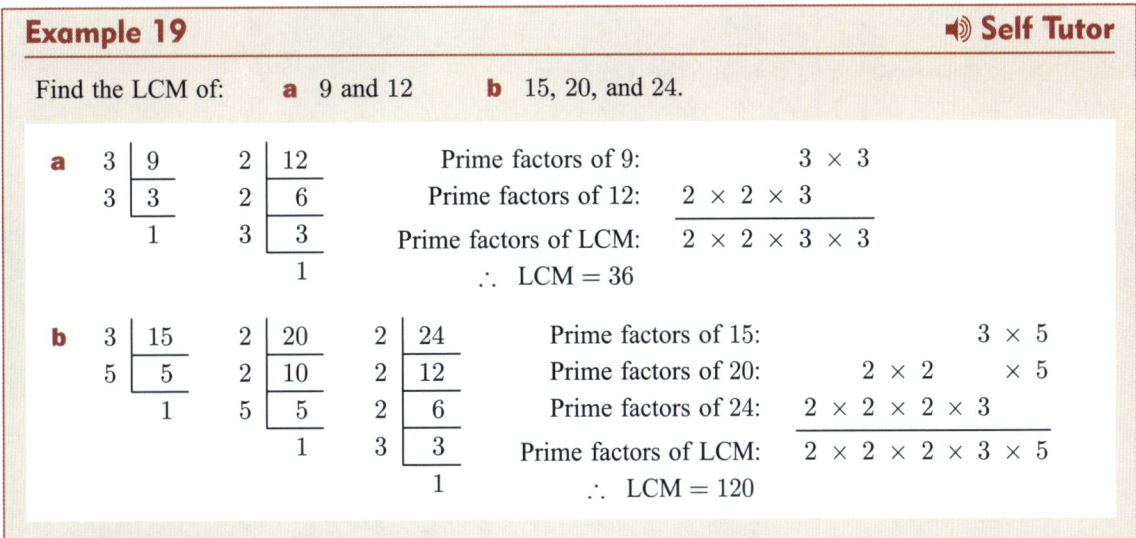

EXERCISE 1K

1 Find the LCM of:
  - **a** 5 and 8
  - **b** 4 and 6
  - **c** 8 and 10
  - **d** 15 and 18
  - **e** 12 and 15
  - **f** 14 and 20
  - **g** 12 and 27
  - **h** 42 and 45

2 Find the LCM of:
  - **a** 2, 3, and 4
  - **b** 5, 7, and 10
  - **c** 4, 6, and 8
  - **d** 5, 8, and 9
  - **e** 8, 10, and 12
  - **f** 9, 15, and 20
  - **g** 14, 18, and 21
  - **h** 20, 25, and 30

3 Find the smallest positive integer which is exactly divisible by 15 and 25.

4 Chris has a piece of rope. It can be cut exactly into either 10 metre or 18 metre lengths. Find the shortest length that Chris' rope could be.

5 Answer the **Opening Problem** on page **12**.

6 Nick has three egg timers. Their sand will last for 4 minutes, 5 minutes, and 6 minutes respectively. If Nick starts the timers together, and turns each one immediately when its sand runs out, how long will it be before they all run out of sand at the same time?

7 Gloria is an avid coffee drinker. Every day, she drinks a coffee at each of these stores:

      Keen Beans: "Buy 5 coffees, get 1 free!"
      The Caffeine Club: "Buy 7 coffees, get 1 free!"
      Expresso Yourself: "Buy 9 coffees, get 1 free!"

Today, Gloria had a free coffee at each of the stores. How long will it be before she next has:

a 2 free coffees in one day

b 3 free coffees in one day?

Review set 1A

1 Write the opposite of each number:
  a 5   **b** -2 **c** $+3$ **d** -11

2 Indicate the position of each point using a number:

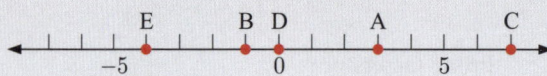

3 **a** Arrange in ascending order: $2, -4, 0, -3, -6, 7, 3$.
    b Find the difference between the greatest and least values in **a**.

4 Find the value of:
  a $4-7$ **b** $-3+6$ **c** 3×-2 **d** -3×-4

5 The minimum temperatures for a week in Beijing were:
$-5°C, -4°C, -5°C, 2°C, 3°C, 3°C, -1°C$.
Calculate the average minimum temperature for that week.

6 A warehouse contains 24 pianos. Each week 3 more are sent to the warehouse from the factory. Over a 4 week period, 1, 6, 3, and 5 pianos are sold. How many pianos are now in the warehouse?

7 Find the value of:
  a $15 - -9$ **b** $-6 + 7 - -3$ **c** $-88 \div 11$
  d $5 - -3 + -7$ **e** $(-8)^2$ **f** $-20 + 15 \div -3$

8 Find the integer equal to:
  a 4^3 **b** $2^3 \times 5^2$ **c** $2^1 \times 5^2 \times 7^1$

9 Calculate:
  a $\sqrt{9}$ **b** $\sqrt{81}$ **c** $\sqrt{100}$

10 Simplify:
  a $24 \div 3 \times 2$ **b** $13 - 5 \times 2 - 4^2$ **c** $5 \times (8 - 2)$
  d $8 \div (5 - 1) \times 3$ **e** $15 - [3 \times (8 - 6) - 2]$

11 Using ×, ÷, +, or − only, replace each □ so that correct equations result.

    a   3 □ 5 □ 4 = 11   **b**   8 □ 6 □ 3 = 6

12 List the factors of:

    a   21   **b**   32   **c**   37

13 Write all of the factor pairs of 66.

14 List the prime numbers between 40 and 50.

15 Express as the product of prime factors in index form:

    a   450   **b**   212

16 Find the lowest common multiple of:

    a   6 and 15   **b**   4 and 11   **c**   5, 8, and 10

17 Find the HCF of:

    a   12 and 14   **b**   24 and 56   **c**   18, 27, and 45

18 In a mathematics competition, students are awarded 3 points for a right answer, and penalised 4 points for a wrong answer. Amy gave 6 wrong answers and 24 right answers. Sean gave 14 wrong answers and 16 right answers.

    a   How many points did each student score?

    b   By how many points did Amy beat Sean?

19 A passenger train goes through a 2-track level crossing every 8 minutes. A freight train goes through the same level crossing every 52 minutes.

A passenger train and a freight train go through the level crossing at the same time. How long will it be before a passenger train and a freight train next pass through the level crossing together?

Review set 1B

1 Insert either < or > in place of □ to make each statement true:

    a   3 □ −1   **b**   −7 □ −4   **c**   −2 □ 5   **d**   0 □ −10

2 Display the numbers on a number line. Hence arrange the numbers from least to greatest.

    a   3, −2, 1, −4, −1   **b**   8, −5, 1, −3, 2

3   **a**   Use a number line to increase −15 by 8.

       b   *Hence* evaluate $3 \times (-15 + 8)$.

4   **a**   Copy and complete: negative × negative = ...

       b   Evaluate -7×-11.

5   **a**   Arrange in descending order: 5, −4, 0, 1, −1, 3, −6.

       b   Find the difference between the largest and smallest values.

6 Find the value of:

    a $4 + -7$ **b** -8×9 **c** $54 \div -6$

    d $(-1)^5$ **e** $-12 \div 4 + 13$ **f** $-2 \times -3 \times -4$

7 Beck, Cathy, Emily, and Ying agreed to meet at a coffee shop. Beck was 9 minutes late, Cathy was 4 minutes early, Emily was 17 minutes late, and Ying was 10 minutes early.

    a Who arrived first?
    b Who arrived closest to the agreed time?
    c Find the difference between the arrival times of:
        i Beck and Ying   **ii** Cathy and Emily
        iii Beck and Emily.

8 List the square numbers between 40 and 90.

9 Evaluate:
    a $(-2)^5$ **b** $2^3 \times 5^1 \times 7^1$ **c** $(-6)^2$

10 List the square numbers between 700 and 800.

11 Simplify:
    a $6^2 + 3 \times 5$ **b** $12 - 8 \div 2 + 1$

12 Insert brackets where necessary to make the following true:
    a $12 \div 6 - 2 = 3$ **b** $6 + 4 \div 2 + 3 = 2$ **c** $18 \div 1 + 2 \times 4 = 2$

13 Use your calculator to evaluate: $200 \div (13 + 12)$

14 List the first five multiples of 8.

15 Write 54 as the product of prime factors in index form.

16 Find three consecutive odd numbers less than 100, which are all composite.

17 For the numbers 18 and 30, find:
    a the highest common factor   **b** the lowest common multiple.

18 A greengrocer sells nectarines by the bag, and each bag contains the same number of nectarines. The greengrocer sold 126 nectarines yesterday, and 198 nectarines today. What is the greatest number of nectarines that could be in each bag?

19 Amy has some lollies in a bag, which she will share with her friends at a party. She knows she can share them equally whether there are four, five, or six children present. What is the smallest number of lollies Amy could have?

2

Real numbers

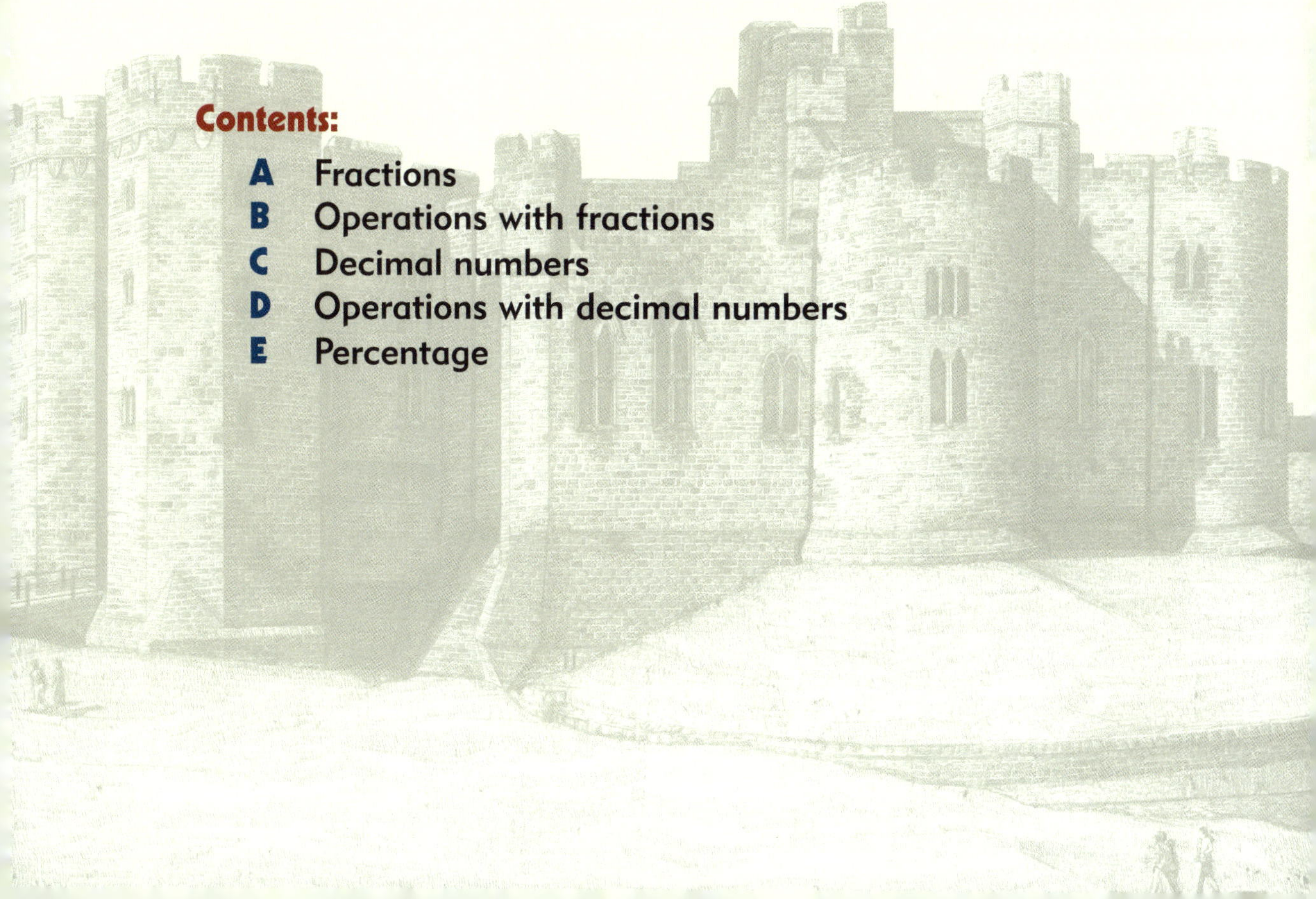

Contents:
- **A** Fractions
- **B** Operations with fractions
- **C** Decimal numbers
- **D** Operations with decimal numbers
- **E** Percentage

Opening problem

An abalone diver has a daily catch limit. He catches $\frac{1}{5}$ of his limit in the first hour, and $\frac{1}{4}$ of his limit in the second hour.

Things to think about:
  a What *fraction* of the limit has he caught so far?
  b What *percentage* of the limit has he caught so far?
  c The diver has caught 18 abalone so far.
   i What is the daily catch limit?
   ii If an average abalone weighs 0.34 kg, can you estimate the total weight of abalone the diver has caught so far?

The set of **real numbers** includes all numbers which can be placed on the number line. It includes the set of whole numbers or **integers**, as well as the **fractions** and **decimals** between them.

In this Chapter we study these numbers and operations with them. We also study how fractions or decimal parts of a whole can be written as **percentages**.

A FRACTIONS

Fractions are obtained when we divide a whole into equal portions.

The fraction $\frac{a}{b}$ means we divide a whole into b equal portions, and then consider a of them.

> The division $a \div b$ can be written as the **common fraction** $\frac{a}{b}$.
>
> the **numerator** is the number of portions considered
> the **bar** indicates division
> the **denominator** is the number of portions we divide a whole into

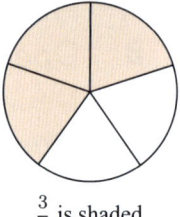

$\frac{3}{5}$ is shaded

TYPES OF FRACTIONS

$\frac{2}{3}$ is called a **proper fraction** as the numerator is less than the denominator.

$\frac{4}{3}$ is called an **improper fraction** as the numerator is greater than the denominator.

$2\frac{1}{2}$ is a **mixed number** which represents $2 + \frac{1}{2}$.

When we perform calculations involving mixed numbers, it is often useful to first convert the mixed number to an improper fraction.

Real numbers (Chapter 2) 37

Example 1 ◀)) Self Tutor

Convert $2\frac{1}{2}$ to an improper fraction.

$2\frac{1}{2} = 2 + \frac{1}{2}$
$= \frac{4}{2} + \frac{1}{2}$ {writing with equal denominators}
$= \frac{5}{2}$

We can write 2 as $\frac{4}{2}$.

PLACING FRACTIONS ON A NUMBER LINE

We can represent fractions on a **number line** by dividing each whole into the number of parts in the denominator. By extending the number line either side of zero, we can represent positive and negative fractions.

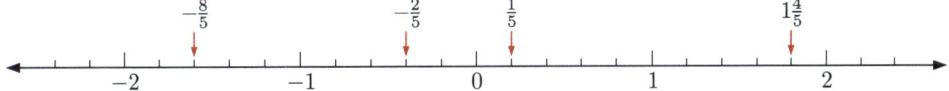

EXERCISE 2A.1

1 Describe the following as proper fractions, improper fractions, or mixed numbers:

a $\frac{2}{7}$ b $\frac{8}{5}$ c $\frac{7}{8}$ d $5\frac{1}{3}$ e $\frac{10}{9}$ f $1\frac{1}{2}$

2 Convert these mixed numbers to improper fractions:

a $1\frac{3}{4}$ b $3\frac{1}{2}$ c $4\frac{1}{5}$ d $2\frac{3}{8}$ e $-1\frac{1}{3}$ f $-3\frac{1}{5}$

3 Plot these fractions on a number line:

a $\frac{1}{3}, -\frac{2}{3}, \frac{5}{3}, -1\frac{1}{3}$ b $\frac{3}{4}, -\frac{1}{4}, 2\frac{1}{4}, -1\frac{1}{4}$ c $\frac{3}{7}, \frac{6}{7}, -\frac{2}{7}, 1\frac{4}{7}, -\frac{10}{7}$

4 Write as a division and hence evaluate:

a $\frac{-6}{3}$ b $\frac{6}{-3}$ c $-\frac{6}{3}$

Discussion

Can you explain why $\frac{-a}{b} = \frac{a}{-b} = -\frac{a}{b}$ for all $b \neq 0$?

EQUAL FRACTIONS

When we multiply or divide the numerator and denominator of a fraction by the same non-zero number, we obtain an **equal** or **equivalent** fraction.

For example, 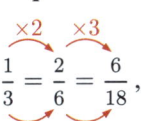 $\frac{1}{3} = \frac{2}{6} = \frac{6}{18}$, and 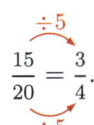 $\frac{15}{20} = \frac{3}{4}$.

Example 2

Express:

a $\frac{3}{4}$ with denominator 32 **b** $\frac{24}{33}$ with denominator 11

a To convert the denominator to 32, we need to multiply by 8.
We must also multiply the numerator by 8.

$$\frac{3}{4} \xrightarrow{\times 8} \frac{24}{32}$$

b To convert the denominator to 11, we need to divide by 3.
We must also divide the numerator by 3.

$$\frac{24}{33} \xrightarrow{\div 3} \frac{8}{11}$$

SIMPLIFYING FRACTIONS

We can **simplify** a fraction by dividing the numerator and denominator by their **highest common factor**. When we have simplified a fraction in this manner, we say it is in **lowest terms**.

Example 3

Write $\frac{32}{40}$ in lowest terms.

The HCF of 32 and 40 is 8.

$$\frac{32}{40} = \frac{32 \div 8}{40 \div 8} = \frac{4}{5}$$

$\therefore$ in lowest terms the fraction is $\frac{4}{5}$.

EXERCISE 2A.2

1 Write with denominator 12:

  a $\frac{1}{4}$ **b** $\frac{2}{3}$ **c** $\frac{5}{6}$ **d** $\frac{15}{36}$

2 Write with denominator 20:

  a $\frac{7}{10}$ **b** $\frac{3}{4}$ **c** $\frac{26}{40}$ **d** $\frac{15}{100}$

3 a Write the fractions $\frac{7}{10}$, $\frac{3}{5}$, and $\frac{2}{3}$ with denominator 30.

  b Hence write the fractions in order from smallest to largest.

4 Write in ascending order:

  a $\frac{3}{8}, -\frac{1}{6}, \frac{5}{6}, \frac{2}{3}, \frac{1}{4}$ **b** $-\frac{1}{3}, \frac{1}{6}, -\frac{2}{9}, \frac{2}{9}, -\frac{5}{18}$

5 Express in lowest terms:

  a $\frac{8}{12}$ **b** $\frac{15}{18}$ **c** $\frac{25}{15}$ **d** $\frac{21}{35}$ **e** $-\frac{2}{12}$

  f $\frac{-5}{15}$ **g** $\frac{36}{48}$ **h** $\frac{8}{-20}$ **i** $\frac{22}{32}$ **j** $\frac{-42}{28}$

6 An enclosure at the zoo contains 16 meerkats. 6 are asleep and 10 are awake.

Find, in lowest terms, the fraction of meerkats that are:
- **a** asleep
- **b** awake.

©iStock.com/pjmalsbury

7 27 beginners, 25 amateurs, and 8 professionals competed in a golf tournament.
- **a** How many players competed in the tournament?
- **b** Find, in lowest terms, the fraction of competitors who were:
  - **i** amateur
  - **ii** professional
  - **iii** either amateur or professional.

Example 4 — Self Tutor

Simplify: $\dfrac{12 + (5 - 7)}{18 \div (6 + 3)}$

$\dfrac{12 + (5 - 7)}{18 \div (6 + 3)}$ {evaluate the brackets first}

$= \dfrac{12 + (-2)}{18 \div 9}$ {simplifying numerator and denominator}

$= \dfrac{10}{2}$

$= 5$

Evaluate the numerator and the denominator first, *then* perform the division.

8 Simplify:
- **a** $\dfrac{72}{4 \times 2}$
- **b** $\dfrac{33}{15 - 4}$
- **c** $\dfrac{32 \div 4}{6 - 4}$
- **d** $\dfrac{13 + 7}{11 - 6}$
- **e** $\dfrac{64 - 16}{17 - 5}$
- **f** $\dfrac{5 \times 2 + 6}{8}$
- **g** $\dfrac{30}{11 - (2 \times 3)}$
- **h** $\dfrac{(4 + 9) - 5}{4 + (9 - 5)}$

9 Evaluate using your calculator:
- **a** $\dfrac{39}{3} - 18$
- **b** $\dfrac{139 - 7}{4 \times 11}$
- **c** $\dfrac{-15 \times 2}{7 - (16 \div 8)}$
- **d** $\dfrac{118 + 8}{3 \times 7}$
- **e** $\dfrac{-240 - 120}{3 \times 4 \times 5}$
- **f** $\dfrac{(10 - 2) \times 3}{32 - (4 \times 5)}$

CALCULATOR INSTRUCTIONS

Game

Click on the icon to practise ordering fractions.

ORDERING FRACTIONS
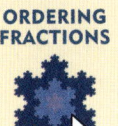

B OPERATIONS WITH FRACTIONS

ADDITION AND SUBTRACTION

To **add** or **subtract** fractions, each fraction must have the same denominator.
- If necessary, convert the fractions so they have the **lowest common denominator** (LCD).
- Add or subtract the new numerators. The denominator stays the same.

Example 5

Find: **a** $\frac{3}{8} + \frac{1}{2}$ **b** $\frac{3}{4} - \frac{2}{3} + \frac{1}{2}$

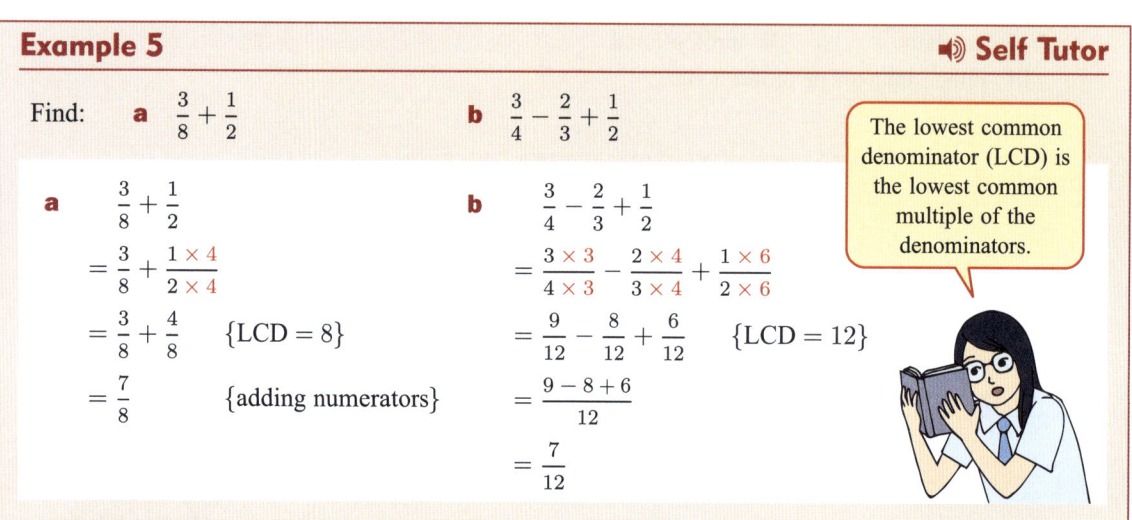

The lowest common denominator (LCD) is the lowest common multiple of the denominators.

Mixed numbers should be written as improper fractions before the addition or subtraction is performed.

Example 6

Find: $2\frac{1}{2} - 1\frac{2}{5}$

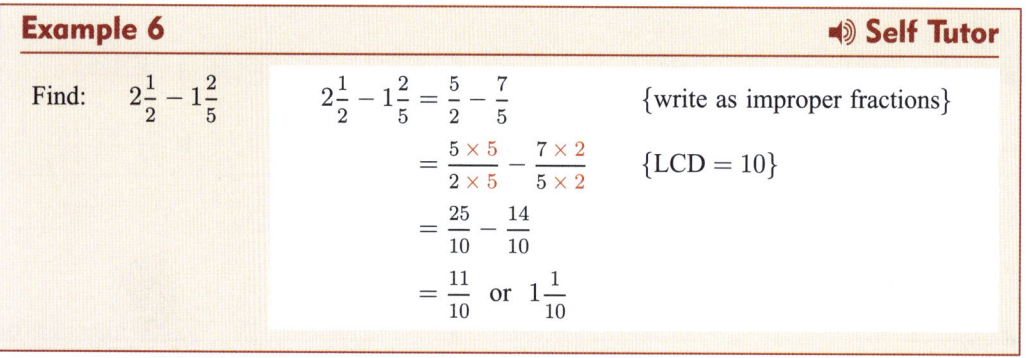

EXERCISE 2B.1

1 Find:

  a $\frac{2}{9} + \frac{5}{9}$ **b** $\frac{6}{7} - \frac{3}{7}$ **c** $\frac{5}{8} + \frac{4}{8}$ **d** $\frac{8}{11} - \frac{3}{11}$

2 Find:

  a $\frac{1}{2} + \frac{3}{4}$ **b** $\frac{5}{6} - \frac{1}{3}$ **c** $\frac{3}{5} + \frac{1}{10}$ **d** $\frac{1}{3} - \frac{1}{4}$

  e $\frac{5}{8} + \frac{3}{4}$ **f** $\frac{5}{6} - \frac{3}{4}$ **g** $\frac{4}{7} + \frac{4}{5}$ **h** $\frac{1}{2} - \frac{5}{9}$

Real numbers (Chapter 2)

3 Find:

a $\dfrac{7}{9} + \dfrac{2}{3} - \dfrac{2}{9}$
b $\dfrac{7}{4} - \dfrac{7}{8} + \dfrac{1}{2}$
c $\dfrac{19}{15} - \dfrac{4}{5} + \dfrac{3}{10}$
d $\dfrac{1}{5} + \dfrac{1}{4} - \dfrac{1}{3}$

4 Find:

a $1 - \dfrac{3}{7}$
b $-\dfrac{1}{2} + \dfrac{3}{4}$
c $1\dfrac{1}{4} + \dfrac{5}{8}$
d $-\dfrac{5}{7} + \dfrac{1}{3}$

e $3 - 1\dfrac{3}{5}$
f $1 - \dfrac{1}{6} - \dfrac{3}{8}$
g $2\dfrac{2}{5} - 1\dfrac{4}{9}$
h $2 - \dfrac{3}{4} - \dfrac{4}{5}$

5 Carter read $\dfrac{1}{5}$ of a book during a flight from Birmingham to Berlin, and another $\dfrac{3}{8}$ of the book during the flight back.

a What fraction of the book has Carter read?
b What fraction of the book remains for Carter to read?

6 A recipe requires $\dfrac{2}{3}$ cup of self-raising flour, and $1\dfrac{1}{2}$ cups of plain flour.

a In total, how much flour is used in the recipe?
b How much more plain flour is used than self-raising flour?

7 Use your calculator to evaluate:

a $\dfrac{1}{5} + \dfrac{1}{6} + \dfrac{1}{7}$
b $\dfrac{1}{8} + \dfrac{7}{10} - \dfrac{29}{30}$

MULTIPLICATION

To **multiply** two fractions, we multiply the two numerators to get the new numerator and multiply the two denominators to get the new denominator.

$$\dfrac{a}{b} \times \dfrac{c}{d} = \dfrac{a \times c}{b \times d}$$

To help make multiplication easier, we can **cancel** any **common factors** in the numerator and denominator *before* we multiply.

Example 7 ◀)) Self Tutor

Find:

a $\dfrac{1}{4} \times \dfrac{2}{3}$
b $1\dfrac{1}{3} \times \dfrac{3}{5}$
c $\left(3\dfrac{1}{2}\right)^2$

a $\dfrac{1}{4} \times \dfrac{2}{3}$
$= \dfrac{1}{{}_2\cancel{4}} \times \dfrac{\cancel{2}^1}{3}$
$= \dfrac{1}{6}$

b $1\dfrac{1}{3} \times \dfrac{3}{5}$
$= \dfrac{4}{{}_1\cancel{3}} \times \dfrac{\cancel{3}^1}{5}$
$= \dfrac{4}{5}$

c $\left(3\dfrac{1}{2}\right)^2$
$= 3\dfrac{1}{2} \times 3\dfrac{1}{2}$
$= \dfrac{7}{2} \times \dfrac{7}{2}$
$= \dfrac{49}{4}$ or $12\dfrac{1}{4}$

Write mixed numbers as improper fractions before multiplying.

DIVISION

> To **divide** by a fraction, we multiply by its **reciprocal**.

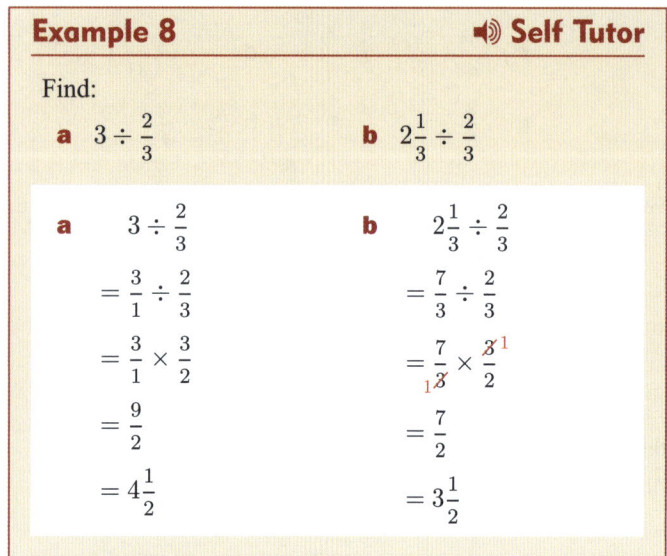

Example 8 — Self Tutor

Find:

a $3 \div \frac{2}{3}$ **b** $2\frac{1}{3} \div \frac{2}{3}$

a $\quad 3 \div \frac{2}{3}$

$= \frac{3}{1} \div \frac{2}{3}$

$= \frac{3}{1} \times \frac{3}{2}$

$= \frac{9}{2}$

$= 4\frac{1}{2}$

b $\quad 2\frac{1}{3} \div \frac{2}{3}$

$= \frac{7}{3} \div \frac{2}{3}$

$= \frac{7}{\cancel{3}_1} \times \frac{\cancel{3}^1}{2}$

$= \frac{7}{2}$

$= 3\frac{1}{2}$

The reciprocal of $\frac{c}{d}$ is $\frac{d}{c}$.
The product of a fraction and its reciprocal is 1.

All of these operations with fractions can be performed on your calculator. Click on the icon for instructions.

CALCULATOR INSTRUCTIONS

EXERCISE 2B.2

1 Find:

a $\frac{1}{3} \times \frac{2}{5}$ **b** $\frac{1}{4} \times \frac{3}{2}$ **c** $\frac{2}{5} \times \frac{2}{3}$ **d** $\frac{2}{3} \times \frac{3}{2}$

e $\frac{2}{3} \times \frac{4}{5}$ **f** $\frac{3}{4} \times \frac{5}{7}$ **g** $4 \times \frac{3}{5}$ **h** $\frac{1}{4} \times 5$

i $1\frac{1}{2} \times \frac{2}{7}$ **j** $1\frac{5}{8} \times \frac{3}{10}$ **k** $\left(\frac{1}{2}\right)^2$ **l** $\frac{1}{2} \times \frac{5}{7} \times \frac{3}{4}$

m $\left(2\frac{1}{5}\right)^2$ **n** $\left(\frac{1}{3}\right)^2 \times 6$ **o** $2\frac{3}{5} \times \frac{5}{4} \times \frac{1}{2}$ **p** $1\frac{1}{2} \times 2\frac{2}{3}$

q $\left(\frac{2}{3}\right)^3$ **r** $1\frac{2}{7} \times 3\frac{1}{3}$ **s** $\frac{3}{4} \times 2\frac{4}{5}$ **t** $8 \times \left(\frac{1}{2}\right)^3$

2 Evaluate:

a $\frac{3}{2} \div \frac{1}{2}$ **b** $\frac{3}{4} \div 2$ **c** $\frac{2}{3} \div \frac{1}{3}$ **d** $\frac{4}{5} \div \frac{1}{2}$

e $\frac{3}{4} \div \frac{5}{7}$ **f** $\frac{1}{3} \div \frac{1}{8}$ **g** $\frac{11}{12} \div \frac{2}{3}$ **h** $\frac{5}{6} \div 3$

i $8 \div \frac{4}{5}$ **j** $1\frac{1}{3} \div \frac{3}{4}$ **k** $\frac{3}{5} \div 1\frac{1}{2}$ **l** $1\frac{2}{3} \div 2\frac{3}{4}$

3 Find:

a $-\frac{4}{5} \times \frac{2}{3}$ **b** $2\frac{1}{4} \times \left(-\frac{1}{3}\right)$ **c** $-\frac{3}{7} \div \frac{5}{6}$ **d** $-2\frac{1}{3} \div \left(-\frac{4}{5}\right)$

4 Calculate:

a $3\frac{3}{7} \times 1\frac{2}{5}$
b $\left(\frac{3}{4}\right)^4$
c $7 - 6 \times \frac{3}{4}$

d $\frac{4}{5} \times 1\frac{1}{2} \div 3$
e $\frac{8 \times 3 \times \frac{1}{3}}{\frac{2}{3}}$
f $1 \div \left(\frac{1}{4} + \frac{2}{3}\right)$

g $1 \div \frac{1}{4} + \frac{2}{3}$
h $\frac{3 - \frac{1}{2}}{3 \times \frac{5}{3}}$
i $\frac{2}{3} + \frac{1}{3} \times 1\frac{1}{2}$

j $\frac{5}{6} \times \frac{4}{5} - \frac{1}{15}$
k $\frac{1}{3} + \frac{1}{3} \div \frac{1}{5} + \frac{3}{5}$
l $1\frac{1}{2} - 2\frac{1}{3} \div 1\frac{2}{3}$

m $12 - \frac{2}{7} \times 3\frac{1}{2}$
n $1\frac{1}{3} + \frac{5}{6} \div \frac{11}{12}$
o $6\frac{2}{5} - \frac{1}{4} \times 1\frac{1}{3} \div \frac{1}{6}$

Check your answers using your calculator.

Remember to use BEDMAS!

Example 9 ◀) Self Tutor

During a season, Joe hit $\frac{2}{5}$ of the home runs for his team.

How many home runs did he hit if there were 40 scored in total?

Joe hit $\frac{2}{5}$ of $40 = \frac{2}{5} \times 40$

$= \frac{2 \times \cancel{40}^{8}}{\cancel{5}_{1}}$

$= 16$ home runs.

The word "of" indicates that we should multiply the numbers.

5 Find:

a $\frac{2}{5}$ of 50
b $\frac{3}{8}$ of 24
c $\frac{5}{7}$ of 35

d $\frac{1}{4}$ of £60
e $\frac{5}{6}$ of 30 m
f $\frac{7}{10}$ of 80 kg

6 Lee is travelling 700 km from Barcelona to Monaco. He stops at Marseille, having travelled $\frac{7}{10}$ of the way. How far has Lee travelled?

7 $\frac{2}{5}$ of the money raised at a charity event is given to the local hospital. The hospital spends $\frac{3}{4}$ of their money on a new X-ray machine. What fraction of the total money raised by the charity was spent on the X-ray machine?

8 240 boys and 300 girls at a school were asked whether they owned a pet. $\frac{1}{3}$ of the children said yes, and $\frac{2}{5}$ of the boys said yes. What fraction of the girls said yes?

9 Trevor ate $\frac{1}{9}$ of a lasagne. Eleanor ate $\frac{1}{6}$ of what remained.

a What fraction of the lasagne did Eleanor eat?
b What fraction of the lasagne now remains?

10 Tina uses $\frac{2}{3}$ tablespoon of butter to make an apricot loaf. How many loaves can she make with 4 tablespoons of butter?

11 Emma buys two identical bottles of shampoo. She uses $\frac{3}{8}$ of one of them at home, and uses $\frac{7}{10}$ of all of the remaining shampoo while she is on holidays. In total, how much shampoo remains?

C DECIMAL NUMBERS

We use **decimal numbers** to display fractions of whole numbers. A decimal number contains a decimal point which separates the whole number part from the fraction.

For example:

- 4.63 means $4 + \frac{6}{10} + \frac{3}{100}$. This number can also be written as the **improper fraction** $\frac{463}{100}$ or as the **mixed number** $4\frac{63}{100}$.

- 14.062 means $14 + \frac{6}{100} + \frac{2}{1000}$.

Expansions of decimals like those above are referred to as **expanded form**.

PLACING DECIMALS ON A NUMBER LINE

To represent decimal numbers on a number line, we choose the scale according to the lowest decimal place value.

For example, when representing $\{-1.8, -1.5, -0.9, -0.2, 0.4, 1.7\}$, the lowest place value is tenths. We divide each whole into tenths.

Example 10

a Write 5.704 in expanded form.

b Write $3 + \frac{2}{10} + \frac{4}{100} + \frac{1}{10\,000}$ in decimal form.

c State the value of the digit 6 in 0.036 24.

d State the numbers A and B on the number line:

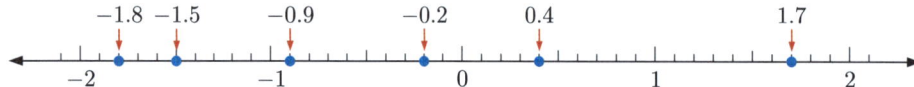

a $5.704 = 5 + \frac{7}{10} + \frac{4}{1000}$

b $3 + \frac{2}{10} + \frac{4}{100} + \frac{1}{10\,000} = 3.2401$

c In 0.036 24, the 6 stands for $\frac{6}{1000}$.

d A is -1.16. B is -0.98.

Real numbers (Chapter 2) 45

Example 11   ◀)) Self Tutor

Write as fractions in simplest form:
a 0.6 **b** 0.045

a $0.6 = \dfrac{6}{10} = \dfrac{6 \div 2}{10 \div 2} = \dfrac{3}{5}$

b $0.045 = \dfrac{45}{1000} = \dfrac{45 \div 5}{1000 \div 5} = \dfrac{9}{200}$

We can convert many fractions to decimals by first writing the fraction so its denominator is a power of 10.

Example 12   ◀)) Self Tutor

Write the following as a decimal:
a $\dfrac{3}{5}$ **b** $\dfrac{7}{25}$ **c** $\dfrac{5}{8}$

a $\dfrac{3}{5} = \dfrac{3 \times 2}{5 \times 2} = \dfrac{6}{10} = 0.6$

b $\dfrac{7}{25} = \dfrac{7 \times 4}{25 \times 4} = \dfrac{28}{100} = 0.28$

c $\dfrac{5}{8} = \dfrac{5 \times 125}{8 \times 125} = \dfrac{625}{1000} = 0.625$

EXERCISE 2C

1 Write in expanded form:
  a 1.2   **b** 1.02   **c** 1.0234   **d** 9.0909   **e** 0.0382

2 Write in decimal form:
  a $4 + \dfrac{5}{10}$ **b** $2 + \dfrac{6}{10} + \dfrac{9}{100}$ **c** $\dfrac{5}{10} + \dfrac{2}{1000}$
  d $\dfrac{3}{100} + \dfrac{8}{1000}$ **e** $2 + \dfrac{5}{1000}$ **f** $4 + \dfrac{1}{100} + \dfrac{4}{10\,000}$

3 State the value of the digit 7:
  a 7295   **b** 571   **c** 0.724   **d** 0.078   **e** 0.000 237

4 Write as fractions in simplest form:
  a 0.2   **b** 0.17   **c** 0.74   **d** 0.04
  e 0.025   **f** 0.008   **g** 0.625   **h** −0.8

5 Place these decimal numbers on a number line:
  a 0.4, 0.7, 1.1, 1.5   **b** 0.5, −0.3, −0.6, 1.2, −1.4   **c** 0.1, 0.35, 0.6, 0.95

6 State the numbers A and B on each number line:

7 Write as a decimal:
   a $\dfrac{9}{10}$ b $\dfrac{17}{20}$ c $\dfrac{2}{5}$ d $\dfrac{3}{25}$ e $\dfrac{31}{50}$ f $\dfrac{3}{8}$

8 Write in ascending order:
   a 0.032, 0.302, 0.32, 0.023, 0.203
   b 1.53, 3.15, 1.35, 1.503, 1.035
   c 0.45, $\dfrac{2}{5}$, 0.6, $\dfrac{1}{2}$, 0.54

9 Write in descending order:
   a 0.604, 0.64, 0.064, 0.406, 0.4006
   b 0.2, $\dfrac{7}{20}$, 0.16, $\dfrac{3}{25}$, 0.261

D OPERATIONS WITH DECIMAL NUMBERS

ADDING AND SUBTRACTING DECIMAL NUMBERS

To add or subtract decimal numbers, we write the numbers under one another so that the decimal points line up. We then add or subtract as we do with whole numbers.

Example 13   ◀) Self Tutor

Find $15.3 + 9.26$.

$$\begin{array}{r} 15.3\,0 \\ +\ \ 9.26 \\ \hline {}^{1} \\ 24.56 \end{array}$$

The zero at the end of 15.3 is added so the numbers have the same number of decimal places.

Example 14   ◀) Self Tutor

Find:
   a $4.632 - 1.507$
   b $8 - 0.706$

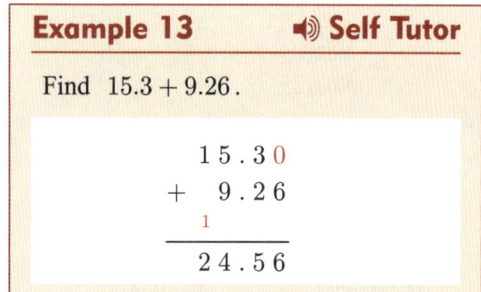

Place the decimal points directly under one another and subtract as for whole numbers.

We insert .000 after the 8 so we have the same number of decimal places in both numbers.

All of the decimal points must line up.

EXERCISE 2D.1

1 Find:
- **a** $2.31 + 4.57$
- **b** $4.9 + 6.3$
- **c** $14.2 + 8.64$
- **d** $19.876 + 5.1$
- **e** $3.2 + 5.1 + 4.8$
- **f** $21.3 + 4.78 + 3.9$
- **g** $3.92 + 4.076$
- **h** $22.019 + 9.3042$
- **i** $37.9 + 1.576 + 0.49$

2 Find:
- **a** $7.84 - 3.22$
- **b** $9.8 - 5.36$
- **c** $11.58 - 3.7$
- **d** $17 - 4.85$
- **e** $14 - 9.227$
- **f** $26.702 - 18.8$
- **g** $47.08 - 3.942$
- **h** $1.9 - 0.137$
- **i** $0.032 - 0.00579$

3 For an overseas holiday, Tony has packed a suitcase weighing 11.6 kg, and a backpack weighing 6.75 kg. Find the total weight of Tony's luggage.

4 In a long jump competition, Samantha jumped 5.23 m. This was 1.37 m further than Theresa's jump. How far did Theresa jump?

5 At the supermarket, Margaret bought a tub of margarine for £3.49, a carton of milk for £1.86, and a bottle of sunscreen for £12.90.
- **a** How much did Margaret spend?
- **b** How much change will she receive from a £20 note?

MULTIPLYING AND DIVIDING BY POWERS OF 10

When **multiplying** by a power of 10, we shift the decimal point to the **right**.

When **dividing** by a power of 10, we shift the decimal point to the **left**.

For example: $0.38 \times 10 = 3.8$ $0.38 \div 10 = 0.038$

$0.38 \times 100 = 38$ $00.38 \div 100 = 0.0038$

MULTIPLICATION AND DIVISION OF DECIMAL NUMBERS

To **multiply** decimal numbers, we multiply the numbers as though they were whole numbers, then divide by the appropriate power of 10.

Example 15 ◀) Self Tutor

Find 0.3×0.15.

$$\begin{aligned} 0.3 \times 0.15 &= (3 \div 10) \times (15 \div 100) \\ &= (3 \times 15) \div (10 \times 100) \\ &= 45 \div 1000 \\ &= 0.045 \quad \text{\{shifting the decimal 3 places left\}} \end{aligned}$$

To **divide** decimal numbers, we write the division as a fraction, then multiply the numerator and denominator by the same power of 10 to make the denominator a whole number. We then perform the division.

48 Real numbers (Chapter 2)

Example 16 — Self Tutor

Find: **a** $2.7 \div 0.3$ **b** $0.002 \div 0.08$

a $2.7 \div 0.3$
$= \dfrac{2.7 \times 10}{0.3 \times 10}$
$= \dfrac{27}{3}$
$= 9$

b $0.002 \div 0.08$
$= \dfrac{0.002 \times 100}{0.08 \times 100}$
$= \dfrac{0.2}{8}$
$= 0.025$

$$8 \overline{) 0.2\,0^{4}0 } \quad 0.0\,2\,5$$

EXERCISE 2D.2

1 Find:
- **a** 5.7×10
- **b** 0.046×100
- **c** $0.06 \times 10\,000$
- **d** $0.000\,01 \times 1000$
- **e** $7 \div 100$
- **f** $3.2 \div 1000$
- **g** $0.022 \div 10$
- **h** $0.091 \div 10\,000$
- **i** $2.005 \div 10\,000$

2 Find:
- **a** 6×0.8
- **b** 0.7×9
- **c** 0.9×0.4
- **d** 1.2×0.7
- **e** 0.6×0.15
- **f** 0.08×0.11
- **g** -0.5×12
- **h** $0.3 \times (-0.4)$
- **i** $(-0.9) \times (-0.8)$

3 Find:
- **a** $1.6 \div 0.2$
- **b** $4 \div 0.8$
- **c** $0.56 \div 0.07$
- **d** $2 \div 0.05$
- **e** $0.03 \div 0.2$
- **f** $0.006 \div 0.03$
- **g** $0.0008 \div 0.05$
- **h** $-1.7 \div 0.1$
- **i** $(-4.2) \div (-0.7)$

4 Dominic drank 2.2 litres of water each day for 8 days. How much water did he drink in this time?

5 A mother duck weighs 2.8 kg. Her baby duckling weighs 0.7 kg. How many times heavier is the mother than her baby?

6 Each kilogram of beef contains 0.8 grams of cholesterol. How many grams of cholesterol are in a 0.250 kg serving of beef?

7 A drinks machine contains 90 litres of water. How many 0.3 litre cups of water can be filled from the machine?

8 Bernard buys 4 bottles of water costing £1.35 each, and 3 fruit bars costing £1.15 each. How much money has Bernard spent altogether?

9 Find, using your calculator:
- **a** $(4.9 + 7.6) \times 12.3$
- **b** $4.9 + 7.6 \times 12.3$
- **c** $93 \div 9 - 6$
- **d** $96 \div (9 - 6)$
- **e** $\dfrac{13.27}{6.4 + 12.8}$
- **f** $\dfrac{11.4}{3.1} - \dfrac{12.5}{7.7}$
- **g** $\dfrac{68.7 - 35.5}{2.8} + 11.3$
- **h** $\dfrac{6.208 - 0.97}{16.41 + 4.232}$

Real numbers (Chapter 2) 49

E PERCENTAGE

We use **percentages** to compare a portion with a whole amount. The whole amount is represented by 100%, which has the value 1.

% reads **per cent**, which means "**in every hundred**".

For example:

- 10% means "10 in every 100", or $\frac{10}{100}$
- 25% means "25 in every 100", or $\frac{25}{100}$.

$$x\% = \frac{x}{100}$$

To convert a percentage into a fraction or a decimal, we divide by 100%.
To convert a fraction or a decimal into a percentage, we multiply by 100%.

EXERCISE 2E.1

1 Copy and complete:

a $6\% = \frac{\square}{100}$ **b** $51\% = \frac{\square}{100}$ **c** $27\% = \frac{\square}{100}$ **d** $86\% = \frac{\square}{100}$

Example 17   ◄)) Self Tutor

Write as a fraction in simplest form:

a 40%   **b** 150%   **c** $12\frac{1}{2}\%$

a 40%
$= \frac{40}{100}$
$= \frac{40 \div 20}{100 \div 20}$
$= \frac{2}{5}$

b 150%
$= \frac{150}{100}$
$= \frac{150 \div 50}{100 \div 50}$
$= \frac{3}{2}$
$= 1\frac{1}{2}$

c $12\frac{1}{2}\%$
$= \frac{12\frac{1}{2} \times 2}{100 \times 2}$
$= \frac{25 \div 25}{200 \div 25}$
$= \frac{1}{8}$

2 Write as a fraction in simplest form:

a 25%   **b** 130%   **c** 65%   **d** 40%
e 210%   **f** 100%   **g** 12%   **h** 2%
i $22\frac{1}{2}\%$ **j** 2.5%   **k** $77\frac{1}{2}\%$ **l** $62\frac{1}{4}\%$

Example 18 — Self Tutor

Write as a decimal:

a 43% **b** $12\frac{1}{2}\%$

a 43%
$= \frac{43}{100}$
$= 0.43$

b $12\frac{1}{2}\%$
$= \frac{12.5}{100}$
$= 0.125$

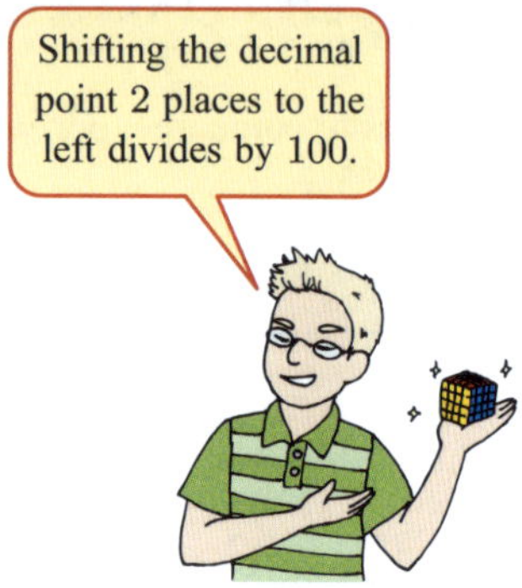

Shifting the decimal point 2 places to the left divides by 100.

3 Write as a decimal:

a 66% **b** 29% **c** 50% **d** 75%
e 180% **f** 205% **g** 300% **h** 128%
i 0.01% **j** 0.3% **k** $10\frac{1}{2}\%$ **l** $56\frac{1}{4}\%$

Example 19 — Self Tutor

Write as a percentage:

a 0.042 **b** $\frac{3}{5}$

a 0.042
$= 0.042 \times 100\%$
$= 4.2\%$

b $\frac{3}{5}$
$= \frac{3}{5} \times 100\%$
$= 60\%$

100% = 1

4 Write as a percentage:

a 0.17 **b** 0.55 **c** 0.09 **d** 0.8
e 0.04 **f** 2 **g** 0.4 **h** 3.5
i 2.05 **j** 3.64 **k** 0.088 **l** 1.409

5 Write as a percentage:

a $\frac{1}{4}$ **b** $\frac{3}{10}$ **c** $\frac{7}{20}$ **d** $\frac{11}{25}$
e $\frac{16}{10}$ **f** $\frac{27}{50}$ **g** $\frac{19}{40}$ **h** $\frac{18}{60}$

Example 20 — Self Tutor

Sarah scored 17 marks out of 20 for her test.

Write her mark as a percentage.

$17 \text{ out of } 20 = \frac{17}{20}$
$= \frac{17}{20} \times 100\%$
$= 85\%$

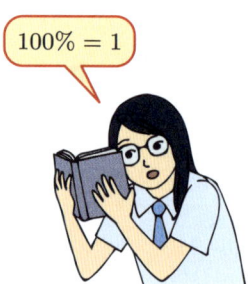

6 Write as a percentage:

  a 30 cm out of 60 cm
  b 14 minutes out of 25 minutes
  c 36 marks out of 40 marks
  d 21 kg out of 30 kg
  e 13 putts out of 20 attempts
  f 66 marks out of 70 marks.

7 Last month Erika was late for her bus 3 days out of 22. On what percentage of the days was she on time?

8 Phyllis sells new cars. Her sales quota for last month was 35 cars. She had a very good month, selling 43 cars. Write this as a percentage of the quota.

Make sure the quantities have the *same units*.

9 Pierre had £5 in his pocket until he spent 85 pence on some sweets. What percentage of his money did Pierre spend?

Example 21 ◀) Self Tutor

Find:
a 10% of 70
b 75% of 80

a 10% of 70
$= \frac{1}{10}$ of 70
$= \frac{1}{10} \times 70$
$= 7$

b 75% of 80
$= \frac{3}{4}$ of 80
$= \frac{3}{4} \times 80$
$= 60$

Some percentages can be written as simple fractions.

10 Find:

  a 10% of 40
  b 50% of 30
  c 25% of 36
  d 20% of 45
  e 70% of 50
  f 75% of 200
  g 60% of 15
  h 150% of 20

Example 22 ◀) Self Tutor

Find:
a 35% of £25 000
b 108% of 5 kg

a 35% of £25 000
$= 0.35 \times £25\,000$
$= £8750$

b 108% of 5 kg
$= 1.08 \times 5$ kg
$= 5.4$ kg

Remember that "of" means "×".

11 Find:

  a 60% of £8
  b 25% of £64
  c 40% of 12 litres
  d 150% of 35 kg

12 Li scored 75% in her test out of 32. What mark did she score?

13 John scored 70% for an examination out of 120. How many marks did he score?

14 A mobile phone manufacturer claims that their new batteries will allow phones to operate for 128% of the current duration. If Lucinda's present battery lasts for 60 hours, for how long should her phone operate with a new battery?

15 Paula receives 5.5% of the income from a book she helped to write. If the income this year was £38 700, how much will she receive?

16 A restaurant charges 7.5% service on the total food bill. Jack and Jill buy food and wine to the value of £54. How much will they have to pay:
  a for service
  b in total?

17 George has played in 250 cricket test matches. In 32% of them his team won, in 24% of them his team lost, and the rest were draws. How many of George's test matches were draws?

18 Julie wants to buy a laptop worth £700. She has already saved £160. Each week she earns £150 from a part-time job, and she saves 30% of her pay. How long will it take Julie to save enough for the laptop?

UNITARY METHOD FOR PERCENTAGE

Jorge and Marcus own a holiday apartment which they rent out. Marcus receives 30% of the income each month. Last month he received £627. How can Marcus work out the total income made from renting out the apartment last month?

In this situation we know a portion of a total amount, and need to work out what the total amount is. To do this, we first work out 1% of the total amount.

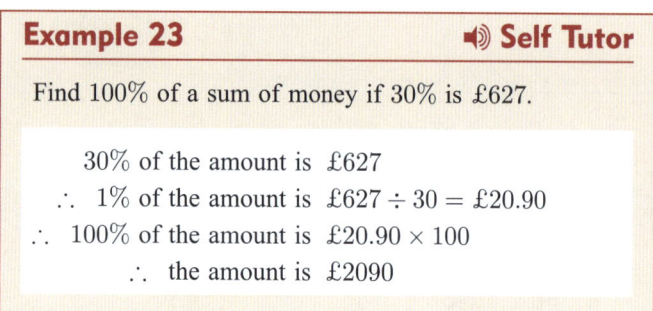

The final answer can be found in one step by multiplying £627 by $\frac{100}{30}$.

EXERCISE 2E.2

1 Find the total amount if:
  a 30% is £48
  b 16% is 2.56 litres
  c 11% is 143 ml
  d 13% is 416 kg
  e 87% is £1131
  f 95% is 399 km
  g 22% is 77 litres
  h 36% is 252 kg
  i 63% is £1323

Example 24 ◆) Self Tutor

Find 60% of a sum of money if 14% is £7280.

14% of the amount is £7280
∴ 1% of the amount is £7280 ÷ 14 = £520
∴ 60% is £520 × 60
∴ 60% is £31 200

2 Find:

    a 40% if 9% is £117   **b** 72% if 6% is 96 kg   **c** 6% if 48% is £630

    d 8% if 95% is 1235 ml   **e** 90% if 6% is 14 kg   **f** 4% if 75% is £465

Example 25   🔊 Self Tutor

82% of the crowd at a basketball match support the Lakers. If there are 24 026 Lakers supporters at the match, how many people are in attendance?

    82% of the crowd is 24 026 people

∴   1% of the crowd is 24 026 ÷ 82 = 293 people

∴ 100% of the crowd is 293 × 100 = 29 300 people

∴ 29 300 people attended the match.

3 A survey showed that 39% of cars passing through a city intersection had only one occupant. If this was 663 cars, how many cars were surveyed?

4 35% of the proceeds from a concert was given to charity. If £26 425 was given to charity, find the total proceeds from the concert.

5 144 girls attend the local school. This is 48% of all of the students. Find the total number of students who attend the school.

Investigation   Terminating and recurring decimals

When we convert a fraction to decimal form, the result may either be a **terminating decimal** or a **recurring decimal**.

> A **terminating decimal** has only a finite number of non-zero digits after the decimal place.

For example, 4.256 is a terminating decimal, as it finishes or terminates after 3 decimal places.

> **Recurring decimals** repeat the same sequence of digits without stopping.

We indicate a recurring decimal with a dot over the repeated digits. If there are more than two repeated digits, we put a dot over the first and last repeated digits.

For example:

- $\frac{8}{11} = 0.727\,272\,72.... = 0.\dot{7}\dot{2}$

- $\frac{88}{111} = 0.792\,792\,792.... = 0.\dot{7}9\dot{2}$

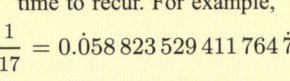

Some decimals take a long time to recur. For example,

$\frac{1}{17} = 0.\dot{0}58\,823\,529\,411\,764\,\dot{7}$

What to do:

1 Write the following decimals in recurring decimal form:
  a 0.2222....
  b 0.353 535....
  c 0.477 77....
  d 1.258 585 8....
  e 0.145 145 145....
  f 0.320 920 920 9....

2 Convert each fraction to a terminating decimal:
  a $\frac{9}{10}$
  b $\frac{1}{4}$
  c $\frac{2}{5}$
  d $\frac{17}{20}$
  e $\frac{3}{4}$
  f $\frac{3}{25}$
  g $\frac{31}{50}$
  h $\frac{3}{8}$
  i $\frac{7}{8}$
  j $\frac{1}{16}$

3 Convert each fraction to a recurring decimal:
  a $\frac{1}{3}$
  b $\frac{7}{9}$
  c $\frac{4}{11}$
  d $\frac{1}{6}$
  e $\frac{4}{7}$

4 Use your calculator to write each fraction as a recurring decimal:
  a $\frac{6}{37}$
  b $\frac{20}{27}$
  c $\frac{13}{30}$
  d $\frac{15}{22}$
  e $\frac{123}{185}$

5 All of the fractions in **2** to **4** are given in lowest terms. Look carefully at the denominators of the fractions.
  a What prime factors do the denominators have, which result in terminating decimals?
  b What prime factors do the denominators have, which result in recurring decimals?
  c Copy and complete:
   When written in lowest terms, a fraction will convert to:
   • a **terminating decimal** if its denominator has no prime factors other than ...... or ......
   • a **recurring decimal** if its denominator has at least one prime factor other than ...... or ......

6 Oliver notices that $\frac{9}{15}$ has a prime factor of 3 in the denominator, so he thinks the fraction will convert to a recurring decimal.
  However, when he enters it into his calculator, he finds that $\frac{9}{15} = 0.6$.
  Can you explain Oliver's mistake?

7 Given that $0.\dot{3} = \frac{1}{3}$, what can we say about $0.\dot{9}$?

Review set 2A

1 Place the fractions $\frac{3}{4}, -\frac{1}{4}, -\frac{7}{4}, 2\frac{1}{4}$ on a number line.

2 Express in lowest terms:
  a $\frac{3}{18}$
  b $\frac{16}{40}$
  c $\frac{49}{28}$
  d $\frac{42}{46}$
  e $\frac{-8}{36}$

3 Find:
  a $\frac{1}{6} - \frac{1}{4}$
  b $7 - 2.36$
  c $1\frac{1}{4} \times \frac{5}{9}$
  d 0.012×0.6
  e $9 \div 2\frac{1}{3}$
  f $0.0028 \div 0.4$

4 State the value of the digit 5 in:

  a 25.7 **b** 1.252 **c** 0.0005 **d** 4.502

5 $\frac{7}{8}$ of Yousef's garden beds are used to grow fruit and vegetables. Of these, $\frac{1}{4}$ of the growing space is taken up by aubergines. What fraction of Yousef's garden beds is taken up by aubergines?

6 At Gary The Greengrocer, watermelons are sold in halves and quarters, as well as whole melons.

  a Gary has $7\frac{1}{2}$ melons that he wishes to cut into quarter melons. How many quarter melons will he have?

  b The sales of watermelons one day were:

whole melons	5
half melons	9
quarter melons	13

  How many watermelons did Gary sell in total?

7 Find:

  a $\frac{3}{8}$ of 32 m **b** $\frac{4}{5}$ of £38 **c** $\frac{7}{12}$ of 20 kg

8 Find:

  a $2 \div \frac{1}{3} - \frac{1}{2}$ **b** $2 \div \left(\frac{1}{3} - \frac{1}{2}\right)$ **c** $2 - \frac{1}{3} \div \frac{1}{2}$ **d** $\left(2 - \frac{1}{3}\right) \div \frac{1}{2}$

9 Write as a fraction in simplest form:

  a 0.31 **b** 0.56 **c** 0.375 **d** −0.45

10 Write as a decimal:

  a $\frac{4}{5}$ **b** $\frac{19}{20}$ **c** $\frac{11}{25}$ **d** $\frac{27}{50}$

11 Write as a decimal:

  a 83% **b** 27.4% **c** 152% **d** 0.4%

12 Write as a percentage:

  a 0.6 **b** $\frac{3}{4}$ **c** 0.08 **d** 2

13 Jillian has repaid £700 of a £1500 debt. What percentage of the debt has been repaid?

14 From a class of 30 students, 6 had visited the Great Wall of China. What percentage of the class had visited the Great Wall?

15 Henri has memorised 60% of a piece of music. If the piece of music takes 5 minutes to play, how long is the section that Henri has memorised?

16 A hospital has 36 doctors, who together make up 24% of the total hospital staff. How many people work at the hospital?

Review set 2B

1 Place the decimals 0.2, 0.6, −0.4, 1.1, −1.3 on a number line.

2 Find:

a $3\frac{2}{5} + 1\frac{2}{3}$ b $0.38 - 0.195$ c 0.64×3.1

3 Simplify:

a $\dfrac{29 - 8}{10 - (6 - 3)}$ b $\dfrac{-7 - 23}{5 - (2 \times 4)}$ c $\dfrac{156 - 24}{(2 - 4) \times 2}$

4 Find the total cost of buying 0.9 kg of bananas priced at £1.30 per kilogram, and 0.75 kg of pears priced at £0.96 per kilogram.

5 Find:

a $3.6 \div 6$ b $3.6 \div 0.6$ c $3.6 \div 0.06$ d $0.36 \div 0.6$

6 Find:

a $2\frac{1}{2} - 1\frac{3}{4}$ b $\frac{7}{4} + \frac{2}{3} - \frac{1}{2}$ c $\frac{5}{3} \times \frac{9}{10}$ d $\frac{3}{8} \div \frac{5}{4}$

7 Ken had $4\frac{1}{2}$ cans of paint left in his shed. After painting his house, he only had $1\frac{2}{3}$ cans of paint left. How much paint did Ken use?

8 Katy baked a lemon slice on Saturday. She ate $\frac{1}{6}$ of it, and her greedy brother ate $\frac{3}{5}$.

a What fraction of the slice has been eaten?

b Express the portion Katy ate as a fraction of the portion her brother ate.

9 Bettina buys 4 entrance tickets to a medieval festival, costing £7.95 each. How much change will she get from £40?

10 Find, using your calculator:

a $(3.21 - 1.7) \div 2.2$ b $\dfrac{4.53 + 2.06}{1.3 \times 3.25}$

11 Write as a fraction in lowest terms:

a 48%   b 15%   c $5\frac{1}{2}$%   d 0.1%

12 Vince brought 25 pumpkins to the market, and sold 18 of them. What percentage of his pumpkins did Vince sell?

13 An elephant eats 5% of its body weight in vegetation each day. Find the weight of vegetation eaten each day by:

a a 3000 kg elephant.   b a 5000 kg elephant.

14 Express as a percentage:

a 75 marks out of 80   b £65 out of £104

15 In a Mathematics competition, the top 0.3% of participants are awarded a prize. Last year 600 000 students took part. How many students were awarded a prize?

16 Anwen scored 82% in her last test. Her friend Bree scored 76%, which was 38 marks.

a How many marks did Anwen score?   b How many marks were in the test?

3

Units

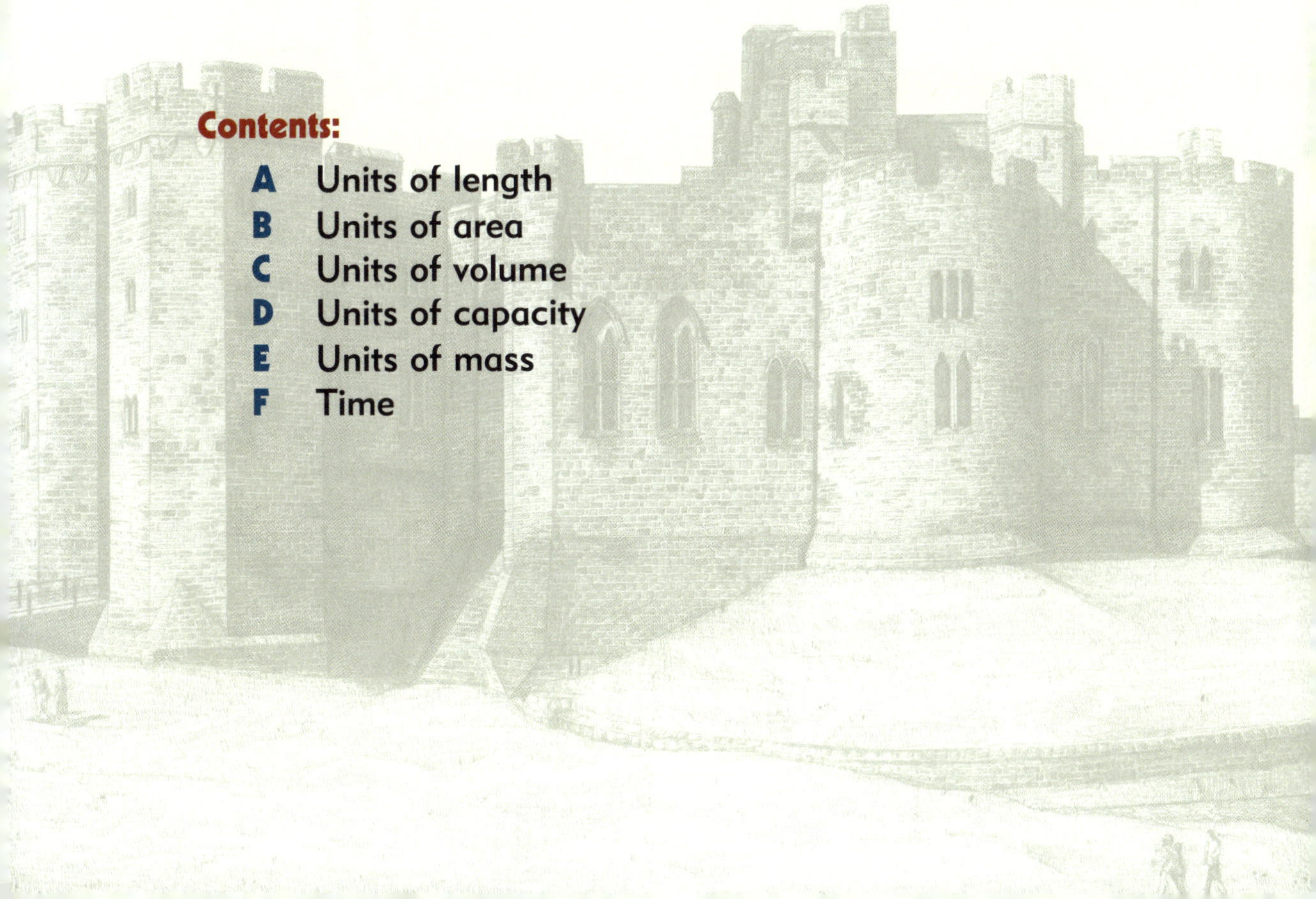

Contents:
- A   Units of length
- B   Units of area
- C   Units of volume
- D   Units of capacity
- E   Units of mass
- F   Time

Opening problem

A fountain in a public square is 2.8 m high. The base of the fountain covers an area of 30 m², and is made from concrete 20 cm thick. The mass of the fountain is 3.6 t, and the fountain can hold a maximum of 1.2 kl of water.

Things to think about:

a What units of measurement were used to describe the fountain?

b Why were different length units used to describe the height of the fountain and the width of the concrete?

c What is the mass of the fountain in kilograms?

d There is currently 800 litres of water in the fountain. How much more water can the fountain hold?

Discussion

Write down five quantities which we commonly *measure*.

What *devices* do we use to measure these quantities?

What *units* are these quantities measured in?

INTERNATIONAL SYSTEM (SI) UNITS

Historical note

The decimal **Metric** system was created at the time of the French Revolution.

Having decided that a new unit of length, the metre, should be one ten millionth of the distance from the North Pole to the Equator, **Pierre Méchain** and **Jean-Baptiste Delambre** set about surveying the 1000 km section of the meridian arc from Dunkirk to Barcelona.

Two platinum bars were deposited in the Archives de la République in Paris in 1799, defining the standard metre and standard kilogram.

Pierre Méchain

Jean-Baptiste Delambre

The **International System of Units**, abbreviated SI from the French *Le Système International d'Unités*, is the world's most widely used system of measurement.

It is founded on seven base units:

Quantity	Name	Symbol
Distance	metre	m
Mass	kilogram	kg
Time	second	s
Electric current	ampere	A

Quantity	Name	Symbol
Temperature	kelvin	K
Intensity of light	candela	cd
Amount of substance	mole	mol

Other SI units, called **derived units**, are defined in terms of the base units by multiplying or dividing them. The result is often given its own special name.

Some of the common SI derived units are:

Quantity	Name	Symbol
Area	square metre	m^2
Volume	cubic metre	m^3
Velocity	metres per second	$m\,s^{-1}$
Angle	radian	rad
Force	newton	N

Quantity	Name	Symbol
Pressure	pascal	Pa
Energy	joule	J
Power	watt	W
Frequency	hertz	Hz

When we multiply one unit by another, we leave a short space between the unit symbols.

When we divide one unit by another, we use an oblique line between the unit symbols, or a negative index. For example, we write metres per second as m/s or $m\,s^{-1}$.

In addition to the base and derived units, the SI allows the use of other units, such as:

Quantity	Name	Symbol	SI equivalent
Time	minute	min	60 s
	hour	h	3600 s
Mass	tonne	t	1000 kg
Capacity	litre	l	$0.001\ m^3$
Area	hectare	ha	$10\,000\ m^2$
Angle	degree	°	$\frac{\pi}{180}$ rad
Temperature	degree Celsius	°C	$K - 273.15$
Pressure	millibar	mb	100 Pa
Distance at sea	Nautical mile	Nm	1.852 km
Speed at sea	Knot	kn	$1.852\ km\,h^{-1}$
Energy	Kilowatt hour	kWh	3.6 MJ

The symbol for litre can be l or L depending on which country you are in. To avoid confusion with the number 1, we will write out the word litre in full.

Smaller or larger multiples of units can be obtained by combining the unit with a particular prefix.

For example:
- "kilo" (k) represents 1000
- "centi" (c) represents $\frac{1}{100}$
- "milli" (m) represents $\frac{1}{1000}$.

For more information on SI units, visit   www.bipm.org/en/publications/si-brochure/

> **Discussion**
>
> - Why is it important to have an international system of units?
> - Which of the units in the tables on the previous page have you seen before? Discuss which units you use that are *not* SI units.
> - For the units you have *not* seen before, discuss what they measure.

In this Chapter we will consider units for length, area, volume, capacity, mass, and time.

A   UNITS OF LENGTH

In the United Kingdom, we commonly use measures of length from both the **SI** or **metric system**, and also the older **imperial system**.

For example, we might talk about:
- the distance between Chester and Conwy in *miles*
- the length and width of a sports ground in *metres* or *yards*
- the length of a dress in *centimetres*
- the length of a desk in *millimetres*
- drill bits, bolts, nails, screws, and pipe threads in both imperial and metric units
- altitude for flying and skydiving in *feet*
- the diameter of a car wheel rim in *inches*.

METRIC LENGTH UNITS

The most common metric length units are connected as follows:

1 kilometre (km) = 1000 metres (m)
1 metre (m) = 100 centimetres (cm)
1 centimetre (cm) = 10 millimetres (mm)

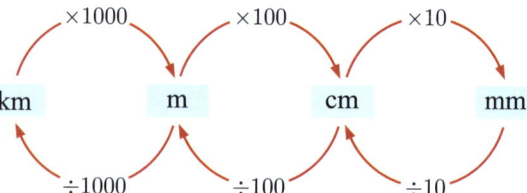

Notice that when we convert from:
- larger units to smaller units we **multiply**
- smaller units to larger units we **divide**.

Units (Chapter 3)

Example 1 ◀)) Self Tutor

Convert:

a 39.4 m to cm   **b** 567 mm to cm   **c** 0.618 m to mm.

a   39.4 m
    $= (39.4 \times 100)$ cm
    $= 3940$ cm

b   567 mm
    $= (567 \div 10)$ cm
    $= 56.7$ cm

c   0.618 m
    $= (0.618 \times 100)$ cm
    $= 61.8$ cm
    $= (61.8 \times 10)$ mm
    $= 618$ mm

EXERCISE 3A.1

1 Suggest an appropriate unit of length for measuring:
    a the length of a baby   **b** the width of an eraser   **c** the height of an oak tree
    d the length of an ant   **e** the length of a pen.

2 Convert:
    a 52 km to m   **b** 115 cm to mm   **c** 1.65 m to cm
    d 6.3 m to mm   **e** 0.625 km to cm   **f** 8.1 km to mm.

3 Convert:
    a 480 cm to m   **b** 54 mm to cm   **c** 5280 m to km
    d 2000 mm to m   **e** 580 000 cm to km   **f** 7 000 000 mm to km.

4 Convert:
    a 42.1 km to m   **b** 210 cm to m   **c** 75 mm to cm
    d 1500 m to km   **e** 1.85 m to cm   **f** 42.5 cm to mm
    g 2.8 km to cm   **h** 16 500 mm to m   **i** 0.25 km to mm.

5 Mike ran 8.5 km while Mal walked 3200 m. How much farther did Mike run than Mal walk?

6 I have 45 coils of fencing wire and each coil is 275 m long. How many kilometres of wire do I have?

7 John has 3.85 km of copper wire. He is setting his machine to cut it into 1.5 cm lengths to be used in electric toasters. How many lengths will be made?

Example 2 ◀)) Self Tutor

Write 7 mm as a percentage of 4 cm.

7 mm out of 4 cm = 7 mm out of 40 mm
                $= \frac{7}{40} \times 100\%$
                $= 17.5\%$

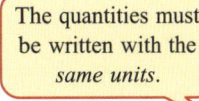

The quantities must be written with the *same units*.

8 Write as a percentage:
    a 48 cm out of 2 m   **b** 6 mm out of 8 cm
    c 800 m out of 7 km   **d** 92 m out of 2.1 km

IMPERIAL LENGTH UNITS

In the imperial system, the most commonly used units for length are **inches**, **feet**, **yards**, and **miles**.

- 1 inch = 2.54 cm

  A paper clip is approximately 1 inch long.

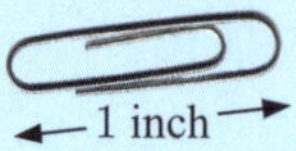

- 1 foot = 30.48 cm

  Your ruler is approximately 1 foot long.

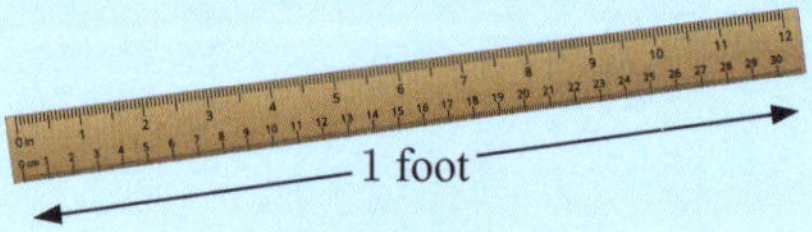

- 1 yard = 91.44 cm or 0.9144 m

  The length of your stride is approximately 1 yard.

- 1 mile ≈ 1609 m or 1.609 km

  An athletics track is 400 m long. So, 4 laps of an athletics track is approximately 1 mile.

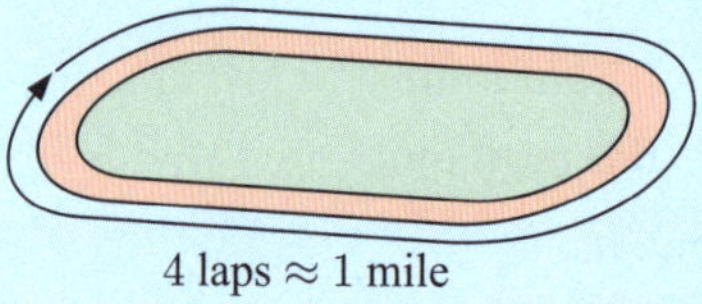

Example 3

Convert:

a 3 inches to cm **b** 5 feet to cm **c** 12 miles to km.

a 3 inches
= (3 × 2.54) cm
= 7.62 cm

b 5 feet
= (5 × 30.48) cm
= 152.4 cm

c 12 miles
≈ (12 × 1.609) km
≈ 19.3 km

EXERCISE 3A.2

1 Convert:

   a 10 inches to cm   **b** 2 feet to cm   **c** 10 feet to m
   d 8 miles to km   **e** 40 yards to m   **f** 76 miles to km.

2 Daniel orders a 6 inch sandwich for lunch. Find the length of his sandwich in centimetres.

3 Brett makes masts for yachts using a new alloy. He received a large order from the USA for masts which are 13 yards long. Find the length of each mast in metres.

4 The distance between Southampton and Brighton is 97 km. Given that 1 km ≈ 0.621 miles, write this distance in miles.

Research

In 1891, **Sir Hugh Munro** published a table of mountains in Scotland with height over 3000 feet. Each mountain of this height is called a **Munro** in his honour.

What to do:

1. What height in metres does a mountain need to be, to be called a Munro?
2. How many Munros are there in Scotland?
3. What is meant by a *Real Munro*? How many *Real Munros* are there?
4. What is meant by a subsidiary top?

Sir Hugh Munro

B UNITS OF AREA

In the metric system, area is measured in **square millimetres**, **square centimetres**, **square metres**, and **square kilometres**. There is also another unit called a **hectare** (ha) which corresponds to a square 100 m × 100 m.

$$1 \text{ mm}^2 = 1 \text{ mm} \times 1 \text{ mm}$$
$$1 \text{ cm}^2 = 10 \text{ mm} \times 10 \text{ mm} = 100 \text{ mm}^2$$
$$1 \text{ m}^2 = 100 \text{ cm} \times 100 \text{ cm} = 10\,000 \text{ cm}^2$$
$$1 \text{ ha} = 100 \text{ m} \times 100 \text{ m} = 10\,000 \text{ m}^2$$
$$1 \text{ km}^2 = 1000 \text{ m} \times 1000 \text{ m} = 1\,000\,000 \text{ m}^2 \text{ or } 100 \text{ ha}$$

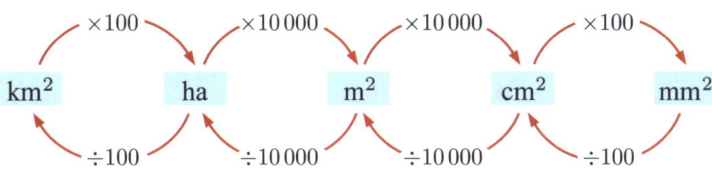

Example 4   ◆) Self Tutor

Convert:
a   6 m² to cm²
b   18 500 m² to ha.

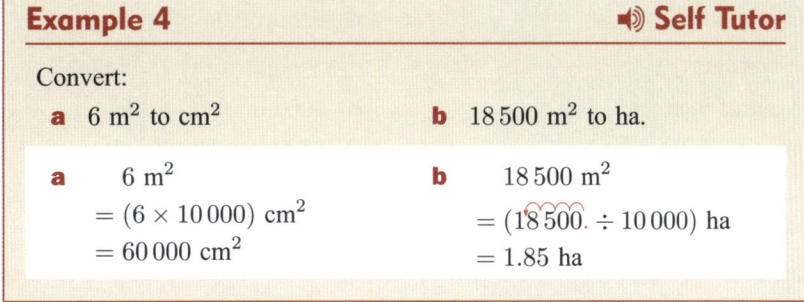

In the imperial system, area is measured using the **acre**.

$$1 \text{ acre} \approx 0.4047 \text{ hectares or } 4047 \text{ m}^2$$

Example 5 ◀) Self Tutor

$1 \text{ m}^2 \approx \dfrac{1}{4047} \text{ acres}$

Convert:

a 76 acres to hectares **b** 965 m² to acres.

a 76 acres
$\approx (76 \times 0.4047)$ hectares
≈ 30.8 ha

b 965 m²
$\approx \dfrac{965}{4047}$ acres
≈ 0.238 acres

Historical note

The Anglo-Saxons defined an acre not just as a particular area, but also with a particular shape. It is about the area which can be ploughed in one day by a pair of yoked oxen pulling a wooden plough.

EXERCISE 3B

1 Suggest an appropriate area unit for measuring:
  a the area of a postage stamp
  b the area of your desktop
  c the area of a vineyard
  d the area of your bedroom floor
  e the area of Wales
  f the area of a toe-nail.

2 Convert:
  a 23 mm² to cm²
  b 3.6 ha to m²
  c 726 cm² to m²
  d 7.6 m² to mm²
  e 8530 m² to ha
  f 0.354 ha to cm²
  g 13.54 cm² to mm²
  h 432 m² to cm²
  i 0.004 82 m² to mm².

3 Convert:
  a 3 km² to m²
  b 0.7 km² to ha
  c 660 ha to m²
  d 660 ha to km²
  e 0.05 m² to cm²
  f 25 cm² to m²
  g 5.2 mm² to cm²
  h 0.72 km² to mm².

4 Convert:
  a 210 acres to hectares
  b 5120 m² to acres
  c 36 hectares to acres
  d 8.2 acres to m².

5 Write as a percentage:
  a 60 mm² out of 5 cm²
  b 3000 m² out of 2 ha
  c 4500 cm² out of 0.8 m²
  d 85 ha out of 1.3 km²

6 Sam purchased 2 m² of material. She needs to cut it into rectangles of area 200 cm² for a patchwork quilt. This can be done with no waste. How many rectangles can Sam cut out?

7 I have purchased a 4.2 ha property. Council regulations allow me to have 5 free range chickens for every 100 m². How many free range chickens am I allowed to have?

8 Erica has just purchased a new property which is a rectangle 128 m by 186 m.
  a Find the area of the property in hectares.
  b The property needs to be sprayed for weeds before Erica can use it for animal grazing. The amount of spray needed is calculated from the land area in acres. Calculate this area for Erica.

C UNITS OF VOLUME

Volume can be measured in **cubic millimetres**, **cubic centimetres**, or **cubic metres**.

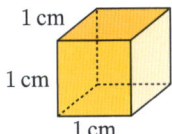

1 cm^3
$= 10 \text{ mm} \times 10 \text{ mm} \times 10 \text{ mm}$
$= 1000 \text{ mm}^3$

Similarly, 1 m^3
$= 100 \text{ cm} \times 100 \text{ cm} \times 100 \text{ cm}$
$= 1\,000\,000 \text{ cm}^3$

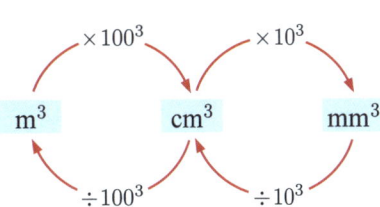

Example 6 ◀) Self Tutor

Convert: **a** 0.481 m^3 to cm^3 **b** $19\,720 \text{ mm}^3$ to cm^3.

a $\quad 0.481 \text{ m}^3$
$= (0.481 \times 100^3) \text{ cm}^3$
$= (0.481 \times 1\,000\,000) \text{ cm}^3$
$= 481\,000 \text{ cm}^3$

b $\quad 19\,720 \text{ mm}^3$
$= (19\,720 \div 10^3) \text{ cm}^3$
$= (19\,720 \div 1000) \text{ cm}^3$
$= 19.72 \text{ cm}^3$

EXERCISE 3C

1 Convert:
  a 0.25 cm^3 to mm^3
  b 0.083 m^3 to cm^3
  c 598 mm^3 to cm^3
  d 9810 cm^3 to m^3
  e $635\,100 \text{ cm}^3$ to m^3
  f 81.5 cm^3 to mm^3.

2 Write as a percentage:
  a 850 mm^3 out of 4 cm^3
  b $30\,000 \text{ cm}^3$ out of 0.7 m^3

3 A pallet of 460 books has total volume 1.2 m^3. Find the volume of each book, in cm^3.

4 A set of 200 dice, each with volume 600 mm^3, fits perfectly inside a box. Find the volume of the box in cm^3.

D UNITS OF CAPACITY

The SI unit for capacity is the **litre**.

1 litre = 1000 millilitres (ml)
1 kilolitre (kl) = 1000 litres
1 megalitre (Ml) = 1000 kilolitres (kl)

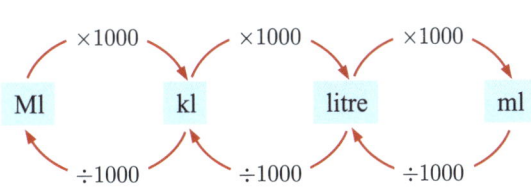

Example 7

Convert:

a 1.56 litres to ml   **b** 790 ml to litres.

a   1.56 litres
$= (1.56 \times 1000)$ ml
$= 1560$ ml

b   790 ml
$= (790 \div 1000)$ litres
$= 0.79$ litres

CONNECTING VOLUME AND CAPACITY

The units for volume and capacity are connected because **1 millilitre (ml)** of fluid fills a container of size 1 cm^3.

We can therefore construct the following table of equivalent units:

$1 \text{ ml} \equiv 1 \text{ cm}^3$
$1 \text{ litre} \equiv 1000 \text{ cm}^3$
$1 \text{ kl} \equiv 1000 \text{ litre} \equiv 1 \text{ m}^3$

$\equiv$ reads "is equivalent to".

EXERCISE 3D

1 Suggest appropriate units of capacity for measuring the amount of water in a:
    a lake   **b** small drink bottle   **c** swimming pool
    d laundry tub   **e** test tube.

2 Convert:
    a 3.76 litres to ml   **b** 47 320 litres to kl   **c** 3.5 kl to litres
    d 0.423 litres to ml   **e** 0.054 kl to ml   **f** 58 340 ml to kl.

3 Write as a percentage:
    a 750 ml out of 4 litres   **b** 600 litres out of 9 kl

4 Rodney has consumed 400 ml of a 1.25 litre soft drink bottle. What percentage of the soft drink remains?

5 A chemist makes up 20 ml bottles of eye drops from a drum of eye drop solution. If the full drum has a capacity of 0.275 kl, how many bottles of eye drop solution can the chemist fill?

6 A shipment of wine is imported from Australia in 1000 litre tanks. There is enough wine to fill 1000 dozen 750 ml bottles. How many tanks are there?

7   **a** State the capacity of a bottle with volume 25 cm^3.
    b Find the volume of a tank with capacity 3200 kl.
    c How many litres are there in a tank with volume 7.32 m^3?

8 The capacity of a car engine is quoted as 1800 cc, meaning 1800 cubic centimetres. Write this capacity in litres.

E UNITS OF MASS

The **mass** of an object is the amount of matter in it.

In the metric system, the base unit of mass is the **kilogram** (kg).

1 kilogram (kg) is the mass of 1 litre of pure water.

The other units we commonly use are the tonne (t), the gram (g), and the milligram (mg).

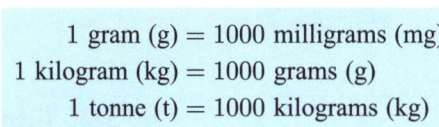

1 gram (g) = 1000 milligrams (mg)
1 kilogram (kg) = 1000 grams (g)
1 tonne (t) = 1000 kilograms (kg)

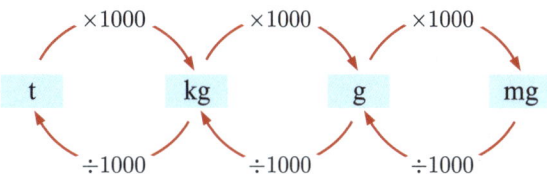

EXERCISE 3E

1 Write down the unit of mass you would choose to measure the mass of:
   a a donkey **b** a grape **c** a textbook
   d a train **e** an ant.

2 Convert:
   a 7 g to mg **b** 7 g to kg **c** 580 g to kg
   d 580 kg to g **e** 56 g to mg **f** 0.45 g to mg
   g 3200 g to kg **h** 1.87 t to kg **i** 47 835 mg to kg
   j 4653 mg to g **k** 2.83 t to g **l** 0.0632 t to g
   m 74 682 g to t **n** 1.7 t to mg **o** 91 275 g to kg.

3 Write 400 kg as a percentage of 2.5 t.

4 Leah ate 250 grams of cake from a cake weighing $1\frac{1}{2}$ kg. What percentage of the cake did she eat?

5 A pack of 500 sheets of paper has mass 2.5 kg. Find the mass of one sheet, giving your answer in the most appropriate unit.

6 A small car has mass 0.96 tonnes. Find the mass, in tonnes, of a car that is 600 kg heavier.

7 In peppermint flavoured lollies, 1 gram of peppermint extract is used per lolly. How many lollies can be made from a drum containing 0.15 t of peppermint extract?

8 A publisher produces a book weighing 856 grams. 6000 of the books are printed and need to be transported to France.
   a How many tonnes of books are to be sent?
   b If the transport costs £450 per tonne, find the total cost of the transport.

F TIME

For thousands of years people measured time by observing the passage of day and night, the stars, and the changes of season. This was important for growing crops and working with animals.

The units of time we use are based on the sun, the moon, and the Earth's rotation.

The base unit of time is the **second**, abbreviated **s**.

The most common units of time are related as follows:

> 1 **minute** = 60 seconds
> 1 **hour** = 60 minutes = 3600 seconds
> 1 **day** = 24 hours
> 1 **week** = 7 days
> 1 **year** = 12 months = $365\frac{1}{4}$ days
> 1 **decade** = 10 years
> 1 **century** = 100 years
> 1 **millennium** = 1000 years

We use h for hours, min for minutes, s for seconds.

Example 8 ◀) Self Tutor

Convert:
- **a** 3 hours 26 minutes 18 seconds into seconds
- **b** 5400 seconds into minutes.

a 3 h 26 min 18 s
$= (3 \times 3600)$ s $+ (26 \times 60)$ s $+ 18$ s
$= 12\,378$ s

b 5400 s
$= (5400 \div 60)$ min
$= 90$ min

EXERCISE 3F.1

1 Convert into seconds:
- **a** 45 minutes
- **b** 38 minutes
- **c** 1 hour 10 minutes
- **d** 3 min 58 s
- **e** 2 h 5 min 28 s
- **f** 5 h 19 min 47 s

2 Convert into minutes:
- **a** $3\frac{1}{2}$ hours
- **b** 1440 seconds
- **c** 5 hours 13 minutes
- **d** 3 days 1 hour 48 minutes

3 Write as a percentage:
- **a** 26 seconds out of 5 minutes
- **b** 9 months out of 2.5 years

4 Find 20% of 3 hours, in minutes.

Example 9 ◀) Self Tutor

Convert 30 240 minutes into days.

30 240 min $= (30\,240 \div 60)$ hours   {60 min in 1 hour}
$= 504$ hours
$= (504 \div 24)$ days   {24 hours in 1 day}
$= 21$ days

5 Convert into days:
- **a** 1248 hours
- **b** 23 040 minutes
- **c** 4 years
- **d** 6 hours

6 Find the number of:
- **a** hours in May
- **b** minutes in December
- **c** seconds in Tuesday.

Units (Chapter 3)

Example 10

Express the following time intervals in years:

a 4 decades   **b** $3\frac{1}{4}$ centuries   **c** 18 millennia

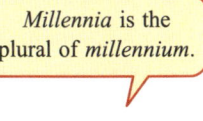
Millennia is the plural of *millennium*.

a 4 decades = 4 × 10 years   {10 years in 1 decade}
            = 40 years

b $3\frac{1}{4}$ centuries = 3.25 × 100 years   {100 years in 1 century}
              = 325 years

c 18 millennia = 18 × 1000 years   {1000 years in 1 millennium}
             = 18 000 years

7 Express in years:

    a 9 decades   **b** 7 centuries   **c** 2 millennia

    d $5\frac{1}{2}$ centuries   **e** $2\frac{1}{2}$ decades   **f** $3\frac{3}{4}$ millennia

8 Which is larger?

    a 38 decades or 4 centuries   **b** 572 centuries or 5 millennia

9 Historical events were first recorded about 5115 years ago. Express this in:

    a centuries   **b** millennia.

Example 11

Calculate the following, expressing your answer in hours, minutes, and seconds:

a 2 h 41 min + 3 h 38 min   **b** 4 h 19 min − 2 h 37 min

a   2 h 41 min
    + 3 h 38 min
    ―――――――
      5 h 79 min   Since 60 min = 1 h, the total is 6 h 19 min.

b $\overset{3}{\cancel{4}}$ h $\overset{79}{\cancel{19}}$ min
    − 2 h 37 min
    ―――――――
      1 h 42 min

10 Calculate the following, expressing your answers in hours, minutes, and seconds:

    a 1 h 19 min + 2 h 42 min + 1 h 7 min   **b** 4 h 51 min 16 s + 2 h 19 min 54 s

    c 12 h − 7 h 55 min   **d** 5 h 23 min − 2 h 48 min

Example 12

What is the time difference between 11:43 am and 3:18 pm?

           11:43 am to 12 noon =   17 min
           12 noon to 3 pm = 3 h
           3 pm to 3:18 pm =   18 min
                                  ――――
    ∴ the time difference is   3 h 35 min

11 Find the time difference between:
   a 4:30 am and 6:55 am
   b 10:08 am and 5:52 pm
   c 3:15 pm and 9:03 pm
   d 7:54 am and 2:29 pm
   e 2:30 am and 7:20 am
   f 10:14 am and 1:51 pm
   g 5:18 pm and 11:32 pm
   h 3:42 pm and 6:08 am the next day.

12 Joseph caught the 7:54 am train into town, arriving in the main station at 8:47 am. It took him 16 minutes to walk from the station to work.
   a How long was Joseph's train journey?
   b At what time did Joseph arrive at work?

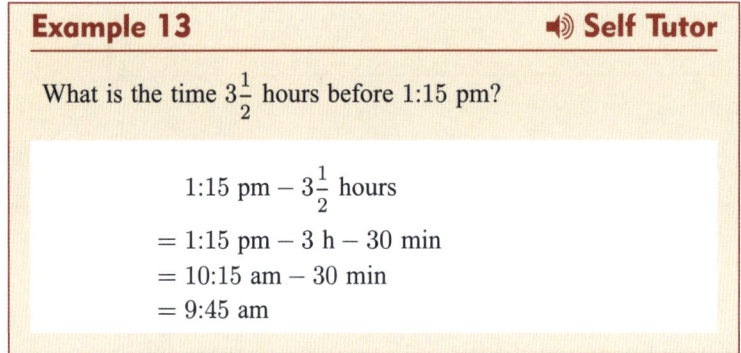

Example 13 ◀) **Self Tutor**

What is the time $3\frac{1}{2}$ hours before 1:15 pm?

$1:15 \text{ pm} - 3\frac{1}{2} \text{ hours}$
$= 1:15 \text{ pm} - 3 \text{ h} - 30 \text{ min}$
$= 10:15 \text{ am} - 30 \text{ min}$
$= 9:45 \text{ am}$

13 Find the time:
   a 3 hours after 7:15 am
   b $5\frac{1}{2}$ hours before 10:26 am
   c $4\frac{1}{2}$ hours after 11:50 am
   d $6\frac{1}{2}$ hours before 2:35 am
   e 3 hours 17 minutes after 9:15 am
   f 4 hours 35 minutes before 1:05 am.

14 I left for work $1\frac{1}{4}$ hours after I woke up. If I left at 8:05 am, at what time did I wake up?

15 My brother is overseas, and he telephoned me at 3:47 am. I was very angry and told him I would ring him back when I woke up in the morning. If I woke up at 7:04 am, how long did my brother have to wait for the return call?

16 Matthew and Ferdinand each captain yachts racing around an island. The race starts and finishes in the same place. The boats leave the start at 2 pm on January 11.
   a Matthew takes 2 days 9 hours 17 minutes to finish the race. When does he finish?
   b Ferdinand finished on January 14 at 1:14 am. How long did it take him to finish the race?

24-HOUR TIME

When a digital clock displays the time ![7:15], this could mean 7:15 am or 7:15 pm. To avoid this problem, we can use **24-hour time**.

> **24-hour time** indicates the amount of time which has passed since midnight.

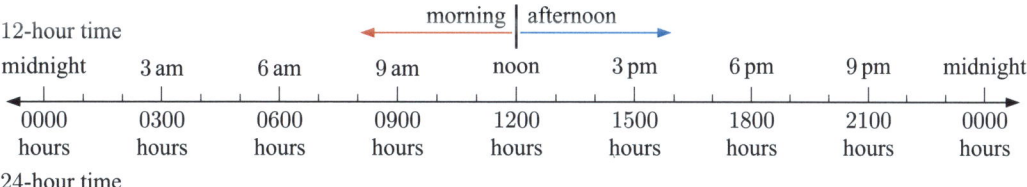

A 24-hour clock would display ![07:15] for 7:15 am, and ![19:15] for 7:15 pm.

We use **four digit notation** when writing 24-hour time.

6:05 am appears as ![06:05], and is written as 0605 hours.

8:35 pm appears as ![20:35], and is written as 2035 hours.

In the table below we compare 12-hour time and 24-hour time.

12-*hour time*	Digital display	24-*hour time*
midnight	00:00	0000 hours
7:42 am	07:42	0742 hours
midday (noon)	12:00	1200 hours
11:29 pm	23:29	2329 hours

24-hour time always uses 4 digits.

Notice that:
- Morning times (am) are from midnight (0000) to midday (1200).
- Afternoon times (pm) are from midday (1200) to midnight (0000).
- Midnight is 0000, not 2400.
- To convert afternoon 24-hour time to 12-hour time, we subtract 1200.

Example 14 ◀)) Self Tutor

Write:
  a 0341 in 12-hour time.   **b** 7:16 pm in 24-hour time.

  a 0341 is 3:41 am.   **b** 7:16 pm is 1916.

EXERCISE 3F.2

1 Write in 24-hour time:
- **a** 9:57 am
- **b** 11:06 am
- **c** 4 o'clock pm
- **d** 2:25 pm
- **e** 8 o'clock am
- **f** 1:06 am
- **g** 8:58 pm
- **h** noon
- **i** 2 minutes past midnight

2 Write in 12-hour time:
- **a** 1140
- **b** 0346
- **c** 1634
- **d** 1900
- **e** 0800
- **f** 2330
- **g** 1223
- **h** 2040

3 Find the time difference between:
- **a** 0930 and 1210
- **b** 0815 and 1740
- **c** 1118 and 1408
- **d** 1326 and 1951
- **e** 1740 and 0505 the next day.

4 Tomorrow, sunrise will be at 0537 and sunset will be at 1918. For how long will the sun be up tomorrow?

5 Explain why 24-hour time is more commonly used than 12-hour time for train and airline timetables.

Historical note

The earliest inventions for measuring time included the sundial, the hourglass, and the waterclock or *clepsydra*.

Over the centuries many different devices were made to measure time more accurately, eventually leading to the watches and clocks we use today. The most accurate clock in the world, the caesium fountain atomic clock, is inaccurate by only one second every 20 million years.

Review set 3A

1 Convert:
- **a** 3.28 km to m
- **b** 755 mm to cm
- **c** 32 cm to m.

2 Convert:
- **a** 1950 mm² to cm²
- **b** 6.4 m² to cm²
- **c** 2 500 000 m² to km².

3 Convert:
- **a** 2600 mm³ to cm³
- **b** 8 000 000 cm³ to m³
- **c** 1.2 m³ to cm³
- **d** 56 cm³ to ml
- **e** 4000 cm³ to litres
- **f** 2.7 litres to cm³.

4 Convert:
  a 56 mg to g
  b 450 g to kg
  c 0.25 t to kg.

5 A staple is made from a piece of wire 3 cm long. How many staples can be made from a roll of wire 1200 m long?

6 A truck with a load limit of 4 tonnes is required to deliver 9000 bricks to a building site. If one brick has a mass of 1.25 kilograms, how many trips will be required?

7 Convert:
  a 2.5 litres to ml
  b 4 Ml to litres
  c 38 litres to kl.

8 Answer the **Opening Problem** on page **58**.

9 Convert into seconds:
  a 30 minutes
  b 4 minutes 47 seconds
  c 2 hours 17 minutes

10 Find the number of hours in August.

11 Find the time difference between:
  a 5:30 am and 10:12 am
  b 11:16 am and 7:41 pm.

12 "Prometheus", one of the world's oldest trees, was believed to be 4844 years old when it was cut down in 1964. Express the age of the tree in:
  a centuries
  b millennia.

13 Kendall's Mathematics exam begins at 1:50 pm, and goes for $2\frac{1}{2}$ hours. At what time does the exam end?

14 Write in 24-hour time:
  a 6:09 am
  b 1:19 pm
  c 10:46 pm

15 Find the time difference between 0845 and 1732.

Review set 3B

1 Convert:
  a 1560 m to km
  b 26.5 cm to mm
  c 1.8 m to cm.

2 Mika ran 5 laps of a 1500 metre circuit. How many kilometres did he run?

3 Sven walks 475 metres to the underground station, then travels 3.5 km on the train. What percentage of the *total* distance Sven travels is on foot?

4 Because he has a cough, Dien has to take 15 ml of cough medicine 3 times a day for 10 days. How much medicine will be left if he buys a 0.5 litre bottle?

5 Convert:
  a 350 mg to g
  b 250 kg to t
  c 16.8 kg to g.

6 Convert:
  a 5.4 cm^2 to mm^2
  b 56 000 cm^2 to m^2
  c 0.8 km^2 to m^2
  d 0.6 cm^3 to mm^3
  e 0.018 m^3 to cm^3
  f 25 000 000 mm^3 to m^3.

7. How many 600 ml cartons of milk could be filled from a tank containing 5.4 kl of milk?

8. How many cubic metres of molten metal would be required to fill 8000 moulds each with 150 cm³ of metal?

9. A winery is situated on a property with area 125 acres. Write this area in:
   a hectares
   b m².

10. A truck was loaded with 300 lawnmowers, each weighing 25 kg. Find the total weight of this load, in tonnes.

11. Mons Meg is a massive cannon built in 1449 by Philip the Good, Duke of Burgundy. It is kept in Edinburgh Castle, Scotland.

    a Mons Meg weighs 6.97 tonnes. Write this mass in kilograms.
    b The length of the cannon is 15 feet. Write this length in metres.
    c The ball diameter (calibre) of the cannon is 20 inches. Write this measurement in centimetres.

12. Convert into days:
    a 360 hours
    b 31 680 minutes
    c 8 years

13. Find the time that is:
    a 7 hours after 10:20 am
    b $3\frac{1}{2}$ hours before 2:07 pm.

14. Aniko left for work at 7:39 am and returned home at 6:43 pm. How long was she away from home?

15. Write in 12-hour time:
    a 0926
    b 1540
    c 2146

16. Elise's friend is coming to visit at 7 pm. The clock in Elise's kitchen currently reads 1624. How long will it be before Elise's friend arrives?

4

Approximation and estimation

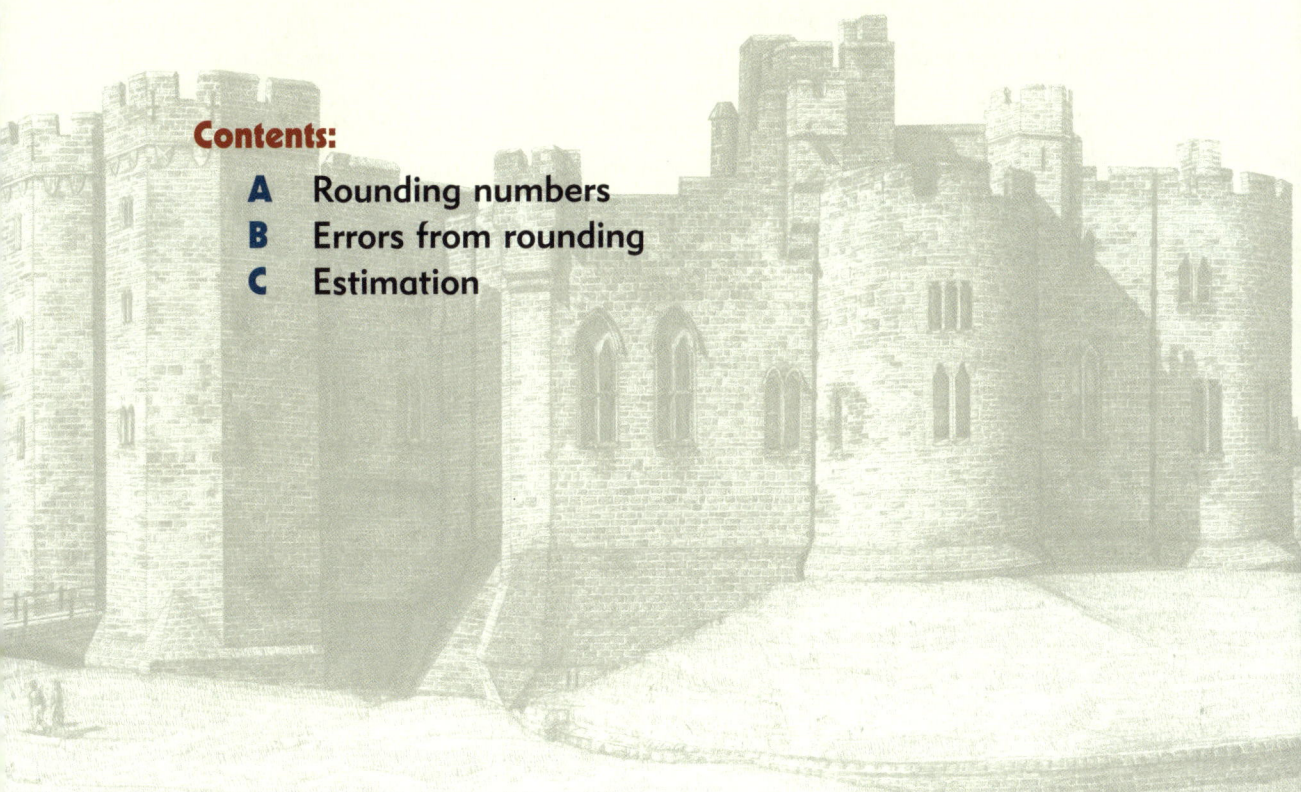

Contents:
- **A** Rounding numbers
- **B** Errors from rounding
- **C** Estimation

Opening problem

In a motorway traffic survey, it was found that 1852 cars carried 4376 people.

Things to think about:

a Can you write the *average* number of people in each car exactly?

b If you were asked to quote the average number of people in each car for a news article, how many *significant figures* would you report? Explain your answer.

c The motorway carries about 12 000 000 cars every year. Estimate the total number of passengers in a year.

Often we do not need to know the exact value of a number, but rather we want a reasonable **estimate** or **approximation** of it.

For example, suppose there are 1517 competitors at the interschools athletics carnival. An estimate of 1500 students gives a good idea of the size of the carnival.

Discussion

What are sensible approximations for:

- 229 students in Year 10
- 2788 students in a school
- 87 066 spectators at the tennis championships this year
- 6 277 168 population at the last census?

A ROUNDING NUMBERS

ROUNDING TO A POWER OF 10

We can round off numbers to the nearest power of ten.

For example, we can round off to the nearest 10, 100, or 1000.

468 is closer to 470 than to 460, so to round to the nearest 10, we **round up** to 470.

463 is closer to 460 than to 470, so to round to the nearest 10, we **round down** to 460.

We use the symbol $\approx$ to mean "is approximately equal to".

So, $468 \approx 470$ and $463 \approx 460$.

When a number is **halfway** between tens we **always round up**. For example, $465 \approx 470$.

ROUNDING TO A NUMBER OF FIGURES

The first **significant figure** of a decimal number is the first (left-most) non-zero digit.

We round to a number of significant figures if we believe this number of digits is important.

For example, to round 34 827 to **two** significant figures, we notice that 34 827 is closer to 35 000 than it is to 34 000. So, 34 827 ≈ 35 000 (to 2 significant figures).

> The rules for rounding off are:
> - If the digit **after** the one being rounded off is **less than 5** (0, 1, 2, 3, or 4), then we **round down**.
> - If the digit **after** the one being rounded off is **5 or more** (5, 6, 7, 8, or 9), then we **round up**.

Example 1 ◀)) Self Tutor

Round off:
- **a** 286 to the nearest 10
- **b** 19 439 to the nearest 100
- **c** 319 to one significant figure
- **d** 3850 to two significant figures.

- **a** 286 ≈ 290   {to the nearest 10}
- **b** 19 439 ≈ 19 400   {to the nearest 100}
- **c** 319 ≈ 300   {to one significant figure}
- **d** 3850 ≈ 3900   {to two significant figures}

EXERCISE 4A.1

1 Round off to the nearest 10:
- **a** 23
- **b** 65
- **c** 68
- **d** 97
- **e** 347
- **f** 561
- **g** 409
- **h** 598

2 Round off to the nearest 100:
- **a** 81
- **b** 671
- **c** 617
- **d** 850
- **e** 349
- **f** 982
- **g** 13 429
- **h** 10 074

3 Round off to the nearest 1000:
- **a** 3015
- **b** 2856
- **c** 8885
- **d** 12 095
- **e** 9995
- **f** 30 905
- **g** 49 895
- **h** 91 500

4 Round off to 3 significant figures:
- **a** 27 461
- **b** 6822
- **c** 704 023
- **d** 21 085 100

5 Round off to the number of significant figures shown in brackets:
- **a** 437   [2]
- **b** 2064   [2]
- **c** 264 183   [3]
- **d** 29 999   [3]
- **e** 48 501   [2]
- **f** 9614   [1]
- **g** 43 188   [3]
- **h** 2 141 097   [4]
- **i** 106 000   [2]
- **j** 370 006   [3]
- **k** 7 206 930   [4]
- **l** 4 008 000   [3]

6 Round off:
- **a** £187 to the nearest £10
- **b** £18 745 to the nearest £1000
- **c** 375 km to the nearest 10 km
- **d** 785 ft to the nearest 100 ft
- **e** a town population of 29 295 to the nearest one thousand
- **f** 995 cm to the nearest metre
- **g** 8945 litres to the nearest kilolitre
- **h** a house price of £274 950 to the nearest £10 000
- **i** the 491 560 sheep on a farm to the nearest 100 000.

7 During the 2017 migration, 187 430 265 swallows flew from Africa to the United Kingdom. Round this number to:

   a 2 significant figures   **b** 3 significant figures.

8 According to a census in 2011, the population in Scotland was 5 295 403. Round this number to:

   a the nearest 1000   **b** 3 significant figures.

ROUNDING DECIMAL NUMBERS

We are often given measurements as decimal numbers. We usually **approximate** the decimal by **rounding off** to the required accuracy.

We have previously seen how to round off whole numbers.

For example: 2185

        ≈ 2190 {to the nearest 10}

        ≈ 2200 {to the nearest 100}

        ≈ 2000 {to the nearest 1000}

We sometimes write 2 d.p. to mean "to 2 decimal places".

We round off decimal numbers in the same way.

For example: 2.1852

        ≈ 2.185 {to 3 decimal places}

        ≈ 2.19 {to 2 decimal places}

        ≈ 2.2 {to 1 decimal place}

        ≈ 2 {to the nearest whole number}

Example 2   ◀)) Self Tutor

Round 39.748 to:

  a the nearest whole number   **b** one decimal place   **c** two decimal places

  a $39.748 \approx 40$ to the nearest whole number   **b** $39.748 \approx 39.7$ to one decimal place

  c $39.748 \approx 39.75$ to two decimal places

Discussion

Suppose we need to round 2.9971 to 2 decimal places to include in a report.

- Suppose you write 3.
  - Will people assume the result was exact?
  - How will they know the accuracy of the approximation?
- Why is the appropriate approximation 3.00?
- What would your approximation be if you rounded the number to 2 significant figures instead?

EXERCISE 4A.2

1 Round to the nearest whole number:
  a 0.813   **b** 7.499   **c** 7.500   **d** 11.674   **e** 128.437

2 Write to 1 decimal place:
  a 2.43   **b** 3.57   **c** 4.92   **d** 6.38   **e** 4.275

3 Write to 2 decimal places:
  a 4.236   **b** 2.731   **c** 5.625   **d** 4.377   **e** 6.5237

4 Write 0.183 75 correct to:
  a 1 decimal place   **b** 2 decimal places   **c** 3 decimal places   **d** 4 decimal places.

5 Round correct to the number of *significant figures* shown in brackets:
  a 42.3   [2]   **b** 6.237   [3]   **c** 0.0462   [2]
  d 0.2461   [2]   **e** 10.27   [3]   **f** 0.999   [2]
  g 0.037 642   [4]   **h** 0.007 639   [2]   **i** 69.7003   [2]
  j 0.005 904   [2]   **k** 3.000 451   [3]   **l** 0.000 345 6   [2]

Example 3   ◀)) Self Tutor

Find $\frac{2}{7}$ correct to 3 decimal places.

$$\begin{array}{r} 0.2\ 8\ 5\ 7 \\ \hline 7 \overline{\smash{\big)}\ 2.0\ ^60\ ^40\ ^50} \end{array} \qquad \therefore \ \frac{2}{7} \approx 0.286$$

6 Find, correct to the number of *decimal places* shown in square brackets:
  a $\frac{17}{4}$ [1]   **b** $\frac{73}{8}$ [2]   **c** 0.3×2.6 [1]
  d 0.12×0.4 [2]   **e** $\frac{8}{11}$ [2]   **f** 0.08×0.03 [3]
  g $(0.7)^2$ [1]   **h** $\frac{37}{6}$ [2]   **i** $\frac{17}{7}$ [3]

7 Use your calculator to find, to 2 decimal places:
  a $\frac{3.675 + 11.291}{5.67}$ **b** $\frac{17.65}{3 - 0.271}$ **c** $\pi \times (5.67)^2$
  d 21.7×18.29 **e** $\frac{21.7}{18.29}$ **f** $(13.29)^2 \times 15.67$
  g $\frac{(16.2)^3}{5.71 \times 3.68}$ **h** $\frac{1}{3} \times \pi \times (6.92)^2$ **i** $\sqrt{\frac{11}{2.77}}$

8 In her maths exam, Julie was asked to round 7.45 cm to one decimal place and to the nearest whole number.
Julie's answer was that 7.45 cm ≈ 7.5 cm (to one decimal place), and that 7.5 cm ≈ 8 cm (to the nearest whole number).
Explain what Julie has done wrong.

B ERRORS FROM ROUNDING

Investigation   Measuring devices

Examine a variety of measuring instruments at school and at home. Make a list of the names of these instruments, what they measure, what their units are, and the degree of accuracy to which they can measure.

For example:

This ruler measures length in centimetres and millimetres, and can measure to the nearest millimetre.

When we take measurements, we are usually reading some sort of scale.

The scale of a ruler may have millimetres marked on it, but when we measure the length of an object, it is likely to fall between two divisions. We **approximate** the length of the object by recording the value at the nearest millimetre mark. In doing so our answer may be inaccurate by up to half a millimetre.

> A measurement is accurate to $\pm \frac{1}{2}$ of the smallest division on the scale.

If we know the accuracy to which a number has been recorded, we can determine the range of values in which the exact value lies. This range of values is sometimes called the **error interval**.

Example 4   Self Tutor

Ling uses a ruler to measure the length l of her pencil case. She records the length as 18.7 cm.
Find the range of values in which the length may lie.

18.7 cm is 187 mm, so the measuring device must be accurate to the nearest half mm.

$\therefore$ the range of values is $187 \pm \frac{1}{2}$ mm

The actual length is in the range $186\frac{1}{2}$ mm to $187\frac{1}{2}$ mm.

$\therefore$ 18.65 cm $\leqslant l < 18.75$ cm

$\leqslant$ means "less than or equal to".

Notice that 18.65 cm is included in the range, but 18.75 cm is not included. This is because 18.65 cm is rounded up to 18.7 cm, and 18.75 cm is rounded up to 18.8 cm.

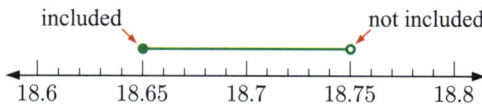

EXERCISE 4B

1 State the accuracy of the following measuring devices:
   a a tape measure marked in cm
   b a measuring cylinder with 1 ml graduations
   c a beaker with 100 ml graduations
   d a set of scales with marks every 500 g.

2 Roni checks his weight every week using scales with 1 kg graduations. This morning he recorded a weight of 68 kg. In what range of values does Roni's actual weight w lie?

3 State the error interval for x given that:
   a when x is rounded to 1 decimal place, the result is 7.3
   b when x is rounded to 2 decimal places, the result is 11.46
   c when x is rounded to 2 significant figures, the result is 280.

4 Tom's digital thermometer said his temperature was 36.4°C. In what range of values did Tom's actual temperature T lie?

5 According to Alex's exercise tracker, Alex walked 7.85 km yesterday. Given that the distance is rounded to 2 decimal places, find the *least* distance Alex could have walked.

6 The newspaper stated that 25 000 people attended a protest march in London.
If this figure had been rounded to two significant figures, what was the largest number of people that could have attended the protest?

7 Four students measured the width of their classroom using the same tape measure. The measurements were 6.1 m, 6.4 m, 6.0 m, 6.1 m.
   a Which measurement is likely to be incorrect?
   b What answer would you give for the width of the classroom?
   c What graduations do you think were on the tape measure?

8 Shirley takes 800 paces to walk from her home to her school. She knows that the length of her pace is 0.7 m, rounded to 1 decimal place. Find the range of possible values for the distance D from Shirley's home to her school.

9 For an overseas flight, Jenny packs a suitcase weighing 12.8 kg, and a backpack weighing 6.7 kg, rounded to 1 decimal place. Find the range of possible values for the total weight w of Jenny's luggage.

10 Dan completed a 400 m race in 52.7 seconds, and Nigel completed the race in 54.3 seconds, rounded to 1 decimal place. Find the range of possible values for the time difference d between the two runners.

11 Ernie needs to fill his fish tank with at least 120 litres of water. He knows that his bucket holds 8.6 litres, rounded to 1 decimal place. What is the minimum number of buckets of water Ernie will need to guarantee his fish tank has enough water?

Example 5

A rectangular block of wood was measured as 78 cm by 24 cm. Find the range of possible values for its perimeter p.

The length of the block could be from $77\frac{1}{2}$ cm to $78\frac{1}{2}$ cm.

The width of the block could be from $23\frac{1}{2}$ cm to $24\frac{1}{2}$ cm.

$\therefore$ the lower bound of the perimeter is $2 \times 77\frac{1}{2} + 2 \times 23\frac{1}{2} = 202$ cm

and the upper bound of the perimeter is $2 \times 78\frac{1}{2} + 2 \times 24\frac{1}{2} = 206$ cm

$\therefore$ 202 cm $\leqslant p <$ 206 cm

12 A rectangular bath mat was measured as 86 cm by 38 cm. Find the range of possible values for its perimeter p.

13 A rectangular garden bed is measured as 252 cm by 143 cm. Find the range of possible values for the total length of edging l required to border the garden bed.

C ESTIMATION

A fast way of estimating a calculation is to perform a **one figure approximation**. We round each number in the calculation to one significant figure, then perform the calculation with these approximations.

Step 1: Leave single digit numbers as they are.
Step 2: Round all other numbers to one significant figure.
Step 3: Perform the calculation.

Example 6

Estimate:
a 57×8 **b** 537×6 **c** 623×69

a 57×8
   $\approx 60 \times 8$
   ≈ 480

b 537×6
   $\approx 500 \times 6$
   ≈ 3000

c 623×69
   $\approx 600 \times 70$
   $\approx 42\,000$

EXERCISE 4C

1 Estimate:
  a 79×4 **b** 47×8 **c** 62×7
  d 88×6 **e** 284×3 **f** 617×7
  g 408×9 **h** 494×6 **i** 2094×7

2 Estimate:
   a 57×42 **b** 73×59 **c** 85×98
   d 275×54 **e** 389×73 **f** 4971×32
   g 3079×29 **h** $40\,989 \times 9$ **i** 880×750

3 Estimate:
   a 4.3×3 **b** 8.9×5 **c** 7×9.6
   d 5.8×2.1 **e** 3.4×6.2 **f** 11.6×28.7

Example 7 — Self Tutor

Estimate $4123 \div 47$.

$$4123 \div 47 \approx 4000 \div 50$$
$$\approx 400 \div 5$$
$$\approx 80$$

4 Estimate:
   a $397 \div 4$ **b** $6849 \div 7$ **c** $79\,095 \div 8$
   d $6000 \div 19$ **e** $80\,000 \div 37$ **f** $18\,700 \div 97$
   g $2780 \div 41$ **h** $48\,097 \div 243$ **i** $798\,450 \div 399$

5 Estimate:
   a $8.3 \div 2$ **b** $9 \div 3.3$ **c** $17.6 \div 3.7$

6 Use estimation only to find which of these calculator answers is reasonable:

   a 489×19 9291   96081   92901
   b 843×74 62382   562382   6238
   c 3907×89 347723   5361243   35723
   d $3132 \div 87$ 3600   36   306

7 a Use your calculator to evaluate $\dfrac{7.96^2 - 4.2 \times 3.86}{2.97}$. Write down your full calculator display.

   b Use an estimation to check that your answer is reasonable.

8 Miki reads 217 words in a minute. Estimate the number of words she can read in one hour.

9 A bricklayer lays 115 bricks each hour. If he works $37\tfrac{1}{2}$ hours each week, estimate how many bricks he will lay in four weeks.

10 Joe can type at 52 words per minute. Estimate the time needed for him to type a document of 3820 words.

11 In a vineyard there are 189 vines in each of 54 rows. Estimate the number of vines in the vineyard.

12 A journey of 1023 km took 19 hours. Estimate the average speed in kilometres per hour.

Example 8

Estimate the cost of 19 pens at £1.95 each.

$$19 \times £1.95 \approx 20 \times £2$$
$$\approx £40$$

13 Estimate the cost of:
- **a** 195 exercise books at 98 pence each
- **b** 27 pot plants at £21 each
- **c** 18 show bags at £8.45 each
- **d** 12 bottles of drink at £2.95 each
- **e** 4 dozen iceblocks at £1.20 each
- **f** 3850 football tickets at £26.50 each.

14
- **a** Estimate the cost of 49 chairs at £19 each.
- **b** Explain whether your answer in **a** is an over-estimate or an under-estimate.

15 To paint and wallpaper a bedroom, Bronwyn needs five rolls of paper at £19.50 each, and one 10 litre tin of paint that costs £68.70. Estimate the total cost of the paint and paper.

16 Andy wants to build a 480 m long post and wire fence. Posts will be 10 m apart and cost £8.75 each. A coil of wire is 100 m long and costs £22.50. Estimate the cost of a four wire fence.

17 Fertiliser costs £250 per tonne to buy and spread. It is spread on fields at a rate of 0.5 tonne per hectare. Estimate the cost to spread fertiliser on a 21.6 ha farm.

Activity

Click on the icon to practise your rounding and estimation skills.

ROUNDING AND ESTIMATION

Review set 4A

1 Round 3579 to the:
- **a** nearest 10
- **b** nearest 100
- **c** nearest 1000.

2 Round off:
- **a** 388 km to the nearest 10 km
- **b** 3501 litres to the nearest kl
- **c** 74 821 students to the nearest 10 000 students.

3 Round:
- **a** £13.68 to the nearest 5 pence
- **b** £13.68 to the nearest pound.

4 Round 28.907 to:
- **a** the nearest whole number
- **b** 1 decimal place
- **c** 2 decimal places.

5 Find $\frac{22}{7}$ correct to:
- **a** 1 decimal place
- **b** 2 decimal places.

6 Use a calculator to find, correct to 2 decimal places:

   a $\pi \times 3.68^2$
   b $\dfrac{48.67}{314.72 \times 0.0766}$
   c $\dfrac{21.66 - 18.79}{21.66 + 18.79}$

7 State the error interval for x given that:

   a when x is rounded to 1 decimal place, the result is 10.3
   b when x is rounded to 3 significant figures, the result is 2.09.

8 A photograph was measured as 15 cm by 10 cm. Find the range of possible values for its perimeter p.

9 **a** How accurate is a tape measure marked in cm?
   b Find the range of possible values for a measurement of 36 cm.

10 Estimate the cost of the following using sensible rounding:

   a 78 notepads at £1.95 each
   b 6 dozen ice creams at £3.95 each.

11 Using 1 figure approximations, find estimates of:

   a 148×6
   b 804×29
   c $3016 \div 58$

12 A tiler can lay 72 tiles each hour and works a 42 hour week.
Estimate the number of tiles laid in a week using 1 figure approximations.

13 Kerosene is sold in drums of capacity 48 litres. Estimate the cost of 97 drums of kerosene costing £1.62 per litre.

Review set 4B

1 Round 4608 to the:

   a nearest 10
   b nearest 100
   c nearest 1000.

2 Round off:

   a 659 km to the nearest 100 km
   b 20 144 tonnes to the nearest 1000 tonnes
   c 156 797 people to the nearest 10 000 people.

3 Round:

   a £69.73 to the nearest 5 pence
   b £172.62 to the nearest pound.

4 Round 55.039 to:

   a the nearest whole number
   b 1 decimal place
   c 2 decimal places.

5 **a** Find $\dfrac{8}{11}$ correct to 3 decimal places.

   b Find 6.8×0.0253 correct to 2 decimal places.

6 Use a calculator to find, correct to 2 decimal places:

   a $\dfrac{\pi}{4}$
   b $\dfrac{12.37 + 63.85}{15.2 \times 1.09}$
   c $\dfrac{1}{3} \times \pi \times (46.73)^2$

7 A set of digital scales displays the weight of Sam's dog as 6.82 kg. Given that the weight is rounded to 2 decimal places, find the range of possible values for the weight w of Sam's dog.

8 Philippa measures the sides of her book as 24.5 cm by 18.5 cm. Find the range of possible values for the perimeter p of the book.

9 Estimate the cost of the following using sensible rounding:
  a 6 magazines at £5.85 each
  b 32 kg of apples at £1.65 per kg.

10 Estimate using 1 figure approximations:
  a 63×9
  b 198×4
  c $1989 \div 42$

11 A company wants to mail advertising to 3065 clients. Each package of advertising takes 2 minutes to prepare for mailing. Estimate how long it would take to prepare the mailing if 3 people were working at once on this task.

12 Laszlo drives 476 miles at an average speed of 65.4 miles h^{-1}. Use your calculator to find the time he takes to drive this distance, correct to the nearest minute.

5

Algebra

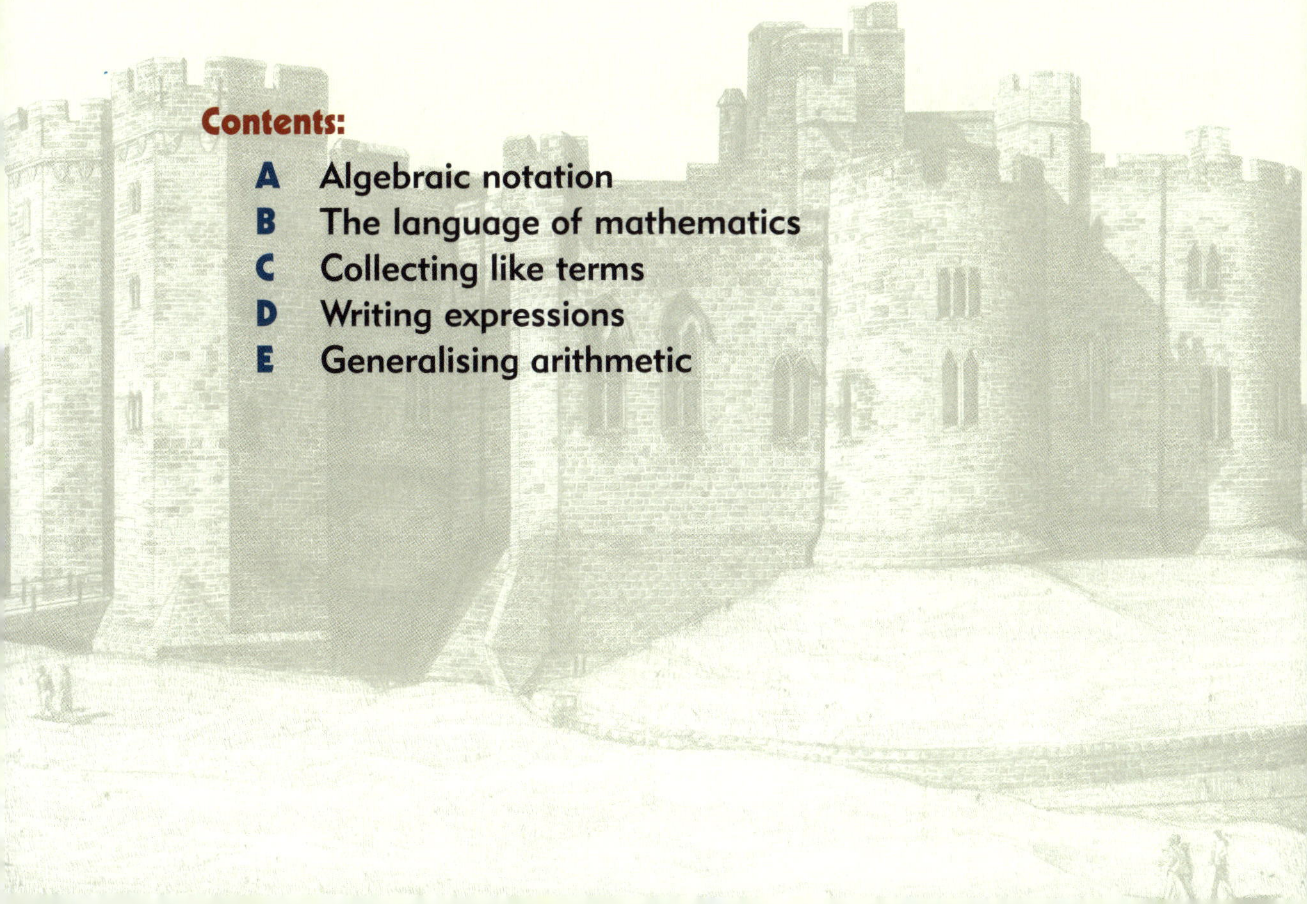

Contents:
- **A** Algebraic notation
- **B** The language of mathematics
- **C** Collecting like terms
- **D** Writing expressions
- **E** Generalising arithmetic

Opening problem

Nadia's brother Tim is 2 years younger than she is. Her mother Melanie is three times Nadia's age. Nadia's father Peter is 4 years older than Melanie.

Things to think about:

a Suppose we use the variable n to represent Nadia's age. How could we represent:
  i Tim's age
  ii Melanie's age
  iii Peter's age?

b If Nadia is 13 years old, how old is Peter?

Algebra is a powerful tool used in mathematics. In algebra we use letters or **pronumerals** to represent unknown values or **variables**. The variables are used in mathematical **expressions**.

For example, the area of any rectangle is found by multiplying its length by its width.

If someone asked you to draw a rectangle, you would probably respond by asking how big the rectangle should be.

The length and width of the rectangle are unknown or variable. We let l represent the length of the rectangle, and w represent its width.

Using algebra, the area of the rectangle can be written as $l \times w$, or lw.

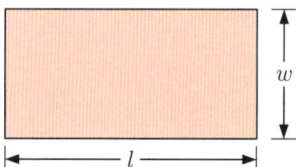

A ALGEBRAIC NOTATION

In algebra the variables are included in expressions as though they were numbers.

However, we obey some rules which help make algebra easier.

In algebra we agree:

- to **leave out** the "×" signs between multiplied quantities
- to write **numerals (numbers) first** in any product
- where products contain two or more letters, we write them in **alphabetical order**.

For example:
- $3b$ is used rather than $3 \times b$ or $b3$
- $3bc$ is used rather than $3cb$.

Example 1   ◀) Self Tutor

Write in product notation:
  a $t \times 6s$
  b $4 \times k + m \times 3$
  c $3 \times (r+s)$

a	$t \times 6s$	**b**	$4 \times k + m \times 3$	**c**	$3 \times (r+s)$
	$= 6st$		$= 4k + 3m$		$= 3(r+s)$

WRITING SUMS AS PRODUCTS

Sums of identical terms can be written using product notation.

For example, $\quad 3 + 3 + 3 + 3 = 4 \times 3 \qquad$ {4 lots of 3}
$\quad \therefore \ b + b + b + b = 4 \times b = 4b \quad$ {4 lots of b}

> **Example 2** ◀)) **Self Tutor**
>
> Simplify:
>
> **a** $\ r + r + r + s + s$ **b** $\ d + d - (a + a + a + a)$
>
> **a** $\quad r + r + r + s + s$
> $\quad = 3r + 2s$
>
> **b** $\quad d + d - (a + a + a + a)$
> $\quad = 2d - 4a$

INDEX NOTATION

We can use **index notation** to simplify algebraic expressions in the same way as we did for numbers.

For example, $\ 3 \times 3 \times 3 \times 3 = 3^4 \ $ and $\ b \times b \times b \times b = b^4$.

> **Example 3** ◀)) **Self Tutor**
>
> Simplify:
>
> **a** $\ 8 \times b \times b \times a \times a \times a$ **b** $\ k + k - 3 \times d \times d \times d$
>
> **a** $\quad 8 \times b \times b \times a \times a \times a$
> $\quad = 8a^3 b^2$
>
> **b** $\quad k + k - 3 \times d \times d \times d$
> $\quad = 2k - 3d^3$

EXERCISE 5A

1 Simplify using product notation:

- **a** $\ 5 \times x$
- **b** $\ c \times 2$
- **c** $\ q \times 7$
- **d** $\ f \times 4g$
- **e** $\ 6q \times p$
- **f** $\ r \times 9s$
- **g** $\ 2a \times 3b$
- **h** $\ m \times 4n$
- **i** $\ a \times 5 \times b$
- **j** $\ q \times 2 \times p$
- **k** $\ j \times k \times l$
- **l** $\ p \times h \times d$

2 Simplify:

- **a** $\ p \times q + r$
- **b** $\ 4 \times x + 5 \times y$
- **c** $\ 2 \times a - b$
- **d** $\ b \times a - c$
- **e** $\ b - a \times c$
- **f** $\ f - g \times 7$
- **g** $\ c \times a + d \times a$
- **h** $\ 12 - r \times s \times 6$
- **i** $\ 3 \times (x + y)$
- **j** $\ 5 \times (d - 1)$
- **k** $\ (w - x) \times 8$
- **l** $\ p \times q \times (r - 2)$

3 Simplify:

- **a** $\ b + b$
- **b** $\ q + q + q$
- **c** $\ x + x + y + y + y + y$
- **d** $\ c + c + c + e$
- **e** $\ 3 + z + y + y$
- **f** $\ a + a + a + a + 7$
- **g** $\ g + g + 2 + g + g$
- **h** $\ 3 - (d + d + d)$
- **i** $\ s - t + t$
- **j** $\ s - (t + t)$
- **k** $\ 4 + r + r + r + 1$
- **l** $\ 2 + a + a + b + b$

Algebra (Chapter 5)

4 Write in expanded form:

a a^4 b f^2 c $4p^3$ d $3t^5$ e $5x^2y$
f $7f^2g^3$ g $(5a)^2$ h $5a^2$ i $p^2 + 2q$ j $p^3 - 3q^2$

5 Write in simplest form:

a $3 \times k \times k$ b $4 \times a \times a \times a$
c $2 \times d \times d \times d \times d$ d $4 \times p \times q \times q$
e $3 \times f \times g \times f \times g$ f $w \times w \times x \times y \times y \times y$
g $m + m \times m$ h $n \times n \times n + n$
i $y \times y - z \times z \times z \times z$ j $a \times a + 7 \times a$
k $8 \times b - b \times b \times b$ l $2 \times p \times q \times q + 6 \times r \times r \times s$
m $h \times h \times 2 - h \times j$ n $3 \times x + 5 \times x \times x \times x$
o $a \times a + 2 \times b \times b \times b - a \times b \times b$

We use indices and product notation to make expressions easier to read.

Example 4

Simplify:

a $2x \times 5$ b $4x \times 3x^2$ c $(6x)^2$

a $\quad 2x \times 5$
$\quad = 2 \times x \times 5$
$\quad = 10x$

b $\quad 4x \times 3x^2$
$\quad = 4 \times x \times 3 \times x \times x$
$\quad = 12x^3$

c $\quad (6x)^2$
$\quad = 6 \times x \times 6 \times x$
$\quad = 36x^2$

6 Simplify the following:

a $2y \times 3$ b $6x \times 2x$ c $3ac \times 4a$
d $(3d)^2$ e $2st \times 3st$ f $a^2 \times 2a^2$
g $4y \times (2y)^2$ h $3g \times g \times 4$ i $3a \times (2a)^2$
j $9b^3 \times 4b^2$ k $(-x) \times 3x$ l $(-2x) \times x^2$

B   THE LANGUAGE OF MATHEMATICS

Some **key words** used in algebra are:

Word	Meaning	Example(s)
variable	an unknown value that is represented by a letter or symbol	$P = 2l + 2w$ has variables P, l, and w.
expression	an algebraic form consisting of numbers, variables, and operation signs	$2x + y - 7$, $\dfrac{2a+b}{c}$
equation	an algebraic form which contains an $=$ sign	$3x + 8 = -1$, $\dfrac{x-1}{2} = -4$

Word	Meaning	Example(s)
terms	algebraic forms which are separated by $+$ or $-$ signs, the signs being included	$3x - 2y + xy - 7$ has four terms. These are $3x$, $-2y$, xy, and -7.
like terms	terms with exactly the same variable form	In $4x + 3y + xy - 3x$: • $4x$ and $-3x$ are like terms • $4x$ and $3y$ are unlike terms • xy and $3y$ are unlike terms.
constant term	a term which does not contain a variable	In $3x - y^2 + 7 + x^3$, 7 is a constant term.
coefficient	the number factor of an algebraic term	In $4x + 2xy - y^3$: • 4 is the coefficient of x • 2 is the coefficient of xy • -1 is the coefficient of y^3.

Example 5 ◀) Self Tutor

Consider $4y^2 - 6x + 2xy - 5 + x^2$.

a Is this an equation or an expression?
b How many terms does it contain?
c State the coefficient of: i x ii x^2.
d State the constant term.

a There is no $=$ sign present, so this is an expression.
b The expression contains five terms: $4y^2$, $-6x$, $2xy$, -5, and x^2.
c i The coefficient of x is -6.   ii The coefficient of x^2 is 1.
d The constant term is -5.

EXERCISE 5B

1 State the coefficient of x in the expression:

  a $3x$
  b $-8x$
  c x
  d $-x$
  e $3 + 4x$
  f $xy - 5x$
  g $3x - 4x^2$
  h $2x^2 - 2x + 1$

2 State the coefficient of y in the expression:

  a $5y$
  b $-5y$
  c $14y$
  d $-7y$
  e $3x - y$
  f $2x + 6y - 3$
  g $y^2 + 2xy + 3y$
  h $3y^2 - 2y + 5$

3 Consider $2x^2 + 5x - 7xy + 5y^2 - 2y + 1$.

  a Is this an equation or an expression?
  b How many terms does it contain?
  c State the coefficient of:
     i x^2 ii y^2 iii xy iv y.
  d State the constant term.

92 Algebra (Chapter 5)

4 State the number of terms in the expression:

    a $4x^2 + 4x + 1$ **b** $p^2 + q^2 - 5pq + 17$ **c** $x^3 - 2x^2 + 5x - \dfrac{1}{x} - 1$

5 Which of the following are equations, and which are expressions?

    a $xy + 2x + 1$ **b** $4x - y = 13$ **c** $4x - 5 = 1$

    d $\dfrac{x}{2} = \dfrac{6}{x}$ **e** $\dfrac{x^2}{3} - 2x^2$ **f** $x^2 + 7x + 10$

6 Identify the like terms in each of these expressions:

    a $3x + 2 + 5x$ **b** $4x + 2y - 3 - 3y$ **c** $2x + 7 - x$

    d $x - 2y + 1 + 3y$ **e** $x^2 - 5x - 8 + x$ **f** $x^2 - 5 + x - 7x^2$

    g $x^2 + 2x - 5x - 10$ **h** $5x - 3x^2 + \dfrac{x}{2} - 1$

C COLLECTING LIKE TERMS

We have seen that **like terms** are algebraic terms which contain the same variables to the same indices.

For example:
- $2xy$ and $-5xy$ are **like terms**
- a^2 and $-3a$ are **unlike terms** because the indices of a are not the same.

Algebraic expressions can often be simplified by adding or subtracting like terms. This is sometimes called **collecting like terms**.

Consider $2a + 4a = \underbrace{a + a}_{\text{"2 lots of } a\text{"}} + \underbrace{a + a + a + a}_{\text{"4 lots of } a\text{"}}$.

In total we have 6 lots of a, and so $2a + 4a = 6a$.

Example 6   ◀)) Self Tutor

Simplify, where possible, by collecting like terms:

    a $3x + 2x$ **b** $7a - 3a$ **c** $-2x + 3 - x$

    d $3bc + bc$ **e** $2x - x^2$

  a $3x + 2x$ **b** $7a - 3a$ **c** $-2x + 3 - x$

       $= 5x$ $= 4a$ $= -3x + 3$

                                                                                        $\{-2x$ and $-x$ are like terms$\}$

  d $3bc + bc$ **e** $2x - x^2$ is in simplest form

       $= 4bc$ $\{2x$ and $-x^2$ are unlike terms$\}$

EXERCISE 5C

1 Simplify, where possible, by collecting like terms:

    a $2 + x + 4$ **b** $q + 5 + 6$ **c** $b + 3 + b + b$

    d $a + a + 7$ **e** $d + d$ **f** $q + 1 + q + 4$

    g $m + 2 - m$ **h** $-k - k + 3$ **i** $5 - p - 7$

2 Simplify, where possible, by collecting like terms:
- **a** $5y - 3y$
- **b** $4z - z$
- **c** $g^2 + g^2$
- **d** $5x + 5$
- **e** $5w^2 - 4w^2$
- **f** $3x - 3x^2$
- **g** $3x - x$
- **h** $3ab + 6ab$
- **i** $m + m + m + m$

3 Simplify, where possible:
- **a** $8p - 8p$
- **b** $8p - p$
- **c** $8p - 8$
- **d** $7pq - pq$
- **e** $ab + 3ab$
- **f** $3q^2 - q^2$
- **g** $3w + 4w + 5w$
- **h** $8xy + 5yx$
- **i** $2z + 5z - 4z$
- **j** $2m + 3m - 5m$
- **k** $5d + 4d - 9$
- **l** $3g + 4g - 7g^2$
- **m** $s + 3s + 4s^2$
- **n** $2x^2 + 2x + 2$
- **o** $2a^2 - b^2$

4 Simplify, where possible:
- **a** $4a + 6a$
- **b** $4a - 6a$
- **c** $-4a + 6a$
- **d** $-4a - 6a$
- **e** $7x + x$
- **f** $7x - x$
- **g** $-7x + x$
- **h** $-7x - x$
- **i** $3n + n^2$
- **j** $-8d - 5d$
- **k** $-8d + 5d$
- **l** $8d - 5d$
- **m** $b + 2 - 3b$
- **n** $2t - 3t - t$
- **o** $2m - 7 - 3m$

5 Simplify, where possible:
- **a** $5x + 4x$
- **b** $5x - 4x$
- **c** $-5x - 4x$
- **d** $p^2 + 2p$
- **e** $k + 4k - 5$
- **f** $n - 6n - 5n$
- **g** $-11m - 4m$
- **h** $4j - 9j + 4$
- **i** $-y - (-8y)$
- **j** $8y - y$
- **k** $8y - (-y) - 5y$
- **l** $y - (-4y) + 3y$

Example 7 ◀)) **Self Tutor**

Simplify, by collecting like terms:
- **a** $2 + 3a - 3 - 2a$
- **b** $x^2 - 2x + 3x - 2x^2$

a
$$2 + 3a - 3 - 2a$$
$$= 3a - 2a + 2 - 3$$
$$= a - 1$$
{$3a$ and $-2a$ are like terms, 2 and -3 are like terms}

b
$$x^2 - 2x + 3x - 2x^2$$
$$= x^2 - 2x^2 - 2x + 3x$$
$$= -x^2 + x$$
{x^2 and $-2x^2$ are like terms, $-2x$ and $3x$ are like terms}

---

6 Simplify, where possible:
- **a** $x + 5 - 3x - 6$
- **b** $8t + 4 - 3t - 1$
- **c** $x + 5y - 6y - 3x$
- **d** $pq + 3 + 5pq - 7$
- **e** $-cd + 2cd + 9cd$
- **f** $2a - 6 + 6 - 3a$
- **g** $12x^2 + 5 - 7x^2 - 7$
- **h** $-5n + 3 + 2n - 6$
- **i** $2v - 7v + w - 6w$
- **j** $-3x^3 - 2x^2 + 3x^3 - x^2$
- **k** $4a - 3b - (-a) - 4b$
- **l** $-2z - 3 - 3z - 4$
- **m** $2p + pq - 3pq - p$
- **n** $-6mn + 3m - mn - 5m$

D WRITING EXPRESSIONS

To convert a statement in words to a mathematical expression, we need to identify key words. The words in the following table tell us what **operations** need to be performed between numbers or variables.

Word	Meaning	Examples
sum	The sum of two or more numbers is obtained by **adding** them.	$4+5$, $x+7$, $r+s+t$ are sums.
difference	The difference between two numbers is the larger one **minus** the smaller one.	$8-3$, $n-11$ (if $n > 11$) are differences.
product	The product of two or more numbers is obtained by **multiplying** them.	2×7, $4b$, pqr are products.
quotient	The quotient of two numbers is the first one mentioned **divided** by the second.	The quotient of a and b is $\frac{a}{b}$.
mean or average	The **mean** or **average** of a set of numbers is their sum divided by the number of numbers.	The mean or average of x, y, and z is $\frac{x+y+z}{3}$.

Example 8   ◀)) Self Tutor

Write an expression for:
- **a** the sum of 6 and a
- **b** the difference between c and d, where $d > c$
- **c** the mean of p, q, and r.

- **a** The sum of 6 and a is $6 + a$.
- **b** Since $d > c$, the difference between c and d is $d - c$.
- **c** The mean of p, q, and r is $\frac{p+q+r}{3}$.

EXERCISE 5D

1 Write an expression for the sum of:
- **a** 9 and 2
- **b** 5 and a
- **c** m and $3n$
- **d** d, e, and f

2 Write an expression for the product of:
- **a** 8 and 6
- **b** 6 and p
- **c** n and $4m$
- **d** b, d, and e

3 Write an expression for the quotient of:
- **a** 6 and 5
- **b** d and 3
- **c** m and $5n$
- **d** $p + q$ and x

> When writing products we leave out the multiplication sign between unknowns and write them in alphabetical order.

4 Write an expression for the mean of:
  a 6 and 10   **b** 9 and d **c** k and $4v$ **d** d, e, and f

5 Write an expression for the difference between:
  a 5 and 8   **b** 6 and s if $6 < s$ **c** 8 and p if $8 > p$

Example 9   ◀)) Self Tutor

Convert into algebraic form:
  a 18 more than a number
  b 7 less than a number
  c double a number
  d double the sum of a number and 7

In each case we let the number be x.
  a 18 more than the number is $x + 18$.
  b 7 less than the number is $x - 7$.
  c double the number is $2 \times x$ or $2x$.
  d The sum of the number and 7 is $x + 7$, so double the sum is $2 \times (x + 7)$ or $2(x + 7)$.
    double ↑ sum of the number and 7

---

6 Write as an algebraic expression:
  a 3 more than a number
  b 5 less than a number
  c one half of a number
  d treble a number
  e one quarter of a number
  f 12 minus a number
  g 1 more than double a number
  h 6 less than five times a number

7 Write an algebraic expression for:
  a 8 more than p
  b g is decreased by 3
  c n is increased by 2
  d the sum of c and 4
  e 3 less than x
  f the product of 4 and f
  g h is divided by 3
  h 4 more than 2 times a
  i double p and add 14
  j the product of x and the square of y
  k the product of the square of 4, and c
  l the square of the product of a and b
  m the sum of the squares of p and q.

8 Copy and complete:
  a Two numbers have a sum of 4. If one of them is s then the other is ......
  b If there are 27 students in a class and b are boys, then there are ...... girls.
  c If the smaller of two consecutive integers is y, then the larger is ......
  d Three consecutive integers in ascending order are x, ..., ...
  e Two consecutive odd integers in ascending order are d and ......
  f Three consecutive integers in descending order are a, ..., ...
  g If the middle of three consecutive integers is m, then the other two are ...... and ......
  h Two numbers differ by 3. If the smaller one is s then the other is ......

E GENERALISING ARITHMETIC

To find algebraic expressions for many real world situations, we first think in terms of numbers or numerical cases. We then proceed to more general cases.

For example, suppose we are asked to find the total cost of x plants which each cost $£y$.

We could start by finding the total cost of 6 plants which each cost £10. This makes it easier to understand that we need to *multiply* the quantities.

The total cost is $6 \times £10 = £60$.

In the same way, the total cost of x plants at $£y$ each is $x \times £y = £xy$.

Example 10 ◀)) Self Tutor

Find: **a** the cost of x bananas at 30 pence each
b the change from £50 when buying y books at £6 each.

a The cost of 7 bananas at 30 pence each is 7×30 pence.
∴ the cost of x bananas at 30 pence each is $x \times 30 = 30x$ pence.

b The total cost of 5 books costing £6 each is 5×6 pounds.
∴ the change from £50 when buying 5 books costing £6 each is $50 - (5 \times 6)$ pounds.
∴ the change from £50 when buying y books costing £6 each is
$50 - (y \times 6)$ pounds $= 50 - 6y$ pounds.

EXERCISE 5E

1 Find the total cost of buying:
   a 5 caps at £20 each   **b** a caps at £20 each   **c** a caps at $£d$ each.

2 Rick is now 14 years old. How old was he:
   a 6 years ago   **b** x years ago?

3 Find the change from £100 when buying:
   a 3 hammers at £15 each   **b** h hammers at £15 each   **c** h hammers at $£p$ each.

4 Patrick decided to go jogging each morning. As a result, he lost 6 kg. If he initially weighed w kg, how much does he weigh now?

5 There were 20 people at a party, then m more people arrived and n people left. How many people are now at the party?

6 Laura buys a apricots and p peaches. Find the total cost (in pounds) if each apricot costs 60 pence and each peach costs 90 pence.

7 Tia is walking to her friend's house, 600 m down the road. Each step she takes is 80 cm long. Tia has walked x steps.
   a How far has she walked?   **b** How far is she from her friend's house?

8 **a** A cyclist travels at an average speed of 15 km per hour for 3 hours. How far has the cyclist travelled?
   b How far would the cyclist travel at an average speed of s km per hour for t hours?

9 **a** Jan has 96 cupcakes to share amongst 8 tables. How many cupcakes does each table receive?

  b If Jan had c cupcakes to share amongst n tables, how many cupcakes would each table receive?

10 Frank buys b balls at £m each and r racquets at £n each. Find the total cost of these items.

Review set 5A

1 Simplify using product notation:
  a $7 \times a$
  b $b \times 3$
  c $b \times a \times 2$
  d $h \times 5 \times g$

2 Simplify:
  a $a + b + b + b$
  b $a + a \times a + a$
  c $2 \times b + b \times b$
  d $6 \times a + a \times 2$
  e $3 \times a - a \times a$
  f $4 \times x \times x - x$

3 Write in expanded form:
  a $7t^4$
  b $(3b)^3$
  c $2b^2 - 3c^3$

4 State the coefficient of x in the expression:
  a $3x + y$
  b $4x^2 - 2x$
  c $\frac{x}{2} - 1$

5 State whether each of the following is an equation or an expression:
  a $3x = 5$
  b $2x^2 - 4x + 7$
  c $(x - 2)^2$
  d $xy + 2y = -7$

6 Consider the expression $5g + 12gh - 2h + 3gh + 4$.
  a How many terms are present?
  b State any like terms.
  c State the coefficient of h.
  d State the constant term.
  e If possible, simplify the expression.

7 Simplify by collecting like terms:
  a $3 + 2a + 5a - 6$
  b $x + 3x + 2x$
  c $3bc + 2cb$
  d $2b - a + 3a + b$
  e $7k + 6k - 6$
  f $9f + 3g - (-8f) - 8g$

8 Write an expression for the average of:
  a 3 and 6
  b 5 and a
  c p and q

9 Write an expression for:
  a 3 more than h
  b 5 less than 2 times g
  c double m less 4
  d a half of the square of a.

10 Gina bought a bag of 18 apples. How many apples will be left in the bag if she eats:
  a 2 apples per day for the next 3 days
  b a apples per day for the next 3 days
  c a apples per day for the next d days?

11 Write as an algebraic expression:
  a the sum of two consecutive whole numbers, the smaller of which is x
  b the total value of x 50 pence coins and $(x + 4)$ 20 pence coins

12 You have a 9 m length of string. If you cut 4 lengths of x m from it, what length remains?

13 Graham buys p pencils and b books. Each pencil costs 60 pence and each book costs 95 pence. Find the total cost of the items, in pence.

Review set 5B

1. Simplify using product notation:

   a $m \times n$ 　　　 b $s \times r$ 　　　 c $m \times p \times n$ 　　　 d $b \times c \times a$

2. Simplify:

   a $2 \times x \times x - 3$ 　　　 b $b \times b \times b \times b$ 　　　 c $a \times a \times a - a \times a$

   d $4 \times y \times y \times 3 \times y$ 　　　 e $5 \times b \times b - 2 \times b$ 　　　 f $s \times s - s \times t$

3. Simplify:

   a $(2x)^3$ 　　　 b $2a^2 \times 3a$ 　　　 c $(-2x) \times (3x)^2$

4. Consider the expression $3x^2 + 4x - 7x + 5$.

   a What are the like terms in this expression?

   b How many terms will the *simplified* expression have?

5. Consider $3x^2 - 2x + 4 - x^2 + x$.

   a Is this an equation or an expression? 　　　 b How many terms does it contain?

   c State the constant term. 　　　 d State any like terms.

6. Identify the like terms in:

   a $3 - x^2 + 2x - 1$ 　　　 b $5 + x - x^2 - (-2x)$

7. Simplify, where possible:

   a $5x - 5$ 　　　 b $3p^2 - p^2$ 　　　 c $1 + b + 5b - 4$

   d $2n^2 + 3n - n^2 + 4n$ 　　　 e $3z - 8z + 2$ 　　　 f $t + 5u - 6t + 7u$

8. Write an expression for:

   a the sum of 3, x, and z 　　　 b the product of $2x$ and $(-3y)$

   c the difference between 2 and x if $x > 2$.

9. Copy and complete:

   a Two numbers differ by a. If the smaller one is 3 then the other is ......

   b Three consecutive even integers in ascending order are x, ..., ...

   c If there are twice as many girls as boys in the club, and there are g girls, then the total number of people in the club is ......

10. Find the change from £20 when buying:

    a 3 ice creams at £2 each 　　　 b 3 ice creams at £p each 　　　 c c ice creams at £p each.

11. A cake stall had 20 muffins. n customers bought 2 muffins each, then a new batch of m muffins arrived. How many muffins does the stall now have?

12. Carlos is now 16 years old. In b years from now, how old will he be?

13. Brian went on a journey to see his friends. He travelled 6 miles to see Jonas, then another k miles to see Susan. He travelled another n miles to see James, then drove the 8 miles directly home. Write an expression for the total distance travelled.

6

Ratio

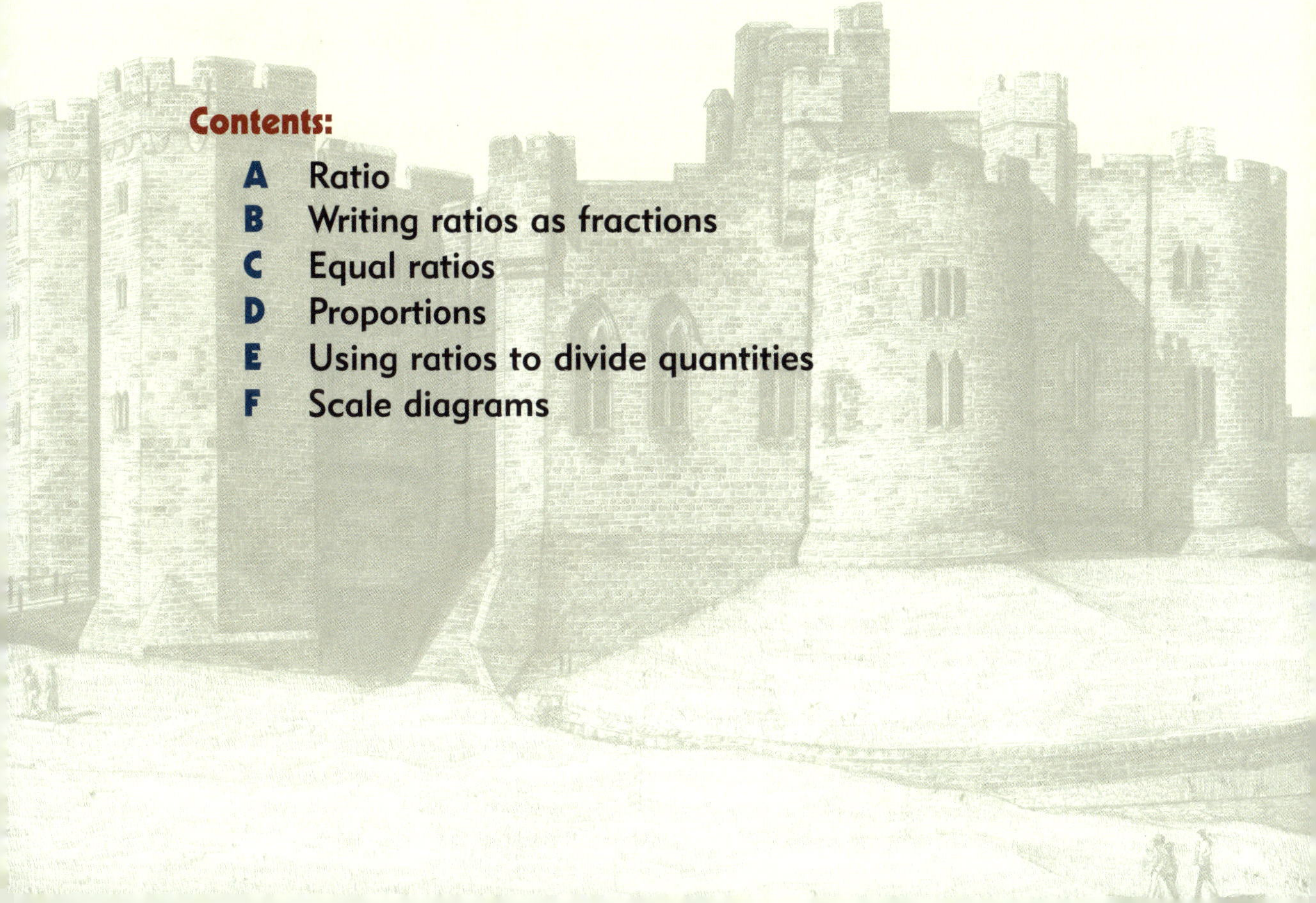

Contents:
- **A** Ratio
- **B** Writing ratios as fractions
- **C** Equal ratios
- **D** Proportions
- **E** Using ratios to divide quantities
- **F** Scale diagrams

Opening problem

Ellen makes guacamole by mixing mashed avocado and yoghurt in the ratio 3 : 2.

Things to think about:
a What does this ratio mean?
b Does the guacamole contain more mashed avocado or yoghurt?
c What *fraction* of the guacamole is:
   i avocado   ii yoghurt?
d Ellen has 150 g of mashed avocado. How much yoghurt should she use?
e How much of each ingredient would Ellen need to make 600 g of guacamole?

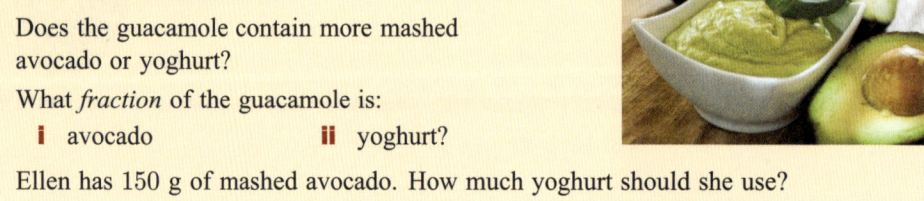

In this Chapter we will study **ratios** and how they are used in **scale diagrams**.

A   RATIO

A **ratio** is an ordered comparison of quantities.

Suppose we have 6 apples and 4 bananas. The ratio of the number of apples to the number of bananas is 6 to 4.

We write this as   apples : bananas = 6 : 4

The numbers in a ratio need to be written in the correct order. The ratio of apples to bananas is 6 : 4, **not** 4 : 6.

Example 1   ◀)) Self Tutor

Find the ratio of the number of squares to the number of triangles.

There are 8 squares and 11 triangles.
∴ squares : triangles = 8 : 11

Although most ratios involve two quantities, ratios may involve more than two quantities. For example, in the picture, the ratio of apples to bananas to oranges is 2 : 5 : 4.

If measurements are involved, we must use the **same units** for each quantity.

For example, the ratio of lengths shown is
20 : 7   {20 mm : 7 mm},   **not**   2 : 7.

2 cm   7 mm

Ratio (Chapter 6) 101

> **Example 2** ◀)) **Self Tutor**
>
> Write as a ratio:
> - **a** Jack has £5 and Jill has 50 pence.
> - **b** Mix 200 ml of cordial with 1 litre of water.
>
> **a** Jack : Jill = £5 : 50 pence   {write in the correct order}
> $\qquad\qquad\quad$ = 500 pence : 50 pence   {write in the same units}
> $\qquad\qquad\quad$ = 500 : 50   {express without units}
>
> **b** cordial : water = 200 ml : 1 litre   {write in the correct order}
> $\qquad\qquad\qquad$ = 200 ml : 1000 ml   {write in the same units}
> $\qquad\qquad\qquad$ = 200 : 1000   {express without units}

EXERCISE 6A

1 Consider the collection of cats and dogs shown.
Which of the following statements is correct?

  A cats : dogs = 3 : 6
  B dogs : cats = 2 : 4
  C cats : dogs = 5 : 4
  D cats : dogs = 4 : 5

2 Find the ratio of:

  a circles to squares

  b ♣s to ♦s.

3 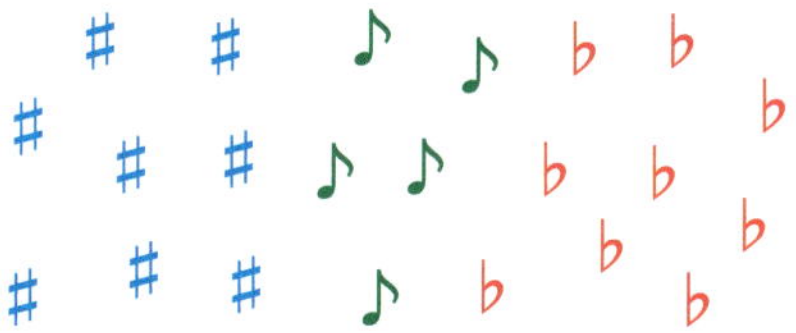 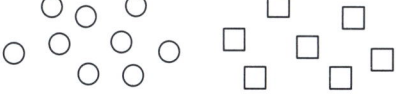   Find the ratio of:

  a ♯s to ♭s
  b ♯s to ♪s
  c ♪s to ♭s
  d ♭s to ♪s
  e ♯s to ♪s to ♭s.

4 Write as a ratio:

  a £8 is to £3
  b 3 litres is to 7 litres
  c 35 kg is to 45 kg
  d £3 is to 50 pence
  e 500 ml is to 3 litres
  f 400 m is to 2.5 km
  g 9 km is to 150 m
  h 12 m is to 8 km
  i 4 h is to 40 min

5 Write as a ratio:

  a For every two laps of the pool that Darryl swims, Toby swims three laps.
  b For every hour Tegan spends working, she spends 10 minutes exercising.
  c Mix 4 cups of flour, 3 cups of sugar, and 1 cup of sultanas.

6 Write as a ratio:

    a 25 ml is to 35 ml is to 50 ml

    b 72 g is to 77 g is to 68 g

    c £35 is to £47.50 is to £52

    d 400 g is to 1.2 kg is to 850 g

B WRITING RATIOS AS FRACTIONS

We can write ratios as fractions by considering the *total number of parts* in the ratio.

For example, vinegar and water are combined in the ratio $2:3$ to make a cleaning liquid. The ratio contains $2 + 3 = 5$ parts in total. For every 5 parts of the cleaning liquid, 2 parts are vinegar and 3 parts are water.

So, $\frac{2}{5}$ of the cleaning liquid is vinegar, and $\frac{3}{5}$ is water.

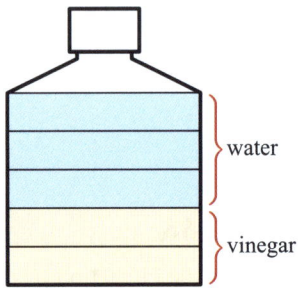

Example 3   ◀)) Self Tutor

The ratio of girls to boys in a class is $3:4$.

What fraction of the class is:

    a girls   **b** boys?

The ratio contains $3 + 4 = 7$ parts in total. Of the 7 parts, 3 parts are girls and 4 parts are boys.

    a $\frac{3}{7}$ of the class are girls.   **b** $\frac{4}{7}$ of the class are boys.

EXERCISE 6B

1 The ratio of adults to children on a bus is $7:2$. What fraction of the passengers are:

    a adults   **b** children?

2 A farmer grows peach trees and pear trees in the ratio $8:3$. What fraction of the trees are:

    a peach trees   **b** pear trees?

3 For each of the following figures, find:

    i the *ratio* of the shaded area to the unshaded area

    ii the *fraction* of the figure which is shaded

    iii the *percentage* of the figure which is shaded.

a 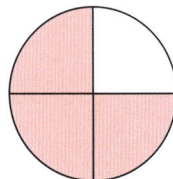   **b** 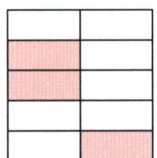   **c**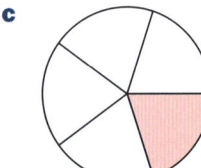

4 A butcher's shop sells chicken, lamb, and beef in the ratio $7:4:3$.
  a What *fraction* of the meat sold is lamb?
  b What *percentage* of the meat sold is chicken?
  c If 168 kg of meat is sold in one day, how much beef is sold?

5 The three children in the Smith family receive a total of £60 pocket money each week. It is divided in the same ratio as their ages. Amy is 8, Brianna is 7, and Caroline is 5.
  a What *fraction* of the pocket money does Amy receive?
  b What *percentage* of the pocket money does Brianna receive?
  c How much pocket money does Caroline receive each week?

6 The ratio of boys to girls in a school choir is $2:3$.
  a What fraction of the choir members are: **i** boys **ii** girls?
  b 30% of the boys are aged under 15, and 50% of the girls are aged under 15. What percentage of all the choir members are aged under 15?

C EQUAL RATIOS

The stamp alongside is 3 cm long and 2 cm wide.

The stamp's length to width ratio is $3:2$.

However, the stamp is also 30 mm long and 20 mm wide.

So, the stamp's length to width ratio is also $30:20$.

We say that $30:20$ and $3:2$ are **equal ratios**.

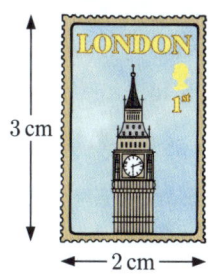

Notice that to get from $30:20$ to $3:2$, we divide each number in the ratio by 10.

$$30:20 = 3:2$$

> If we multiply or divide both parts of a ratio by the same non-zero number, we obtain an **equal ratio**.

Ratios can be expressed in **simplest form** by writing an *equal* ratio with integer parts that are as small as possible.

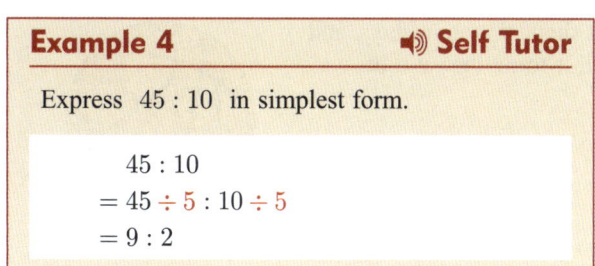

To write a ratio in simplest form, divide each part by their **highest common factor**.

EXERCISE 6C

1 Express each ratio in simplest form:
- **a** 6 : 8
- **b** 8 : 4
- **c** 4 : 12
- **d** 9 : 15
- **e** 3 : 6
- **f** 14 : 8
- **g** 8 : 16
- **h** 18 : 24
- **i** 125 : 100
- **j** 2 : 4 : 6
- **k** 1000 : 50
- **l** 6 : 12 : 24

2 Express each ratio in simplest form:
- **a** 1.2 : 0.7
- **b** 2.1 : 1.4 : 3.5
- **c** $\frac{2}{3} : \frac{1}{3} : \frac{4}{3}$
- **d** $\frac{1}{2} : \frac{1}{3} : \frac{3}{4}$

> **Example 5** ◀)) **Self Tutor**
>
> Express as a ratio in simplest form:
> 20 hours is to 2 days.
>
> 20 hours : 2 days = 20 hours : 48 hours
> $\phantom{20 \text{ hours} : 2 \text{ days}} = 20 : 48$
> $\phantom{20 \text{ hours} : 2 \text{ days}} = 20 \div 4 : 48 \div 4$
> $\phantom{20 \text{ hours} : 2 \text{ days}} = 5 : 12$

3 Express as a ratio in simplest form:
- **a** 70 cm is to 1 m
- **b** 4 days is to 2 weeks
- **c** 2.1 litres is to 600 ml
- **d** 5 mm is to 8 cm
- **e** 1.2 kg is to 800 g
- **f** 50 seconds is to $1\frac{1}{2}$ minutes
- **g** 60 litres is to 1 kl
- **h** 200 kg is to 1.7 t
- **i** 75 mm^2 is to 2 cm^2
- **j** 2.41 km^2 is to 140 ha
- **k** 350 mm^3 is to 4 cm^3
- **l** 1.5 m^3 is to 90 000 cm^3

4 Write the ratio of the first quantity to the second quantity in simplest form:
- **a** a bag of peanuts costing £15 to a bag of pistachios costing £60
- **b** a 40 minute bus trip to a 25 minute train trip
- **c** a tom cat weighing 7.2 kg to a kitten weighing 900 g
- **d** a shelf 30 cm high to a bookcase 1.5 m high
- **e** a 10 km run to a 1500 m swim.

5 Write as a ratio in simplest form:
- **a** the weights of three jugs which are 280 g, 315 g, and 350 g
- **b** the times 600 seconds, 20 minutes, and $\frac{1}{2}$ hour
- **c** the weights of three parcels which are 900 g, 1.2 kg, and 1.6 kg.

6 During a basketball match, Dale scored 3 3-point shots, 9 2-point shots, and 6 free throws worth 1 point each. Find, in simplest form, the ratio of:
- **a** 3-point shots : 2-point shots : free throws
- **b** points earned from 3-point shots to 2-point shots to free throws.

Example 6

Determine whether the ratios $8:10$ and $20:25$ are equal.

We first express the ratios in simplest form:

$8:10$
$= 8 \div 2 : 10 \div 2$
$= 4:5$

$20:25$
$= 20 \div 5 : 25 \div 5$
$= 4:5$

— the same —

So, $8:10 = 20:25$.

Two ratios are equal if they can be expressed in the same simplest form.

7 Determine whether each pair of ratios is equal:
  a $2:6$ and $5:15$
  b $6:9$ and $15:20$
  c $30:18$ and $55:33$
  d $18:28$ and $9:12$
  e $15:21$ and $35:25$
  f $36:52$ and $45:65$

8 Annette, Bert, Claire, and Derek each mixed syrup and water to make a glass of cordial.

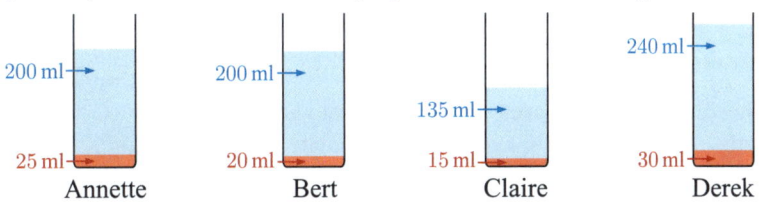

  a Find, in simplest form, the ratio of syrup to water for each person.
  b Which two people mixed the syrup and water in the same ratio?

D PROPORTIONS

A **proportion** is a statement that two ratios are equal.

For example, $6:4 = 9:6$ is a proportion, since both ratios simplify to $3:2$.

If we know three of the values in a proportion, we can always find the fourth value.

Example 7

Find x if $3:5 = 6:x$.

$$3:5 = 6:x \qquad \therefore\ x = 5 \times 2$$
$$\qquad\qquad\qquad \therefore\ x = 10$$

(×2 on both sides)

We need to multiply 3 by 2 to get 6. We therefore multiply 5 by 2 to get x.

EXERCISE 6D

1 Find x if:
  a $2:3 = 8:x$
  b $1:4 = x:12$
  c $3:2 = 15:x$
  d $4:3 = x:21$
  e $5:7 = 25:x$
  f $6:11 = x:77$
  g $5:12 = 40:x$
  h $7:10 = x:80$
  i $4:5 = x:45$

Example 8

The ratio of students to adults on a primary school camp was $9:2$.
If there were 27 students, how many adults were there?

Let the number of adults be x. students : adults $= 27 : x$

$\therefore\ 9 : 2 = 27 : x$ (×3)

$\therefore\ x = 2 \times 3$

$\therefore\ x = 6$

$\therefore$ there were 6 adults.

2 The capacities of two bowls are in the ratio $3:4$. The smaller bowl holds 12 litres. Find the capacity of the larger bowl.

3 Liquid hand soap is a mixture of water to soap in the ratio $5:2$. If I make a 400 g block of soap into liquid hand soap, how much water do I need?

4 A recipe for chocolate ice cream uses melted chocolate and cream in the ratio $3:25$. If 150 ml of melted chocolate is used, how much cream is needed?

5 A photograph has a length to width ratio of $3:2$. When the photograph is enlarged, this ratio must be maintained so the photograph does not appear distorted.
  a If the enlarged photograph is 18 cm long, find its width.
  b If the enlarged photograph is 16 cm wide, find its length.

6 The required ratio of workers to children at a child care centre is changed from $1:8$ to $1:5$. If the child care centre has 120 children, how many more child care workers will be needed?

7 An island contains arctic terns and common terns in the ratio $1 : 3\frac{1}{2}$.
  a Write this ratio in simplest form.
  b Given that there are 600 arctic terns, how many terns live on the island in total?

8 A shop owner likes to stock cola drinks and lemonade drinks in the ratio $5:2$. He currently has 110 cola drinks and 60 lemonade drinks in stock. How many cola drinks should he order to return his stock to the preferred ratio?

9 Ben makes purple paint by mixing red paint and blue paint in the ratio $3:2$. In his shed, Ben has 20 litres of red paint and 12 litres of blue paint. What is the *greatest* amount of purple paint Ben can make?

10 Paula's flower bouquets consist of lilies, roses, and sunflowers in the ratio $8:5:3$. The most recently sold bouquet contained 10 roses. How many:
  a lilies   **b** sunflowers   did the bouquet contain?

11 A fertiliser contains nitrogen, potash, and limestone in the ratio $1 : 3 : 5\frac{1}{3}$.

    a What fraction of the fertiliser is potash?

    b A bag of the fertiliser contains 600 g of nitrogen.

        i How much limestone is in the bag?   **ii** Find the total weight of the bag.

E   USING RATIOS TO DIVIDE QUANTITIES

Quantities can be divided in a particular ratio using fractions.

Example 9   Self Tutor

An inheritance of £60 000 is to be divided between Donny and Marie in the ratio $2 : 3$. How much does each receive?

There are $2 + 3 = 5$ parts.

Donny gets $\frac{2}{5}$ of the inheritance, and Marie gets $\frac{3}{5}$ of the inheritance.

$\therefore$ Donny gets $\quad \frac{2}{5}$ of £60 000   and Marie gets $\quad \frac{3}{5}$ of £60 000

$\qquad\qquad\quad = \frac{2}{5} \times 60\,000 \qquad\qquad\qquad\qquad = \frac{3}{5} \times 60\,000$

$\qquad\qquad\quad = £24\,000 \qquad\qquad\qquad\qquad\quad\; = £36\,000$

EXERCISE 6E

1 Eliza will make 15 litres of green paint by mixing blue paint and yellow paint in the ratio $2 : 1$. How much:

    a blue paint   **b** yellow paint   will Eliza need?

2 The winning pilot from an air race shares his £35 000 prize with his technical crew in the ratio $4 : 3$.

    a What fraction of the money is received by:

        i the pilot   **ii** the technical crew?

    b How much money is received by:

        i the pilot   **ii** the technical crew?

3 The ratio of males to females attending a rugby match was $5 : 3$. Given that 4000 spectators attended the match, how many were:

    a male   **b** female?

4 Pranay and Samar buy a 20 kg bag of rice from the bulk food store. Pranay pays £9 and Samar pays £6.

    a How much did the bag of rice cost?

    b Write down the ratio of Pranay's contribution to Samar's contribution.

    c Pranay and Samar share the rice in proportion to their contribution. How much rice do each of them get?

5 When Josh visits the gym, he uses the treadmill and the weights in the ratio 7 : 3.

  a On Monday Josh spent 30 minutes at the gym. For how long did he use the treadmill?

  b On Wednesday Josh spent $1\frac{1}{2}$ hours at the gym. For how long did he use the weights?

6 Answer the **Opening Problem** on page **100**.

7 Danielle sells small ice creams for £2, and large ice creams for £3. Yesterday she sold small ice creams and large ice creams in the ratio 5 : 3. Given that Danielle sold 320 ice creams in total, how much money did she receive?

8 Katrina and Lee shared some marbles in the ratio 3 : 4. Lee received 8 more marbles than Katrina.

  a How many marbles did they share in total?

  b How many marbles did each person receive?

9 Barry, Robin, and Maurice invest in a house together in the ratio 3 : 2 : 7. The house later sells for £600 000. Fairly divide the revenue amongst the investors.

10 A soil mixture has a peat : compost : vermiculite ratio of 4 : 5 : 1. A gardener needs to make 160 kg of soil mixture.

  How much: **a** peat   **b** compost   **c** vermiculite   is required?

Puzzle

One full glass contains vinegar and water in the ratio 1 : 3. Another glass with twice the capacity of the first, has vinegar and water in the ratio 1 : 4. If the contents of both glasses are mixed together, what is the ratio of vinegar to water?

F SCALE DIAGRAMS

When drawing the floorplan of a house, or the map of a country, it is obviously impractical to make the drawing the same size as the object.

Instead, we draw a smaller diagram which shows the *shape* of the actual object, but with lengths reduced *in proportion*. This diagram is called a **scale diagram**.

The **scale** tells us how many times larger the actual object is than the diagram.

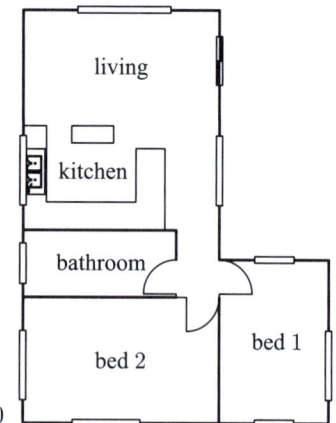

Scale 1 : 200

The scale is often expressed as a ratio. For example, a scale of 1 : 200 tells us that the actual object is 200 times larger than the diagram. We say that the **scale factor** is 200, and so 1 cm on the diagram represents 200 cm or 2 m of actual length.

A divided bar can also be used to show the scale. For example, this scale tells us that 1 cm on the diagram represents 50 m of actual length.

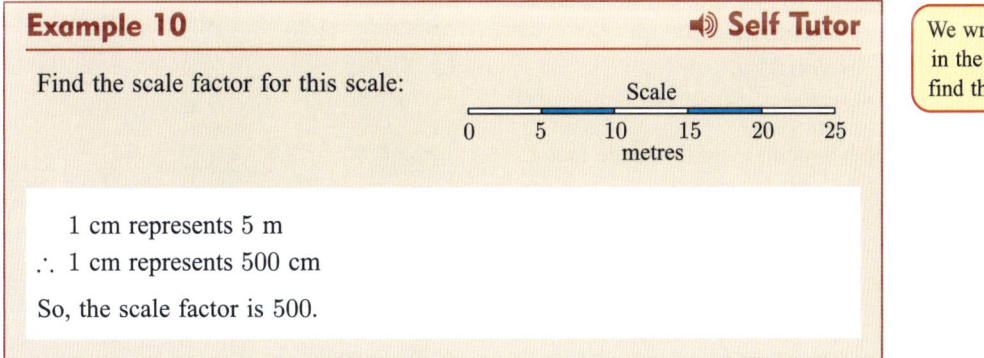

Example 10 ◀) Self Tutor

Find the scale factor for this scale:

We write the lengths in the same units to find the scale factor.

1 cm represents 5 m
∴ 1 cm represents 500 cm

So, the scale factor is 500.

If we are given a scale diagram of an object, we can find the actual measurements for the object using:

$$\text{actual length} = \text{drawn length} \times \text{scale factor}$$

Example 11 ◀) Self Tutor

Alongside is a scale diagram of a building. Use your ruler and the scale given to find:
a the height of the building
b the width of the building.

Scale 1 : 200

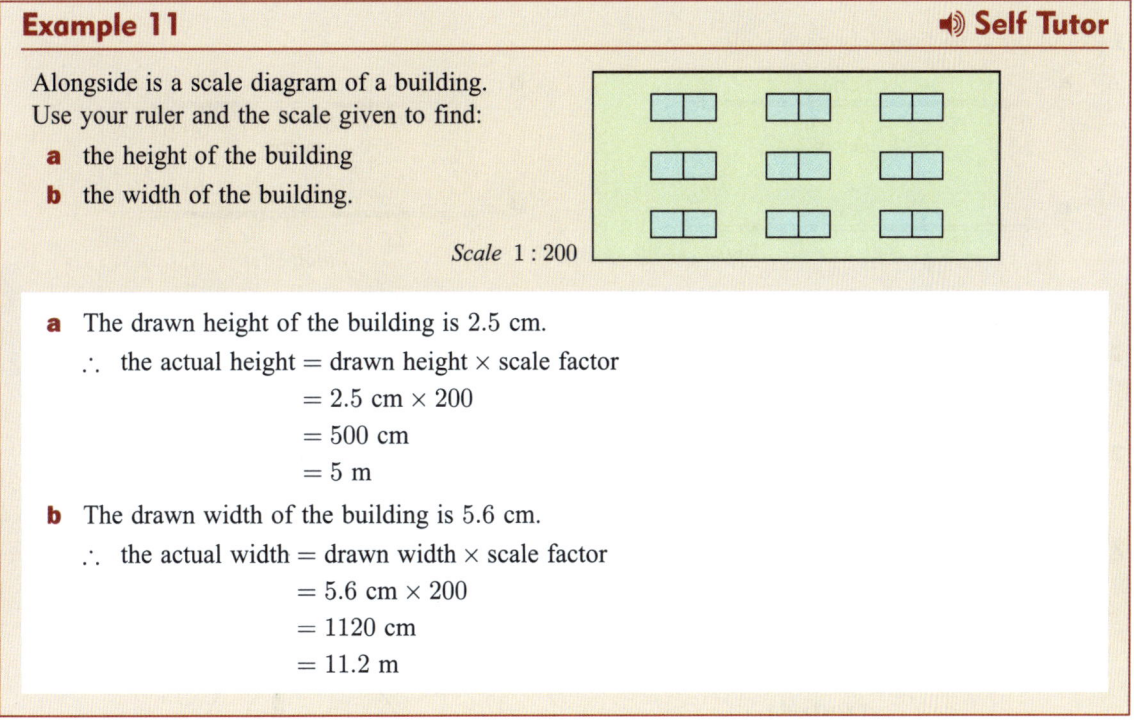

a The drawn height of the building is 2.5 cm.
∴ the actual height = drawn height × scale factor
= 2.5 cm × 200
= 500 cm
= 5 m

b The drawn width of the building is 5.6 cm.
∴ the actual width = drawn width × scale factor
= 5.6 cm × 200
= 1120 cm
= 11.2 m

If we know the actual dimensions of an object, we can determine the required measurements for a scale diagram using:

$$\text{drawn length} = \text{actual length} \div \text{scale factor}$$

Example 12

A football pitch is 100 m long and 70 m wide. Draw a scale diagram of the pitch using the scale 1 : 2000.

The scale factor for the diagram is 2000.

The drawn length = actual length ÷ scale factor
= 100 m ÷ 2000
= 0.05 m
= 5 cm

The drawn width = actual width ÷ scale factor
= 70 m ÷ 2000
= 0.035 m
= 3.5 cm

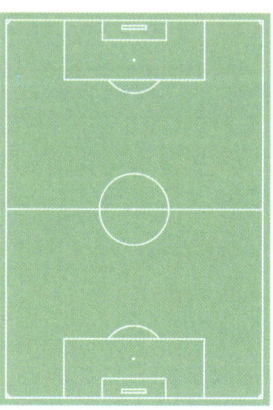

Scale 1 : 2000

EXERCISE 6F

1 Find the scale factor for each scale:

a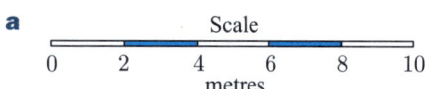

b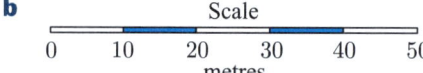

c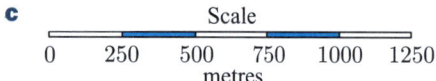

d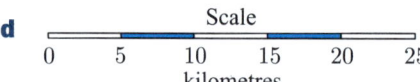

2 For a scale of 1 : 500, find the actual length represented by a drawn length of:
   a 5 mm
   b 2 cm
   c 10 cm
   d 17.5 cm

3 For a scale of 1 : 200, find the length which should be drawn to represent an actual length of:
   a 10 m
   b 35 m
   c 24 m
   d 46.6 m

4 Alongside is a scale diagram of a boat. Use your ruler and the scale given to find:
   a the length of the boat at its longest point
   b the height of the boat's mast.

PRINTABLE DIAGRAMS

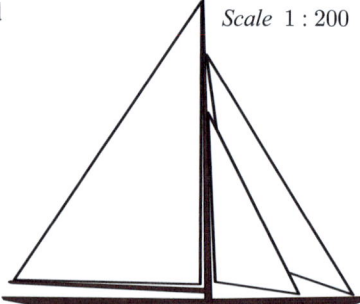

Scale 1 : 200

5 The diagram is a scale drawing of a car park.

a Use your ruler and the scale given to find each side length of the car park.

b Divide the car park into a rectangle and a right angle triangle. Hence find the area of the car park.

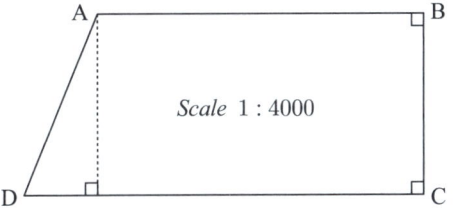

6 Draw a scale diagram of a barn which is 6 m wide and 10 m long. Use a scale of 1 : 200.

7 Draw a scale diagram of a cricket oval which is 150 m long and 100 m wide. Use a scale of 1 : 2000.

8 A block of land is 35 m long and 15 m wide. A building 20 m by 9 m is to be built so it is 15 m from the back fence, and equally spaced between the side fences. Use a scale of 1 : 500 to draw a scale diagram of the situation.

9 Use a scale of your choice to draw a scale diagram of:

a an 8-ball table which is 3 m long and 1.5 m wide

b a triangular field with side lengths 30 m, 40 m, and 50 m.

10 Look at the floorplan alongside.

a What is the scale factor for this diagram?

b Find:

i the length of the house

ii the dimensions of the porch

iii the area of bedroom 2.

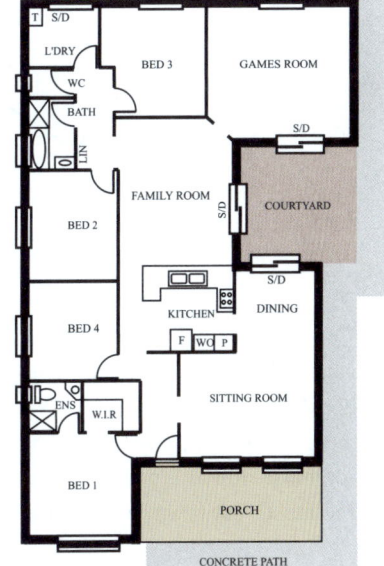

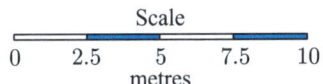

11 The scale for the map alongside is 1 : 2 000 000.

a Find the distance between:

i town A and town B

ii town C and town D.

b Julian's car uses on average 15 pence worth of petrol for each kilometre travelled. Find the cost for Julian to drive from:

i town B to town C

ii town B to town D.

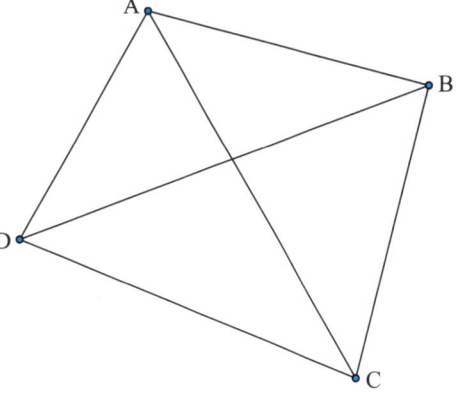

12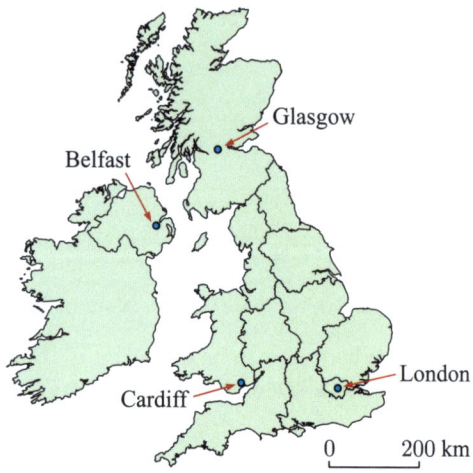
In the map of the United Kingdom alongside, the scale tells us that 200 km of actual length is represented by 1.25 cm on the map.

a What is the scale factor for this diagram?
b Estimate the length of the border between England and Scotland.
c Find the straight line distance between:
  i London and Belfast
  ii Cardiff and Glasgow.

13 In the map below, petrol stations are marked with a Ⓟ.

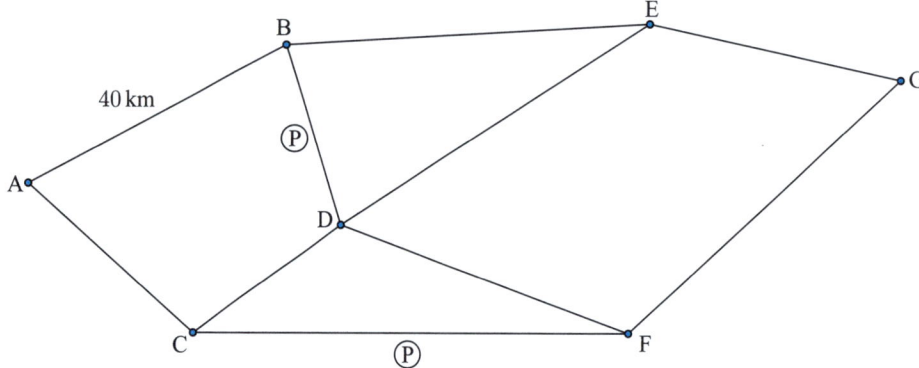

a Find the scale factor for the map.
b Joe wants to travel from town A to town G. Find the shortest route that Joe can take, and the length of that route, if:
  i there are no restrictions
  ii Joe must stop at a petrol station along the way
  iii Joe must visit town D along the way.

14 The formal garden shown is 12 m long and 10 m wide.
  a Find the scale factor for the diagram.
  b Find the diameter of the pond.
  c Find the perimeter of the hedge.

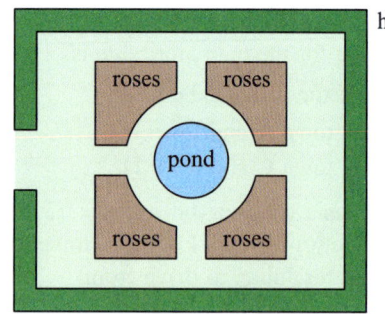

Activity   Scale diagrams

What to do:

1. **a** Measure the dimensions of your bedroom, including the main pieces of furniture.
   b Use an appropriate scale to draw a scale diagram of your bedroom.
   c Compare your diagram with those of your classmates. Did most people use the same scale for their diagram?

2. **a** Use online mapping software to print a map of your school, your street, or a local park.
   b Use the scale given to find the scale factor for the map.
   c Use a tape measure to find the actual length of a feature shown on the map, and compare it with the corresponding distance on the map. Do you think the map is accurate? Discuss your answer.

Review set 6A

1. Find the ratio of triangles to rectangles:

2. A swimming centre offers daytime and evening classes in the ratio $4:7$. What fraction of the classes are:
   a daytime classes
   b evening classes?

3. Express each ratio in simplest form:
   a $10:12$
   b $400:150$
   c $\frac{1}{2}:\frac{1}{5}$

4. During a tennis match, Heather served 16 aces and 10 double faults. Find the ratio of aces to double faults in simplest form.

5. The ratio of cats to dogs at an animal shelter is $3:7$. 50% of the cats are male, and 40% of the dogs are male. What percentage of all the cats and dogs at the shelter are male?

6. Find x if:
   a $4:5 = 12:x$
   b $3:8 = 15:x$
   c $9:2 = x:14$

7. A pride of lions has adults and cubs in the ratio $3:5$. If there are 15 cubs, how many lions are there in total?

8. When James and Lucy go on long car trips, they share the driving in the ratio $4:3$. How far does each person drive on:
   a a 350 km trip
   b a 630 km trip?

9 Two sisters receive an inheritance in the ratio 5 : 4. How much does each receive from £180 000?

10 Eric spends his study time revising Mathematics, Science, and History in the ratio 3 : 5 : 2.
   a Yesterday Eric spent 4 hours revising in total. How long did he spend revising History?
   b Today Eric spent 45 minutes revising Mathematics. How long did he spend revising in total?

11 Alongside is a scale diagram of a building. Use your ruler and the scale given to find:
   a the height of the building
   b the width of the building.

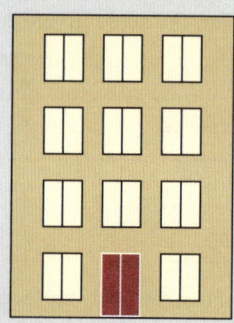

PRINTABLE DIAGRAMS

Scale 1 : 500

12  The actual length of this car is 4.8 m.
   a Find the scale factor for this diagram.
   b Find the diameter of the tyres.

Review set 6B

1 Write as a ratio:
   a Alex drinks 3 glasses of water for every glass of soft drink.
   b For every hour unplugged, my phone requires 7 minutes of recharging.

2 For the figure alongside, find:
   a the ratio of the shaded area to the unshaded area
   b the fraction of the figure which is shaded
   c the percentage of the figure which is shaded.

3 Express as a ratio in simplest form:
   a 8 mm is to 3 cm
   b 350 g is to 1 kg.

4 To make 2-stroke fuel, Tom mixes petrol and oil in the ratio 50 : 1. How much oil should he add to 1500 ml of petrol?

5 Decide whether the following pairs of ratios are equal:
   a 12 : 20 and 21 : 35
   b 18 : 4 and 45 : 12

6 Divide £500 in the ratio 3 : 7.

7 In open water swimming competitions, the ratio of safety personnel to competitors must be 1 : 20.
   a How many safety personnel are required for a competition involving 300 competitors?
   b Only 8 safety personnel are available for a particular competition. What is the maximum number of competitors allowed?

8 Bill sells tropical juice at a school fair. He buys orange juice for £1 per litre, and pineapple juice for £1.40 per litre. He then mixes the orange juice and pineapple juice in the ratio 3 : 2, and sells the resulting tropical juice for £2 per litre. How much profit does he make on each litre of tropical juice?

9 Find the scale factor for each scale:

   a Scale: 0, 50, 100, 150, 200, 250 metres

   b Scale: 0, 10, 20, 30, 40, 50 kilometres

10 Concrete is made by mixing gravel, sand, and cement in the ratio 5 : 3 : 1. How much of each ingredient is required to make 45 tonnes of concrete?

11 Rachael, Shane, and Tim share chocolates in the ratio 2 : 5 : 1. Shane received 15 more chocolates than Rachael. How many chocolates were shared in total?

12 Consider the scale diagram of a house block below.

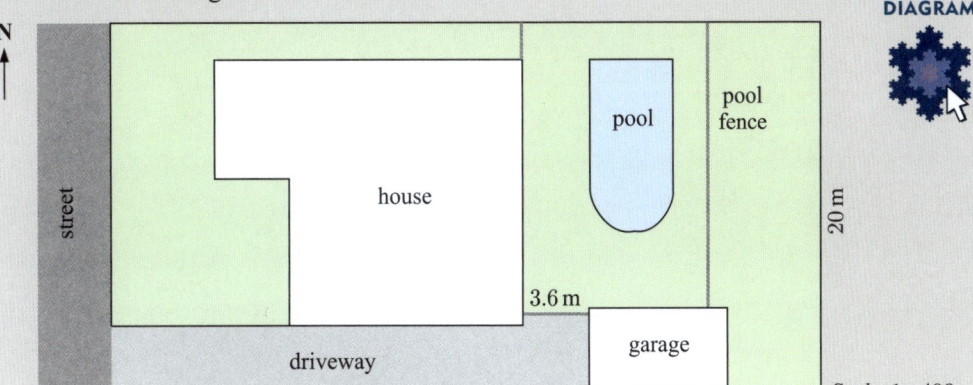

   a What are the dimensions of the block?
   b Find the length of the driveway from the street to the garage.

    c Find the distance between the house and the fence on:
        i the northern boundary   **ii** the southern boundary.
    d How far is the house from the street, at its closest point?
    e How far is the garage wall from the eastern boundary?
    f Find the area of the garage.
    g Find the total length of the pool fences.

7

Indices

Contents:
- **A** Index laws
- **B** Expansion laws
- **C** The zero index law
- **D** The negative index law
- **E** Standard form

118 Indices (Chapter 7)

Opening problem

Amedeo Avogadro (1776 - 1856) first proposed that the volume of a gas, at a given pressure and temperature, is proportional to the number of atoms or molecules present, regardless of the nature of the gas.

In 1865, **Johann Josef Loschmidt** estimated that one gram of hydrogen contains 6.02×10^{23} atoms. This number of atoms is now called one **mole** of a substance.

French physicist **Jean Perrin**, who earned a Nobel Prize in Physics, named the number in honour of Avogadro in 1909, for his work in the field of molecular theory.

Amedeo Avogadro

Things to think about:

a How can we write 6.02×10^{23} as an ordinary number?

b How many atoms would be in one tonne of hydrogen gas?

c Can you find the mass of 10^{30} atoms of hydrogen?

In **Chapter 1** we saw how we can use **index notation** to quickly write a product of factors.

If n is a positive integer, then a^n is the product of n factors of a.

$$a^n = \underbrace{a \times a \times a \times \ldots \times a}_{n \text{ factors}}$$

exponent, power, or index

base

In this Chapter we consider rules for indices which will assist in calculations. The rules will also give meaning to zero and negative indices.

We will also study **standard form** for writing very large and very small numbers.

A INDEX LAWS

Investigation 1   Discovering index laws

We can discover some laws for indices by considering several examples and looking for patterns.

What to do:

1 Copy and complete:

a $2^2 \times 2^3 = (2 \times 2) \times (2 \times 2 \times 2) = 2^5$

b $3^3 \times 3^1 = \boxed{} = \boxed{}$

c $5^4 \times 5^2 = \boxed{} = \boxed{}$

d $a^3 \times a^4 = \boxed{} = \boxed{}$

In general, $a^m \times a^n = \boxed{}$.

WORKSHEET

Indices (Chapter 7)

2 Copy and complete:

a $\dfrac{2^5}{2^3} = \dfrac{2 \times 2 \times \cancel{2} \times \cancel{2} \times \cancel{2}}{\cancel{2} \times \cancel{2} \times \cancel{2}} = 2^2$

b $\dfrac{5^4}{5^1} = \dfrac{5 \times 5 \times 5 \times 5}{5} = \boxed{}$

c $\dfrac{3^6}{3^4} = \boxed{} = \boxed{}$

d $\dfrac{7^4}{7^3} = \boxed{} = \boxed{}$

e $\dfrac{a^5}{a^2} = \boxed{} = \boxed{}$

f $\dfrac{x^7}{x^4} = \boxed{} = \boxed{}$

In general, $\dfrac{a^m}{a^n} = \boxed{}$.

3 Copy and complete:

a $(2^2)^3 = 2^2 \times 2^2 \times 2^2 = (2 \times 2) \times (2 \times 2) \times (2 \times 2) = \boxed{}$

b $(7^3)^2 = 7^3 \times 7^3 \quad = (7 \times 7 \times 7) \times \boxed{} = \boxed{}$

c $(3^2)^4 = \boxed{} = \boxed{} = \boxed{}$

d $(a^4)^3 = \boxed{} = \boxed{} = \boxed{}$

In general, $(a^m)^n = \boxed{}$.

From **Investigation 1** you should have found these **index laws** for **positive indices**:

> If m and n are positive integers, then:
>
> - $a^m \times a^n = a^{m+n}$
>   To **multiply** numbers with the **same base**, keep the base and **add** the indices.
>
> - $\dfrac{a^m}{a^n} = a^{m-n}, \quad a \neq 0$
>   To **divide** numbers with the **same base**, keep the base and **subtract** the indices.
>
> - $(a^m)^n = a^{m \times n}$
>   When **raising** a **power** to a **power**, keep the base and **multiply** the indices.

Example 1   ◀》 Self Tutor

Simplify using the laws of indices:

a $2^3 \times 2^2$

b $x^4 \times x^5$

a $\quad 2^3 \times 2^2$
$= 2^{3+2}$
$= 2^5$
$= 32$

b $\quad x^4 \times x^5$
$= x^{4+5}$
$= x^9$

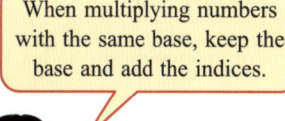

When multiplying numbers with the same base, keep the base and add the indices.

EXERCISE 7A

1 Simplify using the index laws:
- **a** $3^2 \times 3^3$
- **b** $2^2 \times 2^2$
- **c** $5^2 \times 5^4$
- **d** $3^4 \times 3^3$
- **e** 7×7^4
- **f** $(-2)^2 \times (-2)$
- **g** $x \times x$
- **h** $y \times y^2$
- **i** $a \times a^4$
- **j** $n^2 \times n$
- **k** $x^3 \times x^6$
- **l** $y^3 \times y^5$

Example 2 ◀)) Self Tutor

Simplify using the index laws:

a $\dfrac{3^5}{3^3}$ **b** $\dfrac{p^7}{p^3}$

a $\dfrac{3^5}{3^3}$
$= 3^{5-3}$
$= 3^2$
$= 9$

b $\dfrac{p^7}{p^3}$
$= p^{7-3}$
$= p^4$

When dividing numbers with the same base, keep the base and subtract the indices.

2 Simplify using the index laws:
- **a** $\dfrac{2^4}{2^2}$
- **b** $\dfrac{3^3}{3}$
- **c** $\dfrac{7^5}{7^2}$
- **d** $\dfrac{10^4}{10^3}$
- **e** $\dfrac{x^6}{x^2}$
- **f** $\dfrac{y^9}{y^5}$
- **g** $c^6 \div c^4$
- **h** $b^8 \div b^5$

For a power to a power, keep the base and multiply the indices.

Example 3 ◀)) Self Tutor

Simplify using the index laws:

a $(2^3)^2$ **b** $(x^4)^5$

a $(2^3)^2$
$= 2^{3 \times 2}$
$= 2^6$
$= 64$

b $(x^4)^5$
$= x^{4 \times 5}$
$= x^{20}$

3 Simplify using the index laws:
- **a** $(2^2)^3$
- **b** $(10^4)^2$
- **c** $(3^3)^2$
- **d** $(2^4)^3$
- **e** $(x^5)^2$
- **f** $(p^3)^3$
- **g** $(t^3)^4$
- **h** $(z^2)^6$

4 Simplify using the index laws:
- **a** $\dfrac{3^6}{3^2}$
- **b** 5×5^4
- **c** $\dfrac{7^5}{7^3}$
- **d** $(2^4)^2$
- **e** $2^2 \times 2^4$
- **f** $11^3 \div 11^2$
- **g** $(3^2)^3$
- **h** $(5^2)^3$

Indices (Chapter 7)

5 Simplify using the index laws:

- **a** $c^2 \times c^4$
- **b** $b^8 \div b^3$
- **c** $(y^5)^3$
- **d** $y^4 \times y^6$
- **e** $q \times q^6$
- **f** $(z^6)^5$
- **g** $t^{10} \div t^7$
- **h** $a^2 \times a^n$
- **i** $g^4 \div g$
- **j** $n^2 \times n^3 \times n^5$
- **k** $(k^4)^2 \div k$
- **l** $(p^2)^2 \times p^2$

6 Copy and complete, replacing each □ with a number or operation:

- **a** $(7^2)^{\square} = 7^6$
- **b** $2^4 \,\square\, 2 = 2^3$
- **c** $2^2 \,\square\, 2^7 = 2^9$
- **d** $(x^{\square})^4 = x^{12}$
- **e** $a^5 \,\square\, a^5 = a^{10}$
- **f** $5^9 \,\square\, 5^3 = 5^6$

7 Simplify using the index laws:

- **a** $\dfrac{5a^3}{a}$
- **b** $3q^2 \times 5q$
- **c** $8x^2 y \times 2xy^3$
- **d** $\dfrac{21t^3}{3t^2}$

B EXPANSION LAWS

We will now look at index laws for raising a product or quotient to a power.

For example, given $(3a)^4$ or $\left(\dfrac{x}{y}\right)^6$, we need laws which allow us to write the expressions without brackets. We call these **expansion laws**.

Investigation 2 — Discovering expansion laws

Look for any patterns as you complete the following Investigation.

WORKSHEET

What to do:

1 Copy and complete the following:

- **a** $(ab)^4 = ab \times ab \times ab \times ab = a \times a \times a \times a \times b \times b \times b \times b = \boxed{}$
- **b** $(ab)^3 = \boxed{} = \boxed{} = \boxed{}$
- **c** $(2a)^5 = \boxed{} = \boxed{} = \boxed{}$

In general, $(ab)^n = \boxed{}$.

2 Copy and complete:

- **a** $\left(\dfrac{a}{b}\right)^2 = \dfrac{a}{b} \times \dfrac{a}{b} = \dfrac{a \times a}{b \times b} = \boxed{}$
- **b** $\left(\dfrac{a}{b}\right)^3 = \dfrac{a}{b} \times \dfrac{a}{b} \times \dfrac{a}{b} = \boxed{} = \boxed{}$
- **c** $\left(\dfrac{a}{b}\right)^4 = \boxed{} = \boxed{} = \boxed{}$

In general, $\left(\dfrac{a}{b}\right)^n = \boxed{}$ for $b \neq 0$.

From **Investigation 2** you should have found these **expansion laws** for **positive indices**:

If n is a positive integer, then:
- $(ab)^n = a^n b^n$
- $\left(\dfrac{a}{b}\right)^n = \dfrac{a^n}{b^n}$ provided $b \neq 0$.

Example 4 🔊 Self Tutor

Remove the brackets and simplify:
a $(ab)^5$ **b** $(2xy)^3$

a $(ab)^5$
$= a^5 b^5$

b $(2xy)^3$
$= 2^3 \times x^3 \times y^3$
$= 8x^3 y^3$

Raise each factor to the given power.

EXERCISE 7B

1 Remove the brackets and simplify:

a $(pq)^2$ **b** $(xy)^4$ **c** $(ab)^6$ **d** $(abc)^3$
e $(2a)^3$ **f** $(3d)^5$ **g** $(2k)^5$ **h** $(5gh)^2$

Example 5 🔊 Self Tutor

Remove the brackets and simplify:
a $\left(\dfrac{m}{n}\right)^4$ **b** $\left(\dfrac{2}{b}\right)^3$

a $\left(\dfrac{m}{n}\right)^4$
$= \dfrac{m^4}{n^4}$

b $\left(\dfrac{2}{b}\right)^3 = \dfrac{2^3}{b^3}$
$= \dfrac{8}{b^3}$

Raise both the numerator and the denominator to the given power.

2 Remove the brackets and simplify:

a $\left(\dfrac{a}{b}\right)^2$ **b** $\left(\dfrac{b}{2}\right)^3$ **c** $\left(\dfrac{j}{k}\right)^4$ **d** $\left(\dfrac{2}{z}\right)^4$
e $\left(\dfrac{4}{x}\right)^2$ **f** $\left(\dfrac{2}{b}\right)^5$ **g** $\left(\dfrac{q}{2}\right)^4$ **h** $\left(\dfrac{3}{b}\right)^3$

3 Find:

a $\left(\dfrac{2}{5}\right)^2$ **b** $\left(\dfrac{3}{4}\right)^3$ **c** $\left(\dfrac{2}{3}\right)^4$ **d** $\left(\dfrac{1}{2}\right)^5$

Indices (Chapter 7) 123

Example 6

Express in simplest form, without brackets:

a $(3a^2)^2$ **b** $\left(\dfrac{2x}{y}\right)^3$

a $(3a^2)^2 = 3^2 \times (a^2)^2$
$= 9a^4$

b $\left(\dfrac{2x}{y}\right)^3 = \dfrac{2^3 \times x^3}{y^3}$
$= \dfrac{8x^3}{y^3}$

4 Express in simplest form, without brackets:
- **a** $(2a^2)^2$
- **b** $(3b^3)^2$
- **c** $(2c^2)^4$
- **d** $(2d^2)^5$
- **e** $(jk^3)^2$
- **f** $(xy^2)^3$
- **g** $(3g)^2 \times 2g$
- **h** $(5r^2s)^2$

5 Express in simplest form, without brackets:
- **a** $\left(\dfrac{jk}{2}\right)^2$
- **b** $\left(\dfrac{2}{cd}\right)^2$
- **c** $\left(\dfrac{3p}{q}\right)^3$
- **d** $\left(\dfrac{z^2}{5}\right)^2$

C THE ZERO INDEX LAW

For all positive integers n, a^n is defined as the product of n factors of a:

$$a^n = \underbrace{a \times a \times a \times \ldots \times a}_{n \text{ factors}}$$

But what if $n = 0$? In the following **Investigation** we will discover how to define a^0 in a way that preserves the index laws we have already established.

Investigation 3 — The zero index law

What to do:

1 Find the value of:
- **a** $\dfrac{2}{2}$
- **b** $\dfrac{3}{3}$
- **c** $\dfrac{4}{4}$
- **d** $\dfrac{-8}{-8}$
- **e** $\dfrac{-5}{-5}$
- **f** $\dfrac{7}{7}$
- **g** $\dfrac{57}{57}$
- **h** $\dfrac{23}{23}$

2 Copy and complete: When a non-zero value is divided by itself, the result is always ......

For any $a \neq 0$, $\dfrac{a^3}{a^3} = $ ......

3 Use an index law to show that $\dfrac{a^3}{a^3} = a^0$.

4 Hence complete: $a^0 = $ ...... for all $a \neq 0$.
Check your answer by evaluating 2^0 and 5^0 on your calculator.

From **Investigation 3** you should have discovered that $a^0 = 1$ for all $a \neq 0$.

EXERCISE 7C

1. **a** Copy and complete: $\dfrac{3^4}{3^4} = \dfrac{81}{81} = \ldots\ldots$ **b** Show that $\dfrac{3^4}{3^4} = 3^0$.

   c Hence state the value of 3^0.

2. Simplify:
   - **a** 7^0
   - **b** 41^0
   - **c** x^0
   - **d** 5×2^0
   - **e** $8 + 10^0$
   - **f** $6 - 6^0$
   - **g** 11×11^0
   - **h** $p^6 \times p^0$
   - **i** $(2^3)^0$
   - **j** $(2^0)^3$
   - **k** 7×3^0
   - **l** $(7 \times 3)^0$

3. Simplify:
   - **a** $\dfrac{n^2}{n^2}$
   - **b** $\dfrac{k^6}{k^6}$
   - **c** $\dfrac{xy}{y}$
   - **d** $\dfrac{a^3 b^2}{b^2}$

Puzzle

Click on the icon for a printable copy of the solution grid.

PRINTABLE GRID

Across
- 2 $3^8 \times 3^3$
- 5 $10^4 \div 10$
- 7 $7^5 \div 7^3$
- 8 $(11^2)^2$
- 11 $6^6 \div 6^3$
- 12 $11^2 \times 11$
- 14 $3^7 \div 3^2$
- 15 $6^2 - 6^0$
- 16 $2^6 - 5^0$

Down
- 1 $2^5 \times 2^4$
- 2 $(2^2)^5$
- 3 $2^{10} \div 2^6$
- 4 $(3^3)^2$
- 6 7×7^3
- 8 $5^4 \times 5^2$
- 9 $5^6 \div 5^3$
- 10 $5^2 \times 5^3$
- 13 $7^7 \div 7^4$

D THE NEGATIVE INDEX LAW

In the following **Investigation** we consider numbers where the index is negative, such as 7^{-2} and 10^{-4}.

Investigation 4   Negative indices

What to do:

1. Consider the fraction $\dfrac{7^3}{7^5}$.

   a By expanding and then cancelling common factors, show that $\dfrac{7^3}{7^5} = \dfrac{1}{7^2}$.

   b Use an index law to show that $\dfrac{7^3}{7^5} = 7^{-2}$.

   c Hence copy and complete: $7^{-2} = \ldots\ldots$

2. Use the fact that $a^0 = 1$ to copy and complete: $a^{-n} = a^{0-n} = \dfrac{a^0}{\square} = \dfrac{\square}{\square}$.

Indices (Chapter 7) 125

You should have discovered the following law for **negative indices**:

> If a is any non-zero number and n is an integer, then $a^{-n} = \dfrac{1}{a^n}$.
>
> This means that a^n and a^{-n} are **reciprocals** of one another.
>
> In particular, notice that $a^{-1} = \dfrac{1}{a}$.

Example 7 ◀)) Self Tutor

Simplify:
a 3^{-1} **b** 10^{-4}

a 3^{-1}
$= \dfrac{1}{3^1}$
$= \dfrac{1}{3}$

b 10^{-4}
$= \dfrac{1}{10^4}$
$= \dfrac{1}{10\,000}$

The negative index indicates the **reciprocal**.

Using the negative index law,
$\left(\dfrac{2}{3}\right)^{-4} = \dfrac{1}{\left(\dfrac{2}{3}\right)^4}$

$= \dfrac{1^4}{\left(\dfrac{2}{3}\right)^4}$ $\quad \{1^4 = 1\}$

$= \left(\dfrac{1}{\frac{2}{3}}\right)^4$ $\quad$ {expansion law}

$= \left(\dfrac{3}{2}\right)^4$ $\quad$ {simplifying}

So, in general: $\left(\dfrac{a}{b}\right)^{-n} = \left(\dfrac{b}{a}\right)^n$ provided $a \neq 0$, $b \neq 0$.

Example 8 ◀)) Self Tutor

Simplify, giving answers in simplest rational form:
a $\left(\dfrac{2}{3}\right)^{-1}$ **b** $\left(\dfrac{3}{5}\right)^{-2}$ **c** $8^0 - 8^{-1}$

a $\left(\dfrac{2}{3}\right)^{-1}$
$= \left(\dfrac{3}{2}\right)^1$
$= \dfrac{3}{2}$

b $\left(\dfrac{3}{5}\right)^{-2} = \left(\dfrac{5}{3}\right)^2$
$= \dfrac{5^2}{3^2}$
$= \dfrac{25}{9}$

c $8^0 - 8^{-1}$
$= 1 - \dfrac{1}{8}$
$= \dfrac{7}{8}$

"Simplest rational form" means "as a fraction in lowest terms".

EXERCISE 7D

1 Write as a power with a negative index:

 a $\dfrac{3}{3^3}$ **b** $\dfrac{5^2}{5^3}$ **c** $\dfrac{2^3}{2^5}$ **d** $\dfrac{3^2}{3^4}$ **e** $\dfrac{7^4}{7^5}$

2 Simplify:

 a 5^{-1} **b** 4^{-1} **c** 8^{-1} **d** 10^{-1} **e** 3^{-2}
 f 2^{-2} **g** 11^{-2} **h** 7^{-2} **i** 3^{-3} **j** 2^{-7}

3 Simplify:

 a $3^1 - 3^{-1}$ **b** $3^0 + 3^{-1}$ **c** $7^0 - 7^{-1}$ **d** $5^0 + 5^1 - 5^{-1}$

4 Write as a fraction:

 a x^{-1} **b** k^{-1} **c** t^{-3} **d** r^{-5}

5 Simplify, giving answers in simplest rational form:

 a $\left(\dfrac{1}{3}\right)^{-1}$ **b** $\left(\dfrac{1}{5}\right)^{-1}$ **c** $\left(\dfrac{3}{7}\right)^{-1}$ **d** $\left(\dfrac{5}{2}\right)^{-1}$
 e $\left(\dfrac{1}{10}\right)^{-1}$ **f** $\left(\dfrac{5}{6}\right)^{-2}$ **g** $\left(2\dfrac{1}{4}\right)^{-2}$ **h** $\left(\dfrac{2}{3}\right)^{-3}$

Example 9 ◀)) Self Tutor

Write without brackets or negative indices:

 a $8ab^{-1}$ **b** $8(ab)^{-1}$

 a $8ab^{-1} = \dfrac{8a}{1} \times \dfrac{1}{b}$ **b** $8(ab)^{-1} = 8 \times \dfrac{1}{ab}$

 $= \dfrac{8a}{b}$ $= \dfrac{8}{ab}$

6 Write without brackets or negative indices:

 a $2x^{-1}$ **b** $(2x)^{-1}$ **c** $(3q)^{-1}$ **d** $3q^{-1}$
 e $7a^{-2}$ **f** $(7a)^{-2}$ **g** $(5z)^{-2}$ **h** $5z^{-2}$
 i st^{-1} **j** $(st)^{-1}$ **k** gh^{-3} **l** $(gh)^{-3}$
 m $4cd^{-2}$ **n** $(4cd)^{-2}$ **o** $4(cd)^{-2}$ **p** $(7m^{-3})^{-1}$

7 Write as a power of 2, 3, 5, or 7:

 a 4 **b** $\dfrac{1}{4}$ **c** 9 **d** $\dfrac{1}{9}$
 e 16 **f** $\dfrac{1}{16}$ **g** 25 **h** $\dfrac{1}{25}$
 i 81 **j** $\dfrac{1}{81}$ **k** 49 **l** $\dfrac{1}{49}$
 m $\dfrac{1}{1024}$ **n** $\dfrac{1}{125}$ **o** 1 **p** $\dfrac{1}{2^a}$

E STANDARD FORM

There are many situations where we need to describe very large and very small numbers. For example:

- There are about $75\,000\,000\,000\,000$ cells in the human body.
- A glass of tap water contains about $0.000\,04$ grams of chlorine.

Numbers with so many digits can be hard to comprehend and operate with. We can use **standard form** or **scientific notation** to write these numbers in a way that is easier to understand.

> **Standard form** involves writing a given number as *a number between* 1 *and* 10, multiplied by an *integer power of* 10. It has the form
> $$a \times 10^n \text{ where } 1 \leqslant a < 10 \text{ and } n \text{ is an integer.}$$

Consider the pattern alongside.

Notice that each time we divide by 10, the index or exponent of 10 decreases by one.

So, when a number is written in standard form:

- If the original number is greater than or equal to 10, then n is **positive**.
- If the original number is less than 1, then n is **negative**.
- If the original number is between 1 and 10, we write the number as it is and multiply it by 10^0, which is 1.

$$\begin{aligned}
\div 10 \quad 10\,000 &= 10^4 \quad -1\\
\div 10 \quad 1000 &= 10^3 \quad -1\\
\div 10 \quad 100 &= 10^2 \quad -1\\
\div 10 \quad 10 &= 10^1 \quad -1\\
\div 10 \quad 1 &= 10^0 \quad -1\\
\div 10 \quad \tfrac{1}{10} &= 10^{-1} \quad -1\\
\div 10 \quad \tfrac{1}{100} &= 10^{-2} \quad -1\\
\tfrac{1}{1000} &= 10^{-3}
\end{aligned}$$

To write a number larger than 10, we start with a number between 1 and 10, then multiply it by a positive power of 10.

For example:
$$\begin{aligned}
27 &= 2.7 \times 10\\
580 &= 5.8 \times 100 = 5.8 \times 10^2\\
3040 &= 3.04 \times 1000 = 3.04 \times 10^3
\end{aligned}$$

> To multiply a number by 10^n, $n > 0$, shift the decimal point n places to the **right**.

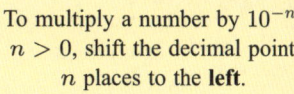

To write a number smaller than 1, we start with a number between 1 and 10, then *divide* it by a power of 10.

For example:
$$\begin{aligned}
0.75 &\qquad 0.0006\\
= 7.5 \div 10 &\qquad = 6 \div 10\,000\\
= 7.5 \times \tfrac{1}{10^1} &\qquad = 6 \times \tfrac{1}{10^4}\\
= 7.5 \times 10^{-1} &\qquad = 6 \times 10^{-4}
\end{aligned}$$

> To multiply a number by 10^{-n}, $n > 0$, shift the decimal point n places to the **left**.

EXERCISE 7E.1

1 Write as powers of 10:
- **a** 100
- **b** 1000
- **c** 10
- **d** 100 000
- **e** 0.1
- **f** 0.01
- **g** 0.0001
- **h** 100 000 000

2 The following values are all equal to 53 000. Which of them is written in standard form?
- **A** 53×10^3
- **B** 0.53×10^5
- **C** 5.3×10^4
- **D** $53\,000 \times 10^0$

3 Copy and complete to write the following numbers in standard form:
- **a** $376 = 3.76 \times 10^{......}$
- **b** $8000 = ... \times 10^3$
- **c** $0.04 = ... \times 10^{-2}$
- **d** $0.005\,07 = 5.07 \times 10^{......}$
- **e** $9\,040\,000 = ... \times 10^6$
- **f** $0.000\,000\,23 = 2.3 \times 10^{......}$

Example 10   ◀)) Self Tutor

Express in standard form:
- **a** 4 500 000
- **b** 0.000 592

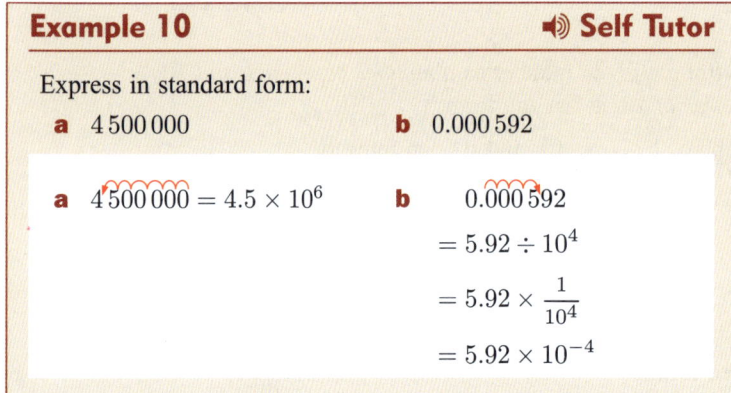

In standard form, there is only one digit before the decimal point, and this digit must be non-zero.

4 Express in standard form:
- **a** 425
- **b** 425 000
- **c** 4.25
- **d** 0.425
- **e** 20.1
- **f** 20 100
- **g** 0.002 01
- **h** 2 010 000
- **i** 3870
- **j** 0.0387
- **k** 38 700 000
- **l** 0.000 387

5 Express in standard form:
- **a** The circumference of the Earth is approximately 40 075 kilometres.
- **b** Bacteria are single cell organisms, some of which have a diameter of 0.0004 mm.
- **c** There are typically 40 million bacteria in a gram of soil.
- **d** The probability that your six numbers will be selected for Lotto on Saturday night is 0.000 000 141 62.
- **e** Superfine sheep have wool fibres as low as 0.01 mm in diameter.
- **f** The central temperature of the Sun is 15 million degrees Celsius.

Example 11   ◀)) Self Tutor

Write as an ordinary decimal number:
- **a** 3.2×10^2
- **b** 5.76×10^{-5}

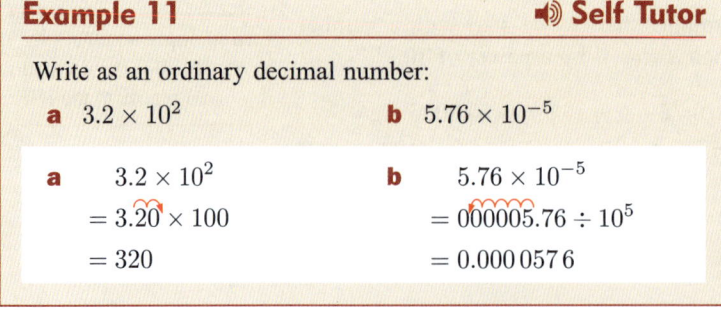

6 Write as an ordinary decimal number:

- **a** 5×10^4
- **b** 3×10^3
- **c** 1.8×10^7
- **d** 8.1×10^2
- **e** 6.5×10^5
- **f** 1.1×10^1
- **g** 2.75×10^8
- **h** 8×10^6

7 Write as an ordinary decimal number:

- **a** 3×10^{-2}
- **b** 9×10^{-5}
- **c** 7×10^{-3}
- **d** 4.1×10^{-4}
- **e** 8.2×10^{-6}
- **f** 7.61×10^{-1}
- **g** 3.25×10^{-7}
- **h** 2×10^{-8}

8 Express as ordinary decimal numbers:

- **a** The wavelength of blue light is about 4.75×10^{-7} m.
- **b** The estimated world population for 2017 was 7.53×10^9 people.
- **c** Physicists in Japan created a model bull which is only 1.2×10^{-5} m long.
- **d** The length of the Earth's equator is approximately 4.01×10^4 km.
- **e** A mosquito weighs about 1.5×10^{-6} kg.

9 Write in standard form:

- **a** 18.17×10^6
- **b** 0.934×10^{11}
- **c** 0.041×10^{-2}

10 Answer the **Opening Problem** on page **118**.

Example 12 ◀) **Self Tutor**

Which number is larger:

a 3.27×10^4 or 1.56×10^5

b 2.6×10^{-4} or 2.79×10^{-4}?

a 1.56×10^5 has a higher power of 10 than 3.27×10^4.
$\therefore 1.56 \times 10^5 > 3.27 \times 10^4$.

b Both numbers have the same power of 10. $2.79 > 2.6$, so $2.79 \times 10^{-4} > 2.6 \times 10^{-4}$.

---

11 Which number is larger:

- **a** 2.2×10^8 or 5.8×10^6
- **b** 7.71×10^{-7} or 3.5×10^{-5}
- **c** 4.9×10^{-3} or 4.9×10^3
- **d** 6.2×10^8 or 6.41×10^8
- **e** 1.006×10^{-6} or 1.01×10^{-6}
- **f** 9×10^{-8} or 0?

The power of 10 indicates the order of magnitude.

12 Electrons and protons are tiny particles present in everything around us.
An electron has mass $0.000\,000\,000\,000\,000\,000\,000\,000\,000\,911$ g.
A proton has mass $0.000\,000\,000\,000\,000\,000\,000\,001\,67$ g.

- **a** Express the mass of an electron in standard form.
- **b** Express the mass of a proton in standard form.
- **c** Which is heavier, an electron or a proton?

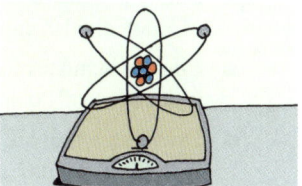

Example 13

Simplify the following, giving your answer in standard form:

a $(5 \times 10^4) \times (4 \times 10^5)$

b $(8 \times 10^5) \div (2 \times 10^3)$

a $(5 \times 10^4) \times (4 \times 10^5)$
$= 5 \times 4 \times 10^4 \times 10^5$
$= 20 \times 10^{4+5}$
$= 2 \times 10^1 \times 10^9$
$= 2 \times 10^{10}$

b $(8 \times 10^5) \div (2 \times 10^3)$
$= \dfrac{8 \times 10^5}{2 \times 10^3}$
$= \dfrac{8}{2} \times 10^{5-3}$
$= 4 \times 10^2$

13 Simplify the following, giving your answer in standard form:

a $(2 \times 10^5) \times (3 \times 10^2)$
b $(4 \times 10^3) \times (6 \times 10^3)$
c $(8 \times 10^5) \times (5 \times 10^6)$
d $(9 \times 10^9)^2$
e $(3 \times 10^4) \times (3 \times 10^{-9})$
f $(7 \times 10^{-3})^2$
g $(8 \times 10^2) \div (2 \times 10^3)$
h $(6 \times 10^7) \div (3 \times 10^{-4})$
i $(2 \times 10^9) \div (8 \times 10^5)$

Example 14

Simplify the following, giving your answer in standard form:

a $7 \times 10^6 + 5 \times 10^6$

b $8.7 \times 10^8 - 3 \times 10^7$

a $7 \times 10^6 + 5 \times 10^6$
$= 12 \times 10^6$
$= 1.2 \times 10^7$

b $8.7 \times 10^8 - 3 \times 10^7$
$= 8.7 \times 10^8 - 0.3 \times 10^8$
$= 8.4 \times 10^8$

To add or subtract numbers in standard form, express the numbers with the same power of 10.

14 Simplify the following, giving your answer in standard form:

a $3 \times 10^8 + 6 \times 10^8$
b $5.2 \times 10^4 - 2 \times 10^4$
c $8 \times 10^7 + 6 \times 10^7$
d $6.4 \times 10^5 - 5.9 \times 10^5$
e $7.1 \times 10^{-3} + 4.2 \times 10^{-3}$
f $2 \times 10^5 + 7 \times 10^4$
g $5 \times 10^9 - 4 \times 10^8$
h $3.6 \times 10^{-4} + 1.7 \times 10^{-5}$
i $5.81 \times 10^4 - 1.4 \times 10^3$
j $7 \times 10^6 + 2.3 \times 10^8$
k $3.54 \times 10^{-7} + 8.5 \times 10^{-9}$
l $1.05 \times 10^{-5} - 8.2 \times 10^{-6}$

15 The table alongside shows the production of cereals for some European regions in 2015.

a Which of the regions listed produced the most cereals?

b Find, in standard form, the cereal production of the United Kingdom in kilograms.

c Find, in standard form, the combined cereal production of Poland and Lithuania.

d How much more cereal did Finland produce than Switzerland?

e How many *times* more cereal did France produce than Luxembourg?

Region	Cereal production (tonnes)
Finland	3.7×10^6
France	7.2×10^7
Lithuania	6.1×10^6
Luxembourg	1.8×10^5
Poland	2.8×10^7
Switzerland	8.9×10^5
United Kingdom	2.5×10^7

Indices (Chapter 7) 131

STANDARD FORM ON A CALCULATOR

Calculators use standard form to display very large and very small numbers. However, if we wish, we can tell the calculator to give all its answers in standard form. Instructions for writing numbers in standard form can be found by clicking on the icon.

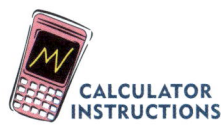

Example 15

Self Tutor

Use your calculator to find:

a $(1.67 \times 10^4) \times (2.3 \times 10^7)$

b $(9.49 \times 10^{-3}) \div (2.6 \times 10^6)$

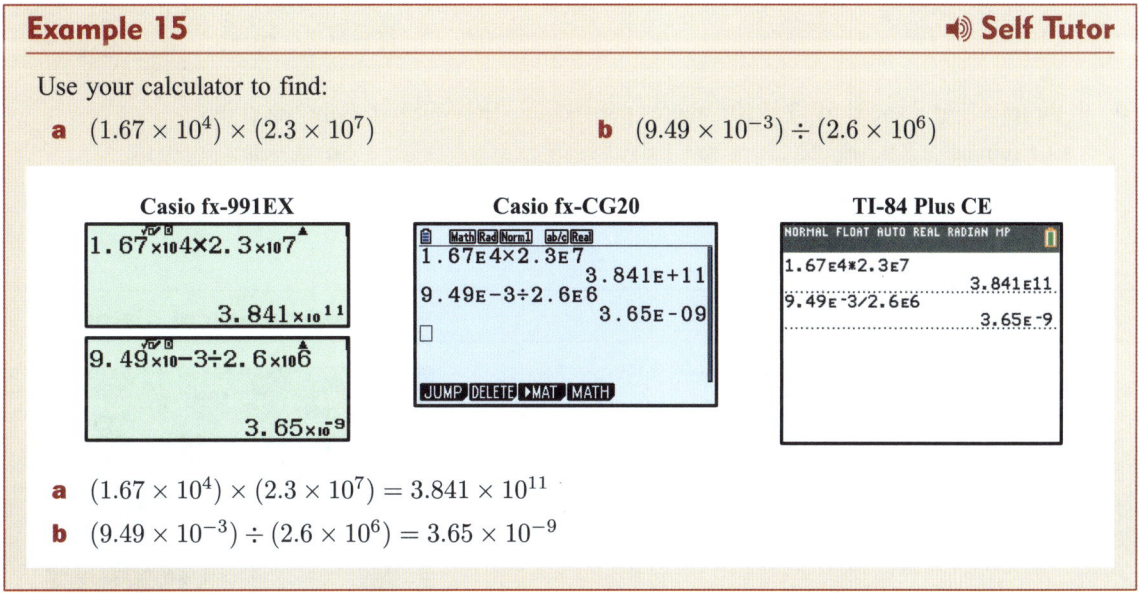

a $(1.67 \times 10^4) \times (2.3 \times 10^7) = 3.841 \times 10^{11}$

b $(9.49 \times 10^{-3}) \div (2.6 \times 10^6) = 3.65 \times 10^{-9}$

EXERCISE 7E.2

1 Calculate the following, giving each answer in standard form. The decimal part should be written correct to 2 decimal places.

a $0.0003 \times 0.01 \div 5000$

b $375 \times 220 \times 290\,000$

c $876\,000 \times 25\,000$

d $800 \times 740 \times 67\,800$

e $0.002\,12 \div 3\,400\,000$

f $0.019 \times 0.000\,27 \times 0.02$

2 Find in standard form, with the decimal part correct to 2 decimal places:

a $(2.81 \times 10^5) \times (3.4 \times 10^4)$

b $(9.81 \times 10^{-4})^2$

c $\dfrac{3.43 \times 10^{-6}}{7 \times 10^7}$

d $(8.66 \times 10^{-3}) \times (7.5 \times 10^{-5})$

e $\dfrac{1}{2.5 \times 10^6}$

f $(7.59 \times 10^4)^3$

3 For the following, give answers in standard form correct to 3 significant figures:

a How many millimetres are there in 479.8 kilometres?

b How many seconds are there in one year?

c How many seconds are there in a millennium?

d How many kilograms are there in 0.5 milligrams?

4 If a rocket travels at 3600 km h^{-1}, how far will it travel in:

   a 1 day   **b** 1 week   **c** 2 years?

Give your answers in standard form with decimal part correct to 2 decimal places. Assume that 1 year = 365 days.

5 Light travels at a speed of 3×10^8 metres per second. How far will light travel in:

   a 1 minute   **b** 1 day   **c** 1 year?

Give your answers in standard form with decimal part correct to 2 decimal places. Assume that 1 year = 365 days.

Activity — Astronomical distances

Astronomers often have to deal with extremely large distances. They measure things like the distance between stars, and the size of galaxies. Because of the large numbers, astronomers have their own set of units.

1 One **astronomical unit** (AU) is the distance between the Earth and the Sun, which is about 150 million kilometres. Write this number in standard form.

2 One **light year** is 9.5×10^{12} km. How many AU is this?

3 For longer distances, astronomers use a unit called a **parsec**, which is about 3.26 light years. Write this number in standard form, in:

   a AU   **b** km.

4 The distance d (in parsecs) from Earth to a star can be calculated using the formula $d = 2 \times p^{-1}$, where p is the *parallax* of the star (found through a series of other measurements).

Copy and complete the following table:

Star	Parallax	Distance from Earth (parsecs)
Alpha Centauri	1.494 46	
Polaris	0.015 12	
Deneb	4.58×10^{-3}	

Review set 7A

1 Simplify using the index laws:

   a $k^3 \times k^6$ **b** $(b^4)^3$ **c** $\dfrac{p^{13}}{p^5}$

2 Express as a power of 5:

   a $5^4 \times 5^3$ **b** $(5^3)^2$ **c** 1

3 Simplify using the index laws:

   a $\dfrac{6b^4}{2b^2}$ **b** $3m^2 \times (-m)$ **c** $2x^2y \times 3xy^2$

4 Express in simplest form, without brackets:

   a $(5x)^2$ **b** $(3mn)^3$ **c** $\left(\dfrac{p}{2q}\right)^4$

5 Simplify:

   a 8^0 **b** 13×13^0 **c** $7 - 5^0$

6 Simplify:

   a 9^{-1} **b** 6^{-2} **c** 10^{-3}

7 Simplify, using the index laws:

   a $5^{-2} \times 5$ **b** $b^7 \div b^{-2}$ **c** $(x^4)^{-2}$

8 Express in simplest form, without brackets or negative indices:

   a $(5c)^{-1}$ **b** $7k^{-2}$ **c** $(4d^2)^{-3}$

9 Write in standard form:

   a 9 **b** $34\,900$ **c** 0.0075

10 Write as an ordinary decimal number:

   a 2.81×10^6 **b** 2.81×10^0 **c** 2.81×10^{-3}

11 Simplify, giving your answer in standard form:

   a $(6 \times 10^3) \times (7.1 \times 10^4)$ **b** $(2.4 \times 10^6) \div (4 \times 10^2)$

   c $1.5 \times 10^5 + 2.8 \times 10^6$ **d** $7.1 \times 10^{-2} - 9.5 \times 10^{-3}$

12 The Earth orbits around the Sun at a speed of approximately 1.07×10^5 km h^{-1}. How far does the Earth move, relative to the Sun, in:

   a 1 day   **b** 1 week   **c** 1 year?

Give your answers in standard form with decimal part correct to 2 decimal places. Assume that 1 year = 365 days.

Review set 7B

1 Simplify using the index laws:

   a $x^3 \times x^3$ **b** $\dfrac{c^{12}}{c^7}$ **c** $(d^{11})^3$

2 Simplify, using the index laws:

   a $3^2 \times 3^6$ **b** $2^5 \div 2^5$ **c** $(y^3)^{-1}$

3 Copy and complete, replacing each $\square$ and $\triangle$ with a number or operation:

   a $3^5 \,\square\, 3^2 = 3^3$ **b** $(x^3)^{\square} = x^6$ **c** $(4a)^{\square} \times a^3 = 64a^{\triangle}$

4 Express in simplest form, without brackets:

   a $(-2x)^3$ **b** $(3m^2)^2$ **c** $\left(\dfrac{m}{4n}\right)^3$

5 Simplify:

   a $21^0 - 3$ **b** y^0 **c** $\left(\dfrac{3}{5}\right)^3$

6 Remove the brackets and simplify:

   a $\left(\dfrac{c}{d}\right)^2$ **b** $\left(\dfrac{q}{4}\right)^3$ **c** $\left(\dfrac{ab}{8}\right)^2$

7 Simplify, giving answers in simplest rational form:

   a $\left(\dfrac{4}{5}\right)^{-1}$ **b** $\left(\dfrac{2}{7}\right)^{-2}$ **c** $\left(3\dfrac{1}{3}\right)^{-2}$

8 Write without brackets or negative indices:

   a $a^{-2}b$ **b** $(5a^{-2})^2$ **c** $(5p^{-3})^{-1}$

9 Write in standard form:

   a 263.57 **b** $0.000\,511$ **c** $863\,400\,000$

10 Write as an ordinary decimal number:

   a 2.78×10^0 **b** 3.99×10^7 **c** 2.081×10^{-3}

11 Simplify, giving your answer in standard form:

   a $(8 \times 10^3)^2$ **b** $(3.6 \times 10^5) \div (6 \times 10^{-2})$

   c $6.8 \times 10^4 + 7.03 \times 10^4$ **d** $5.8 \times 10^{-3} - 8 \times 10^{-5}$

12 How many kilometres are there in 0.21 millimetres? Give your answer in standard form.

Algebra: Expansion

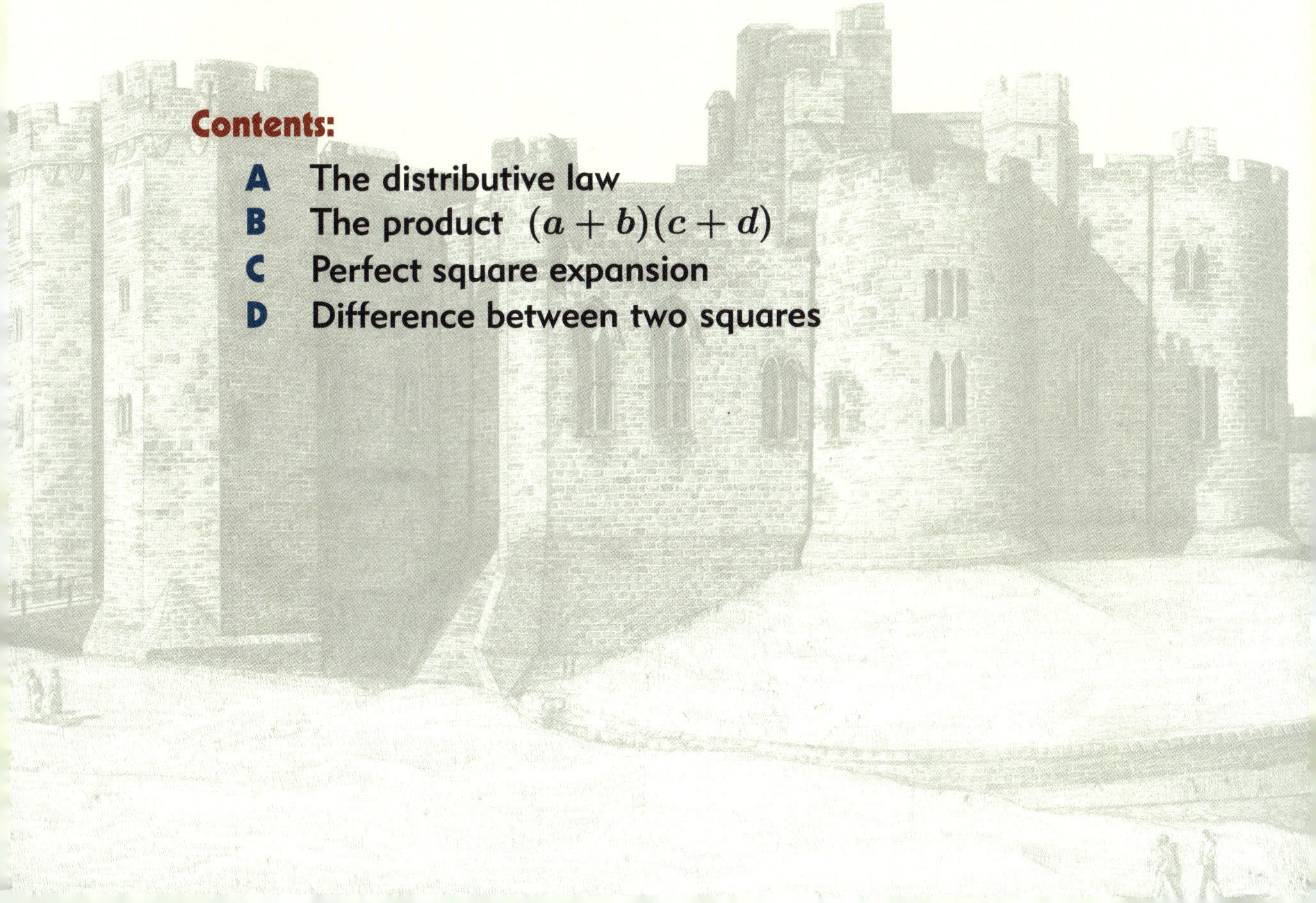

Contents:
- **A** The distributive law
- **B** The product $(a+b)(c+d)$
- **C** Perfect square expansion
- **D** Difference between two squares

Opening problem

Anton thinks that to find the square of the sum of two numbers, you can just square each of the numbers, then add the results.

Things to think about:

a Does $(5+3)^2 = 5^2 + 3^2$?

b Can you write a statement in words which describes Anton's expression $5^2 + 3^2$?

c Can you explain why Anton is incorrect?

When we write real life problems in terms of algebra, we often obtain expressions containing brackets. To solve equations containing these expressions, we may need to **expand** the brackets and **simplify** the result. In this Chapter we will study the expansion of algebraic expressions.

A THE DISTRIBUTIVE LAW

Over the summer holidays, Jasmin is able to do some jobs. The holidays are 11 weeks long.

Each week, Jasmin earns £5 from her father for washing the family car, and £7 from her grandfather for mowing lawns.

Over the 11 week period, Jasmin earns a total of $11 \times £5$ from her father, and $11 \times £7$ from her grandfather.

We could also say that Jasmin earns 11 lots of $(£5 + £7)$, which is $11 \times (5 + 7)$ pounds.

Consequently, $11(5 + 7) = 11 \times 5 + 11 \times 7$.

Notice that the factor 11 outside the brackets is multiplied by each term inside the brackets. We say that 11 is the **coefficient** of the expression in the brackets.

$\overset{\frown}{a(b+c)} = ab + ac$ is called the **distributive law**.

The distributive law says that we must multiply the coefficient by each term within the brackets, then add the results.

GEOMETRIC DEMONSTRATION

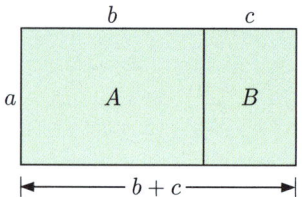

The area of the large rectangle is $a(b+c)$.

However, this area could also be found by adding the areas of the two small rectangles. This is area A + area $B = ab + ac$.

So, $a(b+c) = ab + ac$. {equating areas}

Algebra: Expansion (Chapter 8) 137

Example 1   Self Tutor

Expand and simplify:

a $5(x+4)$ **b** $4(y-3)$

a $\quad 5(x+4)$
$= 5 \times x + 5 \times 4$
$= 5x + 20$

b $\quad 4(y-3)$
$= 4(y+ -3)$
$= 4 \times y + 4 \times (-3)$
$= 4y - 12$

> Multiply each term inside the brackets by the factor outside the brackets.

EXERCISE 8A

1 Expand and simplify:

  a $2(x+7)$ **b** $3(x-2)$ **c** $4(a+3)$ **d** $5(a+c)$
  e $6(b-3)$ **f** $7(m+4)$ **g** $2(n-p)$ **h** $4(p-q)$
  i $3(5+x)$ **j** $5(y-x)$ **k** $8(t-8)$ **l** $6(d+e)$
  m $4(10-j)$ **n** $7(y+n)$ **o** $2(n-12)$ **p** $8(11-d)$

Example 2   Self Tutor

Expand and simplify:

a $3(2a+7)$ **b** $x(5-4x)$

a $\quad 3(2a+7)$
$= 3 \times 2a + 3 \times 7$
$= 6a + 21$

b $\quad x(5-4x)$
$= x(5+ -4x)$
$= x \times 5 + x \times (-4x)$
$= 5x - 4x^2$

2 Expand and simplify:

  a $9(2x+1)$ **b** $3(1-3x)$ **c** $5(2a+3)$ **d** $11(1-2n)$
  e $6(3x+y)$ **f** $5(x-2y)$ **g** $4(3b+c)$ **h** $2(a-2b)$
  i $7(a-5b)$ **j** $12(2+3d)$ **k** $8(3-4y)$ **l** $6(5b+3a)$
  m $11(2x-y)$ **n** $7(c-9d)$ **o** $6(m+7n)$ **p** $8(8a-c)$

3 Expand and simplify:

  a $x(x+2)$ **b** $x(5-x)$ **c** $a(2a+4)$ **d** $b(5-3b)$
  e $a(b+2c)$ **f** $a(a^2+1)$ **g** $x(3-4x)$ **h** $3x(6-x)$
  i $5x(x-4)$ **j** $4a(1-a)$ **k** $7b(b+2)$ **l** $a(a+b)$
  m $b(3-8b)$ **n** $m(m-3n)$ **o** $c(c-4a)$ **p** $6p(4-7p)$

Example 3

Expand and simplify:

a $-4(x+3)$ **b** $-(2x-4)$ **c** $-a(a+7)$

a $\quad -4(x+3)$
$= -4 \times x + -4 \times 3 \qquad \{-4 \text{ is multiplied by } x \text{ and by } 3\}$
$= -4x - 12$

b $\quad -(2x-4)$
$= -1(2x-4)$
$= -1 \times 2x + -1 \times (-4) \qquad \{-1 \text{ is multiplied by } 2x \text{ and by } -4\}$
$= -2x + 4$

c $\quad -a(a+7)$
$= -a \times a + -a \times 7 \qquad \{-a \text{ is multiplied by } a \text{ and by } 7\}$
$= -a^2 - 7a$

4 Expand and simplify:

a $-2(x+2)$ **b** $-3(x+4)$ **c** $-4(x-2)$
d $-5(5-x)$ **e** $-(a+2)$ **f** $-(x-3)$
g $-(5-x)$ **h** $-(2x+1)$ **i** $-3(4-x)$
j $-4(5x-2)$ **k** $-5(3-4c)$ **l** $-2(7-5x)$

5 Expand and simplify:

a $-a(a+1)$ **b** $-b(b+4)$ **c** $-c(5-c)$
d $-x(2x+4)$ **e** $-2x(1-x)$ **f** $-3y(y+2)$
g $-4a(5-a)$ **h** $-6b(3-2b)$ **i** $-xy(2y-x)$

Example 4

Expand and simplify:

a $4 + 2(x+3)$ **b** $8 - 3(2y-1)$

a $\quad 4 + 2(x+3)$
$= 4 + 2 \times x + 2 \times 3$
$= 4 + 2x + 6$
$= 2x + 10$

b $\quad 8 - 3(2y-1)$
$= 8 + -3(2y-1)$
$= 8 + -3 \times 2y + -3 \times (-1)$
$= 8 - 6y + 3$
$= 11 - 6y$

With practice you will not need all of these steps.

6 Expand and simplify:

a $3(x+2)+5$
b $3x+2(2x+1)$
c $7-6(2x-3)$
d $11x-(2+x)$
e $6+5(1-2x)$
f $11-(3-2x)$
g $16-7(1-3x)$
h $x+6+3(4+x)$
i $8x+1+2(3-2x)$
j $7-(1-2x)$
k $2x-(8+7x)+3$
l $8-5(11-3x)$
m $5x+x(x+2)$
n $8x+x(x-1)$
o $7x-x(x+3)$
p $x^2-x(2-x)$
q $4x-x(x-3)+2x^2$
r $3x^2-2x(x-5)-6x$

Example 5 ◀)) Self Tutor

Expand and simplify:

a $2(3x-1)+3(5-x)$
b $x(2x-1)-2x(5-x)$

a $\quad 2(3x-1)+3(5-x)$
$= 2 \times 3x + 2 \times -1 + 3 \times 5 + 3 \times -x$
$= 6x - 2 + 15 - 3x$
$= 3x + 13$

b $\quad x(2x-1)-2x(5-x)$
$= x \times 2x + x \times -1 + -2x \times 5 + -2x \times -x$
$= 2x^2 - x - 10x + 2x^2$
$= 4x^2 - 11x$

In **b**, the minus sign in front of $2x$ affects *both* terms inside the following bracket.

7 Expand and simplify:

a $2(x-3)+3(x+4)$
b $4b+(a-b)$
c $4b-(a-b)$
d $3(x+2)+5(4-x)$
e $6(m-2)-3(2m+1)$
f $7n-5(3-2n)$
g $5(y-x)+6(x-y)$
h $a(a+2)+5(a-3)$
i $x(x+5)-3(x-4)$
j $a^2+a(a+3)$
k $-a^2-a(a-1)$
l $x(x+y)-y(x+y)$
m $-3(x-6)-(2-x)$
n $4(3x-2)-(3x+1)$
o $2x(x-5)-3x(2-x)$

Discussion

Can you explain why:
- an even number has the form $2m$, where m is an integer
- an odd number has the form $2n+1$, where n is an integer
- the product of an even number and an odd number is always even?

Activity   "Think of a number" games

Algebra is a powerful tool in mathematical problem solving. It can help us describe problems in general terms, and often gives us an insight into *why* something works.

What to do:

1 Play the following "think of a number" game with a partner:

> Think of a number.
> Double it.
> Subtract 4.
> Halve the result.
> Add 3.
> Subtract your original number.

Repeat the game choosing different starting numbers.

2 We can use algebra to explain why the answer to the game above is always 1. Let x represent the starting number.

Copy and complete the following argument by writing down each step in terms of x:

Think of a number.		x
Double it.	gives	$2x$
Subtract 4.	gives	$2x - 4$
Halve the result.	gives	$\frac{1}{2}(\ldots\ldots - \ldots\ldots)$ or $\ldots\ldots - \ldots\ldots$
Add 3.	gives	$\ldots\ldots - \ldots\ldots + 3$ or $\ldots\ldots + \ldots\ldots$
Subtract your original number.	gives	$\ldots\ldots + \ldots\ldots - x$ or $\ldots\ldots$

3 Try the following "think of a number" game:

> Think of a number.
> Treble it.
> Add 9.
> Divide the result by 3.
> Subtract 3.

What is your answer? Repeat the game using different numbers.

4 For the game above, let x be the starting number. Use algebra to explain how the game works.

5 Make up your own "think of a number" game. Test it with algebra before you try it with others.

B THE PRODUCT $(a+b)(c+d)$

Consider the large rectangle alongside. It has side lengths $(a+b)$ and $(c+d)$, so the overall area is $(a+b)(c+d)$.

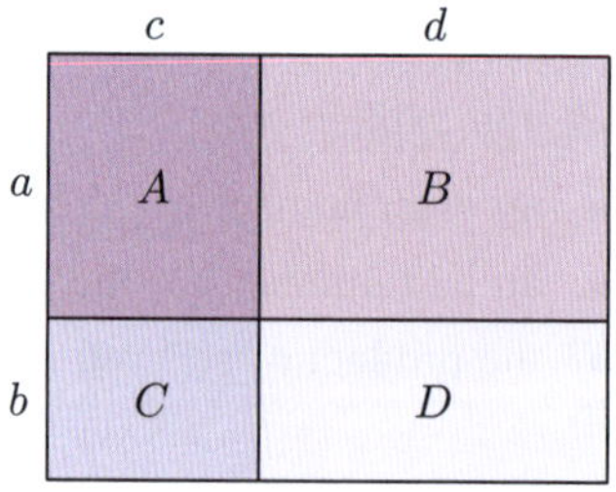

However, this area can also be obtained by adding the areas of the four smaller rectangles:

Overall area $= $ area $A \, + \,$ area $B \, + \,$ area $C \, + \,$ area D
$\phantom{\text{Overall area }} = ac + ad + bc + bd$

Equating areas gives

$$(a+b)(c+d) = ac + ad + bc + bd$$

Algebra: Expansion (Chapter 8) 141

This expansion rule is sometimes called the **FOIL** rule as:

$$(a+b)(c+d) = ac + ad + bc + bd$$

with inners and outers indicated, and labelled **F**irsts, **O**uters, **I**nners, **L**asts.

We can establish this result algebraically by using the distributive law several times:

$$(a+b)(c+d) = a(c+d) + b(c+d)$$
$$= ac + ad + bc + bd$$

Example 6 ◀) Self Tutor

Expand and simplify:

a $(x+3)(x+2)$ **b** $(2x+1)(3x-2)$

a $(x+3)(x+2)$
$= x \times x + x \times 2 + 3 \times x + 3 \times 2$
$= x^2 + 2x + 3x + 6$
$= x^2 + 5x + 6$

b $(2x+1)(3x-2)$
$= 2x \times 3x + 2x \times -2 + 1 \times 3x + 1 \times -2$
$= 6x^2 - 4x + 3x - 2$
$= 6x^2 - x - 2$

EXERCISE 8B

1 Expand and simplify:

a $(x+1)(x+2)$
b $(x+4)(x+2)$
c $(x+5)(x-1)$
d $(x+3)(x+1)$
e $(y+3)(y+2)$
f $(a+3)(a+7)$
g $(x+2)(x-2)$
h $(x-4)(x+2)$
i $(x+7)(x-3)$
j $(x-9)(x+2)$
k $(x-4)(x+3)$
l $(x+6)(x-2)$
m $(x-4)(x-3)$
n $(x-5)(x-8)$
o $(x-11)(x-4)$
p $(2x+3)(x-1)$
q $(x-4)(3x+2)$
r $(2x+3)(2x-4)$
s $(3x+2)(4x+1)$
t $(1-3x)(2x+1)$
u $(6-x)(2x+5)$
v $(4x-3)(1+3x)$
w $(4-x)(4+5x)$
x $(8-x)(2x+5)$

Example 7 ◀) Self Tutor

Expand and simplify:

a $(x+7)^2$ **b** $(3x-2)^2$

a $(x+7)^2$
$= (x+7)(x+7)$
$= x^2 + 7x + 7x + 49$
$= x^2 + 14x + 49$

b $(3x-2)^2$
$= (3x-2)(3x-2)$
$= 9x^2 - 6x - 6x + 4$
$= 9x^2 - 12x + 4$

The middle two terms are identical.

2 Expand and simplify:

a $(x+1)^2$ b $(x+4)^2$ c $(x-2)^2$ d $(x-5)^2$
e $(3+y)^2$ f $(3-y)^2$ g $(2x+1)^2$ h $(2x-1)^2$
i $(1+4a)^2$ j $(1-4a)^2$ k $(a+b)^2$ l $(a-b)^2$

Example 8   ◉) Self Tutor

Expand and simplify:

a $(x+3)(x-3)$ b $(2x-5)(2x+5)$

a $(x+3)(x-3)$
$= x^2 - 3x + 3x - 9$
$= x^2 - 9$

b $(2x-5)(2x+5)$
$= 4x^2 + 10x - 10x - 25$
$= 4x^2 - 25$

The middle two terms add to zero!

3 Expand and simplify:

a $(x+2)(x-2)$ b $(y-5)(y+5)$ c $(a+7)(a-7)$
d $(b-4)(b+4)$ e $(3+x)(3-x)$ f $(6-y)(6+y)$
g $(1+a)(1-a)$ h $(8-b)(8+b)$ i $(2x+1)(2x-1)$
j $(3a-2)(3a+2)$ k $(3+5b)(3-5b)$ l $(5-4y)(5+4y)$

C PERFECT SQUARE EXPANSION

An expression of the form $(a+b)^2$ is called a **perfect square**.

The large square alongside has area $(a+b)^2$.

We can also write the total area as
$a^2 + ab + ab + b^2$
$= a^2 + 2ab + b^2$

So, $(a+b)^2 = a^2 + 2ab + b^2$

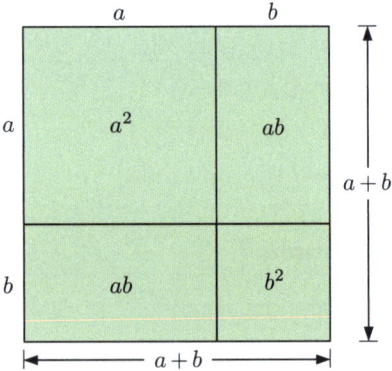

The following is a useful way of remembering the perfect square rule:

$(a+b)^2$ = a^2 + $2ab$ + b^2

↑ square of first term   ↑ twice the product of the two terms   ↑ square of second term

Algebra: Expansion (Chapter 8) 143

Example 9 ◆)) Self Tutor

Expand and simplify: **a** $(x+6)^2$ **b** $(2x+5)^2$

a $(x+6)^2$
$= x^2 + 2 \times x \times 6 + 6^2$
$= x^2 + 12x + 36$

b $(2x+5)^2$
$= (2x)^2 + 2 \times 2x \times 5 + 5^2$
$= 4x^2 + 20x + 25$

EXERCISE 8C

1 Expand and simplify:

  a $(x+2)^2$ **b** $(a+3)^2$ **c** $(x+5)^2$ **d** $(x+11)^2$

  e $(2x+1)^2$ **f** $(2x+4)^2$ **g** $(3x+2)^2$ **h** $(4x+3)^2$

  i $(7x+1)^2$ **j** $(2x+y)^2$ **k** $(6+x)^2$ **l** $(3+5x)^2$

2 Consider the **Opening Problem** on page **136**. Anton thinks that $(5+3)^2 = 5^2 + 3^2$.

  a By evaluating each side of the equation, show that Anton is incorrect.

  b Check that $(5+3)^2 = 5^2 + 2 \times 5 \times 3 + 3^2$.

Example 10 ◆)) Self Tutor

Expand and simplify:

a $(5-x)^2$ **b** $(3x-7)^2$

a $(5-x)^2$
$= (5+-x)^2$
$= 5^2 + 2 \times 5 \times (-x) + (-x)^2$
$= 25 - 10x + x^2$

b $(3x-7)^2$
$= (3x+-7)^2$
$= (3x)^2 + 2 \times 3x \times (-7) + (-7)^2$
$= 9x^2 - 42x + 49$

3 Expand and simplify:

  a $(x-4)^2$ **b** $(x-1)^2$ **c** $(x-6)^2$ **d** $(d-3)^2$

  e $(4-a)^2$ **f** $(7-x)^2$ **g** $(3x-1)^2$ **h** $(6-d)^2$

  i $(2x-5)^2$ **j** $(3-4a)^2$ **k** $(3a-2b)^2$ **l** $(3-5x)^2$

4 Expand and simplify:

  a $(2x+3)^2$ **b** $(4a-1)^2$ **c** $(3y+5)^2$

  d $(3a-4)^2$ **e** $(2x-7)^2$ **f** $(8+3a)^2$

  g $(2+5b)^2$ **h** $(6-5x)^2$ **i** $(4-5y)^2$

Discussion

Can you explain why:
- when an even number is squared, the result is always even
- when an odd number is squared, the result is always odd?

D DIFFERENCE BETWEEN TWO SQUARES

Consider the product $(a+b)(a-b)$.

Using the FOIL rule to expand this product,
$$(a+b)(a-b)$$
$$= a^2 - ab + ab - b^2$$
$$= a^2 - b^2$$

The middle two terms add to zero.

Thus, $\quad \boxed{(a+b)(a-b) = a^2 - b^2}$

This is called the **difference between two squares** expansion, because the expression on the right hand side is the difference between the two squares a^2 and b^2.

GEOMETRIC DEMONSTRATION

In the figure alongside,
shaded area = area of large square − area of small square
$$= a^2 - b^2$$

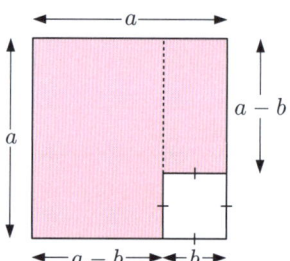

If the rectangle on the right hand side is rotated and placed on top of the remaining shaded area, we form a new rectangle.

$\therefore$ shaded area $= (a+b)(a-b)$

$\therefore$ $(a+b)(a-b) = a^2 - b^2$ {equating areas}

DEMO

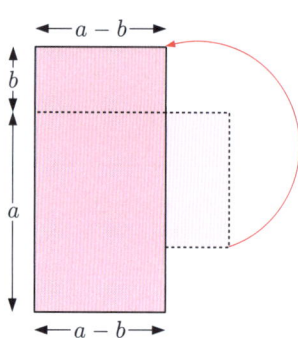

Example 11 ◀ Self Tutor

Expand and simplify:

a $(x+5)(x-5)$ **b** $(3-y)(3+y)$

a $(x+5)(x-5)$
$= x^2 - 5^2$
$= x^2 - 25$

b $(3-y)(3+y)$
$= 3^2 - y^2$
$= 9 - y^2$

Algebra: Expansion (Chapter 8)

EXERCISE 8D

1 Expand and simplify:
- **a** $(x+3)(x-3)$
- **b** $(x-3)(x+3)$
- **c** $(3+x)(3-x)$
- **d** $(3-x)(3+x)$
- **e** $(x+2)(x-2)$
- **f** $(2-x)(2+x)$
- **g** $(x+6)(x-6)$
- **h** $(a+4)(a-4)$
- **i** $(b-1)(b+1)$
- **j** $(p-q)(p+q)$
- **k** $(5+n)(5-n)$
- **l** $(7-y)(7+y)$

Example 12 ◀)) Self Tutor

Expand and simplify:
a $(2x-3)(2x+3)$ **b** $(5-3y)(5+3y)$

a $(2x-3)(2x+3)$
$= (2x)^2 - 3^2$
$= 4x^2 - 9$

b $(5-3y)(5+3y)$
$= 5^2 - (3y)^2$
$= 25 - 9y^2$

2 Expand and simplify:
- **a** $(2x+1)(2x-1)$
- **b** $(5x+2)(5x-2)$
- **c** $(4a+3)(4a-3)$
- **d** $(3b+5)(3b-5)$
- **e** $(4x+1)(4x-1)$
- **f** $(1-4x)(1+4x)$
- **g** $(7-2y)(7+2y)$
- **h** $(3-2x)(3+2x)$
- **i** $(3x+2)(3x-2)$

3 Use the difference between two squares expansion to find the value of 31×29.

4 Expand and simplify:
- **a** $(x+4)(x-4) + (x+1)^2$
- **b** $(x+2)(x-2) - (x+1)^2$
- **c** $(2+x)(2-x) - (x+2)^2$
- **d** $(x-1)^2 - (1-2x)(1+2x)$

Review set 8A

1 Expand:
- **a** $2(x+11)$
- **b** $5x(7-x)$
- **c** $-4(3-2x)$

2 Expand and simplify:
- **a** $(x-6)(x+4)$
- **b** $(x-9)(x-4)$
- **c** $(2x+5)(x-2)$

3 Expand and simplify:
- **a** $(x+8)^2$
- **b** $(x-4)^2$
- **c** $(7-2x)^2$

4 Expand and simplify:
- **a** $(x+7)(x-7)$
- **b** $(1+a)(1-a)$
- **c** $(4x+5)(4x-5)$

5 Expand and simplify:
- **a** $3(x-4)$
- **b** $-5(2x-3)$
- **c** $x(1-x)$

6 Expand and simplify:
   a $7 + 4(x - 2)$
   b $3 - 2(x + 4)$
   c $-4(2x - 3) - (x + 1)$

7 Expand and simplify:
   a $(x + 4)(x + 7)$
   b $(x + 8)(x - 2)$
   c $(1 - x)(5x + 6)$
   d $(2a - 3)(a - 9)$

8 Expand and simplify:
   a $(x + 3)(x + 2) + 3(x - 4)$
   b $(n - 3)(n + 5) - (n - 2)(n + 3)$

9 Expand and simplify:
   a $(x + 4)^2$
   b $(x - 10)^2$
   c $(2x + 5)^2$
   d $(3 - 4x)^2$

10 Expand and simplify:
   a $(x + 9)(x - 9)$
   b $(3x + 2)(3x - 2)$
   c $(4 + 3x)(4 - 3x)$

Review set 8B

1 Expand and simplify:
   a $3(x - 8)$
   b $-2(6 - 3x)$
   c $2x^2(x + 1)$

2 Expand and simplify:
   a $(x + 7)(x - 1)$
   b $(y - 8)(y + 3)$
   c $(3x - 1)(x + 3)$

3 Expand and simplify:
   a $4 + 5(x + 2)$
   b $13 - 3(x - 4)$
   c $5x + x(x - 8)$

4 Expand and simplify:
   a $(x + 6)^2$
   b $(2x - 5)^2$
   c $(3x + 7)(3x - 7)$

5 Expand and simplify:
   a $-3(x + 4)$
   b $x(y + 3)$
   c $2a(a - 5)$

6 Expand and simplify:
   a $(x + 9)(x - 4)$
   b $(2x - 7)(x + 6)$
   c $(2a - 3)(a - 4)$

7 Expand and simplify:
   a $(y + 3)^2$
   b $(3x + 2)^2$
   c $(4a - b)^2$

8 Expand and simplify:
   a $(5 - x)(5 + x)$
   b $(3y + 4)(3y - 4)$
   c $(6a + 5)(6a - 5)$

9 Expand and simplify:
   a $-4(x + 2) - (x + 1)^2$
   b $5(y - 3) + (y + 5)(y - 5)$

10 Expand and simplify:
   a $(x + 4)^2 - (x - 2)(x + 2)$
   b $(2x + 3)(2x - 3) - (x - 3)^2$

Linear equations and inequalities

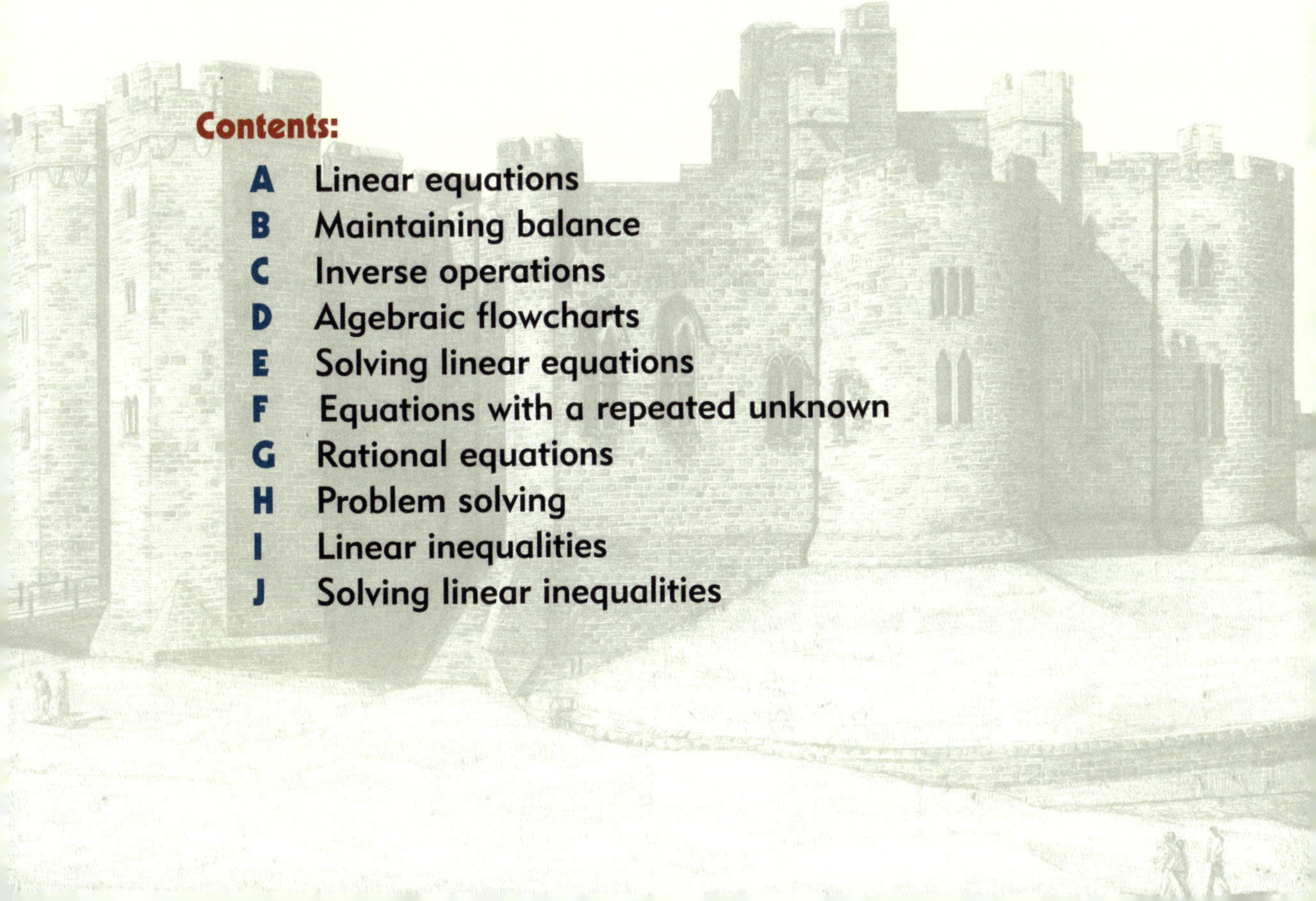

Contents:

- **A** Linear equations
- **B** Maintaining balance
- **C** Inverse operations
- **D** Algebraic flowcharts
- **E** Solving linear equations
- **F** Equations with a repeated unknown
- **G** Rational equations
- **H** Problem solving
- **I** Linear inequalities
- **J** Solving linear inequalities

148 Linear equations and inequalities (Chapter 9)

Opening problem

We are often faced with problems where we need to work out the value of an unknown quantity.

For example, consider the following questions:

- **i** When 6 is added to a number, the result is 8. What is the number?
- **ii** When 3 is subtracted from a number, the result is 11. What is the number?
- **iii** When a number is multiplied by 4, the result is 28. What is the number?
- **iv** When a number is divided by 5, the result is 3. What is the number?

Things to think about:

- **a** Can you write each of the problems above using symbols?
- **b** What techniques can you use to *solve* these problems?

Equations are a fundamental part of mathematics, and an important tool for problem solving. The use of equations dates back to Ancient Egypt, where the scribe **Ahmes** recorded a series of problems which were solved using equations.

An **equation** is a mathematical sentence which indicates that two expressions have the same value. The expressions are connected by an *equal* sign $=$.

The **left hand side** (LHS) of an equation is on the left of the $=$ sign.

The **right hand side** (RHS) of an equation is on the right of the $=$ sign.

For example: $\underbrace{5x - 3}_{\text{LHS}} = \underbrace{2x + 6}_{\text{RHS}}$

We can often convert a worded problem into an equation, then follow a formal procedure to **solve** the equation and hence the problem.

In other problems, we may be considering when one quantity is greater than or less than another quantity. The problem may then be written as a mathematical **inequality**.

An **inequality** is a mathematical sentence which compares the values of two expressions.
It contains one of the symbols $>$, $<$, $\geqslant$, or $\leqslant$.

A LINEAR EQUATIONS

Linear equations are equations in which the variable is raised only to the power 1.

All linear equations can be written in the form $ax + b = 0$ where a and b are constants, $a \neq 0$, and x is the variable.

For example:
- $3x + 6 = 15$, $\frac{x}{5} - 4 = -2$, and $10 - 7.5y = -5$ are linear equations
- $x^2 + 4x + 3 = 0$, $x - \sqrt{x} = 2$, and $x^5 = 32$ are not linear equations.

A **solution** of an equation is a value of the variable which makes the equation true.

Consider the equation $5x - 3 = 2x + 6$.

When $x = 3$, LHS $= 5(3) - 3$ and   RHS $= 2(3) + 6$
                     $= 15 - 3$ $= 6 + 6$
                     $= 12$ $= 12$ also.

In general, a linear equation will have one solution.

So, $x = 3$ is a solution of the equation.

SOLUTION BY INSPECTION

Some simple equations are easily solved by **inspection**.

For example, consider $x + 7 = 20$. We know that $13 + 7 = 20$, so $x = 13$ must be a solution.

EXERCISE 9A

1 Solve by inspection:

  a $x + 2 = 3$ **b** $x - 7 = 2$

  c $10 + x = 15$ **d** $3 - x = -2$

  e $x - 4 = -6$ **f** $2x = 8$

  g $-7x = 28$ **h** $4x = 52$

  i $\frac{x}{8} = 5$ **j** $\frac{x}{5} = -35$

  k $\frac{10}{x} = -2$ **l** $\frac{-12}{x} = 3$

The correct solution makes LHS = RHS.

2 One of the numbers in brackets is the correct solution to the given equation. Find the correct solution.

  a $3x + 8 = 14$ $\{0, 1, 2, 3\}$ **b** $5 - 3x = -4$ $\{0, 1, 2, 3\}$

  c $7x + 3 = -11$ $\{-4, -3, -2, -1\}$ **d** $2x - 5 = x$ $\{-5, 0, 5, 10\}$

  e $\frac{k}{4} = k + 3$ $\{-8, -4, 4, 8\}$ **f** $7 - 2m = m + 4$ $\{-2, -1, 0, 1, 2\}$

B MAINTAINING BALANCE

For any equation, the LHS must always equal the RHS. We can therefore think of an equation as a set of scales that must always be in **balance**.

The balance of an equation is maintained provided we perform the same operation on **both sides** of the equals sign.

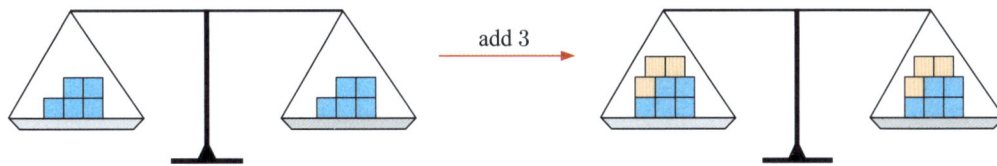

add 3

150   Linear equations and inequalities   (Chapter 9)

> **Discussion**
>
> Will the balance of an equation be maintained if we:
> - add the same amount to both sides
> - subtract the same amount from both sides
> - multiply both sides by the same amount
> - divide both sides by the same amount?

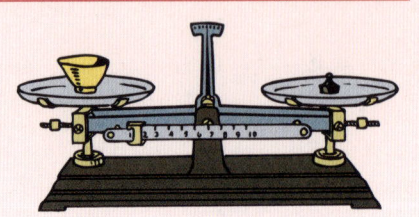

To maintain the balance, whatever operation we perform on one side of the equation, we must also perform on the other.

Example 1   ◀) Self Tutor

Write down the equation which results when:

a   3 is added to both sides of $x - 3 = 8$
b   4 is taken from both sides of $2x + 4 = 18$
c   both sides of $5x = 15$ are divided by 5
d   both sides of $\dfrac{x}{4} = -7$ are multiplied by 4.

a $x - 3 = 8$
$\therefore\ x - 3 + 3 = 8 + 3$
$\therefore\ x = 11$

b $2x + 4 = 18$
$\therefore\ 2x + 4 - 4 = 18 - 4$
$\therefore\ 2x = 14$

c $5x = 15$
$\therefore\ \dfrac{5x}{5} = \dfrac{15}{5}$
$\therefore\ x = 3$

d $\dfrac{x}{4} = -7$
$\therefore\ \dfrac{x}{4} \times 4 = -7 \times 4$
$\therefore\ x = -28$

EXERCISE 9B

1 Write down the equation which results when we add:
   a 3 to both sides of $x - 3 = 2$
   b 9 to both sides of $x - 9 = 0$
   c 4 to both sides of $5x - 4 = 11$
   d 5 to both sides of $7x - 5 = x + 1$.

2 Write down the equation which results when we subtract:
   a 1 from both sides of $x + 1 = 5$
   b 6 from both sides of $2x + 6 = 10$
   c 5 from both sides of $3x + 5 = 2$
   d 9 from both sides of $4x + 9 = 3x + 11$.

3 Write down the equation which results when we multiply both sides of:
   a $\dfrac{x}{2} = 8$ by 2
   b $\dfrac{x - 1}{5} = 1$ by 5
   c $\dfrac{3x}{7} = 2$ by 7
   d $\dfrac{3x - 4}{4} = -10$ by 4.

4 Write down the equation which results when we divide both sides of:
   a $4x = -40$ by 4
   b $-2x = 18$ by -2
   c $3(2 - x) = 15$ by 3
   d $-5(2x - 1) = -55$ by -5.

C INVERSE OPERATIONS

When I woke up this morning there were 4 eggs in my refrigerator.

I found my chickens had already laid 2 eggs, so I then had $4 + 2 = 6$ eggs in total.

I fried 2 eggs to have with my breakfast, so I now have $6 - 2 = 4$ eggs left.

I now have the same number of eggs that I started with. Adding 2 eggs and subtracting 2 eggs cancel each other out.

> Addition and subtraction are **inverse operations**.

Over the next week my chickens lay very well. The number of eggs in my fridge doubles, so I now have $4 \times 2 = 8$ eggs.

On the weekend I bake some cakes. Only half of the eggs now remain. This is $8 \div 2 = 4$ eggs, which is the same number that I started with.

The operations of multiplying by 2 and dividing by 2 cancel each other out.

> Multiplication and division are **inverse operations**.

Example 2   ◀⟩ Self Tutor

State the inverse of:

  a $\times 4$ **b** $\div 7$ **c** $+6$ **d** -3

  a The inverse of $\times 4$ is $\div 4$.   **b** The inverse of $\div 7$ is $\times 7$.
  c The inverse of $+6$ is -6.   **d** The inverse of -3 is $+3$.

We can use inverse operations to solve simple equations. To keep the equation balanced, we must perform the same operation on both sides of the equation.

Example 3   ◀⟩ Self Tutor

Solve for x using a suitable inverse operation:

  a $x + 6 = 13$ **b** $y - 4 = -1$ **c** $4g = 20$ **d** $\dfrac{h}{7} = -6$

  a $x + 6 = 13$
     $\therefore \ x + 6 - 6 = 13 - 6$
     $\therefore \ x = 7$

  b $y - 4 = -1$
     $\therefore \ y - 4 + 4 = -1 + 4$
     $\therefore \ y = 3$

  c $4g = 20$
     $\therefore \ \dfrac{4g}{4} = \dfrac{20}{4}$
     $\therefore \ g = 5$

  d $\dfrac{h}{7} = -6$
     $\therefore \ \dfrac{h}{7} \times 7 = -6 \times 7$
     $\therefore \ h = -42$

EXERCISE 9C

1 State the inverse of:
- **a** $+3$
- **b** -8
- **c** $\times 2$
- **d** $\div 5$
- **e** $+12$
- **f** $\div 6$
- **g** -5
- **h** $\times 9$
- **i** $+\frac{2}{3}$
- **j** $\div 13$
- **k** $\times 15$
- **l** $-\frac{4}{5}$

2 Simplify:
- **a** $x - 3 + 3$
- **b** $x + 5 - 5$
- **c** $x \div 12 \times 12$
- **d** $x \times 9 \div 9$
- **e** $p + 4 - 4$
- **f** $3q \div 3$
- **g** $\frac{8r}{7} \times 7$
- **h** $\frac{2s}{3} \div \frac{2}{3}$

3 Solve for x using a suitable inverse operation:
- **a** $x + 4 = 10$
- **b** $\frac{x}{10} = 1$
- **c** $x - 9 = -2$
- **d** $2x = 12$
- **e** $x - 5 = 1$
- **f** $x + 2 = 0$
- **g** $\frac{x}{7} = -5$
- **h** $5x = -55$
- **i** $\frac{x}{8} = -4$
- **j** $-20x = 60$
- **k** $x + 5 = -2$
- **l** $x - 7 = -16$
- **m** $\frac{x}{-3} = 3$
- **n** $x - 4 = 0$
- **o** $-9x = -81$
- **p** $x + 9 = -9$

4 Solve for x using a suitable inverse operation:
- **a** $12x = -48$
- **b** $x + 9 = -18$
- **c** $x - 8 = -12$
- **d** $\frac{x}{15} = -6$
- **e** $x + 13 = 49$
- **f** $13x = 0$
- **g** $\frac{x}{-11} = -4$
- **h** $x - 21 = -7$

5 Solve using a suitable inverse operation:
- **a** $\frac{a}{5} = -2$
- **b** $3b = 4$
- **c** $-c = 9$
- **d** $4 + d = 1$
- **e** $e - 7 = 3$
- **f** $\frac{1}{4}f = 15$
- **g** $z - 7 = -9$
- **h** $\frac{w}{6} = -12$

D ALGEBRAIC FLOWCHARTS

To solve harder equations, we need to know how algebraic expressions are "built up". We can then use appropriate inverse operations to "undo" the expression and isolate the variable. An **algebraic flowchart** is useful to help us do this.

For example, to "build up" the expression $3x + 2$, we start with x, multiply it by 3, then add on 2.

$$\boxed{x} \xrightarrow{\times 3} \boxed{3x} \xrightarrow{+2} \boxed{3x+2}$$

We know that the inverse operation of $\times 3$ is $\div 3$, and the inverse operation of $+2$ is -2.

Linear equations and inequalities (Chapter 9) 153

To "undo" the expression $3x + 2$, we perform **inverse operations** in the **reverse order**.

$$\boxed{3x+2} \xrightarrow{-2} \boxed{3x} \xrightarrow{\div 3} \boxed{x}$$

Example 4 ◀)) Self Tutor

Use flowcharts to show how each expression can be "built up", and how it can be "undone" using inverse operations:

a $4x - 7$ **b** $4(x - 7)$

a *Building up*:

$$\boxed{x} \xrightarrow{\times 4} \boxed{4x} \xrightarrow{-7} \boxed{4x-7}$$

Undoing:

$$\boxed{4x-7} \xrightarrow{+7} \boxed{4x} \xrightarrow{\div 4} \boxed{x}$$

b *Building up*:

$$\boxed{x} \xrightarrow{-7} \boxed{x-7} \xrightarrow{\times 4} \boxed{4(x-7)}$$

Undoing:

$$\boxed{4(x-7)} \xrightarrow{\div 4} \boxed{x-7} \xrightarrow{+7} \boxed{x}$$

EXERCISE 9D

1 Use flowcharts to show how to "build up" and "undo":

a $7x + 3$ **b** $7(x + 3)$ **c** $5(x - 2)$

d $5x - 2$ **e** $\dfrac{x}{3} + 1$ **f** $\dfrac{x+1}{3}$

g $\dfrac{x}{8} - 5$ **h** $\dfrac{x-5}{8}$ **i** $2x - 6$

j $\dfrac{x}{-3} + 10$ **k** $8(x - 7)$ **l** $\dfrac{x-3}{4}$

The order of operations is very important!

Example 5 ◀)) Self Tutor

Use flowcharts to show how to "build up" and "undo":

a $\dfrac{3x+4}{7}$ **b** $1 - \dfrac{x}{2}$

a *Building up*:

$$\boxed{x} \xrightarrow{\times 3} \boxed{3x} \xrightarrow{+4} \boxed{3x+4} \xrightarrow{\div 7} \boxed{\dfrac{3x+4}{7}}$$

Undoing:

$$\boxed{\dfrac{3x+4}{7}} \xrightarrow{\times 7} \boxed{3x+4} \xrightarrow{-4} \boxed{3x} \xrightarrow{\div 3} \boxed{x}$$

b *Building up*:

$$\boxed{x} \xrightarrow{\div -2} \boxed{-\dfrac{x}{2}} \xrightarrow{+1} \boxed{1-\dfrac{x}{2}}$$

Undoing:

$$\boxed{1-\dfrac{x}{2}} \xrightarrow{-1} \boxed{-\dfrac{x}{2}} \xrightarrow{\times -2} \boxed{x}$$

2 Use flowcharts to show how to "build up" and "undo":

a $\dfrac{3x+2}{5}$
b $\dfrac{3x}{5}+2$
c $\dfrac{3(x+2)}{5}$
d $\dfrac{7x-1}{6}$
e $\dfrac{7x}{6}-1$
f $\dfrac{7(x-1)}{6}$
g $\dfrac{5x}{6}-3$
h $\dfrac{5(x-3)}{6}$
i $\dfrac{5x-3}{6}$
j $1-\dfrac{2x}{3}$
k $\dfrac{1-2x}{3}$
l $\dfrac{2(1-x)}{3}$

Activity 1 — Expression invaders

Click on the icon to practise building and undoing expressions.

EXPRESSION INVADERS

E SOLVING LINEAR EQUATIONS

To solve equations like $4x-5=25$, we first consider how the expression on the LHS has been "built up". We then isolate the unknown by using **inverse operations** in the **reverse order**.

Activity 2 — Solving by scales

Click on the icon to solve equations using a set of scales.

PRACTICE

Example 6   Self Tutor

Solve for x: $4x-5=25$

$4x-5=25$
$\therefore\ 4x-5+5=25+5$ {adding 5 to both sides}
$\therefore\ 4x=30$ {simplifying}
$\therefore\ \dfrac{4x}{4}=\dfrac{30}{4}$ {dividing both sides by 4}
$\therefore\ x=7\tfrac{1}{2}$ {simplifying}

Check: LHS $=4\left(7\tfrac{1}{2}\right)-5=30-5=25=$ RHS ✓

Check your answer by substituting back into the original equation.

EXERCISE 9E

1 Solve for x:

a $2x+1=5$
b $4x+7=27$
c $3x+7=19$
d $3x+1=-23$
e $5x-9=11$
f $8x-3=0$
g $2x-7=-4$
h $2x-11=23$
i $7+8x=-9$
j $6-3x=0$
k $8+13x=34$
l $11+4x=-6$

Linear equations and inequalities (Chapter 9) 155

Example 7

Solve for x: $\quad \dfrac{x}{4} + 5 = -8$

$\dfrac{x}{4} + 5 = -8$

$\therefore \dfrac{x}{4} + 5 - 5 = -8 - 5 \quad$ {subtracting 5 from both sides}

$\therefore \dfrac{x}{4} = -13 \quad$ {simplifying}

$\therefore \dfrac{x}{4} \times 4 = -13 \times 4 \quad$ {multiplying both sides by 4}

$\therefore x = -52 \quad$ {simplifying}

Check: LHS $= \dfrac{-52}{4} + 5 = -13 + 5 = -8 =$ RHS ✓

2 Solve for x:

a $\dfrac{x}{2} + 1 = 3$ 　　　　b $\dfrac{x}{2} - 5 = 6$ 　　　　c $\dfrac{x}{8} + 3 = 5$

d $\dfrac{x}{3} - 4 = -1$ 　　　e $\dfrac{x}{3} - 4 = -11$ 　　f $\dfrac{x}{6} + 2 = -2$

g $\dfrac{x}{9} - 7 = 0$ 　　　　h $\dfrac{x}{7} + 5 = -3$ 　　　i $\dfrac{x}{11} + 31 = 33$

Example 8

Solve for x: $\quad 32 - 5x = 8$

$32 - 5x = 8$

$\therefore 32 - 5x - 32 = 8 - 32 \quad$ {subtracting 32 from both sides}

$\therefore -5x = -24$

$\therefore \dfrac{-5x}{-5} = \dfrac{-24}{-5} \quad$ {dividing both sides by -5}

$\therefore x = 4\dfrac{4}{5}$

Check: LHS $= 32 - 5\left(4\dfrac{4}{5}\right) = 32 - 5\left(\dfrac{24}{5}\right) = 32 - 24 = 8 =$ RHS ✓

3 Solve for x:

a $15 - 2x = 7$ 　　　　b $2 - 3x = 8$ 　　　　c $1 - 4x = -15$

d $4 - 5x = -21$ 　　　e $16 - 8x = 0$ 　　　　f $22 - 3x = 1$

g $14 - x = -1$ 　　　　h $19 - 4x = -9$ 　　　i $-5x + 12 = -8$

Example 9

Solve the equation: $\dfrac{2x-3}{3} = -2$

$\dfrac{2x-3}{3} = -2$

$\therefore \dfrac{2x-3}{3} \times 3 = -2 \times 3$ {multiplying both sides by 3}

$\therefore 2x - 3 = -6$

$\therefore 2x - 3 + 3 = -6 + 3$ {adding 3 to both sides}

$\therefore 2x = -3$

$\therefore \dfrac{2x}{2} = \dfrac{-3}{2}$ {dividing both sides by 2}

$\therefore x = -\dfrac{3}{2}$

Check: LHS $= \dfrac{2(-\tfrac{3}{2}) - 3}{3} = \dfrac{-6}{3} = -2 =$ RHS ✓

4 Solve for x:

a $\dfrac{x+1}{3} = 4$ **b** $\dfrac{4x-1}{5} = 7$ **c** $\dfrac{2x-5}{2} = 1$

d $\dfrac{3x+1}{4} = -5$ **e** $\dfrac{5x+6}{-2} = 7$ **f** $\dfrac{2x+1}{-5} = 11$

g $\dfrac{11x-1}{8} = -7$ **h** $\dfrac{6x-2}{-5} = -2$ **i** $\dfrac{11+4x}{3} = -11$

Example 10

Solve the equation: $3(2x - 1) = -21$

$3(2x - 1) = -21$

$\therefore \dfrac{3(2x-1)}{3} = \dfrac{-21}{3}$ {dividing both sides by 3}

$\therefore 2x - 1 = -7$

$\therefore 2x - 1 + 1 = -7 + 1$ {adding 1 to both sides}

$\therefore 2x = -6$

$\therefore \dfrac{2x}{2} = \dfrac{-6}{2}$ {dividing both sides by 2}

$\therefore x = -3$

Check: LHS $= 3(2(-3) - 1) = 3(-7) = -21 =$ RHS ✓

5 Solve for x:

a $2(x - 1) = 18$ **b** $3(2x + 1) = 15$ **c** $5(2x - 7) = 10$

d $4(3x - 5) = -28$ **e** $-4(3x - 2) = 44$ **f** $7(3x - 7) = -49$

g $6(3x - 2) = 12$ **h** $-5(4x + 1) = -15$ **i** $-6(3 + 8x) = -18$

6 Solve the following equations:

a $3a + 5 = 14$
b $\frac{x}{8} - 1 = 55$
c $\frac{3x-1}{2} = 7$
d $4(x + 5) = 24$
e $6(n - 2) = 12$
f $5a + 9 = -31$
g $\frac{2x-5}{4} = 0$
h $\frac{x}{5} - 3 = 12$
i $\frac{x+15}{3} = 6$
j $5(2n - 1) = -35$
k $\frac{3k+5}{2} = 13$
l $-8(5z + 1) = 24$

LEARNING ALGEBRA

Activity 3 — Real world problems

We are often faced with real world problems where we need to work out the value of an unknown quantity.

For example, consider these problems:

A Each of my cats has 4 kittens. There are 28 kittens in all. How many cats do I have?

B My training course runs for 6 hours. It finishes at 8 pm. What time does it start?

C This morning I bought some plants from the nursery. When I divided them equally into 5 large pots, I found I had 3 plants in each. How many plants did I buy this morning?

D Bernadette tells me that 3 years ago she was 11 years old. How old is she now?

What to do:

1 Answer the **Opening Problem** on page **148**.

2 Match each of the real world problems above to the corresponding "find the number" question in the **Opening Problem**. Hence solve the real world problems.

F EQUATIONS WITH A REPEATED UNKNOWN

If the unknown or variable appears more than once in the equation, we need to take extra steps in its solution.

DEMO

For example, consider the equation $3x + 1 = x + 7$.

In this case the unknown appears twice, once on each side of the equation.

158   Linear equations and inequalities   (Chapter 9)

We can represent the equation $3x + 1 = x + 7$ using the set of scales shown.

The unknown x is the number of blocks in each bag.

We can add or subtract bags or blocks on both sides of the scales to maintain the balance.

Removing a bag is like subtracting x from both sides. This gives us the equation $2x + 1 = 7$.

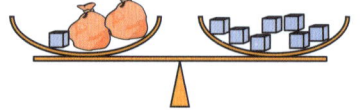

We can then use inverse operations to solve for x.

In general, we follow these steps to solve equations:

> *Step 1:* If necessary, expand any brackets and collect like terms.
>
> *Step 2:* If necessary, remove the unknown from one side of the equation. Remember to balance the other side.
>
> *Step 3:* Use inverse operations to isolate the unknown and solve the equation.

Example 11   ◀)) Self Tutor

Solve for d: $2d + 3(d - 1) = 7$

$$2d + 3(d - 1) = 7$$
$$\therefore \ 2d + 3d - 3 = 7 \quad \text{\{expanding brackets\}}$$
$$\therefore \ 5d - 3 = 7 \quad \text{\{collecting like terms\}}$$
$$\therefore \ 5d = 10 \quad \text{\{adding 3 to both sides\}}$$
$$\therefore \ d = 2 \quad \text{\{dividing both sides by 5\}}$$

Check: LHS $= 2(2) + 3(2 - 1) = 4 + 3 = 7 =$ RHS ✓

EXERCISE 9F

1 Solve:

a $2x + 3x = 10$

b $5x + 4 + 3x = -36$

c $3x - 6 - 2x + 7 = 16$

d $4y - 9 + 3y - 5 = -14$

e $5x + 3(x + 2) = 30$

f $2m + 4(m - 3) = 3$

g $7x + 2(x + 1) = 28$

h $-3(p + 2) + 4(p - 2) = 1$

i $t + \dfrac{t}{3} = 8$

j $d - \dfrac{d}{5} = -8$

k $2x + \dfrac{x}{3} = -14$

l $\dfrac{x}{2} - \dfrac{x}{3} = 2$

Remember to check your solution by substituting it into the equation.

Example 12

Solve for x: $\quad 3x + 2 = x + 14$

$3x + 2 = x + 14$
$\therefore \quad 2x + 2 = 14 \qquad$ {subtracting x from both sides}
$\therefore \quad 2x = 12 \qquad$ {subtracting 2 from both sides}
$\therefore \quad x = 6 \qquad$ {dividing both sides by 2}

Check: LHS $= 3(6) + 2 = 18 + 2 = 20$
RHS $= 6 + 14 = 20$ ✓

2 Solve:
    a $\;2x - 1 = 5 + x$ **b** $\;3x + 1 = x + 5$ **c** $\;2x - 6 = 7x + 14$
    d $\;4x - 5 = 2x + 1$ **e** $\;7x + 7 = 3x + 23$ **f** $\;x - 6 = 5x + 4$

Example 13

Solve for x: $\quad 4x + 3 = 23 - x$

$4x + 3 = 23 - x$
$\therefore \quad 5x + 3 = 23 \qquad$ {adding x to both sides}
$\therefore \quad 5x = 20 \qquad$ {subtracting 3 from both sides}
$\therefore \quad x = 4 \qquad$ {dividing both sides by 5}

Check: LHS $= 4(4) + 3 = 16 + 3 = 19$
RHS $= 23 - 4 = 19$ ✓

3 Solve:
    a $\;x + 2 = 4 - x$ **b** $\;2x + 5 = 10 - 3x$ **c** $\;2 + 7x = 3 - 3x$
    d $\;3x - 11 = 9 - 2x$ **e** $\;8 - 9x = 3 - 4x$ **f** $\;9 - 3x = 5 - 6x$

Example 14

Solve for x: $\quad 3(x + 2) = x - 1$

$3(x + 2) = x - 1$
$\therefore \quad 3x + 6 = x - 1 \qquad$ {expanding the brackets}
$\therefore \quad 2x + 6 = -1 \qquad$ {subtracting x from both sides}
$\therefore \quad 2x = -7 \qquad$ {subtracting 6 from both sides}
$\therefore \quad x = -\dfrac{7}{2} \qquad$ {dividing both sides by 2}

Check: LHS $= 3\left(\left(-\dfrac{7}{2}\right) + 2\right) = 3\left(-\dfrac{3}{2}\right) = -\dfrac{9}{2}$
RHS $= \left(-\dfrac{7}{2}\right) - 1 = -\dfrac{9}{2}$ ✓

4 Solve:
   a $2(x+1) = x+5$
   b $3(t-2) = t+4$
   c $6(2x-3) = x-7$
   d $4(3y-1) = 2y+1$
   e $2(a-5) = 5-3a$
   f $8(p+2) = 1+3p$

5 **a** Try to solve $7(a+3) = 21 + 7a$. What do you notice?
   b How many values of a satisfy the equation?

6 **a** Try to solve $3(a+1) = 4 + 3a$. What do you notice?
   b How many values of a satisfy the equation?

7 Solve the following equations:
   a $4 - x = 2(x+1) + 1$
   b $3x + 1 = 2(1-3x) + 19$
   c $10 - x = 5(x-3) + 7$
   d $7 + 6p = 3 + 2(1-p)$

8 Solve for x:
   a $3(x-5) = 5x+1$
   b $x - 2 + 4(x-1) = 2$
   c $4 - 3x - (x+5) = 3$
   d $3x - 2(x+3) - 4x = 0$
   e $3(x-6) = 2(2-x) + 3$
   f $2(x-8) = 3(x-5)$
   g $4(2x+3) - 10 = 5(3-x)$
   h $7 - (2-3x) = 12 - (5-4x)$

LEARNING ALGEBRA

G RATIONAL EQUATIONS

Rational equations are equations involving fractions. To solve rational equations, we write all the fractions in the equation with the same **lowest common denominator (LCD)**, and then equate the numerators.

For fractions whose denominators involve the variable, the lowest common denominator is found in the same way as for numerical fractions.

For example:
- in $\dfrac{x}{4} = \dfrac{x+1}{6}$ the LCD is 12
- in $\dfrac{5}{3x} = \dfrac{7}{9}$ the LCD is $9x$

Example 15 ◀) Self Tutor

Solve for x: $\dfrac{x}{2} = \dfrac{3+x}{5}$

$\dfrac{x}{2} = \dfrac{3+x}{5}$ has LCD $= 10$

$\therefore \dfrac{x}{2} \times \dfrac{5}{5} = \dfrac{2}{2} \times \left(\dfrac{3+x}{5}\right)$ {creating a common denominator}

$\therefore 5x = 2(3+x)$ {equating numerators}
$\therefore 5x = 6 + 2x$ {expanding brackets}
$\therefore 5x - 2x = 6 + 2x - 2x$ {subtracting $2x$ from both sides}
$\therefore 3x = 6$
$\therefore x = 2$ {dividing both sides by 3}

Notice the insertion of brackets here.

EXERCISE 9G

1 Solve for x:

   a $\dfrac{x}{2} = \dfrac{5}{9}$ **b** $\dfrac{x-1}{5} = \dfrac{x}{10}$ **c** $\dfrac{x}{3} = \dfrac{x+1}{2}$

   d $\dfrac{x+2}{5} = \dfrac{3x-1}{4}$ **e** $\dfrac{3x}{4} = \dfrac{6-x}{5}$ **f** $\dfrac{2x-5}{4} = \dfrac{3x-1}{-2}$

Example 16   ◀)) Self Tutor

Solve for x: $\dfrac{1}{2x} = -3$

$\dfrac{1}{2x} = -3$

$\therefore \dfrac{1}{2x} = -3 \times \dfrac{2x}{2x}$ {creating a common denominator}

$\therefore 1 = -6x$ {equating numerators}

$\therefore x = -\dfrac{1}{6}$ {dividing both sides by -6}

2 Solve for x:

   a $\dfrac{1}{x} = 3$ **b** $\dfrac{1}{x} = -4$ **c** $\dfrac{3}{2x} = 1$ **d** $-\dfrac{1}{3x} = 2$

Example 17   ◀)) Self Tutor

Solve for x: $\dfrac{4}{x} = \dfrac{3}{4}$

$\dfrac{4}{x} = \dfrac{3}{4}$ has LCD $= 4x$

$\therefore \dfrac{4}{x} \times \dfrac{4}{4} = \dfrac{3}{4} \times \dfrac{x}{x}$ {creating a common denominator}

$\therefore 16 = 3x$ {equating numerators}

$\therefore x = \dfrac{16}{3}$ {dividing both sides by 3}

3 Solve for x:

   a $\dfrac{3}{x} = \dfrac{2}{7}$ **b** $\dfrac{4}{x} = \dfrac{5}{9}$ **c** $\dfrac{6}{5} = \dfrac{4}{x}$ **d** $\dfrac{5}{6} = -\dfrac{2}{x}$

   e $\dfrac{4}{3x} = \dfrac{8}{7}$ **f** $\dfrac{3}{2x} = \dfrac{5}{4}$ **g** $\dfrac{7}{2x} = -\dfrac{1}{8}$ **h** $\dfrac{9}{2x} = -\dfrac{1}{4}$

Example 18

Solve for x: $\dfrac{3}{x-2} = 2$

$\dfrac{3}{x-2} = 2$

$\therefore \ \dfrac{3}{x-2} = 2 \times \dfrac{(x-2)}{(x-2)}$ {creating a common denominator}

$\therefore \ 3 = 2(x-2)$ {equating numerators}

$\therefore \ 3 = 2x - 4$ {expanding brackets}

$\therefore \ 2x = 7$ {adding 4 to both sides}

$\therefore \ x = \dfrac{7}{2}$ {dividing both sides by 2}

4 Solve for x:

a $\dfrac{1}{x+3} = 1$ **b** $\dfrac{2}{x-2} = 3$ **c** $\dfrac{3}{1-x} = 2$

Example 19

Solve for x: $\dfrac{5}{1-x} = -\dfrac{1}{3}$

$\dfrac{5}{1-x} = -\dfrac{1}{3}$

$\therefore \ \dfrac{5}{1-x} \times \dfrac{3}{3} = -\dfrac{1}{3} \times \dfrac{(1-x)}{(1-x)}$ {creating a common denominator}

$\therefore \ 15 = -(1-x)$ {equating numerators}

$\therefore \ 15 = -1 + x$ {expanding brackets}

$\therefore \ x = 16$

5 Solve for x:

a $\dfrac{2}{3-x} = \dfrac{1}{2}$ **b** $\dfrac{-1}{x+2} = \dfrac{2}{3}$ **c** $\dfrac{-3}{4-x} = -\dfrac{4}{5}$

Activity 4   Identities

An **identity** is an equation which is always true no matter what value the variable takes.

For example, $2x + 3x = 5x$ is true for all values of x, so this equation is an identity.

What to do:

1 Decide whether the following equations are identities:

    a $x + 2 = 2 + x$ **b** $x - x = 0$ **c** $3 + x = 3x$

    d $x \times 1 = x$ **e** $0 \times x = 0$ **f** $4 - x = x - 4$

    g $\dfrac{x}{-1} = -x$ **h** $\dfrac{x}{5} = 5x$ **i** $\dfrac{x}{x} = 1$

2 For the equations in **1** which are *not* identities, find the value(s) of x which make the equation true.

Linear equations and inequalities (Chapter 9) 163

H PROBLEM SOLVING

Many problems can be translated into **algebraic equations**. To solve problems using algebra, we follow these steps:

Step 1: Decide on the unknown quantity and allocate it a variable such as x.
Step 2: Translate the problem into an equation.
Step 3: Solve the equation by isolating the variable.
Step 4: Check that the solution satisfies the original problem.
Step 5: Write the answer in sentence form, describing how the solution relates to the original problem.

Example 20   ◀) Self Tutor

The sum of 3 consecutive even integers is 132. Find the smallest integer.

Let x be the smallest even integer
$\therefore$ the next is $x + 2$, and the largest is $x + 4$.

So, $\quad x + (x + 2) + (x + 4) = 132 \qquad$ {their sum is 132}
$\qquad \therefore \quad 3x + 6 = 132$
$\qquad \therefore \quad 3x + 6 - 6 = 132 - 6 \qquad$ {subtracting 6 from both sides}
$\qquad \therefore \quad 3x = 126$
$\qquad \therefore \quad \dfrac{3x}{3} = \dfrac{126}{3} \qquad$ {dividing both sides by 3}
$\qquad \therefore \quad x = 42$

$\therefore$ the smallest integer is 42.

Example 21   ◀) Self Tutor

If twice a number is subtracted from 11, the result is 4 more than the number. What is the number?

Let x be the number.
$\qquad \therefore \quad 11 - 2x = x + 4$
$\qquad \therefore \quad 11 - 2x + 2x = x + 4 + 2x \qquad$ {adding $2x$ to both sides}
$\qquad \therefore \quad 11 = 3x + 4$
$\qquad \therefore \quad 11 - 4 = 3x + 4 - 4 \qquad$ {subtracting 4 from both sides}
$\qquad \therefore \quad 7 = 3x$
$\qquad \therefore \quad \dfrac{7}{3} = \dfrac{3x}{3} \qquad$ {dividing both sides by 3}
$\qquad \therefore \quad x = 2\tfrac{1}{3}$

So, the number is $2\tfrac{1}{3}$.

EXERCISE 9H

1. When a number is doubled, the result is 18. Find the number.
2. When 6 is added to a number, the result is 11. Find the number.
3. When a number is trebled and the result is decreased by 5, the answer is 19. Find the number.
4. Two consecutive integers have a sum of 173. Find the numbers.
5. Three consecutive integers add to 108. Find the smallest of them.
6. When a number is decreased by 1 and the resulting number is halved, the answer is 45. Find the number.
7. Three times a number is equal to 17 minus the number. Find the number.
8. Cat food tins are sold in packs of 6. Rachel bought 42 tins in total. How many packs did she buy?
9. Julian's sister owns 3 more shirts than he does. Between them they own 15 shirts. How many shirts does Julian own?
10. A plane flying from Liverpool to Madrid is carrying 30 rows of passengers, as well as 12 crew members. There are 222 people on board the plane, and every seat is taken. How many passengers are in each row?
11. Drew, Paige, and Henry went to a birthday party. Drew ate 7 more sweets than Paige, and Henry ate 2 fewer sweets than Paige. Between them they ate 20 sweets. How many sweets did Paige eat?
12. During a volleyball training session, Keela drank twice as much water as Carol, and Xavier drank 100 ml more water than Keela. Between them they drank 3 litres of water. How much water did Carol drink?
13. In a cricket match, Emma scored 5 fewer runs than Alex, and Alex scored three times as many runs as Toni. Between them they scored 93 runs. How many runs did each player score?

Example 22 ◀◣ Self Tutor

Cans of sardines are sold in two sizes. Small cans cost £2 each, and large cans cost £3 each. If 15 cans of sardines were bought for a total of £38, how many small cans were bought?

Size	Cost per can	Number bought	Value
small	£2	x	£$2x$
large	£3	$15 - x$	£$3(15 - x)$
		15	£38

$2x + 3(15 - x) = 38$

$\therefore\ 2x + 45 - 3x = 38$ {expanding brackets}

$\therefore\ 45 - x = 38$

$\therefore\ -x = -7$ {subtracting 45 from both sides}

$\therefore\ x = 7$

So, 7 small cans were bought.

14 Isaac is going to boarding school. He buys school shirts at £35 each and trousers at £49 each. Altogether he buys 9 items, and their total cost is £357. How many shirts does he buy?

15 I have 36 coins in my pocket, all of which are 5-cent or 10-cent coins. If their total value is €3.20, how many 5-cent coins do I have?

16 Oranges cost 25 pence each and apples cost 30 pence each. I bought 5 more oranges than apples, and the total cost was £4.55. How many apples did I buy?

17 Ellie is now four times as old as her son. In 5 years' time she will be three times as old as her son. How old is Ellie's son now?

18 Four years ago, Adrian was one quarter of his brother's age. In two years' time, his age doubled will equal his brother's age. How old is Adrian now?

I LINEAR INEQUALITIES

For safety reasons, many rides at theme parks have height restrictions.

For example, it is common to see signs which say:

"You must be at least 120 cm tall to go on this ride."

We could write this requirement as height $\geqslant 120$ cm or $H \geqslant 120$ cm.

This means that if a person has height $H = 120$, 120.5, 125, or 150 cm, then he or she can go on the ride.

All of these values satisfy the *inequality* $H \geqslant 120$ cm.

> An **algebraic inequality** is a mathematical sentence which compares the values of two expressions. It contains one of the symbols $>$, $<$, $\geqslant$, or $\leqslant$.

Symbol	Meaning	Examples
$>$	is greater than	$5 > 3$, $a > 7$, $2\frac{1}{2} > x$
$<$	is less than	$3 < 5$, $7 < a$, $x < 2\frac{1}{2}$
$\geqslant$	is greater than or equal to	$5 \geqslant 4$, $b \geqslant 20$, $\frac{4}{5} \geqslant y$
$\leqslant$	is less than or equal to	$4 \leqslant 5$, $20 \leqslant b$, $y \leqslant \frac{4}{5}$

For example, $5x - 2 > 7$ is a *linear inequality* which indicates that the value of the expression $5x - 2$ is greater than 7.

Notice in the examples above that if we interchange the LHS and RHS, then the **inequality sign** needs to be **reversed**.

Example 23

Rewrite the following inequality with the variable on the LHS:

a $8 < x$
b $1.5 \geqslant y$

a $8 < x$ is written as $x > 8$
b $1.5 \geqslant y$ is written as $y \leqslant 1.5$

SOLUTIONS TO LINEAR INEQUALITIES

While a linear *equation* usually only has one solution, a linear *inequality* is usually true for many values of the variable.

For example:

$x = 2$ indicates that the only value which satisfies the equation is 2

$x > 2$ indicates that any real number greater than 2 will satisfy the inequality, and so there are infinitely many solutions.

We can display the solutions to an inequality using a **number line**.

For example, the solutions to $x > 2$ are shown alongside. The arrow indicates that all values of x to the right of 2 satisfy the inequality. The open circle indicates that 2 is not included.

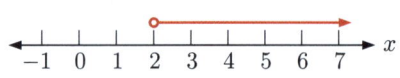

- If the value of the variable may be infinitely large, we use a ray → or ← written above the number line.
- If the sign is $>$ or $<$, the ray has an open circle ○→ or ←○ .
  This indicates that the value at the ○ is **not included** in the solution.
- If the sign is $\geqslant$ or $\leqslant$, the ray has a closed circle ●→ or ←● .
  This indicates that the value at the ● is **included** in the solution.

Example 24

Draw separate number lines to illustrate the following inequalities:

a $a > 1$ b $a < -1$ c $a \geqslant 1$ d $a \leqslant -1$

a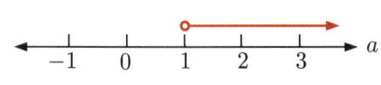
b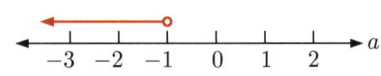
c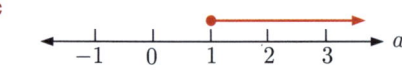
d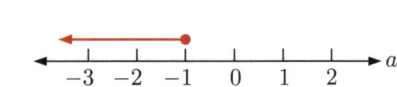

EXERCISE 9I

1 Write a mathematical sentence for:

    a the speed S must not exceed 40 km h^{-1}
    b the age A must be at least 18 years
    c a is greater than 3
    d b is less than or equal to -3
    e d is less than 5
    f -20 is greater than or equal to x
    g 4 is less than y
    h z is greater than or equal to 0.

2 Rewrite the following inequalities with the variable on the LHS:

   a $2 < x$ **b** $5 > b$ **c** $2\frac{1}{2} \leqslant c$

   d $-7 \geqslant d$ **e** $-19 > a$ **f** $-3 < p$

It is usual to write the variable on the LHS of an inequality.

3 Draw separate number lines to illustrate the following inequalities:

   a $x > 2$ **b** $a > -3$ **c** $b \leqslant 2$

   d $m \geqslant -1$ **e** $3 < x$ **f** $2 > a$

   g $x > 3\frac{1}{2}$ **h** $x \leqslant -1\frac{1}{4}$

4 Given that $a > 5$, decide if the following statements are true or false. Draw a number line to help you if necessary.

   a 7 is a possible value for a **b** -5.3 is a possible value for a

   c 5 is a possible value for a **d** 12.2 is *not* a possible value for a

   e -100.8 is *not* a possible value for a **f** 0 is a possible value for a

5 **a** On the same number line, illustrate the solutions to the inequalities $x > 3$ and $x \leqslant \frac{5}{2}$.

   b State the only integer value of x which does not satisfy either of these inequalities.

6 **a** Suppose p and q are whole numbers, where $p > 25$ and $q \leqslant 10$. Find the *smallest* possible value of $p - q$.

   b Suppose w and x are whole numbers, where $w \leqslant 36$ and $x < 20$. Find the *largest* possible value of $w + x$.

J   SOLVING LINEAR INEQUALITIES

We have previously compared a mathematical equation to a balanced set of scales. An inequality is like an **unbalanced** set of scales.

For example, consider the inequality $3x + 2 > 7$. We let a square represent x and a circle represent 1.

To solve a linear inequality, we need to maintain the *imbalance*.

We can carry out the same operation on both sides of the inequality sign, but we need to make sure this will not change its solutions.

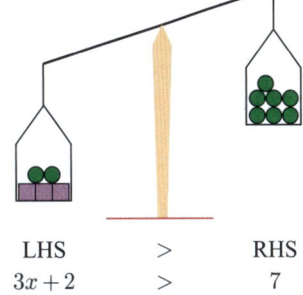

LHS   >   RHS
$3x + 2$ > 7

Investigation   Operations with inequalities

In this Investigation we will begin with inequalities which we know are true, and then perform various operations to both sides of the inequalities. By checking whether the resulting inequalities are true or false, we will establish some rules for operating with inequalities.

PRINTABLE WORKSHEET

Adding or subtracting a positive or negative number

What to do:

1 Copy and complete the following table where T means *true* and F means *false*:

True inequality	Operation	Working	Simplify	T or F
$5 > 1$	add 3 to both sides	$5 + 3 > 1 + 3$	$8 > 4$	T
$11 > 7$	add 6 to both sides			
$8 > -3$	add 4 to both sides			
$-7 < -4$	add 5 to both sides			
$-6 < 5$	add 11 to both sides			
$8 > 6$	subtract 2 from both sides			
$9 > 7$	subtract 4 from both sides			
$2 > -1$	subtract 2 from both sides			
$-7 < -5$	subtract 5 from both sides			

$a > b$ if a is to the right of b on the number line.

2 If I *add* a number to both sides of an inequality, will it still be true?

3 If I *subtract* a number from both sides of an inequality, will it still be true?

Multiplying or dividing by a positive or negative number

What to do:

1 Copy and complete the following table, where T means *true* and F means *false*:

True inequality	Operation	Working	Simplify	T or F
$5 > 2$	multiply by 3	$5 \times 3 > 2 \times 3$	$15 > 6$	T
$-3 < -2$	multiply by 4			
$4 > 3$	multiply by (-3)			
$-2 > -5$	multiply by (-4)			
$7 > 3$	multiply by 2			
$-4 < -2$	multiply by 3			
$2 > -6$	multiply by (-2)			
$-5 < -3$	multiply by (-4)			
$6 > 3$	divide by 3			
$-6 > -8$	divide by 2			
$6 > 3$	divide by (-3)	$\frac{6}{-3} > \frac{3}{-3}$	$-2 > -1$	F
$-6 > -8$	divide by (-2)			
$7 < 14$	divide by 7			
$-10 < -5$	divide by 5			
$7 < 14$	divide by (-7)			

2 If I multiply or divide both sides of an inequality by a *positive* number, will it still be true?

3 If I multiply or divide both sides of an inequality by a *negative* number, will it still be true?

You should have found that, when we multiply or divide both sides of an inequality by a *negative* number, the inequality is no longer true.

To illustrate why this occurs, consider the values 5 and 2 on this number line. 5 is to the *right* of 2 on the number line, so $5 > 2$.

Suppose we multiply each value by -1, giving -5 and -2. -5 is now to the *left* of -2. Therefore, to keep the inequality correct, we must **reverse** the inequality sign, giving $-5 < -2$.

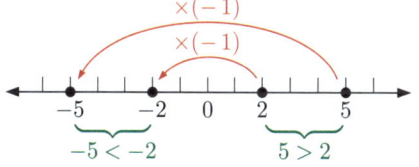

RULES FOR SOLVING INEQUALITIES

- If we **swap** the LHS and RHS, we **reverse** the inequality sign.
- If we **add to** or **subtract from** both sides, we **keep** the inequality sign.
- If we **multiply** or **divide** both sides by:
  - a **positive** number we **keep** the inequality sign
  - a **negative** number we **reverse** the inequality sign.

We reverse the sign when multiplying or dividing by a negative number! The reverse of $>$ is $<$, and the reverse of $\geqslant$ is $\leqslant$.

Example 25   ◀) Self Tutor

Solve the following inequalities, and show their solutions on separate number lines:

a $a - 4 > 5$ **b** $3b \leqslant 9$ **c** $4 - 2x > 0$

a $a - 4 > 5$
$\therefore\ a - 4 + 4 > 5 + 4$ {adding 4 to both sides}
$\therefore\ a > 9$

Check: We choose a value of a which satisfies $a > 9$, say $a = 10$.
If $a = 10$, we have $10 - 4 > 5$
$\therefore\ 6 > 5$ which is true. ✓

b $3b \leqslant 9$
$\therefore\ \dfrac{3b}{3} \leqslant \dfrac{9}{3}$ {dividing both sides by 3}
$\therefore\ b \leqslant 3$

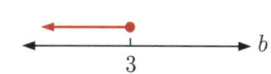

Check: If $b = 2$, we have $3 \times 2 \leqslant 9$
$\therefore\ 6 \leqslant 9$ which is true. ✓

c $4 - 2x > 0$
$\therefore\ 4 - 4 - 2x > 0 - 4$ {subtracting 4 from both sides}
$\therefore\ -2x > -4$
$\therefore\ \dfrac{-2x}{-2} < \dfrac{-4}{-2}$ {reversing the sign}
$\therefore\ x < 2$

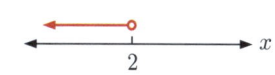

Check: If $x = 0$, we have $4 - 2 \times 0 > 0$
$\therefore\ 4 > 0$ which is true. ✓

EXERCISE 9J

1 Solve the following inequalities, and show their solutions on separate number lines:

a $a + 4 > 6$
b $3b \leqslant -9$
c $s - 4 < 2$
d $\dfrac{c}{5} < 2$
e $x - 8 \geqslant -3$
f $-4b > 16$
g $5 + t > 0$
h $5k < -30$
i $-\dfrac{m}{5} \geqslant 12$

2 Solve the following inequalities, and show their solutions on separate number lines:

a $2x + 5 > 11$
b $5m - 3 \leqslant 17$
c $3a - 2 \geqslant 8$
d $4a + 9 < 1$
e $7 - 2b < -3$
f $16 + 7s > 2$
g $2a - 4 \leqslant 0$
h $12 - 5b > -3$
i $3b - 1 \geqslant 0$
j $5n + 7 < -3$
k $18 - x > 5$
l $11 - 4b \leqslant 4$

Example 26   ◀)) **Self Tutor**

Solve $\dfrac{x}{4} - 2 \geqslant 3$, and show its solution on a number line.

$\dfrac{x}{4} - 2 \geqslant 3$

$\therefore \dfrac{x}{4} - 2 + 2 \geqslant 3 + 2$ {adding 2 to both sides}

$\therefore \dfrac{x}{4} \geqslant 5$

$\therefore \dfrac{x}{4} \times 4 \geqslant 5 \times 4$ {multiplying both sides by 4}

$\therefore x \geqslant 20$

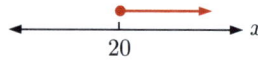

Check: If $x = 32$, we have $\dfrac{32}{4} - 2 \geqslant 3$

$\therefore 6 \geqslant 3$ which is true. ✓

3 Solve the following inequalities, and show their solutions on separate number lines:

a $\dfrac{x}{3} + 1 > 4$
b $\dfrac{b}{5} - 2 \leqslant -3$
c $\dfrac{c}{4} + 4 \geqslant 8$
d $\dfrac{2x}{3} - 2 < 4$
e $\dfrac{3x}{2} + 5 \geqslant -2$
f $1 - \dfrac{x}{2} < 4$
g $2 - \dfrac{x}{4} \geqslant -3$
h $3 - \dfrac{2x}{5} < 2$
i $5 - \dfrac{3x}{4} > -2$

4 Solve the following inequalities, and show their solutions on separate number lines:

a $\dfrac{a - 3}{2} < 6$
b $\dfrac{b + 5}{4} \geqslant 1$
c $\dfrac{4c + 3}{5} \leqslant -1$
d $\dfrac{4 - a}{2} > 5$
e $\dfrac{5 - 3x}{2} \leqslant -6$
f $\dfrac{3 - 2x}{3} \geqslant -1$

5 Solve the following inequalities by first interchanging the LHS and RHS:
  a $5 > 4 + x$
  b $2 \leqslant 6c + 14$
  c $-4 > 2b$
  d $7 \leqslant \frac{a}{3}$
  e $3 < \frac{d-2}{4}$
  f $6 \geqslant 3(p+2)$

6 Solve for x, and show the solution on a number line:
  a $5x - 3 > 3x + 1$
  b $2x + 1 \geqslant 4x + 7$
  c $8x + 6 < 3x + 1$
  d $2x + 7 > 7x + 3$
  e $6x + 2 \leqslant 3x - 7$
  f $x - 11 \leqslant 6x - 1$

7 a Find the largest integer which satisfies:
  i $2x - 5 \leqslant 15$
  ii $\frac{x+2}{4} < -1$
  iii $5 - \frac{2x}{3} \geqslant 2$

  b Find the smallest integer which satisfies:
  i $3x + 4 \geqslant 30$
  ii $\frac{x}{4} - 2 > 5$
  iii $3x - 1 < 6x + 12$

Activity 5 — Sir Cool and the inequality crusade

Click on the icon to help Sir Cool rid the world of inequality!

THE ADVENTURES OF SIR COOL

Review set 9A

1 State the inverse of $\div 6$.

2 Find the equation which results when:
  a 2 is added to both sides of $3x - 2 = -11$
  b 9 is subtracted from both sides of $4x + 9 = -1$.

3 Solve $x + 5 = 2$ by inspection.

4 One of the numbers 6, 10, -6, or -4 is the solution to the equation $8x + 3 = 3(x - 9)$. Find the solution.

5 Solve using a suitable inverse operation:
  a $x - 11 = 4$
  b $6x = 42$
  c $a + 4 = -9$
  d $\frac{t}{-3} = 7$

6 Use a flowchart to show how the following expressions are "built up" from x:
  a $\frac{x}{6} + 1$
  b $4(3x - 4)$
  c $\frac{2 - 4x}{3}$

7 Use a flowchart to show how to isolate x from the following expressions:
  a $\frac{x}{7} - 3$
  b $4(x + 1)$
  c $1 + 3x$

8 Solve for x:
 a $10x - 7 = 13$
 b $5 + 4x = 29$
 c $3 - 2x = 9$
 d $4x + 7 = -1$
 e $4x - 5 = 5x - 6$
 f $\dfrac{5x + 4}{3} = -2$

9 Solve for x:
 a $7x - 6 = 6x - 1$
 b $4(5x + 1) = 14$
 c $3x + 2(3 - x) = -3$
 d $4x - 2(3x - 1) = 5 - 7x$
 e $3(x + 6) - 4(4 - 2x) = 7x + 6$
 f $9 - 5(x - 1) = 2(x + 4)$

10 Solve for x:
 a $\dfrac{x + 3}{5} = \dfrac{x - 4}{2}$
 b $\dfrac{4}{3x} = \dfrac{10}{7}$
 c $\dfrac{3}{2 - x} = \dfrac{2}{3}$

11 When a number is increased by 11 and the result is doubled, the answer is 48. Find the number.

12 The sum of three consecutive integers is 63. Find the smallest of the integers.

13 When 7 times a certain number is decreased by 11, the result is 31 more than the number. Find the number.

14 I have 25 coins consisting of 5 p and 50 p pieces. If the total value is £7.10, how many 5 p coins do I have?

15 Draw separate number lines to illustrate:
 a $x \geqslant 5$
 b $3 > x$

16 Solve the following inequalities, and show their solutions on separate number lines:
 a $3 - 2x \geqslant -4$
 b $\dfrac{2x + 3}{2} < -3$
 c $3(x + 2) - 1 \leqslant 1 - 4x$

17 a Solve the inequality $\dfrac{x}{5} - 3 > -5$ and display the solution on a number line.
  b Solve the inequality $8 - 3(x + 2) \geqslant 1$ and display the solution on a number line.
  c Find the values of x which satisfy *both* inequalities.

Review set 9B

1 State the inverse of:
 a multiplying by 5
 b subtracting 7.

2 Solve using a suitable inverse operation:
 a $a - 3 = 4$
 b $-5b = 45$
 c $c + 17 = 7$
 d $\dfrac{d}{8} = -12$

3 Use a flowchart to show how the following expressions are "built up" from x:
 a $\dfrac{x + 3}{2}$
 b $5x - 9$
 c $\dfrac{x}{7} - 2$

4 Use a flowchart to show how to isolate x from the following expressions:

a $3x - 1$ **b** $\dfrac{8x + 10}{3}$ **c** $5 - \dfrac{2x}{3}$

5 Solve for x:

a $3x + 5 = 17$ **b** $\dfrac{x}{4} + 1 = -11$ **c** $5 + 2x = 3$

6 Solve for x:

a $\dfrac{1 + 2x}{3} = 1$ **b** $\dfrac{1 + 2x}{3} = 0$ **c** $\dfrac{1 + 2x}{3} = -1$

7 Solve for x:

a $2x + 1 = x + 8$ **b** $x - 4 = 5x - 1$ **c** $3(x - 3) = 8 - x$

8 Solve for x:

a $3(4 - x) - 2x = -13$ **b** $3x - 5 = 3 - x$
c $2(4x - 3) + x = 3(2x - 1) + 2$ **d** $2(x - 3) - 3(4 - x) = 4(2x - 5)$

9 Ms Maxwell wrote the equation $\dfrac{x}{3} + 2 = 7$ on the board. She told her class there were *two* good ways to solve the equation.

  a The first method is to first multiply *each* term in the equation by 3.
   i Find the equation that results when *each* term is multiplied by 3.
   ii Hence solve the equation.
  b The second method is to first subtract 2 from each side of the equation.
   i Find the equation that results when 2 is subtracted from both sides of $\dfrac{x}{3} + 2 = 7$.
   ii Hence solve the equation.
  c Do the two methods give the same answer?

10 Solve for x:

a $\dfrac{x - 3}{4} = \dfrac{2x}{5}$ **b** $\dfrac{3}{5x} = -\dfrac{7}{8}$ **c** $\dfrac{2}{x + 5} = \dfrac{6}{7}$

11 Four times a number is equal to the number plus 15. Find the number.

12 The sum of two consecutive odd integers is 36. Find the larger integer.

13 Five more than a certain number is nine less than three times the number. Find the number.

14 Writing pads cost £1.35 each and pens cost £0.85 each. I bought twice as many pens as pads, and the total cost was £18.30. How many pads did I buy?

15 Sadao likes collecting action figures. In total he has 44 figures belonging to three categories. He has 2 more transformers than chogokin, and he has 1 more anime figure than transformers. How many chogokin does Sadao have?

16 Draw separate number lines to illustrate:

    a $a \geqslant -2$ **b** $b < -1$ **c** $c > 4\frac{1}{2}$

17 Solve the following inequalities:

    a $3 - 2x \geqslant 0$ **b** $\frac{x}{5} - 3 < -1$ **c** $6(2x + 1) \geqslant 6 - 13x$

10

Geometry

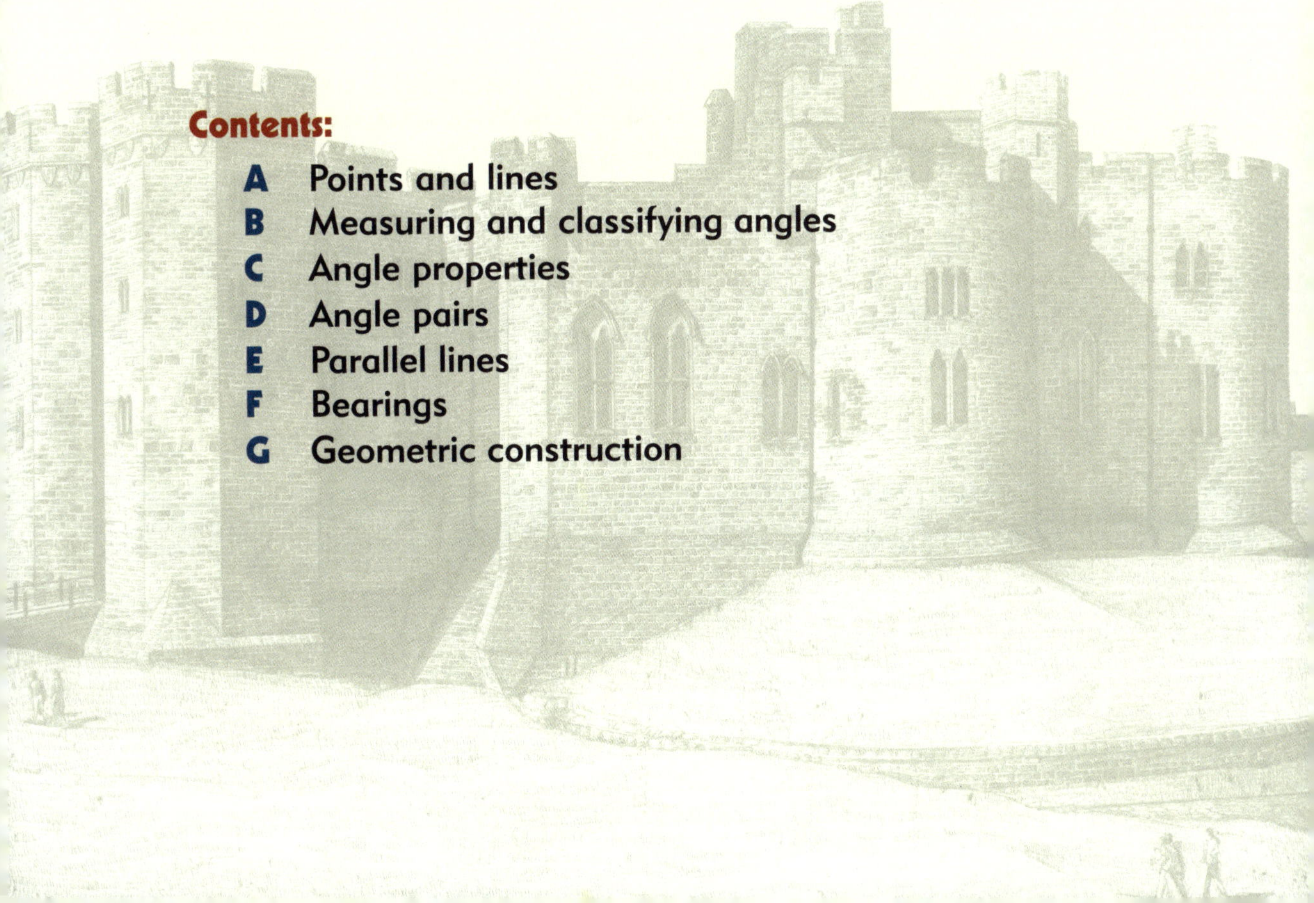

Contents:

- **A** Points and lines
- **B** Measuring and classifying angles
- **C** Angle properties
- **D** Angle pairs
- **E** Parallel lines
- **F** Bearings
- **G** Geometric construction

Opening problem

This diagram shows three lines. We can see points of **intersection** where line 1 meets the other lines.

Things to think about:

a How can we describe the point where:
  i line 1 meets line 2
  ii line 1 meets line 3?

b Lines 2 and 3 do not meet in this diagram.
  i If we were to extend the lines, do you think they would eventually meet?
  ii By measuring angles in the diagram, can we test whether the lines will eventually meet?

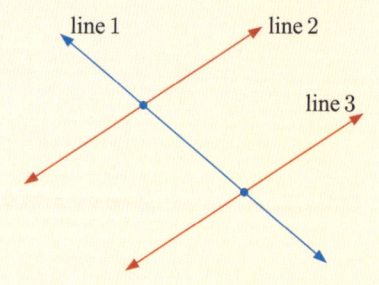

If we look carefully, we can see **angles** in many objects and situations. We see them in the framework of buildings, the pitches of roof structures, the steepness of ramps, and the positions of boats from a harbour and aeroplanes from an airport.

The measurement of angles dates back more than 2500 years and is still very important today in architecture, building, surveying, engineering, navigation, space research, and many other industries.

Research   Degree measure

The Babylonian Empire was founded in the 18th century BC by Hammurabi in lower Mesopotamia, which is today in southern Iraq. It lasted over a thousand years, finally being absorbed into the Persian Empire of Darius in the 6th century BC.

1 The Babylonians invented the **astrolabe**.
  Find out what an astrolabe measures.

2 How many degrees did the Babylonians decide should be in one full turn? Why did they choose this number?

A POINTS AND LINES

POINTS

We use a **point** to mark a location or position.

Examples of points are:

- the corner of your desk
- the tip of your compass needle.

Points do not have size. We say they are **infinitely small**. In geometry, however, a point is represented by a small dot so we can see it. To help identify the point, we label it with a capital letter.

Geometry (Chapter 10) 177

For example: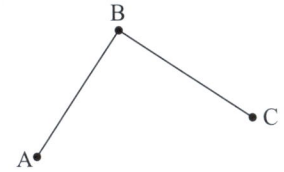

The letters A, B, and C identify the points.

The letters allow us to make statements like:
"the distance from A to B is ...." or
"the angle at B measures ....".

STRAIGHT LINES

A **straight line**, usually just called a **line**, is a continuous infinite collection of points which lie in a particular direction. A line has no beginning or end.

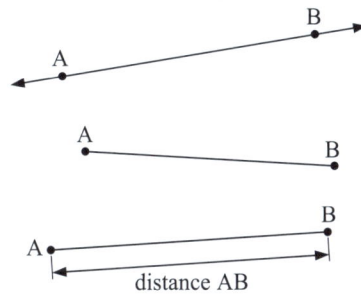

Line AB is the endless straight line which passes through points A and B. We can call it "line AB" or "line BA". There is only one straight line which passes through both A and B.

Line segment AB is the part of the line AB that connects A with B.

The **distance AB** is the length of the line segment AB.

If three or more *points* lie on a single straight line, we say that the points are **collinear**.

For example, in the diagram the points A, B, C, and D are collinear.

If three or more *lines* meet or intersect at the same point, we say that the lines are **concurrent**.

For example, the lines shown are concurrent at point B.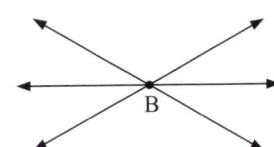

PARALLEL AND INTERSECTING LINES

In mathematics, a **plane** is a flat surface like a table top or a sheet of paper. It goes on indefinitely in all directions, so it has no boundaries.

Two straight lines in the same plane may either be **parallel** or **intersecting**.

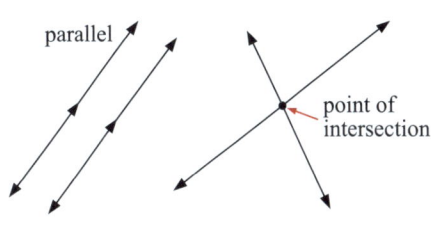

Parallel lines are lines which are always a fixed distance apart and never meet.

Arrowheads at the ends show the line continues forever. Arrowheads in the middle show parallel lines.

Discussion

1 Give *two* examples in the classroom of:
  a a point
  b a line segment
  c a flat surface
  d an angle
  e parallel lines.

2 How many different lines can you draw through:
  a two distinct points A and B
  b all three distinct collinear points A, B, and C
  c one point A
  d all three distinct non-collinear points A, B, and C?

3 You may have heard shapes such as rectangles and triangles being described as "two-dimensional". How many dimensions are there in:
  a a line
  b a point?

EXERCISE 10A

1 Describe, with a sketch, the meaning of:
  a a line segment
  b a point of intersection
  c parallel lines
  d collinear points
  e concurrent lines.

2 Give *all* ways of naming the following lines:

  a

  b

3 PQR is a triangle.
  a Name the three sides of the triangle.
  b Which sides intersect at point P?

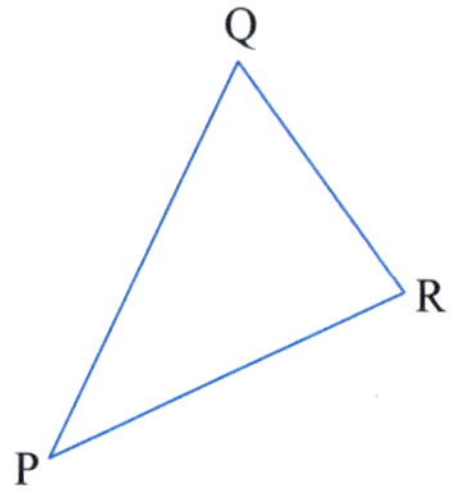

4 Name the point of intersection between:
  a line 2 and line 3
  b line 1 and line 3
  c line AB and line segment DE
  d line AC and line DF.

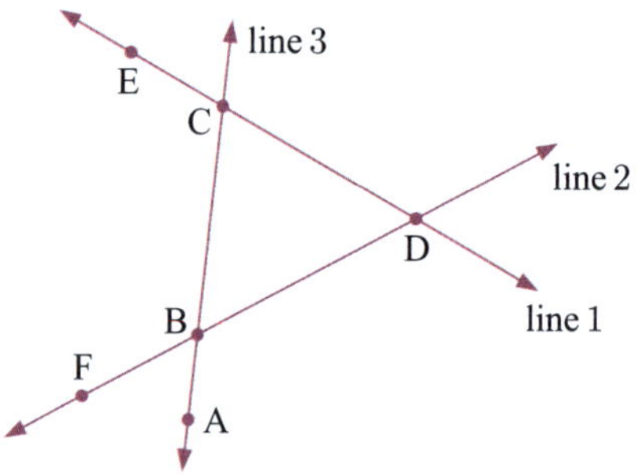

5 Draw a diagram for each statement:
  a X is a point on line segment PQ.
  b Lines EF and GH meet at point M.
  c S, T, U, and V are collinear.
  d Lines JK and MN are parallel.
  e Line segment AB, line CD, and line EF are concurrent at G.

Geometry (Chapter 10) 179

6 **a** Name the line AB in three other ways.
   b How many lines go through point D?
   c What can be said about:
      i lines EF and AD
      ii points A, D, and F
      iii lines CD and EG?

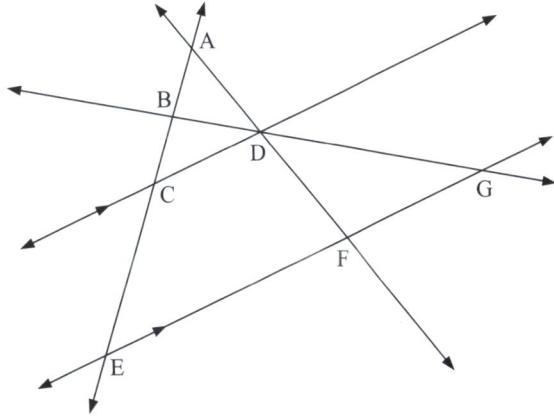

B MEASURING AND CLASSIFYING ANGLES

Whenever two lines or edges meet, an **angle** is formed between them. In mathematics, an angle is made up of two **arms** which meet at a point called the **vertex**.

The **size** or **measure** of the angle is the amount of turning or rotation from one arm to the other.

> Two angles are **equal** if they have the same size or degree measure.

THE PROTRACTOR

Alongside is a **protractor** placed with its centre at B and its base line on AB. The amount of turning from AB to BC is 110 degrees.

We write $\widehat{ABC} = 110°$ which reads "the angle ABC measures 110 degrees".

$\widehat{ABC}$ is called **three point notation**. We use it to make it clear which angle we are referring to.

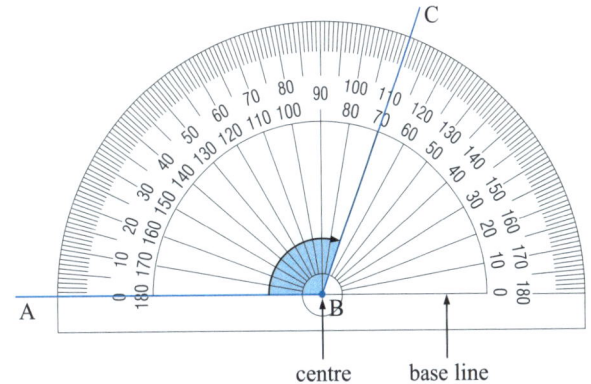

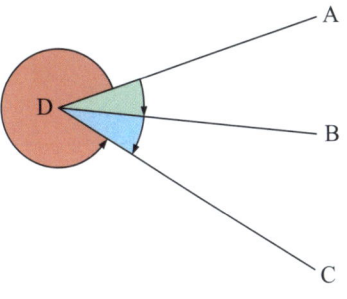

For example, if we want to talk about the angle shaded green in this figure, we cannot just say the angle at D. This could refer to many angles, including the blue one and the red one.

The green angle is $\widehat{ADB}$ or $\widehat{BDA}$.

The blue angle is $\widehat{BDC}$ or $\widehat{CDB}$.

$\widehat{ADC}$ is made up of the green angle *and* the blue angle.

The red angle is called the *reflex* $\widehat{ADC}$, since its size is more than 180°.

CLASSIFYING ANGLES

Angles are classified according to their **size**.

Revolution	Straight Angle	Right Angle
 One complete turn. One revolution = 360°.	 $\frac{1}{2}$ turn. 1 straight angle = 180°.	 $\frac{1}{4}$ turn. 1 right angle = 90°.
Acute Angle	**Obtuse Angle**	**Reflex Angle**
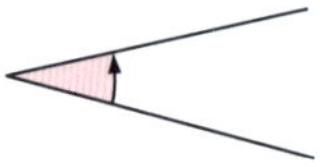 Less than a $\frac{1}{4}$ turn. An acute angle has size between 0° and 90°.	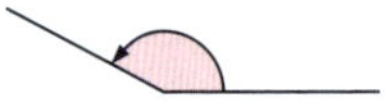 Between $\frac{1}{4}$ turn and $\frac{1}{2}$ turn. An obtuse angle has size between 90° and 180°.	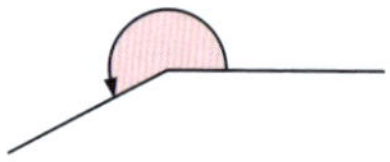 Between $\frac{1}{2}$ turn and 1 turn. A reflex angle has size between 180° and 360°.

EXERCISE 10B

1 Match the names to the correct angles:

    a $\widehat{ABC}$ **b** $\widehat{CAB}$ **c** $\widehat{BCA}$ **d** $\widehat{CBD}$

    A   **B** 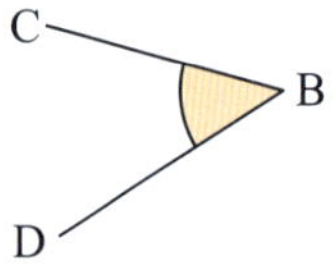   **C** 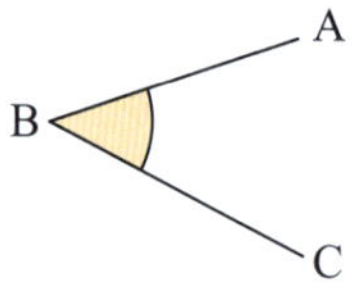   **D**

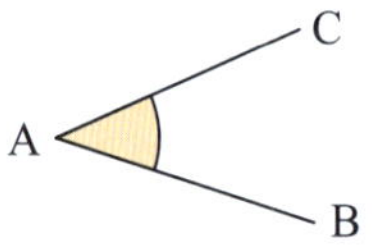

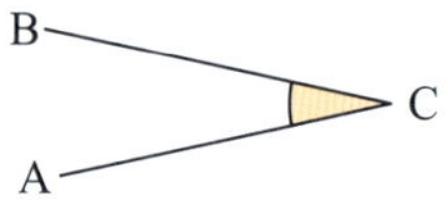

2 Draw and label each of the following angles:

    a $\widehat{PQR}$ **b** $\widehat{RQP}$ **c** reflex $\widehat{EFG}$ **d** $\widehat{CAB}$

3 a Find the sizes of these angles without using your protractor:

    i $\widehat{STU}$ **ii** $\widehat{WTU}$

    iii $\widehat{XTV}$ **iv** $\widehat{STW}$

   b Classify each angle in **a** as acute, right, or obtuse.

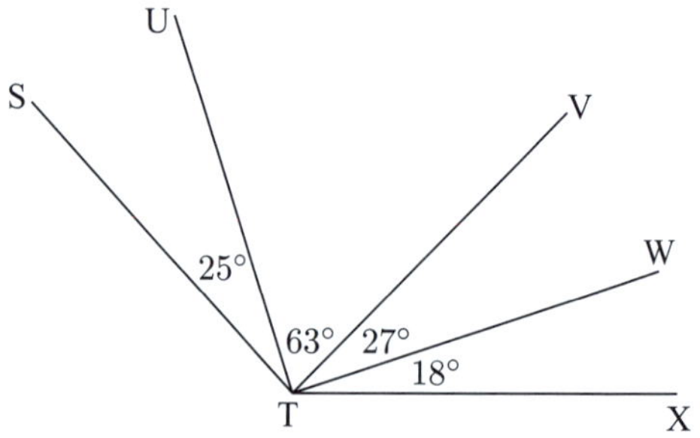

4 Use your ruler and protractor to draw angles with the following sizes:

    a 38°   **b** 89°   **c** 120°

Ask a friend to check the accuracy of your angles.

5 Consider the figure alongside.

a Find the angles corresponding to:

  i BÂD   ii DB̂C   iii AD̂B

b Classify the following angles as acute, obtuse, or reflex:

  i f ii a iii h

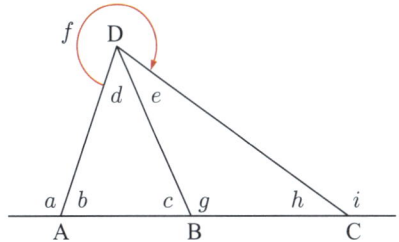

6 Draw a free-hand sketch of:

a obtuse XŶZ   b revolution B   c reflex YŜP
d right CẐR   e acute JK̂L   f straight EF̂G.

7 Use a protractor to measure the named angles:

a i BÂD   b i CB̂D
   ii CB̂A   ii DX̂E
   iii XÂB   iii AD̂C

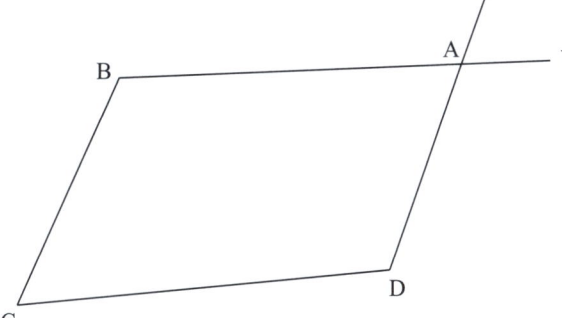

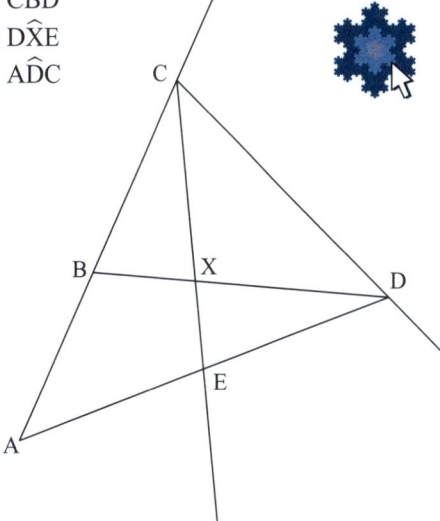

PRINTABLE DIAGRAMS

8 Kit hits the billiard ball so that it follows the path shown. What *acute* angle will it make with the edge of the table?

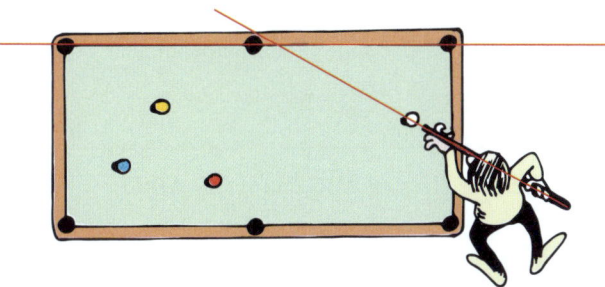

9 An awning is extended out the front of a café as shown. Measure the angle between the top of the awning and the support post.

C ANGLE PROPERTIES

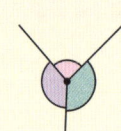
Angles at a point add to 360°.

There are 360° in one complete turn.

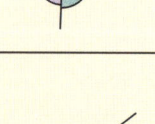
Angles on a line add to 180°.
Angles which add to 180° are called **supplementary angles**.

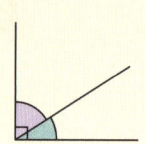
Angles in a right angle add to 90°.
Angles which add to 90° are called **complementary angles**.
Lines or line segments which meet at 90° are said to be **perpendicular**.

A small square is used to indicate a right angle.

For example:

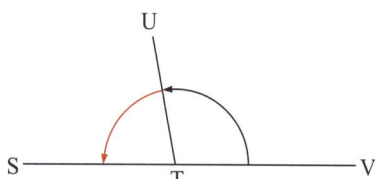

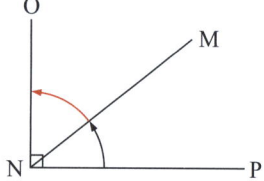

$S\widehat{T}U$ and $U\widehat{T}V$ are supplementary because they are angles on a line.

$M\widehat{N}O$ and $M\widehat{N}P$ are complementary because they are angles in a right angle.

ON and NP are perpendicular.

Discussion

For any angle ABC, what is the relationship between the sizes of $A\widehat{B}C$ and reflex $A\widehat{B}C$?

Example 1   ◀)) Self Tutor

a Are angles with sizes 37° and 53° complementary?
b What angle size is supplementary to 48°?

a $37° + 53° = 90°$. So, the angles are complementary.
b The angle size supplementary to 48° is $180° - 48° = 132°$.

EXERCISE 10C

1 Add the following pairs of angles, and state whether they are complementary, supplementary, or neither:
- **a** 109°, 71°
- **b** 67°, 117°
- **c** 62°, 28°
- **d** 155°, 31°
- **e** 25°, 55°
- **f** 64°, 116°

2 Find the size of the angle complementary to:
- **a** 15°
- **b** 87°
- **c** 43°

3 Find the size of the angle supplementary to:
- **a** 129°
- **b** 57°
- **c** 90°

4 Classify the following angle pairs as complementary, supplementary, or neither:
- **a** $\hat{COA}$ and $\hat{COE}$
- **b** $\hat{AOD}$ and $\hat{EOC}$
- **c** $\hat{BOC}$ and $\hat{COD}$
- **d** $\hat{COE}$ and $\hat{DOB}$

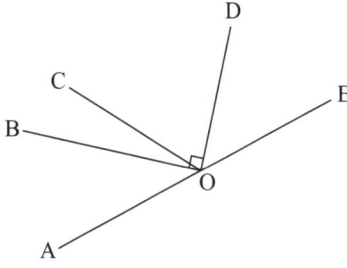

5 Copy and complete:
- **a** The size of the angle complementary to $x°$ is ......
- **b** The size of the angle supplementary to $y°$ is ......
- **c** Two lines are perpendicular if they meet at ......

Example 2 ◀) **Self Tutor**

Find the value of the unknown:

a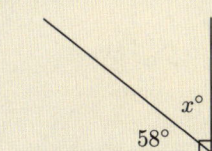

b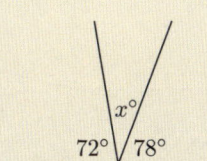

c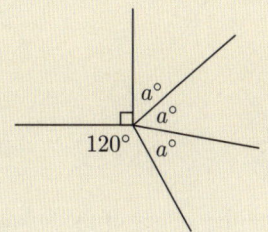

a The angles 58° and $x°$ are complementary.
∴ $x + 58 = 90$
∴ $x = 32$

b The three angles are supplementary, so they add to 180°.
∴ $72 + x + 78 = 180$
∴ $x + 150 = 180$
∴ $x = 30$

c We have five angles at a point, so the sum of the five angles is 360°.
∴ $3a + 90 + 120 = 360$
∴ $3a + 210 = 360$
∴ $3a = 150$
∴ $a = 50$

6 Find the value of the unknown:

a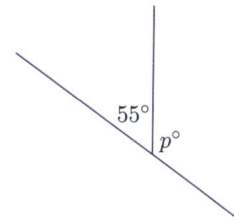

b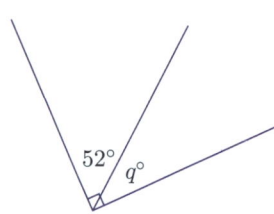

c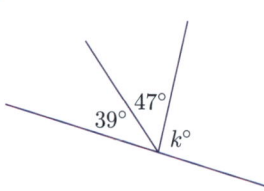

d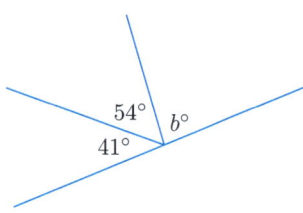

e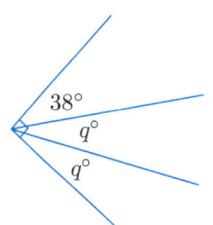

f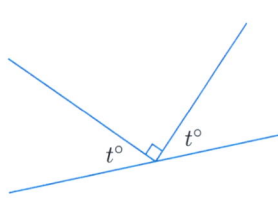

g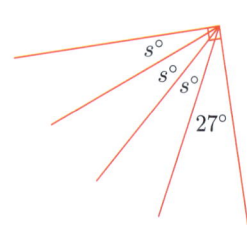

h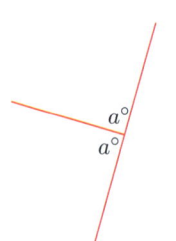

i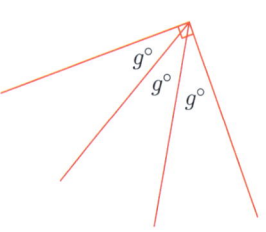

7 Find the sizes of the unknown angles:

a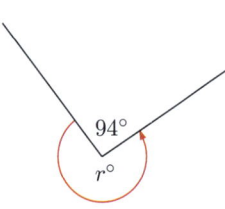

b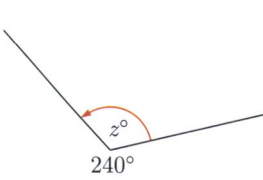

c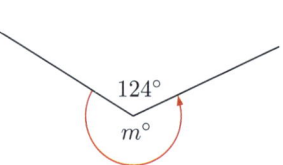

8 Find the values of the unknowns:

a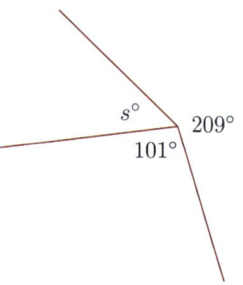

b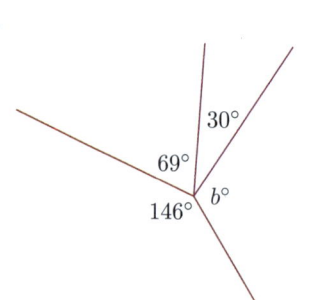

c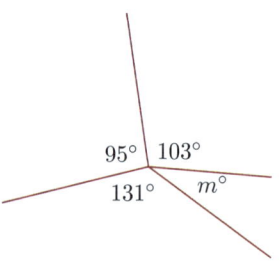

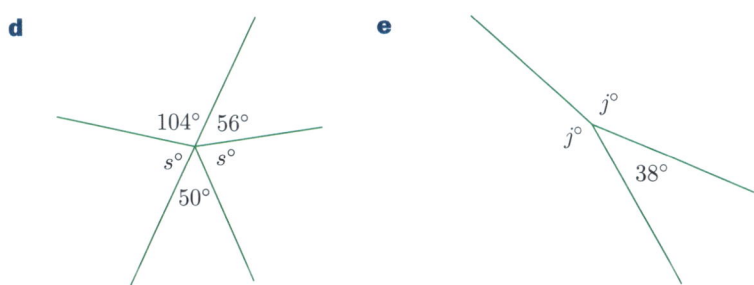

D | ANGLE PAIRS

We have already seen how lines drawn in a plane are either **parallel** or **intersecting**.

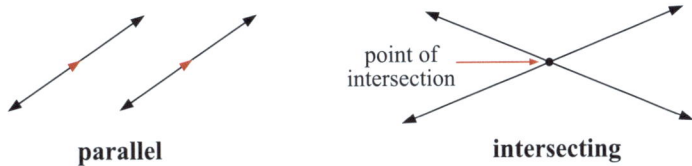

When we are dealing with several lines in a plane, we can identify a number of **angle pairs**.

VERTICALLY OPPOSITE ANGLES

> **Vertically opposite angles** are formed when two straight lines intersect. The two angles are directly opposite each other through the vertex.

For example:

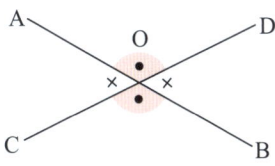

$A\hat{O}C$ and $D\hat{O}B$ are vertically opposite.
$A\hat{O}D$ and $C\hat{O}B$ are vertically opposite.

Measure the pairs of vertically opposite angles carefully. You should find that:

> When two straight lines intersect, **vertically opposite** angles are *equal* in size.

Proof:

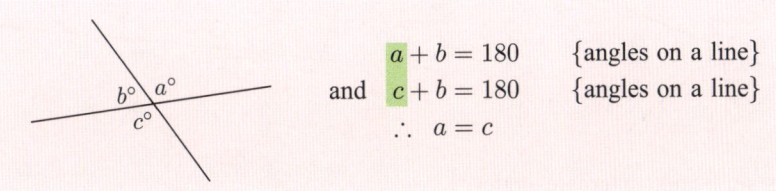

$a + b = 180$ {angles on a line}
and $c + b = 180$ {angles on a line}
$\therefore\ a = c$

GEOMETRY PACKAGE

CORRESPONDING, ALTERNATE, AND CO-INTERIOR ANGLES

A third line that crosses two other straight lines is called a **transversal**.

When two or more straight lines are cut by a transversal, three different angle pairs are formed:

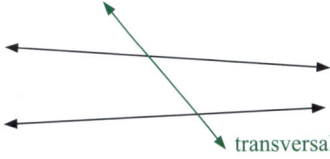

transversal

Corresponding angle pairs	Alternate angle pairs	Co-interior angle pairs
The angles marked • and × are **corresponding angles** because they are both in the *same position*. They are on the *same side* of the transversal and the *same side* of the two straight lines.	The angles marked • and × are **alternate angles**. They are on *opposite sides* of the transversal and *between* the two straight lines.	The angles marked • and × are **co-interior angles**. They are on the *same side* of the transversal and *between* the two straight lines. Co-interior angles can also be called **allied angles**.

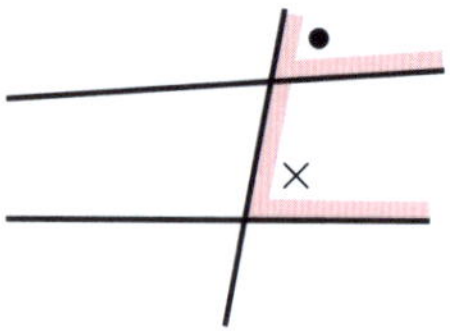

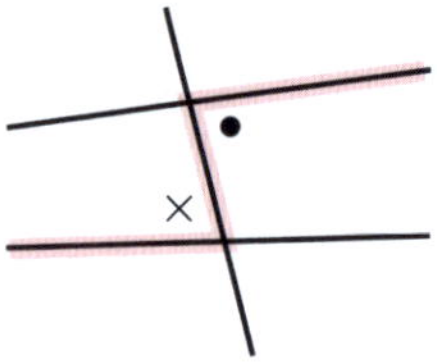

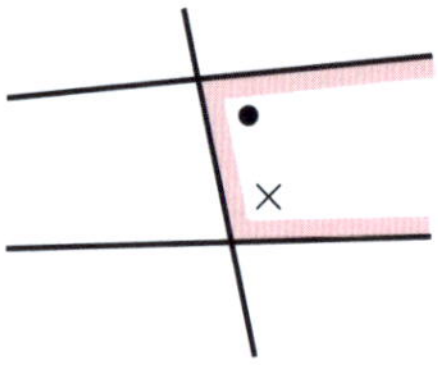

Example 3

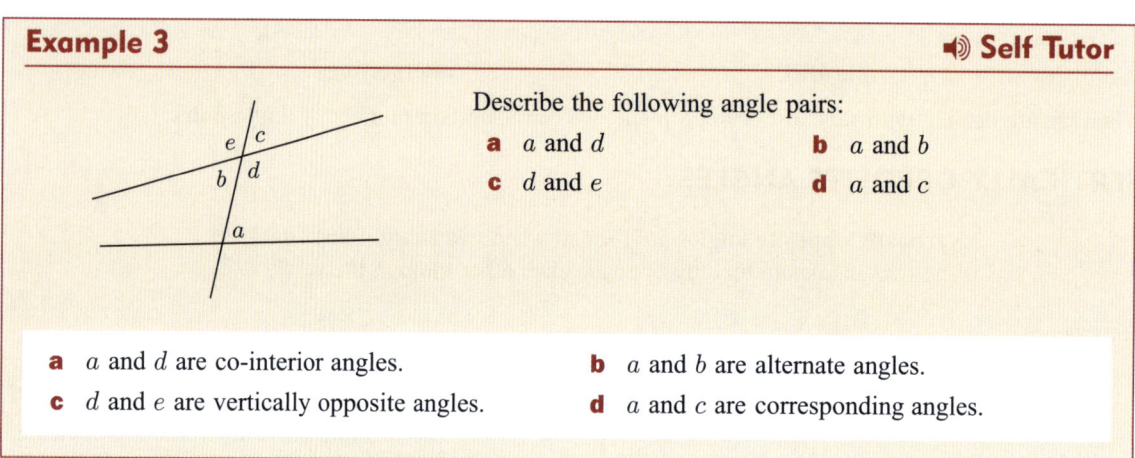

Describe the following angle pairs:

a a and d **b** a and b
c d and e **d** a and c

a a and d are co-interior angles.
b a and b are alternate angles.
c d and e are vertically opposite angles.
d a and c are corresponding angles.

EXERCISE 10D

1 In each diagram, list the pairs of vertically opposite angles:

a

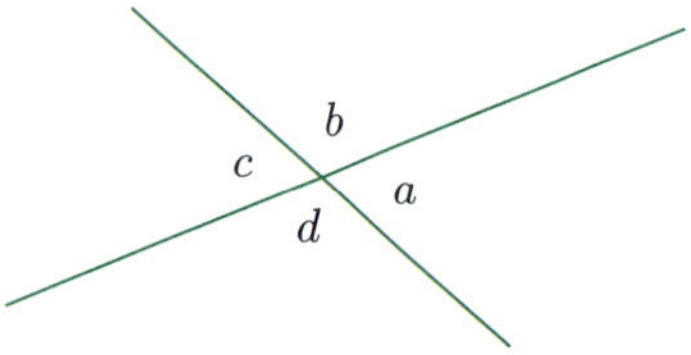

b

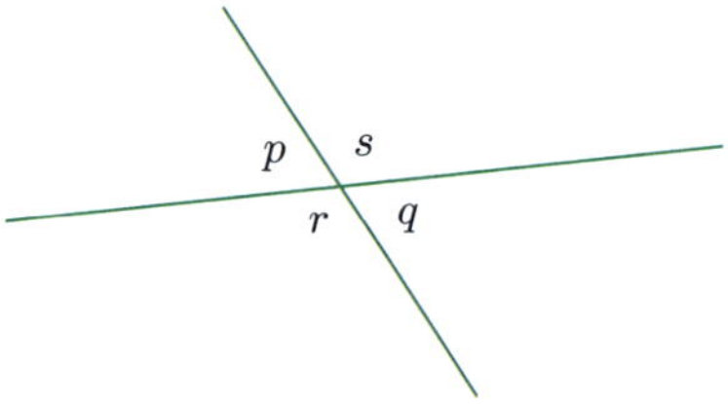

2 In which diagrams are s and t alternate angles?

A 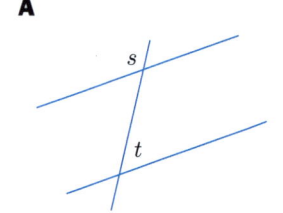 **B** 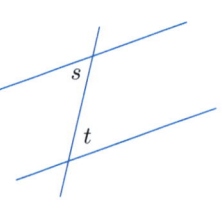 **C** 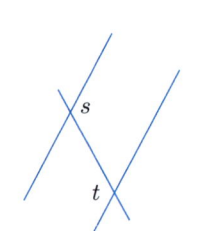 **D**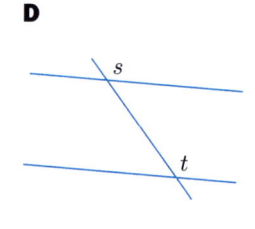

3 Which angle is alternate to angle q?

a 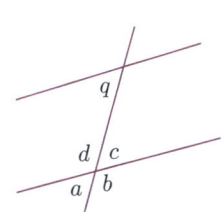 b 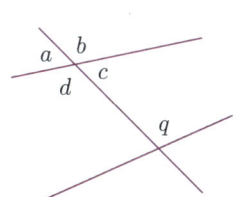 c 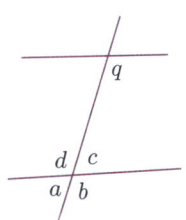 d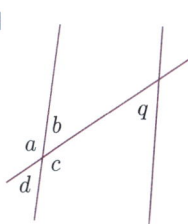

4 In which diagrams are the marked angles corresponding?

A 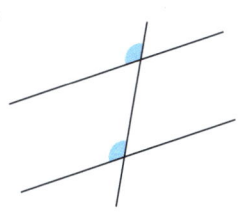 B 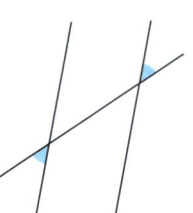 C 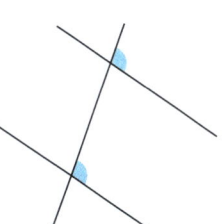 D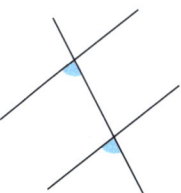

5 Which angle is corresponding to angle k?

a 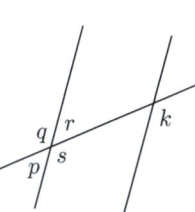 b 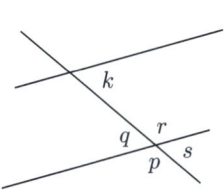 c 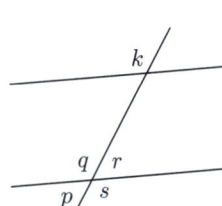 d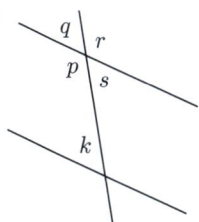

6 In which diagrams are the marked angles co-interior?

A 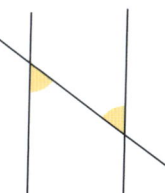 B 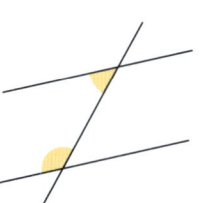 C 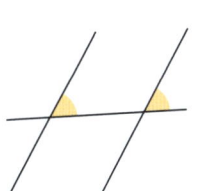 D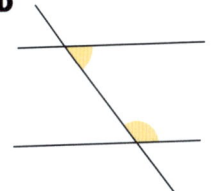

7 Which angle is co-interior with angle f?

a 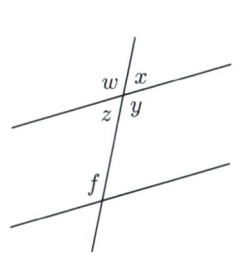 b 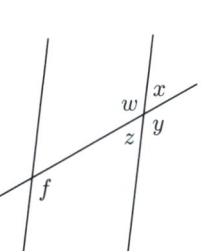 c 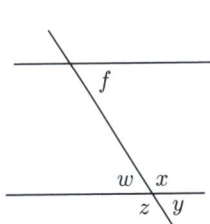 d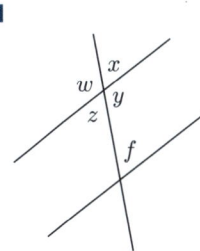

8 Classify the following angle pairs as either corresponding, alternate, co-interior, or vertically opposite:

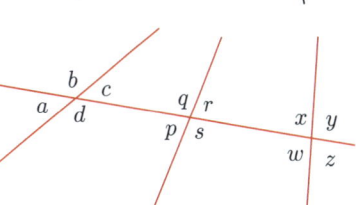

- **a** a and p
- **b** r and w
- **c** r and x
- **d** z and s
- **e** b and q
- **f** a and c
- **g** x and z
- **h** w and s
- **i** c and p

188   Geometry   (Chapter 10)

E   PARALLEL LINES

If the two lines cut by a transversal are parallel, then corresponding, alternate, and co-interior angle pairs have special properties. We will discover these properties in the following **Investigation**.

Investigation   Angle pairs on parallel lines

What to do:

1 Print this worksheet so you can write directly onto it.

2 In each diagram, measure the angles marked.

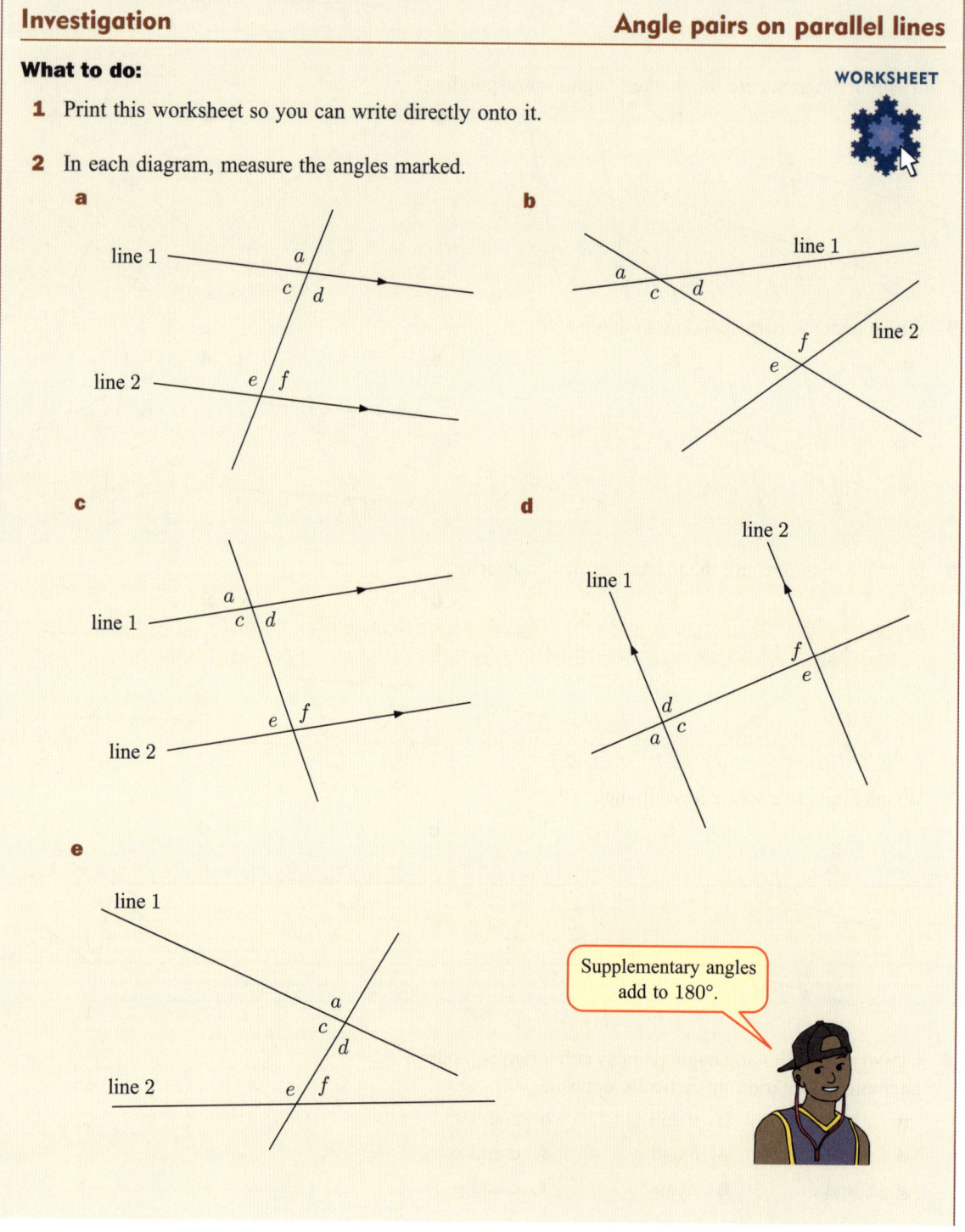

Supplementary angles add to 180°.

Geometry (Chapter 10)   189

Diagram	Are lines 1 and 2 parallel?	Are the corresponding angles a and e equal?	Are the alternate angles c and f equal?	Are the co-interior angles d and f equal? Are they supplementary?	
				equal	supplementary
a					
b					
c					
d					
e					

3 What can you conclude from your results?

From the **Investigation** you should have discovered the following important facts:

> When **parallel lines** are cut by a **transversal**:
> - corresponding angles are equal in size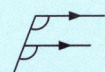
>
> - alternate angles are equal in size
>
> - co-interior angles are supplementary, which means they add up to 180°.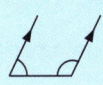

Using these geometrical facts, we can find unknown values for angles on parallel lines.

Example 4   ◀) Self Tutor

Find the value of the unknown, giving a brief reason for your answer:

a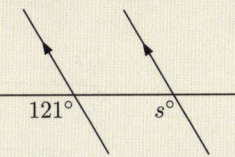

b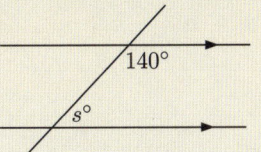

The special properties only apply if the lines cut by the transversal are *parallel*.

a Corresponding angles on parallel lines are equal.
∴ $s = 121$

b Co-interior angles on parallel lines are supplementary.
∴ $s + 140 = 180$
∴ $s = 40$

Geometry (Chapter 10)

TESTS FOR PARALLELISM

Suppose two lines are cut by a transversal.
- If pairs of corresponding angles are equal in size then the lines are parallel.
- If pairs of alternate angles are equal in size then the lines are parallel.
- If pairs of co-interior angles are supplementary then the lines are parallel.

GEOMETRY PACKAGE

Example 5   ◀) Self Tutor

Decide if the figure contains parallel lines, giving a brief reason for your answer:

a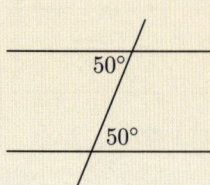

b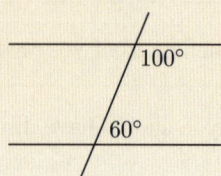

a These alternate angles are equal, so the lines are parallel.

b These co-interior angles add to $160°$, so they are not supplementary.
∴ the lines are *not* parallel.

EXERCISE 10E

1 Find, giving brief reasons, the values of the unknowns:

a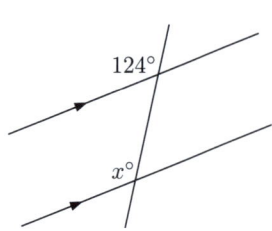

b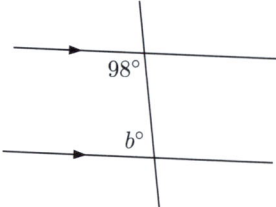

c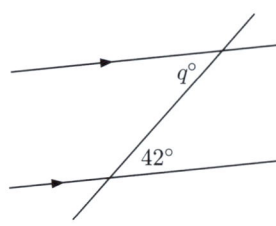

d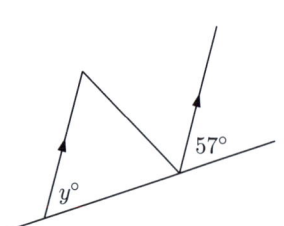

e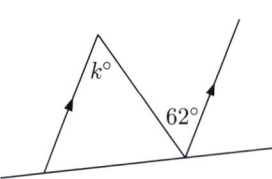

f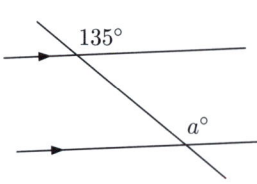

g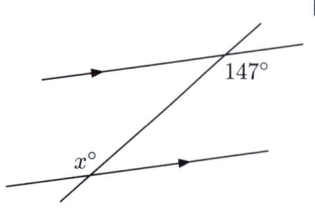

h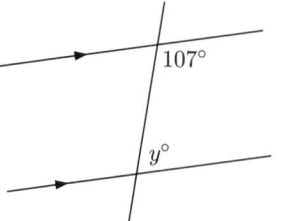

i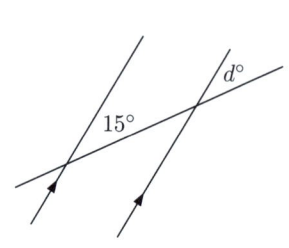

2 Find the values of the unknowns in alphabetical order, giving brief reasons:

a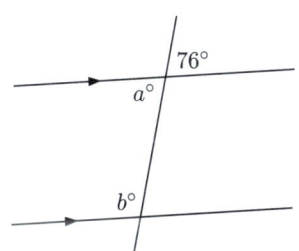

b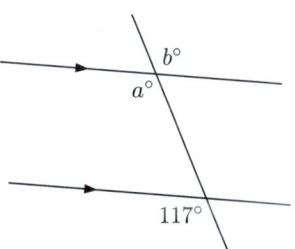

c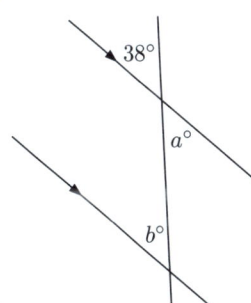

d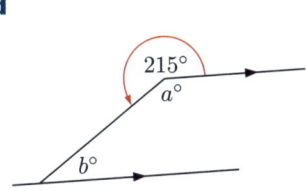

e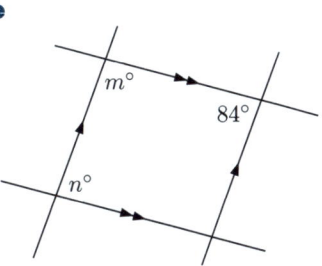

f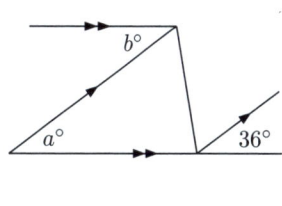

3 Write a statement connecting the unknowns, giving a brief reason:

a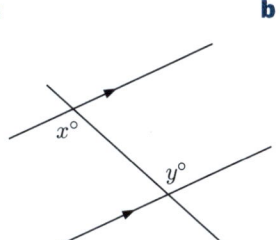

b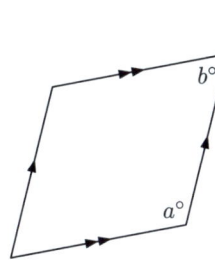

c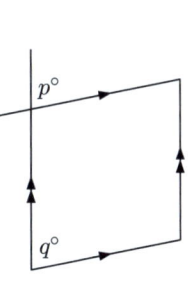

d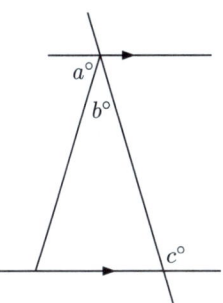

4 The following figures are not drawn to scale.
Decide if each figure contains a pair of parallel lines, giving brief reasons for your answers.

a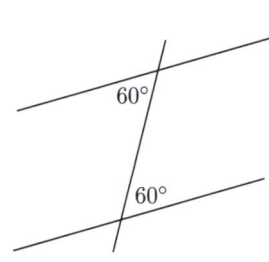

b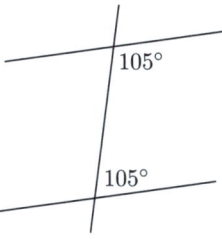

c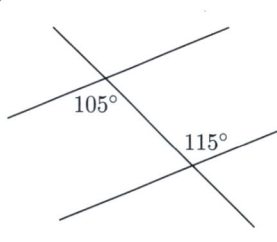

d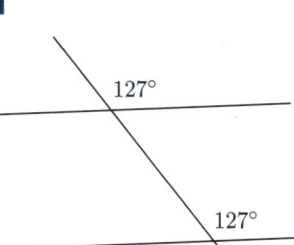

e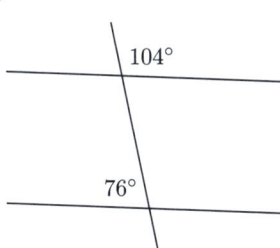

f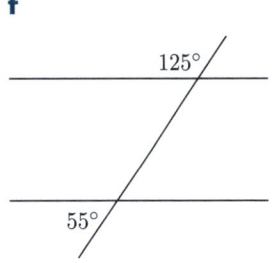

192 Geometry (Chapter 10)

5 In the diagram alongside, Troy decides that $y = 112$.
Explain why Troy is incorrect.

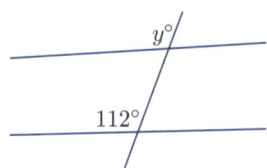

6 The following figures are not drawn to scale. Find the value of a, giving reasons for your answer.

a 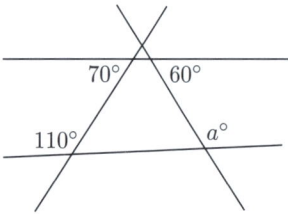   **b**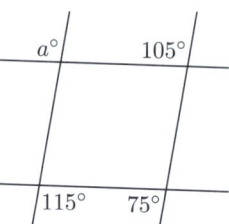

Puzzle

1 When a transversal intersects two parallel lines, ... angles are on the same side of the transversal and on the same side of the parallel lines.

2 Points that lie in a straight line are ......

3 Angles between parallel lines on the same side of a transversal are ...... angles.

4 A line which intersects two parallel lines is a ......

5 An angle which measures between 90° and 180° is ......

6 Lines which intersect at right angles are ......

7 Two angles on a straight line are ...... angles.

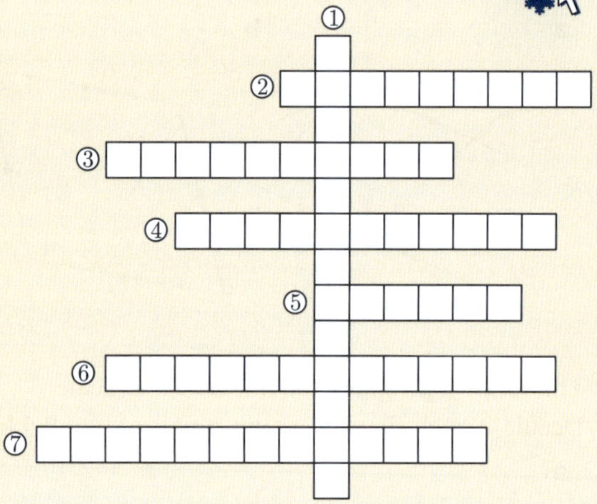

PRINTABLE

F BEARINGS

COMPASS BEARINGS

On a compass we will usually see:
- the **cardinal directions** north, south, east, and west
- the **ordinal directions** northeast, southeast, southwest, and northwest.

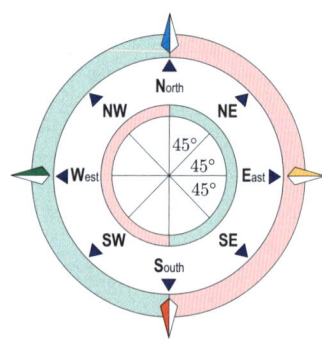

TRUE BEARINGS

We can measure a direction by comparing it with the **true north direction**. We call this a **true bearing**.

Imagine you are standing at point A, facing north. You turn **clockwise** through an angle until you face B. The **bearing of B from A** is the angle through which you have turned.

So, the bearing of B from A is the clockwise measure of the angle between the "north" line through A, and the line AB.

In the diagram alongside, the bearing of B from A is 120° from true north. We write this as 120°T or 120°.

To find the **bearing of A from B**, we place ourselves at point B, face north, then turn clockwise until we face A. The true bearing of A from B is 300°.

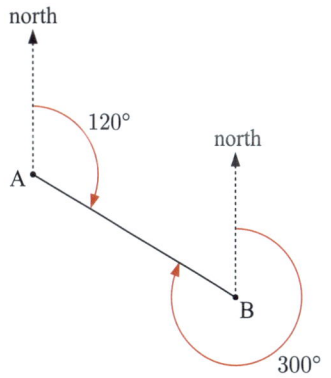

Note:
- A true bearing is always written using three digits. For example, we write 070° rather than 70°.
- The bearing of A from B, and the bearing of B from A always differ by 180°.
- True north lines are parallel, so we can use angle pair properties to find unknown angles in bearing problems.

EXERCISE 10F

1 Draw diagrams to represent bearings from O of:
   a 055°   **b** 140°   **c** 330°   **d** 255°

2 Find the bearing of Q from P if the bearing of P from Q is:
   a 124°   **b** 068°   **c** 244°   **d** 321°

A true bearing must be from 000° to 360°.

3 A, B, and C are checkpoints in an orienteering course. For each of the following, find the bearing of:
   i B from A   **ii** C from B   **iii** B from C
   iv C from A   **v** A from B   **vi** A from C.

a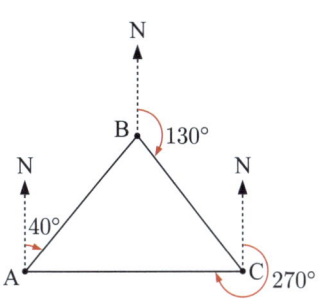

b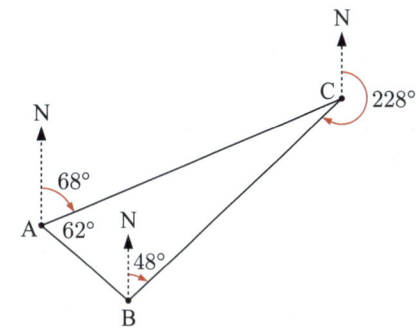

4 For each of the following, find the bearing of:

    i B from A   **ii** A from B   **iii** C from A

    iv A from C   **v** C from B   **vi** B from C.

a 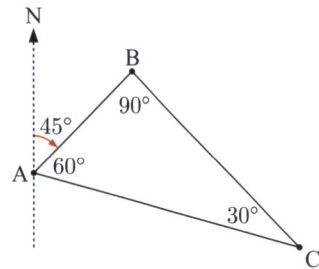   **b**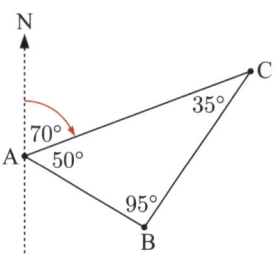

5 The centre of a large city is a square oriented as shown. State the compass bearing and true bearing of:

    a B from A   **b** C from A   **c** D from A

    d A from B   **e** D from B   **f** A from C

    g B from C   **h** B from D.

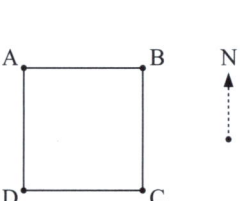

6 The locations of three islands are shown on the scale diagram alongside.

    a Find the distance and bearing of:

      i island B from island A

      ii island C from island B.

    b Island D is located 7 km from island B, on the bearing 200° from island B.

      i Copy the scale diagram, and show the location of island D.

      ii Find the distance and bearing of island D from island A.

PRINTABLE DIAGRAM

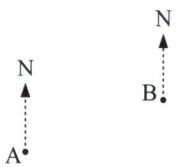

Scale: 1 cm = 4 km

G   GEOMETRIC CONSTRUCTION

In a **geometric construction** we use a ruler and compass to accurately draw a geometric figure.

Historical note — Geometric construction

The **Ancient Greeks** were the first to attempt ruler and compass constructions, over 2000 years ago. They used a ruler without any markings, known as a **straight edge**. They discovered how to bisect an angle, duplicate a line segment, and construct shapes such as equilateral triangles, squares, and regular pentagons.

Euclid outlined these constructions and many others in his book titled *Elements*.

However, there were other constructions the Greeks were unable to perform. These constructions included trisecting an angle, constructing a square with the same area as a given circle, and constructing a regular heptagon. It was not until the 18th and 19th centuries that these constructions were proved to be impossible using a straight edge and compass only.

CONSTRUCTING A PERPENDICULAR BISECTOR

The red line on this figure is **perpendicular** to the line segment AB.

It passes through M which is midway between A and B, so we say it **bisects** the line segment AB.

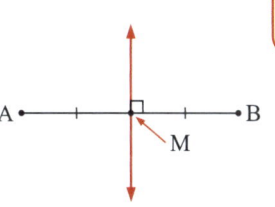

> Identical tick marks indicate equal length.

We therefore say the red line is the **perpendicular bisector** of line segment AB.

We can use a ruler and compass to construct the perpendicular bisector of a line segment.

Example 6   ◀◉ Self Tutor

The line segment AB has length 4 cm. Construct the perpendicular bisector of AB.

Step 1: With centre A, and radius more than 2 cm but less than 4 cm, draw an arc of a circle to cut AB as shown.

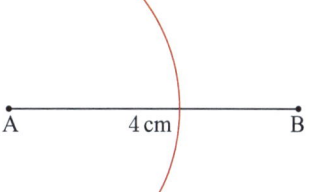

Step 2: Repeat *Step 1*, but with centre B. Make sure that the first arc is crossed twice at C and D.

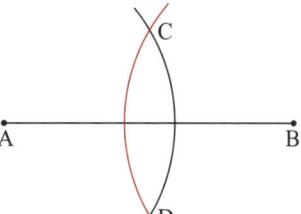

Step 3: With pencil and ruler, join C and D. CD and AB are perpendicular, and meet at M, the midpoint of line segment AB.

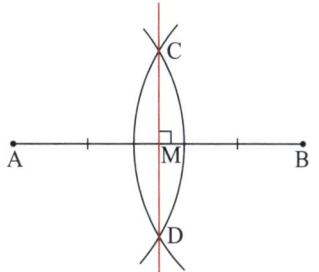

CD is the perpendicular bisector of the line segment AB.

Any point P on the perpendicular bisector of the line segment AB is **equidistant** from A and B. This means that P is the same distance from A and B.

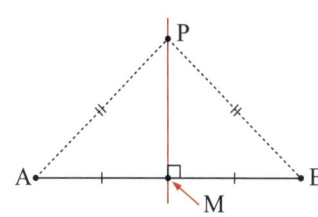

CONSTRUCTING A 90° ANGLE TO A LINE

A **right angle** or **90° angle** can be constructed without a protractor or set square. This allows us to construct a line which is **perpendicular** to another line.

Example 7
Self Tutor

Construct an angle of 90° at P on the line segment XY.

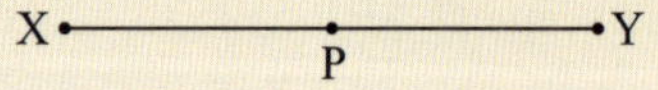

Step 1: On a line segment XY, draw a semi-circle with centre P and convenient radius which cuts XY at M and N.

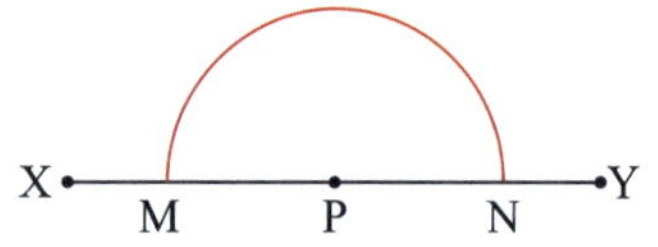

Step 2: With centre M and convenient radius larger than length MP, draw an arc above P.

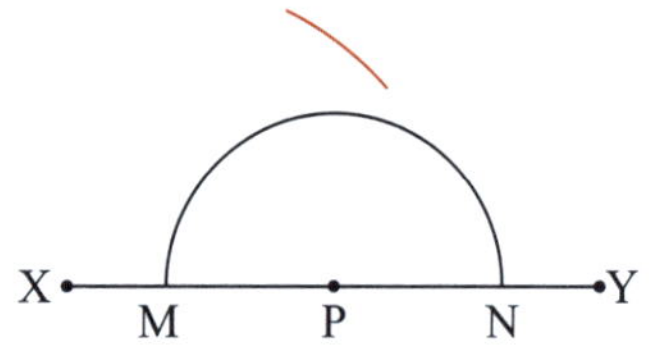

Step 3: With centre N and the *same* radius, draw an arc to cut the first one at W.

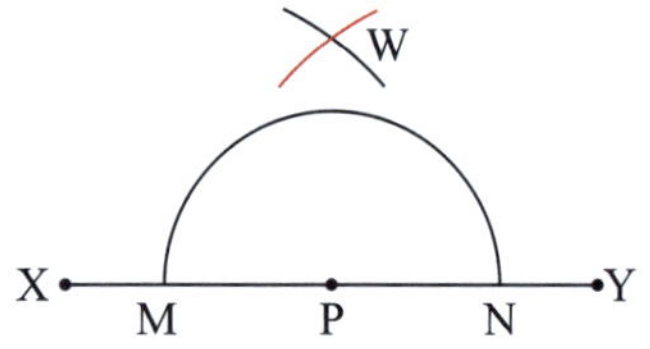

Step 4: Draw the line from P through W. WP̂Y and WP̂X are both 90°.

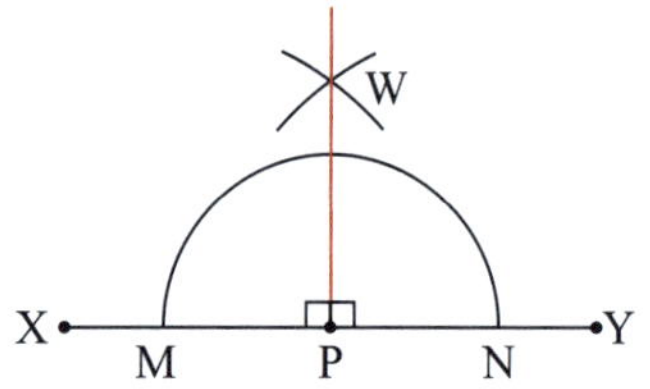

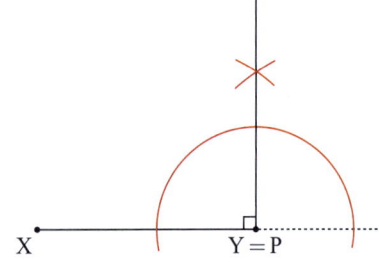

In the Example above, in some cases the point P may be close to one end of the line segment, or indeed be the end of the line segment. In these cases you may need to extend the line segment.

For example, in the construction alongside, Y and P are the same point. We say they **coincide**. We can still construct a right angle at P, but we need to extend the line segment first.

CONSTRUCTING A PERPENDICULAR TO AN EXTERNAL POINT

If we are given a line AB and a point P not on the line, we can construct the perpendicular to P from the line. This is useful because PN is the shortest distance from P to the line AB.

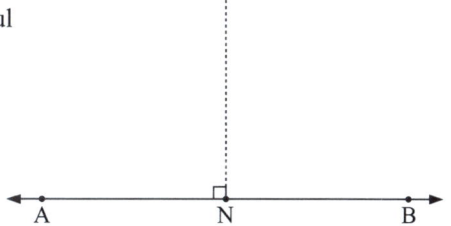

Example 8

◄)) **Self Tutor**

Construct a perpendicular from the line segment XY to the external point P.

Step 1: With centre P, draw an arc to cut the line at A and B.

Step 2: With centre A, draw an arc below P.

Step 3: With centre B and the *same* radius, draw an arc to cut the previous one at C.

Step 4: Join P to C. We let N be the point of intersection with the original line.
P$\hat{N}$X = P$\hat{N}$Y = 90°, and PN is the shortest distance from the line to P.

BISECTING ANGLES

When we **bisect** an angle with a straight line, we divide it into two angles of equal size.

Example 9

Self Tutor

Bisect $A\widehat{B}C$.

A
B
C

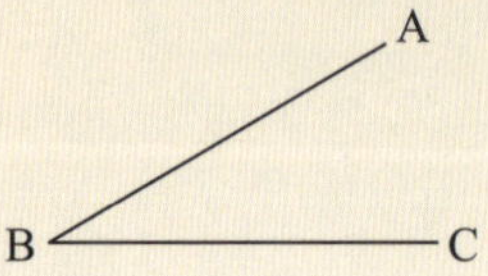

Step 1: With centre B, draw an arc which cuts BA and BC at P and Q respectively.

A
P
B
C
Q

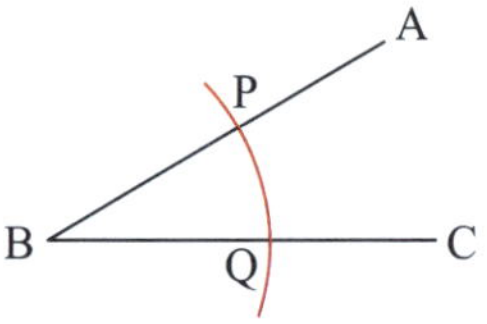

Step 2: With Q as the centre, draw an arc within $A\widehat{B}C$.

A
P
B
C
Q

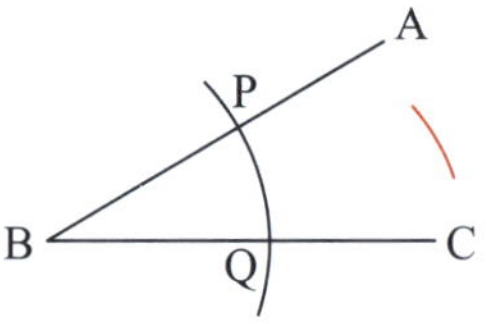

Step 3: Keeping the *same* radius and with centre P, draw another arc to intersect the previous one at point M.

A
P
M
B
C
Q

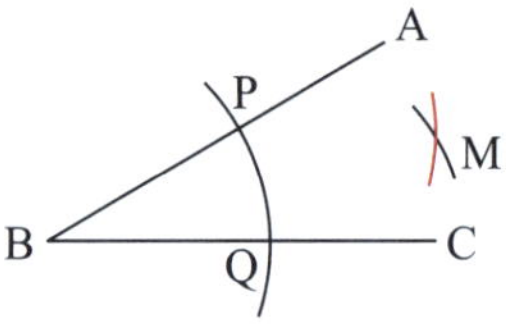

Step 4: Join B to M.
BM bisects $A\widehat{B}C$,
so $A\widehat{B}M = C\widehat{B}M$.

A
P
M
B
C
Q

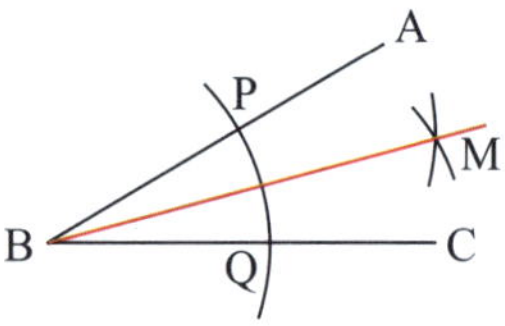

Any point P on the angle bisector of $A\widehat{B}C$ is equidistant from the lines BA and BC.

A
P
B
C

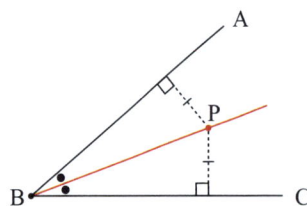

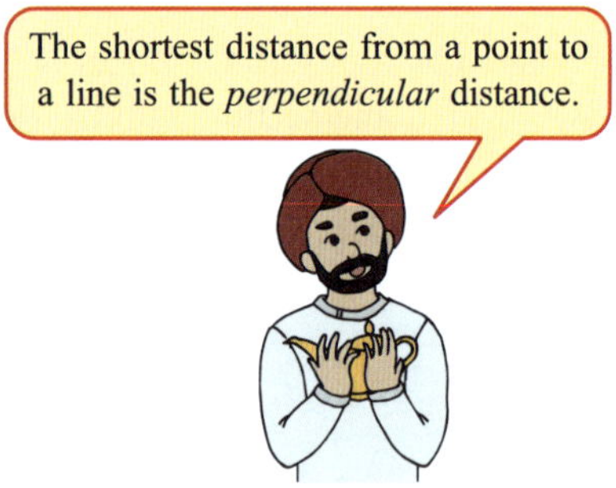

The shortest distance from a point to a line is the *perpendicular* distance.

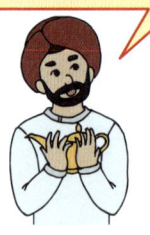

Geometry (Chapter 10)

EXERCISE 10G

1 **a** Draw a line segment PQ of length 5 cm.

    b Use a compass and ruler to construct the perpendicular bisector of PQ.

    c Let the perpendicular bisector of PQ meet PQ at Y. Check your perpendicular bisector by measuring the lengths of PY and QY.

2 **a** Accurately draw the line segment AB with the dimensions shown. Mark on it the points A, B, and C.

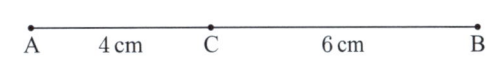

    b Use a compass and ruler to construct a right angle at C below the line segment AB.

3 **a** Draw a line segment PQ of length 6 cm.

    b Construct a right angle at Q.

4 **a** Use a ruler and protractor to accurately draw $A\widehat{B}C$ of size $60°$, with AB = BC = 6 cm.

    b Construct a right angle from line segment BC to A. Let the perpendicular meet BC at X.

    c Hence find the distance between B and X.

5 **a** Use your protractor to accurately draw $A\widehat{B}C$ of size $80°$.

    b Use a compass and ruler only to bisect $A\widehat{B}C$.

    c Use a protractor to check the accuracy of your construction.

6 **a** Draw any triangle ABC and carefully bisect its three angles.

    b Repeat with another triangle DEF of different shape.

    c Check with other students in your class for any observations about the three angle bisectors.

    d Copy and complete: "The three angle bisectors of a triangle ......".

DEMO

7 AB is a line segment. The point P is 3 cm from it.

    a Construct the perpendicular from AB to P.

    b Explain why the length of the perpendicular must be the shortest distance from AB to P.

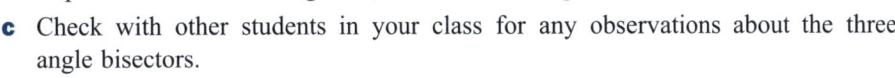

8 **a** Draw a line segment AB of length 8 cm.

    b Construct the perpendicular bisector of AB, meeting AB at C.

    c Locate a point D on the perpendicular bisector which is 3 cm from C.

    d Construct a line segment DE which is perpendicular to CD.

    e Explain why AB and DE are parallel.

9 **a** Draw any triangle, and construct the perpendicular bisectors of its three sides.

    b Repeat **a** with a different triangle.

    c Hence, copy and complete: "The three perpendicular bisectors of the sides of a triangle are ......".

10 Copy triangle ABC alongside, and locate:

a point P such that P lies on BC, and AP is perpendicular to BC

b point Q such that Q lies on AB, and is equidistant from A and C.

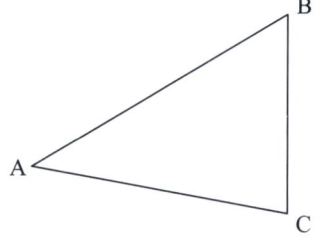

PRINTABLE DIAGRAMS

11 a Draw a line segment PQ of length 6 cm.

b Illustrate the set of points which are:
- equidistant from P and Q, *and*
- 4 cm or less from P.

12 A scale diagram of a field is shown alongside. A gold coin has been hidden in the field.

The coin is:
- equidistant from AD and CD
- 40 m from B.

Copy the scale diagram, and locate the position of the coin.

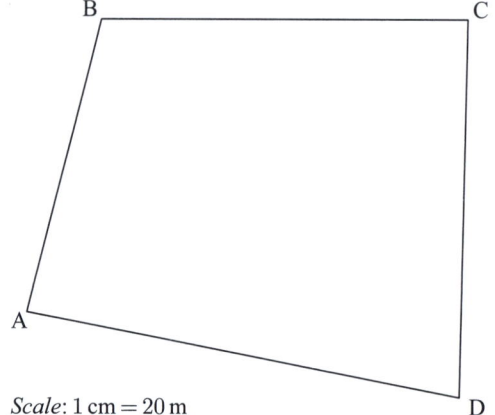

Scale: 1 cm = 20 m

Activity   Constructing shapes

We can use geometric construction to construct various shapes.

VIDEO

For example, we can construct a triangle ABC with sides 4 cm, 3 cm, and 2 cm long using the following procedure:

Step 1: Draw a line segment of length 4 cm. We will call this line segment AB, and use it as the base of the triangle.

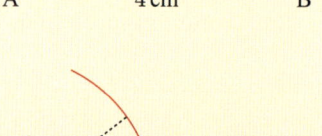

Step 2: Open your compass to a radius of 2 cm. Using this radius, draw an arc from A.

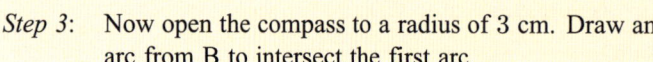

Step 3: Now open the compass to a radius of 3 cm. Draw an arc from B to intersect the first arc.

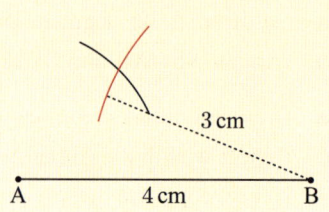

Step 4: The point of intersection of the two arcs is the third vertex C of the triangle ABC. Draw line segments AC and BC to complete the triangle.

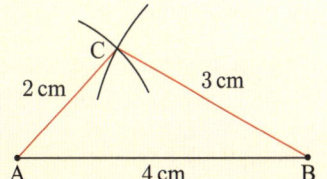

What to do:

1 Construct a triangle with side lengths:
   a 3 cm, 4 cm, and 5 cm
   b 4 cm, 6 cm, and 7 cm.

2 **a** Draw a line segment AB of length 5 cm.
   b Construct a square with AB as one side.
   c Use a ruler and protractor to check that the figure is a square.

3 Construct the circle which passes through points A, B, and C by following these steps:
   Step 1: Construct the perpendicular bisectors of line segments AB and BC.
   Step 2: Let the perpendicular bisectors intersect at O. This is the centre of the circle.
   Step 3: With centre O and radius OA, draw a circle. This circle should pass through A, B, and C.

a 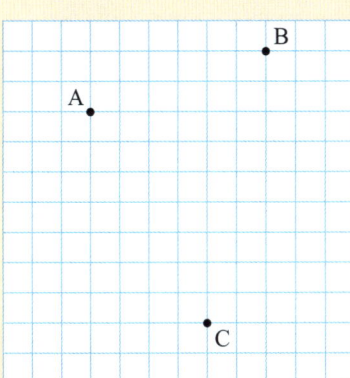   **b**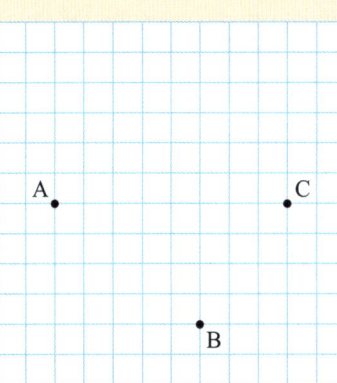

VIDEO CLIP

PRINTABLE DIAGRAMS

Review set 10A

1 Find:
   a the angle complementary to 53°
   b the angle supplementary to 130°.

2 Find the value of the unknown:

   a 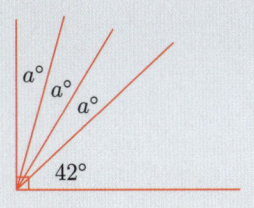   **b** 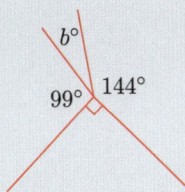   **c**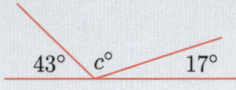

3 How many points are needed to determine the position of a line?

4 Draw a diagram to illustrate the following statement:
   "Line segments AB and CD intersect at P."

5 Find, giving a reason, the value of x:

   a 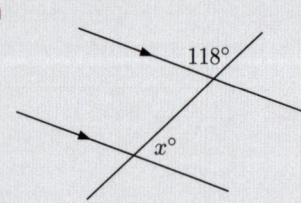 **b** 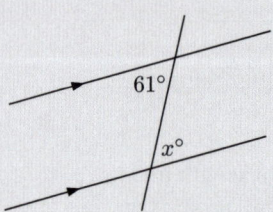 **c**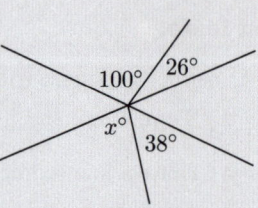

6 Find, giving a reason, the value of m:

   a 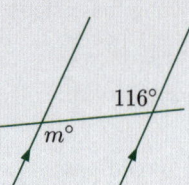 **b** 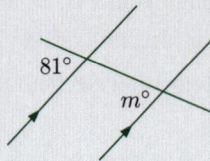 **c**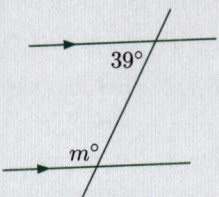

7 Write down an equation connecting the unknowns. Give reasons for your answers.

   a 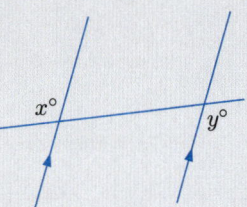 **b**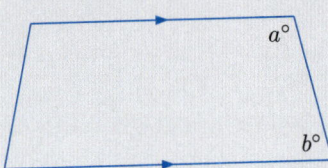

8 Find the bearing of:
   a B from A   **b** A from B
   c C from B   **d** B from C
   e A from C   **f** C from A.

   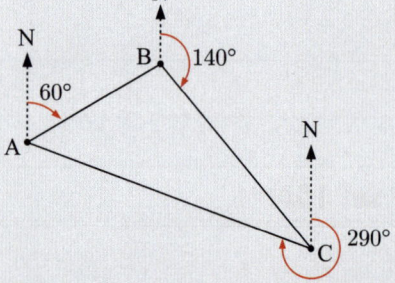

9 State whether each figure contains a pair of parallel lines. Give reasons for your answers.

   a 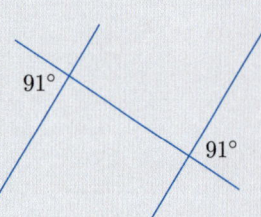 **b**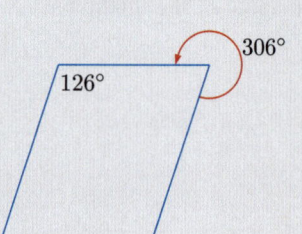

10 **a** Draw a line segment PQ of length 6 cm.
  b Construct the perpendicular bisector of PQ, meeting PQ at X.
  c Check your perpendicular bisector by measuring the lengths of PX and QX.

11 In any triangle, a line from a vertex, perpendicular to the side opposite, is called an **altitude** of the triangle.

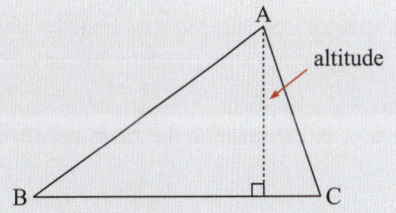

  a Draw any triangle ABC, and construct the three altitudes of the triangle.
  b What do you notice about the three altitudes?

Review set 10B

1 Consider the diagram alongside.
  a Name the line AB in two other ways.
  b What can be said about:
    i points A, B, and C
    ii lines AD and BD?

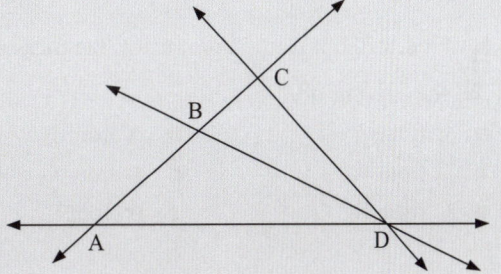

2 Find:
  a the angle complementary to 65°
  b the angle supplementary to 88°.

3 Find, giving a reason, the value of the unknown:
  a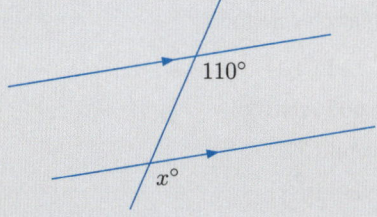
  b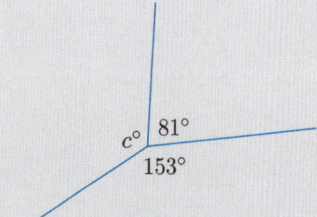

4 Find, giving a reason, the value of the unknown:
  a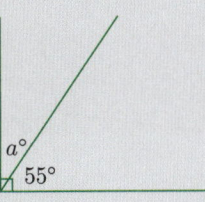
  b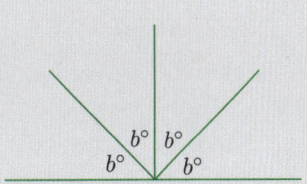

5 Draw a line segment AB of length 8 cm. Construct an angle of 90° at B using a compass and ruler only. Hence draw BC of length 6 cm which is perpendicular to AB.

6 Draw and label the following angles:
  a reflex $B\hat{A}C$
  b acute $P\hat{Q}R$
  c obtuse $T\hat{R}S$

204 Geometry (Chapter 10)

7 **a** Find the angle corresponding to:
  i BD̂A   **ii** DĈB   **iii** BÂC
b Classify the following angles as acute, obtuse, or reflex:
  i c   **ii** a   **iii** d

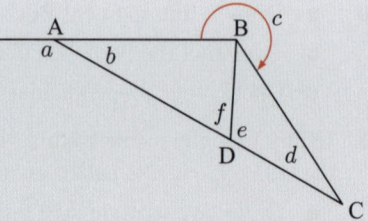

8 **a** Determine the measure of reflex PQ̂R.

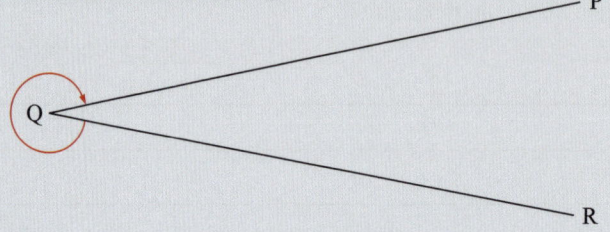

b Find, without using a protractor, the measure of acute PQ̂R. Justify your answer.

9 Find the bearing of:
  a B from A   **b** A from B
  c C from A   **d** A from C
  e C from B   **f** B from C.

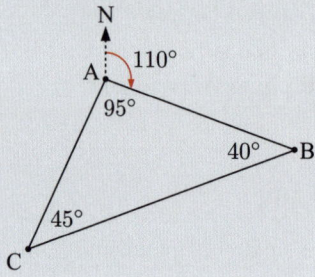

10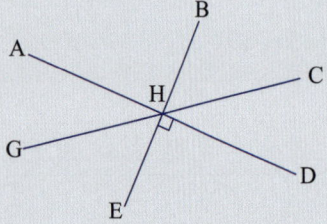

Classify the following angles as complementary, supplementary, or neither:
  a CĤA and CĤD
  b AĤG and AĤB
  c BĤC and CĤD
  d BĤC and EĤG

11 Decide if WX is parallel to YZ, giving reasons for your answer.

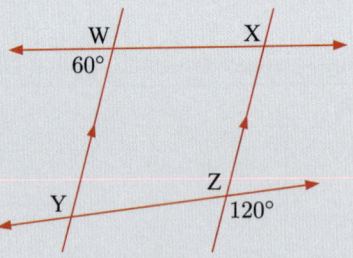

12 **a** Draw a line segment AB of length 4 cm.
  b Show the possible positions of point C such that:
   • BC is perpendicular to AB, *and*
   • BC = 2 cm.

:::

11

Shape

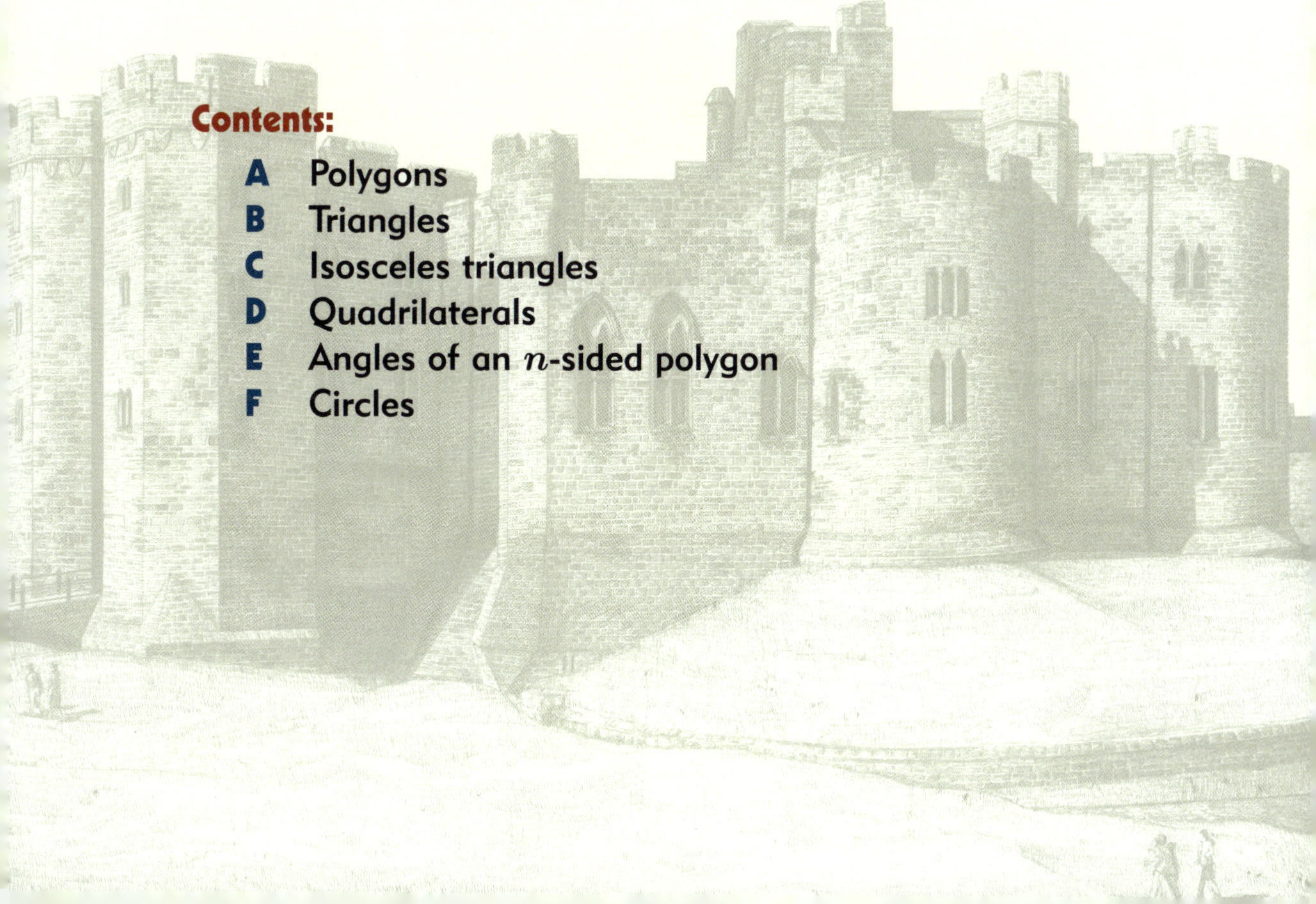

Contents:

- **A** Polygons
- **B** Triangles
- **C** Isosceles triangles
- **D** Quadrilaterals
- **E** Angles of an n-sided polygon
- **F** Circles

Opening problem

The figure alongside contains two pairs of parallel lines.

Things to think about:
a   What name is given to this shape?
b   Which angles in the figure are equal?
c   What is the sum of the sizes of:
   i   the blue angles
   ii the green angles?
d   What is the total sum of the angles of this figure?

A POLYGONS

A shape that is drawn on a flat surface or plane is called a **plane figure**.

If the boundary of a shape has no beginning or end, it is said to be **closed**.

A **polygon** is a closed plane figure with straight line sides which do not cross.

Some simple examples of polygons are:

triangle   quadrilateral   pentagon
3 sides   4 sides   5 sides

Polygons are named according to the number of sides they have. For example, a 9-sided polygon can be called a 9-gon. However, many polygons are known by more familiar names, such as those in the table.

Number of sides	Polygon name
3	**Tri**angle
4	**Quad**rilateral
5	**Penta**gon
6	**Hexa**gon
7	**Hepta**gon
8	**Octa**gon
9	**Nona**gon
10	**Deca**gon

A **vertex** of a polygon is a point where two sides meet.

The plural of vertex is **vertices**.

In any polygon, the number of sides equals the number of vertices.

REGULAR POLYGONS

A **regular polygon** has all sides of equal length **and** all angles of equal measure.

This is a regular hexagon:

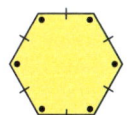

This pentagon is not regular even though its sides are equal in length. Its angles are not all equal.

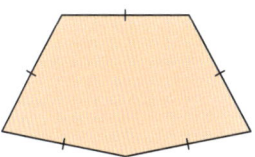

This polygon is not regular even though its angles are equal. Its sides are not all equal in length.

DIAGONALS OF A POLYGON

A **diagonal** of a polygon is a straight line segment which joins a pair of vertices across the polygon.

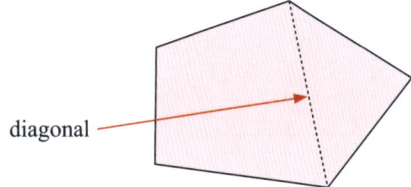

EXERCISE 11A

1 Which of these figures is a polygon?
   Give a reason if the figure is not a polygon.

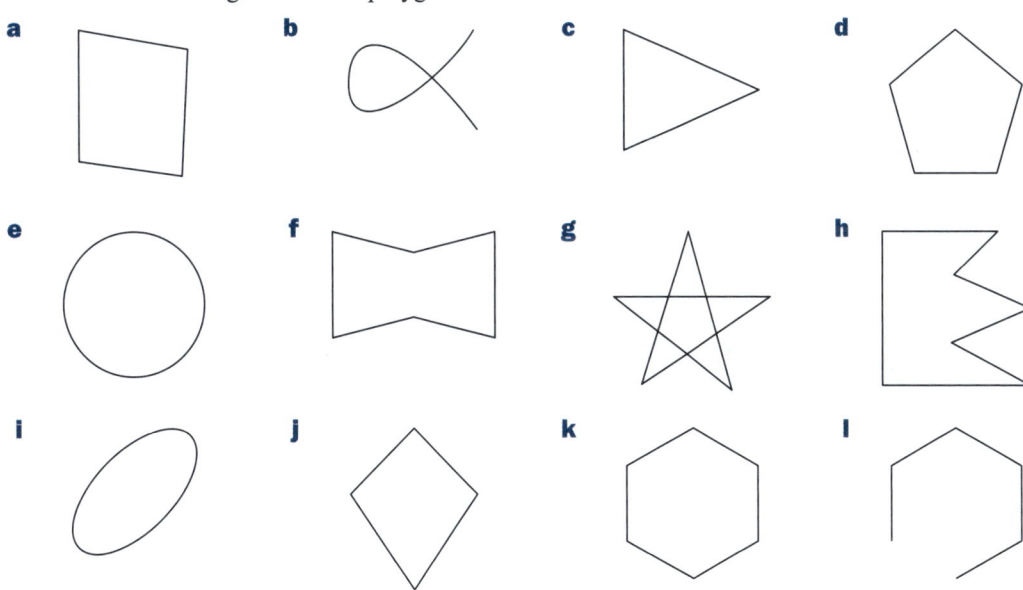

2 Write down the name given to a polygon with:

    a three sides   **b** four sides   **c** six sides

    d seven sides   **e** eight sides   **f** nine sides.

3 Name these polygons according to their number of sides:

a    **b** 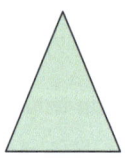   **c**    **d**

e 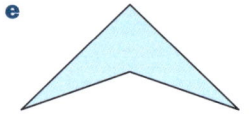   **f**    **g**    **h**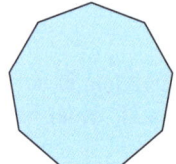

4 Explain why these figures are not regular polygons:

a 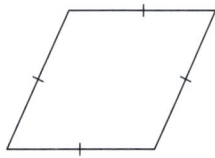   **b**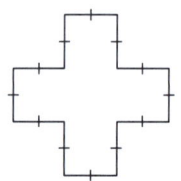

c    **d**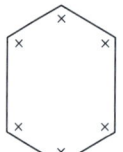

Tick marks show sides of equal length.

Angle markings show angles of the same size.

5 Draw each of the following polygons, marking equal side lengths and equal angles:

    a regular triangle   **b** regular pentagon   **c** regular octagon.

6 Sketch each of the following polygons, and draw all of their diagonals:

    a a quadrilateral   **b** a pentagon   **c** an octagon.

B   TRIANGLES

A **triangle** is a polygon which has three sides.

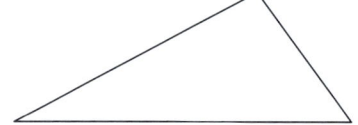

CLASSIFICATION BY SIDES

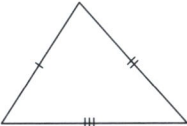
Scalene triangle
no equal sides

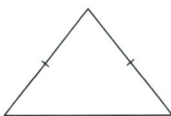
Isosceles triangle
two equal sides

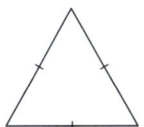
Equilateral triangle
three equal sides

CLASSIFICATION BY ANGLES

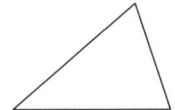
Acute angled triangle
all acute angles

Obtuse angled triangle
one obtuse angle

Right angled triangle
one right angle

PROPERTIES OF TRIANGLES

All triangles have the following properties:

- The sum of the interior angles of a triangle is $180°$.

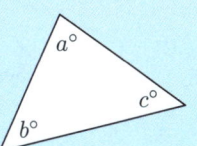 $a + b + c = 180$ **GEOMETRY PACKAGE**

- Any exterior angle is equal to the sum of the interior opposite angles.

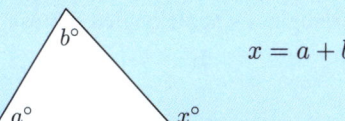 $x = a + b$ **GEOMETRY PACKAGE**

- The longest side is opposite the largest angle.

Proof that the sum of the angles of a triangle is $180°$:

Draw a triangle ABC with angles $a°$, $b°$, and $c°$.

Draw a line segment DE through B which is parallel to AC.

Using equal alternate angles,

$\widehat{ABD} = a°$ and $\widehat{CBE} = c°$.

But $\quad a + b + c = 180 \quad$ {angles on a line}

$\therefore \quad a + b + c = 180$

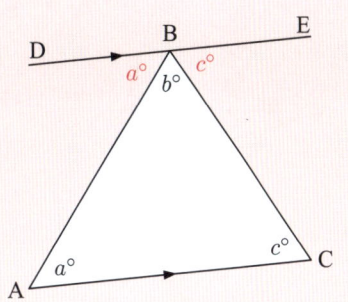

Example 1

Find the value of the unknown:

a 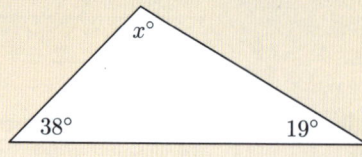 **b**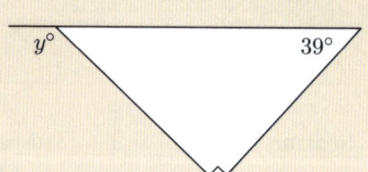

a $x + 38 + 19 = 180$ {angle sum of a triangle}
$\therefore\ x + 57 = 180$
$\therefore\ x + 57 - 57 = 180 - 57$ {subtracting 57 from both sides}
$\therefore\ x = 123$

b $y = 39 + 90$ {exterior angle of a triangle}
$\therefore\ y = 129$

EXERCISE 11B

1 Classify the following triangles as scalene, isosceles, or equilateral:

a 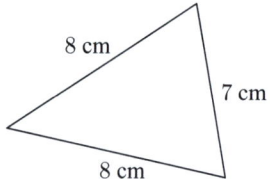 **b** 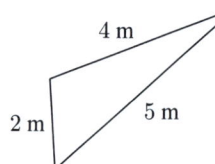 **c**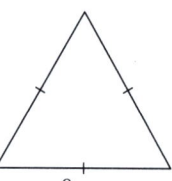

2 Classify the following triangles as acute angled, obtuse angled, or right angled:

a 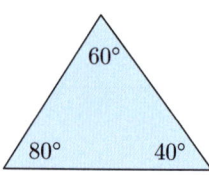 **b** 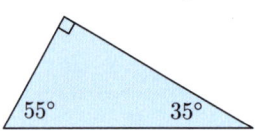 **c**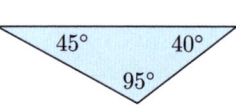

3 These diagrams are not drawn to scale, but the information on them is correct. Find the values of the unknowns:

a 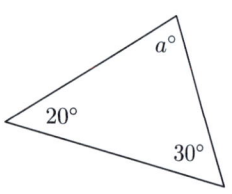 **b** 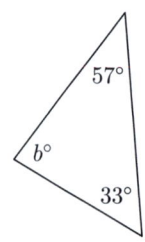 **c**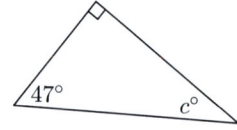

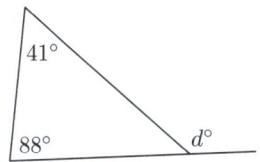
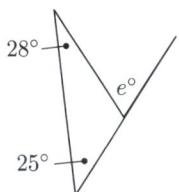

4 Two of the angles in Nancy's triangular pizza slice are 72° and 58°. Find the measure of the third angle.

5 State whether the following statements are *true* or *false*:
  a The sum of the angles of a triangle is equal to two right angles.
  b A right angled triangle can contain an obtuse angle.
  c The sum of two angles of a triangle is always greater than the third angle.
  d The two smaller angles of a right angled triangle are supplementary.

6 The following triangles are *not* drawn to scale. State the longest side of each triangle.

> The longest side is opposite the largest angle.

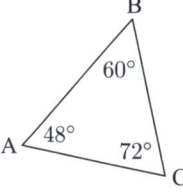
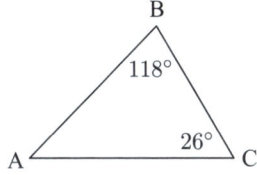
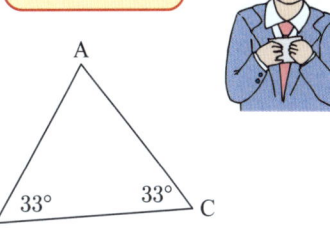
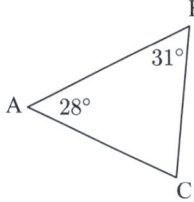
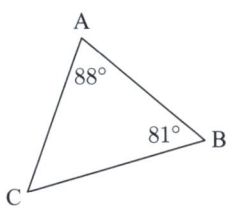
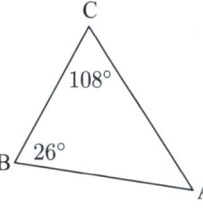
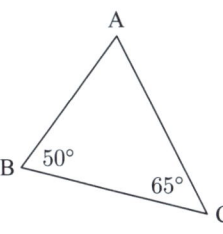
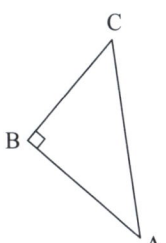
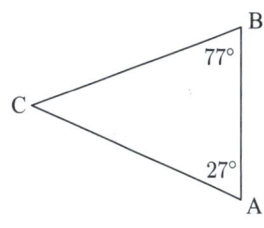

Example 2

Find the values of the variables:

a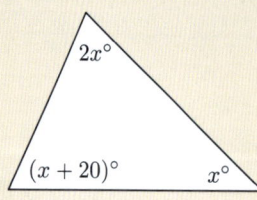

b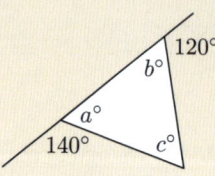

a $2x + x + (x + 20) = 180$ {angle sum of a triangle}
$\therefore 4x + 20 = 180$ {collecting like terms}
$\therefore 4x = 160$ {subtracting 20 from both sides}
$\therefore x = 40$

b $a = 180 - 140 = 40$ {angles on a line}
Likewise $b = 180 - 120 = 60$
But $a + b + c = 180$ {angle sum of a triangle}
$\therefore 40 + 60 + c = 180$
$\therefore 100 + c = 180$
$\therefore c = 80$

7 Find the values of the variables:

a 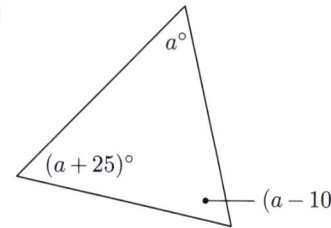 **b** 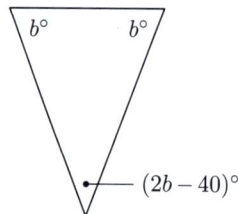 **c**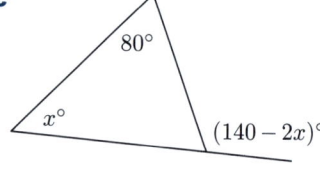

d 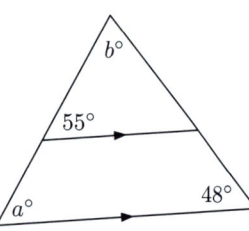 **e** 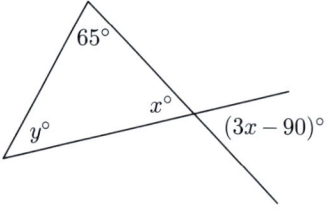 **f**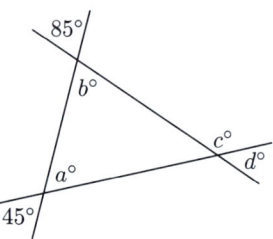

8 For the figure alongside:

a Explain why $\widehat{CBD} = 90°$.

b Find the size of $\widehat{BDC}$.

c Hence find the size of $\widehat{ADB}$.

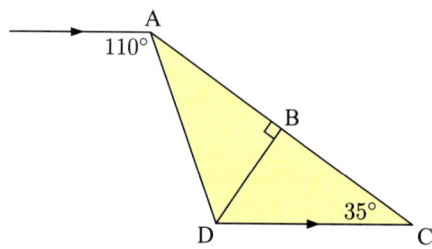

Shape (Chapter 11) 213

Research

Research the use of triangles in the construction of bridges.

Explain what is meant by the statement *"the triangle is the only rigid polygon"* and how this helps the bridge structure.

C ISOSCELES TRIANGLES

An **isosceles triangle** is a triangle in which two sides are equal in length.

The angles opposite the two equal sides are called the **base angles**.

The vertex where the two equal sides meet is called the **apex**.

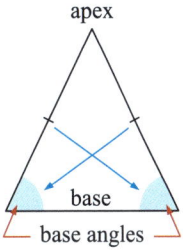

GEOMETRY PACKAGE

THE ISOSCELES TRIANGLE THEOREM

In any isosceles triangle:
- the base angles are equal
- the line joining the apex to the midpoint of the base bisects the vertical angle and meets the base at right angles.

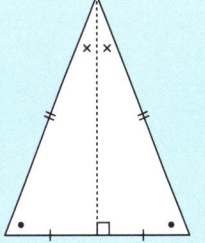

CONVERSES

With many theorems there are *converses* which we can use in problem solving.

Converse 1: If a triangle has two equal angles, then it is isosceles.
Converse 2: The angle bisector of the apex of an isosceles triangle bisects the base at right angles.

Discussion

What does the word *converse* mean?

Activity — Converses of the isosceles triangle theorem

What to do:

Click on the icon to run the interactive software. Use the software to decide which of the following are also converses of the isosceles triangle theorem:

INTERACTIVE TRIANGLES

1. If the line joining one vertex to the midpoint of the opposite side is perpendicular to that side, then the triangle is isosceles.
2. If the line joining one vertex to the midpoint of the opposite side bisects the angle at the vertex, then the triangle is isosceles.
3. If a perpendicular to one side of the triangle passes through a vertex and bisects the angle at that vertex, then the triangle is isosceles.

Example 3 ◀) Self Tutor

Find x:

a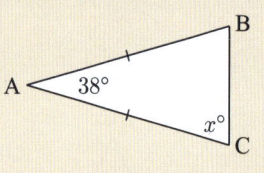

b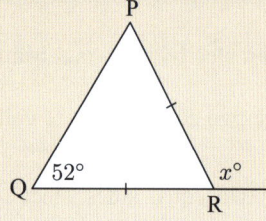

a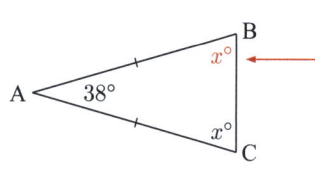

Since $AB = AC$, the triangle is isosceles.
$\therefore \ \widehat{ABC} = x°$ {isosceles triangle theorem}
Now $x + x + 38 = 180$ {angle sum of a triangle}
$\therefore \ 2x + 38 = 180$
$\therefore \ 2x = 142$
$\therefore \ x = 71$

b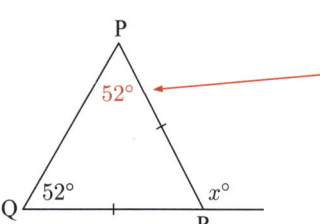

Since $PR = QR$, the triangle is isosceles.
$\therefore \ \widehat{QPR} = 52°$ {isosceles triangle theorem}
$\therefore \ x = 52 + 52$ {exterior angle of a triangle}
$\therefore \ x = 104$

EXERCISE 11C

1 Find x:

a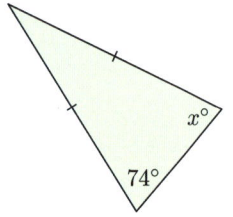

b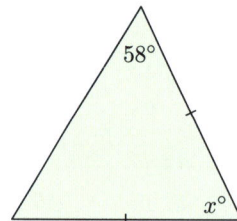

c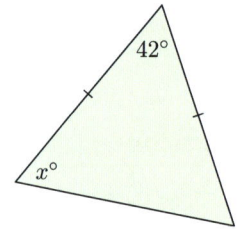

Shape (Chapter 11) 215

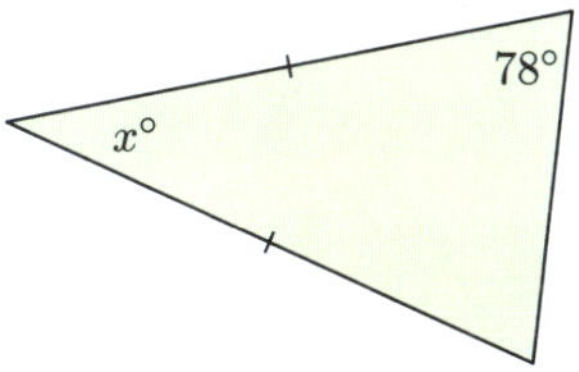

d 78°, x°

e x°, (90 − x)°

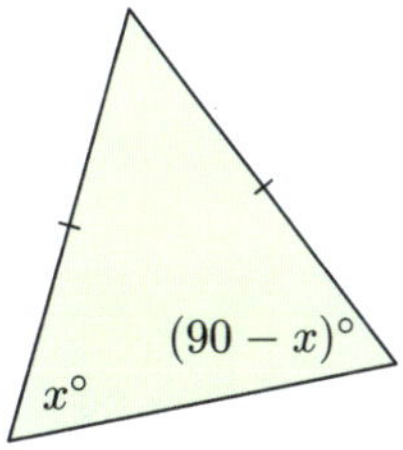

f 3x°, 2x°

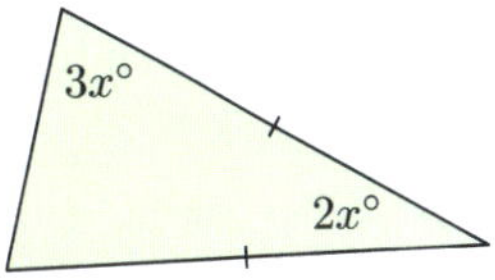

g x°, 70°

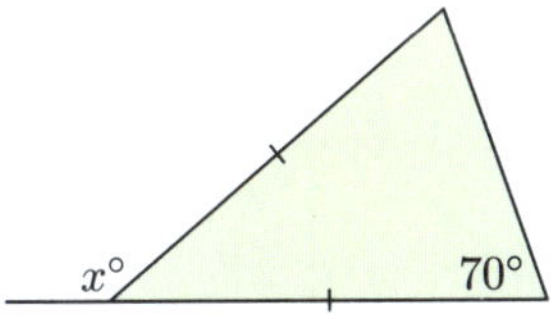

h 26°, x°

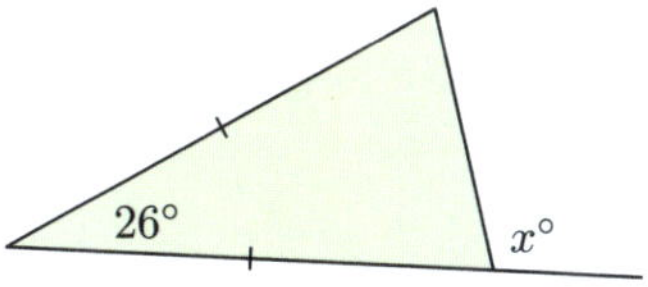

i x°, 66°

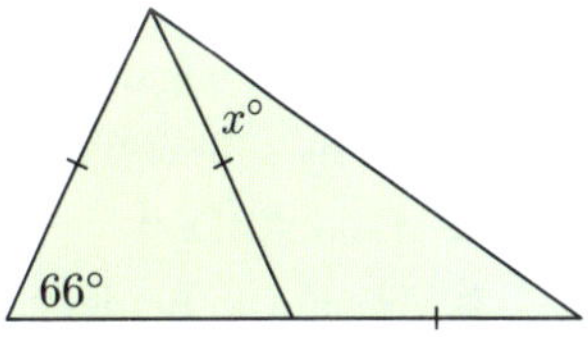

2 Find x, giving brief reasons:

a Triangle ABC: B 70°, C 70°, AB = x cm, AC = 12 cm

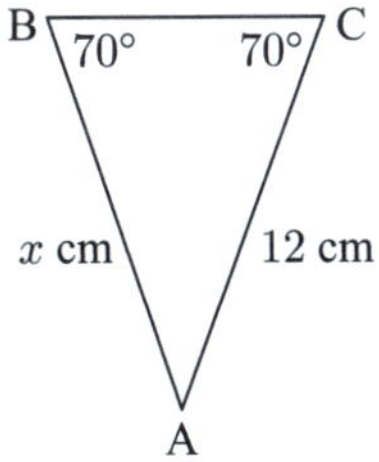

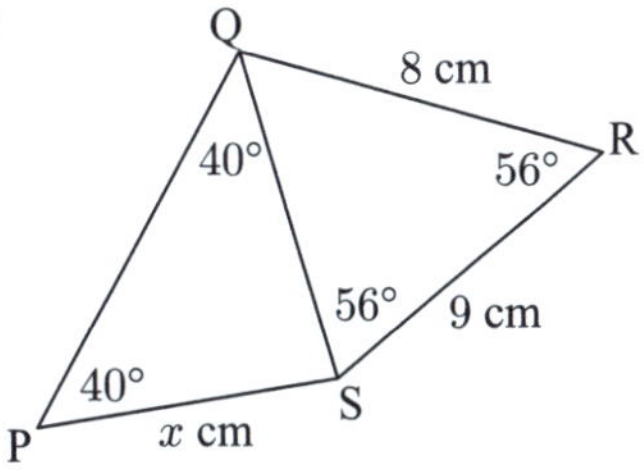

b P 40°, Q 40°, QR = 8 cm, angle at R = 56°, RS = 9 cm, angle at S = 56°, PS = x cm

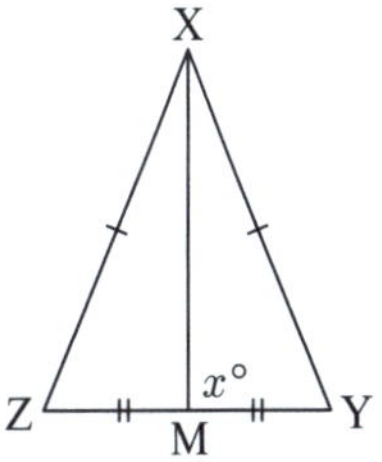

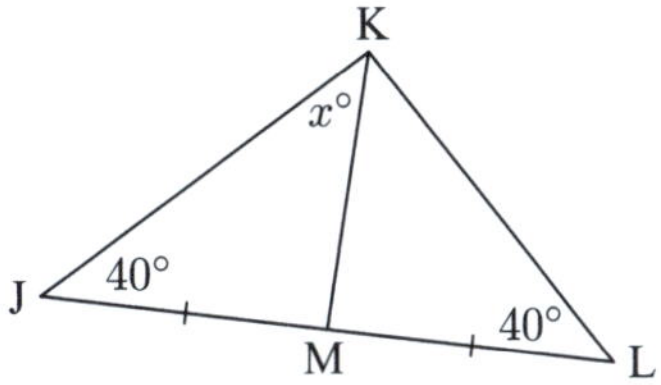

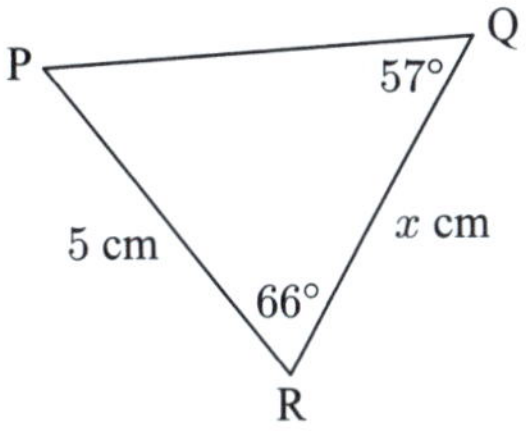

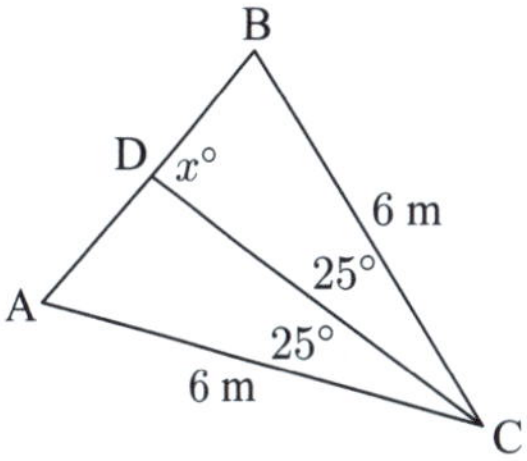

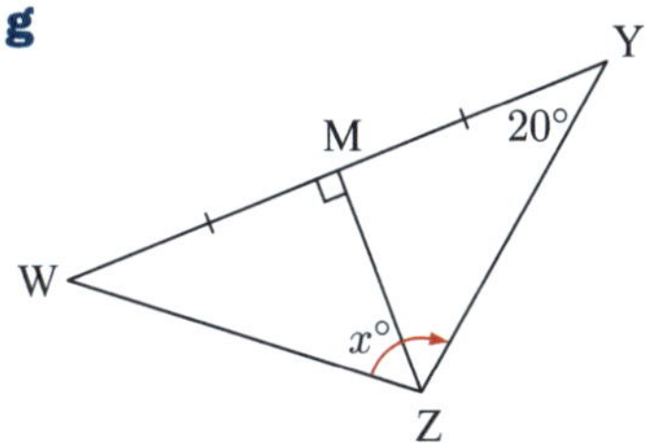

c Triangle XZY with M on ZY: ZM = MY, XZ = XY, angle at M = $x°$

d J 40°, angle KML = 40°, angle JKM = $x°$

e P, angle Q = 57°, PR = 5 cm, QR = x cm, angle R = 66°

f Triangle with D on AC: angle at D = $x°$, DB and DC = 6 m, angles 25° and 25°

g W, M (right angle), Y 20°, angle at Z = $x°$

h DE = 4 m, EF marked equal, angle D = 60°, DF = x m

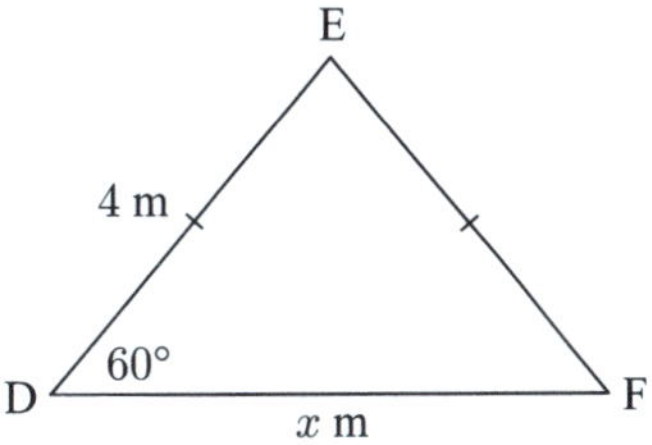

3 The triangular control frame of a hang glider has an angle of 46° between the two equal sides. Find the measure of the other two angles.

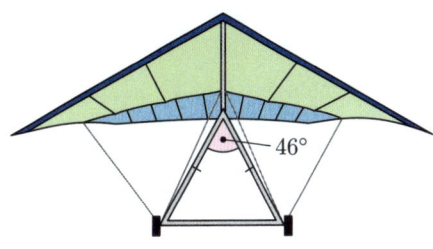

4 The figure alongside has not been drawn to scale, but the information given is correct.

  a Find x.

  b What can be deduced about the triangle?

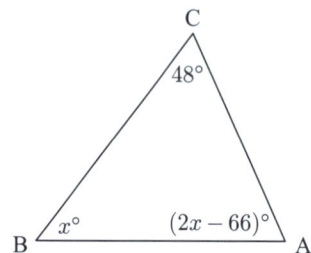

5 The figure alongside has not been drawn to scale, but the information given is correct.

  a Find $\widehat{ABD}$.

  b What can be deduced about triangle ABD?

  c Find x.

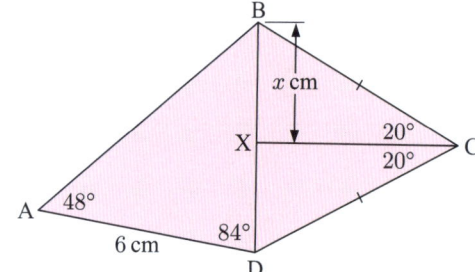

6 For the figure alongside, determine the measure of $\widehat{AEB}$.

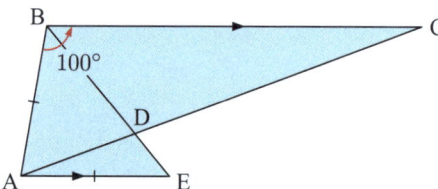

7 a Explain why each angle of an equilateral triangle measures $60°$.

  b Hence describe how to construct a $60°$ angle to a line segment AB using only a compass and ruler.

D QUADRILATERALS

A **quadrilateral** is a polygon which has four sides.

ANGLES OF A QUADRILATERAL

Suppose a quadrilateral is drawn on a piece of paper.

If the four angles are torn off and reassembled at a point, we notice that the angle sum is always $360°$.

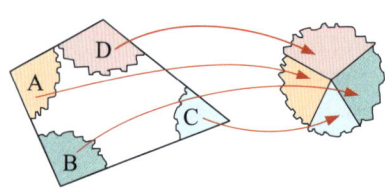

> The sum of the interior angles of a quadrilateral is $360°$.

Proof:

Suppose we divide the quadrilateral into two triangles. Each triangle has an angle sum of 180°, so the angle sum of the quadrilateral is $2 \times 180° = 360°$.

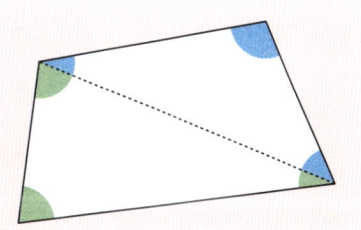

GEOMETRY PACKAGE

Example 4   ◀) Self Tutor

Find the value of x:

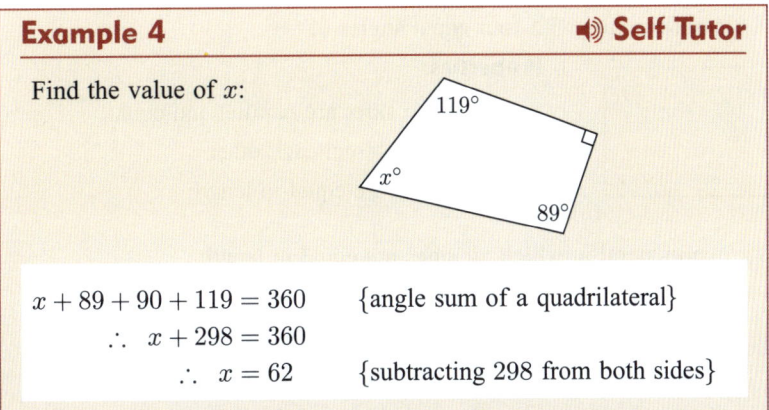

$x + 89 + 90 + 119 = 360$ {angle sum of a quadrilateral}

$\therefore \ x + 298 = 360$

$\therefore \ x = 62$ {subtracting 298 from both sides}

EXERCISE 11D.1

1 Find the value of x:

a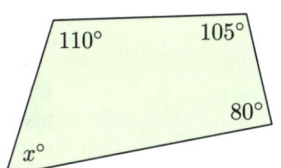

b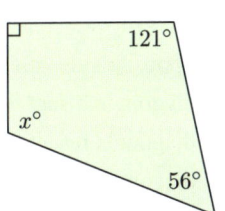

c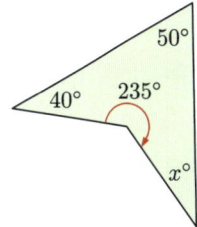

d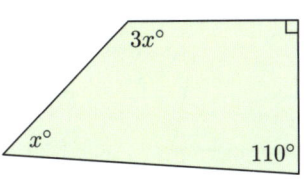

e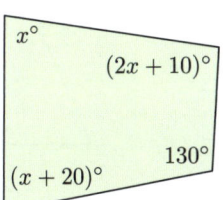

f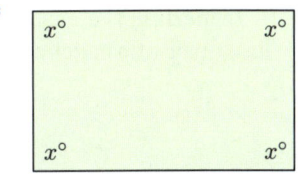

2 Find the values of the unknowns:

a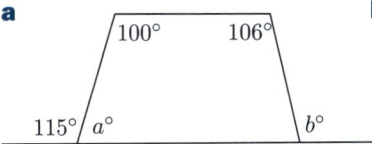

b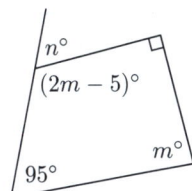

c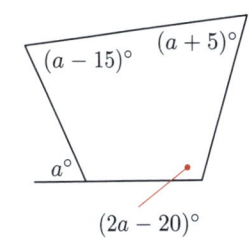

SPECIAL QUADRILATERALS

There are six special quadrilaterals.

1 A **parallelogram** is a quadrilateral which has opposite sides parallel.

Properties:
- opposite sides are equal in length
- opposite angles are equal in size
- diagonals bisect each other.

GEOMETRY PACKAGE

2 A **rectangle** is a parallelogram with four equal angles of 90°.

Properties:
- opposite sides are parallel and equal
- diagonals bisect each other
- diagonals are equal in length.

GEOMETRY PACKAGE

3 A **rhombus** is a quadrilateral in which all sides are equal in length.

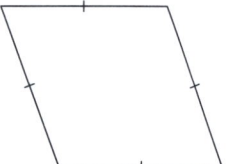

Properties:
- opposite sides are parallel
- opposite angles are equal in size
- diagonals bisect each other at right angles
- diagonals bisect the angles at each vertex.

GEOMETRY PACKAGE

4 A **square** is a rhombus with four equal angles of 90°.

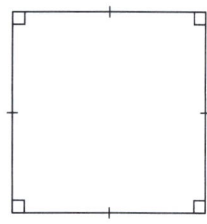

Properties:
- opposite sides are parallel
- diagonals bisect each other at right angles
- diagonals bisect the angles at each vertex
- diagonals are equal in length.

GEOMETRY PACKAGE

5 A **trapezium** is a quadrilateral which has a pair of parallel opposite sides.

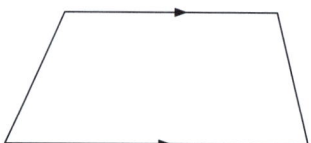

6 A **kite** is a quadrilateral which has two pairs of adjacent sides equal in length.

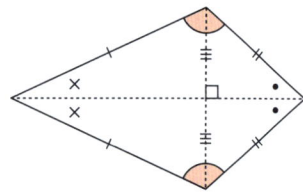

Properties:
- one diagonal is a line of symmetry
- one pair of opposite angles are equal
- diagonals cut each other at right angles
- **one** diagonal bisects **one** pair of angles at the vertices
- one of the diagonals bisects the other.

Example 5

Draw three diagrams to show all the properties of a parallelogram.

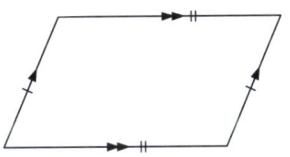
opposite sides are equal

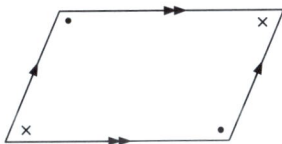
opposite angles are equal

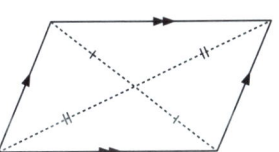
diagonals bisect each other

Example 6

Find x, giving brief reasons for your answer:

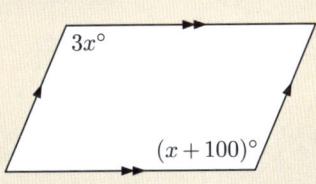

The figure is a parallelogram.

$\therefore \ 3x = x + 100$ {opposite angles of a parallelogram}
$\therefore \ 2x = 100$ {subtracting x from both sides}
$\therefore \ x = 50$

EXERCISE 11D.2

1 Draw four or more diagrams which show all the properties of:

  a a square   **b** a kite   **c** a rhombus.

2 Find the values of the variables:

a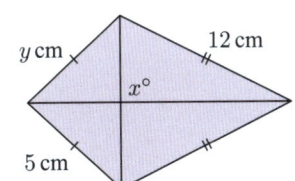

b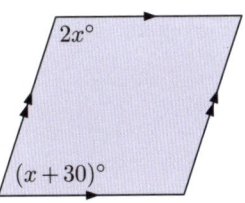

c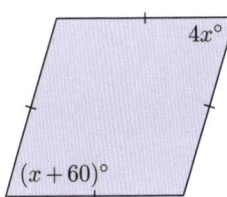

d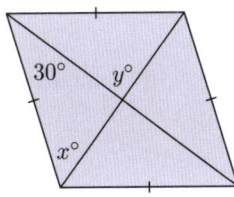

e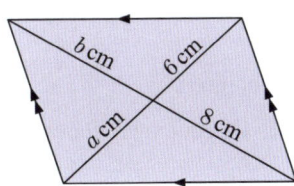

f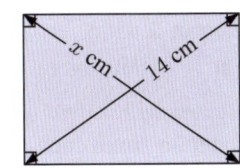

3 True or false?

  a A square is a quadrilateral in which all sides are equal.
  b A quadrilateral in which all sides are equal is a square.
  c The diagonals of a parallelogram are equal in length.
  d The diagonals of a kite intersect at right angles.

4 Jarrod draws a quadrilateral ABCD and its diagonals AC and BD. He notices that AC and BD intersect at right angles.
   a What types of quadrilateral could ABCD be? Explain your answer.
   b Jarrod measures the diagonals, and notices that one diagonal is twice the length of the other. He also notices that the diagonals bisect each other. What type of quadrilateral must ABCD be? Explain your answer.
   c If the shorter diagonal AC is 4 cm long, sketch and label quadrilateral ABCD.

5 Find the values of the unknowns:
   a 130°, $b°$, $a°$, $(2a - 40)°$
   b 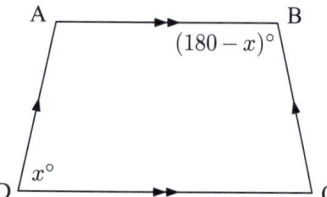 $b°$, $a°$, 42°
   c 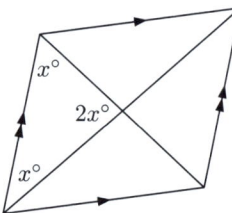 40°, $(3x+2)$ cm, $(10-x)$ cm, 30°, $a°$

6 These figures are not drawn to scale, but the information on them is correct. What can be deduced about the quadrilateral?
   a Quadrilateral ABCD with $(180-x)°$ at B and $x°$ at D.
   b 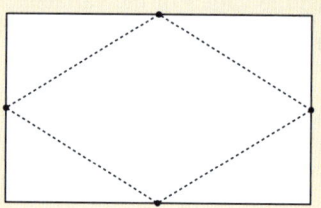 $x°$, $2x°$, $x°$

7 True or false? "A square is a special case of both a parallelogram and a rhombus."

Investigation 1   The midpoints of a quadrilateral

When the midpoints of adjacent sides of a rectangle are joined, the resulting figure appears to be a parallelogram.

What to do:
1 Draw your own rectangle. Find the midpoints of the sides, and join them.
2 Repeat **1** with:
    a a parallelogram   **b** a rhombus   **c** a kite   **d** a trapezium.
3 Repeat with a few quadrilaterals of your own choosing, including ones such as this:

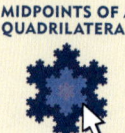
MIDPOINTS OF A QUADRILATERAL

4 Copy and complete: "When the midpoints of adjacent sides of a quadrilateral are joined, the resulting figure is always a ......".

Research

Research the use of rhombuses, parallelograms, and kites in art and architecture.

Moorish tiles at La Alhambra, Granada.

E ANGLES OF AN n-SIDED POLYGON

We have seen that a quadrilateral can be divided into 2 triangles, each with an angle sum of 180°, so the sum of the angles in a quadrilateral is $2 \times 180° = 360°$.

We can likewise find the sum of the angles of *any* polygon by dividing it into triangles.

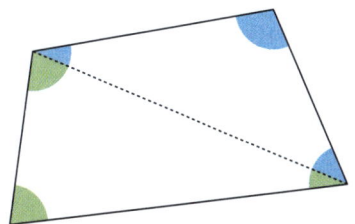

Investigation 2   Angles of an n-sided polygon

What to do:

1. Draw any pentagon, and label one of its vertices A. Draw in all of the diagonals from A. Notice that 3 triangles are formed.

2. Repeat with a hexagon, a heptagon (7-gon), and an octagon, drawing diagonals from one vertex only.

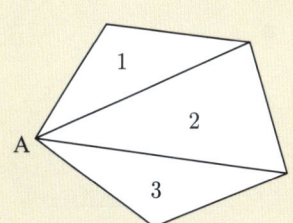

3. Copy and complete the following table:

Polygon	Number of sides	Number of triangles	Angle sum of polygon
quadrilateral	4	2	$2 \times 180° = 360°$
pentagon	5	3	
hexagon			
heptagon			
octagon			
20-gon			

4. Copy and complete:
   "The sum of the sizes of the interior angles of any n-sided polygon is ...... $\times 180°$."

From the **Investigation** you should have discovered that:

> The sum of the sizes of the interior angles of any n-sided polygon is $(n-2) \times 180°$.

Example 7   ◆) Self Tutor

Find x:

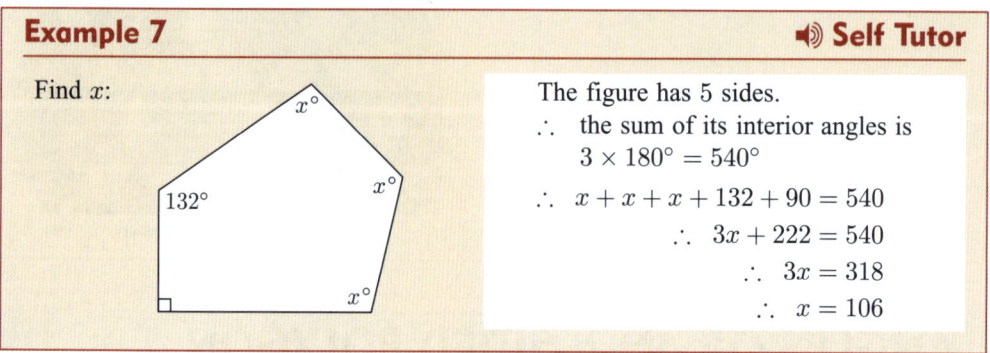

The figure has 5 sides.
∴ the sum of its interior angles is
$3 \times 180° = 540°$

∴ $x + x + x + 132 + 90 = 540$
∴ $3x + 222 = 540$
∴ $3x = 318$
∴ $x = 106$

EXERCISE 11E

1 Find the sum of the angles of:

  a   **b**   **c**

  d a polygon with 12 sides   **e** a 15-gon.

2 Find x:

  a 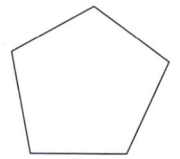   **b** 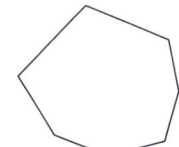   **c**

  d 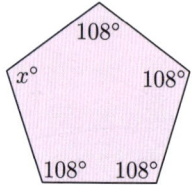   **e** 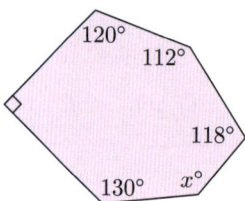   **f**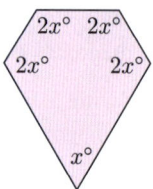

  g 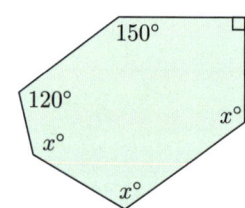   **h** 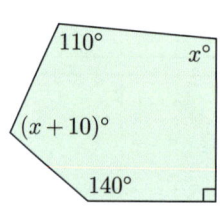   **i**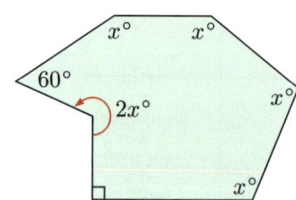

Shape (Chapter 11) 223

3 A pentagon has three right angles and two other equal angles. What is the size of each of the two equal angles?

4 **a** Find x.
b Find the size of the largest angle of the hexagon.

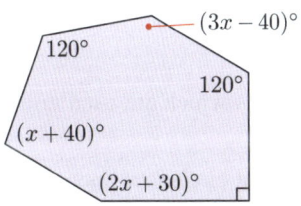

5 The sum of the angles of a polygon is $1980°$. How many sides has the polygon?

6 **a** What is the maximum number of reflex angles that a hexagon can have?
b Draw a hexagon with this number of reflex angles.

7 **a** Copy and complete the following table:

Regular polygon	Number of sides	Sum of angles	Size of each angle
triangle			
quadrilateral			
pentagon			
hexagon			
octagon			
decagon			

A **regular** polygon has all sides of equal length and all angles of equal size.

b Copy and complete:
 i the sum of the angles of an n-sided polygon is ......
 ii the size of each angle of a regular n-sided polygon is ......
c Find the size of each angle of a regular 12-sided polygon.

8 In the regular hexagon alongside, CF is parallel to DE.
Find the measure of:
 a BÂF
 b AB̂F
 c CB̂F
 d FĈD
 e CF̂E
 f BF̂C

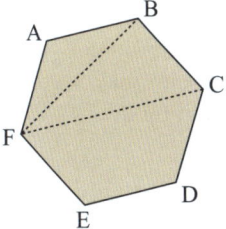

F CIRCLES

A **circle** is a two-dimensional shape. All points on the circle are the same distance from a fixed point called the **centre** of the circle.

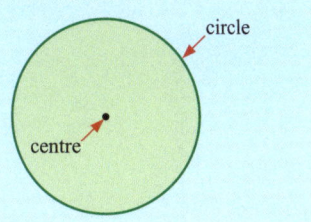

Builders in ancient Egypt constructed circles using a rope fixed at one end to a spike in the ground. Keeping the rope taut, they used a stick to draw the circle around the spike.

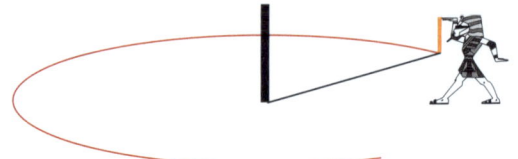

These are some words which are used to describe different parts of a circle:

- A **chord** of a circle is a line which joins any two points of the circle.
- A **diameter** of a circle is a chord which passes through the circle's centre.
- A **radius** of a circle is a straight line segment which joins the circle's centre to any point on the circle. **Radii** is the plural of radius.
- A **semi-circle** is a half of a circle.
- An **arc** is a part of a circle. It joins any two different points on the circle.
  For any two (non-opposite) points, we can define a **minor arc** and a **major arc** which are the shorter and longer arcs around the circle respectively.

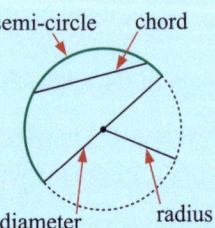

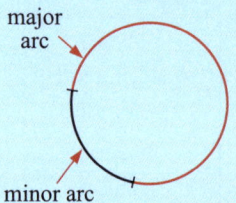

- A **segment** of a circle is the region between a chord and the circle.

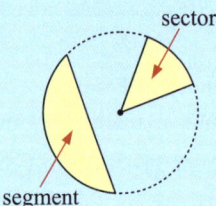

A **sector** of a circle is the region between two radii and the circle.

We can define minor and major segments and sectors just as we did for arcs.

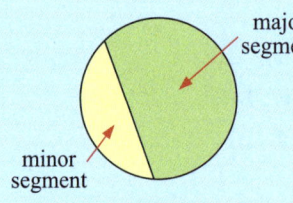

- A **tangent** to a circle is a line which *touches* the circle but does not enter it. A tangent is always at right angles to the radius at that point.

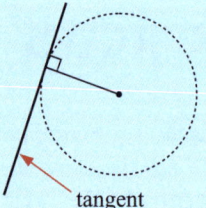

EXERCISE 11F

1 Match the part of the figure indicated to the phrase which best describes it:

a 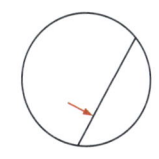 b 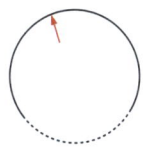 c 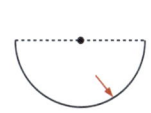 d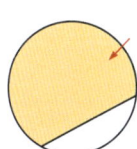

e 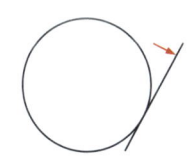 f 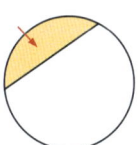 g 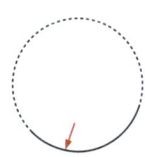 h

i 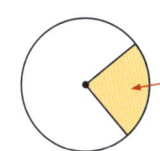 j 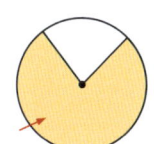 k 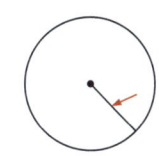 l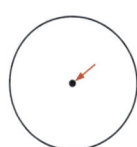

- **A** semi-circle
- **B** radius
- **C** minor arc
- **D** major arc
- **E** diameter
- **F** chord
- **G** minor segment
- **H** centre
- **I** major segment
- **J** major sector
- **K** minor sector
- **L** tangent

2 What name can be given to the longest chord that you can draw in a circle?

3 a Explain why the diameter of a circle is always twice as long as its radius.
  b Find:
   i the diameter of a circle with radius 4 cm
   ii the radius of a circle with diameter 12 cm.

> We commonly use **radius** to refer to the length of any radius of a circle, and **diameter** to refer to the length of any diameter.

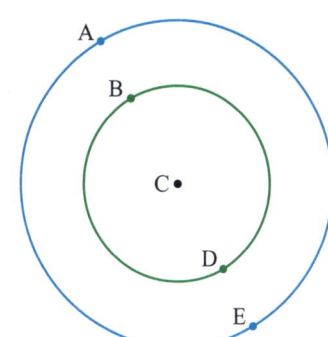

4 The circles shown both have centre C. The larger circle has radius 5 cm, and the smaller circle has radius 3 cm. Points A, B, C, D, and E are collinear.
Find the distance between:

- **a** C and B
- **b** C and A
- **c** B and D
- **d** A and E
- **e** A and B
- **f** E and B.

5 **a** Use a compass to draw a circle with radius 23 mm.
   b Find the diameter of the circle.
   c On the circle, draw a chord AB with length 4 cm.
   d Label the major arc of the circle with endpoints A and B.
   e Shade the minor segment of the circle which can be formed using points A and B.

Investigation 3   The angle in a semi-circle

In this Investigation we consider the *size* of an angle in a semi-circle.

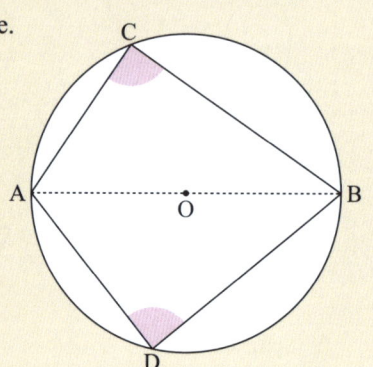

What to do:

1. Draw a circle of radius greater than 5 cm.
2. Draw any diameter AB of the circle, which divides the circle into two semi-circles.
3. Choose any point C on one of the semi-circles.
4. Measure the angle ACB.
5. Now choose any point D on the second semi-circle.
6. Measure the angle ADB.
7. What do you suspect about the angle in a semi-circle?
8. Click on the icon to run software for measuring the angle in a semi-circle.

   GEOMETRY PACKAGE

9. Comment on the statement: *The angle in a semi-circle is always a right angle.*
10. Draw the radius OC. What type of triangles are AOC and BOC?
11. Explain why:
    a $O\hat{A}C = O\hat{C}A$ **b** $O\hat{B}C = O\hat{C}B$
12. Explain why $A\hat{C}B = O\hat{A}C + O\hat{B}C$.
13. Hence explain why $A\hat{C}B$ is a right angle.

Review set 11A

1 Name these polygons according to their number of sides:

  a   **b** 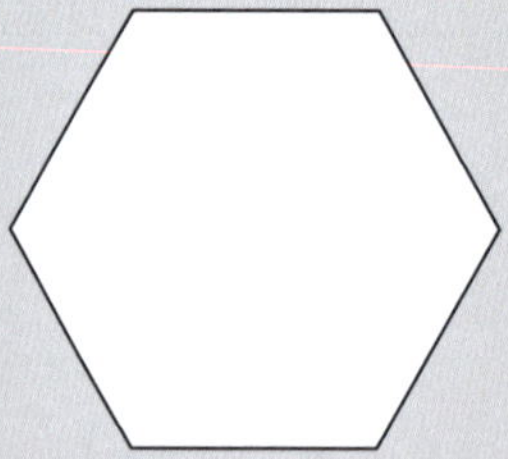   **c**

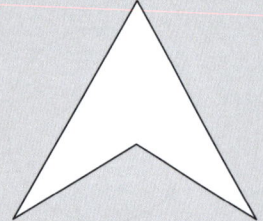      

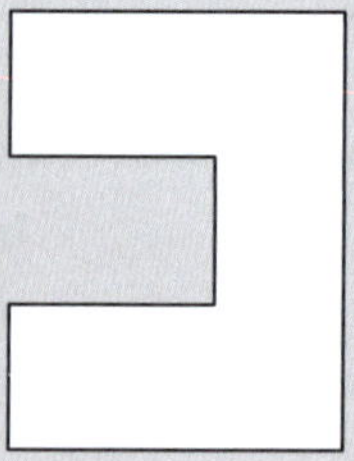

2 Use a protractor to classify the following triangles as acute, obtuse, or right angled.

a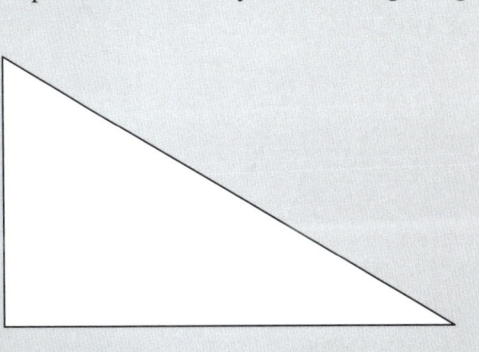

b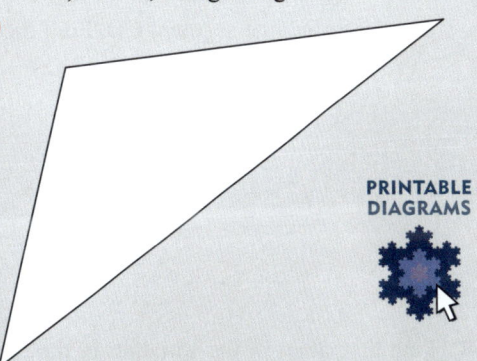

3 Classify each triangle by side lengths and by angles:

a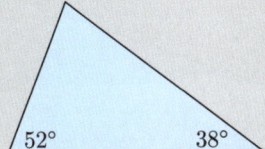

b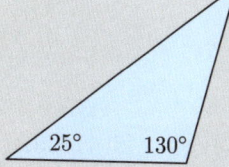

c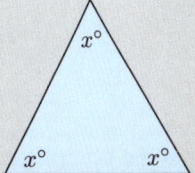

4 Find the values of the variables, giving brief reasons for your answers:

a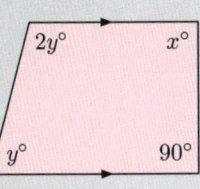

b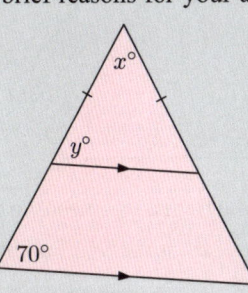

c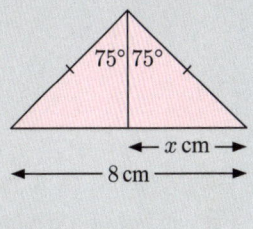

5 The quadrilaterals below are not drawn to scale, but the information on them is correct. Classify each quadrilateral.

a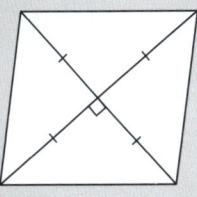

b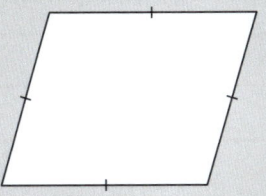

c

6 Find the values of the variables:

a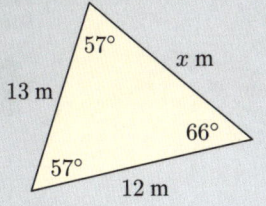

b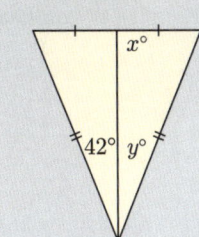

c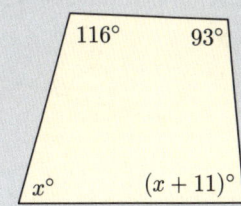

7 In the figure alongside, BC = CD and the line segment AB is parallel to the line segment CD. Find the size of $A\hat{B}C$.

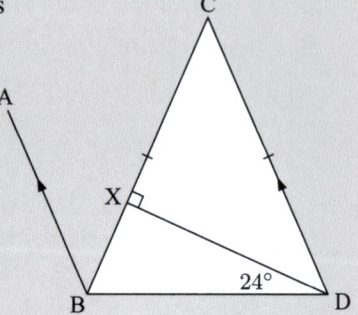

8 Find the values of the variables in the figure alongside:

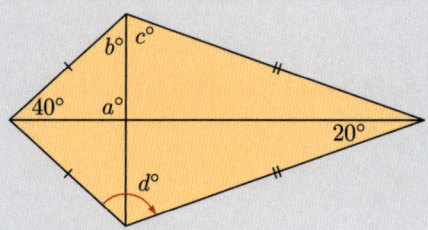

9 Find the sum of the angles of a polygon with 11 sides.

10 Find the value of x:

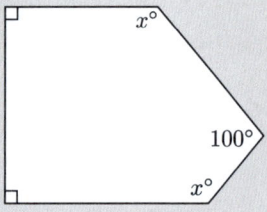

11 Clearly define, with the aid of diagrams, the meaning of:
   a an *arc* of a circle
   b a *sector* of a circle
   c a *chord* of a circle.

12 Are the following statements true or false? Explain your answers.
   a A minor arc of a circle is always shorter than a semi-circle.
   b A chord of a circle is always longer than the radius of the circle.

Review set 11B

1 Sketch:
   a a regular pentagon
   b an irregular quadrilateral.

2 Use a ruler to classify the following triangles as equilateral, isosceles, or scalene:
   a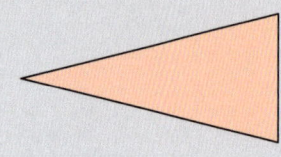
   b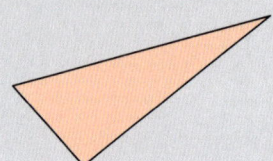

PRINTABLE DIAGRAMS

3 Classify this triangle as:
  a scalene, isosceles, or equilateral
  b acute angled, obtuse angled, or right angled.

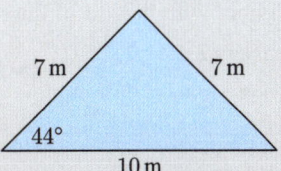

4 Draw two diagrams to illustrate the properties of a rectangle.

5 Find the values of the variables:

  a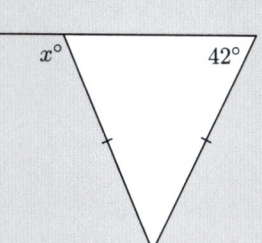

  b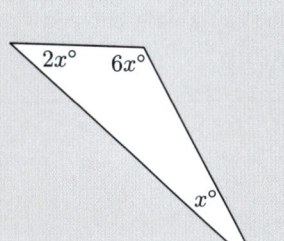

  c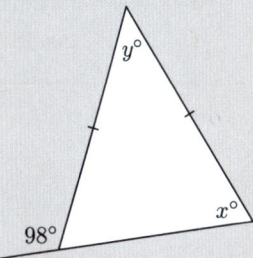

6 True or false?
  a The diagonals of a square are equal in length.
  b The diagonals of a rhombus intersect at right angles.

7 Find the values of the variables, giving brief reasons for your answers.

  a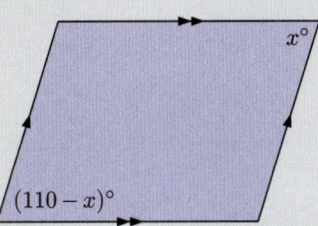

  b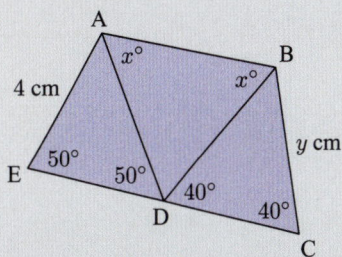

8 The quadrilaterals below are not drawn to scale, but the information on them is correct. Classify each quadrilateral.

  a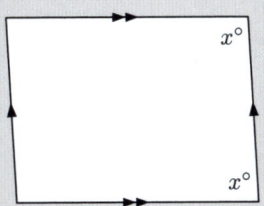

  b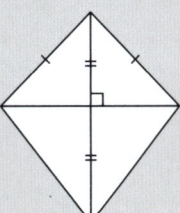

  c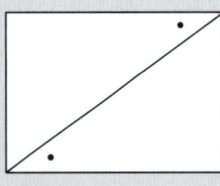

9 A hexagon has two right angles and four other equal angles. What is the size of each of the four equal angles?

10 a What is the maximum number of reflex angles a pentagon can have?
  b Draw a pentagon with this number of reflex angles.

11 Clearly define, with the aid of diagrams:
  a a *semi-circle*
  b a *major arc* of a circle
  c a *minor segment* of a circle
  d a *tangent* to a circle.

12
  a Use a compass to draw a circle with radius 32 mm.
  b Draw a diameter AB of the circle.
  c Draw a point C which lies on the circle such that AC is 16 mm.
  d Use a protractor to measure $A\hat{C}B$.
  e Draw the diameter of the circle with one end C. Label its other end D.
  f Classify quadrilateral ACBD.

12

Solids

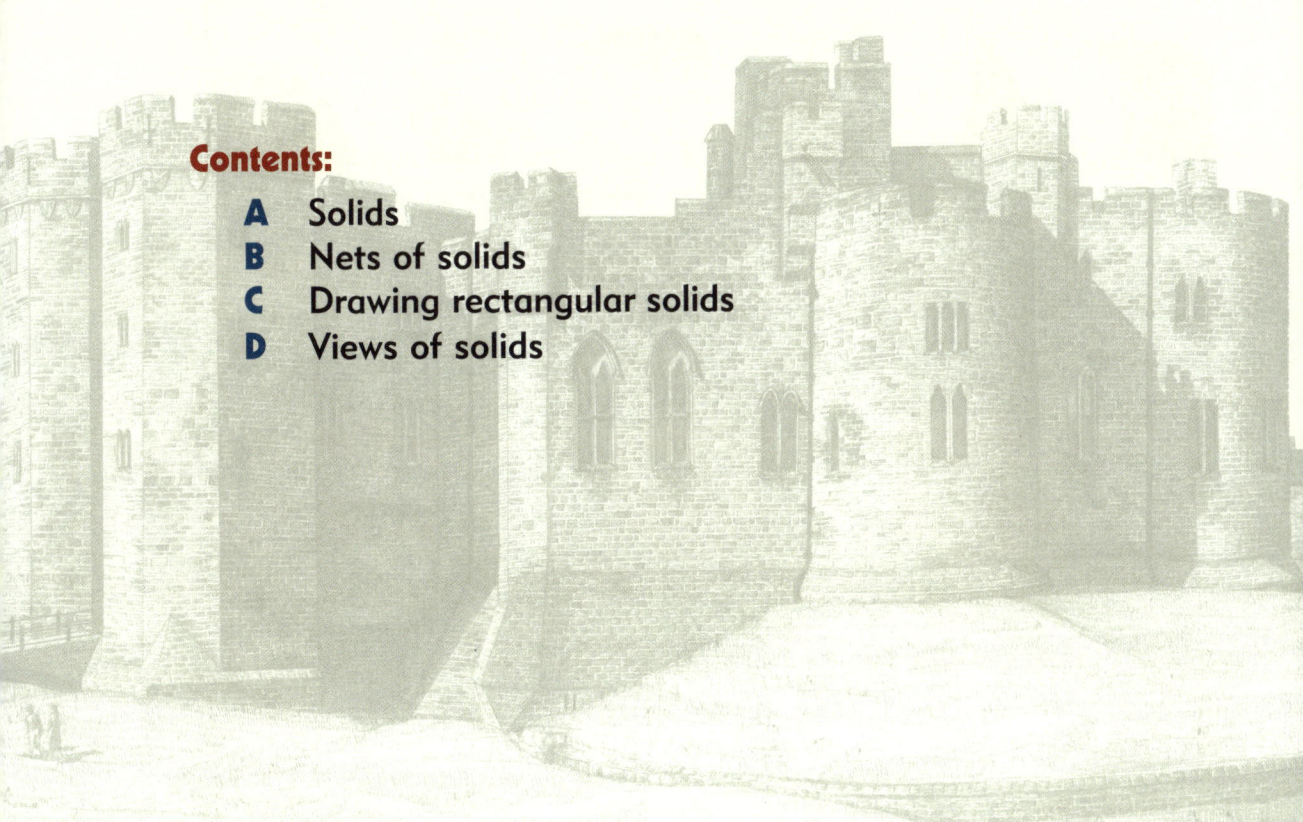

Contents:
- **A** Solids
- **B** Nets of solids
- **C** Drawing rectangular solids
- **D** Views of solids

Opening problem

Consider the diagram alongside.

Things to think about:
- **a** What object does the diagram represent?
- **b** Is the object 2-dimensional or 3-dimensional?
- **c** Is the *diagram* 2-dimensional or 3-dimensional?
- **d** Can you draw a 2-dimensional shape which could be folded to create this object?
- **e** How can we illustrate how the object looks from different directions?

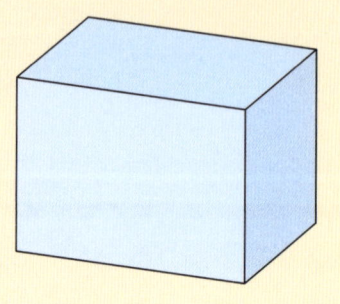

In this Chapter we will study three-dimensional **solids**. We will consider how they can be represented on a two-dimensional page, and two-dimensional **nets** which can be folded to create them.

A SOLIDS

A **solid** is a three-dimensional body which occupies space.

The diagrams below show a collection of solids. Each solid has three dimensions: *length*, *width*, and *height*.

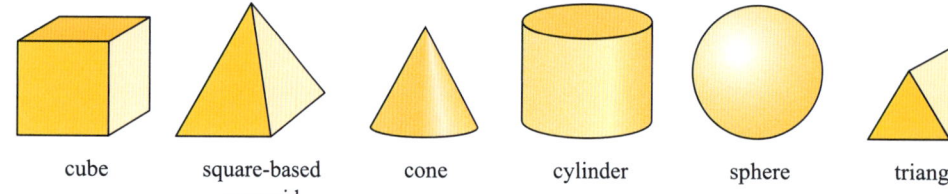

cube   square-based pyramid   cone   cylinder   sphere   triangular prism

The boundaries of a solid are called **surfaces**. Surfaces may be flat or curved. Flat surfaces with straight edges are also called **faces**. A **vertex** is a point where three or more edges meet to form a corner.

For example:

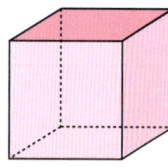 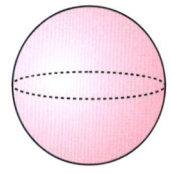 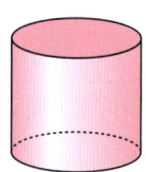

A cube is bounded by six faces. It has eight vertices.   A sphere is bounded by one curved surface.   A cylinder is bounded by two flat surfaces and one curved surface.

When we draw solids, we often use dashed lines to show edges which are hidden at the back of the solid. The dashed lines remind us these edges are there, even if we cannot normally see them. Dashed lines can also help us to appreciate the three-dimensional nature of the solids.

DEMO

Solids (Chapter 12) 233

PRISMS

A **prism** is a solid with a uniform cross-section that is a polygon.

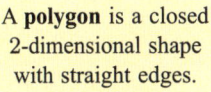
A **polygon** is a closed 2-dimensional shape with straight edges.

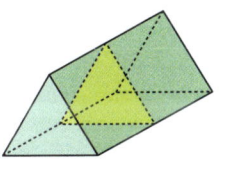
triangular prism

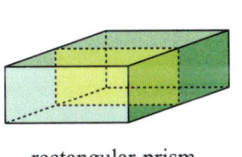
rectangular prism or cuboid

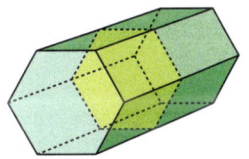
hexagonal prism

CYLINDERS

A **cylinder** is a solid with a uniform cross-section that is a circle.

PYRAMIDS

A **pyramid** is a solid with a polygonal base, and triangular faces which come from the edges of the base to meet at a point called the **apex**.

A triangular-based pyramid is also called a **tetrahedron**.

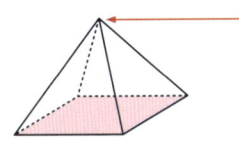
square-based pyramid

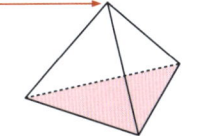
triangular-based pyramid

apex

CONES

A **cone** is a solid with a circular base, and a curved surface from the base to a point called the **apex**.

apex

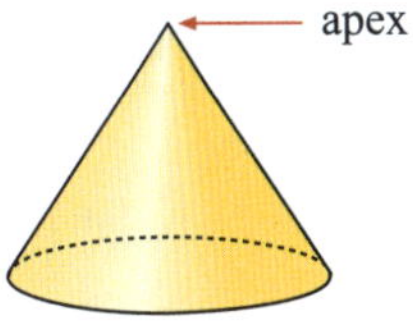

Discussion

- Is a cylinder a prism?
- Is a cone a pyramid?

EXERCISE 12A

1 Name the following solids:

a 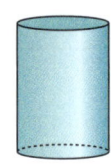   b 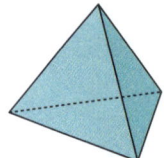   c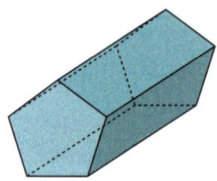

2 Draw a diagram to represent:
- **a** a cone
- **b** a rectangular-based pyramid
- **c** a sphere
- **d** an octagonal prism
- **e** a hexagonal-based pyramid.

3 Name the solid which best resembles:
- **a** a can of soup
- **b** a marble
- **c** a cereal box
- **d** a witch's hat
- **e** a four-sided die
- **f** a coin.

4 For each solid below, state the number of vertices, edges, and faces.

a 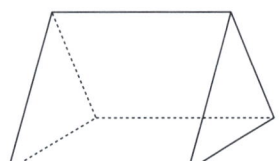   b 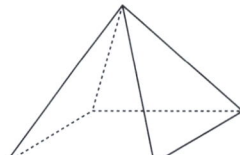   c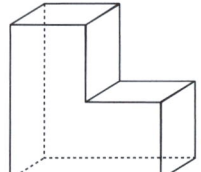

5 What shape are the side faces of a:
- **a** prism
- **b** pyramid?

6 Draw a solid which has:
- **a** only a curved surface
- **b** a curved and a flat surface
- **c** two flat surfaces and one curved surface
- **d** four faces
- **e** 12 edges
- **f** six faces and six vertices.

B NETS OF SOLIDS

A **net** is a two-dimensional shape which may be folded to form a solid.

For example, the following nets may be cut out and folded along the dotted lines to form common solids:

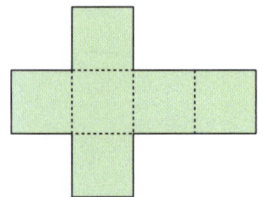   becomes   a **cube**

PRINTABLE NETS

Solids (Chapter 12) 235

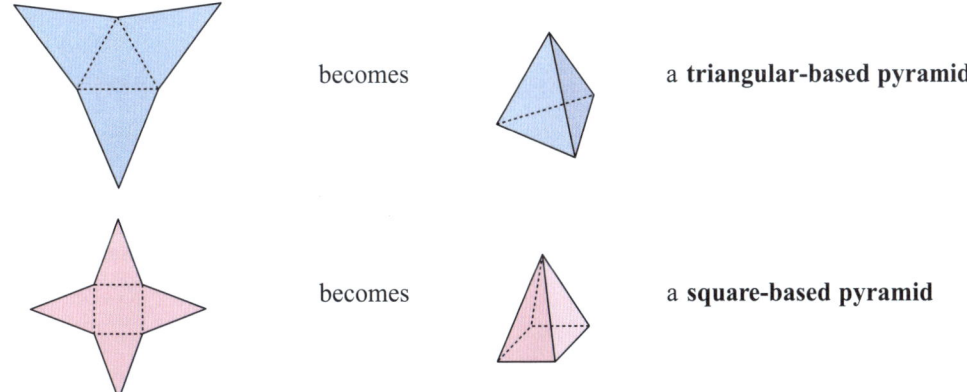

becomes — a **triangular-based pyramid**

becomes — a **square-based pyramid**

Click on the icon to view demonstrations of how the nets form the solids.

DEMO

EXERCISE 12B

1 For each of the following nets, draw and name the corresponding solid:

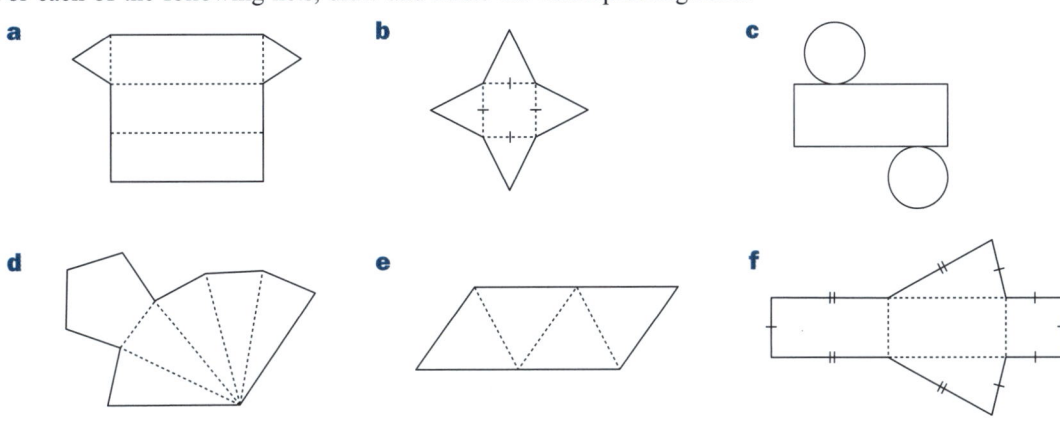

2 Draw nets for each of the following three-dimensional solids, clearly marking the lengths of the sides:

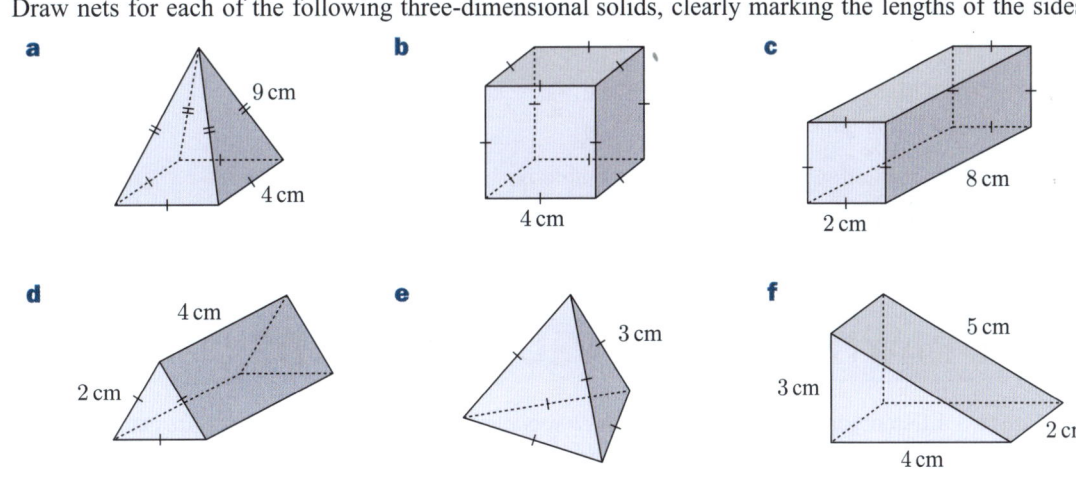

3 Copy this net of a rectangular prism. Place tick marks on the remaining lines to indicate the sides of equal length.

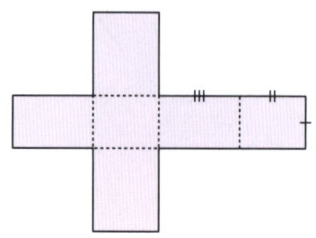

4 Which of the following nets can be used to make this cube?

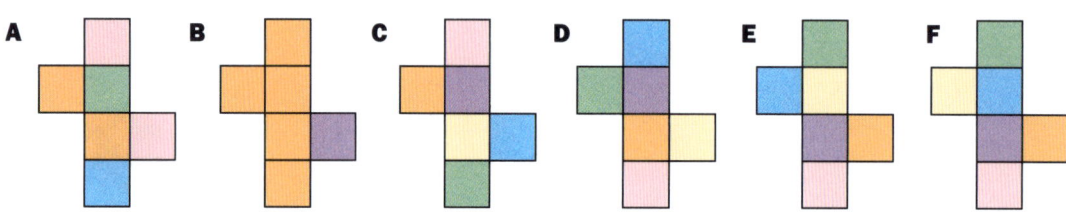

5 Draw the net for this prism, clearly marking the lengths of the sides.

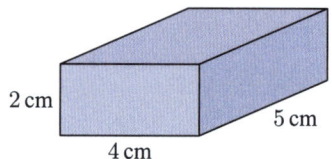

Investigation   Triangular-based pyramids

In this Investigation we will use an acute angled triangle as a net to form a triangular-based pyramid.

PRINTABLE NETS

What to do:

1 Click on the icon and print the different nets. Fold the nets along the dashed lines to form triangular-based pyramids.

2 On separate sheets of paper, draw acute angled triangles of different shapes. Cut them out with scissors.

3 For each triangle, use three folds of the paper to attempt to construct a triangular-based pyramid.

4 Will your method work if the original triangle is:
  a right angled   **b** obtuse angled?

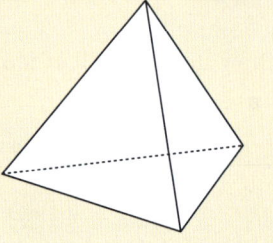

C DRAWING RECTANGULAR SOLIDS

There are two different methods we can use to draw rectangular solids. These methods are called **projections** because we *project* the image of the three-dimensional solid onto the two-dimensional paper.

OBLIQUE PROJECTIONS

To draw a cube using an **oblique projection** we use the following steps:

Step 1: Draw a square for the front face.

Step 2: Draw edges back from the front face at 45°, and shorter than those of the front face.

Step 3: Complete the cube.

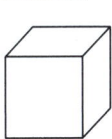

Step 4: If appropriate, draw in dashed lines to show the hidden edges.

ISOMETRIC PROJECTIONS

When drawing a rectangular solid using an **isometric projection**, we use **isometric graph paper** which is made up of equilateral triangles.

We start with a vertical edge of the solid. The horizontal edges are drawn inclined at 30°.

The diagram alongside shows the isometric projection of a cube. Notice that all the edges drawn have the same length. The edge AB appears closest to us. This is often the **starting edge** of the figure, or first edge drawn.

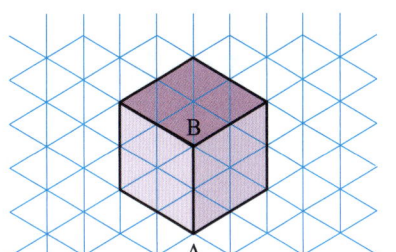

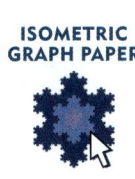

ISOMETRIC GRAPH PAPER

EXERCISE 12C

1 Draw an oblique projection of a box which has sides 2 units by 2 units by 1 unit. Start with a 2 unit by 1 unit rectangle as the front face.

2 Draw the following solids on isometric paper. Use the darker lines as the starting edges.

a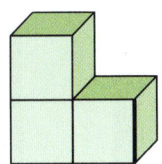

b

c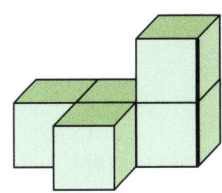

238   Solids   (Chapter 12)

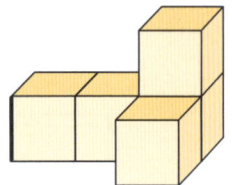 d   e   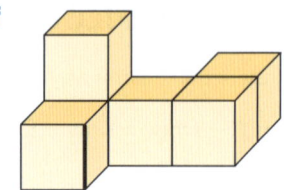 f

3 Redraw these isometric projections as oblique projections.

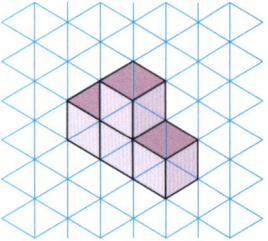 a   b   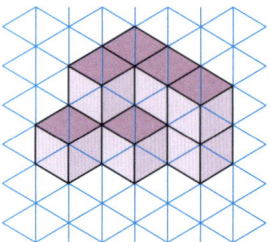 c

Activity 1	Isometric solids
Click on the icon to access this Activity.	 ISOMETRIC SOLIDS

D   VIEWS OF SOLIDS

When drawing a three-dimensional solid, we cannot show the details on all of the faces at the same time, because many of the faces are hidden from view.

Instead we can create several drawings of the solid from different angles.

The top view is called the **plan** of the object.

The front, back, left, and right views are called **elevations**.

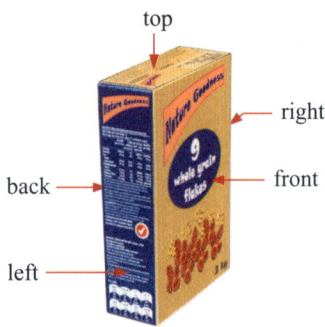

plan   front elevation   back elevation   left elevation   right elevation

Solids (Chapter 12)

EXERCISE 12D.1

1 A rectangular prism is 10 cm long, 5 cm wide, and 4 cm high. Sketch, including dimensions, the:

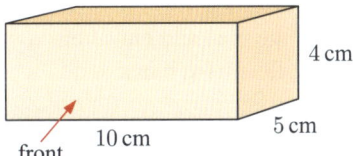

- **a** plan
- **b** front elevation
- **c** back elevation
- **d** left elevation
- **e** right elevation.

2 The numbers on a die are arranged so that the sum of each pair of opposite faces is seven.

- **a** Which number is on the bottom of the die?
- **b** Sketch the die alongside from the:
  - **i** front
  - **ii** top
  - **iii** right
  - **iv** left
  - **v** back.

3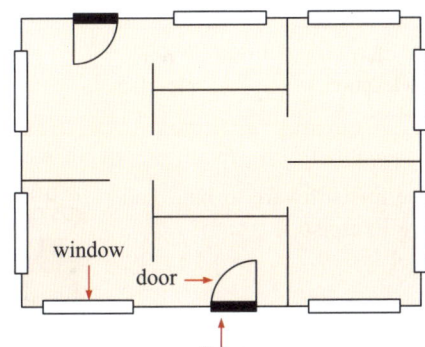

The diagram shows an architect's plan for a building. The doors and windows are indicated on the plan. Sketch the:

- **a** front elevation
- **b** back elevation
- **c** left elevation
- **d** right elevation.

4 A cube has the letters A, B, C, D, E, and F painted on its faces. Three different views of the cube are shown alongside.

Using the bottom view of the cube as a basis, draw the:

- **a** front elevation
- **b** right elevation
- **c** plan
- **d** left elevation
- **e** back elevation.

BLOCK SOLIDS

Drawings of block solids on isometric graph paper can also be viewed from different angles.

We assume that on the isometric graph paper, the view from the bottom left corner is the front view.

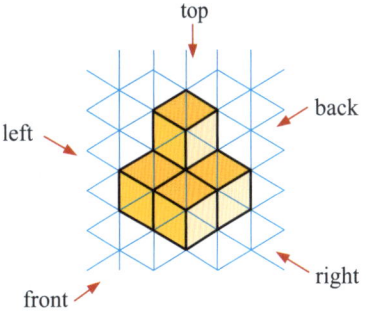

Example 1

Draw the plan and elevations of the illustrated block solid.

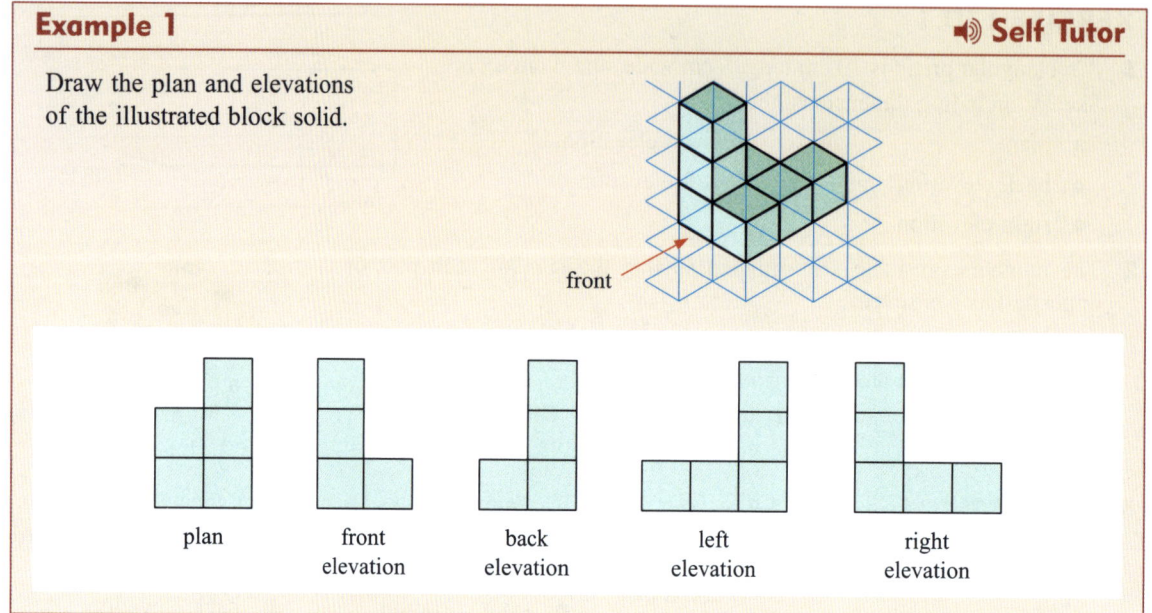

Click on the icon to run the Blockbuster software. You can use this software to help you visualise block solids and answer the questions in the following Exercise.

BLOCK BUSTER

EXERCISE 12D.2

1 Draw the plan and elevations of these block solids. Assume that all blocks are visible.

a   b   c

d   e   f

Example 2

These diagrams show the different views of a block solid. Draw the object on isometric paper.

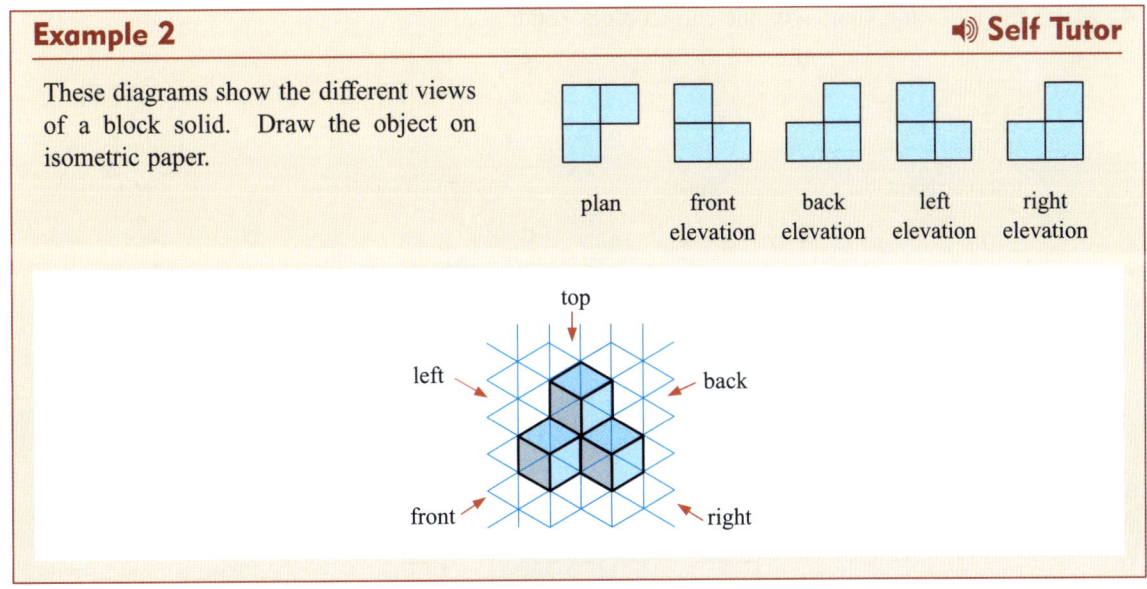

2 Draw the block solid which has these views:

a

b

c

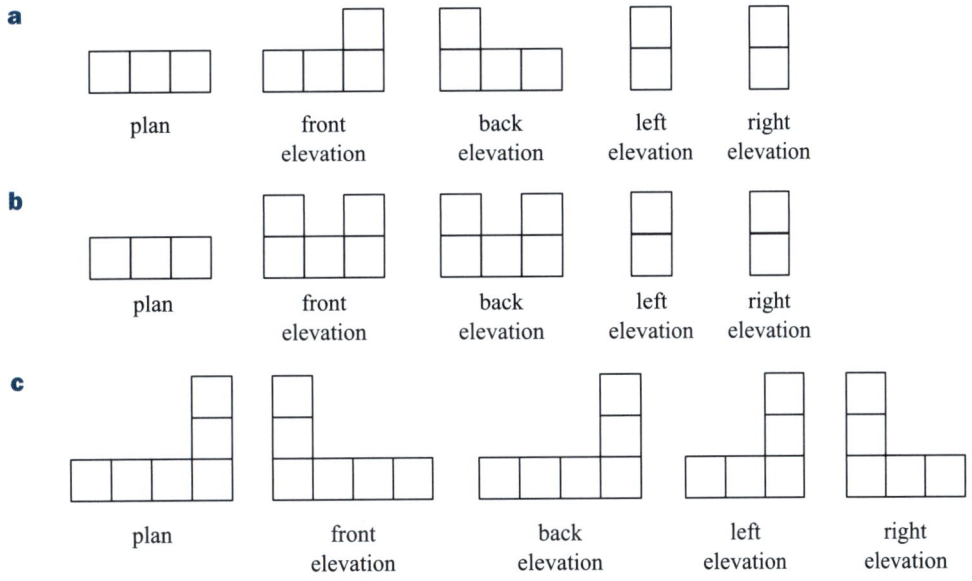

3 Explain why it is impossible for a block solid to have these views:

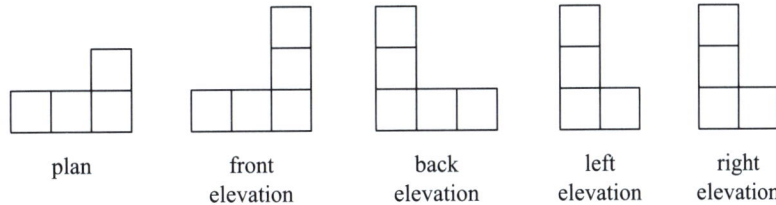

4 Match the following views with the correct block solid:

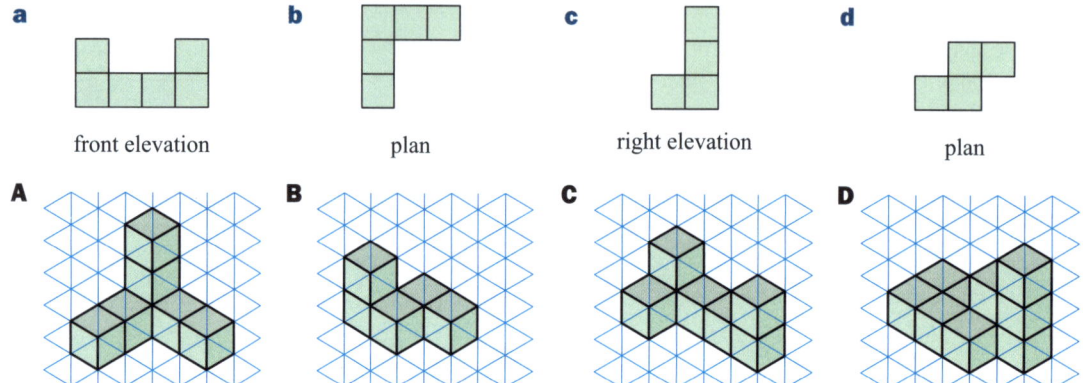

5 Draw *every* block solid which has these views. The individual blocks cannot be glued together.

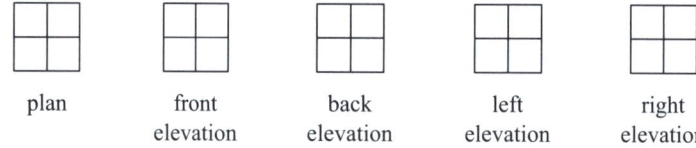

Activity 2   Humble houses

The Humble House factory manufactures cubic living quarters for countries where the conditions are mostly dry and hot. Heat enters every roof and exposed wall at the same rate.

For a house made of **one cube**, heat enters in equal amounts from 5 sides, but not the floor.

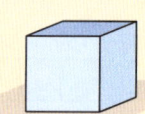

There are two possible house designs made from **two cubes** placed together with faces touching:

A 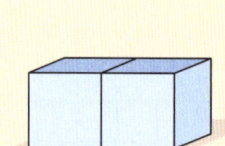   **B**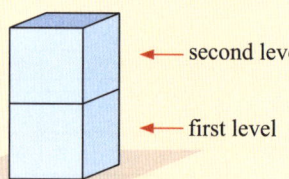

What to do:

1 How many exposed faces are there in each of the designs **A** and **B**? Which one would be more suitable for hot conditions?

2 Draw the four different possible housing arrangements using three adjacent cubes. The blocks must touch face to face and the building must be free-standing. For example, columns for support are not acceptable, since we do not want heat to come in through the floor.

3 From your models, determine the two "best" 3-cube structures which would allow the least amount of heat to come in.

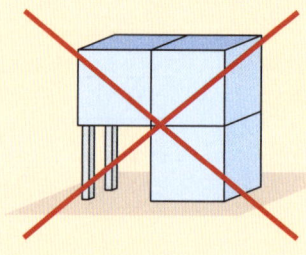

4 Investigate the possible 4-cube buildings, and determine the model which would allow the least amount of heat to come in.

5 How many different possible 5-cube buildings are there? Which one is "best"?

6 Write some general conclusions about how these buildings should be designed to minimise the amount of heat coming in.

Activity 3 — Painted cubes

A cube is painted and then cut into 8 smaller cubes.

On dismantling the $2 \times 2 \times 2$ cube, we see that all 8 cubes have paint on exactly 3 faces.

In this Activity we consider how many cubes are painted the same when the cube is cut into a $3 \times 3 \times 3$ cube and a $4 \times 4 \times 4$ cube.

What to do:

1 Copy and complete:

Cube cut	3 faces painted	2 faces painted	1 face painted	No faces painted
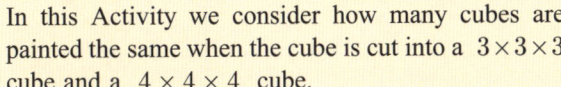	8	0	0	0

2 From the results in your table, what patterns do you notice?

Review set 12A

1. Name the following solids:
   a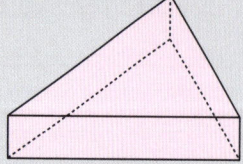
   b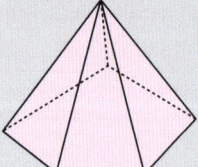

2. Draw a net for a triangular-based pyramid.

3. For the net shown:
   a Name the corresponding solid.
   b State the number of vertices, edges, and faces of the corresponding solid.

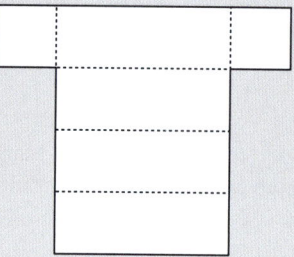

4. Draw an oblique projection of a rectangular prism which is 5 cm long, 3 cm wide, and 2 cm high. Start with a 5 cm by 2 cm rectangle as the front face.

5. Draw the following as isometric projections. Use the darker lines as the starting edges.
   a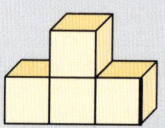
   b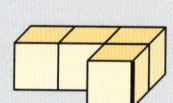
   c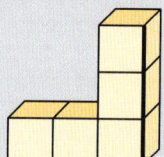

6. Draw the block solid with these views:

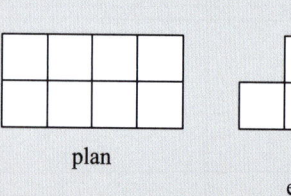 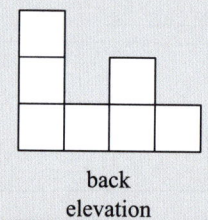 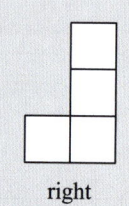

plan   front elevation   back elevation   left elevation   right elevation

7. The plan of a house is given alongside. The house is 3 metres high. Sketch, including dimensions, the:
   a front elevation
   b back elevation
   c left elevation
   d right elevation.

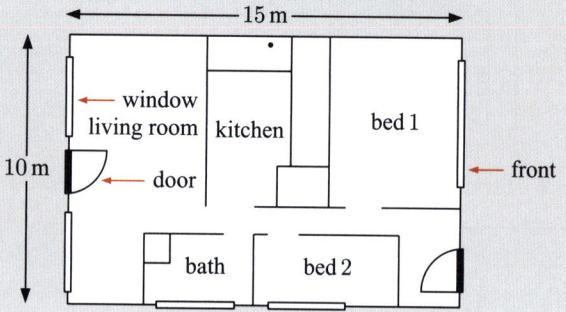

8 Draw the plan and elevations for:

a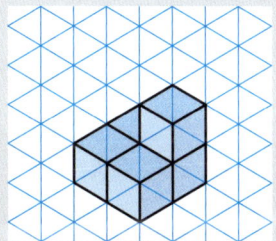

b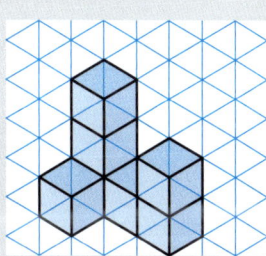

Review set 12B

1 Draw a diagram to represent:

   a a cylinder   **b** a square-based pyramid.

2 Draw a net for making a 5 cm by 3 cm by 1 cm rectangular prism.

3 Name the solid which best resembles a six-sided die.

4 Draw these isometric projections as oblique projections:

a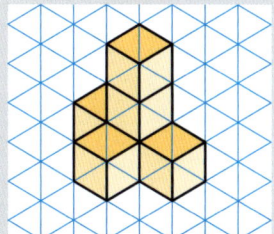

b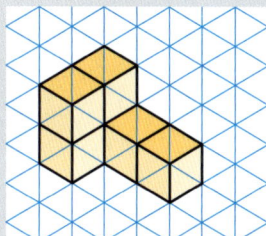

5 Sketch, including dimensions, the following views of this rectangular prism:

   a front elevation   **b** plan
   c left elevation   **d** right elevation
   e back elevation.

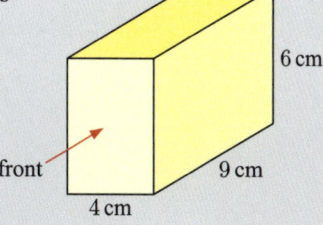

6 Draw the net for this triangular prism, clearly marking the lengths of the sides.

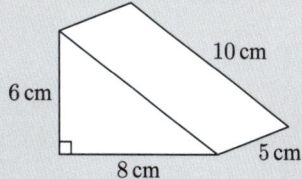

7 Draw the plan and elevations for:

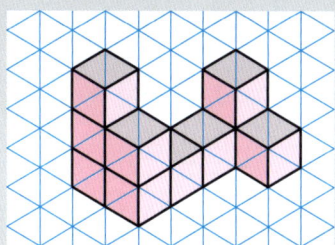

8 Draw the block solid with these views:

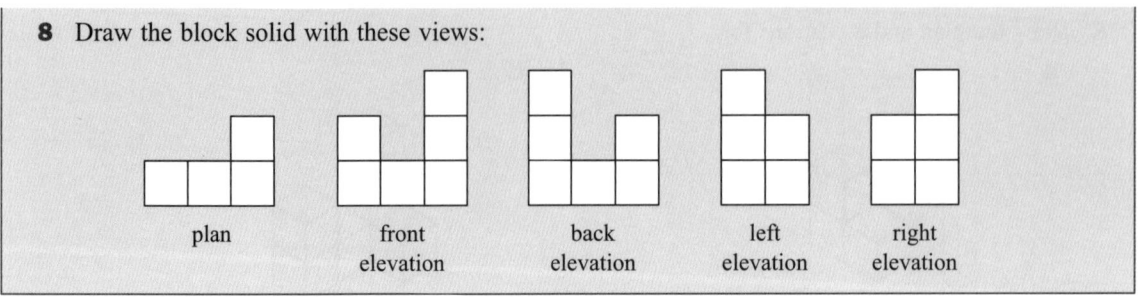

13

Algebraic factorisation

Contents:
- **A** Common factors
- **B** Factorising with common factors
- **C** Difference between two squares
- **D** Perfect square factorisation
- **E** Factorising quadratic trinomials
- **F** Miscellaneous factorisation

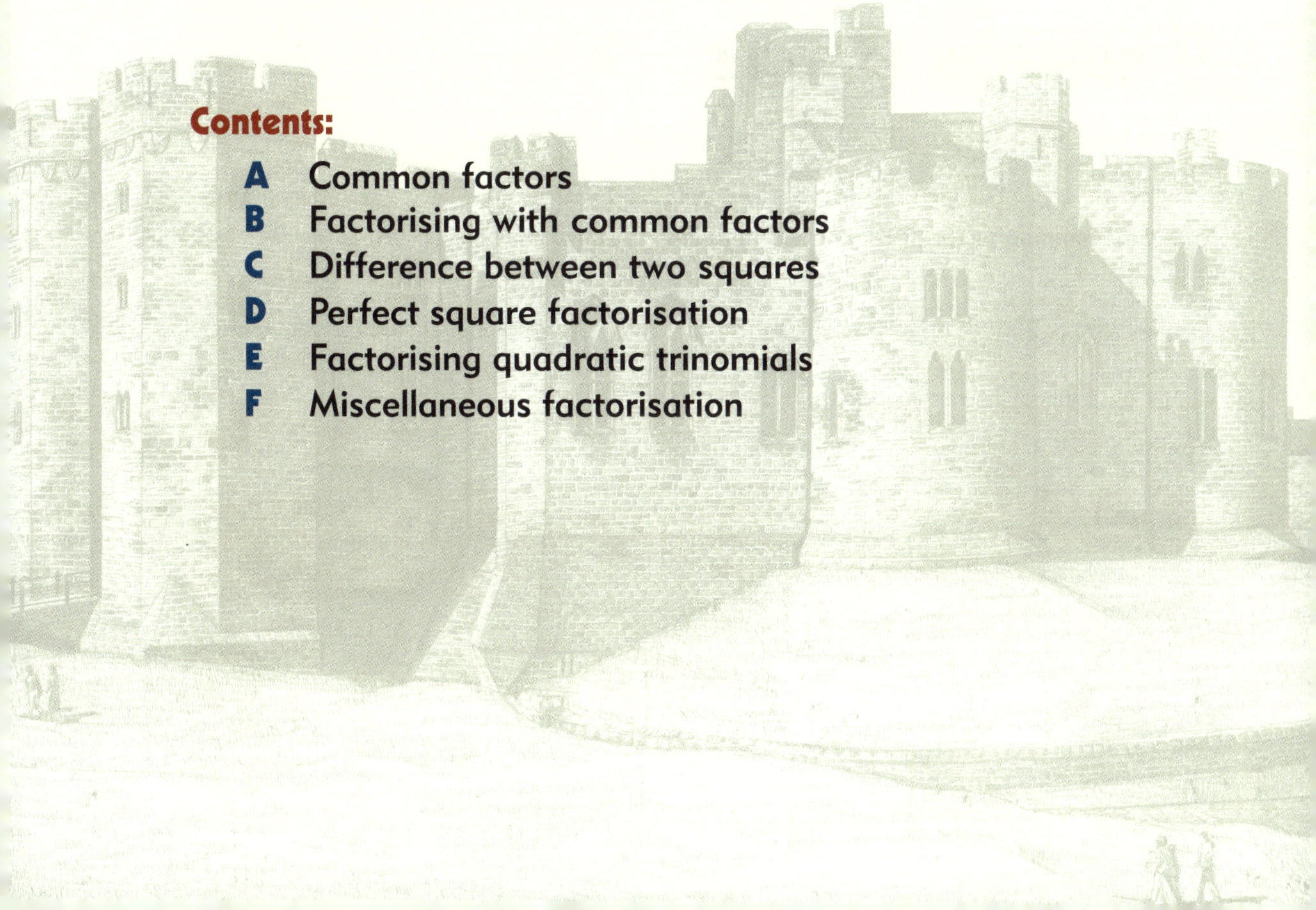

Opening problem

A square pool will be surrounded by a path as shown. The landscaper needs to know the area of the path so he knows how many pavers to buy.

Things to think about:

a Can you explain why the area of the path is $(L^2 - l^2)$ m²?

b Can you show that this area can be written as:
  i $(L+l)(L-l)$ m²
  ii $4x(x+l)$ m²?

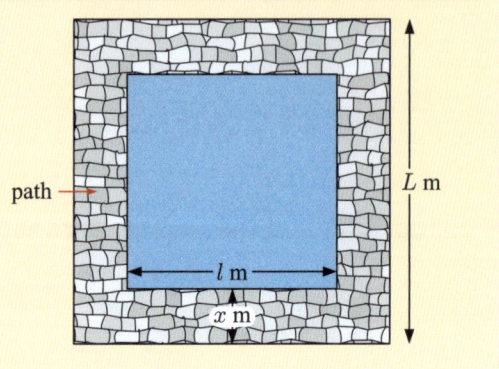

In **Chapter 8** we learnt how to perform algebraic **expansions** such as writing $3(x+2)$ as $3x+6$. In this Chapter we will learn how to **factorise** algebraic expressions.

Factorisation is the reverse process of expansion.

$$3(x+2) \xrightleftharpoons[\text{factorisation}]{\text{expansion}} 3x+6$$

A COMMON FACTORS

In previous years we have seen how numbers can be expressed as the product of factors.

A number which is a factor of two or more numbers is called a **common factor** of these numbers.

The **highest common factor** or **HCF** of a set of numbers is the largest factor that is common to all of them.

For example, $30 = 2 \times 3 \times 5$ and $140 = 2 \times 2 \times 5 \times 7$. We see that 2 and 5 are the common factors of 30 and 140.

∴ the HCF of 30 and 140 is $2 \times 5 = 10$.

We can find the HCF of **algebraic products** in the same way.

Example 1 ◀) Self Tutor

Find the highest common factor of:
a $8a$ and $12b$
b $4x^2$ and $6xy$

a
$8a = 2 \times 2 \times 2 \times a$
$12b = 2 \times 2 \times 3 \times b$
∴ HCF $= 2 \times 2$
  $= 4$

b
$4x^2 = 2 \times 2 \times x \times x$
$6xy = 2 \times 3 \times x \times y$
∴ HCF $= 2 \times x$
  $= 2x$

Write each term as a product of its **factors**!

EXERCISE 13A

1 Find the highest common factor of:
- **a** $3x$ and 6
- **b** $2b$ and b
- **c** $7b$ and 28
- **d** $4q$ and $5q$
- **e** $9c$ and $27c$
- **f** $8d$ and $12d$
- **g** $11x$ and 12
- **h** $16z$ and $20z$
- **i** $42x$ and $70x$

2 Find the HCF of:
- **a** $15ab$ and $13ba$
- **b** st and $3st$
- **c** $12x$ and $18xy$
- **d** f and f^2
- **e** c^3 and $5c$
- **f** r^3 and r^2
- **g** $8d$ and $2d^2$
- **h** $5a^3$ and $25a$
- **i** $7b^2$ and $14b^3$
- **j** $26gh$ and $39hg$
- **k** $6jk$ and $12jk^2$
- **l** $8x^2y$ and $18xy^2$
- **m** $10x$, $20xy$, and $30y^2$
- **n** $5abc$, $10a^2b$, and $15bc$
- **o** $22p^2q$ and $48pq^2$

Example 2   ◀)) Self Tutor

Find the HCF of $3(x+3)$ and $(x+3)(x+1)$.

$$3(x+3) = 3 \times (x+3)$$
$$(x+3)(x+1) = (x+3) \times (x+1)$$
$$\therefore \text{ HCF} = (x+3)$$

3 Find the HCF of:
- **a** $3(b+6)$ and $(b+5)(b+6)$
- **b** $7(1+b)^2$ and $7(2+b)(1+b)$
- **c** $y^2(y-3)$ and $y(y-3)$
- **d** $16(x-4)^2$ and $8(x-4)(x-2)$
- **e** $9(x-8)^2$ and $12(x-7)(x-8)$
- **f** $5d(d+1)$ and $15d(d+1)^2$

4 Find the HCF of:
- **a** $x+1$ and x^2+1
- **b** x^2 and $x-1$
- **c** $x(1-x)$ and $(x+1)(x-1)$

If there are no common factors, the HCF is 1.

Activity

Click on the icon to run a game involving algebraic common factors.

GAME

B FACTORISING WITH COMMON FACTORS

Factorisation is the process of writing an expression as a **product** of its **factors**.

Factorisation is the reverse process of **expansion**.

When we expand an expression, we remove its brackets.

When we factorise an expression, we insert brackets.

Notice that $5(x-1)$ is the *product of the two factors*, 5 and $x-1$.

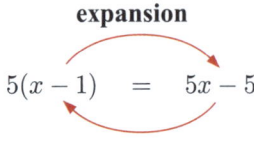

$$5(x-1) = 5x - 5$$

To factorise an algebraic expression involving a number of terms, we look for the HCF of the terms. We write it in front of a set of brackets. We then use the reverse of the distributive law to complete the factorisation.

> Reversing the distributive law gives $ab + ac = a(b+c)$.

For example: $6x^2$ and $2xy$ have HCF $= 2x$,

$$\therefore \quad 6x^2 + 2xy = 2x \times 3x + 2x \times y$$
$$= 2x(3x + y)$$

Example 3   ◀) Self Tutor

> With practice, the middle line is not necessary.

Fully factorise:

a $3a + 6$
b $ab - 2bc$

a $\quad 3a + 6$
$\quad = 3 \times a + 3 \times 2$
$\quad = 3(a+2) \quad$ {HCF is 3}

b $\quad ab - 2bc$
$\quad = a \times b - 2 \times b \times c$
$\quad = b(a - 2c) \quad$ {HCF is b}

EXERCISE 13B

1 Fully factorise:

a $3x - 6$ **b** $8d + 40$ **c** $10x - 30$ **d** $4 + 2x$
e $5p + 5q$ **f** $ab + ad$ **g** $42 - 35x$ **h** $gh + h$
i $y + yz$ **j** $p - pq$ **k** $3e + ef$ **l** $9x - xy$
m $jk - 3hk$ **n** $rs - st$ **o** $3xy - 2yz$ **p** $4mn + lm$

Example 4 Self Tutor

Fully factorise:

a $8x^2 + 12x$ **b** $3y^2 - 6xy$

a $8x^2 + 12x$
$= 2 \times 4 \times x \times x + 3 \times 4 \times x$
$= 4x(2x + 3)$ {HCF is $4x$}

b $3y^2 - 6xy$
$= 3 \times y \times y - 2 \times 3 \times x \times y$
$= 3y(y - 2x)$ {HCF is $3y$}

2 Fully factorise:

a $x^2 + 7x$ **b** $3x^2 + 9x$ **c** $8x - 4x^2$
d $5x - 25x^2$ **e** $6x^2 + 15x$ **f** $x^3 + 2x^2$
g $x^3 + 3x$ **h** $8x^3 - x^2$ **i** $x^2y - xy^2$
j $2xy^2 + 8x^2y$ **k** $4x^2 - 8x + 12$ **l** $9x^2 - 45x - 27x^3$

3 Write an expression for the total area of the shape in:

a factorised form **b** expanded form.

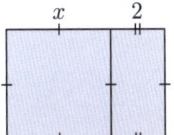

Example 5 Self Tutor

Fully factorise: $-2a + 6ab$

$-2a + 6ab$
$= 6ab - 2a$ {Rewrite with $6ab$ first.}
$= 2 \times 3 \times a \times b - 2 \times a$
$= 2a(3b - 1)$ {HCF is $2a$}

4 Fully factorise:

a $-5a + 10$ **b** $-20 + 12b$ **c** $-8c + 16d$
d $-ab + 3a$ **e** $-p + pq$ **f** $-y^2 + y$
g $-14y + 7y^2$ **h** $-cd + 2d^2$ **i** $-z^2 + z$

Example 6 Self Tutor

Fully factorise: $-2x^2 - 4x$

$-2x^2 - 4x$
$= -2 \times x \times x + -2 \times 2 \times x$
$= -2x(x + 2)$ {HCF is $-2x$}

5 Fully factorise:

  a $-6x - 24$ **b** $-8 - 8x$ **c** $-2x - 2y$
  d $-10p - 5q$ **e** $-ab - a$ **f** $-7a - 21a^2$
  g $-16g^2 - 8g$ **h** $-18t^2 - 12st$ **i** $-22e^2 - 33e$

Example 7 ◀)) **Self Tutor**

Fully factorise:

  a $2(x+3) + x(x+3)$ **b** $x(x+4) - (x+4)$

  a $2(x+3) + x(x+3)$ {HCF $= (x+3)$}
  $= (x+3)(2+x)$

  b $x(x+4) - (x+4)$
  $= x(x+4) - 1(x+4)$ {HCF $= (x+4)$}
  $= (x+4)(x-1)$

---

6 Fully factorise:

  a $a(a+2) + 3(a+2)$ **b** $4(x-1) + x(x-1)$
  c $5(x+4) - x(x+4)$ **d** $x(x-9) + (x-9)$
  e $a(c+d) - b(c+d)$ **f** $x(x-3) - 3(x-3)$
  g $t(r+s) - (r+s)$ **h** $y(y-6) + y - 6$

Example 8 ◀)) **Self Tutor**

Fully factorise $(x-1)(x+2) + 3(x-1)$

$(x-1)(x+2) + 3(x-1)$ {HCF $= (x-1)$}
$= (x-1)[(x+2) + 3]$
$= (x-1)(x+5)$

We use square brackets in the second line to help distinguish them.

---

7 Fully factorise:

  a $(x+4)(x+5) + 3(x+4)$ **b** $2(x+1) + (x+1)(x-3)$
  c $(x-6)(x+2) - 2(x+2)$ **d** $(x-10)^2 + 4(x-10)$
  e $(d+7)^2 + (d+6)(d+7)$ **f** $(x-y)(y-1) - 5(y-1)$
  g $15(a+1)^2 - 10(a+1)$ **h** $3(n-3) - 7(n-3)^2$
  i $n(n-3) - 2(n-3)(n+1)$ **j** $(x-14)^2 - 3(x-14)(x+8)$

8 Jake has factorised $12x^2 - 18xy$ as $6(2x^2 - 3xy)$.

  a Explain why Jake has not factorised the expression fully.
  b Fully factorise $12x^2 - 18xy$.

Algebraic factorisation (Chapter 13) 253

9 All of the regions in the diagram are rectangles. Write an expression for the area of each of the following regions, fully factorising your answers.
  a blue   **b** red   **c** total

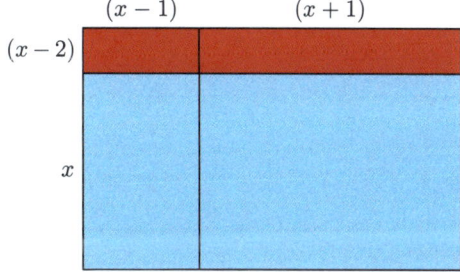

C DIFFERENCE BETWEEN TWO SQUARES

Investigation   The difference between two squares

In the diagram alongside, a square with side length b has been cut from a square with side length a.

 WORKSHEET

What to do:

1 Explain why the green shaded area is given by $a^2 - b^2$.

2 Copy the above diagram, or print the **worksheet** by clicking on the icon. Cut along the dotted line.

3 Rearrange the two trapezia to form a rectangle as shown.

4 Find, in terms of a and b, the lengths AB and BC.

5 Hence find the area of the rectangle in the form $(......)(......)$.

6 What can be deduced by comparing the areas in **1** and **5**?

Using the FOIL rule, $(a+b)(a-b) = a^2 - ab + ab - b^2$
$\qquad\qquad\qquad\qquad\qquad\;\; = a^2 - b^2$

We can factorise expressions of the form $a^2 - b^2$ by reversing this process:

$$a^2 - b^2 = (a+b)(a-b)$$

Example 9   ◀)) Self Tutor

Fully factorise:

a $x^2 - 4$ **b** $1 - 25y^2$

a $\quad x^2 - 4$
$\quad = x^2 - 2^2$
$\quad = (x+2)(x-2)$

b $\quad 1 - 25y^2$
$\quad = 1^2 - (5y)^2$
$\quad = (1+5y)(1-5y)$

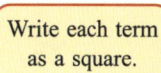
Write each term as a square.

EXERCISE 13C

1 Fully factorise:

- **a** $d^2 - e^2$
- **b** $p^2 - q^2$
- **c** $q^2 - p^2$
- **d** $x^2 - y^2$
- **e** $x^2 - 9$
- **f** $x^2 - 100$
- **g** $y^2 - 49$
- **h** $16x^2 - 9$
- **i** $4b^2 - 1$
- **j** $36y^2 - 49$
- **k** $81 - d^2$
- **l** $121 - 9y^2$

Example 10 ◀) Self Tutor

Always look for common factors first.

Fully factorise:
- **a** $2x^2 - 18$
- **b** $x^3 - xy^2$

a $\quad 2x^2 - 18$
$= 2(x^2 - 9)$
$= 2(x + 3)(x - 3)$

b $\quad x^3 - xy^2$
$= x(x^2 - y^2)$
$= x(x + y)(x - y)$

2 Fully factorise:

- **a** $3x^2 - 75$
- **b** $8x^2 - 32$
- **c** $5a^2 - 20$
- **d** $5x^2 - 5$
- **e** $9b^2 - 900$
- **f** $80 - 5t^2$
- **g** $6k^2 - 24$
- **h** $15 - 15y^2$
- **i** $r^3 - 49r$
- **j** $z^3 - z$
- **k** $x^5 - x^3$
- **l** $x^3y - xy^3$

Example 11 ◀) Self Tutor

Fully factorise:
- **a** $4a^2 - 9b^2$
- **b** $x^2y^2 - 16$

a $\quad 4a^2 - 9b^2$
$= (2a)^2 - (3b)^2$
$= (2a + 3b)(2a - 3b)$

b $\quad x^2y^2 - 16$
$= (xy)^2 - 4^2$
$= (xy + 4)(xy - 4)$

3 Fully factorise:

- **a** $25a^2 - b^2$
- **b** $x^2 - 81y^2$
- **c** $16x^2 - 49y^2$
- **d** $25p^2 - 36q^2$
- **e** $81j^2 - k^2$
- **f** $s^2t^2 - 9$
- **g** $49x^2 - y^2z^2$
- **h** $4d^2 - 25e^2$

Example 12 ◀) Self Tutor

Notice the use of the square brackets.

Fully factorise:
- **a** $(x + 2)^2 - 9$
- **b** $25 - (x - 2)^2$

a $\quad (x + 2)^2 - 9$
$= (x + 2)^2 - 3^2$
$= [(x + 2) + 3][(x + 2) - 3]$
$= [x + 2 + 3][x + 2 - 3]$
$= (x + 5)(x - 1)$

b $\quad 25 - (x - 2)^2$
$= 5^2 - (x - 2)^2$
$= [5 + (x - 2)][5 - (x - 2)]$
$= [5 + x - 2][5 - x + 2]$
$= (x + 3)(7 - x)$

4 Fully factorise:

    a $(x+2)^2 - 4$ **b** $(x-3)^2 - 36$ **c** $100 - (x-11)^2$

    d $81 - (x-8)^2$ **e** $(x+7)^2 - 1$ **f** $9 - (x+4)^2$

5   **a** Write 391 as the difference between two squares.

    b Hence show that 17 is a factor of 391.

6 Answer the **Opening Problem** on page **248**.

D   PERFECT SQUARE FACTORISATION

We have seen that

$$(a+b)^2 = (a+b)(a+b) = a^2 + ab + ab + b^2 = a^2 + 2ab + b^2$$

and

$$(a-b)^2 = (a-b)(a-b) = a^2 - ab - ab + b^2 = a^2 - 2ab + b^2$$

We can reverse this process to factorise expressions of the form $a^2 + 2ab + b^2$ or $a^2 - 2ab + b^2$.

$$a^2 + 2ab + b^2 = (a+b)^2$$
$$a^2 - 2ab + b^2 = (a-b)^2$$

*$a^2 + 2ab + b^2$ and $a^2 - 2ab + b^2$ are called **perfect squares** because they factorise into the product of two identical factors.*

IDENTIFYING PERFECT SQUARES

Notice that $(a+b)^2 = a^2 + 2ab + b^2$ and $(a-b)^2 = a^2 - 2ab + b^2$.
                         squares   squares

A perfect square must contain two squares a^2 and b^2, and a middle term which is $\pm 2ab$. The sign of the middle term indicates whether the perfect square is $(a+b)^2$ or $(a-b)^2$.

Example 13   ◀)) Self Tutor

Determine whether the following expressions are perfect squares:

    a $x^2 + 12x + 36$ **b** $x^2 - 4x + 9$

    a $x^2 + 12x + 36 = x^2 + 12x + 6^2$

        The expression contains two squares x^2 and 6^2, and the middle term $12x = 2 \times x \times 6$.

        $\therefore$ $x^2 + 12x + 36$ is a perfect square.

    b $x^2 - 4x + 9 = x^2 - 4x + 3^2$

        The expression contains two squares x^2 and 3^2, but the middle term $-4x \neq -2 \times x \times 3$.

        $\therefore$ $x^2 - 4x + 9$ is *not* a perfect square.

Example 14 Self Tutor

Factorise:

a $x^2 + 20x + 100$ **b** $x^2 - 8x + 16$

a $x^2 + 20x + 100$
$= x^2 + 2 \times x \times 10 + 10^2$
$= (x + 10)^2$

b $x^2 - 8x + 16$
$= x^2 - 2 \times x \times 4 + 4^2$
$= (x - 4)^2$

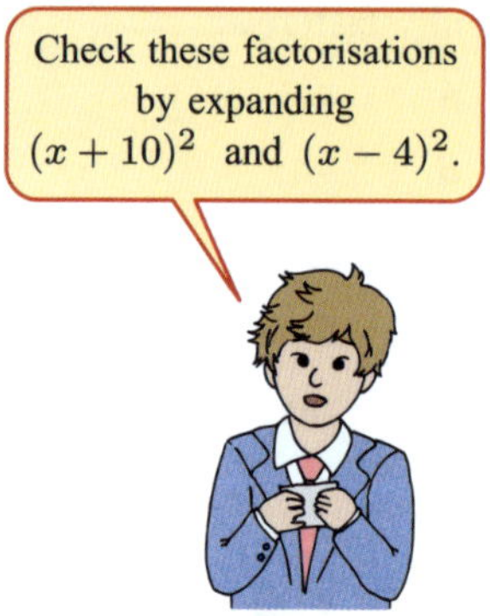

Check these factorisations by expanding $(x + 10)^2$ and $(x - 4)^2$.

EXERCISE 13D

1

a
b
A_1
A_2
A_3
A_4

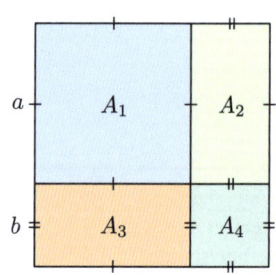

a Write expressions for the areas A_1, A_2, A_3, and A_4.

b Hence explain why $(a + b)^2 = a^2 + 2ab + b^2$.

2 Determine whether the following expressions are perfect squares:

a $x^2 + 16x + 64$ **b** $x^2 + 14x + 36$ **c** $a^2 - 6a + 9$
d $x^2 - 12x + 49$ **e** $t^2 - 2t + 1$ **f** $x^2 + 10x + 100$
g $1 - 2x + x^2$ **h** $4 + 8x + x^2$ **i** $9 - 12x + 4x^2$

3 Factorise:

a $x^2 - 2x + 1$ **b** $x^2 + 10x + 25$ **c** $x^2 + 8x + 16$
d $x^2 + 16x + 64$ **e** $x^2 - 14x + 49$ **f** $x^2 + 22x + 121$
g $x^2 - 20x + 100$ **h** $x^2 - 18x + 81$ **i** $x^2 - 8x + 16$
j $x^2 + 12x + 36$ **k** $x^2 - 6x + 9$ **l** $x^2 + 24x + 144$
m $1 + 2x + x^2$ **n** $4 - 4x + x^2$ **o** $9 + 6x + x^2$

Example 15 Self Tutor

Factorise:

a $16x^2 + 24x + 9$ **b** $4x^2 - 20x + 25$

a $16x^2 + 24x + 9$
$= (4x)^2 + 2 \times 4x \times 3 + 3^2$
$= (4x + 3)^2$

b $4x^2 - 20x + 25$
$= (2x)^2 - 2 \times 2x \times 5 + 5^2$
$= (2x - 5)^2$

4 Factorise:

a $4x^2 - 4x + 1$ **b** $25x^2 + 20x + 4$ **c** $64x^2 - 48x + 9$
d $9x^2 + 42x + 49$ **e** $36x^2 - 60x + 25$ **f** $16x^2 - 40x + 25$

5 Factorise:

a $2x + x^2 + 1$
b $x^2 + 9 + 6x$
c $-4x + 4 + x^2$
d $16 + x^2 - 8x$
e $1 + 4x^2 + 4x$
f $4y^2 + 9 - 12y$

6 Determine whether each expression is a perfect square, and factorise those which are:

a $x^2 + 18x + 81$
b $x^2 - 8x + 25$
c $4x^2 + 6x + 1$
d $9x^2 - 24x + 16$
e $49x^2 + 14x + 1$
f $x^2 - 4xy + 4y^2$

Example 16 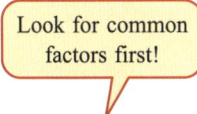 Self Tutor

Fully factorise:

a $3x^2 - 18x + 27$
b $-2x^2 + 8x - 8$

a $3x^2 - 18x + 27$
$= 3(x^2 - 6x + 9)$
$= 3(x^2 - 2 \times x \times 3 + 3^2)$
$= 3(x - 3)^2$

b $-2x^2 + 8x - 8$
$= -2(x^2 - 4x + 4)$
$= -2(x^2 - 2 \times x \times 2 + 2^2)$
$= -2(x - 2)^2$

Look for common factors first!

7 Fully factorise:

a $2x^2 + 12x + 18$
b $2x^2 - 4x + 2$
c $3x^2 + 12x + 12$
d $4x^2 - 32x + 64$
e $5x^2 + 50x + 125$
f $-x^2 + 12x - 36$
g $-x^2 - 16x - 64$
h $-2x^2 + 40x - 200$
i $-3x^2 + 30x - 75$

8 Evaluate $30^2 + 2 \times 30 \times 1 + 1^2$, and hence find the square root of 961.

E FACTORISING QUADRATIC TRINOMIALS

A **quadratic trinomial** is an algebraic expression of the form $ax^2 + bx + c$ where x is a variable and a, b, c are constants, $a \neq 0$.

We have previously used the FOIL rule to expand expressions of the form $(x + a)(x + b)$.

For example, $(x + 5)(x + 3) = x^2 + 3x + 5x + 15 = x^2 + 8x + 15$

    Firsts Outers Inners Lasts

We now consider the reverse process. If we are given the expression $x^2 + 8x + 15$, how can we deduce that it can be factorised as $(x + 5)(x + 3)$?

To do this, notice that the coefficient of x, which is 8, is the **sum** of 5 and 3. The constant term, which is 15, is the **product** of 5 and 3.

So, to factorise $x^2 + 8x + 15$, we need two numbers whose sum is 8 and whose product is 15. The numbers are 5 and 3, so $x^2 + 8x + 15 = (x + 5)(x + 3)$.

258 Algebraic factorisation (Chapter 13)

$$x^2 + px + q = (x+a)(x+b)$$
where a and b are two numbers whose sum is p, and whose product is q.

This process is sometimes called **sum and product** factorisation.

Example 17   ◀)) Self Tutor

Factorise: $x^2 + 11x + 24$

We need to find two numbers with sum 11 and product 24.
We list pairs of numbers with product 24, and find their sum.

Number pair	1, 24	2, 12	3, 8	4, 6
Sum	25	14	11	10

↑ this one

Most of the time we can find these two numbers mentally. We then do not need to show all of the working.

The numbers we want are 3 and 8.
$\therefore \; x^2 + 11x + 24 = (x+3)(x+8)$

If we try all of the number pairs with a particular product but none of the pairs have the correct sum, then we cannot use this method for factorisation.

EXERCISE 13E

1 Find two numbers which have:
   a sum 8 and product 7
   b sum 11 and product 30
   c sum 17 and product 60
   d sum 2 and product -8
   e sum -1 and product -20
   f sum -15 and product 56
   g sum -22 and product -48
   h sum -9 and product -10

2 Factorise:
   a $x^2 + 7x + 10$
   b $x^2 + 14x + 33$
   c $x^2 + 19x + 34$
   d $x^2 + 24x + 23$
   e $x^2 + 16x + 48$
   f $x^2 + 9x + 20$
   g $x^2 + 21x + 54$
   h $x^2 + 29x + 100$
   i $x^2 + 16x + 63$

Example 18   ◀)) Self Tutor

Factorise: $x^2 - 7x + 12$

sum $= -7$ and product $= 12$
$\therefore$ the numbers are -3 and -4
$\therefore \; x^2 - 7x + 12 = (x-3)(x-4)$

The sum is negative but the product is positive, so both numbers must be negative.

3 Factorise:
- **a** $x^2 - 9x + 20$
- **b** $x^2 - 18x + 56$
- **c** $x^2 - 13x + 42$
- **d** $x^2 - 11x + 28$
- **e** $x^2 - 3x + 2$
- **f** $x^2 - 16x + 63$
- **g** $x^2 - 20x + 96$
- **h** $x^2 - 17x + 30$
- **i** $x^2 - 8x + 15$

4 Explain why the following quadratic trinomials cannot be factorised using the "sum and product" method:
- **a** $x^2 + 3x - 2$
- **b** $x^2 + x + 2$

Example 19 ◀)) Self Tutor

Factorise: **a** $x^2 - 2x - 15$ **b** $x^2 + x - 6$

a sum $= -2$ and product $= -15$
∴ the numbers are -5 and $+3$
∴ $x^2 - 2x - 15 = (x - 5)(x + 3)$

b sum $= 1$ and product $= -6$
∴ the numbers are -2 and $+3$
∴ $x^2 + x - 6 = (x - 2)(x + 3)$

Since the product is negative, the numbers must be opposite in sign.

5 Factorise:
- **a** $x^2 - 5x - 6$
- **b** $x^2 - 7x - 18$
- **c** $x^2 + 11x - 80$
- **d** $x^2 + x - 72$
- **e** $x^2 + 8x - 33$
- **f** $x^2 - 4x - 77$
- **g** $x^2 + 16x - 57$
- **h** $x^2 - x - 90$
- **i** $x^2 - 17x - 84$
- **j** $x^2 - 3x - 10$
- **k** $x^2 + 4x - 45$
- **l** $x^2 + 6x - 72$

Example 20 ◀)) Self Tutor

Fully factorise by first removing a common factor: $3x^2 + 6x - 72$

$3x^2 + 6x - 72$ {look for a common factor}
$= 3(x^2 + 2x - 24)$ {sum $= 2$, product $= -24$
$= 3(x + 6)(x - 4)$ ∴ the numbers are 6 and -4}

6 Fully factorise by first removing a common factor:
- **a** $2x^2 + 6x + 4$
- **b** $4x^2 + 28x + 40$
- **c** $3x^2 + 12x + 9$
- **d** $4x^2 - 8x - 12$
- **e** $9x^2 + 27x - 36$
- **f** $5x^2 - 20x + 15$
- **g** $7x^2 - 28x - 35$
- **h** $5x^2 - 45x - 110$
- **i** $6x^2 - 72x + 120$
- **j** $3x^2 - 15x - 18$
- **k** $2x^2 - 14x - 36$
- **l** $10x^2 - 10x - 200$

F MISCELLANEOUS FACTORISATION

In the following Exercise you will need to determine which factorisation method to use.
The flowchart below may be useful:

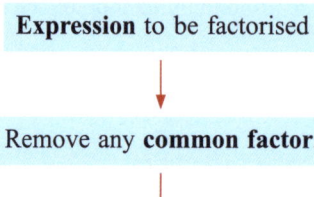

Look for:
- difference between two squares
  $a^2 - b^2 = (a+b)(a-b)$
- perfect squares
  $a^2 + 2ab + b^2 = (a+b)^2$
  $a^2 - 2ab + b^2 = (a-b)^2$
- sum and product type
  $x^2 + px + q = (x+a)(x+b)$
  where $a+b = p$ and $ab = q$.

EXERCISE 13F

1 Fully factorise:

- **a** $4a^2 + 8a$
- **b** $3b^2 + 15$
- **c** $6x - 36y$
- **d** $p^2 + 12p + 35$
- **e** $x^2 - 17x - 18$
- **f** $x^2 - 3x - 4$
- **g** $-x^2 - 49x$
- **h** $h^2 + 2h^3$
- **i** $st^2 - 2st$
- **j** $-2x^2 - 18$
- **k** $r^2 + r - 42$
- **l** $20x - 5x^3$
- **m** $3z - 21yz$
- **n** $a^2 + 16ab$
- **o** $y^2 - 7y + 12$
- **p** $9x^2 - 9x - 18$
- **q** $6y^2 - 36y + 54$
- **r** $x^3 + 4x^2 + 4x$

2 Fully factorise:

- **a** $x^2 - 10x + 25$
- **b** $x^2 - 169$
- **c** $4c^2 - 16$
- **d** $10 - 10y^2$
- **e** $5x^2 - 125$
- **f** $12f^2 + 24f + 12$
- **g** $81y^2 - 49x^2$
- **h** $d^2 - 11d + 24$
- **i** $2x^2 - 32x + 128$

3 Fully factorise:

- **a** $pq^2 - p^2q - pq$
- **b** $4b^3 - 4b^2$
- **c** $c^2d^2 - 8cd$
- **d** $16y - 9y^3$
- **e** $(x+1)^2 - 2(x+1)$
- **f** $e^2f - fg^2$
- **g** $2(x-5) - x(x-5)$
- **h** $9(x+y) - y(x+y)$
- **i** $s(s+t) - t(s+t)$
- **j** $k^2 - 100k$
- **k** $6x^2 + 30x - 84$
- **l** $49mn^2 - m^3$

LEARNING ALGEBRA

Review set 13A

1. Find the HCF of:
   a. $3a^2b$ and $6ab$
   b. $3(x+1)$ and $6(x+1)^2$

2. Fully factorise:
   a. $x^2 - 3x$
   b. $3mn + 6n^2$
   c. $ax^3 + 2ax^2$

3. Fully factorise:
   a. $-2x^2 - 32x$
   b. $d(t+2) - 4(t+2)$
   c. $(x-1)^2 - (x-1)$
   d. $2x(x+3) - 5(x+3)$
   e. $3(g+1)^2 - 9(g+1)$
   f. $b(b-c) - c(b-c)$

4. Determine whether the following are perfect squares:
   a. $x^2 - 4x + 9$
   b. $16x^2 + 24x + 9$

5. Fully factorise:
   a. $x^2 - 25$
   b. $100 - k^2$
   c. $9 - 16x^2$
   d. $y^2 - x^2$
   e. $9a^2 - 4b^2$
   f. $6x^2 - 24$

6. Fully factorise:
   a. $x^2 + 4x + 4$
   b. $x^2 - 10x + 25$
   c. $x^2 + 14x + 49$

7. Fully factorise:
   a. $4x^2 + 20x + 25$
   b. $9x^2 - 6x + 1$
   c. $5x^2 - 20x + 20$

8. Fully factorise:
   a. $x^2 + 10x + 21$
   b. $x^2 + 4x - 21$
   c. $x^2 - 4x - 21$
   d. $6 - 5x + x^2$
   e. $4x^2 + 8x - 12$
   f. $x^2 + 13x + 36$
   g. $20 + 9x + x^2$
   h. $2x^2 - 2x - 60$
   i. $3x^2 - 30x + 48$

Review set 13B

1. Find the HCF of:
   a. $6y^2$ and $8y$
   b. $4(x-2)$ and $2(x-2)(x+3)$

2. Fully factorise:
   a. $2x^2 + 6x$
   b. $-2xy - 4x$
   c. $(x+1)(x+3) - 2(x+3)$

3. Fully factorise:
   a. $xy^3 - 16xy$
   b. $3x^2 - 60x + 300$
   c. $p(a+2) - q(a+2)$
   d. $4cd^2 - 6c^2d$
   e. $(k-3) + (k-3)^2$
   f. $x(x-1) + 3x$

4. Explain why $x^2 - 2x + 4$ cannot be factorised by the "sum and product" method.

5. Fully factorise:
   a. $25 - x^2$
   b. $4a^2 - 9b^2$
   c. $49 - 9z^2$

6. Fully factorise:
   a. $x^2 - 12x + 36$
   b. $n^2 - 6n + 9$
   c. $16 + 8x + x^2$
   d. $-6x + 9 + x^2$
   e. $25 + x^2 - 10x$
   f. $x^2 + 35 - 12x$

7 Fully factorise:
   a $2x^2 + 4x + 2$ **b** $x^3 - 16x$ **c** $-12x + 18x^2 + 2$

8 Fully factorise:
   a $x^2 + 12x + 35$ **b** $x^2 + 2x - 35$ **c** $x^2 - 12x + 35$
   d $2x^2 - 4x - 70$ **e** $30 - 11x + x^2$ **f** $x^2 - 8x - 20$
   g $x^2 - 14x + 33$ **h** $4x^2 - 36x - 88$ **i** $2x^2 + 10x - 72$

14

Tables, graphs, and charts

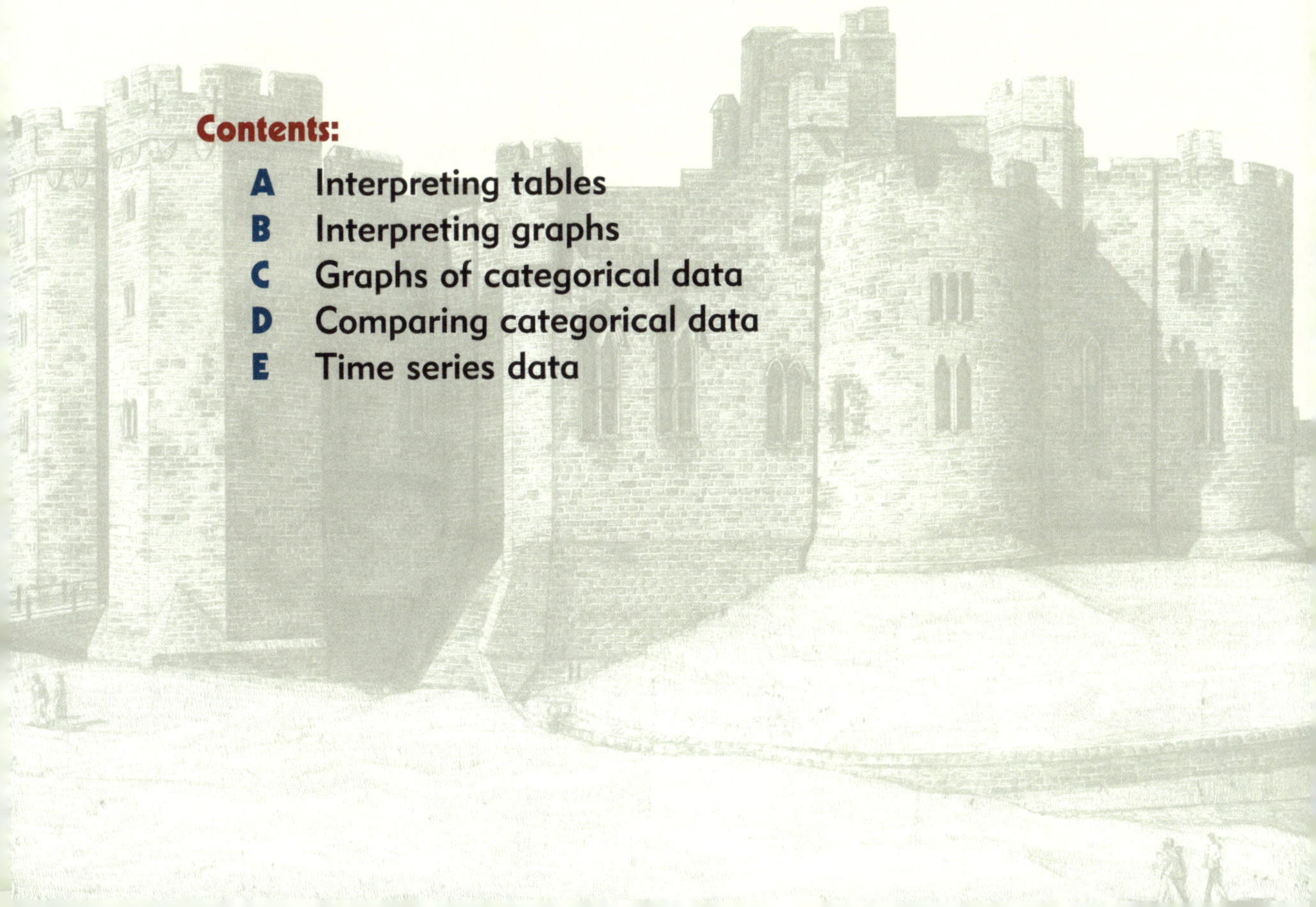

Contents:
- **A** Interpreting tables
- **B** Interpreting graphs
- **C** Graphs of categorical data
- **D** Comparing categorical data
- **E** Time series data

Opening problem

Petrice is a potter who lives in Cheshire. She sells her ceramics on the internet. The multiple bar chart shows her sales over a 4 year period, both within the United Kingdom and internationally.

Things to think about:

a What is the *trend* of the sales within the United Kingdom?

b How would you compare the international sales with those within the United Kingdom?

c Petrice's work was featured in a European ceramics magazine. Which year do you think this was? Explain your answer.

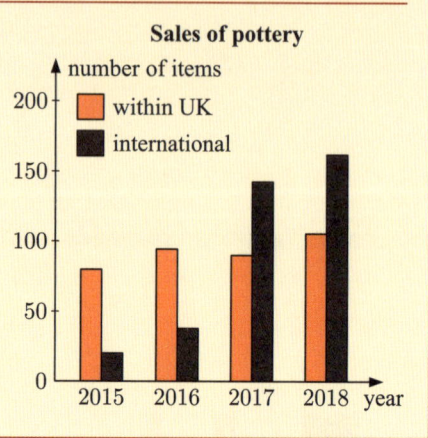

When we are presented with **tables** and **graphs** containing data, it is important that we interpret and analyse the data correctly. We need to understand the **variables** presented in the table or graph, and what they mean in the real world.

TYPES OF GRAPHS

BAR CHART

A **bar chart** is a popular method of displaying statistical data. The information may be displayed either vertically or horizontally. The height (if vertical) or length (if horizontal) of each bar indicates the frequency of the category it represents. All bars have the same width.

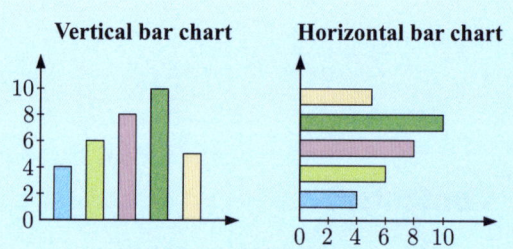

PICTOGRAM

A **pictogram** is set out in the same way as a bar chart, but pictures are used instead of bars to represent the quantities involved.

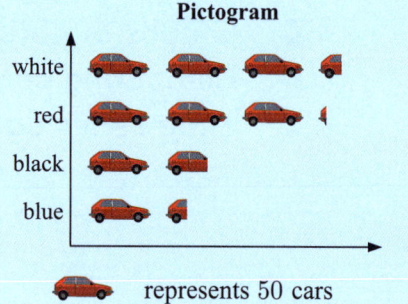

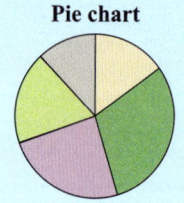 represents 50 cars

PIE CHART

In a **pie chart**, a circle is divided into sectors which represent the categories. The size of each sector is proportional to the frequency of the category it represents, so its sector angle can be found as a fraction of 360°.

LINE GRAPH

A **line graph** consists of a curve, or a series of straight line segments, which shows the relationship between two quantities.

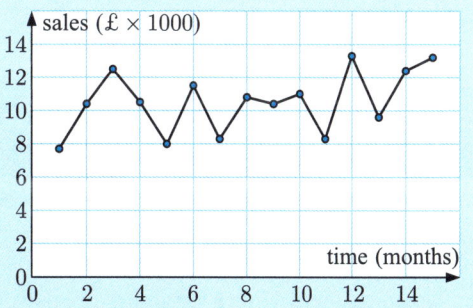

A INTERPRETING TABLES

Tables are often used to display large amounts of data. When answering questions about tables, we need to be able to identify the data which is relevant to the question. We locate the data by using the correct **row** and **column** of the table.

Example 1   ◀)) Self Tutor

The timetable below shows the Eurostar trains from London to Brussels.

a What time does the 12:58 train from London arrive in Brussels?
b On which days of the week does train 9110 run?
c On Thursday, how many trains stop at Calais?
d How long is the journey of train 9116 expected to take on Saturday?

Mon	Tue	Wed	Thur	Fri	Sat	Sun	LONDON St Pancras Intl	EBBSFLEET International	ASHFORD International	CALAIS Fréthun	LILLE Europe	BRUSSELS Midi/Zuid	Train no.
✓	-	-	-	✓	-	-	06:13	06:30	06:52	-	-	09:22	9108
-	✓	✓	✓	-	-	-	06:47	07:04	07:25	08:59	09:30	10:07	9110
-	-	-	-	-	✓	-	06:57	-	07:28	-	09:26	10:05	9110
✓	-	-	-	✓	-	-	08:04	-	-	-	10:26	11:05	9114
-	✓	✓	✓	-	-	-	08:54	09:14	-	-	11:23	12:02	9116
✓	-	-	-	✓	-	-	08:54	09:14	-	10:56	11:27	12:05	9116
-	-	-	-	-	✓	✓	08:58	09:15	-	10:59	11:30	12:08	9116
✓	-	✓	✓	✓	✓	-	10:58	11:15	-	-	13:26	14:05	9126
-	-	-	-	-	-	✓	11:04	-	-	-	13:26	14:05	9126
✓	✓	✓	✓	✓	✓	✓	12:58	13:15	-	14:59	15:30	16:08	9132
✓	✓	✓	✓	-	-	✓	15:04	-	-	-	17:26	18:05	9140
-	-	-	-	-	✓	-	16:04	-	-	-	18:26	19:05	9144
✓	✓	✓	✓	✓	-	✓	17:04	-	-	-	19:26	20:05	9148
-	-	-	-	-	-	✓	17:55	-	-	18:28	20:26	21:05	9152
-	-	-	✓	✓	-	-	18:04	-	-	-	20:26	21:05	9152
-	-	-	-	-	✓	✓	19:04	-	-	20:59	21:30	22:08	9156
✓	✓	✓	✓	✓	-	-	19:34	-	-	-	21:29	22:00	22:38	9158
-	-	-	-	-	-	✓	20:03	-	-	-	22:26	23:05	9162

a The 12:58 train from London arrives in Brussels at 16:08.
b The 9110 runs on Tuesday, Wednesday, Thursday, and Saturday.

c On Thursday, 3 trains (9110, 9132, and 9158) stop at Calais.

d On Saturday, the 9116 departs London at 08:58 and arrives in Brussels at 12:08.

$08:58 + 2$ minutes $= 09:00$

$09:00 + 3$ hours $\ = 12:00$

$12:00 + 8$ minutes $= 12:08$

$\therefore$ the journey is expected to take 3 hours 10 minutes.

EXERCISE 14A

1 Use the timetable in **Example 1** to answer the following questions:
   a What time does the Wednesday 17:04 train from London arrive in Brussels?
   b Which train arrives in Brussels at 12:02 on Tuesday?
   c On which days does train 9110 stop at Ebbsfleet?
   d How long does the 9126 train take to reach Brussels on Sunday?
   e How many minutes earlier does the 9152 arrive at Brussels on Thursday and Friday than on Sunday?
   f Jeremy arrives at London St Pancras at 06:20 Saturday and boards the next train for Lille. At what time should he arrive?

2 The table below indicates the days that certain places open during the end of year holiday period:

	Dec 24	Dec 25	Dec 26	Dec 27	Dec 28	Dec 29	Dec 30	Dec 31	Jan 1	Jan 2
Banks	✓	✗	✗	✗	✓	✓	✓	✓	✗	✗
Supermarkets	✓	✗	✗	✓	✓	✓	✓	✓	✗	✓
Libraries	✓	✗	✗	✗	✗	✗	✗	✗	✗	✓
Department stores	✓	✗	✓	✓	✓	✓	✓	✓	✗	✓

   a Are the libraries open on December 29?
   b Are the department stores open on December 26?
   c On how many days during this period are the banks open?
   d On which days are: **i** all of the places open **ii** none of the places open?

3 A furniture store offers a delivery service for its products. The fee depends on the weight of the furniture to be delivered, and the distance the furniture must be transported. The delivery costs are shown in the table below.

50 - 100 kg includes 50 kg but not 100 kg.

		Distance		
		0 - 20 km	20 - 40 km	40 - 60 km
Weight	0 - 50 kg	£19	£34	£50
	50 - 100 kg	£35	£62	£85
	100+ kg	£50	£90	£105

   a Suzette bought a cupboard weighing 60 kg. She lives 12 km from the furniture store. How much will it cost Suzette to have the cupboard delivered to her home?
   b Jason lives 25 km from the furniture store. He bought a chair weighing 20 kg, a table weighing 35 kg, and a desk weighing 50 kg.
   i Find the total weight of the furniture Jason bought.
   ii Find the cost of delivering these items.

c Lillian lives 47 km from the furniture store. She bought a 70 kg chest of drawers, and had it delivered to her home. A month later, she bought a 45 kg bookshelf, which was also delivered to her home.
  i Find the total cost of the two deliveries.
  ii How much would Lillian have saved if the items had been delivered together?

4 The table below shows the number of internet users per 100 inhabitants in various countries from 2007 to 2016.

	2007	2008	2009	2010	2011	2012	2013	2014	2015	2016
Indonesia	5.79	7.92	6.92	10.92	12.28	14.52	14.94	17.14	21.98	25.37
Japan	74.30	75.40	78.00	78.21	79.05	79.50	88.22	89.11	91.06	92.00
South Korea	78.80	81.00	81.60	83.70	83.76	84.07	84.77	87.56	89.65	92.72
New Zealand	69.76	72.03	79.70	80.46	81.23	81.64	82.78	85.50	88.22	88.47
Philippines	5.97	6.22	9.00	25.00	29.00	36.24	48.10	49.60	53.70	55.50
Russia	24.66	26.83	29.00	43.00	49.00	63.80	67.97	70.52	73.41	76.41
Singapore	69.90	69.00	69.00	71.00	71.00	72.00	80.90	79.03	79.01	81.00
Thailand	20.03	18.20	20.10	22.40	23.67	26.46	28.94	34.89	39.32	47.51
United States of America	75.00	74.00	71.00	71.69	69.73	74.70	71.40	73.00	74.55	76.18
Vietnam	20.76	23.92	26.55	30.65	35.07	36.80	38.50	41.00	43.50	46.50

a How many inhabitants out of 100 used the internet in Singapore in:
  i 2009   **ii** 2014?
b In which years did more than 40 out of 100 Russian inhabitants use the internet?
c Which of the listed countries had the most internet users per 100 inhabitants in:
  i 2011   **ii** 2015?
d Which listed country had the largest increase of internet users per 100 inhabitants from 2007 to 2016?

5 A group of hotel guests were asked to rate the hotel's performance in a number of categories. The results are presented in the table below:

	Excellent	Good	Fair	Poor
Cleanliness	20%	31%	35%	14%
Staff	32%	27%	15%	26%
Restaurant	9%	19%	37%	35%
Value for money	12%	16%	28%	44%
Location	33%	39%	18%	10%
Facilities	27%	34%	26%	13%

a What percentage of people rated the hotel's cleanliness as "Fair"?
b What percentage of people rated the hotel's staff as "Good" or "Excellent"?
c What was the most common response when rating the hotel's value for money?
d 130 people rated the hotel's facilities as "Fair". How many hotel guests were surveyed?

6 The table below shows the distances in miles between some of the places along the Wales Coast Path in the north of Wales.

	Flint	Mostyn	Talacre	Rhyl	Pensarn	Colwyn Bay	Conwy	Llanfairfechan
Flint		9	13	22	26	32	47	55
Mostyn	9		5	13	17	23	38	46
Talacre	13	5		8	13	19	33	42
Rhyl	22	13	8		5	11	25	34
Pensarn	26	17	13	5		6	21	29
Colwyn Bay	32	23	19	11	6		15	23
Conwy	47	38	33	25	21	15		8
Llanfairfechan	55	46	42	34	29	23	8	

a Why are some of the boxes shaded in?
b Find the distance between:
  i Mostyn and Colwyn Bay
  ii Rhyl and Conwy
  iii Llanfairfechan and Flint
  iv Talacre and Pensarn.
c Describe how the places are ordered.
d Redraw the table with places in alphabetical order.

7 The table below shows some of the 2011/12 student numbers for higher education qualifications in the United Kingdom, as well as the total percentage changes in student numbers from 2006/07.

Subject area	Female	Male	Total	Total % change from 2006/07
Medicine & dentistry	10 650	7555	18 205	30.4
Subjects allied to medicine	68 470	17 280	85 750	1.6
Biological sciences	34 450	20 970	55 420	28.9
Veterinary science	875	255	1130	35.1
Agriculture & related subjects	3635	2255	5890	23.9
Physical sciences	11 110	15 100	26 210	23.6
Mathematical sciences	4720	6765	11 485	30.8
Computer science	5750	24 765	30 515	−2.4
Engineering & technology	8595	42 085	50 680	31.2
Architecture, building & planning	7340	14 405	21 745	28.4
Social studies	46 255	27 485	73 740	22.1
Law	19 585	13 480	33 065	9
Business & administrative studies	69 655	70 370	140 025	43.3
Languages	25 345	11 495	36 840	17.2
Historical & philosophical studies	15 000	13 170	28 170	11.6
Creative arts & design	37 755	23 535	61 290	31.5
Education	61 430	18 915	80 345	11.2

a How many female students studied Law in 2011/12?
b Did more males study Computer science or Education?

c Which subject listed had the **i** highest **ii** lowest total number of students?

d What percentage of Biological sciences students were female?

e How many of these subject areas increased their student intake from 2006/07?

f Estimate, to three significant figures, the total number of students studying Social studies in 2006/07.

B INTERPRETING GRAPHS

Many newspapers and magazines use graphs to display information. Graphs are used to make the information more visually appealing and easier to understand.

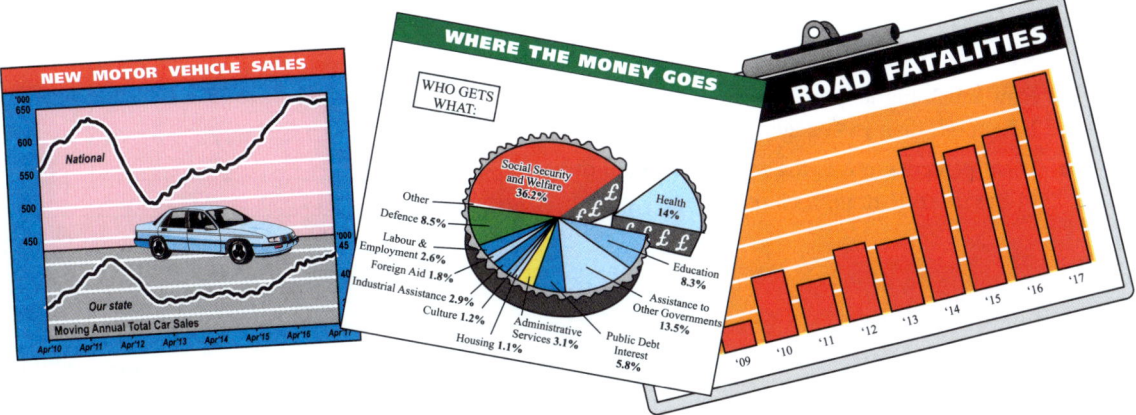

Example 2   ◀)) Self Tutor

This bar chart shows the number of different items sold in an electrical goods store in a given month.

a How many computers were sold for the month?

b How many more television sets were sold than DVD/Blu-ray players?

c What percentage of the items sold were tablets?

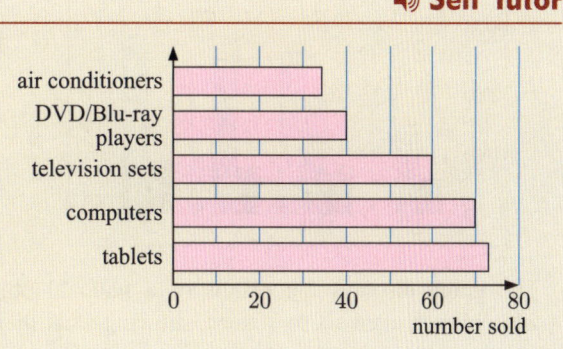

a 70 computers were sold.

b 60 television sets were sold. 40 DVD/Blu-ray players were sold.
∴ 20 more television sets than DVD/Blu-ray players were sold.

c 73 tablets were sold.
∴ the total number of items sold was $34 + 40 + 60 + 70 + 73 = 277$
∴ $\frac{73}{277} \times 100\% \approx 26.4\%$ of the items sold were tablets.

EXERCISE 14B

1. The graph shows the attendances at various Friday night events at an Arts festival.
   a. Which event was most popular?
   b. How many more people attended the Drama than attended the Modern Dance?
   c. What percentage of people went to see the Jazz group?

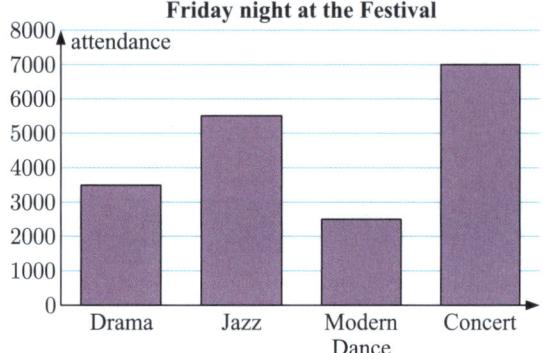

2. a. Estimate the mean house price in 2017 in:
      i. Liverpool   ii. Edinburgh.
   b. How much higher was the mean house price in Edinburgh than in Glasgow?
   c. What percentage was the mean Liverpool price of the mean Birmingham price?

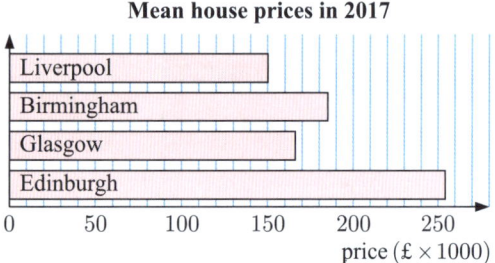

3.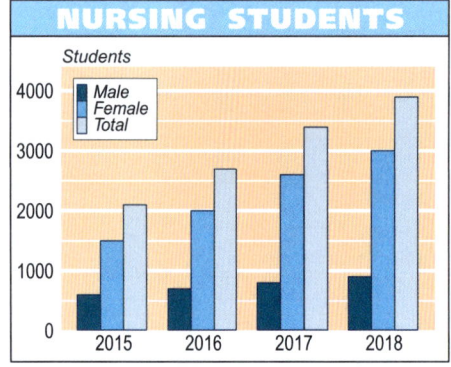

   The graph shows the number of students studying nursing in London over a period of four years.
   a. Estimate the total number of nursing students in 2016.
   b. Estimate the percentage of the total nursing students in 2016 who were male.
   c. In which year was the proportion of male students the highest?

4. The graph displays the results of a national poll of 240 children aged six to sixteen, showing what they do at home between 3 pm and 6 pm.
   a. Estimate the percentage of children who:
      i. watch TV, use the internet, or listen to music
      ii. spend time with friends, siblings, or pets.
   b. Of the 240 children surveyed in the poll:
      i. how many do schoolwork
      ii. how many more spend time on the phone than do chores?

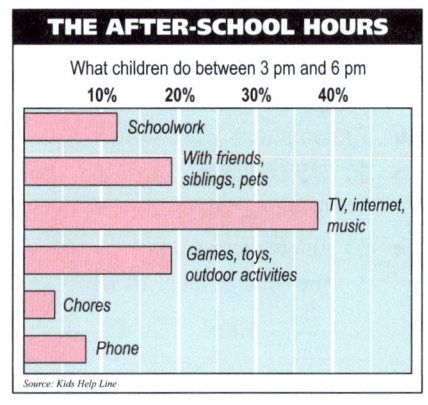

5 This pictogram shows the number of flights leaving an airport each day for a week.

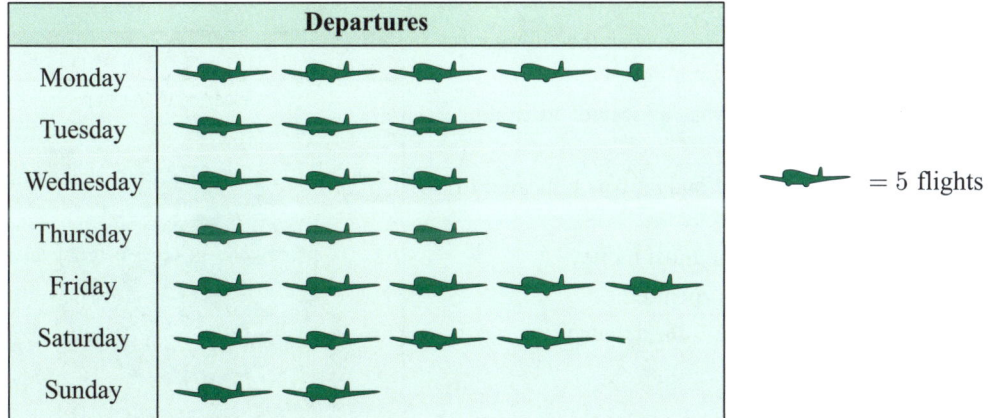

  a How many flights left on Thursday?
  b What day shows the greatest number of departures?
  c How many flights left over the weekend?

6 Ian counted the vehicles that passed as he waited for his bus. Each icon on the pictogram represents two vehicles.

  a How many:
    i trucks
    ii bicycles went past?
  b Was the number of cars greater than the sum of all the other vehicles?

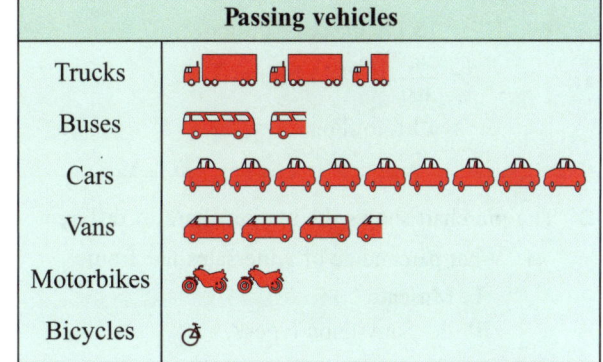

7 The graph alongside indicates the number of ticketed arts performances in Manchester from 2007 to 2017.

  a Find the number of ticketed arts performances in 2007.
  b Find the increase in ticketed arts performances from 2016 to 2017.
  c Which year showed the greatest increase in performances from the previous year?

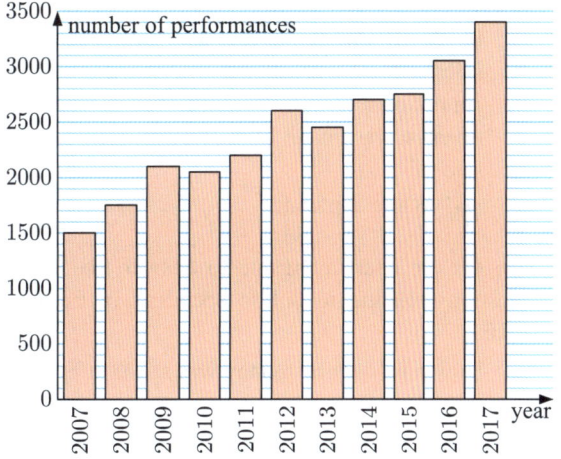

Example 3

Self Tutor

Use the pie chart alongside to answer the following questions regarding public spending.

a How much money was allocated to public spending in 2018?

b What percentage of the budget was allocated to education?

c Which area received the most money?

d How much money was spent on:

i the NHS **ii** defence?

Source: www.ukpublicspending.co.uk

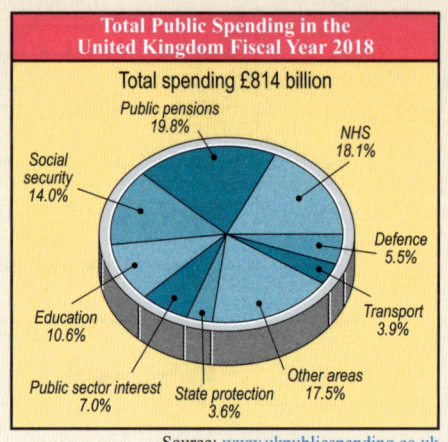

a £814 billion was allocated to public spending.

b 10.6% of the budget was allocated to education.

c Public pensions received the most money, accounting for 19.8% of the spending.

d **i** 18.1% of £814 billion
$= \frac{18.1}{100} \times$ £814 billion
$\approx$ £147 billion

ii 5.5% of £814 billion
$= \frac{5.5}{100} \times$ £814 billion
$\approx$ £44.8 billion

8 The pie chart shows the market share of different wines in the United States of America.

a What percentage of wine sales are from:

i Muscato

ii the Sauvignon types?

b What type of wine has the highest market share?

c In 2010, the retail revenue from wine sales in the United States of America was approximately \$30 billion. How much revenue was made in that year from the sale of:

i White Zinfandel **ii** Merlot?

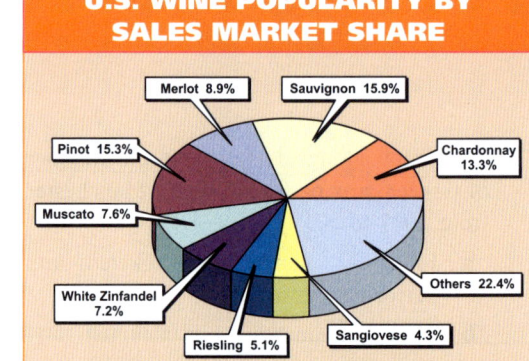

9 Use the pie chart alongside to answer the following questions relating to public spending in England in 2016.

a How much public money was spent in England in 2016?

b What percentage of public money was spent on welfare?

c How much money was spent on:

i health care **ii** education?

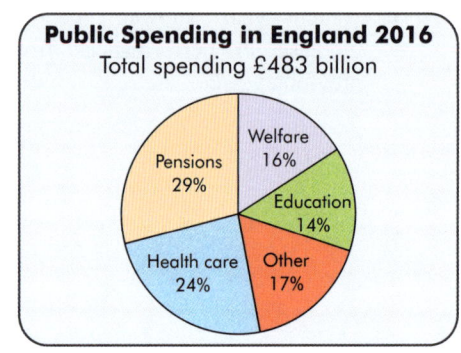

10 On a spring day in Brighton, Ebony used a thermometer to measure the temperature in her bedroom. Her results are shown in the line graph below:

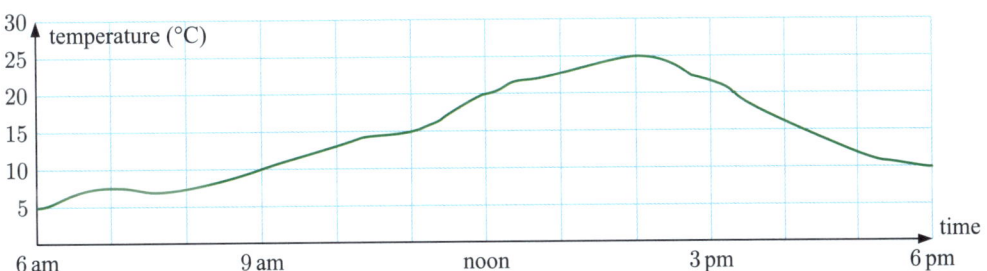

a Find the temperature at:
  i 9 am   **ii** noon.
b **i** What was the maximum temperature during the day?
  ii At what time did this maximum temperature occur?
c Was the temperature increasing or decreasing at 4 pm? Explain your answer.
d By how much did the temperature vary during the 12-hour period.

11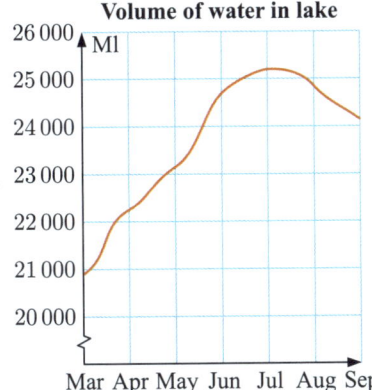

a What information is contained in the line graph alongside?
b How much water was in the lake at the start of:
  i April   **ii** August?
c Over which month did the water volume:
  i increase the most
  ii decrease the most?
d What was the maximum water volume during this period?
e What was the percentage increase in water volume from the start of March to the start of June?

C GRAPHS OF CATEGORICAL DATA

Categorical data is data which can be placed in categories.

For example, suppose the students in Alan's class are asked to name their favourite subject. The data collected is categorical data. The possible categories may include Mathematics, Art, Science, Music, and English.

We organise categorical data using a **tally and frequency table**, and display the data using a bar chart, pie chart, or pictogram.

Example 4

The students in Alan's class were asked their favourite subject. The responses were:

Art, Music, Mathematics, Music, English, Mathematics, Science, Art,
Music, Science, Music, Mathematics, Art, Music, Mathematics, Music

a Construct a tally and frequency table to organise the data.
b Draw a vertical bar chart to display the data.

a

Favourite subject	Tally	Frequency
Mathematics	||||	4
Art	|||	3
Science	||	2
Music	⦀|	6
English	|	1
Total		16

b Vertical bar chart showing frequency of favourite subjects: Mathematics 4, Art 3, Science 2, Music 6, English 1.

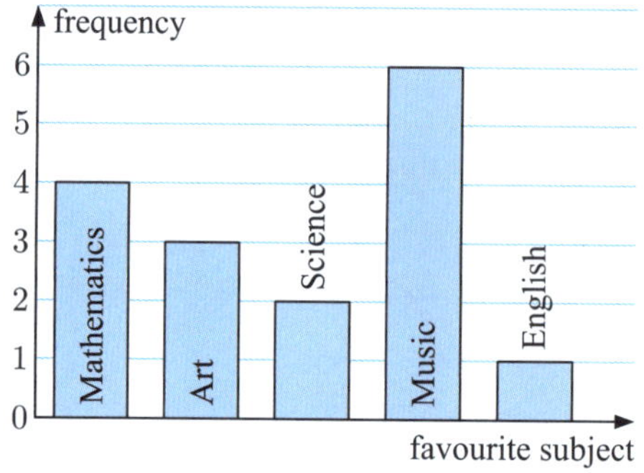

EXERCISE 14C

1 Students in a science class obtained the following levels of achievement:

D C C A A C D C B C C D
B C C C C E B A C C B C B C

a Draw a tally and frequency table to organise the data.
b How many students obtained a C?
c What fraction of students obtained a B?
d Draw a horizontal bar chart to display the data.

2 People visiting the local show were asked whether they preferred the side shows (S), the farm animals (F), the ring events (R), the dogs and cats (D), or the wood chopping (W).

The results were: S R W S S   W F D D S   R S F W S   R S R W S   S R R R F

  a Draw a tally and frequency table for the data.

  b What percentage of people preferred the wood chopping?

  c Draw a vertical bar chart to display the data.

3 The 20 players in a football team voted to decide who should be their captain. The results are given in the table alongside.

  a Draw a horizontal bar chart to display the data.

  b Which candidate received the most votes?

  c What percentage of the team voted for:

    i Luke   **ii** Greg or Steve?

Candidate	Votes
Cameron	3
Greg	7
Luke	4
Steve	6

4 At a school camp, the students selected their favourite ice cream flavour out of chocolate (C), strawberry (S), vanilla (V), and lime (L).

The results were:

  C V C S S   V L S C V   C V S L V   S C C V V   C S L C V
  V C L S C   C C V L S   S L V C V   C L C S C   L C V L C

The most common response is called the *mode*.

  a Organise this data into a tally and frequency table.

  b How many students chose vanilla?

  c What percentage of the students chose lime?

  d Find the most common response.

  e Draw a vertical bar chart to display the data.

Example 5   🔊 Self Tutor

The table shows the results when the Year 10 students at a school were asked "What is your favourite fruit?"

Construct a pie chart to display this data.

Fruit	Frequency
Orange	13
Apple	21
Banana	10
Pineapple	7
Pear	9
Total	60

There are 60 students in the sample, so each student is represented by $\frac{1}{60}$th of $360°$ or $6°$ on the pie chart.

We calculate the sector angles:

  $13 \times 6° = 78°$ for Orange
  $21 \times 6° = 126°$ for Apple
  $10 \times 6° = 60°$ for Banana
  $7 \times 6° = 42°$ for Pineapple
  $9 \times 6° = 54°$ for Pear.

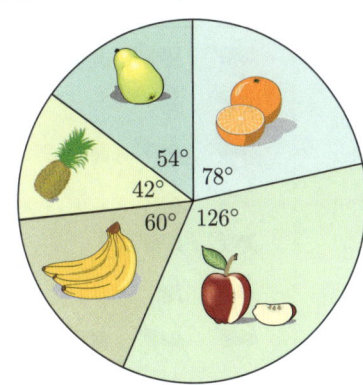

5 A survey of eye colour in a group of 30 teenagers revealed these results:

Eye colour	Blue	Brown	Green	Grey
Number of students	9	12	2	7

   a Illustrate the results on a pie chart.

   b What percentage of the group have: **i** green eyes   **ii** blue or grey eyes?

6 This bar chart shows the different types of traffic fines handed out by a police officer over one week.

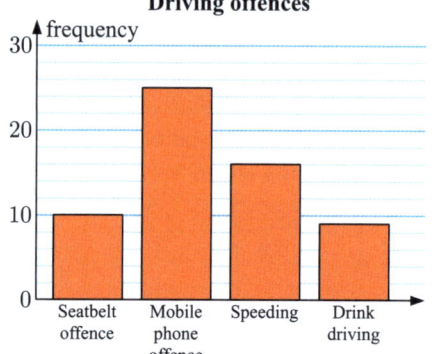

   a How many fines did the police officer hand out in total?

   b Draw a pie chart to illustrate this data.

   c Determine whether each statement is true or false:

      i The most common fine was for drink driving.

      ii Fines for speeding were about one quarter of all fines.

      iii More than half of the fines were either for seatbelt or mobile phone offences.

7 90 people were asked their opinions of a new sports stadium. The table alongside gives some of the results.

Given that "Fair" would occupy 100° of a pie chart of the responses:

Response	Frequency
Excellent	8
Good	30
Fair	
Poor	

   a Complete the table.

   b Draw a pie chart to display the data.

Example 6   ◀ㅇ Self Tutor

This table shows the number of dogs competing in each category of a show.

Draw a pictogram to represent the data.

Category	Number of dogs
Terrier	30
Toy	10
Working	13
Sporting	22
Hound	11
Non-sporting	8
Herding	19

Dog show competitors

Terrier	🐕 🐕 🐕 🐕 🐕 🐕
Toy	🐕 🐕
Working	🐕 🐕 🐕
Sporting	🐕 🐕 🐕 🐕 🐕
Hound	🐕 🐕 🐕
Non-sporting	🐕 🐕
Herding	🐕 🐕 🐕 🐕

🐕 = 5 dogs

8 Jessica recorded the activities of people using a path for one hour on Sunday morning.

Draw a pictogram to represent the results, using

👤 to represent 4 people.

Activity	Number of people
Walking	36
Jogging	19
Cycling	26
Rollerblading	13

9 This table shows the types of cards sold by a stationery store in one day.

Given that the store sold 64 cards in total:

a Find the number of "get well" cards sold.

b Draw a pictogram to represent the data, using

▭ to represent 5 cards.

c What percentage of cards sold were wedding cards?

Type of card	Number sold
Anniversary	8
Birthday	27
Get Well	
Thank You	10
Wedding	12

D COMPARING CATEGORICAL DATA

To understand the significance of the results we collect, we often need to compare two data sets.

For example, in **Example 4** we studied the favourite subjects of students in Alan's class. We now also consider the students in Bill's class, whose favourite subjects are shown in the table alongside.

Favourite subject	Frequency
Mathematics	3
Art	5
Science	2
Music	2
English	4
Total	16

MULTIPLE BAR CHART

To compare the results from Alan's class and Bill's class, we can draw a bar chart for each data set on the same axes. This is known as a **multiple bar chart**. A different colour is used for each data set, and a legend is included so we can see clearly which data set is which.

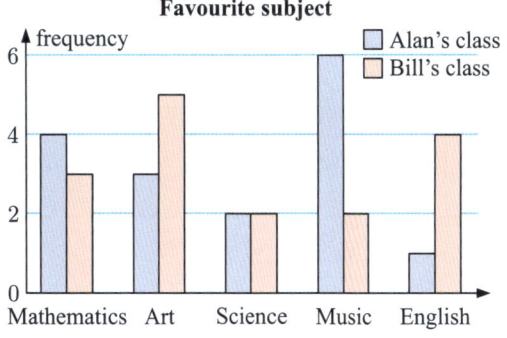

COMPOSITE BAR CHART

Alternatively, we can place the results for Bill's class on top of those for Alan's class, so that each subject is represented by a single bar split into segments. This is known as a **composite bar chart**.

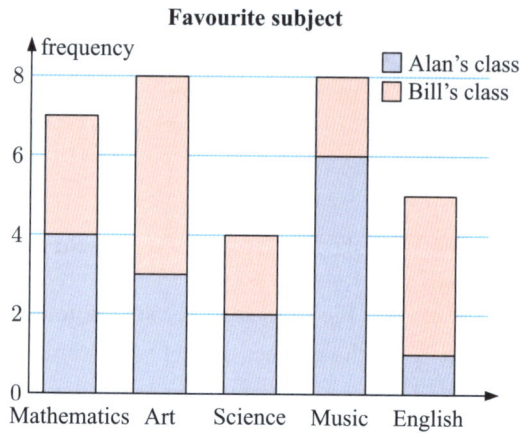

Example 7

This composite bar chart shows the number of girls and boys visiting a playground on Friday, Saturday, and Sunday.

a How many:
  i girls visited the playground on Friday
  ii boys visited the playground on Saturday
  iii children visited the playground on Sunday?

b On what day was the percentage of girls visiting the playground the highest?

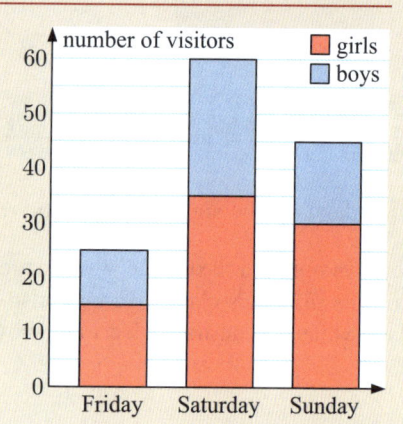

a **i** 15 girls visited the playground on Friday.
  ii $60 - 35 = 25$ boys visited the playground on Saturday.
  iii 45 children visited the playground on Sunday.

b Percentage of girls visiting on Friday $= \dfrac{15}{25} \times 100\% = 60\%$

Percentage of girls visiting on Saturday $= \dfrac{35}{60} \times 100\% \approx 58.3\%$

Percentage of girls visiting on Sunday $= \dfrac{30}{45} \times 100\% \approx 66.7\%$

The percentage of girls visiting the playground was highest on Sunday.

EXERCISE 14D

1 The students in classes A and B were asked whether they live north, east, south, or west of their school. This multiple bar chart shows their responses.

   a How many students from class A live south of the school?
   b How many students from class B live west of the school?
   c Find the most common response for:
      i class A
      ii class B.
   d In which class are there more students who live north of the school?
   e What percentage of students who live east of the school are from class B?

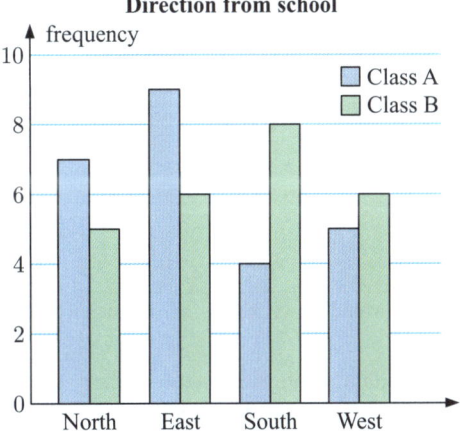

2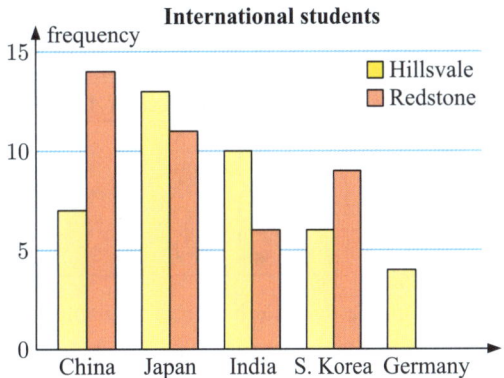

This composite bar chart shows the number of tennis, squash, and badminton members at a local sports club.

   a State the number of:
      i junior tennis members
      ii adult badminton members.
   b Does squash have more junior members or adult members?
   c For which sport is the percentage of adult members highest?
   d The sports club charges £15 for each junior member, and £25 for each adult member. Find the total amount received by the club for memberships.

3 Hillsvale School and Redstone School each have 40 international students. This multiple bar chart shows the countries that these international students come from.

   a How many of Hillsvale's international students come from India?
   b Which school does not have any students from Germany?
   c Which school has more students from:
      i China
      ii Japan?

4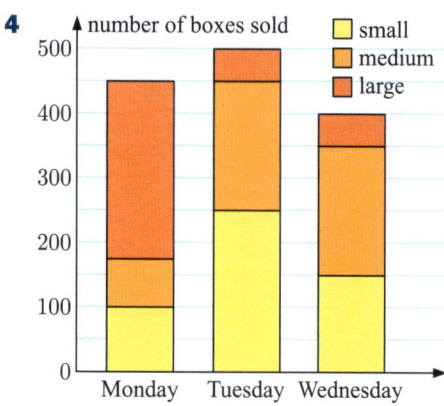

This composite bar chart shows the number of small, medium, and large boxes of popcorn sold at a cinema over three days.

a On which day was the most boxes of popcorn sold?

b What percentage of popcorn boxes sold on Wednesday were small or medium?

c Which size box was most popular on Tuesday?

d On which day do you think the cinema discounted the price of large popcorn? Explain your answer.

5 The data alongside shows the drinks purchased by students at recess and lunch time.

Recess	
Drink	Frequency
Orange juice	8
Soft drink	12
Milkshake	5
Water	6

Lunch	
Drink	Frequency
Orange juice	20
Soft drink	23
Milkshake	15
Water	5

Frank drew this composite bar chart to represent the data.

a Describe two mistakes Frank has made.

b Redraw the chart so that it is correct.

c Were more orange juices sold during recess or lunch?

d What percentage of drinks purchased at lunch time were milkshakes?

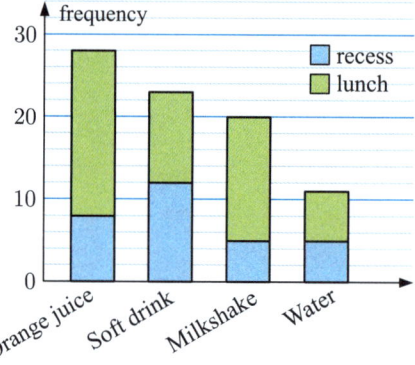

6 30 children and 30 adults were asked which section of the newspaper they enjoyed most.

Section	Children	Adults
News	5	10
Sport	7	9
Comics	10	4
Puzzles	8	7

a Draw a multiple bar chart to display the data.

b Which newspaper section was most popular with:
  i children
  ii adults?

c Which section has the most difference in popularity between children and adults? Discuss your answer.

7 On a particular day, a fire truck and an ambulance each received 20 call-outs. The data alongside shows the location of each call-out.

Location	Fire truck	Ambulance
House	6	9
Apartment	2	4
Office	5	2
Factory	7	5

a Draw a composite bar chart to display the data.
b Which type of location was most common for:
  i the fire truck   **ii** the ambulance?
c Which vehicle was called out to more offices?

E TIME SERIES DATA

The table below shows the number of customers in a city restaurant at hourly intervals:

Time	10 am	11 am	12 pm	1 pm	2 pm	3 pm	4 pm	5 pm	6 pm	7 pm	8 pm
Number of customers	12	22	20	29	25	16	10	13	27	35	20

When a variable is recorded at various times like this, the resulting data is called **time series data**.

We can use a **line graph** to display time series data. We plot each data point on a grid, then join the points with straight line segments.

Displaying time series data enables us to see **trends** in the data.

Example 8   ◀)) Self Tutor

The circumference of Warren's head was measured every 2 years from birth.

Age (years)	0	2	4	6	8	10
Head circumference (cm)	35	40	42	44	45	46

a Draw a line graph of Warren's head circumference over time.
b Describe the trend of the graph.

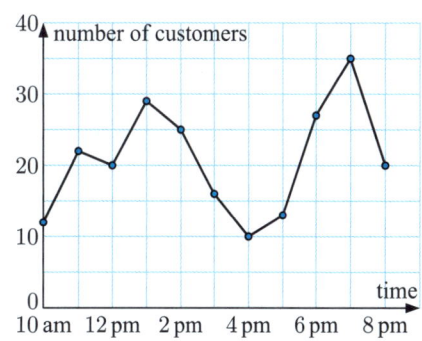

a

b Warren's head circumference increased quickly at first. The rate slowed down over time, and we would expect that once Warren reaches adulthood it would stop increasing altogether.

EXERCISE 14E

1 The number of squirrels in a region of forest was recorded at yearly intervals.

Year	2011	2012	2013	2014	2015	2016	2017	2018
Number of squirrels	300	378	455	482	490	488	491	490

- **a** Draw a line graph to display the data.
- **b** Describe how the squirrel population has changed over time.

2 Peter recorded the distance he walked each day during February:

Day	1	2	3	4	5	6	7	8	9	10	11	12	13	14
Distance (km)	3.82	4.68	4.69	2.41	3.22	3.25	4.09	5.90	3.76	5.70	2.82	1.89	2.50	4.25

Day	15	16	17	18	19	20	21	22	23	24	25	26	27	28
Distance (km)	4.93	4.67	3.48	2.92	4.25	3.42	3.42	3.84	4.77	2.77	3.08	2.28	4.38	5.01

- **a** Draw a line graph to display the data.
- **b** On how many days during February did Peter walk more than 4 km?
- **c** On which day was Peter most active?

3 The table below shows Jana's heart rate during a 60 minute gym session.

Time (minutes)	0	5	10	15	20	25	30	35	40	45	50	55	60
Heart rate (beats per minute)	50	80	95	112	120	142	130	145	158	152	165	160	135

- **a** Draw a line graph to display the data.
- **b** At what time was Jana's heart rate highest?
- **c** Describe the trend of the data.

4 The water consumption for a household is recorded each quarter for two years.

Year	2017				2018			
Quarter	1	2	3	4	1	2	3	4
Usage (kl)	59.4	63.2	75.8	70.5	60.3	62.2	79.5	74.3

A **quarter** is 3 months.

- **a** Draw a line graph to display the data.
- **b** In which quarter of the year does the household use:
  - **i** the most water
  - **ii** the least water?

  Give an explanation for your answers.

5 Leah owns a homewares shop. The table below shows the quarterly profit made by the shop over three years, in thousands of pounds.

	2014				2015				2016			
Quarter	1	2	3	4	1	2	3	4	1	2	3	4
Profit (£ × 1000)	27	20	18	35	31	23	24	42	34	32	29	47

a Draw a line graph to display the data.

b In which quarter does the shop generally make the most profit? Can you explain why this may occur?

c Describe the long-term trend of the data.

Discussion   Misleading graphs

Some people may try to trick or mislead others by the way they draw their graphs.

For example, Kelly owns two shops. One of them is managed by John, and the other by Wei Li. Last year John's shop earned a profit of £45 000, whereas Wei Li's profit was £38 000.

John draws this graph to show the profits earned by the two shops, and gives it to Kelly.

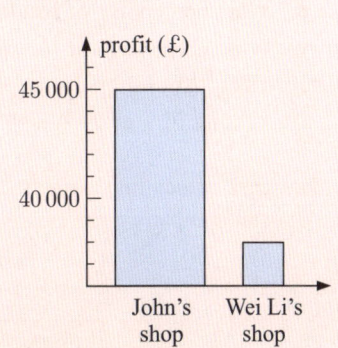

What to do:

1 Discuss the misleading features of John's graph.

2 Why do you think John has drawn the graph like this?

3 Discuss the misleading features of these graphs:

a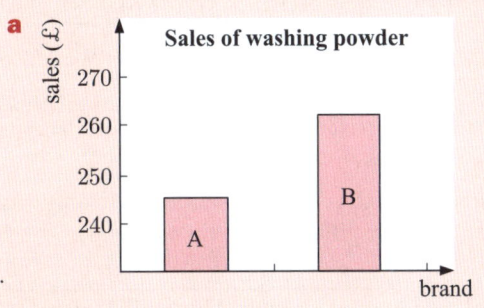

b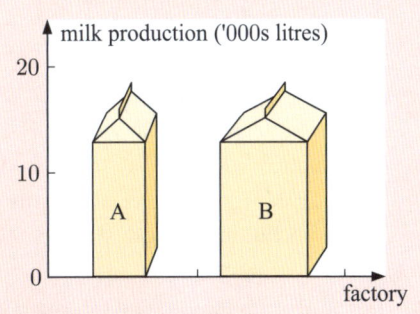

c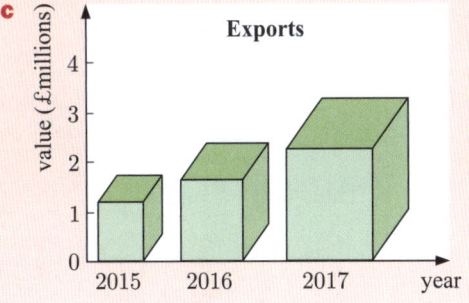

d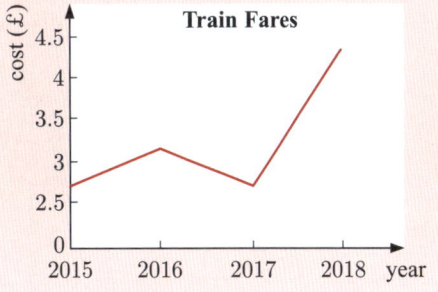

Review set 14A

1. A local council conducted a survey. They asked their residents to rate, out of five, the standard of different facilities and services in the area. The results are given alongside.

	1	2	3	4	5
Libraries	11%	12%	19%	26%	32%
Public transport	16%	32%	25%	14%	13%
Schools	17%	28%	25%	16%	14%
Shops	7%	16%	20%	35%	22%
Parks	11%	23%	24%	29%	13%
Health services	9%	14%	27%	26%	24%

   a What percentage of residents rated health services as 5 out of 5?
   b What percentage of residents rated the standard of schools as 4 or above?
   c The council will upgrade the facilities and services for which 30% or more of residents gave a rating of 1 or 2. Which facilities and services will the council be upgrading?

2. The cost of sending parcels overseas depends on the weight of the parcel and the destination. A postal company's prices are given in the table below:

		Weight			
		< 0.5 kg	0.5 to 1 kg	1 to 1.5 kg	1.5 to 2 kg
	Europe	£6	£9	£11	£13
Destination	World Zone 1	£8	£13	£17	£20
	World Zone 2	£9	£14	£18	£22

   a Danni has a parcel weighing 1.2 kg. She wants to send it to Canada, which is in World Zone 1. How much will this cost her?
   b Owen has 3 parcels which need to be sent to different locations within Europe. The parcels weigh 0.2 kg, 0.9 kg, and 1.6 kg. How much will Owen spend on postage?
   c Martin needs to send some gifts to New Zealand, which is in World Zone 2. He only has £20 to spend on postage. What is the maximum weight in gifts that Martin can send?

3. a In the pictogram, what does 🥛 represent?
   b On which day were the milk sales:
      i greatest   ii least?
   c How much milk was sold on:
      i Thursday   ii Friday?

4

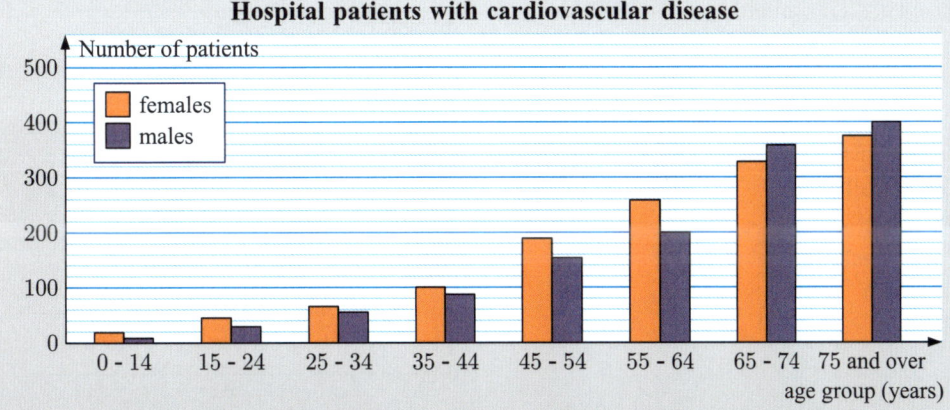

The multiple bar chart above shows the number of cardiovascular patients admitted to a hospital over 3 months. Use the graph to find:

a how many 55 - 64 year old males were admitted to hospital

b the percentage of the total patients aged 45 - 54 who were female

c which age groups had more male patients than female patients.

5 A chemical engineer is performing a reaction using large quantities of liquid. The line graph shows the volume of liquid in the collection tank over a 24 hour period. During the day, three samples of the liquid were taken for quality control.

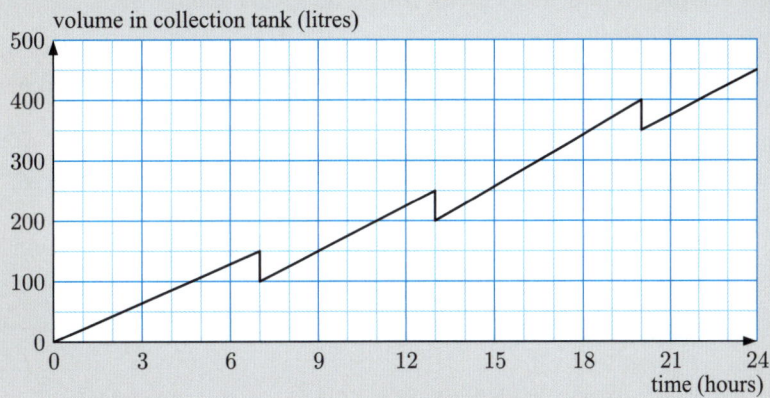

a How much liquid was produced in the first 7 hours?

b How much liquid was sampled each time for quality control?

c What was the total amount of liquid produced by the reaction over the whole day?

6 A survey of hair colour in a class of 40 students revealed the results in the table.

a Construct a horizontal bar chart to display this data.

b Which was the most common hair colour?

c What percentage of students in the class have black or blond hair?

Hair colour	Frequency
Red	4
Brown	17
Black	11
Blond	8

7 The size of Neil's monthly phone bill is displayed alongside.

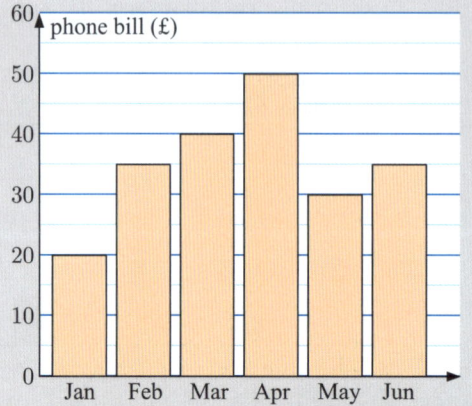

- **a** How much was Neil's phone bill in March?
- **b** In which month did Neil receive the highest bill?
- **c** Find the increase in Neil's phone bill from January to February.

8 The table below shows the results of a Mathematics competition for students at School A.

Result	Frequency
High distinction	52
Distinction	89
Credit	118
Participation	101

- **a** How many students from School A took part in the competition?
- **b** Draw a pie chart to display the data.
- **c** This pie chart shows the competition results for students at School B.
  Comment on the validity of this statement:
  "School B had more students who scored a Distinction than School A."

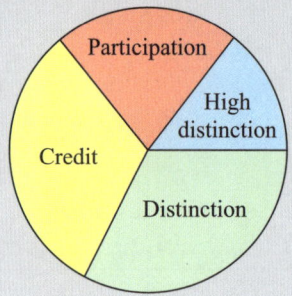

9 This multiple bar chart shows the leg injuries received by two rugby teams during a season.

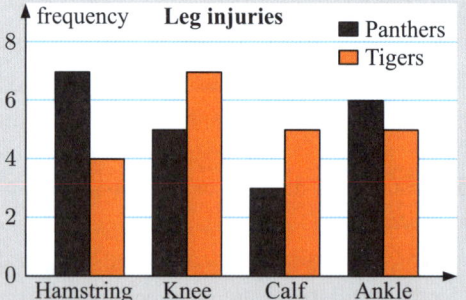

- **a** How many calf injuries were received by:
  - **i** the Panthers
  - **ii** the Tigers?
- **b** Which team suffered the most ankle injuries?
- **c** Which injury was most common for:
  - **i** the Panthers
  - **ii** the Tigers?

10 A group of children at a summer camp were asked which sport they wanted to play. The choices were tennis, swimming, cricket, basketball, and athletics.

Girls

Sport	Frequency
Tennis	8
Swimming	6
Cricket	5
Basketball	4
Athletics	7

Boys

Sport	Frequency
Tennis	6
Swimming	9
Cricket	6
Basketball	5
Athletics	4

a Draw a composite bar chart to display the data.
b Was tennis more popular with girls or boys?
c What percentage of children who chose athletics were boys?
d Which sport was most popular overall?

11 The attendances at a football team's home games are shown below.

Game	1	2	3	4	5	6	7	8	9	10	11	12
Attendance (× 1000)	37	34	48	30	33	24	30	18	26	13	17	16

a Draw a line graph to display the data.
b Describe the trend in the data.

Review set 14B

1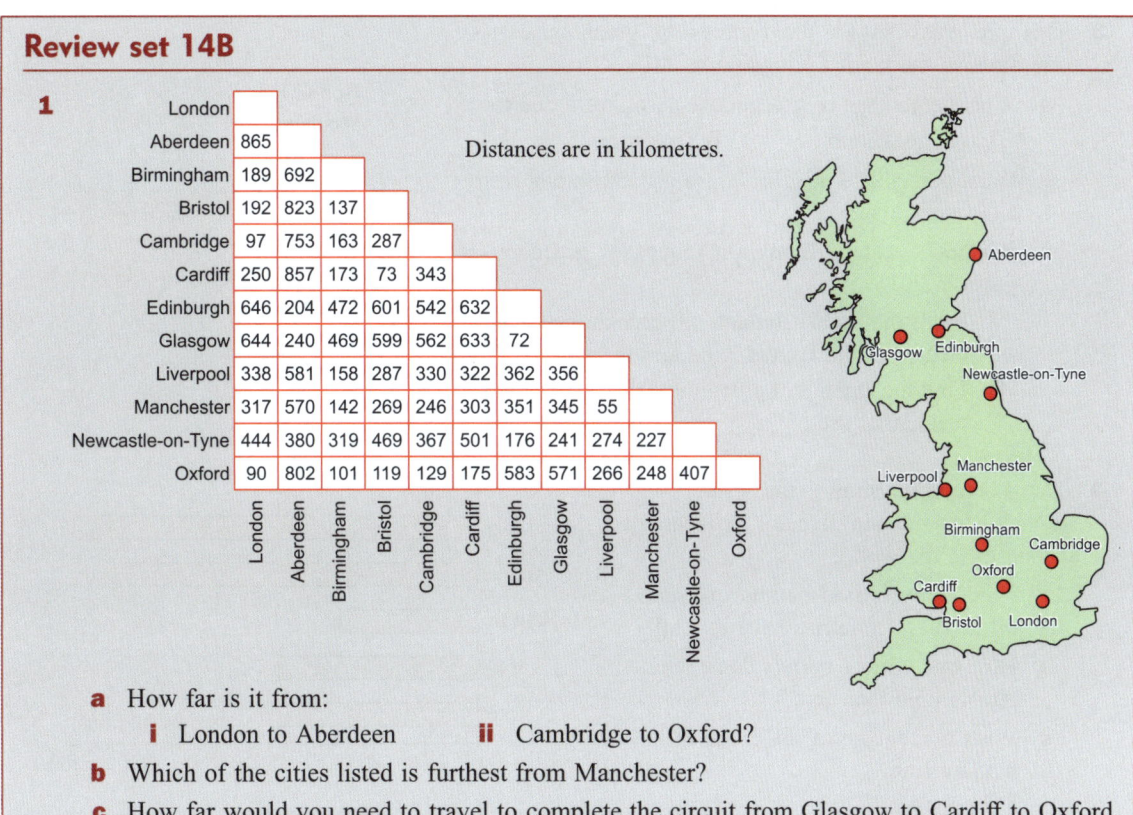

Distances are in kilometres.

	London	Aberdeen	Birmingham	Bristol	Cambridge	Cardiff	Edinburgh	Glasgow	Liverpool	Manchester	Newcastle-on-Tyne	Oxford
Aberdeen	865											
Birmingham	189	692										
Bristol	192	823	137									
Cambridge	97	753	163	287								
Cardiff	250	857	173	73	343							
Edinburgh	646	204	472	601	542	632						
Glasgow	644	240	469	599	562	633	72					
Liverpool	338	581	158	287	330	322	362	356				
Manchester	317	570	142	269	246	303	351	345	55			
Newcastle-on-Tyne	444	380	319	469	367	501	176	241	274	227		
Oxford	90	802	101	119	129	175	583	571	266	248	407	

a How far is it from:
  i London to Aberdeen   **ii** Cambridge to Oxford?
b Which of the cities listed is furthest from Manchester?
c How far would you need to travel to complete the circuit from Glasgow to Cardiff to Oxford and back to Glasgow?

2 The nutritional information panel from a bottle of flavoured milk is shown below.

NUTRITIONAL INFORMATION			
SERVINGS PER PACK: 1			
SERVING SIZE: 500 ml	PER 500 ml SERVE	%RDI PER SERVE	PER 100 ml
ENERGY	1035 kJ	12%	207 kJ
PROTEIN	16.5 g	34%	3.3 g
FAT, TOTAL	4.5 g	7%	0.9 g
- SATURATED	3.0 g	13%	0.6 g
CARBOHYDRATE	34.5 g	11%	6.9 g
- SUGARS	31.0 g	34%	6.2 g
DIETARY FIBRE	0.5 g	2%	0.1 g
SODIUM	225 mg	10%	45 mg
CALCIUM	625 mg	78%	125 mg

a How much sugar is in the bottle?
b How much sodium is in 100 ml of the milk?
c The milk provides 2% of the Recommended Daily Intake (RDI) of a particular nutrient.
  i What nutrient is it?
  ii What is the RDI of this nutrient in grams?

3 This pie chart shows the sources of greenhouse gas emissions in the United Kingdom in 2015.

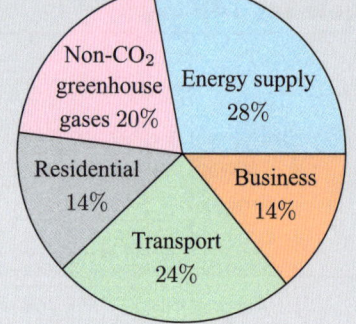

  a What percentage of greenhouse gases were due to:
    i energy supply   **ii** transport?
  b What percentage of greenhouse gas emissions were CO_2 emissions?
  c Business created 70 million tonnes of greenhouse gases in 2015.
    i Find the total amount of greenhouse gases created in the United Kingdom in 2015.
    ii Find the amount of greenhouse gases created by residential use.

4 The bar graph shows the crude oil reserves for the top five oil producing countries in the world.

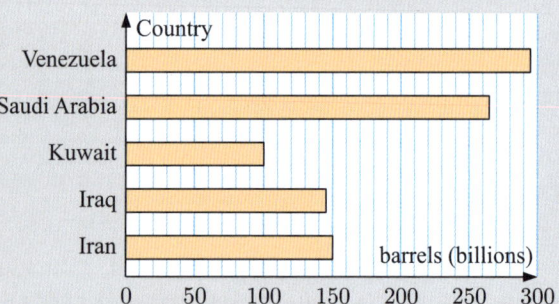

  a How many billion barrels of crude oil do these countries have in total?
  b How many more barrels does Saudi Arabia hold than Iraq?
  c What percentage of the crude oil is held by Iran?

5 Tian is a keen fisherman. He has recorded his weekend fish catches over a 7-week period.

  a What was Tian's catch in:
   i week 4   **ii** week 5?
  b Given that Tian caught 60 fish in total over the 7-week period, fill in the result for week 3.

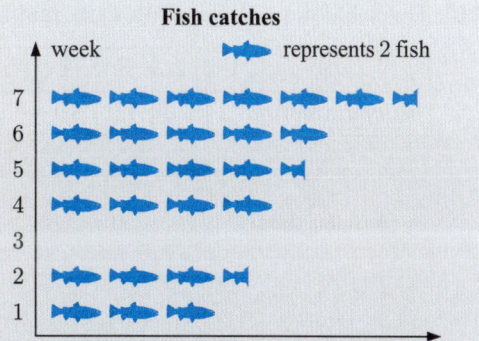

6 Sixty people whose houses had been burgled were asked where they were at the time of the burglary. The responses are shown alongside.

  a Draw a vertical bar chart to display the data.
  b What percentage of people were shopping when they were burgled?

Response	Frequency
At home	12
At work	20
Shopping	5
On holidays	10
Visiting friends	13
Total	60

7 Answer the **Opening Problem** on page 264.

8

Month	House sales
April	40
May	15
June	30
July	28
August	13
September	30

This table shows the number of houses sold by Bill Black Real Estate in a six month period.

Draw a pictogram to display the data, where 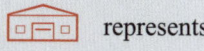 represents 10 houses.

9 This composite bar chart shows the time Megan spent in different stages of sleep over 3 nights.

  a How much sleep did Megan get on Monday?
  b How much more light sleep did Megan get on Tuesday than on Wednesday?
  c What type of sleep did Megan get the least of on Tuesday?
  d On which night did Megan get the greatest percentage of deep sleep?

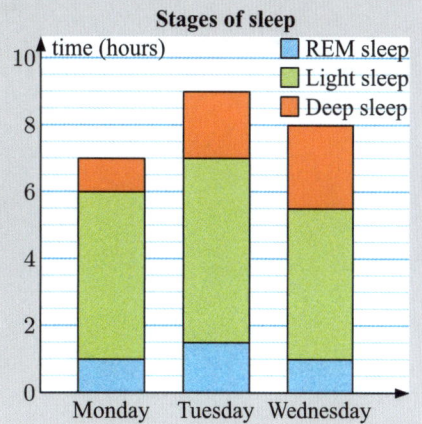

10 This table shows Liverpool's daily rainfall over a two week period.

Day	1	2	3	4	5	6	7	8	9	10	11	12	13	14
Rainfall (mm)	3.4	2.3	4.0	3.1	6.5	5.2	0	4.7	10.3	12.5	11.6	13.5	15.0	11.1

Lewis constructed this line graph to represent the data:

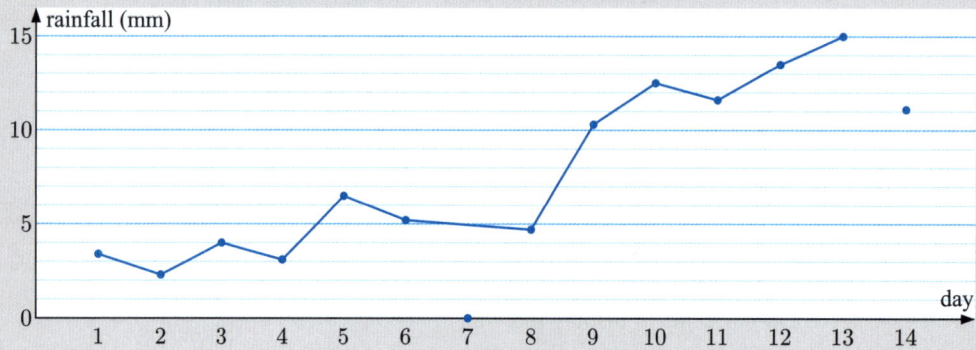

a Describe two mistakes Lewis has made.
b Redraw the line graph correctly.
c Describe the trend of the data.

15

Pythagoras' theorem

Contents:

- **A** Solving $x^2 = k$
- **B** Pythagoras' theorem
- **C** The converse of Pythagoras' theorem
- **D** Problem solving

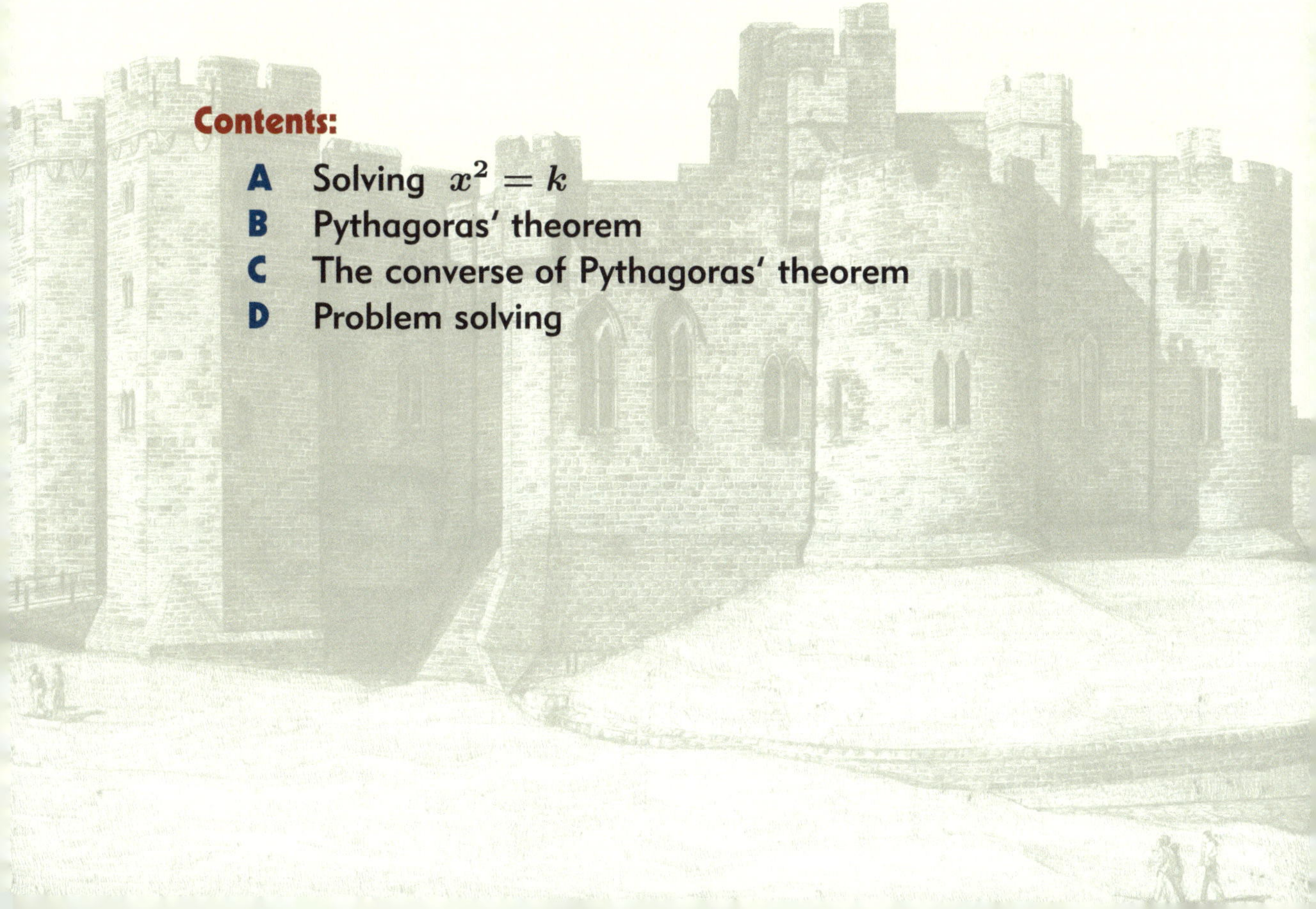

Opening problem

Phil is concerned that the street lamp outside his house is not quite at right angles to the ground.

He marks a point A on the lamp 1.5 m from its base, and a point B on the ground 2 m from the lamp's base. Using a tape measure, he finds that the distance between A and B is 2.40 m.

Things to think about:

a What assumptions has Phil made?

b Is the street lamp at right angles to the ground?

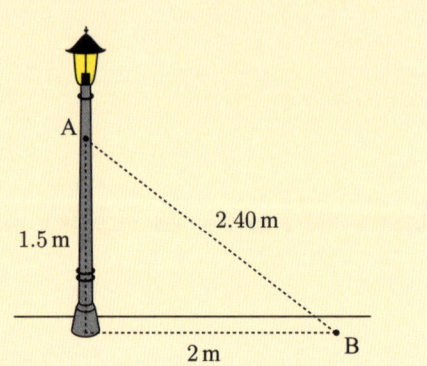

Historical note

For many centuries people have used right angled corners to construct buildings and to divide land into rectangular fields. They have done this quite accurately by relatively simple means.

Over 3000 years ago the Egyptians knew that a triangle with sides in the ratio $3:4:5$ was right angled. They used a loop of rope with 12 knots equally spaced along it to make corners in their building construction.

Around 500 BC, the Greek mathematician **Pythagoras of Samos** proved a rule which connects the sides of a right angled triangle. According to legend, he discovered the rule while studying the tiled palace floor as he waited for an audience with the ruler Polycrates.

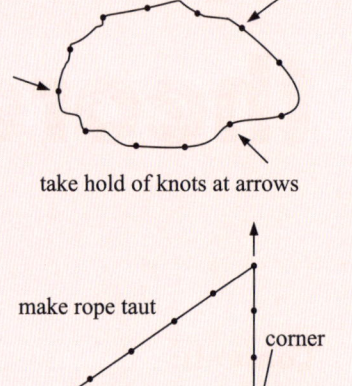

The discovery of **Pythagoras' theorem** led to the classification of a different type of number which does not have a terminating or recurring decimal value, but which does have a distinct place on the number line. These numbers are called **surds** and are **irrational numbers**.

A SOLVING $x^2 = k$

In **Chapter 1** we saw square roots such as $\sqrt{7}$.

Notice that $\sqrt{7} \times \sqrt{7} = 7$ and $(-\sqrt{7}) \times (-\sqrt{7}) = 7$ {as negative × negative = positive}.

So, if we were asked to solve the equation $x^2 = 7$, it is clear that x could equal $\sqrt{7}$ or $-\sqrt{7}$. The squares of both of these numbers are 7.

We write the solutions as $x = \pm\sqrt{7}$, which reads "plus or minus the square root of 7".

Consider $x^2 = k$.
If $k > 0$, then $x = \pm\sqrt{k}$.
If $k = 0$, then $x = 0$ is the only solution.
If $k < 0$, then there are **no real solutions**.

A **real** number is a number which can be placed on the number line.

Example 1

Solve for x:

a $x^2 = 4$ **b** $x^2 = 11$ **c** $x^2 = -5$

a $x^2 = 4$
$\therefore x = \pm\sqrt{4}$
$\therefore x = \pm 2$ $\{x = 2$ or $x = -2\}$

b $x^2 = 11$
$\therefore x = \pm\sqrt{11}$ $\{x = \sqrt{11}$ or $x = -\sqrt{11}\}$

c $x^2 = -5$ has no real solutions, since x^2 cannot be negative.

If $x^2 = k$ where $k > 0$, then there are **two** solutions.

Example 2

Solve for x:

a $x^2 + 4 = 9$

b $x^2 + (2x)^2 = 20$

a $x^2 + 4 = 9$
$\therefore x^2 = 5$ {subtracting 4 from both sides}
$\therefore x = \pm\sqrt{5}$

b $x^2 + (2x)^2 = 20$
$\therefore x^2 + 4x^2 = 20$ {index law}
$\therefore 5x^2 = 20$
$\therefore x^2 = 4$ {dividing both sides by 5}
$\therefore x = \pm\sqrt{4}$
$\therefore x = \pm 2$

EXERCISE 15A

1 If possible, solve for x:

a $x^2 = 9$ **b** $x^2 = 49$ **c** $x^2 = 36$ **d** $x^2 = 0$
e $x^2 = 1$ **f** $x^2 = 17$ **g** $x^2 = 23$ **h** $x^2 = 100$
i $x^2 = -4$ **j** $x^2 = -7$ **k** $x^2 = 27$ **l** $x^2 = -27$

2 Solve for x:

a $x^2 + 5 = 9$ **b** $x^2 + 16 = 25$ **c** $x^2 + 2 = 27$
d $x^2 + 7 = 23$ **e** $8 + x^2 = 44$ **f** $x^2 + 14 = 39$
g $10 + x^2 = 60$ **h** $x^2 + 5 = 49$ **i** $6 + x^2 = 18$

3 Solve for x:

a $2x^2 = 18$
b $3x^2 = 48$
c $5x^2 = 20$
d $x^2 + x^2 = 32$
e $x^2 + 2x^2 = 90$
f $x^2 + 5x^2 = 120$
g $x^2 + (2x)^2 = 45$
h $x^2 + (3x)^2 = 70$
i $x^2 + (2x)^2 = 1$

B PYTHAGORAS' THEOREM

A **right angled triangle** is a triangle which has a right angle as one of its angles.

The side **opposite** the right angle is called the **hypotenuse**. It is the **longest** side of the triangle.

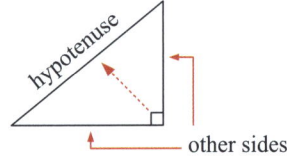

If a triangle is right angled, the special relationship connecting the lengths of its sides is called **Pythagoras' theorem**.

Investigation 1 — Discovering Pythagoras' theorem

Consider a right angled triangle which has a hypotenuse of length c cm, and the other two sides have lengths a cm and b cm.

We are looking for an equation which connects a, b, and c.

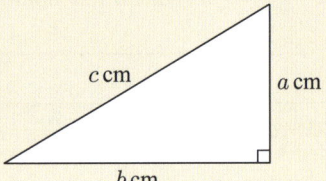

What to do:

1 Draw a horizontal line of length 4 cm. At one end, draw a vertical line of length 3 cm. The lines you have drawn form a right angle.

2 Complete a right angled triangle by drawing in the hypotenuse. In this case $a = 3$ and $b = 4$. Find c by measuring this hypotenuse.

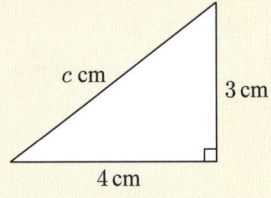

3 **a** Copy the table and complete the second row using the information from **2**.
   b Complete the rest of the table using three more right angled triangles for which a and b are specified.

a	b	c	a^2	b^2	c^2	$a^2 + b^2$
3	4		9			
6	8					
5	12					
4	7					

4 State any conclusions you draw from the information in this table.

5 Construct two more right angled triangles with lengths a and b of your choosing. Does your conclusion hold for these triangles?

PYTHAGORAS SIMULATION

6 Click on the icon to further explore the side lengths of right angled triangles.

PYTHAGORAS' THEOREM

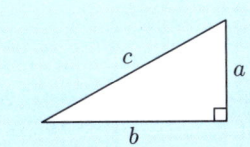

In a right angled triangle with hypotenuse of length c, and other sides of length a and b,

$$c^2 = a^2 + b^2.$$

In geometric form, Pythagoras' theorem states:

In any right angled triangle, the area of the square on the hypotenuse is equal to the sum of the areas of the squares on the other two sides.

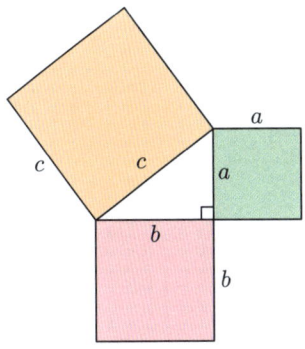

Look back at the tile pattern on page **292**. Can you see this figure in the pattern?

WORKSHEET

Discussion — Pythagoras' theorem

How did Pythagoras prove his theorem? He probably did not use algebra, since there is no evidence algebra was invented until well after his lifetime.

PYTHAGORAS' PROOF

Click on the icon for a possible answer.

There are over 400 different proofs of Pythagoras' theorem. One of them is presented below.

Proof:

On a square we draw 4 identical right angled triangles, as illustrated. A smaller square is formed in the centre.

Suppose the hypotenuse of each triangle has length c, and the other two side lengths are a and b.

The total area of the large square
$= 4 \times$ area of one triangle $+$ area of smaller square

$\therefore (a+b)^2 = 4 \times \frac{1}{2}ab + c^2$

$\therefore a^2 + 2ab + b^2 = 2ab + c^2$

$\therefore a^2 + b^2 = c^2$

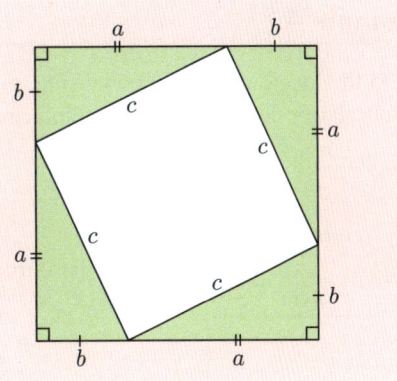

We can use Pythagoras' theorem to find side lengths in right angled triangles.

Example 3

Self Tutor

Find the length of the hypotenuse in the given right angled triangle:

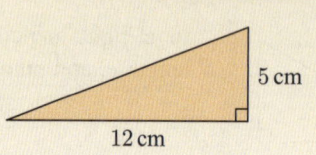

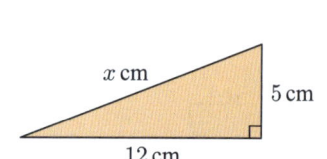

Let the hypotenuse have length x cm.

$\therefore\ x^2 = 12^2 + 5^2$ {Pythagoras}

$\therefore\ x^2 = 144 + 25$

$\therefore\ x^2 = 169$

$\therefore\ x = \pm\sqrt{169}$

$\therefore\ x = 13$ {as $x > 0$}

The hypotenuse has length 13 cm.

We reject the negative answer as the length of a side must be positive!

EXERCISE 15B

1 Find the length of the hypotenuse in each right angled triangle. Give your answers to 3 significant figures where necessary.

a 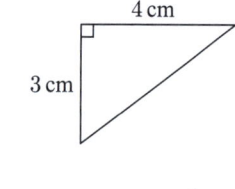 **b** 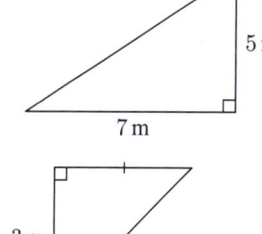 **c**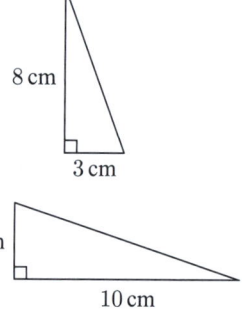

d   **e**   **f**

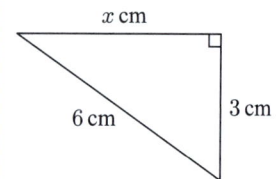 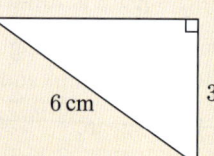

Example 4

Self Tutor

Find the length of the unknown side, giving your answer correct to 2 decimal places.

Let the third side have length x cm.

$\therefore\ x^2 + 3^2 = 6^2$ {Pythagoras}

$\therefore\ x^2 + 9 = 36$

$\therefore\ x^2 = 27$

$\therefore\ x = \pm\sqrt{27}$

$\therefore\ x = \sqrt{27}$ {as $x > 0$}

The third side has length ≈ 5.20 cm.

Casio fx-991EX

$\sqrt{27}$

5.196152423

2 Find, correct to 2 decimal places where necessary, the length of the unknown side in each right angled triangle.

a 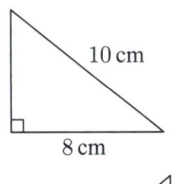   b 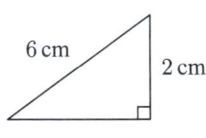   c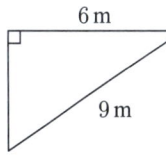

d 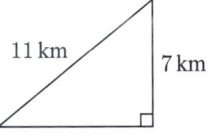   e 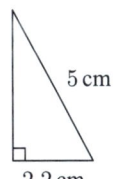   f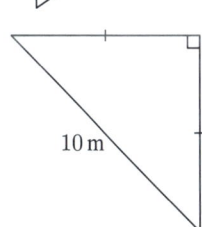

3 Find the length of the unknown side in each right angled triangle.

a 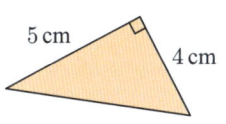   b 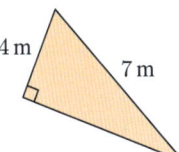   c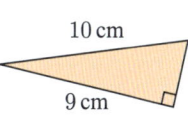

Make sure you identify the hypotenuse.

d 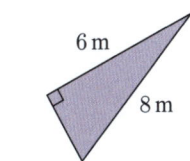   e 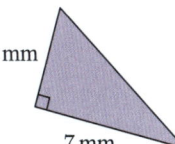   f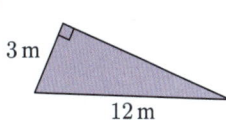

Example 5 ◀) Self Tutor

Find the value of y:

a 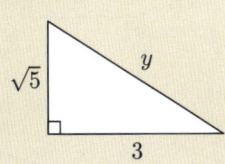   b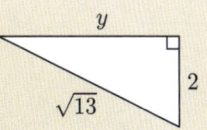

a $y^2 = 3^2 + (\sqrt{5})^2$ {Pythagoras}
∴ $y^2 = 9 + 5$
∴ $y^2 = 14$
∴ $y = \pm\sqrt{14}$
∴ $y = \sqrt{14}$ {as $y > 0$}

b $y^2 + 2^2 = (\sqrt{13})^2$ {Pythagoras}
∴ $y^2 + 4 = 13$
∴ $y^2 = 9$
∴ $y = \pm\sqrt{9}$
∴ $y = 3$ {as $y > 0$}

4 Find the value of y:

a 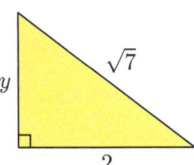   b 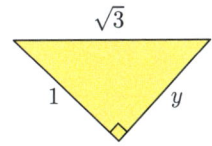   c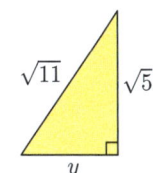

Example 6

Find the unknown lengths:

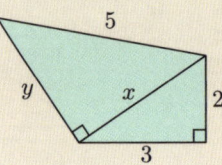

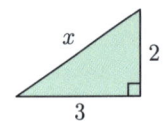

$x^2 = 3^2 + 2^2$ {Pythagoras}
$\therefore\ x^2 = 9 + 4$
$\therefore\ x^2 = 13$
$\therefore\ x = \sqrt{13}$ {as $x > 0$}

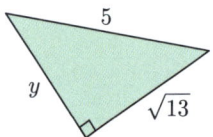

$y^2 + (\sqrt{13})^2 = 5^2$ {Pythagoras}
$\therefore\ y^2 + 13 = 25$
$\therefore\ y^2 = 12$
$\therefore\ y = \sqrt{12}$ {as $y > 0$}

5 Find the unknown lengths:

a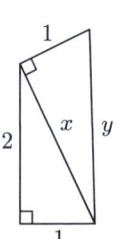

b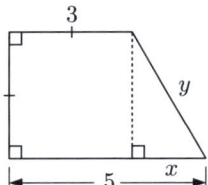

c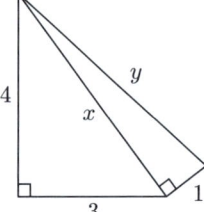

d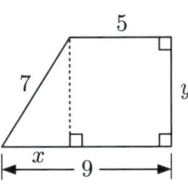

e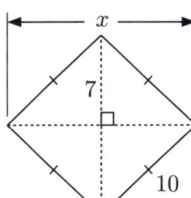

f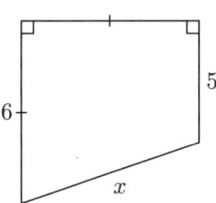

Example 7

Find x:

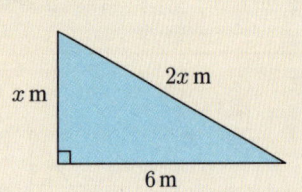

$(2x)^2 = x^2 + 6^2$ {Pythagoras}
$\therefore\ 4x^2 = x^2 + 36$
$\therefore\ 3x^2 = 36$
$\therefore\ x^2 = 12$
$\therefore\ x = \sqrt{12}$ {as $x > 0$}

6 Find the value of x:

a 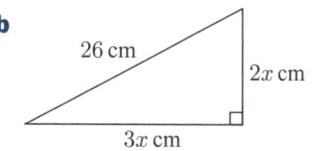 **b**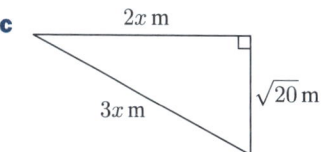

c

7 Find the length AC:

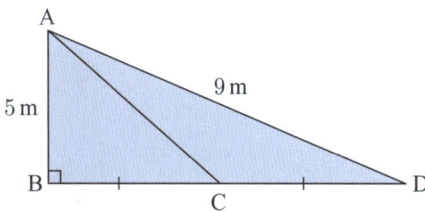

8 Use the figure below to show that $\sqrt{2} + \sqrt{8} = \sqrt{18}$.

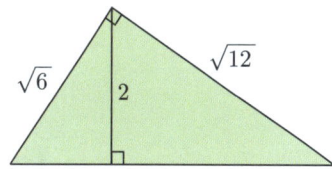

9 Find the distance AB in each of the following figures:

a **b** 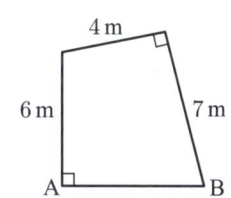 **c**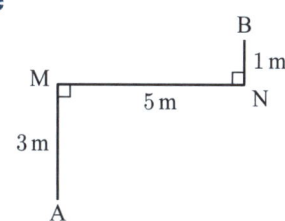

C THE CONVERSE OF PYTHAGORAS' THEOREM

If we are given the lengths of three sides of a triangle, the **converse of Pythagoras' theorem** gives us a simple **test** to determine whether the triangle is right angled.

THE CONVERSE OF PYTHAGORAS' THEOREM

If a triangle has sides of length a, b, and c units where $a^2 + b^2 = c^2$, then the triangle is right angled.

PYTHAGORAS SIMULATION

Example 8   ◀)) Self Tutor

Is the triangle with sides 8 cm, 9 cm, and 12 cm right angled?

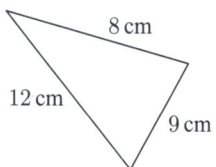

The two shorter sides have lengths 8 cm and 9 cm.

Now $8^2 + 9^2 = 64 + 81 = 145$

whereas $12^2 = 144$

∴ $8^2 + 9^2 \neq 12^2$

∴ the triangle is not right angled.

EXERCISE 15C

1 The following figures are not drawn to scale. Which of the triangles are right angled?

a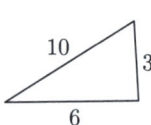

b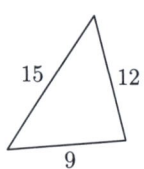

> The right angle would be opposite the hypotenuse or longest side.

c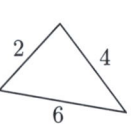

d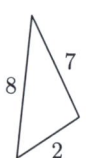

e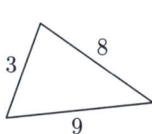

f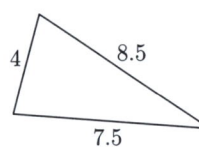

2 The following figures are not drawn to scale. Which of the triangles are right angled? For those triangles that are, indicate which angle is the right angle.

a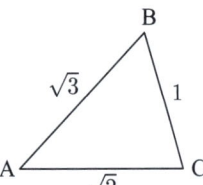

b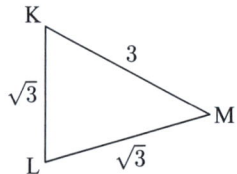

c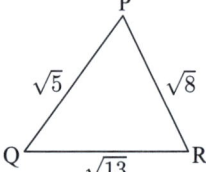

3 Meredith is trying to make a scarf by cutting some cloth into a right angled triangle. The triangle she makes has side lengths 40 cm, 42 cm, and 58 cm. Is Meredith's triangle right angled?

4 Answer the **Opening Problem** on page **292**.

Activity — Testing for right angles

What to do:

1. Select an object around your school which appears to be at right angles to the ground. It could be a telegraph pole, a table leg, or a wall.

2. Mark a point on your object, and measure its height a above the ground.

3. Mark a point on the ground, and measure its distance b from the base of the object.

4. Measure the distance c between the two points. Make sure that your tape measure is taut.

5. Use the converse of Pythagoras' theorem to determine whether the object is at right angles to the ground.

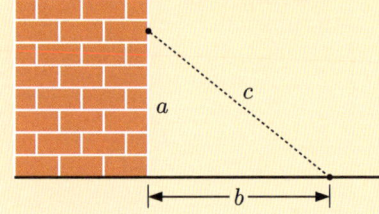

D PROBLEM SOLVING

Right angled triangles occur in many practical problems. In these situations we can apply Pythagoras' theorem to help find unknown side lengths.

The problem solving approach involves the following steps:

Step 1: Draw a neat, clear diagram of the situation.
Step 2: Mark known lengths and right angles on the diagram.
Step 3: Use a symbol such as x to represent the unknown length.
Step 4: Write down Pythagoras' theorem for the given information.
Step 5: Solve the equation.
Step 6: Where necessary, write your answer in sentence form.

The following special geometrical figures contain right angled triangles:

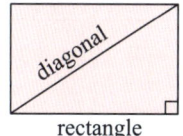
rectangle

Each corner of a **rectangle** is a right angle. We can construct a diagonal to form a right angled triangle.

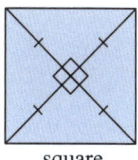
square

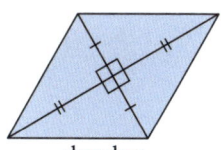
rhombus

In a **square** and a **rhombus**, the diagonals bisect each other at right angles.

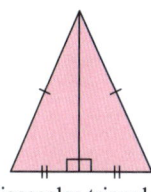
isosceles triangle

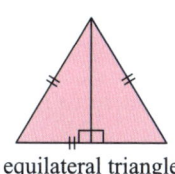
equilateral triangle

In an **isosceles triangle** and an **equilateral triangle**, the altitude bisects the base at right angles.

Example 9   ◀)) Self Tutor

The rectangular frame of a gate is 3 m by 5 m.
Find, to the nearest centimetre, the length of the diagonal support across the frame.

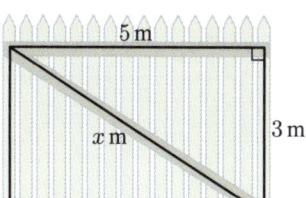

Let the diagonal support have length x m.

Now $x^2 = 3^2 + 5^2$ {Pythagoras}
$\therefore \ x^2 = 9 + 25$
$\therefore \ x^2 = 34$
$\therefore \ x = \sqrt{34}$ {as $x > 0$}

$\therefore$ the support is $\sqrt{34} \approx 5.83$ m long.

EXERCISE 15D

1 A rectangle has sides of length 5 cm and 8 cm. Find the length of its diagonals.

2 Find the length of the diagonal of a square with side length 6.8 cm. Give your answer as a decimal to the nearest mm.

3 What is the longest iron rod which can be placed flat across the diagonal of a 4 m by 2.5 m garden shed floor?

4 A rhombus has diagonals 6 cm and 10 cm. Find the length of one side of the rhombus.

5 Three roads AB, BC, and CA form a right angled triangle. AC is 9 km long and BC is 5 km long. Liam rides his bicycle from A to B to C. What extra distance does he travel compared with going directly from A to C?

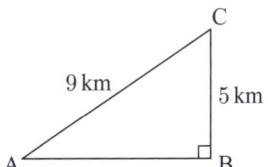

Example 10

An 8 m long ladder has its feet placed 3 m out from a vertical wall. How far up the wall will the ladder reach, to the nearest cm?

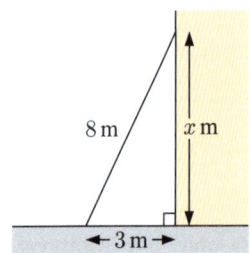

Suppose the ladder reaches x m up the wall.

Now $x^2 + 3^2 = 8^2$ {Pythagoras}

$\therefore x^2 + 9 = 64$

$\therefore x^2 = 55$

$\therefore x = \sqrt{55}$ {as $x > 0$}

$\therefore$ the ladder reaches $\sqrt{55} \approx 7.42$ m up the wall.

6 When the feet of a ladder are placed 2.5 m from a wall, the ladder just reaches a window which is 8 m from the ground. How long is the ladder?

7 A car travels 9 km due north and then 12 km due west. How far is the car from its starting point?

8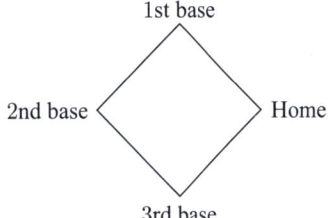

A baseball "diamond" is a square whose sides are 27 m long. Find, to the nearest $\frac{1}{10}$ metre, the distance from the home plate to second base.

9 To check that his set square was right angled, Roger measured its sides. The two shorter sides were 8 cm and 11.55 cm long, and the longest side was 14.05 cm long. Is the set square right angled?

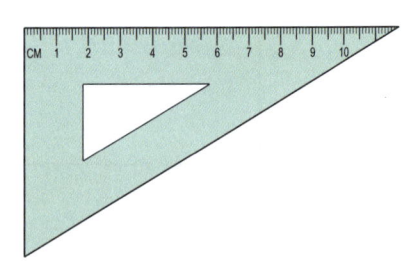

10 Two trains A and B leave the station at the same time. Train A travels north at a constant speed of 45 km per hour. Train B travels east at a constant speed of 70 km per hour.
  a How far will each train have travelled after 3 hours?
  b Find the distance between A and B after 3 hours.

11 A large flagpole is held to the ground by six cables, as illustrated. If the cables have to be replaced, what length of cabling must be purchased?

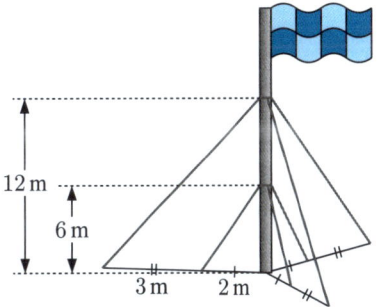

Example 11   ◀)) Self Tutor

An equilateral triangle has sides of length 8 cm. Find the height of the triangle, to the nearest mm.

We draw an altitude which *bisects* the base.

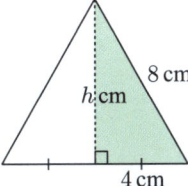

$h^2 + 4^2 = 8^2$ {Pythagoras}
$\therefore h^2 + 16 = 64$
$\therefore h^2 = 48$
$\therefore h = \sqrt{48}$ {as $h > 0$}
$\therefore$ the height of the triangle ≈ 6.9 cm.

12 Find the height of an equilateral triangle with sides of length 10 cm.

13 An isosceles triangle has equal sides measuring 10 cm, and a base which is 12 cm long. Find the length of the altitude of the triangle from the apex to the base.

14 Find the length of the truss AB in the roof structure shown:

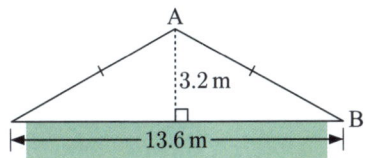

15 How high is the roof above the walls in the roof structure shown?

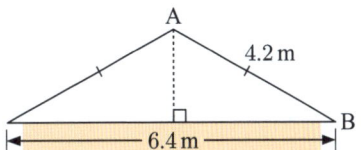

Example 12   ◀)) Self Tutor

A square has diagonals of length 10 cm. Find the length of a side, to the nearest mm.

Let the sides have length x cm.

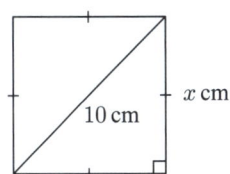

$\therefore x^2 + x^2 = 10^2$ {Pythagoras}
$\therefore 2x^2 = 100$
$\therefore x^2 = 50$
$\therefore x = \sqrt{50}$ {as $x > 0$}
$\therefore$ the sides have length $\sqrt{50} \approx 7.1$ cm.

16 A square has diagonals of length 15 cm. Find the length of a side, to the nearest mm.

17 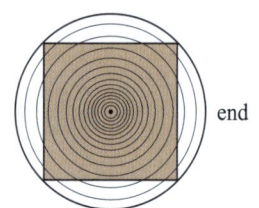 A log is 60 cm in diameter. Find the dimensions of the largest square section beam which can be cut from the log.

18 The longer side of a rectangle is three times the length of the shorter side. Each diagonal of the rectangle is 22 cm long. Find the dimensions of the rectangle.

19 An equilateral triangle has an altitude of length 16 cm. Find the length of each side.

Review set 15A

1 Solve for x:
   a $x^2 = 81$ **b** $x^2 + 5 = 30$ **c** $3x^2 = -12$

2 Find x, giving your answer to 2 decimal places where necessary:

  a 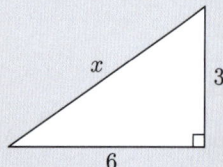   **b** 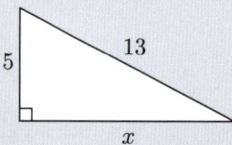   **c**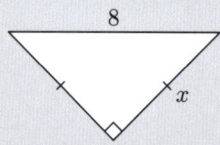

3 A rectangle has sides of length 6 cm and 7 cm. Find the length of the diagonal, to the nearest mm.

4 The given triangles are not drawn to scale. Are either of the triangles right angled? Explain your answers.

  a 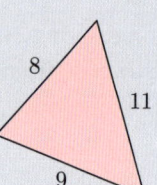   **b**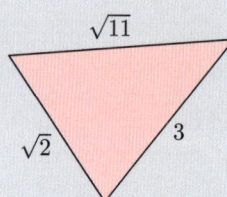

5 Find the unknown lengths:

  a 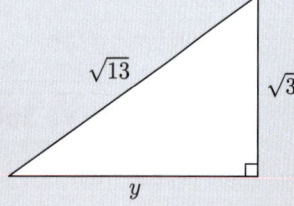   **b**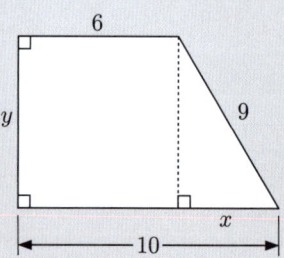

6 An 8 m long ladder leans against a vertical wall, with its feet 2.5 m away from the wall. Find how far up the wall the ladder reaches, to the nearest cm.

7 Two brothers leave their house at the same time. Alex runs due east at a constant speed of 10 km per hour, and Boris walks due south at a constant speed of 4 km per hour.
  a How far has each brother travelled after 30 minutes?
  b Find the distance between the two brothers after 30 minutes.

8 A garden gate is 2.4 metres wide and 1.2 metres high. The gate is strengthened by a diagonal strut.
   a How long is the strut?
   b Calculate the length of steel needed for the frame of the gate, including the strut.

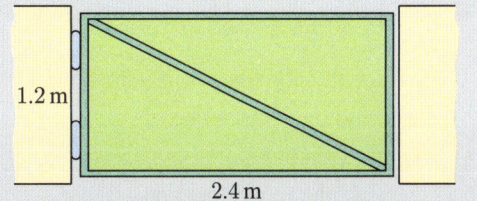

9 Use the figure alongside to show that $\sqrt{2} + \sqrt{18} = \sqrt{32}$.

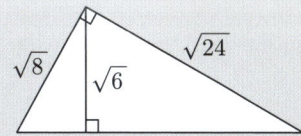

10 A young tree has a 2 m support rope tied to a peg in the ground 1.2 m from its base. How high up the tree is the rope tied?

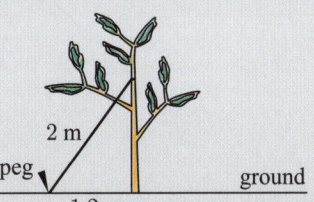

11 Find the length of the truss AB for the roof structure shown.

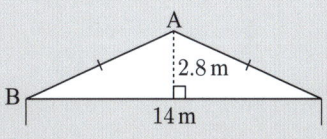

Review set 15B

1 Solve for x:
   a $6x^2 = 30$
   b $2x^2 + 2x^2 = 48$
   c $x^2 + (2x)^2 = 65$

2 Find x, giving your answer to 3 significant figures where necessary:
   a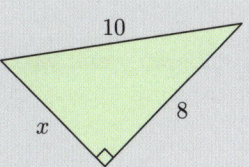
   b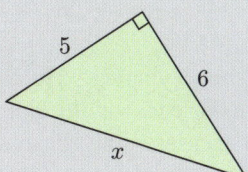
   c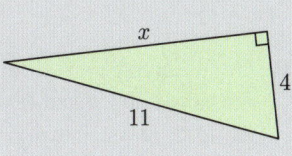

3 A rectangle measures 4 m by 5 m. Find the length of its diagonals, to 1 decimal place.

4 Find the values of x and y:

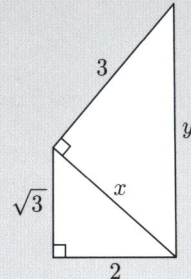

5 The triangle below is not drawn to scale. State whether or not the triangle is right angled, and if it is, which angle is the right angle.

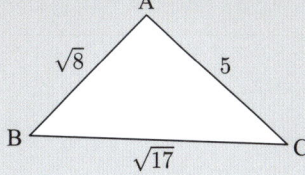

6 When viewed from above, a clothesline has the design shown.
Each of the arms is 1.6 m long.
  a Find the value of x.
  b Calculate the *total* length of:
    i the innermost line
    ii the outermost line.
  c Find the total length of cord needed for the clothesline.

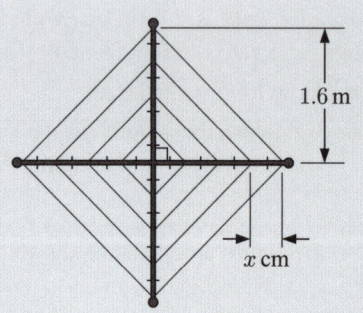

7 Find the distance AB in the following figures:

a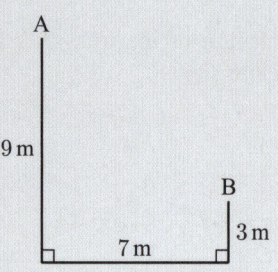

b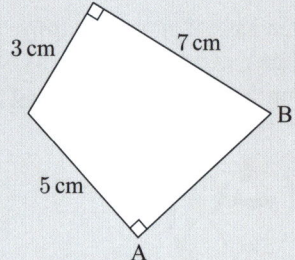

8 A 1.5 m ladder reaches three times as far up a vertical wall as the base is out from the wall. How far up the wall does the ladder reach?

9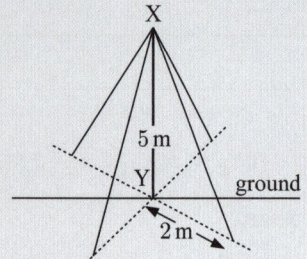

A pole XY is 5 metres tall. Four wires from the top of the pole X connect it to the ground.
Each wire is pegged 2 metres from the base of the pole. Find the total length of the four wires.

10 Gui is laying out a course for show jumping. He has measured the distances between the ends of the poles as 11.2 m as shown. He then measures the diagonal distances x and y, and finds they are equal.
Explain the significance of this result.

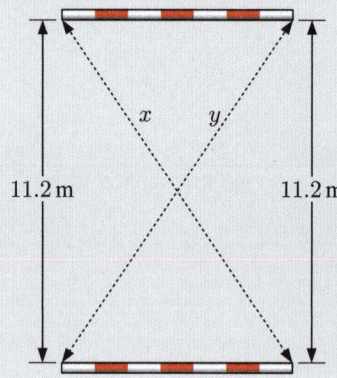

16

Measurement

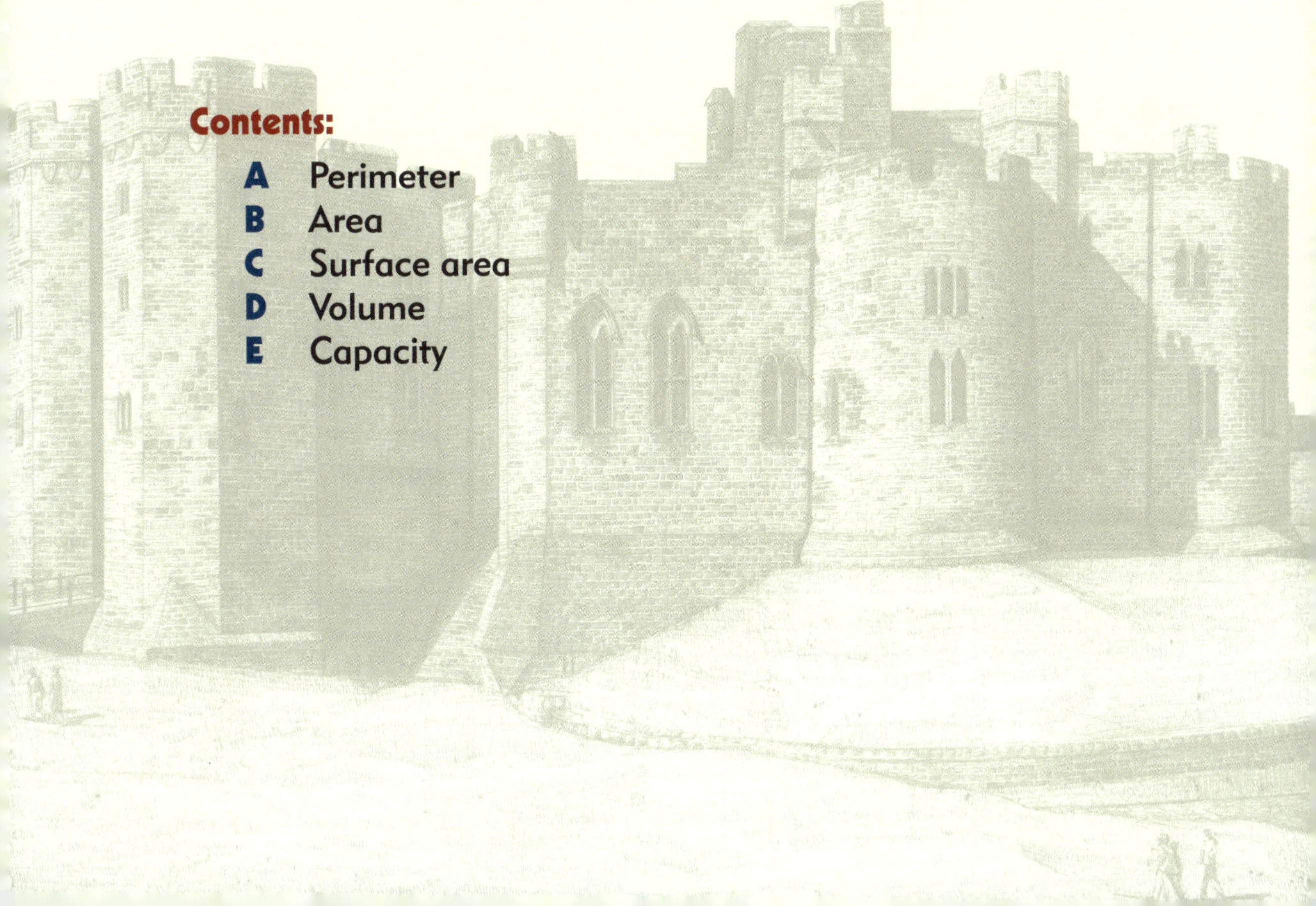

Contents:
- **A** Perimeter
- **B** Area
- **C** Surface area
- **D** Volume
- **E** Capacity

308 Measurement (Chapter 16)

Opening problem

Jacob wants to throw a water balloon at his brother. It is a sphere containing 1 litre of water.

Things to think about:

a What is the *volume* of the filled balloon?
b What is the *radius* of the water balloon?
c What is the outer *surface area* of the water balloon?

Constructing a building or a bridge, joining the circuits of a microchip, or rendezvousing in space to repair a satellite, all require the precise use of **measurement**.

A PERIMETER

The **perimeter** of a closed figure is the total distance around its **boundary**.

POLYGONS

The perimeter of a **polygon** is the sum of the lengths of its sides.

For example:

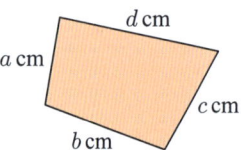

Perimeter $= (a + b + c + d)$ cm

Example 1 ◉ Self Tutor

Find the perimeter of the following figures:

a
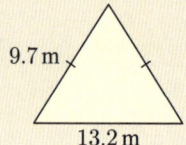

b
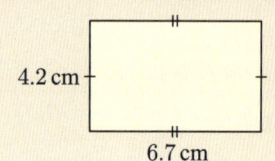

a Perimeter $= (2 \times 9.7 + 13.2)$ m
            $= 32.6$ m

b Perimeter $= (2 \times 4.2 + 2 \times 6.7)$ cm
            $= (8.4 + 13.4)$ cm
            $= 21.8$ cm

CIRCLES

In the case of the circle, the perimeter is given the special name **circumference**.

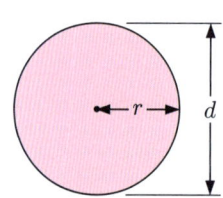

π is an irrational number. It is approximately 3.141 59 ....

Circumference $C = 2\pi r$ or $C = \pi d$

Example 2

Find the perimeter of the following figures:

a circle with radius 8 cm

b stadium shape with straight side 40 m

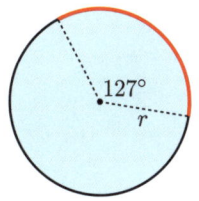
CALCULATOR INSTRUCTIONS

a Circumference
$= 2\pi r$
$= 2\pi \times 8$ cm
$= 16\pi$ cm
≈ 50.3 cm

b Perimeter
$=$ circumference of circle $+ 2 \times$ length of each straight side
$= ((\pi \times 40) + (2 \times 40))$ m
$= (40\pi + 80)$ m
≈ 206 m

ARC LENGTH

Consider the red arc shown. The angle at the centre of the circle is $127°$.

Since there are $360°$ in the whole circle, the arc makes up $\dfrac{127}{360}$ of the circle's circumference.

So, the length of the arc is $\left(\dfrac{127}{360}\right) \times 2\pi r$.

$$\text{Arc length} \quad l = \left(\dfrac{\theta}{360}\right) \times 2\pi r$$

Example 3

Find the perimeter of the sector: (135°, 4 mm)

Perimeter
$= (4 + 4)$ mm $+$ length of arc
$= 8$ mm $+ \left(\dfrac{135}{360}\right) \times 2 \times \pi \times 4$ mm
≈ 17.4 mm

EXERCISE 16A

1 Find the perimeter of each figure:

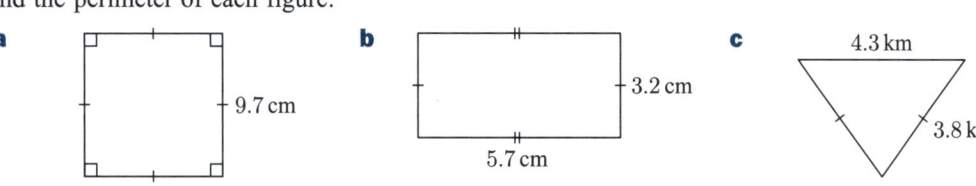

a square, 9.7 cm   **b** rectangle, 3.2 cm by 5.7 cm   **c** triangle, 4.3 km and 3.8 km

d 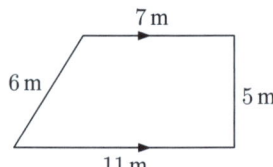 **e** 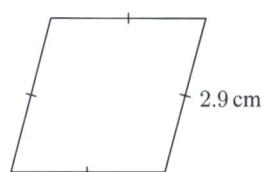 **f**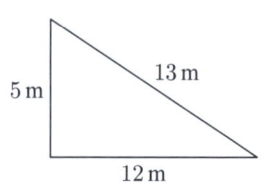

g 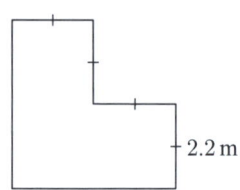 **h** 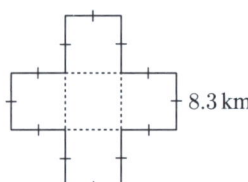 **i**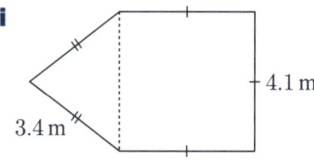

2 Find the perimeter of each figure:

a 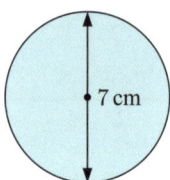 **b** 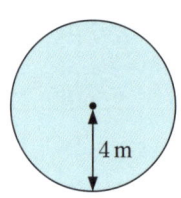 **c**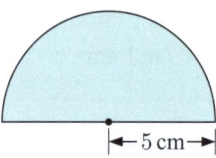

d 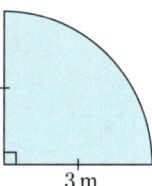 **e** 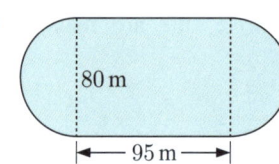 **f**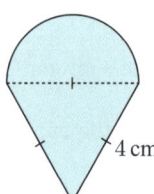

g 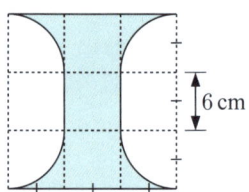 **h** 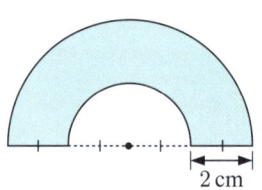 **i**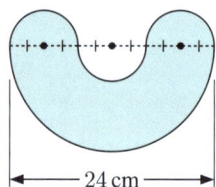

3 Find the perimeter of the following figures:

a 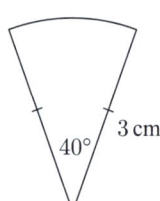 **b** 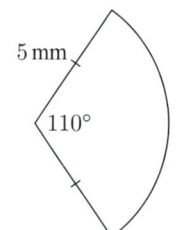 **c**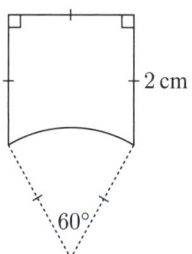

4 An Olympic boxing ring is 6.1 m by 6.1 m. It has four ropes around each side, as shown. Find:
  a the perimeter of the boxing ring
  b the total length of the rope required
  c the total cost of the rope if it costs £9.90 per metre.

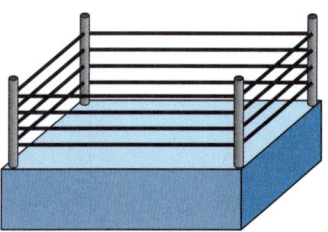

5 A rectangular room is 3.5 m by 4.8 m. It has a door 80 cm wide, and a window 1.3 m wide. A wallpaper frieze runs around the room just under the cornice, so it goes over the door and window. The frieze is repeated at eye level, and so is broken for the door and window at this height. Find the total length of frieze used in the room.

6 The horses on a merry-go-round complete 20 laps during each ride.
   a Find, correct to 1 decimal place, the total distance travelled during a ride by:
      i the black horse   **ii** the white horse.
   b How much further does the black horse travel?

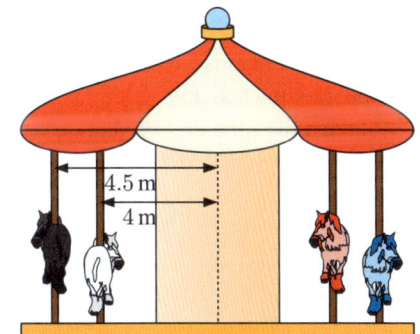

7 A tennis court has the dimensions shown.
   a What is the perimeter of the court?
   b Find the total length of all the marked lines, not including the net.

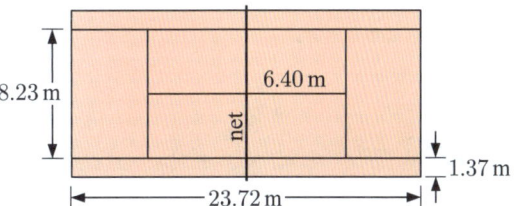

8 A football has diameter 24 cm. How many times must it roll along the ground, to travel the length of a 100 metre football field?

9 A lighting company produces conical lampshades from sectors of circles as illustrated. When the lampshades are made, lace is stitched around the circular base. Determine the total cost of the lace for 1500 lampshades if the lace costs £0.75 per metre.

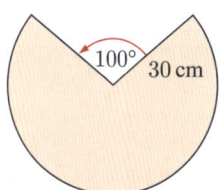

10  Find the total length of ribbon used to tie the box illustrated. 25 cm of ribbon is required for the bow.

11 At Bushby Park there is a 5 m diameter circular pond which is surrounded by a 1 m wide garden bed, and then a 3 m wide lawn. A safety fence is placed around the lawn with posts every 3 m and a gateway 1.84 m wide. The gate is wrought iron.
   a How many metres of safety fence are needed?
   b How many posts are needed?
   c If the posts cost £15.75 and the safety fence costs £18.35 per metre, calculate the total cost of the fence (excluding the gate).

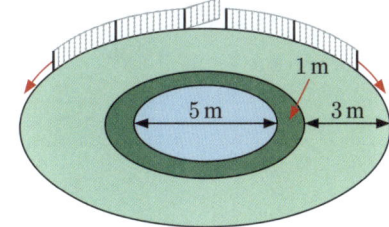

Puzzle — Rope around the Earth

Consider these two scenarios:

1 A loop of rope is placed tightly around a circular table.
1 metre of rope is then added to the loop, and the rope is stretched into a circle so there is a gap between the rope and the table.

2 A loop of rope is placed tightly around the Earth. Again, 1 metre of rope is then added to the loop, and the rope is stretched into a circle so there is a gap between the rope and the Earth.

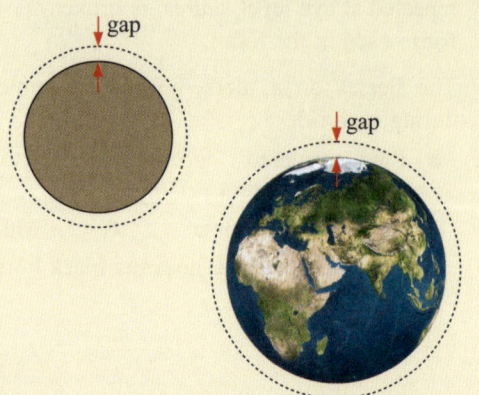

Do you think the gap will be larger in scenario **1** or scenario **2**? Perform some calculations to find out whether you are correct!

B AREA

All around us we see surfaces such as walls, ceilings, paths, and sporting fields. All of these surfaces have boundaries which define their shape.

> The **area** of a region is the amount of **surface** within its boundaries.
> The **area** of surface of a closed figure is measured in terms of the number of square units it encloses.

In previous years you should have seen the following area formulae:

Shape	Figure	Formula
Rectangle	(rectangle with length and width)	Area = length × width
Triangle	(triangles with base and height)	Area = $\frac{1}{2}$ × base × height
Parallelogram	(parallelogram with base and height)	Area = base × height
Trapezium	(trapezium with parallel sides a and b, height h)	Area = $\left(\dfrac{a+b}{2}\right) \times h$

DEMO

DEMO

DEMO

Example 4

Find the area of:

a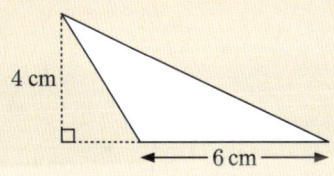

b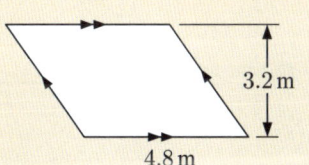

c
14 mm
12 mm
26 mm

a Area
$= \frac{1}{2}(\text{base} \times \text{height})$
$= \frac{1}{2} \times 6 \times 4$ cm^2
$= 12$ cm^2

b Area
$= \text{base} \times \text{height}$
$= 4.8 \times 3.2$ m^2
$= 15.36$ m^2

c Area
$= \left(\frac{a+b}{2}\right) \times h$
$= \left(\frac{14+26}{2}\right) \times 12$ mm^2
$= 240$ mm^2

EXERCISE 16B.1

1 Find the area of the following figures:

a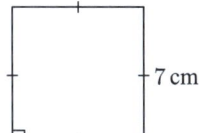

b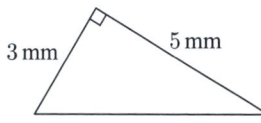

c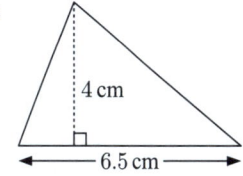

d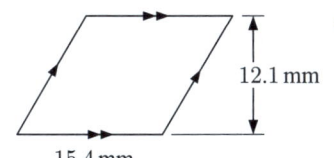

e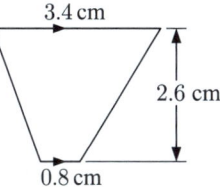

f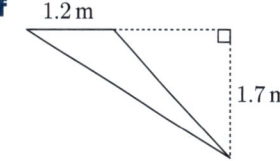

Example 5

An equilateral triangle has sides of length 6 cm. Find its area.

The altitude bisects the base at right angles.

$\therefore \ a^2 + 3^2 = 6^2$ {Pythagoras}
$\therefore \ a^2 + 9 = 36$
$\therefore \ a^2 = 27$
$\therefore \ a = \sqrt{27}$ {as $a > 0$}

Area $= \frac{1}{2} \times \text{base} \times \text{height}$
$= \frac{1}{2} \times 6 \times \sqrt{27}$
$= 3\sqrt{27}$ cm^2
≈ 15.6 cm^2

So, the area is about 15.6 cm^2.

2 Find the area of the following figures:

a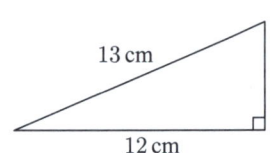

b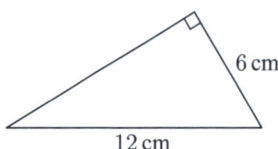

c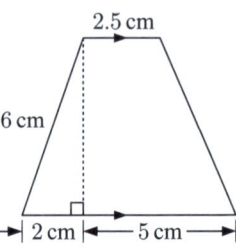

d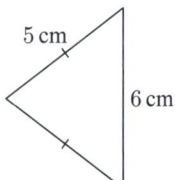

e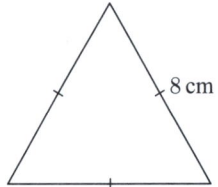

f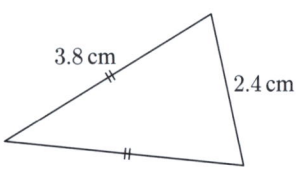

AREA OF A CIRCLE

Consider cutting a circle of radius r into 16 equal sectors and arranging them as shown:

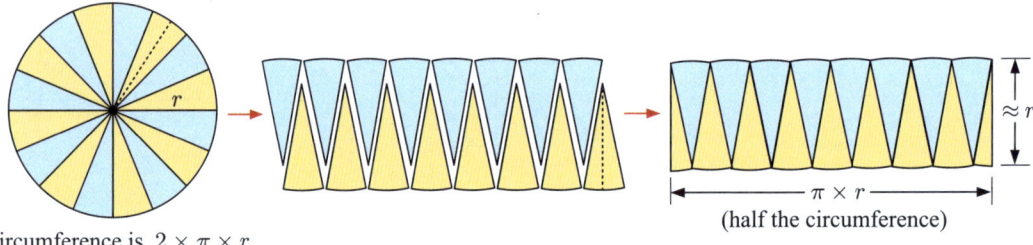

circumference is $2 \times \pi \times r$

(half the circumference)

The figure obtained closely resembles a rectangle. The height of the rectangle is the radius of the circle.

The top "edge" is the sum of all the arc lengths of the blue sectors. This is half the circumference of the circle, which is $\frac{1}{2} \times 2\pi r = \pi \times r$.

The bottom "edge" is made up from the arcs of the yellow sectors in a similar way.

If the original circle is cut into thousands of equal sectors and arranged in the same way, the resulting figure is indistinguishable from a rectangle.

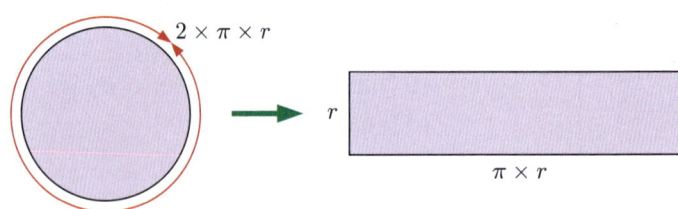

So, the area of the circle $A = $ length $\times$ width of the rectangle
$\therefore A = \pi \times r \times r$
$\therefore A = \pi r^2$

DEMO

The area of a circle of radius r is given by **Area** $= \pi r^2$.

Measurement (Chapter 16)

AREA OF A SECTOR

The area of a **sector** is a fraction of the area of the circle it is taken from.

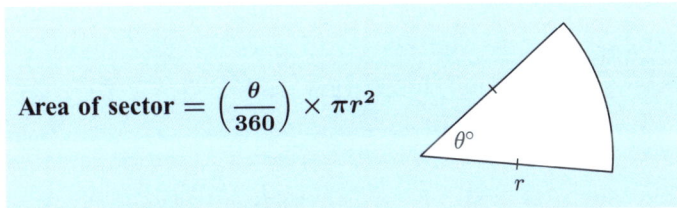

$$\text{Area of sector} = \left(\frac{\theta}{360}\right) \times \pi r^2$$

Example 6

Find, to 1 decimal place, the shaded area:

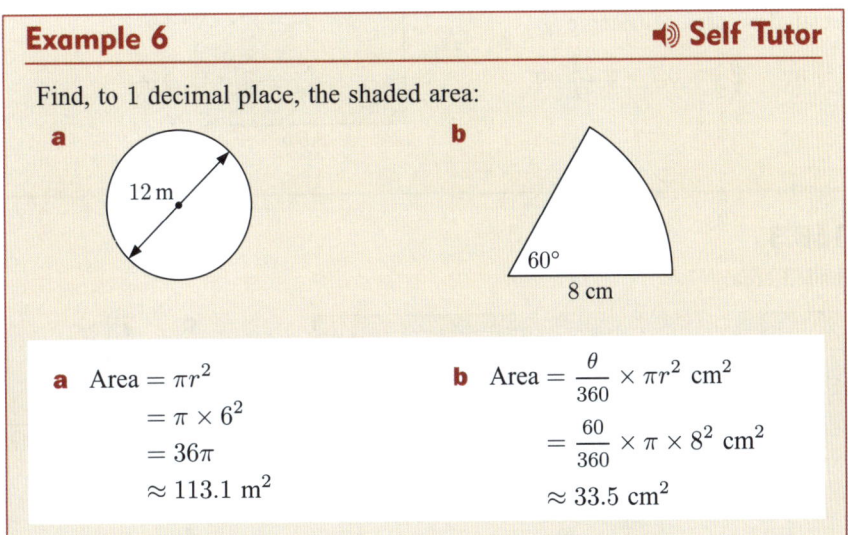

a Area $= \pi r^2$
$= \pi \times 6^2$
$= 36\pi$
≈ 113.1 m^2

b Area $= \dfrac{\theta}{360} \times \pi r^2$ cm^2
$= \dfrac{60}{360} \times \pi \times 8^2$ cm^2
≈ 33.5 cm^2

EXERCISE 16B.2

1 Find the area of a circle with: **a** radius 3 m **b** diameter 7 cm.

2 Find the shaded area:

a 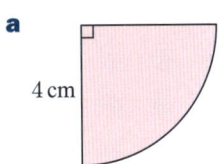   **b** 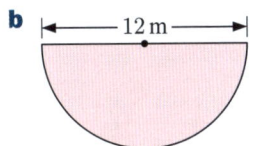   **c** 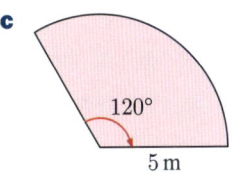   **d**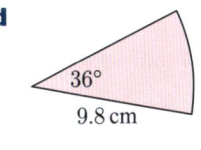

COMPOSITE SHAPES

If we have an unusual shape for which we do not have an area formula, we can often find its area by considering it as the sum or difference between several other shapes. In this case, we call it a **composite shape**.

For example,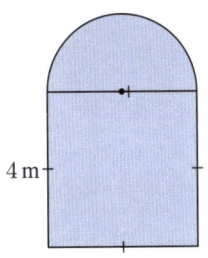

shaded area = area of square + area of semi-circle
$= (4 \times 4)$ m$^2 + \dfrac{1}{2}(\pi \times 2^2)$ m^2
$= (16 + 2\pi)$ m^2
≈ 22.3 m^2

Example 7

Find the shaded area:

Shaded area
= area of rectangle − area of triangle
= (8.6×5.4) m² − $\left(\dfrac{1}{2} \times 3.7 \times 2.2\right)$ m²
= 42.37 m²

EXERCISE 16B.3

1 Find the shaded area:

a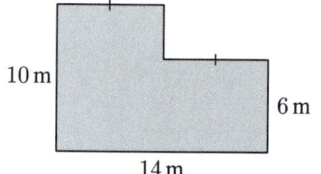

b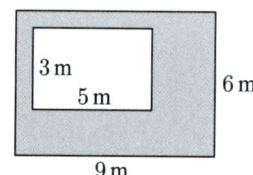

c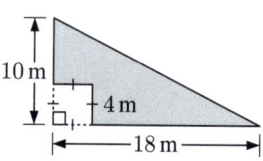

d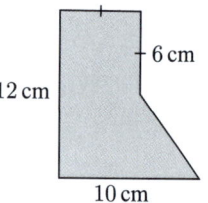

e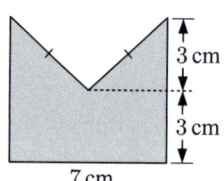

f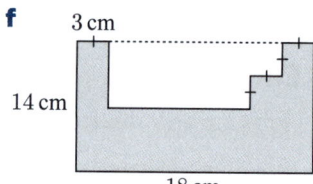

2 Find the shaded area:

a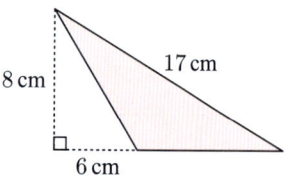

b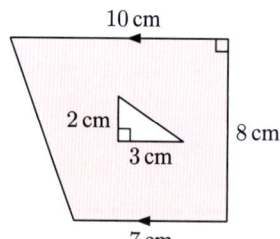

c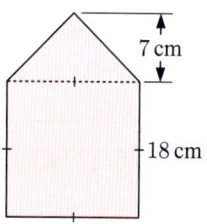

3 Find the shaded area:

a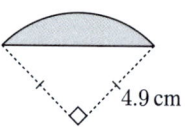

b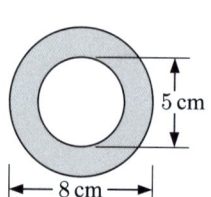

c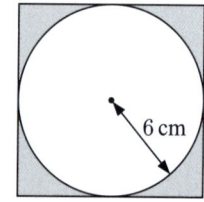

Measurement (Chapter 16) 317

d 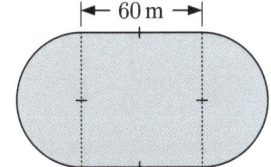 e 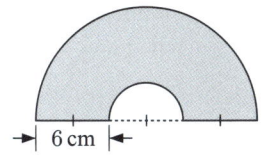 f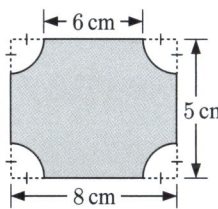

PROBLEM SOLVING

Example 8 ◀)) Self Tutor

The Highways Department orders 25 road signs that warn motorists to watch out for squirrels.
a By assuming the signs are triangular, find the area of metal sheeting (in m^2) required for the 25 signs.
b To allow for wastage when the signs are cut, an extra 20% of the metal is needed. How much metal needs to be purchased?
c The sheet metal costs £28.40 per m^2. What will its cost be?

a Area of one sign $= \frac{1}{2} \times 0.75 \times 0.65$ m^2
≈ 0.244 m^2
∴ the area of 25 signs $\approx 25 \times 0.244$ m^2
≈ 6.1 m^2

b Area needed to be purchased
$= 6.1$ m$^2 \times 120\%$
$= 7.32$ m^2

c Cost of metal
$= 7.32$ m$^2 \times$ £28.40 per m^2
$\approx$ £208

EXERCISE 16B.4

1  A farmer wishes to fertilise his paddock using 150 kg of superphosphate per hectare. The paddock is 550 m × 300 m. What amount of superphosphate will he need to spread?

2 The diagram shows the dimensions of a courtyard. It is to be paved with 60 cm square tiles costing £9.40 each.
  a How many tiles will be needed?
  b How much will the tiles cost?

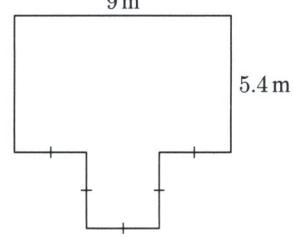

3 A gravel path 1 m wide is placed around a circular garden bed of diameter 2 m. The gravel costs £7.90 per m^2.
  a Find the area of the path.
  b Find the cost of the gravel.

4 A door is in the shape of a rectangle surmounted by a semi-circle. The width of the door is 1.2 m, and the height of the door is 2.5 m. Find the total area of the door.

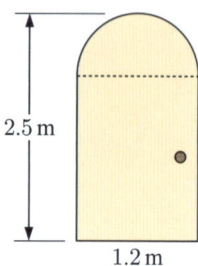

5 A restaurant uses square tables with sides of length 1.3 m, and round tablecloths with diameter 2 m. Determine the percentage of each tablecloth which overhangs its table.

6 6 identical metal discs are stamped out of an 18 cm by 12 cm sheet of copper as illustrated. What percentage of the copper is wasted?

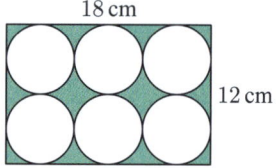

7 A 15 cm by 20 cm rectangle has the same perimeter as a square. Which figure has the greater area? Explain your answer.

8 Connie, Damien, Nick, and Uyen are sharing a pizza.
  a Find, correct to 2 decimal places, the area of:
   i Connie's slice   **ii** Damien's slice
   iii Nick's slice   **iv** Uyen's slice.
  b Find the sum of the areas in **a**. Check your answer by finding the area of the whole pizza.

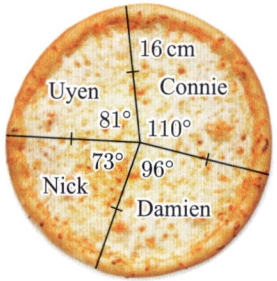

9 A chess board consists of 5 cm squares of blackwood for the black squares, and maple for the white squares. The squares are surrounded by an 8 cm wide blackwood border. Determine the percentage of the board which is made of maple.

10 The diagram shows the dimensions of a table-top. A protective cloth is cut from a roll 1.6 m wide to exactly fit the table-top. The cloth costs £18.40 per metre of length.
  a What length of cloth must be purchased?
  b Calculate the cost of the fabric.
  c Find the area of the table-top.
  d Calculate the percentage of cloth that is wasted.

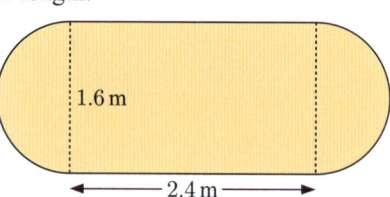

Activity 1   Sam the sheep

Click on the icon to obtain this Activity finding areas of composite shapes.

SAM THE SHEEP

Measurement (Chapter 16) 319

C SURFACE AREA

SOLIDS WITH PLANE FACES

The **surface area** of a three-dimensional figure with plane faces is the sum of the areas of the faces.

A *plane* face is one which is flat.

The surface area is therefore the same as the area of the **net** required to make the figure.

Example 9 ◆) Self Tutor

Find the surface area of the rectangular prism:

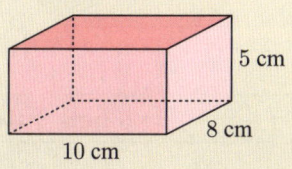

The figure has 2 faces which are 10 cm × 8 cm
                2 faces which are 10 cm × 5 cm
            and 2 faces which are   8 cm × 5 cm.

Total surface area
$= 2 \times 10 \times 8 \text{ cm}^2 + 2 \times 10 \times 5 \text{ cm}^2 + 2 \times 8 \times 5 \text{ cm}^2$
$= (160 + 100 + 80) \text{ cm}^2$
$= 340 \text{ cm}^2$

Example 10 ◆) Self Tutor

Find the total surface area of this wedge.

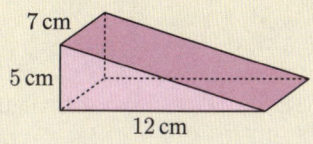

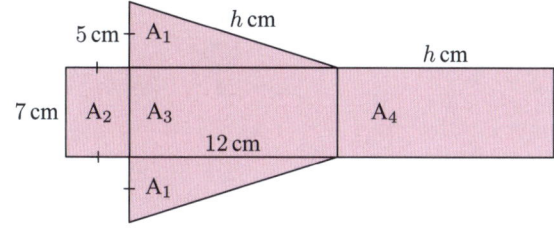

Let the slant edge have length h cm.
$\therefore h^2 = 12^2 + 5^2$ {Pythagoras}
$\therefore h^2 = 169$
$\therefore h = \sqrt{169} = 13$ {as $h > 0$}

$A_1 = \frac{1}{2}bh$ $\quad A_2 = 7 \times 5$ $\quad A_3 = 12 \times 7$ $\quad A_4 = 13 \times 7$
$ = \frac{1}{2} \times 12 \times 5$ $ = 35 \text{ cm}^2$ $ = 84 \text{ cm}^2$ $ = 91 \text{ cm}^2$
$ = 30 \text{ cm}^2$

The total surface area $= 2 \times A_1 + A_2 + A_3 + A_4$
$= 2 \times 30 + 35 + 84 + 91$
$= 270 \text{ cm}^2$

EXERCISE 16C.1

1. Find the surface area of a cube with sides:
   - **a** 3 cm
   - **b** 4.5 cm
   - **c** 9.8 mm

2. Find the surface area of the following rectangular prisms:

   a 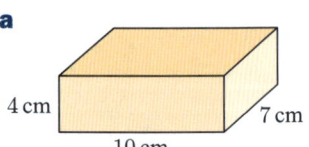   **b** 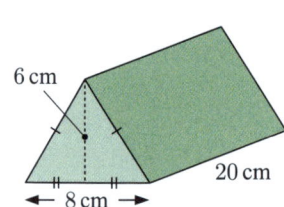   **c**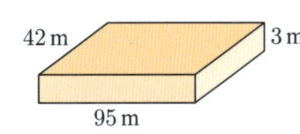

3. Find the surface area of the following triangular prisms:

   a 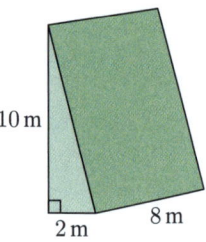   **b** 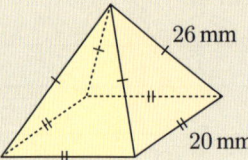   **c**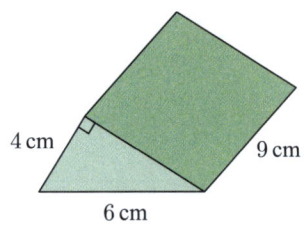

Example 11 ◀) Self Tutor

Find the surface area of the square-based pyramid:

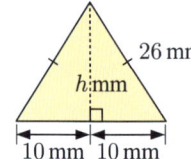

The figure has:
- 1 square base
- 4 triangular faces

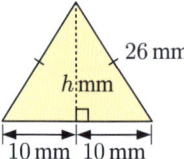

$h^2 + 10^2 = 26^2$ {Pythagoras}
$\therefore h^2 + 100 = 676$
$\therefore h^2 = 576$
$\therefore h = 24$ {as $h > 0$}

Total surface area $= 20 \times 20 \text{ mm}^2 + 4 \times \left(\frac{1}{2} \times 20 \times 24\right) \text{ mm}^2$
$= (400 + 960) \text{ mm}^2$
$= 1360 \text{ mm}^2$

4. Find the surface area of the following square-based pyramids:

   a 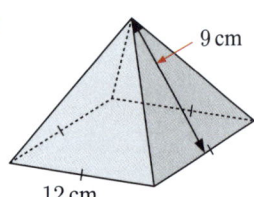   **b** 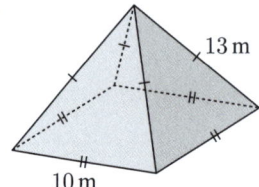   **c**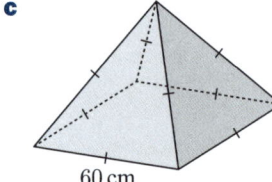

5 Find the surface area of each solid:

a 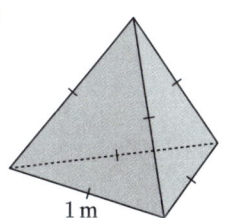 1 m

b 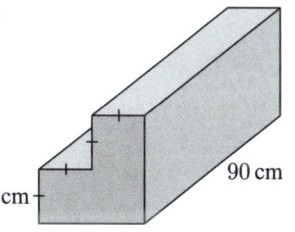 16 cm, 90 cm

c 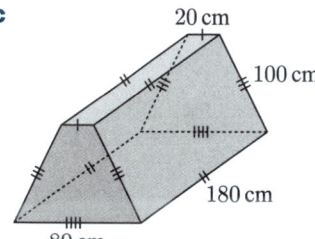 20 cm, 100 cm, 180 cm, 80 cm

6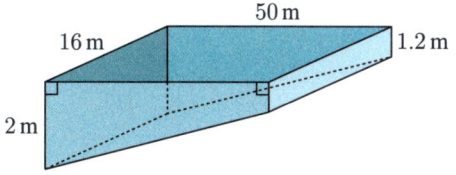

The base and walls of the swimming pool shown are tiled. The tiles cost £25 per m².

a Find the total area of tiles.

b Find the value of the tiles.

7 A marquee with the dimensions as shown is made from canvas. The marquee has no floor. Find the total cost of the canvas if it costs £31.50 per square metre.

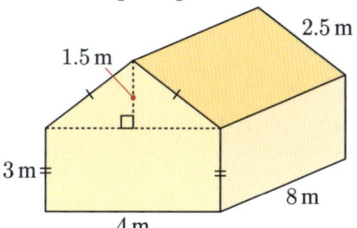

8 An aviary has the shape of a hexagonal prism. Find the area of netting required to cover the aviary, including the floor.

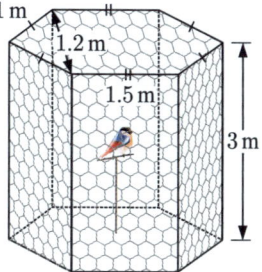

CYLINDERS

The cylinder shown has no top or bottom. If the cylinder is cut, opened out, and flattened, it takes the shape of a rectangle.

The length of the rectangle is the circumference of the cylinder.

The width of the rectangle is the height of the cylinder.

∴ for a hollow cylinder, the outer surface area

$$A = \text{area of rectangle}$$
$$= \text{length} \times \text{width}$$
$$= 2\pi rh$$

DEMO

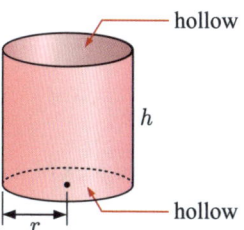

Hollow cylinder (no ends)	Open can (one end)	Solid cylinder (two ends)
$A = 2\pi rh$	$A = 2\pi rh + \pi r^2$	$A = 2\pi rh + 2\pi r^2$

322 Measurement (Chapter 16)

CONES

The curved surface of a cone is made from a sector of a circle. The radius of the sector is equal to the slant height s of the cone. The arc length AB of the sector is equal to the circumference of the base.

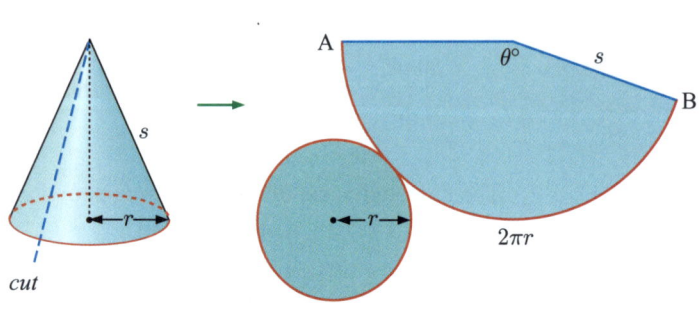

arc AB $= 2\pi r$

$\therefore \left(\dfrac{\theta}{360}\right) 2\pi s = 2\pi r$

$\therefore \dfrac{\theta}{360} = \dfrac{r}{s}$

$\therefore$ the area of the curved surface
$=$ the area of the sector

$= \left(\dfrac{\theta}{360}\right) \pi s^2$

$= \dfrac{r}{s} \times \pi s^2$

$= \pi r s$

Hollow cone (no end)	Solid cone (closed end)
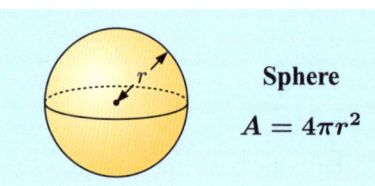 $A = \pi r s$	 $A = \pi r s + \pi r^2$

SPHERES

Sphere
$A = 4\pi r^2$

SURFACE AREA OF A SPHERE

Example 12 ◀)) Self Tutor

Find the surface area of each solid:

a 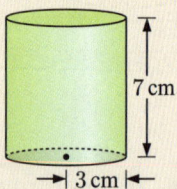 7 cm, 3 cm

b 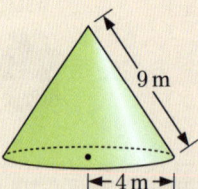 9 m, 4 m

c 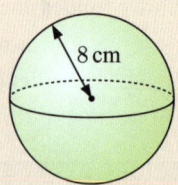 8 cm

a Surface area $= 2\pi r h + 2\pi r^2$
$= (2 \times \pi \times 3 \times 7 + 2 \times \pi \times 3^2)$ cm^2
≈ 188 cm^2

b Surface area $= \pi r s + \pi r^2$
$= (\pi \times 4 \times 9 + \pi \times 4^2)$ m^2
≈ 163 m^2

c Surface area $= 4\pi r^2$
$= 4 \times \pi \times 8^2$ cm^2
≈ 804 cm^2

EXERCISE 16C.2

1 Find the outer surface area of each cylinder:

a solid

b can (no top)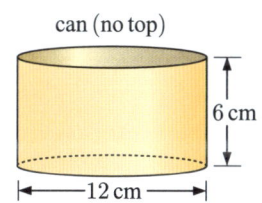

c solid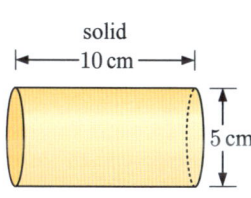

d well (no top)

e solid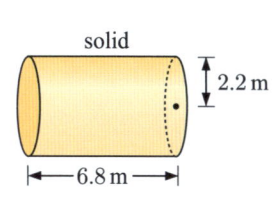

f hollow throughout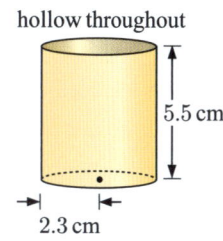

2 Find the outer surface area of each cone:

a solid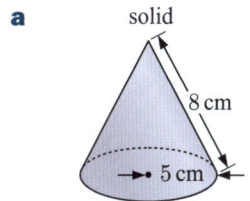

b hollow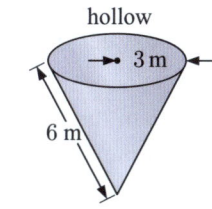

c solid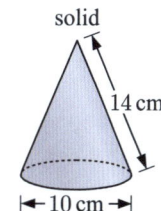

d hollow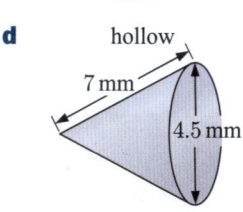

e solid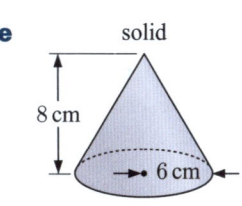

f hollow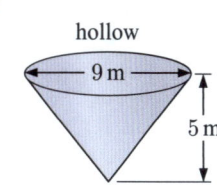

3 Find the surface area of each solid:

a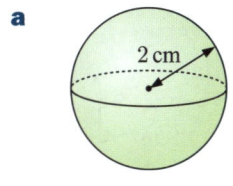

b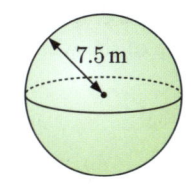

c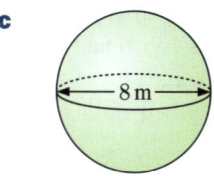

d

e solid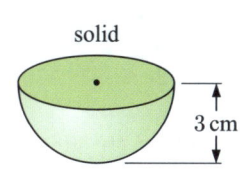

f solid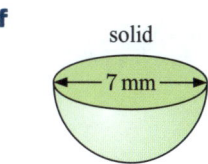

4 A ball bearing has a radius of 1.2 cm. Find the surface area of the ball bearing.

5 A conical piece of filter paper has a base radius of 2 cm, and is 5 cm high.

  a Find the slant height l.

  b Hence find the outer surface area of the filter paper.

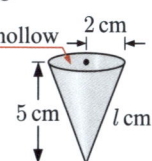

6 Find, correct to 1 decimal place, the surface area of each solid:

a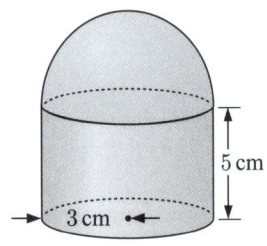

b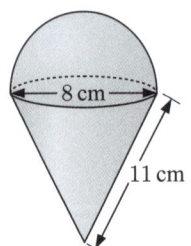

c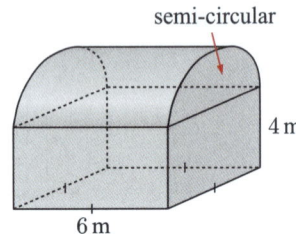

7 A cylindrical tank is 6 m long and has diameter 8 m. Its outer surface (including both ends) is to be painted bright red. Each litre of paint covers 5 m². It is purchased in 5 litre cans costing £52.50 each.
   a Find the surface area to be painted.
   b Find the number of cans of paint which must be purchased.
   c What is the cost of the paint?

8 We commonly use a sphere to model the Earth, even though it is not a *perfect* sphere. The Earth has a radius of approximately 6400 km.
   a Estimate the surface area of the Earth.
   b 71% of the Earth's surface is covered by water. Estimate this area.
   c China has a land area of 9 706 961 km².
      i What percentage of the surface area of the Earth is China?
      ii What percentage of the *land* area of the Earth is China?

9 The outside of this observatory needs cleaning. The cost of cleaning the hemispherical glass roof is £4 per square metre, and the cost of cleaning the cylindrical walls is £2.50 per square metre. Find, to the nearest pound, the total cost of cleaning the observatory.

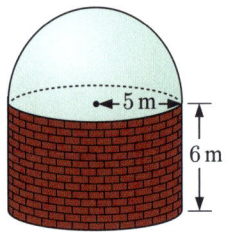

10 A wedge with angle 20° is cut from the centre of a cylindrical cake of radius 13 cm and height 6 cm. Each side of the wedge, excluding the bottom, is to be covered with icing.
Find the total surface area of cake to be iced.

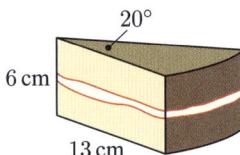

D VOLUME

The **volume** of a solid is the amount of space it occupies.

RECTANGULAR PRISM

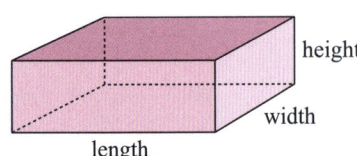

Volume = length × width × height

SOLIDS OF UNIFORM CROSS-SECTION

In the triangular prism alongside, any vertical slice parallel to the front triangular face will be the same size and shape as that face. Solids like this are called *solids of uniform cross-section*. The cross-section in this case is a triangle.

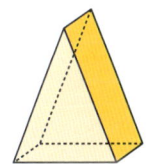
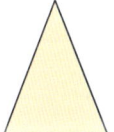

Another example is this hexagonal prism:

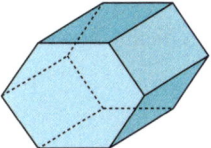
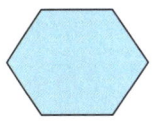

For any solid of uniform cross-section:

Volume = area of cross-section × length

For a **cylinder**, the cross-section is a circle.

Volume = area of circle × height
$= \pi r^2 \times h$

Volume $= \pi r^2 h$

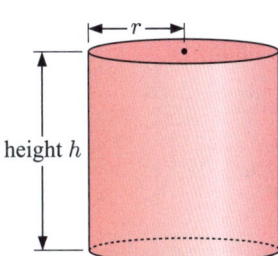

Example 13   ◀) Self Tutor

Find the volume of each solid:

a
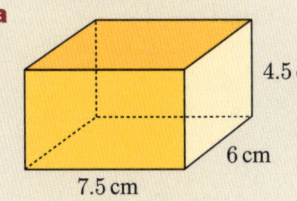

b area = 30 cm²
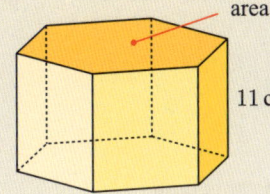

c
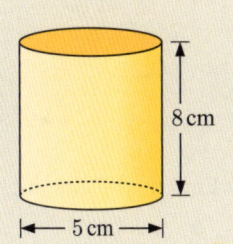

a   Volume
= length × width × height
= 7.5 cm × 6 cm × 4.5 cm
= 202.5 cm³

b   Volume
= area of cross-section × height
= 30 cm² × 11 cm
= 330 cm³

c   The base has diameter 5 cm, so the radius is 2.5 cm.
$V = \pi r^2 h$
$= \pi \times 2.5^2 \times 8$ cm³
≈ 157 cm³

EXERCISE 16D.1

1 Find the volume of each rectangular prism:

a 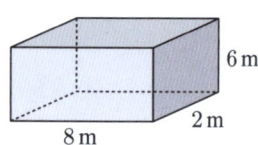 b 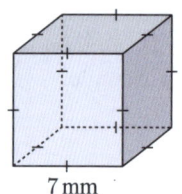 c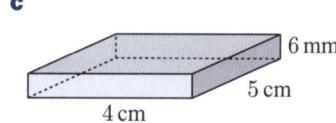

2 Find, rounded to 3 significant figures, the volume of each cylinder:

a 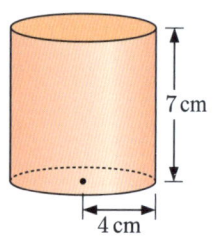 b 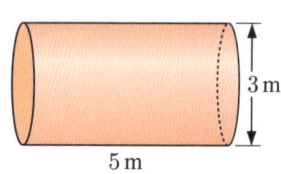 c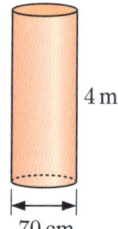

3 Find the volume of each solid:

a 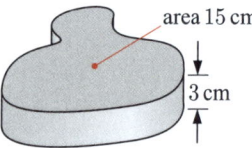 b 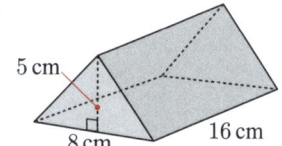 c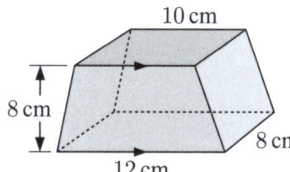

d 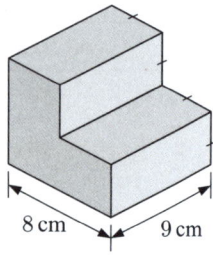 e 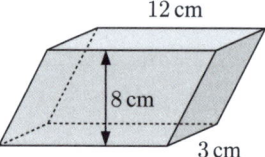 f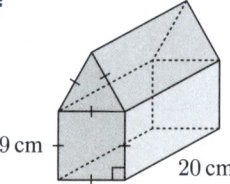

4 Find the volume of:

a this brick 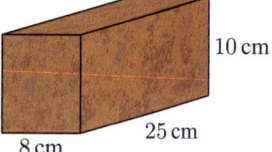 b this coin 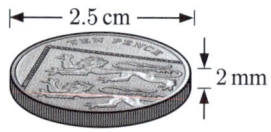 c this wedge of cheese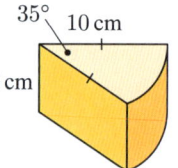

d this picket 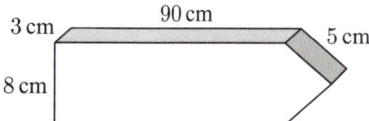 e this dumbbell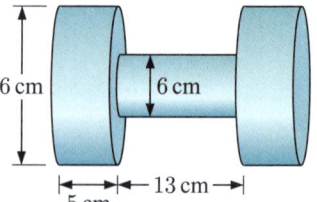

5 1 m³ of brass is melted down and cast into solid door handles with the shape shown. How many handles can be cast?

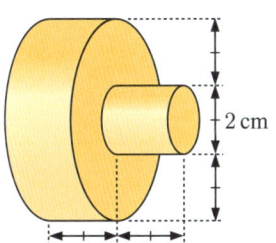

6 A concrete tank has an external diameter of 5 m and an internal height of 3 m. The walls and base of the tank are 20 cm thick. The concrete costs £142 per m³.
  a Find the volume of concrete required to make:
    i the base   **ii** the walls.
  b Find the total volume of concrete required.
  c Find the cost of the concrete required to make the tank.

7 A swimming pool has the dimensions shown alongside.
  a Find the area of a trapezium-shaped side.
  b Determine the volume of water required to fill the pool.

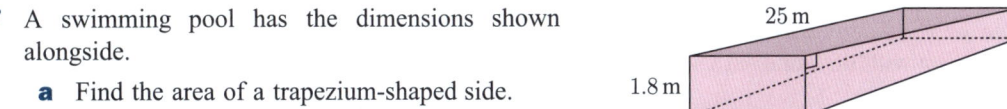

8 In the town square, there is a fountain in the middle of a circular pond. The pond is 6 metres in diameter. A concrete wall 30 cm wide and 60 cm high is built around the edge of the pond.
  a Find the area of the top of the wall.
  b Find the volume of concrete required for the wall.

9 A rubber tube has the dimensions shown.
  a Find the cross-sectional area of one end of the tube.
  b Hence find the volume of rubber used to make the tube.
  c If the rubber weighs 1500 kg per m³, find the weight of the tube.

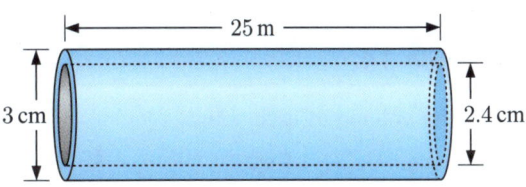

Activity 2   Packing boxes

Click on the icon to obtain this Activity.

PACKING BOXES

PYRAMIDS AND CONES

Pyramids and cones are known as **tapered solids**. They have a flat base, and come to a point called the **apex**.

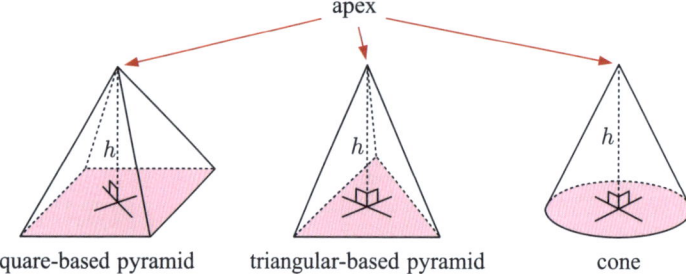

square-based pyramid   triangular-based pyramid   cone

Tapered solids do **not** have a uniform cross-section. For example, we can see that the perpendicular cross-section of a cone is always a circle, but its radius decreases as we move up the cone.

The volume of a square-based pyramid is one third of the volume of a prism with the same base area and height.

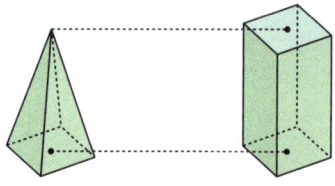

Similarly, the volume of a cone is one third of the volume of a cylinder with the same base area and height.

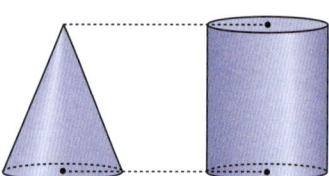

$$\text{Volume of a tapered solid} = \tfrac{1}{3} \times \text{area of base} \times \text{height}$$

Example 14   ◀⑨ Self Tutor

Find the volume of each solid:

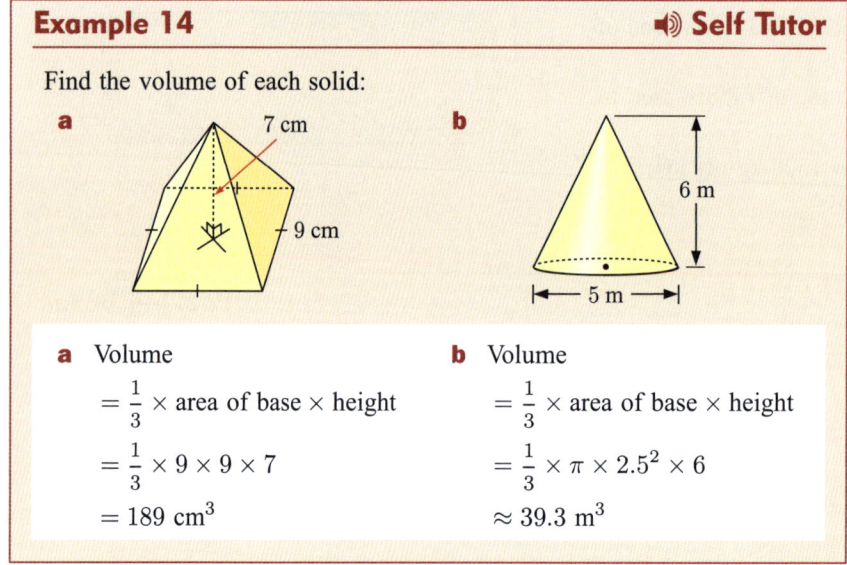

a Volume
$= \tfrac{1}{3} \times \text{area of base} \times \text{height}$
$= \tfrac{1}{3} \times 9 \times 9 \times 7$
$= 189$ cm^3

b Volume
$= \tfrac{1}{3} \times \text{area of base} \times \text{height}$
$= \tfrac{1}{3} \times \pi \times 2.5^2 \times 6$
≈ 39.3 m^3

The volume of a cone is $V = \tfrac{1}{3}\pi r^2 h$.

EXERCISE 16D.2

1 Find the volume of each solid:

a 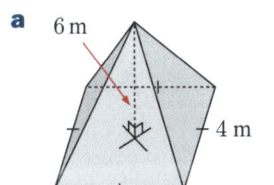   b 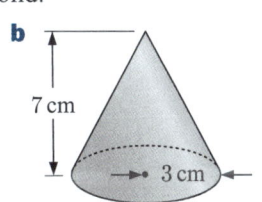   c 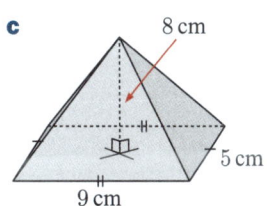   d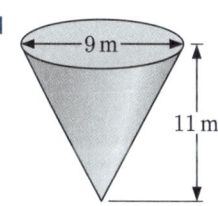

2 Use Pythagoras' theorem to help find the volume of each solid:

a 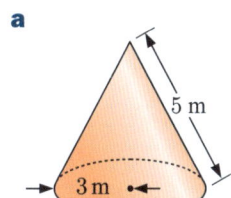   b 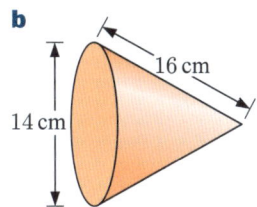   c 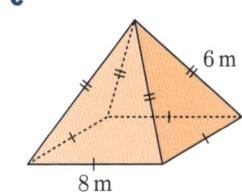   d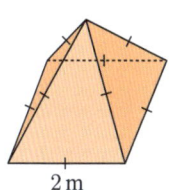

3 The *Pyramide du Louvre* in Paris is a square-based pyramid with sides 35 metres long and height 21.6 metres.
Calculate the volume of air in this pyramid.

4 A conical heap of gravel is 1.4 m high and has a diameter of 2.6 m. Find the volume of gravel in the heap.

5 A wax crayon has the dimensions shown. Find the total volume of wax required to make a set of 24 crayons.

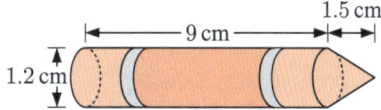

6 0.6 cubic metres of metal is melted down and cast into garden stakes with the dimensions shown. How many garden stakes can be made?

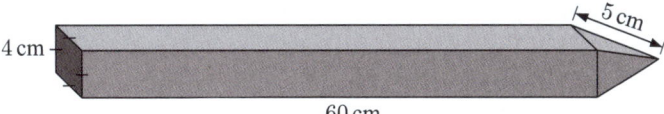

SPHERES

The volume of a sphere with radius r is given by

$$V = \tfrac{4}{3}\pi r^3$$

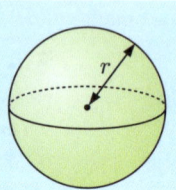

Example 15
◀)) **Self Tutor**

Find the volume of this sphere:

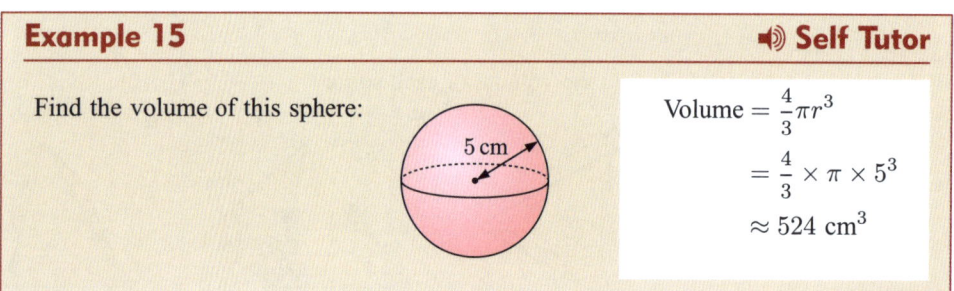

Volume $= \tfrac{4}{3}\pi r^3$

$= \tfrac{4}{3} \times \pi \times 5^3$

≈ 524 cm^3

EXERCISE 16D.3

1 Find the volume of each solid:

a 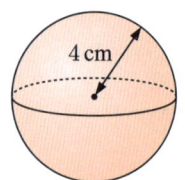 **b** 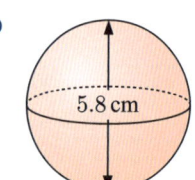 **c** 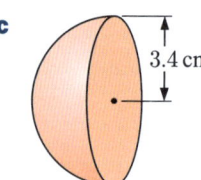 **d**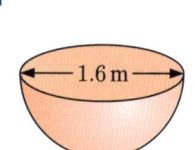

2 Find the volume of:
   a a sphere with radius 6 mm
   b a hemisphere with radius 7 cm.

3 Find the volume of each solid:

a 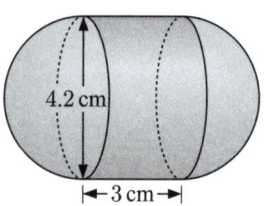 **b** 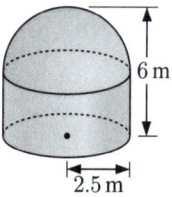 **c**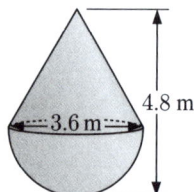

4 A beach ball has a diameter of 1.2 m. Find the volume of air inside the ball.

5 A garden ornament is shaped like a mushroom. The base is a cylinder 8 cm high and 6 cm in diameter. The top of the mushroom is 20 cm in diameter.
What volume of concrete, in m^3, is needed to make 50 mushrooms?

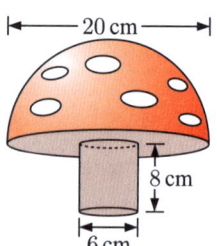

Measurement (Chapter 16) 331

6 A hollow spherical glass bauble contains a winter snowman scene.
The bauble has internal diameter 6.8 cm and external diameter 7.0 cm. What volume of glass was used to make it?

7 A block of lead 3.2 cm × 2.1 cm × 4.6 cm is melted and used to craft solid spheres with radius 6 mm.
  a Find the volume of the block of lead.
  b Find the volume of each sphere.
  c How many spheres can be made?
  d What percentage of the lead will be wasted?

E CAPACITY

The **capacity** of a container is the quantity of fluid or gas used to fill it.

You should remember that the units of volume and capacity are connected:

$$1 \text{ ml} \equiv 1 \text{ cm}^3$$
$$1 \text{ litre} \equiv 1000 \text{ cm}^3$$
$$1 \text{ kl} = 1000 \text{ litres} \equiv 1 \text{ m}^3$$

Example 16   ◀》 Self Tutor

Find the capacity of a 3 m by 2.4 m by 1.8 m tank.

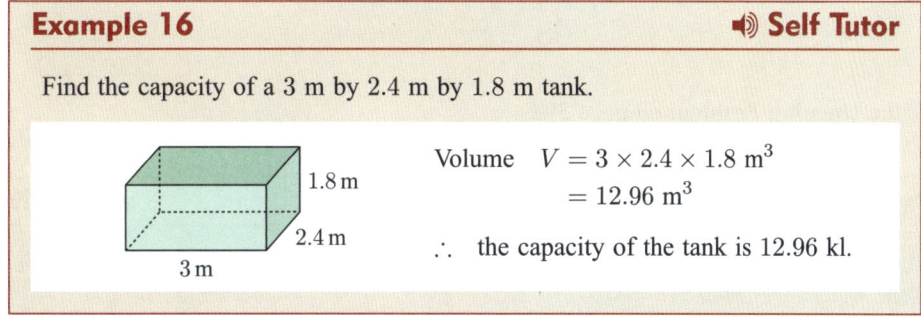

Volume $V = 3 \times 2.4 \times 1.8 \text{ m}^3$
             $= 12.96 \text{ m}^3$

∴ the capacity of the tank is 12.96 kl.

EXERCISE 16E

1 Find the capacity of each container:

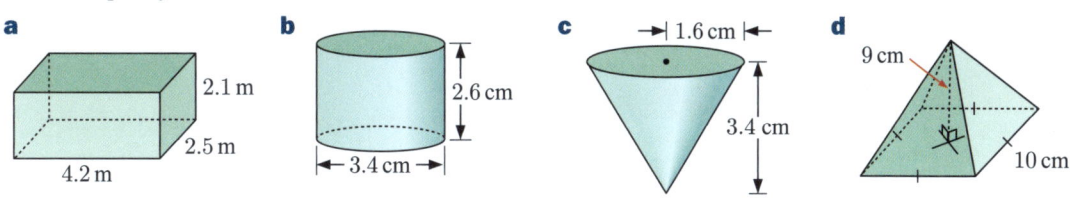

2 How many cylindrical bottles 12 cm high and with 6 cm diameter could be filled from a tank containing 125 litres of detergent?

3 A bottle of perfume is spherical with a diameter of 36 mm. Calculate the capacity of the bottle, in ml.

4 The average depth of water in a lake is 1.7 m. It is estimated that the total surface area of the lake is 1.35 ha.
   a Convert 1.35 ha to m².
   b How many kilolitres of water does the lake contain?

5 A car has a rectangular prism petrol tank with dimensions 48 cm by 56 cm by 20 cm.
   a Find the capacity of the petrol tank.
   b The car consumes petrol at an average rate of 8.7 litres per 100 km. How far could it travel on a full tank of petrol?

6 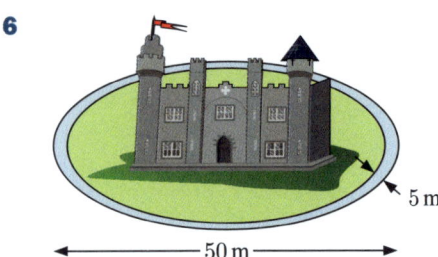 A castle is surrounded by a circular moat which is 5 m wide and 2 m deep. The diameter of the outer edge of the moat is 50 m. Find, in kilolitres, the quantity of water in the moat.

7 The inside of a glass vase consists of a cylinder with diameter 8 cm and depth 15 cm, with a hemisphere below it.
Find the quantity of water that the vase can hold.

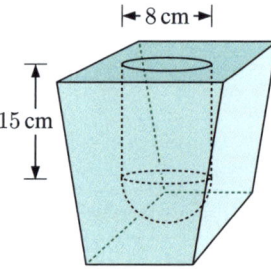

8 Answer the **Opening Problem** on page 308.

Example 17 ◀) Self Tutor

14.4 kl of water is pumped into an empty cylindrical tank with base radius 2 m. How high up the tank will the water level rise?

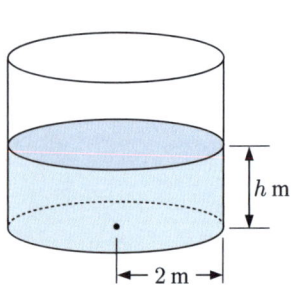

Since $1 \text{ kl} \equiv 1 \text{ m}^3$, the volume of water pumped into the tank is 14.4 m³.

If the water in the cylindrical tank is h m deep,
its volume $= \pi \times 2^2 \times h$ $\{V = \pi r^2 h\}$
$\phantom{\text{its volume }} = 4\pi h \text{ m}^3$
$\therefore \ 4\pi h = 14.4$ {equating volumes}
$\therefore \ h = \dfrac{14.4}{4\pi} \approx 1.15$

$\therefore$ the water level will rise to 1.15 m.

---

9 20 kl of water is pumped into an empty cylindrical tank with base radius 3 m. How high up the tank will the water level rise?

10 Water enters a cylindrical rainwater tank at 80 litres per minute. The base diameter of the tank is 2.4 m and the height is 4 m.

  a Find the capacity of the full tank.

  b How high up the tank will the water level rise after 10 minutes?

Review set 16A

1 Find the perimeter and area of each figure:

  a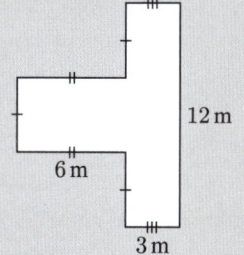

  b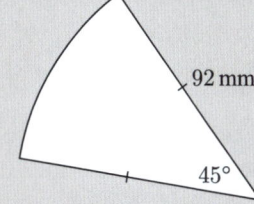

  c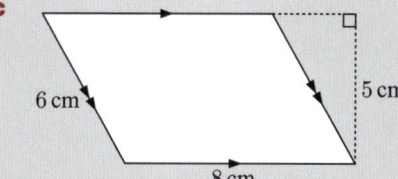

  d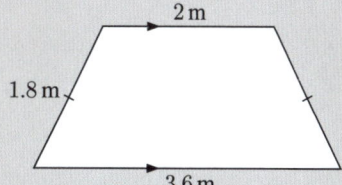

2 A rectangular bathroom measuring 3 m by 2 m is to be decorated on all walls with a single row of patterned tiles. There is a doorway measuring 90 cm wide. Each patterned tile is 15 cm long and costs £5. Find:

  a the total length of patterned tiles required   **b** the total cost of the tiles.

3 A circular playing field has radius 80 metres. The field is surrounded by a fence 10 metres from the edge of the field. Determine the length of the fence.

4 Find the area of:

  a a sector with radius 10 cm and angle 120°

  b a right angled triangle with base 5 cm and hypotenuse 13 cm

  c a circle with diameter 15 cm.

5 Find the outer surface area of each solid:

  a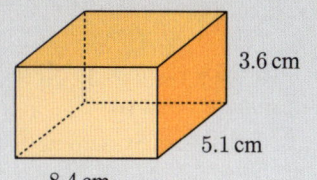

  b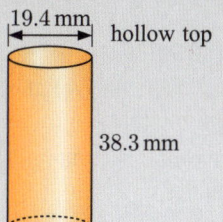

  c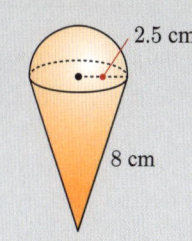

6 Find the volume of each solid:

a 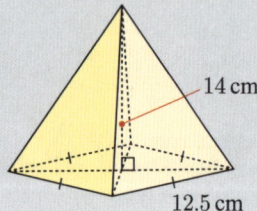   b 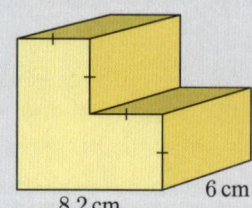   c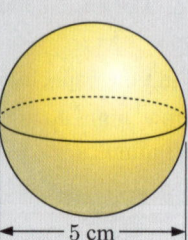

7 A solid cone has radius 5 cm and perpendicular height 8 cm. Find its:
   a volume
   b surface area.

8 Soup cans have a base diameter of 7 cm and a height of 10 cm.
   a Exactly how many such cans can be filled from a vat containing 2000 litres of soup?
   b Calculate the total surface area of metal required to make the cans in **a**.

9 A semi-circular tunnel with the dimensions shown is made of concrete. The tunnel is 220 m long, and the concrete costs £256 per m^3.

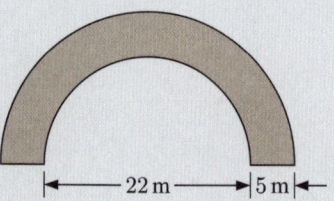

   a Find the cross-sectional area of the tunnel.
   b Find the volume of concrete used in the tunnel.
   c Find the cost of the concrete.

10 a Find the capacity of a swimming pool with the dimensions shown.
   b If the pool was filled to a depth 10 cm from the top, how much water would it contain?

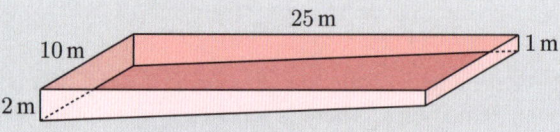

11 The Pyramid of Khufu in Egypt has a square base with sides 230.4 m and a height of 138.8 m.

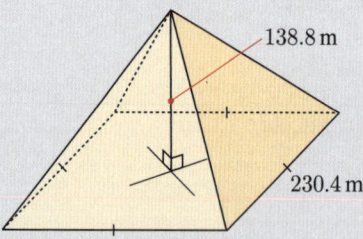 

   a Find the total volume of the pyramid.
   b If each cubic metre of stone weighs 2.67 tonnes, find the total mass of stone used. Give your answer in standard form. Assume that the pyramid is solid stone for this part.
   c The King's Chamber in the pyramid is rectangular and measures 10.47 m by 5.23 m by 5.97 m. Find the capacity of the chamber.

Review set 16B

1 Find the perimeter and area of each figure:

a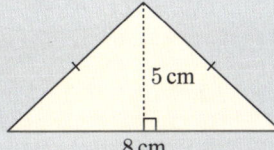

b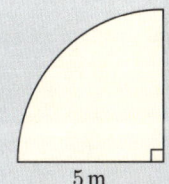

c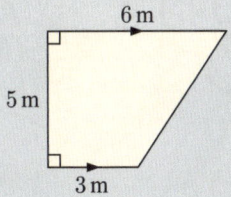

d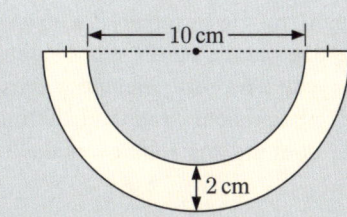

2 Find the perimeter of:

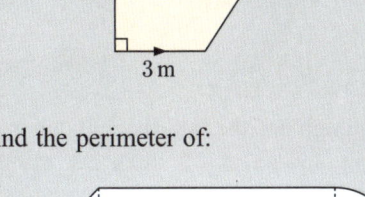

3 Find the shaded area, rounded to 2 decimal places.

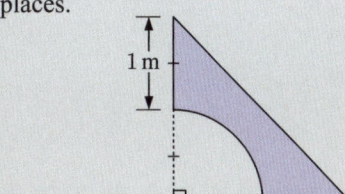

4 Find the outer surface area of each solid:

a 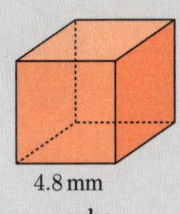 cube

b 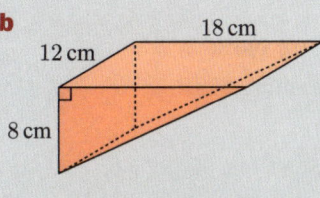 triangular prism

c 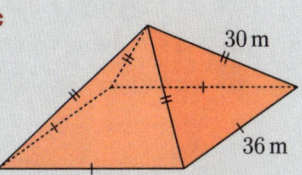 square-based pyramid

d 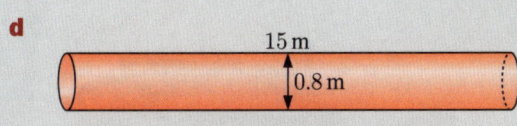 hollow throughout

5 The moon is approximately spherical with radius 1737 km. Estimate:
  a the distance around the equator of the moon
  b the surface area of the moon.

6 The diagram shows the cross-section of a railway cutting that needs to be excavated. The cutting will be 56 m long. Find the volume of soil that needs to be excavated.

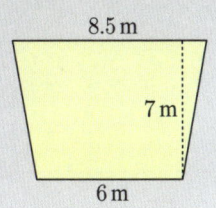

7 Find the volume of each solid:

a 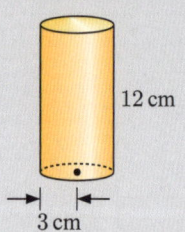   b   c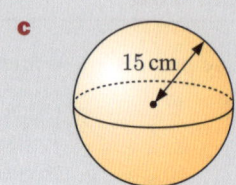

8 A security-conscious man builds himself a round fort of radius 10 m. He surrounds it with a moat 3 m wide. A large dog patrols on the other side of the moat to deter anyone who may attempt to cross it. The dog makes 150 complete circuits of the moat every day. How far does the dog walk every day?

9 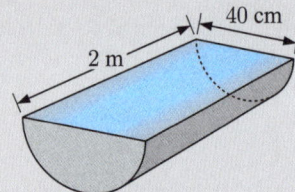 A horse's drinking trough has the dimensions shown. Water flows into the empty trough at 12 litres per minute for 10 minutes. Will the trough overflow?

10 A block of wood is 30 cm by 40 cm by 20 cm. Its corner is chopped by an axe, and a section removed as shown. Find, to 1 decimal place:

  a the remaining volume of the block
  b the surface area of the resulting block.

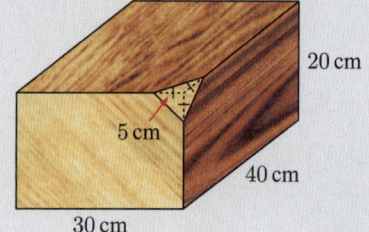

11 The Pantheon in Rome was built during the reign of emperor Augustus around 27 BC, and rebuilt by the emperor Hadrian around 126 AD. Its centre is designed as a hemisphere on a cylinder so that a sphere with diameter 43.3 m will fit exactly inside it.

Find the capacity of this chamber in kl.

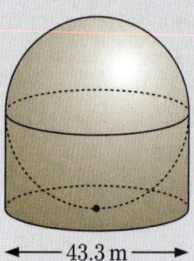

17

Formulae and functions

Contents:
- **A** Number machines
- **B** Formula construction
- **C** Substituting into formulae
- **D** Rearranging formulae
- **E** Rearrangement and substitution
- **F** Predicting formulae

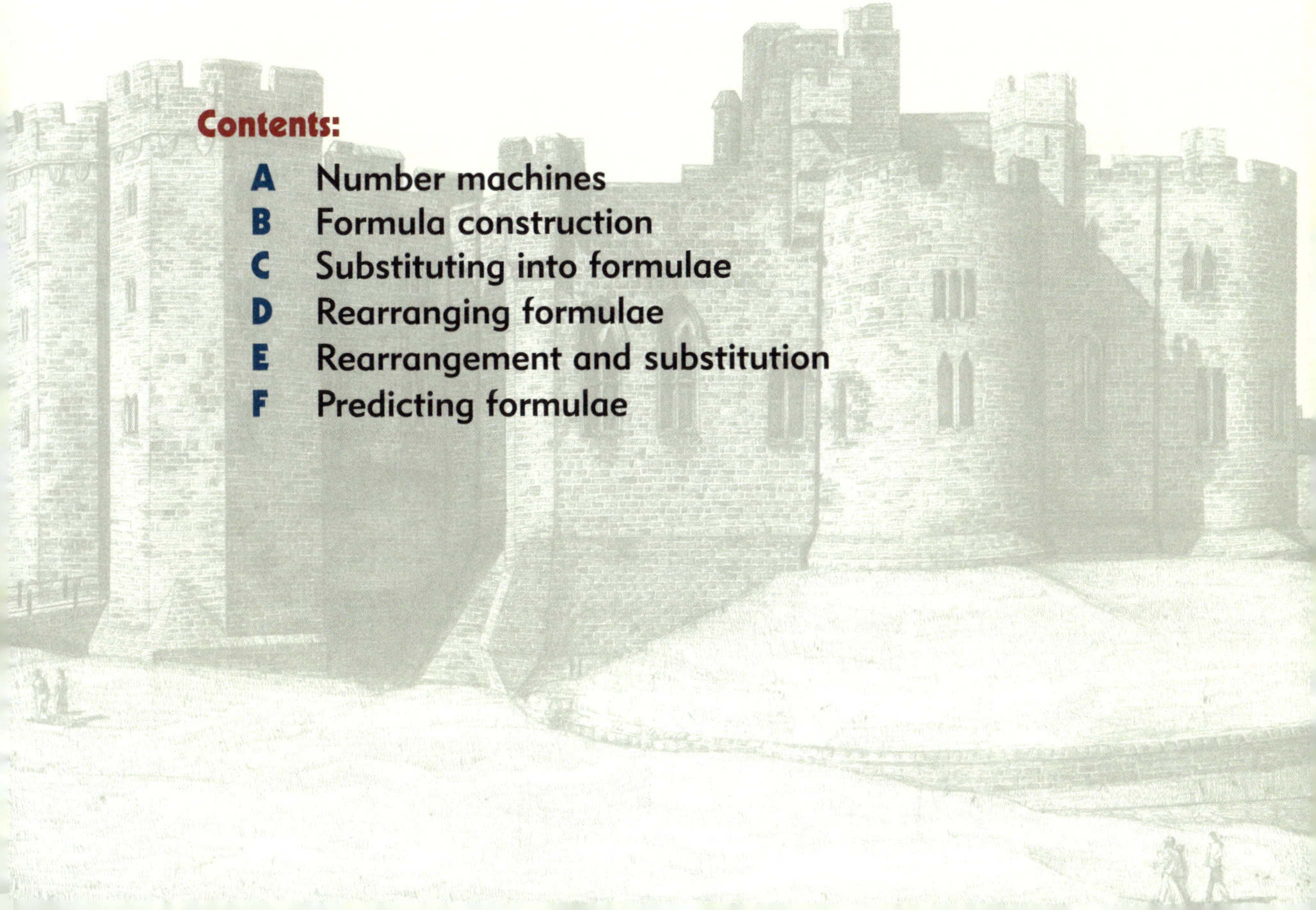

Opening problem

In Gaelic football games, teams can score goals worth 3 points each, as well as individual points.

While watching a football game, Josh noticed something unusual about Mayo's score.

Mayo had scored 2 goals and 6 points, which is a total of 12 points, but he also recognised that $2 \times 6 = 12$.

Josh wondered whether there were other football scores with this property.

Things to think about:

a For a score of g goals and p points to have this property, can you explain why $3g + p = gp$?

b Can you *rearrange* this formula to make p the subject?

c By *substituting* different values for g, can you find other scores which have this property?

A **formula** is an equation which connects two or more variables.

For example, the formula $s = \dfrac{d}{t}$ relates the three variables *speed* (s), *distance travelled* (d), and *time taken* (t).

We usually write a formula with one variable on its own on the left hand side. The other variable(s) and constants are written on the right hand side.

The variable on its own is called the **subject** of the formula. We say this variable is written *in terms of* the other variables.

A NUMBER MACHINES

Consider a machine which operates on numbers.

For any **input number** put into the machine, the machine calculates an **output number** according to a rule or formula.

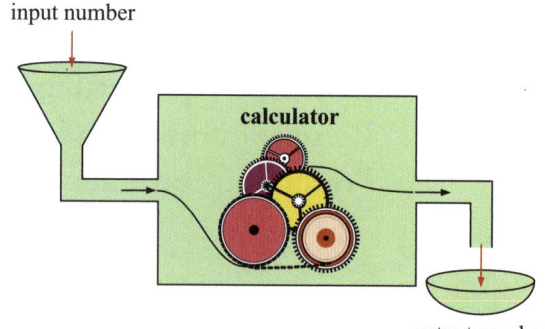

For example, consider the rule "the output number M is three times the input number n, plus seven".

We can write this rule in symbols as the formula $M = 3 \times n + 7$
or $M = 3n + 7$.

Input number (n)	Calculation	Output number (M)
1	$3 \times 1 + 7$	10
2	$3 \times 2 + 7$	13
3	$3 \times 3 + 7$	16
⋮	⋮	⋮

Formulae and functions (Chapter 17) 339

Example 1 ◀) Self Tutor

Consider the rule *three times the input number, then subtract one*.

a Write the rule in symbols using input number n and output number M.
b Calculate the output numbers corresponding to input numbers 1, 2, 5, and 10.

a $M = 3 \times n - 1$ or $M = 3n - 1$

b
Input number (n)	Calculation	Output number (M)
1	$3 \times 1 - 1$	2
2	$3 \times 2 - 1$	5
5	$3 \times 5 - 1$	14
10	$3 \times 10 - 1$	29

The number machine will perform operations in a rule or formula in a particular order. We can use a **flowchart** to show the calculations.

Example 2 ◀) Self Tutor

Consider the rule *three times the input number, plus one*.

Use a flow chart to find the output numbers corresponding to input numbers 1, 2, 4, 7.

Input				Output
1	$\xrightarrow{\times 3}$	3	$\xrightarrow{+1}$	4
2	$\xrightarrow{\times 3}$	6	$\xrightarrow{+1}$	7
4	$\xrightarrow{\times 3}$	12	$\xrightarrow{+1}$	13
7	$\xrightarrow{\times 3}$	21	$\xrightarrow{+1}$	22

We multiply the input number by 3 and *then* add 1.

EXERCISE 17A.1

1 For the following input numbers and rules, calculate the output numbers:

a *Four times the input number.*

Input number	Calculation	Output number
1		
2		
3		
4		

b *The input number plus three.*

Input number	Calculation	Output number
4		
6		
12		
26		

c *Add two then multiply by 3.*

Input number	Calculation	Output number
0		
1		
2		
5		

d *Subtract 3 then multiply by 2.*

Input number	Calculation	Output number
3		
4		
10		
15		

2 Copy and complete this flowchart for the rule *double the input number then subtract* 2.

Input → ×2 → □ → −2 → Output
1 → □ → □
2 → □ → □
3 → □ → □
4 → □ → □

3 Use a flowchart to find the output numbers for each rule and given input numbers:
  a *Add* 3 *then double the result,* input numbers 0, 1, 3, 5
  b *Halve the input number, plus four,* input numbers 2, 6, 10, 18
  c *Add* 10 *then divide by* 3, input numbers 2, 8, 14, 23

4 Consider these number machines:

A Input → ×2 → □ → −1 → Output

B Input → +7 → □ → ÷3 → Output

  a For each number machine, write the rule using input number n and output number M.
  b Find the input number that produces the same output number in both number machines.

INPUTS FROM OUTPUTS

To find the input number needed to give a particular output number, we move in reverse along the flowchart, performing **inverse operations** at each step.

Example 3   ◀) Self Tutor

Consider the rule *triple the input number and add one.* Calculate the input numbers for the output number:

a 16   **b** 28

a Input   Output
  □ →×3→ □ →+1→ 16
  □ →×3→ 15 ⇄ +1/−1 ⇄ 16
  5 ⇄ ×3/÷3 ⇄ 15 ⇄ +1/−1 ⇄ 16
  ∴ the input number was 5.

b Input   Output
  □ →×3→ □ →+1→ 28
  □ →×3→ 27 ⇄ +1/−1 ⇄ 28
  9 ⇄ ×3/÷3 ⇄ 27 ⇄ +1/−1 ⇄ 28
  ∴ the input number was 9.

EXERCISE 17A.2

1. **a** Copy and complete the flowchart for the rule *double the input number, plus two*. Use it to find the input number for the output number 10.

   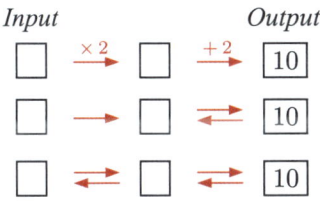

   b Construct your own flowchart to find the input number corresponding to the output number:

   i 2   **ii** 4

2. For each rule, calculate the input numbers needed to produce the given output numbers:

		Output numbers:
a	*The input number plus four.*	$\{5, 7, 15\}$
b	*Five times the input number, minus three.*	$\{7, 12, 17\}$
c	*Add one to the input number then double the result.*	$\{2, 6, 12\}$

3. Consider the rule *square the input number then add one*.

   a Explain why the inverse operation for "square the input number" is "plus or minus the square root of the number".

   b Hence find the input numbers which would produce the output number:

   i 2   **ii** 5   **iii** 17

B FORMULA CONSTRUCTION

When we try to construct a formula to connect related variables, we often start with numerical examples. They are useful to help us understand the situation before we generalise the result.

Example 4   ◀)) Self Tutor

Write a formula for the amount £A in a person's bank account if initially the balance was:

a £5000, and £200 was withdrawn each week for 10 weeks
b £5000, and £200 was withdrawn each week for w weeks
c £5000, and £x was withdrawn each week for w weeks
d £B, and £x was withdrawn each week for w weeks.

a $A = 5000 - 200 \times 10$

b $A = 5000 - 200 \times w$
$\therefore A = 5000 - 200w$

c $A = 5000 - x \times w$
$\therefore A = 5000 - xw$

d $A = B - x \times w$
$\therefore A = B - xw$

> We do not simplify the amount in **a** because we want to see how the formula is put together.

EXERCISE 17B

1 Write a formula for the amount £A earnt for working:
  a 5 hours at £15 per hour
  b 5 hours at £p per hour
  c t hours at £p per hour.

2 Write a formula for the amount £A in a bank account if the initial balance was:
  a £2000, and then £150 was deposited each week for 8 weeks
  b £2000, and then £150 was deposited each week for w weeks
  c £2000, and then £d was deposited each week for w weeks
  d £P, and then £d was deposited each week for w weeks.

3 Write a formula for the total cost £C of hiring a plumber given a fixed call-out fee of:
  a £40, plus £60 per hour for 5 hours of work
  b £40, plus £60 per hour for t hours of work
  c £40, plus £x per hour for t hours of work
  d £F, plus £x per hour for t hours of work.

4 In a multiple choice mathematics competition, students are awarded 3 points for each question answered correctly, and penalised 1 point for each question answered incorrectly. Write a formula for the number of points P scored by a student who:
  a answers 15 questions and gets 10 of them correct
  b answers 20 questions and gets c of them correct
  c answers a questions and gets c of them correct.

5 A musical recital consists of performances by a number of musicians, with a short break between each performance. Write a formula for the duration D minutes of a recital consisting of:
  a 4 performances of 6 minutes each, with a 2 minute break between performances
  b 5 performances of m minutes each, with a 3 minute break between performances
  c 8 performances of m minutes each, with a b minute break between performances
  d p performances of m minutes each, with a b minute break between performances.

6 A rectangular paddock is fenced into a rectangular array of yards so that each yard is connected by a gate to each adjacent yard. A 2×3 arrangement of yards is shown alongside. Write a formula for the number of gates G for:
  a a 2×3 arrangement
  b a 3×5 arrangement
  c a 4×4 arrangement
  d an $m \times n$ arrangement.

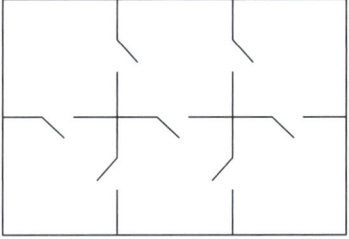

Example 5

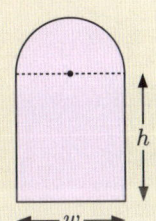

The illustrated door consists of a semi-circle and a rectangle. Find a formula for the area of the door in terms of the width w and height h of the rectangular part.

The area of a rectangle = height × width
$$= hw$$

The radius of the semi-circle is $\frac{w}{2}$.

$\therefore$ the area of the semi-circle $= \frac{1}{2} \times$ (area of full circle)

$$= \frac{1}{2} \times \pi r^2$$

$$= \frac{1}{2} \times \pi \times \left(\frac{w}{2}\right)^2$$

$$= \frac{1}{2} \times \pi \times \frac{w^2}{4}$$

$$= \frac{1}{8}\pi w^2$$

$\therefore$ the total area is $A = hw + \frac{1}{8}\pi w^2$

We can use known geometric formulae to help construct formulae for more complicated shapes.

7 Write a formula for the perimeter P of the following figures:

a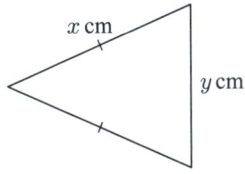
b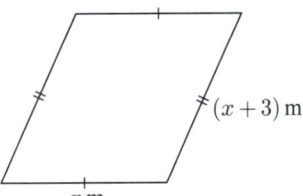
c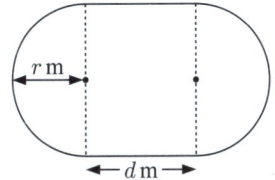

8 Find a formula for the area A of each of the shaded regions:

a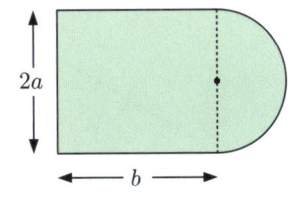
b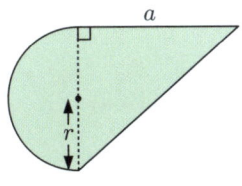
c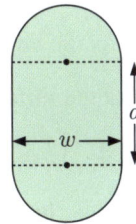

d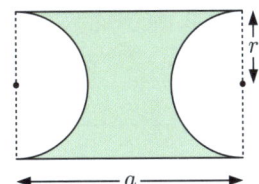
e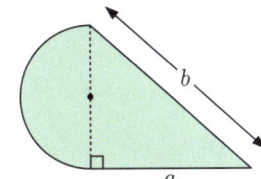

9 Find a formula for the volume V of each of the following objects:

a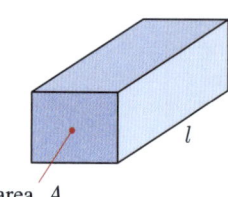

b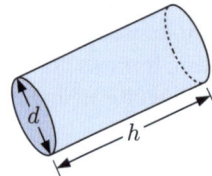

c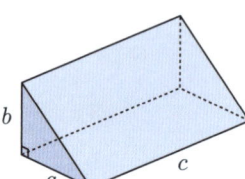

d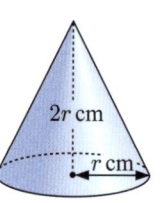

e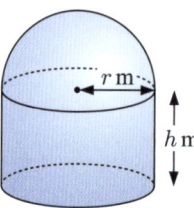

10 Find a formula for the surface area A of each of the following:

a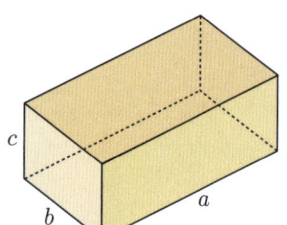

b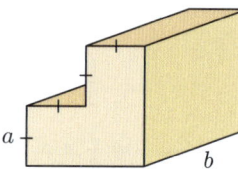

c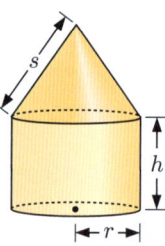

d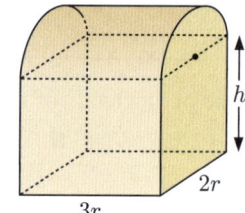

11 Consider the net for a cone shown alongside.

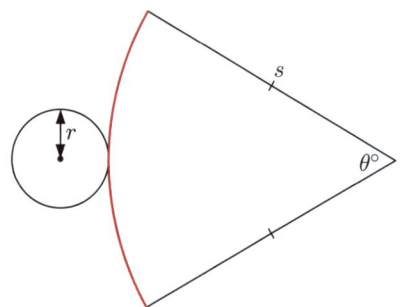

 a Explain why the red arc has length:

  i $\dfrac{\theta}{360} \times 2\pi s$ ii $2\pi r$

 b Hence explain why $s = \dfrac{360}{\theta} r$.

 c Show that the surface area of the cone is given by $A = \pi r^2 \left(1 + \dfrac{360}{\theta}\right)$.

12 A cylindrical pipe has outside radius R, inside radius r, and length l. Show that the volume of concrete used to make the pipe is given by
$$V = \pi l (R+r)(R-r).$$

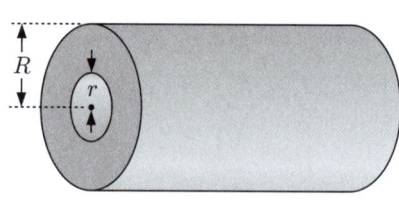

C SUBSTITUTING INTO FORMULAE

Suppose a formula contains two or more variables, and we know the value of all but one of them. We can **substitute** the known values into the formula to find the corresponding value of the unknown variable.

Step 1: Write down the formula.
Step 2: State the values of the known variables.
Step 3: Substitute the known values into the formula to form a one variable equation.
Step 4: Solve the equation for the unknown variable.

Example 6 ◀) Self Tutor

When a stone is dropped from a cliff, the total distance fallen after t seconds is given by the formula $D = \frac{1}{2}gt^2$ metres, where $g = 9.8$ m/s^2. Find:

a the distance fallen after 4 seconds

b the time, to the nearest $\frac{1}{100}$th second, taken for the stone to fall 200 metres.

a $\quad D = \frac{1}{2}gt^2 \quad$ where $\quad g = 9.8 \quad$ and $\quad t = 4$

$\therefore \ D = \frac{1}{2} \times 9.8 \times 4^2 = 78.4$

$\therefore$ the stone has fallen 78.4 metres.

b $\quad D = \frac{1}{2}gt^2 \quad$ where $\quad D = 200 \quad$ and $\quad g = 9.8$

$\therefore \ \frac{1}{2} \times 9.8 \times t^2 = 200$

$\therefore \ 4.9t^2 = 200$

$\therefore \ t^2 = \frac{200}{4.9}$

$\therefore \ t = \sqrt{\frac{200}{4.9}} \quad$ {t must be positive}

$\therefore \ t \approx 6.39$

$\therefore \ $ the time taken is about 6.39 seconds.

EXERCISE 17C

1 The formula for finding the circumference C of a circle with radius r is $C = 2\pi r$. Find:

   a the circumference of a circle of radius 4.2 cm
   b the radius of a circle with circumference 112 cm
   c the diameter of a circle with circumference 400 metres.

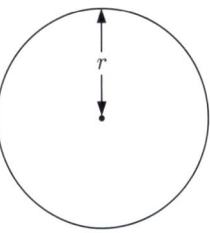

2 When a stone is dropped from the top of a cliff, the distance fallen after t seconds is given by the formula $D = \frac{1}{2}gt^2$ metres, where $g = 9.8$ m/s². Find:

   a the distance fallen in the first 2 seconds

   b the time taken for the stone to fall 100 metres.

3 The area A of a circle with radius r is $A = \pi r^2$. Find:

   a the area of a circle with radius 6.4 cm

   b the radius of a circular swimming pool which has an area of 160 m².

4 The volume of a cylinder with radius r and height h is given by $V = \pi r^2 h$. Find:

   a the volume of a cylindrical tin can with radius 8 cm and height 21.2 cm

   b the height of a cylinder with radius 6 cm and volume 120 cm³

   c the radius, in mm, of a copper pipe with volume 470 cm³ and length 6 m.

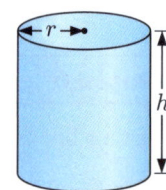

5 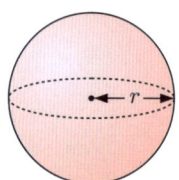 The formula for the surface area A of a sphere with radius r is $A = 4\pi r^2$. Find:

   a the surface area of a sphere with radius 7.5 cm

   b the radius, in cm, of a spherical balloon which has a surface area of 2 m².

6 The perimeter of each of the following figures is 12 cm. For each figure, write a *formula* for the perimeter P, and hence find the value of x.

   a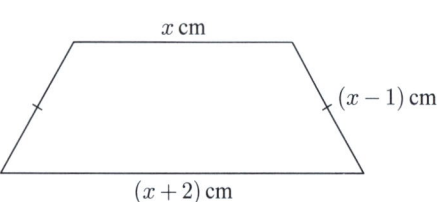

   b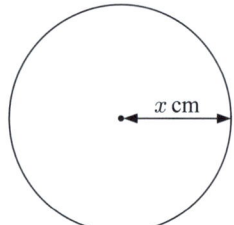

7 **a** Write a formula for the area of an equilateral triangle with sides of length x cm.

   b An equilateral triangle has area $16\sqrt{3}$ cm². Find the length of its sides.

8 The *period* or time taken for one complete swing of a simple pendulum is given approximately by $T = \frac{1}{5}\sqrt{l}$ seconds, where l is the length of the pendulum in centimetres. Find:

   a the time for one complete swing of a pendulum with length 45 cm

   b the length of a pendulum which has a period of 1.8 seconds.

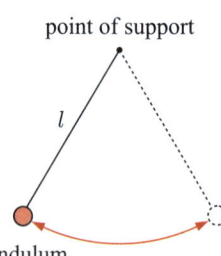

Activity   Pizza pricing

Luigi's Pizza Parlour has a "Seafood Special" pizza advertised this week.

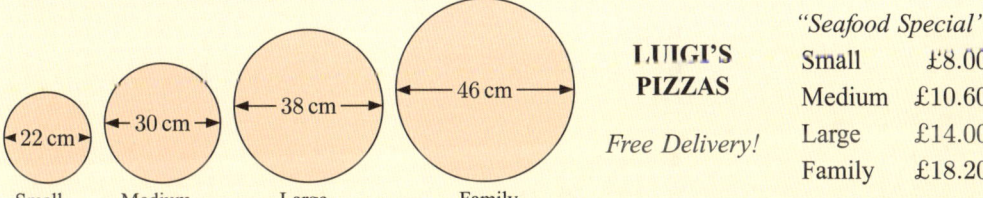

"Seafood Special"	
Small	£8.00
Medium	£10.60
Large	£14.00
Family	£18.20

Sasha, Enrico, and Bianca each attempted to find Luigi's formula for the price £P of each pizza size. The formulae they worked out for a pizza of radius r cm were:

Sasha: $P = \dfrac{17r - 27}{20}$ Enrico: $P = \sqrt{\dfrac{33r - 235}{2}}$ Bianca: $P = 5 + \dfrac{r^2}{40}$.

What to do:

1. Investigate the suitability of each formula.
2. Luigi is introducing a Party size pizza of diameter 54 cm. What do you think his price will be?

D REARRANGING FORMULAE

In the formula $D = xt + p$, D is expressed in terms of the other variables, x, t, and p. We say that D is the **subject** of the formula.

We can rearrange a formula to make one of the other variables the subject. However, we must do this carefully to ensure that the formula is still true.

> We **rearrange** formulae using the same processes which we use to solve equations. Anything we do to one side we must also do to the other.

Example 7   ◀》 Self Tutor

Make y the subject of:

a $2x + 3y = 12$ **b** $x = 5 - cy$

a $2x + 3y = 12$
$\therefore \; 3y = 12 - 2x$ {subtracting $2x$ from both sides}
$\therefore \; y = \dfrac{12 - 2x}{3}$ {dividing both sides by 3}
$\therefore \; y = 4 - \dfrac{2}{3}x$

b $x = 5 - cy$
$\therefore \; x + cy = 5$ {adding cy to both sides}
$\therefore \; cy = 5 - x$ {subtracting x from both sides}
$\therefore \; y = \dfrac{5 - x}{c}$ {dividing both sides by c}

EXERCISE 17D

1 Make y the subject of:
- **a** $x + y = 7$
- **b** $x - y = 3$
- **c** $x + 2y = 1$
- **d** $2x + 5y = 10$
- **e** $3x + 4y = 20$
- **f** $2x - y = 8$
- **g** $2x + 7y = 14$
- **h** $5x + 2y = 20$
- **i** $2x - 3y = -12$

2 Make x the subject of:
- **a** $p + x = r$
- **b** $xy = z$
- **c** $3x + a = d$
- **d** $5x + 2y = d$
- **e** $ax + by = p$
- **f** $y = mx + c$
- **g** $2 + tx = s$
- **h** $p + qx = m$
- **i** $6 = a + bx$

3 Make y the subject of:
- **a** $z = t - 5y$
- **b** $c - 2y = p$
- **c** $a - 3y = t$
- **d** $n - ky = 5$
- **e** $a - by = n$
- **f** $p = a - ny$
- **g** $4 - xy = c$
- **h** $w = 6 - ay$
- **i** $-k = m - ty$

Example 8   ◆) Self Tutor

Make z the subject of $c = \dfrac{m}{z}$.

$$c = \dfrac{m}{z}$$
$$\therefore \ cz = m \qquad \{\text{multiplying both sides by } z\}$$
$$\therefore \ z = \dfrac{m}{c} \qquad \{\text{dividing both sides by } c\}$$

4 Make z the subject of:
- **a** $az = \dfrac{b}{c}$
- **b** $p = \dfrac{q}{z}$
- **c** $\dfrac{a}{z} = d$
- **d** $\dfrac{3}{d} = \dfrac{2}{z}$
- **e** $\dfrac{7}{z} = \dfrac{k}{n}$
- **f** $\dfrac{p}{z} = -\dfrac{q}{t}$
- **g** $\dfrac{z}{2} = \dfrac{a}{z}$
- **h** $\dfrac{b}{z} = \dfrac{z}{n}$
- **i** $\dfrac{m}{z} = \dfrac{z}{a-b}$

5 Make:
- **a** a the subject of $F = ma$
- **b** r the subject of $C = 2\pi r$
- **c** d the subject of $V = ldh$
- **d** K the subject of $A = \dfrac{b}{K}$
- **e** h the subject of $A = \dfrac{bh}{2}$
- **f** T the subject of $I = \dfrac{PRT}{100}$

6 The surface area of a cylinder with radius r and height h is given by $A = 2\pi r^2 + 2\pi rh$.
Rearrange this formula to make h the subject.

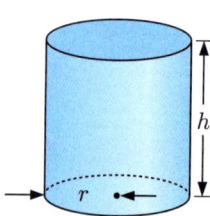

Example 9

Make t the subject of $s = \frac{1}{2}gt^2$, given that $t > 0$.

$\frac{1}{2}gt^2 = s$ {rewriting with t^2 on the LHS}

$\therefore\ gt^2 = 2s$ {multiplying both sides by 2}

$\therefore\ t^2 = \frac{2s}{g}$ {dividing both sides by g}

$\therefore\ t = \sqrt{\frac{2s}{g}}$ {as $t > 0$}

7 Make:
- **a** r the subject of $A = \pi r^2$, given that $r > 0$
- **b** x the subject of $N = \frac{x^2}{a}$
- **c** k the subject of $M = 5k^2$
- **d** x the subject of $D = \frac{n}{x^3}$
- **e** x the subject of $y = 4x^2 - 7$, given that $x < 0$
- **f** Q the subject of $P^2 = Q^2 + R^2$

Example 10

Make x the subject of $T = \frac{a}{\sqrt{x}}$.

$T = \frac{a}{\sqrt{x}}$

$\therefore\ T^2 = \left(\frac{a}{\sqrt{x}}\right)^2$ {squaring both sides}

$\therefore\ T^2 = \frac{a^2}{x}$

$\therefore\ T^2 x = a^2$ {multiplying both sides by x}

$\therefore\ x = \frac{a^2}{T^2}$ {dividing both sides by T^2}

8 Make:
- **a** a the subject of $d = \frac{\sqrt{a}}{n}$
- **b** l the subject of $T = \frac{1}{5}\sqrt{l}$
- **c** a the subject of $c = \sqrt{a^2 - b^2}$
- **d** d the subject of $\frac{k}{a} = \frac{5}{\sqrt{d}}$
- **e** l the subject of $T = 2\pi\sqrt{\frac{l}{g}}$
- **f** b the subject of $A = 4\sqrt{\frac{a}{b}}$

Example 11

Make x the subject of $ax + 3 = bx + d$.

$ax + 3 = bx + d$
$\therefore ax - bx = d - 3$ {writing terms containing x on the LHS}
$\therefore x(a - b) = d - 3$ {x is a common factor on the LHS}
$\therefore x = \dfrac{d - 3}{a - b}$ {dividing both sides by $(a - b)$}

If the variable we wish to make the subject appears more than once, we will need factorisation or expansion.

9 Make x the subject of:
- **a** $3x + a = bx + c$
- **b** $ax = c - bx$
- **c** $mx + a = nx - 2$
- **d** $8x + a = -bx$
- **e** $a - x = b - cx$
- **f** $rx + d = e - sx$

10 Make x the subject of:
- **a** $4(x + y) = x + 1$
- **b** $5x - z = 2(3 - x)$
- **c** $a(x + 5) = b(x - 1)$
- **d** $k(2x - 1) = x - 7$
- **e** $3(xy + 2) = 2x$
- **f** $m(1 - 4x) = nx - m$

Example 12

Make x the subject of $T = \dfrac{a}{x - b}$.

$T = \dfrac{a}{x - b}$
$\therefore T(x - b) = a$ {multiplying both sides by $(x - b)$}
$\therefore Tx - Tb = a$
$\therefore Tx = a + Tb$ {adding Tb to both sides}
$\therefore x = \dfrac{a + Tb}{T}$ {dividing both sides by T}

11 Make:
- **a** a the subject of $P = \dfrac{2}{a + b}$
- **b** r the subject of $T = \dfrac{8}{q + r}$
- **c** q the subject of $A = \dfrac{B}{p - q}$
- **d** x the subject of $A = \dfrac{3}{2x + y}$

E REARRANGEMENT AND SUBSTITUTION

In the Section on formula substitution, the known variables were replaced by numbers, and we then solved an equation to find the unknown.

In situations when we need to perform this process several times, it is quicker to **rearrange** the formula first, and then **substitute**.

Example 13

The volume of a cone is given by $V = \frac{1}{3}\pi r^2 h$, where r is the base radius and h is the height.

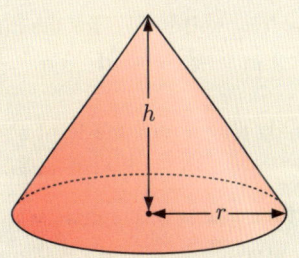

a Rearrange this formula to make r the subject.
b Hence, find the base radius of a cone with:
   i height 6 cm and volume 100 cm^3
   ii height 10 cm and volume 200 cm^3
   iii height 15 cm and volume 150 cm^3.

a $\quad V = \frac{1}{3}\pi r^2 h$

$\therefore\ 3V = \pi r^2 h \qquad$ {multiplying both sides by 3}

$\therefore\ \dfrac{3V}{\pi h} = r^2 \qquad$ {dividing both sides by πh}

$\therefore\ r = \sqrt{\dfrac{3V}{\pi h}} \qquad$ {as r must be positive}

b i When $h = 6$ and $V = 100$, $\ r = \sqrt{\dfrac{3 \times 100}{\pi \times 6}} = \sqrt{\dfrac{50}{\pi}} \approx 3.99$

So, the base radius is approximately 3.99 cm.

ii When $h = 10$ and $V = 200$, $\ r = \sqrt{\dfrac{3 \times 200}{\pi \times 10}} = \sqrt{\dfrac{60}{\pi}} \approx 4.37$

So, the base radius is approximately 4.37 cm.

iii When $h = 15$ and $V = 150$, $\ r = \sqrt{\dfrac{3 \times 150}{\pi \times 15}} = \sqrt{\dfrac{30}{\pi}} \approx 3.09$

So, the base radius is approximately 3.09 cm.

EXERCISE 17E

1 The area of a sector with radius r and angle θ is given by the formula $A = \dfrac{\theta}{360} \times \pi r^2$.

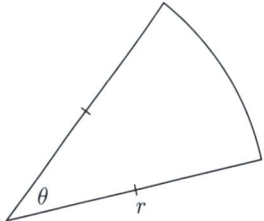

   a Rearrange this formula to make θ the subject.
   b Hence, find the angle of a sector with:
      i radius 3 cm and area 5 cm^2
      ii radius 7 cm and area 45 cm^2
      iii radius 8.5 cm and area 135 cm^2.

2 a Make a the subject of the formula $K = \dfrac{d^2}{2ab}$.

   b Find the value of a when:
      i $K = 112$, $d = 24$, $b = 2$ **ii** $K = 400$, $d = 72$, $b = 0.4$

3 The height of a bush after t years is given by the formula $H = 1 + \sqrt{t}$ metres.
  a Rearrange this formula to make t the subject.
  b How long will it take for the bush to reach a height of:
    i 2 m **ii** 3 m **iii** 3.5 m?

4 The formula for the volume V of a sphere with radius r is $V = \frac{4}{3}\pi r^3$.
  a Make r the subject of the formula.
  b Find the radius of a sphere which has volume:
    i 40 cm^3 **ii** 800 cm^3 **iii** 1 000 000 cm^3.

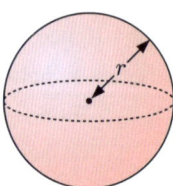

5 An object with constant acceleration a m/s^2 travels s m. Its initial speed is u m/s and its final speed is v m/s. The variables are connected by the formula $v^2 - u^2 = 2as$.
  a Rearrange the formula to make v the subject, where $v \geqslant 0$.
  b Find the final speed of an object which travels:
    i 100 m with initial speed 5 m/s and constant acceleration 2 m/s^2
    ii 1.5 km with initial speed 10 m/s and constant acceleration 0.9 m/s^2.

6 The *winning percentage* of a tennis player who has won w matches and lost l matches is given by the formula $P = \dfrac{w}{w+l} \times 100\%$.
  a Find the winning percentage of a player who has won 10 matches and lost 7 matches.
  b Rearrange the formula to make w the subject.
  c This year Mary has lost 15 matches, with a winning percentage of 37.5%. How many matches has she won?
  d Over his career, Claude has won 84 matches and lost 49 matches. His aim is to increase his winning percentage to 65%. How many consecutive matches must he win to reach his target?

7 Consider two objects with masses m_1 kg and m_2 kg, which are d m apart. The gravitational force between the objects is given by the formula

$$F = G\frac{m_1 m_2}{d^2} \text{ Newtons}$$

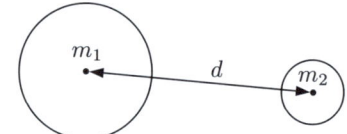

where $G \approx 6.67 \times 10^{-11}$ is the universal gravitational constant.
  a The Earth has mass 5.97×10^{24} kg, and the Moon has mass 7.35×10^{22} kg. Given that the Earth and the Moon are approximately 3.82×10^8 m apart, find the gravitational force between them. Give your answer in standard form.
  b Rearrange the formula so that d is the subject.
  c **i** The Sun has mass 1.99×10^{30} kg. Given that the gravitational force between the Sun and the Earth is 3.54×10^{22} N, find the distance between the Sun and the Earth.
  ii Two planets each have mass 2.32×10^{26} kg, and the gravitational force between them is 1.76×10^{14} N. Find the distance between the planets.

8 Answer the **Opening Problem** on page **338**.

Formulae and functions (Chapter 17)

F PREDICTING FORMULAE

We can often predict a formula for a general situation by examining simple cases and looking for a pattern.

For example, the set of even numbers is $\{2, 4, 6, 8, 10, 12, ...\}$.

We observe that: the 1st term is 2×1
the 2nd term is 2×2
the 3rd term is 2×3, and so on.

We see from the pattern that the 13th term will be 2×13.

So, we generalise by saying that "the nth even number is $2n$".

The coefficient 2 in $2n$ indicates that the terms increase by 2 each time n is increased by 1.

> To *generalise* from a pattern, we replace the term number by n.

Example 14

Examine the matchstick pattern: △ , △▽ , △▽△ , ...

How many matches are needed to make:
- **a** the first diagram
- **b** the second diagram
- **c** the third diagram
- **d** the 4th diagram
- **e** the nth diagram?

a 3 matches
b 5 matches
c 7 matches

d The fourth diagram is △▽△▽, which contains 9 matches.

e So far, the set of numbers is $\{3, 5, 7, 9, ...\}$. We are adding 2 matches each time, so the formula must involve $2n$.
The expression $2n$ would generate the set $\{2, 4, 6, 8, ...\}$, whereas our set of numbers is always 1 more than these values.
$\therefore$ there are $2n + 1$ matches in the nth diagram.
Check: If $n = 1$, $2(1) + 1 = 3$ ✓
If $n = 2$, $2(2) + 1 = 5$ ✓
If $n = 3$, $2(3) + 1 = 7$ ✓

EXERCISE 17F

1 Examine the matchstick pattern: □ , □□ , □□□ , □□□□ , ...

How many matchsticks make up the:
- **a** first diagram
- **b** second diagram
- **c** third diagram
- **d** 4th diagram
- **e** 5th diagram
- **f** nth diagram?

2 Examine the matchstick pattern:

How many matchsticks make up the:
- **a** first diagram
- **b** second diagram
- **c** third diagram
- **d** 4th diagram
- **e** 5th diagram
- **f** nth diagram?

3 Examine the matchstick pattern:

How many matchsticks make up the:
- **a** first diagram
- **b** second diagram
- **c** third diagram
- **d** 4th diagram
- **e** 5th diagram
- **f** nth diagram
- **g** 10th diagram
- **h** 100th diagram?

4 Examine the matchstick pattern:

How many matchsticks make up the:
- **a** first diagram
- **b** second diagram
- **c** third diagram
- **d** 4th diagram
- **e** 5th diagram
- **f** nth diagram
- **g** 10th diagram
- **h** 100th diagram?

5 a Find:
  - **i** $1 + 3$
  - **ii** $1 + 3 + 5$
  - **iii** $1 + 3 + 5 + 7$
  - **iv** $1 + 3 + 5 + 7 + 9$

  b If S_n is the sum of the first n positive odd numbers, then $S_1 = 1$, $S_2 = 4$, and $S_3 = 9$. Predict a formula for S_n.

6 a Find:
  - **i** $1 + 2$
  - **ii** $1 + 2 + 4$
  - **iii** $1 + 2 + 4 + 8$
  - **iv** $1 + 2 + 4 + 8 + 16$

  b Predict a formula for the sum of the first n terms of the number set $\{1, 2, 4, 8, 16, ...\}$.

Review set 17A

1 Consider the rule *halve the input number then square the result*.
  - **a** Write the rule in symbols using input number n and ouput number M.
  - **b** Use a table to calculate the output number corresponding to the input number:
    - **i** 2
    - **ii** 4
    - **iii** 8
    - **iv** 12

2 For the rule *subtract three from the input number then double the result*, find the input number needed to produce the output number:
  - **a** 2
  - **b** 8
  - **c** 0
  - **d** -4

3 a A trough is initially empty. Write a formula for the volume of water V in the trough if:
  - **i** six 8-litre buckets of water are poured into it
  - **ii** n 8-litre buckets of water are poured into it
  - **iii** n l-litre buckets of water are poured into it.

  b A trough initially contains 25 litres of water. Write a formula for the volume of water V in the trough if n buckets of water, each containing l litres, are poured into it.

4 The average speed of an object which travels d km in t hours is given by the formula $s = \dfrac{d}{t}$ km/h.

   a Find the average speed of a truck which travels 540 km in 6 hours.

   b Find the distance travelled by an aeroplane which flies for $6\tfrac{1}{2}$ hours at an average speed of 600 km/h.

5 Find a formula for the surface area A of the solid alongside.

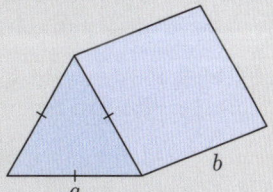

6 Make x the subject of:

   a $mx + n = 3p$ **b** $\dfrac{7}{y} = \dfrac{5}{x}$

7 Make k the subject of:

   a $T = \sqrt{k - l^2}$ **b** $P = 2k^2 - r$, $k < 0$

8 1.5 m³ of garden soil is dumped on a flat surface. It forms a cone whose diameter equals its height.

   a Suppose the diameter is x m. Write a formula for the volume of the heap in terms of x.

   b Find the diameter of the heap of soil.

9 The electric current in a circuit with voltage E volts, resistance r ohms, and load resistance R ohms, is given by the formula $I = \dfrac{E}{r + R}$ amperes.

   a Find the current in a circuit with voltage 24 V, resistance 0.5 ohms, and load resistance 2.5 ohms.

   b Rearrange the formula to make r the subject.

   c Find the resistance of a circuit with current 1.5 amperes, voltage 7.725 V, and load resistance 5 ohms.

10 Examine the matchstick pattern:

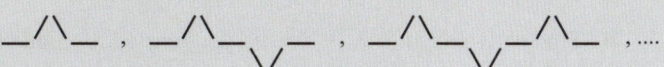

   How many matchsticks make up the:

   a first diagram   **b** second diagram   **c** third diagram

   d 4th diagram   **e** 5th diagram   **f** nth diagram?

11 To convert temperatures from degrees Fahrenheit (°F) to Kelvin (K), we use the formula $K = \dfrac{5}{9}(F - 32) + 273.15$.

   a Convert the following temperatures to Kelvin, correct to 1 decimal place:

      i 50°F   **ii** −130°F   **iii** 150°F

   b Rearrange the formula to make F the subject.

   c Convert the following temperatures to degrees Fahrenheit:

      i 313.15 K   **ii** 0 K   **iii** 200 K

Review set 17B

1 Consider the rule *multiply by* 4 *then subtract* 7. Use a flowchart to find the output number produced by input number:
   a 2   **b** 4   **c** 1

2 Write a formula for the bill £B at a restaurant if there is a charge of:
   a £15 for corkage, plus £25 per person for 5 people
   b £c for corkage, plus £25 per person for p people
   c £c for corkage, plus £m per person for p people.

3 Write a formula for the number of edge pieces E (excluding corner pieces) in a:
   a 3×5 jigsaw puzzle
   b 4×8 jigsaw puzzle
   c $m \times n$ jigsaw puzzle.

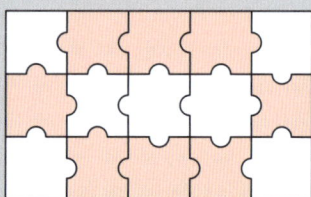

4 Consider the formula $M = p - qr$. Find:
   a M when $p = 19$, $q = -3$, and $r = 6$
   b r when $M = -2$, $p = 14$, and $q = 2$.

5 Find a formula for the volume V of the solid of uniform cross-section shown.

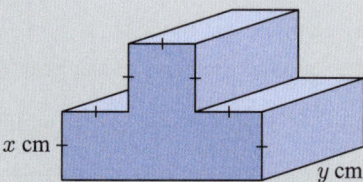

6 Make a the subject of:
   a $B = ad - f$ **b** $\dfrac{Q}{\sqrt{a}} = \dfrac{t}{3}$ **c** $G = \sqrt{\dfrac{5}{a+1}}$

7 Amy is trying to find pairs of numbers which have the same sum and product. In other words, she is looking for number pairs a and b such that $ab = a + b$.
   a Rearrange this formula to make b the subject.
   b Find b given that $a = 3$. Check your answer by finding the sum and product of the numbers.

8 Examine the matchstick pattern:

    , 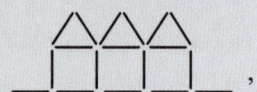 , ...

   How many matchsticks make up the nth diagram?

9 **a** Find: **i** $2 + 4$ **ii** $2 + 4 + 6$ **iii** $2 + 4 + 6 + 8$ **iv** $2 + 4 + 6 + 8 + 10$
   b Hence write a formula for the sum of the first n positive even numbers.

10 The kinetic energy of an object with mass m kg which is moving with speed v m/s, is given by the formula $E = \dfrac{1}{2}mv^2$ joules, $v \geqslant 0$.
   a Find the kinetic energy of a person with mass 80 kg moving at 5 m/s.
   b Rearrange the formula to make v the subject.
   c A running wombat with mass 25 kg has 800 joules of kinetic energy. Find the speed of the wombat.

18

Applications of percentage

Contents:

- **A** Percentage increase and decrease
- **B** Business calculations
- **C** Chain percentage problems
- **D** Appreciation and depreciation
- **E** Simple interest
- **F** Compound interest

Opening problem

Michelle bought an antique clock for £1500. It appreciated in value by 4% each year.

Michelle sold the clock after 5 years, and invested the money in an account earning 8% per annum compound interest. She left the money in the account for 3 years.

Things to think about:
- **a** How much did Michelle sell the clock for?
- **b** What was the final value of the investment?
- **c** Would Michelle have been better off investing the £1500 in the account initially, instead of buying the clock?

We have previously seen that a **percentage** is a fraction out of 100. For example, $20\% = \frac{20}{100} = 0.2$.

In the world around us, we often talk about **percentage change**.

We see percentage changes most often with money, when we talk about discounts, mark-up, interest, and tax.

A PERCENTAGE INCREASE AND DECREASE

If we *increase* an amount by 20% then we have 　$(100\% + 20\%)$ of the amount
　　　　　　　　　　　　　　　　　　　　　　　　　$= 120\%$ of the amount
　　　　　　　　　　　　　　　　　　　　　　　　　$= 1.2 \times$ the amount.

In this case we say 1.2 is the **multiplier**, since we multiply by 1.2 to get the final amount.

If we *decrease* an amount by 20% then we have 　$(100\% - 20\%)$ of the amount
　　　　　　　　　　　　　　　　　　　　　　　　　$= 80\%$ of the amount
　　　　　　　　　　　　　　　　　　　　　　　　　$= 0.8 \times$ the amount.

In this case the multiplier is 0.8.

Example 1　　　　　　　　　　　　　　　　　　　　　　　　　　　　　　◀)) Self Tutor

A fruit grower picked 1830 kg of apples last year. This year she expects her crop to be 30% bigger. How many kilograms of apples does she expect to pick this year?

The expected new crop $= (100\% + 30\%)$ of 1830 kg
　　　　　　　　　　　$= 130\% \times 1830$ kg
　　　　　　　　　　　$= 1.3 \times 1830$ kg
　　　　　　　　　　　$= 2379$ kg

She expects to pick 2379 kg.

EXERCISE 18A.1

1 Write down the multiplier corresponding to:
- **a** an increase of 30%
- **b** a decrease of 10%
- **c** an increase of 15%
- **d** a decrease of 35%
- **e** an increase of 12%
- **f** a decrease of 7.5%.

2 The price of vegetables has risen by 20% because of dry weather. How much will Justine need to pay for tomatoes that usually cost £3.50 per kilogram?

3 Kurt's business is expanding. He has increased the number of staff by 25%. If he previously employed 64 people, how many does he employ now?

4 In 2015 Su-Lin's salary was £48 000. In 2016 it increased by 35% when she was promoted to manager. What was her salary in 2016?

Example 2   ◀)) Self Tutor

Stefan expected to harvest 2000 kg of cherries this year, but bad storms damaged 60% of his crop. What weight of cherries can he expect to harvest now?

The expected new weight = (100% − 60%) of 2000 kg
$$= 40\% \times 2000 \text{ kg}$$
$$= 0.4 \times 2000 \text{ kg}$$
$$= 800 \text{ kg}$$

Stefan can expect to harvest 800 kg.

5 Marius found that travelling on the new freeway decreased his travelling time to work by 12%. He used to take 50 minutes to get to work. How long does he take now?

6 In a clearance sale, the price of a new car was decreased by 35%. If the car normally cost £14 960, what was the new price?

7 If Claudia walked to school following the footpaths, she would walk 920 metres. If she walked across the park she could reduce this distance by 16%. How far would she walk then?

FINDING A PERCENTAGE CHANGE

When the size of a quantity changes, the multiplier for the change is calculated by:

$$\text{multiplier} = \frac{\text{new amount}}{\text{original amount}}$$

We can then use the multiplier to determine the percentage change.

Example 3

Determine the percentage change when:

a 50 kg is increased to 70 kg

b £160 is decreased to £120.

a multiplier = $\dfrac{\text{new amount}}{\text{original amount}}$

$= \dfrac{70 \text{ kg}}{50 \text{ kg}}$

$= 1.4$

This corresponds to a 40% increase.

b multiplier = $\dfrac{\text{new amount}}{\text{original amount}}$

$= \dfrac{£120}{£160}$

$= 0.75$

This corresponds to a 25% decrease.

EXERCISE 18A.2

1 Find the percentage change when:
   a £20 is increased to £22
   b 80 ml is decreased to 68 ml
   c 45 g is decreased to 27 g
   d 90 cm is increased to 1.35 m.

*A multiplier greater than 1 shows an **increase**. A multiplier less than 1 shows a **decrease**.*

2 Describe the percentage change in the following situations:
   a The price of a haircut last month was £30. It has since risen to £34.50.
   b 150 people attended a community picnic last year. It rained this year, so only 108 people attended.
   c Arthur bought a house for £420 000. It is now worth £490 000.
   d At his school's sports day, Casey threw the javelin 56.33 m, breaking the previous school record of 52.40 m.
   e John completed a half-marathon in 1 hour and 52 minutes, improving on his previous best time of 2 hours and 8 minutes.

3 Harriot really loves snakes. For her birthday in 2014, she was given a pet carpet python by her uncle. Since then, she has measured its length on her birthday each year.

Year	2014	2015	2016	2017
Length	58 cm	79 cm	94 cm	1.07 m

   a Calculate the percentage by which the snake's length increased from:
      i 2014 to 2015
      ii 2015 to 2016
      iii 2016 to 2017.
   b Calculate the overall percentage increase for the 3-year period.

4 The following table shows the populations of some European countries, in millions of people, in 2006 and 2016:

	Germany	Italy	France	Spain	Greece	Portugal	United Kingdom
2006	82.4	58.1	63.6	44.7	11.0	10.5	60.8
2016	82.7	60.6	66.9	46.6	10.7	10.3	65.6

a Calculate the percentage change in population from 2006 to 2016 for each country.

b Which country had the:

  i largest percentage increase **ii** largest percentage decrease in population?

FINDING THE ORIGINAL AMOUNT

Given an original amount and a certain percentage change, we have seen how to obtain a new amount.

It is also useful to be able to solve the reverse problem, so that if we know the percentage change and the new amount, we can calculate the original amount.

Example 4 ◀) Self Tutor

In one season, Josephine's camellia grew 25%. It is now 1.68 m tall. How high was the camellia at the start of the season?

$$\text{multiplier} = \frac{\text{new height}}{\text{original height}}$$

$$\therefore\ 1.25 = \frac{1.68 \text{ m}}{\text{original height}}$$

$$\therefore\ \text{original height} = \frac{1.68 \text{ m}}{1.25}$$

$$\approx 1.34 \text{ m}$$

Josephine's camellia was 1.34 m high at the start of the season.

EXERCISE 18A.3

1 Find the original amount given that:

  a after an increase of 20%, the length was 24 cm

  b after a decrease of 15%, the mass was 51 kg

  c after an increase of 3.6%, the amount was £129.50

  d after an increase of 130%, the capacity was 9200 litres

  e after a decrease of 0.8%, the attendance was 49 600 people.

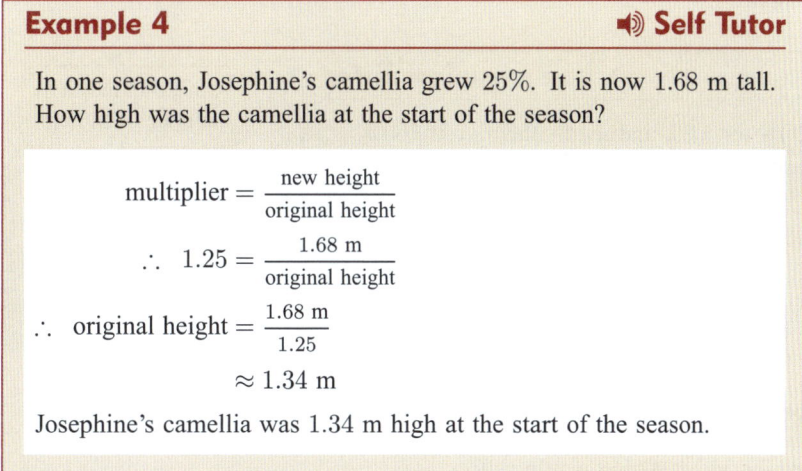

We *divide* the new amount by the multiplier to find the original amount.

2 In 2014, 88 nations participated in the Winter Olympics. This was a 10% increase from the 2006 Winter Olympics. How many nations participated in the 2006 Winter Olympics?

3 From 1993 to 2013, the recorded number of black rhinoceroses increased by 97.2% to 4880. Estimate the black rhinoceros population in 1993.

4 Joan has just received an electricity bill of £283.50. This is 32.5% less than her previous bill. How much was Joan's previous bill?

Discussion

If an amount is *increased* by 18%, and the resulting amount is then *decreased* by 18%, do we return to the original amount?

B BUSINESS CALCULATIONS

Many people choose to run their own business. A successful small business owner needs to have a good understanding of percentages in order to calculate mark-up, discounts, and tax.

MARK-UP

To make a profit, a retailer needs to sell goods at a price *greater* than the cost price which he or she paid for the goods. Starting with the cost price, the retailer applies a percentage increase called a **mark-up** to calculate the selling price.

Example 5   ◉ Self Tutor

Alfred buys shirts for £20, and marks them up by 60%. At what price does he sell them?

Alfred sells the shirts for   £20 × 160%
                            = £20 × 1.6
                            = £32

A mark-up increases the price of an item!

EXERCISE 18B.1

1 Find the price for which Alfred would sell:
   a a shirt costing him £24, if he applies a 40% mark-up
   b a skirt costing him £50, if he applies a 55% mark-up
   c a jacket costing him £120, if he applies a 35% mark-up
   d a suit costing him £320, if he applies a 30% mark-up
   e a pair of trousers costing him £45, if he applies a 28% mark-up.

2 A baker puts a mark-up of 30% on the bread sold to storekeepers. On average, it costs the baker 90 pence per loaf to make the bread. What price does the baker charge storekeepers?

3 Jacinta sells fruit and vegetables. Find the selling price of:
   a apples costing her £2.30 per kg, if she marks them up by 60%
   b pineapples costing her £0.80 each, if she marks them up by 80%
   c cabbages costing her £1.20 each, if she marks them up by 120%.

4 A butcher buys chicken breast fillets for £2.84 per kg, then adds a 65% mark-up. What price per kilogram will the customer pay?

5 A sports store buys footballs for £15 each, and applies an 80% mark-up.
  a At what price does the sports store sell the footballs?
  b How much profit does the store make on each football sold?

Example 6 ◀) **Self Tutor**

A carton of brussels sprouts costs a farmer £3.35 to produce. He sells the carton to a supermarket for £4.35. Find the percentage mark-up.

$$\text{multiplier} = \frac{\text{selling price}}{\text{cost price}} = \frac{£4.35}{£3.35} \approx 1.3$$

$\therefore$ the farmer applied a 30% mark-up.

---

6 Kane bought a lawnmower for £180 and marked it up to £240 to sell in his shop. Find the percentage mark-up.

7 A jewellery store buys a bracelet for £70, and sells it for £126. What percentage mark-up has been applied?

8 Louise bought a 30-can pack of soft drinks for £8 to sell in her shop. She sells each can for £1.50.
  a What percentage mark-up has been applied to each can?
  b Find the total profit Louise will receive from selling all 30 cans.

Example 7 ◀) **Self Tutor**

An electrical goods store buys a TV set at wholesale price. The price of the TV is increased by 25% for sale by the store. Its selling price is now £550.
For what price did the store buy the TV set?

cost price × multiplier = selling price
$\therefore$ cost price × 1.25 = £550   {100% + 25% = 125% = 1.25}
$\therefore$ cost price = $\frac{£550}{1.25}$ = £440

So, the store bought the television for £440.

---

9 A shop applies a 45% mark-up on fresh fish, and sells it for £25.50 per kg.
Find the cost price of the fish.

10 A litre of milk retails for £1.47 after being marked up by 40%. What was the cost price?

DISCOUNT

To encourage a sale, a retailer may offer a **discount**. This means that the **marked price** is **reduced** or **discounted** by a certain amount or percentage.

Example 8   Self Tutor

A dress marked at £150 is discounted by 20% for the end of season sale. Find the sale price of the dress.

Sale price = £150 × 80%   {100% − 20% = 80%}
= £150 × 0.8
= £120

A discount reduces the price of an item!

EXERCISE 18B.2

1 Find the sale price of a dress which was originally marked at:
   a £80, and is discounted 30%
   b £160, and is discounted 25%
   c £240, and is discounted 35%
   d £300, and is discounted 12.5%.

2 A bicycle marked at £560 is to be discounted by 30%. Calculate the sale price.

3 A CD is marked at £20, and is discounted by 15%. What is its sale price?

4 A refrigerator is marked at £850, and is discounted by 18%. What is its sale price?

5 A department store is offering the following discounts:
   • 5% off for customers who spend more than £100
   • 7.5% off for customers who spend more than £200
   • 10% off for customers who spend more than £500.

   Find the final price paid by a customer who buys:
   a a television set marked at £180
   b a washing machine marked at £320
   c a dishwasher marked at £890.

6 Paul and Emily are planning a holiday. They have £500 to spend on caravan hire. The caravan hire costs £70 per day, but there is a 25% discount if they book early.
   For how many extra days can Paul and Emily hire the caravan if they book early?

7 A wok is discounted from £75 to £69 in a sale. Find the percentage discount.

8 A vacuum cleaner has recommended retail price £160. In a clearance sale, it is discounted to £124. Calculate the percentage discount.

9 A fan was discounted by 20% and sold for £23.20. What was the marked price of the fan?

10 A pair of bathers sold for £44 after being discounted by 45%. What was the marked price of the bathers?

11 A clothing store is having a 15% off sale. The price of a coat has been reduced to £119. How much does the coat usually cost?

Discussion

> £10 off your next purchase!   10% off your next purchase!

- If you could use one of these vouchers for your next purchase in a store, which would you choose?
- If you could use *both* vouchers, which would you use *first*?

VALUE ADDED TAX (VAT)

The **value added tax (VAT)** is a tax which is applied to most goods and services that you buy. In the United Kingdom, this tax is applied at a rate of 20%.

Example 9   ◉ Self Tutor

A shoe store wants to receive £60 from selling a pair of shoes.

a How much VAT must be added to this price?
b How much will the customer pay in total?

a VAT added = £60 × 20%
              = £60 × 0.2
              = £12

b The customer will pay in total
       £60 + £12 = £72.

EXERCISE 18B.3

1 A sports store wants to receive £50 from the sale of a cricket bat. Find:
  a the VAT to be added
  b the selling price of the bat.

2 A hairdresser charges £45 for a haircut, plus VAT. Find:
  a the VAT to be paid
  b the total cost of a haircut.

3 A mechanic charges £85 per hour for parts and labour, plus VAT. If a job takes three hours to complete, find the total cost of the job.

4 Brett charges £38.20 per hour, plus VAT, for customer support. If he spends 1 hour 40 minutes helping a customer, find the total cost.

Discussion

If you know the final price of an item, how can you work out how much of the amount is VAT?

C CHAIN PERCENTAGE PROBLEMS

Sometimes we need to use a multiplier more than once within a problem.

Example 10   ◀) Self Tutor

Greeting cards are bought for £1.40 by a newsagent. He marks up the price by 50%, then adds on 20% VAT. What price does the customer pay?

Final price = £1.40 × 150% × 120%
            = £1.40 × 1.5 × 1.2
            = £2.52

The £1.40 is increased by 50%, **and then** increased by a further 20%.

EXERCISE 18C

1 Kirsty buys a skirt for £22. She applies a 55% mark-up, then adds 20% VAT. For what price will Kirsty sell the skirt?

2 On his 5th birthday, Sean was 110 cm tall. He grew by 8% the next year, and by 5% the year after that. How tall was Sean on his 7th birthday?

3 A surfboard manufactured for £300 is sold to a retail shop for a 40% profit. The shop then applies a 20% mark-up. What price does the customer pay?

4 The population of a town was 25 000 in 2016. The population increased by 15% in 2017, and then decreased by 6% in 2018. Find the population of the town in 2018.

5 A fisherman sells a crayfish to a wholesaler for £35. The wholesaler sells it to a fresh fish shop for 60% profit. However, the fish shop is forced to sell the crayfish at a 25% loss. What price does the customer pay?

6 In a particular city it is estimated that, for each additional kilometre from the city centre, the average house price decreases by 4%.

   If the average house price 2 km from the city centre is £350 000, find the average house price:

   a 3 km from the city centre   b 5 km from the city centre.

7 An art supplies store buys canvases for £40, and applies a 40% mark-up. During a sale, a discount is offered of 12.5% for non-members, and 15% for members. Find the difference in the sale price for non-members and members.

Example 11   ◀) Self Tutor

Over 3 consecutive years, Heather's salary increased by 3%, 6%, and 4%. What is the overall percentage increase in Heather's salary over this period?

Suppose Heather's original salary was £x.

Heather's salary after 3 years = £x × 1.03 × 1.06 × 1.04
                              ≈ £x × 1.135

∴ overall, Heather's salary increased by ≈ 13.5%.

8 Over 2 consecutive years, a company's sales increased by 6% and then increased by 3%. Find the overall percentage increase in sales over this period.

9 A rare book is bought by an antiques dealer, and he marks its price up by 50% for sale. One of his regular customers is interested in buying the book, so the dealer offers the customer a 20% discount. Find the overall percentage mark-up in the price of the book.

10 A baby's weight increased by 28% in the first month then 17% in the second month. The baby weighed 5.4 kg after 2 months. Find the weight of the baby when it was born.

11 Over 2 years, the yield of apples from an apple tree decreased by 13%, then increased by 13%.
Does this mean that the yield returned to its original level after 2 years? Explain your answer.

12 Julia's parents have agreed to increase her allowance by 20% each year. Julia thinks her allowance when she is 12 will be 60% higher than when she is 9. Is Julia correct? Explain your answer.

D APPRECIATION AND DEPRECIATION

When the value of an investment or an item such as a house increases, we say it **appreciates** in value.

You may have noticed that the prices of everyday goods and services also increase over time. This is known as **inflation**, and is a form of appreciation.

Example 12   ◆) Self Tutor

The inflation rate over the next three years is predicted to be 2%, then 3%, then 3.5%. If an item currently costs £140, and its cost rises in line with the predicted inflation, find the item's cost in 3 years' time.

Cost in 3 years = £140 × 1.02 × 1.03 × 1.035
= £152.23

Electrical equipment, furniture, vehicles, and machinery all lose value over time. This may be because they become damaged or worn, or because their technology is no longer the latest and best available. We say the items **depreciate** in value.

Example 13   ◆) Self Tutor

A computer is bought for £1500. It depreciates by 25% in the first year, then 30% in the second year. Find the value of the computer after 2 years.

Value after 2 years = £1500 × 0.75 × 0.7
= £787.50

EXERCISE 18D

1. The inflation rate is predicted to be 3% and 3.5% over the next two years. What would you expect a bicycle costing £500 today to cost in two years' time?

2. Three years ago, an ornate vase was valued at £2000. Over the next three years, it appreciated by 7%, 5%, and 8%. Find the present value of the vase.

3. Sophie bought a car for £5000. Over the next two years it depreciated by 20% and 15%. Find the value of the car after two years.

4. A photocopier is bought for £3000, and depreciates by 15% each year.
    a Find the value of the photocopier after 3 years.
    b How many years will it take for the value of the photocopier to fall below £1500?

5. Kevin and Don each bought a motorbike for £8000. Kevin's motorbike depreciated at 22% per year, and Don's motorbike depreciated at 12% per year. Find the difference in value of the motorbikes after 3 years.

6. A share fund reported a 9% increase in value for year 1, a 13% decrease in value for year 2, and a 4% increase in value for year 3. What was the overall percentage increase or decrease of the share fund over the 3 years?

7. Jamie bought a piece of cricket memorabilia. Over the next 4 years the piece appreciated by 5%, depreciated by 4%, depreciated by 8%, and appreciated by 6%. Find the overall percentage appreciation or depreciation of the item over the 4 years.

8. An apartment was valued at £270 000. It appreciated by 5% each year for 3 years, then depreciated by 5% each year for 3 years. Find:
    a the value of the apartment after 6 years
    b the overall percentage appreciation or depreciation of the apartment.

E SIMPLE INTEREST

If you **borrow** money from a bank, you must repay the loan in full, and also pay an additional charge called **interest**.

One method for calculating interest is called **simple interest**.

> **Simple interest** is interest that is calculated each year as a fixed percentage of the original amount borrowed.
>
> The fixed percentage is called the **interest rate**, and is usually written as a percentage **per annum**, which means "per year".

For example, suppose £4000 is borrowed at 10% per annum simple interest.

Each year, the simple interest charge is 10% of £4000 = £400.

Applications of percentage (Chapter 18) 369

Example 14   🔊 Self Tutor

Find the simple interest payable on a loan of £60 000 borrowed at 9% p.a. for:
  a 4 years **b** 5 months

p.a. means "per annum".

The simple interest charge each year = 9% of £60 000
$$= 0.09 \times £60\,000$$
$$= £5400$$

a The simple interest for 4 years
$$= £5400 \times 4$$
$$= £21\,600$$

b 5 months is $\frac{5}{12}$ of a year
∴ the simple interest for 5 months
$$= \frac{5}{12} \times £5400$$
$$= £2250$$

Example 15   🔊 Self Tutor

Find the total amount needed to repay a loan of £40 000 borrowed at 9% p.a. simple interest for 5 years.

The simple interest charge each year = 9% of £40 000
$$= 0.09 \times £40\,000$$
$$= £3600$$
∴ the simple interest for 5 years $= £3600 \times 5$
$$= £18\,000$$
∴ the total to be repaid $= £40\,000 + £18\,000$
$$= £58\,000$$

EXERCISE 18E

1 Find the simple interest charged when:
  a £5000 is borrowed for 1 year at 12% per annum simple interest
  b £2500 is borrowed for 2 years at 8% p.a. simple interest
  c £40 000 is borrowed for 5 years at 11% p.a. simple interest
  d £250 000 is borrowed for 9 months at 20% p.a. simple interest.

Total to be repaid = original amount + interest.

2 Find the total amount needed to repay a loan of:
  a £2400 borrowed for 3 years at 10% p.a. simple interest
  b £8000 borrowed for 7 years at 12% p.a. simple interest
  c £7500 borrowed for $2\frac{1}{2}$ years at 8% p.a. simple interest
  d £23 000 borrowed for 4 months at 15% p.a. simple interest.

3 Kyle borrows £25 000 at 6% p.a. simple interest for 4 years.
  a Find the total amount needed to repay the loan.
  b Calculate the monthly repayment required to pay this loan off in 48 equal instalments.

4 Alice borrows £4700 from a finance company to buy her first car. The rate of simple interest is 17% and she borrows the money over a 5 year period. Find:
  a the total amount Alice must repay the finance company
  b her equal monthly repayments.

5 Richard borrowed £6000 at 4% p.a. simple interest. By the end of the loan, Richard had paid £7200 to repay the loan in full. For how long did Richard borrow the money?

6 Jenny took out a loan at 8% p.a. simple interest for 6 years. She paid a total of £1680 interest on the loan. How much money did she borrow?

F COMPOUND INTEREST

A more common method for calculating interest is **compound interest**.

If you leave your money in the bank for a period of time, the interest is automatically added to your account.

With compound interest, any interest that is added to your account will also earn interest in the next time period.

Compound interest allows you to earn interest on interest!

> **Compound interest** is calculated as a percentage of the total amount at the end of the previous compounding period.

In this course we will only consider interest which compounds each year.

Suppose £1000 is placed in an account earning interest at a rate of 10% p.a. The interest is allowed to compound itself for three years. We say it is earning "10% p.a. compound interest".

We can show this in a table:

Year	Amount at beginning of year	Compound interest	Amount at end of year
1	£1000	10% of £1000 = £100	£1000 + £100 = £1100
2	£1100	10% of £1100 = £110	£1100 + £110 = £1210
3	£1210	10% of £1210 = £121	£1210 + £121 = £1331

After 3 years there is a total of £1331 in the account. We have earned £331 in compound interest.

Notice that the amount in the account increases by 10% each year. This means that we can use chain percentage calculations to find the balance of a compound interest account.

Example 16

a What will £5000 invested at 8% p.a. compound interest amount to after 3 years?

b How much interest is earned?

a The amount increases by 8% each year.
∴ the total amount after 3 years = £5000 × 1.08 × 1.08 × 1.08
                                 = £6298.56

b Interest earned = £6298.56 − £5000
                     = £1298.56

EXERCISE 18F

1 Sunil invested £4000 at 5% p.a. compound interest. Find the value of the investment after 2 years.

2 Cassandra invested £7000 at 3.5% p.a. compound interest for 2 years. Find:
   a the value of the investment after 2 years
   b the interest earned.

3 Donna borrows £6500 for 3 years at 4% p.a. compound interest.
   a Find the total amount Donna must repay after 3 years.
   b In total, how much interest will Donna be charged?

4 How much compound interest is earned by investing £9000 for 3 years at 4.5% p.a.?

5 George invests £6200 at 6% p.a. compound interest.
   a Find the balance of the account after 4 years.
   b Find the overall percentage increase in value of the investment.

6 Shelley invests £8500 in an account which pays 3.7% p.a. compound interest.
   a Find the value of the investment after 3 years.
   b How many years will it take for the investment to be worth at least £10 000?

7 You have £8000 to invest for 3 years. You have been offered two investment options:
   Option 1: Invest at 9% p.a. simple interest.
   Option 2: Invest at 8% p.a. compound interest.
   a Calculate the amount accumulated at the end of the 3 years for both options, and decide which option to take.
   b Would you change your decision if you were investing for 5 years?

8 Helen invests £2000 in an account which offers 6% p.a. compound interest in the first 2 years, and then 3.5% p.a. compound interest for each subsequent year.
   a Find the value of the investment after 4 years.
   b Would Helen have been better off investing in an account which paid a constant compound interest rate of 5% p.a.?

9 Warren invests £5000 in an account which pays 4% p.a. compound interest.

   a Explain why the amount in the account after n years is given by the formula $A = 5000 \times 1.04^n$ pounds.

   b Find the amount in the account after:

      i 3 years   **ii** 7 years.

   c How many years will it take for the investment to be worth at least £8000?

Activity   Compound interest

We can use the graphing package to investigate the growth of a compound interest investment.   **GRAPHING PACKAGE**

Suppose £1000 is invested at 9% p.a. compound interest.

After x years, the value of the investment is 1000×1.09^x pounds.

What to do:

1   **a** Use the graphing package to graph the function $Y = 1000 \times 1.09\wedge X$.

   b Use your graph to find the value of the investment after:

     i 1 year   **ii** 2 years   **iii** 5 years
     iv 10 years   **v** 15 years   **vi** 20 years.

   c Use the graph to find the amount of interest paid in the fifth year.

2 Suppose we want to know how long it will take for the investment to reach £2000.

   a Draw the graphs of $Y = 1000 \times 1.09\wedge X$ and $Y = 2000$ on the same set of axes.

   b Find the intersection point of the graphs. Hence, determine how long it will take for the investment to reach £2000.

   c How long will it take for the investment to reach:

     i £5000   **ii** £10 000?

3 Another bank offers an interest rate of 15% p.a. *simple interest*.

The investment would earn 15% of £1000 = £150 interest each year. The value of the investment after x years would be $1000 + 150x$ pounds.

   a Draw the graphs of $Y = 1000 \times 1.09\wedge X$ and $Y = 1000 + 150X$ on the same set of axes.

   b Find the value of the simple interest investment after:

     i 1 year   **ii** 2 years   **iii** 5 years
     iv 10 years   **v** 15 years   **vi** 20 years.

   c Find the points of intersection of the graphs.

   d Over what time period would it be better to choose:

     i the compound interest account   **ii** the simple interest account?

Review set 18A

1 Find the percentage change when:

   a £80 is increased to £85   **b** 3.5 litres is decreased to 2.5 litres.

2 A football stadium normally has a maximum capacity of 60 000 people. However, one stand is under renovation, so the stadium's capacity has been reduced by 8.5%. How many people can the stadium currently hold?

3 A toaster is sold to a retailer for £38, and the retailer marks it up by 40%. What price does the customer pay?

4 I bought a house for £420 000. Over the next four years it increased in value by 9%, decreased by 3%, increased by 7%, and increased by 12%. Find the value of the house after these four years.

5 As a result of a redevelopment, the area of a city park was reduced by 15% to 3.4 hectares. Find the original area of the park.

6 A lounge suite was bought for a basic cost of £1780. It was marked up for a 35% profit, then 20% VAT was added. Determine the final selling price of the lounge suite.

7 When Luke completed his university degree 3 years ago, he had a student loan of £22 000 to repay. Over the next 3 years, the inflation rate was 2.2%, 1.9%, and 2.6%.
  a If Luke's loan rose in line with inflation, find the amount he has to repay now.
  b Luke will receive a 10% discount on his loan if he repays the loan in a single payment. How much will he need to pay to repay the loan in a single payment?

8 Find the simple interest payable on a loan of £4000 at 7% p.a. for 3 years.

9 Raj borrows £5000 from a bank at $8\frac{1}{2}$% p.a. simple interest. He will repay the money over a 4 year period, with equal monthly repayments.
  a Find the total amount that Raj must repay the bank.
  b Find Raj's equal monthly repayments.

10 **a** What will an investment of £8000 at 6% p.a. compound interest amount to after 4 years?
  b What part of this is interest?

11 A caravan costing £15 000 depreciates by 16% p.a. each year.
  a Explain why the value of the caravan after n years is given by the formula $V = 15\,000 \times 0.84^n$ pounds.
  b Find the value of the caravan after 5 years.
  c How many years will it take for the value of the caravan to fall below £5000?

Review set 18B

1 As part of a road safety campaign, fines for all traffic offences will increase by 5%.
Copy and complete the table alongside, showing the changes to each fine.

Offence	Old fine	New fine
Speeding	£200	
Drink driving	£840	
Not wearing seatbelt		£273
Illegal parking		£52.50

2 An item of clothing is marked at £49, but is discounted 15% in a sale. Find the sale price.

3 The annual rate of inflation is predicted to be 3% next year, then 3.5% in the year after that. If an item currently costs £50, estimate its cost in two years' time.

4 A blender was discounted by 20% and sold for £64. Find the marked price of the blender.

5 A softball bat was sold for £120 including 20% VAT.
  a How much will the seller receive once the tax is passed on to the government?
  b How much is the VAT?

6 A soufflé increases in height by 125% while in the oven. If the finished height is 13.5 cm, how tall was the soufflé when it was placed in the oven?

7 A new café opened in January in the city centre. The number of customers visiting the café increased by 8% each month.
  a Find the overall percentage increase in visitors from April to July.
  b Given that the café had 2106 customers in March, how many customers did they have in February?

8 Answer the **Opening Problem** on page 358.

9 Darren borrows £15 000 from the bank to extend his house. The bank will charge him simple interest of 6.8% p.a. for a 5 year loan. Find the total amount Darren will need to repay.

10 You have £5000 to invest for 4 years. Which of these options should you take? Explain your answer.
  A 7% p.a. simple interest   **B** 6% p.a. compound interest

11 Denise invests £30 000 in an account paying 4.6% p.a. compound interest.
  a Find the value of the investment after 3 years.
  b How many years will it take for the investment to be worth at least £40 000?

# 19

Rates

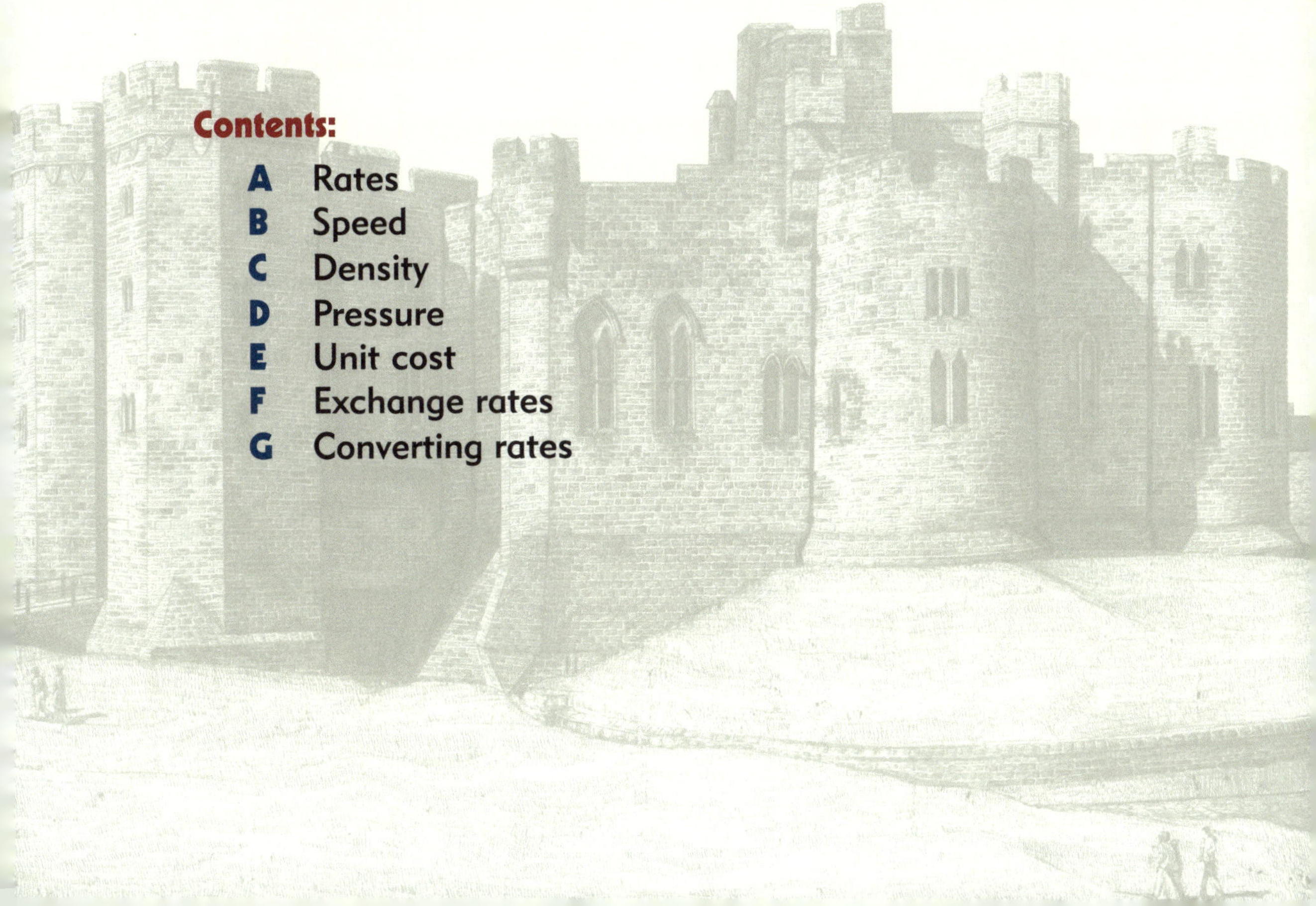

Contents:
- **A** Rates
- **B** Speed
- **C** Density
- **D** Pressure
- **E** Unit cost
- **F** Exchange rates
- **G** Converting rates

Opening problem

A greyhound runs 515 m in 29.5 seconds.

A horse gallops 1650 m in 1 minute 45 seconds.

A cheetah sprints 380 m in 12.9 seconds.

Things to think about:

a How can we compare the speeds of the three animals?

b Which animal is fastest?

A RATES

We have seen that a **ratio** is an ordered comparison of quantities of the **same** kind. For example, we can have a ratio of lengths or a ratio of times.

> A **rate** is an ordered comparison of quantities of **different** kinds.

For example, a person's *heart rate* is a comparison between the *number of heart beats* and the *time*.

When we write a rate, we do not use a ratio sign " : ", but instead we divide one quantity by another.

Since we are comparing quantities of *different* kinds, units are very important. We must always include units in our answer. We use the word *per* which means "for every", or a slash /, to separate the units.

For example, if a person's heart beats 65 times every minute, we write their heart rate as 65 beats per minute, or 65 beats/minute.

The slash / indicates division.

Activity 1   Measuring your heart rate

One way to determine your fitness level is to measure your heart rate. It is usually measured in beats per minute.

What to do:

1. Find your pulse on your wrist, or the side of your neck.
2. Count how many times you can feel your pulse in one minute.
3. Compare your heart rate with those of your classmates.
4. What happens to your heart rate when you exercise?

Other common examples of rates are:

	Examples of units
Rates of pay	pounds per hour
Petrol consumption	litres per 100 km or km per litre
Annual rainfall	mm per year
Unit cost	pounds per kg
Population density	people per square kilometre

Example 1   ◄)) Self Tutor

A tap fills a 9 litre bucket in 3 minutes. Express this as a rate in simplest form.

$$\text{rate} = \frac{9 \text{ litres}}{3 \text{ minutes}}$$

$$\therefore \text{ rate} = \frac{9}{3} \text{ litres per minute}$$

$$\therefore \text{ rate} = 3 \text{ litres per minute}$$

Where necessary, round the rates to 2 decimal places.

EXERCISE 19A

1 Write down the meaning of each rate:
- **a** 5 miles per h
- **b** 15 pounds per h
- **c** 7 litres per s
- **d** 99 pence per litre
- **e** 30 kg per h
- **f** 14 g per min
- **g** 96 euros per day
- **h** 66 m per s
- **i** 21 ml per h

2 Suggest units which could be used to measure:
- **a** a person's rate of pay
- **b** an aeroplane's speed
- **c** the price of petrol
- **d** the typing speed of a secretary
- **e** the change in an oven's temperature over time.

3 Copy and complete:
- **a** A car uses 10 litres of petrol every 160 km. The rate of petrol consumption is ...... km per litre.
- **b** A train travels 416 miles over 8 hours. This is a rate of ...... miles per hour.
- **c** 28 litres of water drains from a tank in 8 seconds. This is a rate of ...... litres per s.
- **d** A carton of milk costs £2.18 for 2 litres. This is a rate of £...... per litre.
- **e** A driver works for 3 hours and receives £51. His rate of pay is £...... per hour.

4 Jennifer's heart beats 375 times in 5 minutes. Express this rate in beats per minute.

5 The Peterson household used 1170 megajoules of gas during April. Express this rate of energy use in megajoules per day.

6 Annie travels 25 miles by train to school. Her journey takes 45 minutes.
Victoria travels 20 miles by car, a journey which takes 40 minutes.
- **a** Find the rate of travel for each girl in miles per min.
- **b** Which mode of transport is more efficient?

7 Xinsong works 8 hours a week as a waiter, earning £64.
Jay works 6 hours each week in the kitchen, earning £54.

  a Find the rate of pay for each person in pounds per hour.

  b Who is paid at a higher rate?

Example 2   ◀)) Self Tutor

Henry eats 240 peanuts every 3 minutes.

  a Find Henry's rate of eating peanuts.
  b How many peanuts will Henry eat in 10 minutes?

  a   Henry's rate of eating peanuts
      $= \dfrac{240 \text{ peanuts}}{3 \text{ minutes}}$
      $= 80$ peanuts per minute
  b In 10 minutes Henry will eat
      $80 \times 10 = 800$ peanuts.

8 A family of four uses 2800 litres of water each week.

  a Find the rate of water usage in litres per day.

  b How much water will the family use in 20 days?

9 Judy works part-time at a local café. She earned £29.60 for working 4 hours last week.

  a Find Judy's rate of pay.

  b This week Judy worked 19 hours. How much will she earn this week if she is paid the same hourly rate?

10 A milk truck takes 5 minutes to discharge 6750 litres of milk.
At this rate, how much milk would the truck discharge in 18 minutes?

11 Jeff is a gardener who can plant 5 trees each hour.

  a Today Jeff planted trees for 3 hours. How many trees did he plant?

  b Jeff thinks that he will plant 80 trees tomorrow. For how long is Jeff expecting to work?

12 It costs £96 to buy a 32 m length of fibre optic cable.

  a Find the cost of each metre of cable.

  b Find the cost of a cable of length 27 m.

  c Find the length of cable that could be bought for £450.

13 To travel 518 km, a car uses 28 litres of petrol.

    a Find the rate at which the petrol is used in:

        i km per litre   **ii** litres per 100 km.

    b At this rate, how many litres of fuel would be needed to travel 1480 km?

    c If fuel costs £1.35 per litre, how much would the journey in **b** cost?

14 This table shows the cost of holding an event at a function centre.

Number of people	Cost
1 - 50	£30 per person
51 - 70	£28 per person
71 - 90	£26 per person
91 - 100	£25 per person

    a Find the total cost of holding an event for:

        i 40 people   **ii** 80 people.

    b Karen has £1800 to spend on an event. Find the maximum number of people who can attend the event.

Puzzle

It takes Melanie 12 minutes to cut a log into 4 pieces. In total, how long will it take her to cut the log into 10 pieces?

B   SPEED

The most common rate that we use is **speed**, which is a comparison between the *distance travelled* and the *time taken*.

The **instantaneous speed** of an object refers to how fast the object is travelling at a particular instant.

In the metric system, speed is commonly measured in km/h or m/s. However, in the United Kingdom we still usually measure the speed of vehicles in miles/hour.

For example, when you are in a car, the speedometer might say that you are travelling at 50 miles per hour.

However, when we go on a journey, we do not always travel at a constant speed. We need to slow down for other cars, and stop at traffic lights. For the whole journey, therefore, we calculate an **average speed** by comparing the total distance travelled with the total time taken.

$$\text{average speed} = \frac{\text{total distance travelled}}{\text{total time taken}}$$

This formula can be rearranged as: distance = speed × time

                        or   time = $\frac{\text{distance}}{\text{speed}}$

You can use the triangle alongside to help you remember these.

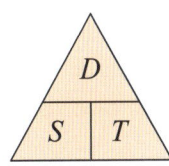

Example 3

Erica cycled 36.8 miles in 2 hours.
  a Find her average speed.
  b Cycling at the same rate, how long would it take Erica to cycle 86 miles?

a average speed

$= \dfrac{\text{distance travelled}}{\text{time taken}}$

$= \dfrac{36.8 \text{ miles}}{2 \text{ hours}}$

$= 18.4 \text{ miles/h}$

b time

$= \dfrac{\text{distance}}{\text{speed}}$

$= \dfrac{86 \text{ miles}}{18.4 \text{ miles/h}}$

$\approx 4.67 \text{ hours}$

$\approx 4 \text{ h } 40 \text{ min}$

EXERCISE 19B.1

1 Find, in miles per hour, the average speed of:
  a a cyclist who travels 100 miles in 4 hours
  b a boat which travels 150 miles in 5 hours
  c an athlete who runs 18 miles in 1.5 hours
  d an aeroplane which takes 50 minutes to fly 750 miles.

2 The speed limit on a freeway is 70 miles/h. Jason drives 174 miles along the freeway in 2.4 hours. Has he broken the law?

3 Bernadette drives her car at an average speed of 32 miles/h.
  a If Bernadette drives for 3 hours, how far does she travel?
  b How long would it take Bernadette to travel 54 miles at this speed?

4 A model train travels around a circular track with radius 5 m as shown. The train takes 20 seconds to complete a lap of the track. Find the average speed of the train. Give your answer in metres per second, correct to 1 decimal place.

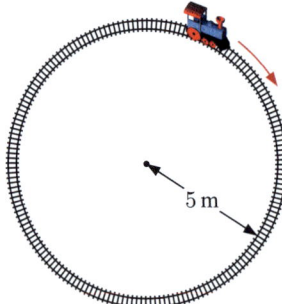

5 **a** How far does Liam travel if his aeroplane flies at 210 km/h for 1 hour and 40 minutes?
  b When Liam makes the return journey he is now flying against the wind, and his plane averages only 175 km/h. How long does the return flight take him?

6 Yiren walks 60 metres in 22.5 seconds, while Sean walks 150 metres in 1 minute.
  a Find the average speed of each person.
  b Yiren and Sean each walk 2000 m at their normal speed. Who will finish first, and by how much?

DISTANCE-TIME GRAPHS

A **distance-time graph** or **travel graph** for a journey shows the relationship between *distance travelled* and the *time taken*.

We will see how distance-time graphs can be used to calculate the speeds of travel at different stages of a journey.

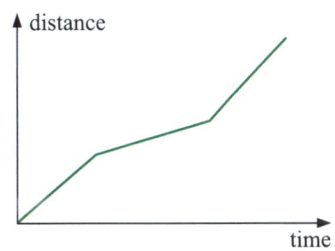

Investigation — Travel graphs

Brian is riding his bicycle along a flat stretch of road. He travels 100 metres every 10 seconds.

What to do:

1. Copy and complete the table alongside, showing the total distance Brian has travelled over the first 50 seconds.

Time (seconds)	Distance (metres)
0	0
10	100
20	200
30	
40	
50	

2. Plot these points on a graph, and join the points with a line. What feature of the graph indicates that Brian is travelling at a constant speed?

3. For Brian's journey so far, find:
   a the total distance travelled
   b the time taken
   c the average speed, in metres per second.

4. Brian then encounters a steep downhill section. He travels 200 metres in the next 10 seconds.
   a Extend your table to record the total distance travelled after 60 seconds.
   b Extend the graph to include this new point. What feature of the graph indicates that Brian's speed has changed?

5. For the period when Brian was travelling downhill, find:
   a the distance travelled
   b the time taken
   c the average speed, in metres per second.

6. For the *whole journey*, find:
   a the total distance travelled
   b the total time taken
   c the average speed, in metres per second.

From the **Investigation** you should have observed that when the speed is constant, the distance-time graph will be a straight line.

If the distance-time graph is not a straight line, we can still find the **average speed** between any two points on the graph.

Example 4

The graph shows the progress of a train travelling between cities.

a How far does the train travel in the first 3 hours?
b Find the speed of the train for the first 3 hours.
c Find the speed of the train during the final hour of the journey.
d Find the average speed of the train for the entire journey.

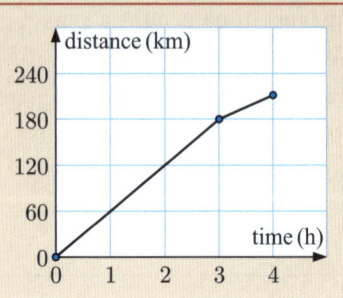

a The train travels 180 km in the first 3 hours.

b Speed = $\dfrac{\text{distance travelled}}{\text{time taken}} = \dfrac{180 \text{ km}}{3 \text{ hours}} = 60$ km/h.

c The speed of the train was 30 km/h.

d Average speed = $\dfrac{\text{total distance travelled}}{\text{total time taken}}$

$= \dfrac{210 \text{ km}}{4 \text{ hours}}$

$= 52.5$ km/h

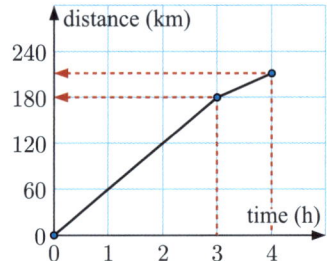

EXERCISE 19B.2

1 This distance-time graph shows the progress of a truck travelling between two cities.

  a Is the truck travelling at constant speed? Explain your answer.
  b How far does the truck travel in the first 2 hours?
  c Find the speed of the truck.

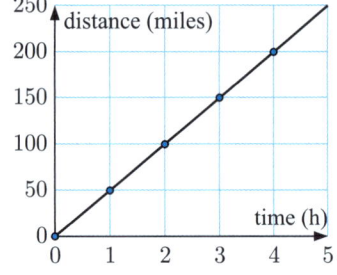

2 Every 2 hours, a cyclist travels 40 km.

  a Copy and complete the distance-time graph opposite.
  b How far will the cyclist travel in 3 hours?
  c How long will it take the cyclist to travel 100 km?
  d Find the speed of the cyclist.

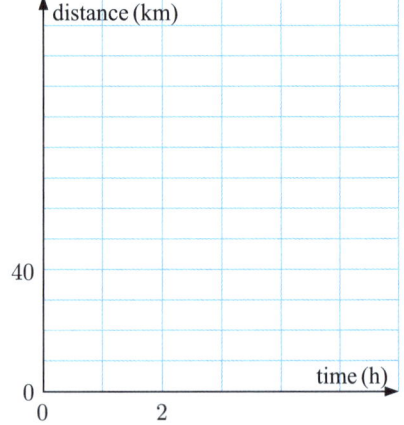

3 This travel graph shows the progress of a car travelling from town A to B to C.

   a How far is it from A to B?

   b How long did the car take to get from A to B?

   c What was the speed of the car while travelling from A to B?

   d How far is it from B to C?

   e How long did the car take to get from B to C?

   f What was the speed of the car while travelling from B to C?

   g How far is it from A to C?   **h** How long did the car take to get from A to C?

   i Find the *average speed* of the car from A to C.

C DENSITY

Puzzle

Which is heavier, 1 tonne of lead, or 1 tonne of feathers?

The answer to this Puzzle is that the objects are as heavy as each other, since both objects have mass 1 tonne. However, many people guess that the lead is heavier, since a *certain volume* of lead will be much heavier than the *same volume* of feathers. They have in fact compared the lead and feathers using a rate called **density**.

Lead is much more dense than feathers.

> The **density** of an object is its mass per unit of volume.

Density can be found using the formula:

$$\text{density} = \frac{\text{mass}}{\text{volume}}$$

Density is usually measured in grams per cubic centimetre.

For example, the density of pure gold is 19.3 grams per cm^3. This means that every cubic centimetre of pure gold weighs 19.3 grams.

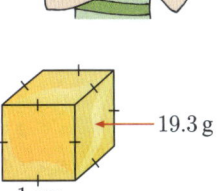

THE DENSITY OF WATER

1 ml or 1 cm^3 of pure water at 4°C weighs 1 gram.

> The density of pure water is 1 gram per cm^3.

If an object has density less than 1 gram per cm^3, then it will float on water. If its density is greater than 1 gram per cm^3, then it will sink.

This table lists the densities of some common materials in g per cm³.

Material	Density	Material	Density
carbon dioxide	0.002	aluminium	2.7
petrol	0.70	iron	7.8
ice	0.92	lead	11.3
water	1.00	gold	19.3
milk	1.03	platinum	21.4

Example 5 ◀)) Self Tutor

Find the density of a piece of timber which is 60 cm by 10 cm by 3 cm and weighs 1.62 kg.

Mass of timber = 1.62 kg
 = 1620 g

Volume of timber = 60 × 10 × 3 cm³
 = 1800 cm³

∴ density of timber = $\frac{\text{mass}}{\text{volume}}$

= $\frac{1620 \text{ g}}{1800 \text{ cm}^3}$

= 0.9 g per cm³

EXERCISE 19C

1 Find the density, in g per cm³, of:
  a an object with mass 20 g and volume 5 cm³
  b a stone with mass 315 g and volume 35 cm³
  c a metal disc which weighs 1.13 kg and has volume 50 cm³.

2 A block of ebony is 1.1 m × 3 cm × 4 cm, and weighs 1.4 kg. Find the density of the block in g per cm³.

3 A pair of dice and their weights are shown alongside. Which die is made from the denser material?

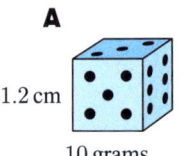

1.2 cm

10 grams

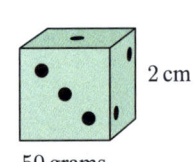

2 cm

50 grams

4 Petrol and water do not mix. If the two liquids are poured into a container, they will separate into two layers. Which is the upper layer? Explain your answer.

5 The doorstop shown weighs 200 grams. If it was dropped into water, would it sink or float? Explain your answer.

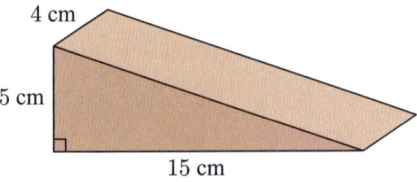

Example 6

The density of silver is 10.5 g per cm³. Find:

a the mass of 50 cm³ of silver

b the volume of 700 g of silver.

a density = $\frac{\text{mass}}{\text{volume}}$

∴ mass = density × volume
 = 10.5 × 50
 = 525 g

b density = $\frac{\text{mass}}{\text{volume}}$

∴ volume = $\frac{\text{mass}}{\text{density}}$
 = $\frac{700}{10.5}$
 ≈ 66.7 cm³

6 The density of zinc is 7.14 g per cm³. Find:
  a the mass of 60 cm³ of zinc
  b the volume of 400 g of zinc.

7 Find the mass of:
  a 200 cm³ of aluminium
  b 80 ml of milk.

8 The density of glass is 2.6 g per cm³. Find the mass of this glass marble.

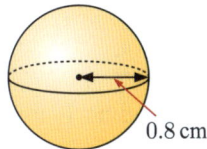

0.8 cm

9 Find the volume of:
  a 60 g of lead
  b 1.3 kg of iron.

10 300 g of copper and 100 g of tin are combined to make a bronze alloy. The density of copper is 8.96 g per cm³, and the density of tin is 7.31 g per cm³.
  a Find the volume of tin used.
  b Find the density of the bronze alloy.

Activity 2   Will it sink or float?

You will need: ruler, set of scales, container of water

What to do:

1. Gather several solid objects from around your classroom or your home. The objects should have a shape that you can calculate the volume of, such as a rectangular prism or a cylinder.
   Before performing any calculations, predict whether each object will sink or float in water.

2. Measure the dimensions of each object. Use your measurements to find the volume of each object.

3. Use the scales to find the mass of each object.

4. Calculate the density of each object. If you wish, you may now change your predictions about whether each object will sink or float.

5. Place each object in the container of water. Were your predictions correct?

D PRESSURE

When an object is placed on a surface, the object exerts a **force** on the surface.

To find the **pressure** exerted by the object, we divide the force by the **area** over which the force is applied.

$$\text{pressure} = \frac{\text{force}}{\text{area}}$$

Discussion

Is Gary more likely to break his kitchen table if he stands on it, or lies down on it?

Can you use the formula above to help explain your answer?

Force is measured in newtons (N), so pressure is usually measured in N per cm^2 or N per m^2.

The **newton** is a unit of force named after **Sir Isaac Newton**.

Example 7   ◉ Self Tutor

a Find the pressure when a force of 235 N is applied to an area of 5 m^2.

b A box exerts a force of 30 N on a table. The pressure on the table is 0.2 N/cm^2. Find the area over which the force is applied.

a pressure $= \dfrac{\text{force}}{\text{area}}$

$= \dfrac{235 \text{ N}}{5 \text{ m}^2}$

$= 47 \text{ N/m}^2$

b area $= \dfrac{\text{force}}{\text{pressure}}$

$= \dfrac{30 \text{ N}}{0.2 \text{ N/cm}^2}$

$= 150 \text{ cm}^2$

EXERCISE 19D

1 Find the pressure when a force of 75 N is applied to an area of 1.5 m^2.

2 A block with base area 500 cm^2 produces 0.04 N/cm^2 of pressure when placed on the floor. Find the force exerted by the block on the floor.

3 A pot plant exerts a force of 8 N on a window sill. The pressure on the window sill is 0.1 N/cm². Find the area over which the force is applied.

4 A bicycle applies a force of 160 N to the road. At any given time, 2.4 cm² of the surface of each tyre makes contact with the road. Find the pressure exerted by the bicycle on the road.

5 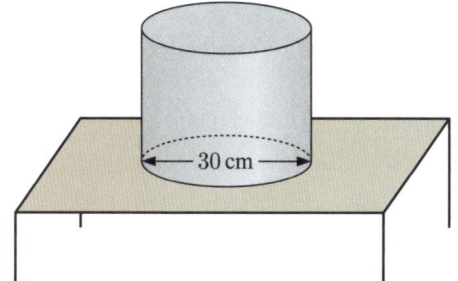 This cylinder exerts a force of 40 N on the table.
  a Find the pressure applied by the cylinder on the table.
  b What force would the cylinder need to exert, to produce 0.3 N/cm² of pressure?

6 This container exerts a force of 3500 N on the floor.
  a Find the pressure applied by the container on the floor.
  b Suppose the container is tipped over, so the shaded face is on the floor.
  Find the pressure now applied by the prism on the floor.

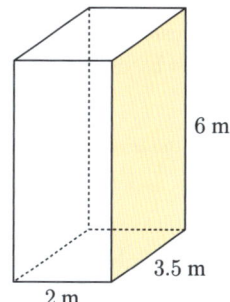

E UNIT COST

When shopping, it is important to get good value for money. However, it is often not obvious which item represents the best value for money, because the same item can come in several different sized packages.

To properly compare prices, we need to convert the cost of an item into a rate. This rate is called the **unit cost**. It might be written as the cost per gram, the cost per 100 grams, the cost per kilogram, or the cost per litre. We then compare the unit costs for packages of different sizes.

Activity 3   Unit pricing

Most supermarkets include unit pricing on the price tags of their items.

Next time you are at a supermarket, find out how the unit prices for the following items are measured:

- milk
- batteries
- flour
- dishwashing liquid
- steak
- paper towels

LEMONADE £0.79
1.5 litre
£0.53 per litre

Example 8

By comparing the cost per 100 g, decide which box of cereal is better value for money.

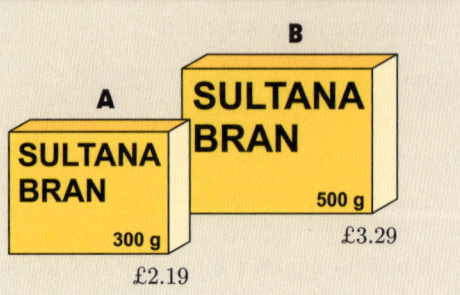

A 300 g = 3 lots of 100 g

$\therefore$ cost per 100 g $= \dfrac{£2.19}{3}$

$= £0.73$ per 100 g

B 500 g = 5 lots of 100 g

$\therefore$ cost per 100 g $= \dfrac{£3.29}{5}$

$= £0.658$ per 100 g

So, **B** is better value for money.

EXERCISE 19E

1 Use your calculator to find the unit cost for each of the following items. Express your answer using the units in brackets.

  a packet of 3 tennis balls for £11.40   (£ per ball)
  b 5 kg potatoes for £8.45   (£ per kg)
  c 250 g packet of crisps for £1.40   (pence per g)
  d 1.25 litres soft drink for £0.99   (pence per litre)
  e 4.2 m of ribbon for £8.40   (£ per m)
  f 35 litres of petrol for £43.40   (pence per litre)

2 Consider the following grocery items and decide which is the better value for money:

  a compare cost per 100 g

  b compare cost per 100 ml

  c compare cost per box

  d compare cost per tablet

e compare cost per 10 m **f** compare cost per 10 g

50 m £3.65   20 m £1.55

420 g £3.85   250 g £2.15

3 A supermarket sells 110 g tubes of toothpaste for £3.19, and 160 g tubes of toothpaste for £3.99.

  a Calculate the price per 10 g for each size of toothpaste.
  b Which size is better value for money?
  c The supermarket offers a "3 for 2" deal where if you buy two 110 g tubes of toothpaste, you receive a third one free. Does this represent better value for money than buying the 160 g tubes?

F EXCHANGE RATES

If you have travelled to other countries, you may have noticed that in different places people use different types of money. We call these **currencies**. For example, the United States of America uses the US dollar, most European countries use the euro, and Japan uses the yen.

If you visit a place which uses a different currency, you will need to sell some of your money and buy some of theirs in **exchange**. An **exchange rate** is used to work out how much your money is worth in the other currency.

For example, suppose that the exchange rate between the British pound and the US dollar is 1 pound = 1.3 US dollars. This means that 1 pound can be exchanged for 1.3 US dollars.

- To convert pounds into US dollars, we **multiply** by 1.3.
- To convert US dollars into pounds, we **divide** by 1.3.

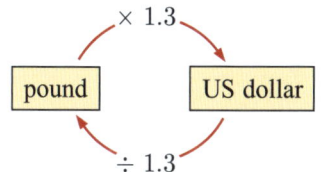

Exchange rates change constantly, so it is a good idea to exchange money at a time when the exchange rate will give you a higher return.

Example 9   ◀) Self Tutor

Suppose the exchange rate between the pound and the US dollar is 1 pound = 1.3 US dollars.
  a Convert 200 pounds into US dollars.   **b** Convert 650 US dollars into pounds.

  a 200 pounds = (200 × 1.3) US dollars
              = 260 US dollars

  b 650 US dollars = (650 ÷ 1.3) pounds
              = 500 pounds

EXERCISE 19F

1 Suppose the exchange rate between the British pound and the Australian dollar is
1 pound = 1.8 Australian dollars. Convert:
   - **a** 100 pounds into Australian dollars
   - **b** 250 pounds into Australian dollars
   - **c** 55 Australian dollars into pounds
   - **d** 1100 Australian dollars into pounds.

2 Suppose the exchange rate between the euro and the pound is 1 euro = 0.89 pounds. Convert:
   - **a** 40 euros into pounds
   - **b** 270 euros into pounds
   - **c** 900 pounds into euros
   - **d** 1500 pounds into euros.

3 Suresh has travelled from India to England. He wants to convert 20 000 Indian rupees into pounds. The current exchange rate is 1 pound = 85.9 rupees. How many pounds will Suresh receive?

4 Estelle lives in Paris, and has 700 euros to spend on accommodation for a 5 night trip to London. She looks online, and sees a hotel which costs 120 British pounds per night. The current exchange rate is 1 euro = 0.91 British pounds. Will Estelle be able to afford the hotel?

5 Steve is travelling from England to New Zealand for a holiday. The current exchange rate is 1 British pound = 1.79 New Zealand dollars.
   - **a** When Steve arrives in New Zealand, he converts 2000 pounds into New Zealand dollars. How many New Zealand dollars does he receive?
   - **b** During his holiday, Steve spends 1380 New Zealand dollars. How many New Zealand dollars does he have left at the end of his holiday?
   - **c** When he returns to England, Steve exchanges his New Zealand dollars back into pounds. How many pounds does he receive?

G CONVERTING RATES

It is often useful to convert a rate into different units so it is easier to understand for the situation we are dealing with.

Discussion

2 metres per second is the same rate as 7.2 kilometres per hour. Which rate makes it easier to understand the situation if you are:
- walking 300 m to the bus stop
- hiking for 5 hours?

Example 10   ◀)) Self Tutor

A petrol bowser pumps petrol at the rate of 600 litres per hour. Write this rate in litres per minute.

In 1 hour, the bowser pumps 600 litres.

There are 60 minutes in 1 hour, so in 1 minute the bowser pumps $\frac{600}{60} = 10$ litres.

This is a rate of 10 litres per minute.

Rates (Chapter 19) 391

EXERCISE 19G.1

1 A fire hose discharges water at the rate of 180 litres per minute. Write this rate in litres per hour.

2 Kelly's heart rate is 60 beats per minute. Write her heart rate in:
   a beats per second
   b beats per hour
   c beats per day.

3 A shower head has a flow rate of 7.5 litres per minute. Write this rate in:
   a ml per second
   b litres per hour.

4 A bamboo plant grows 18 m in 60 days. Write this growth rate in:
   a m per day
   b m per hour
   c mm per hour.

5 Reg eats 175 g of potato chips per day. Write this rate in:
   a grams per week
   b kilograms per week.

6 A book on a desk exerts 0.03 N/cm² of pressure. Write this pressure in N/m².

7 The density of a material is 6.8 g per cm³. Write this density in kg per m³.

SPEED CONVERSIONS

Roger rides his bicycle at $36 \text{ km/h} = \dfrac{36 \text{ km}}{1 \text{ hour}}$

$= \dfrac{36\,000 \text{ m}}{3600 \text{ s}}$ {1 h = 60 min = 60 × 60 s = 3600 s}

$= 10 \text{ m/s}$

So, travelling at 10 m/s is the same as travelling at 36 km/h. We say that these are **equivalent** rates.

Notice that travelling at 1 m/s is the same as travelling at 3.6 km/h.

- To convert m/s into km/h, we **multiply by 3.6**.
- To convert km/h into m/s, we **divide by 3.6**.

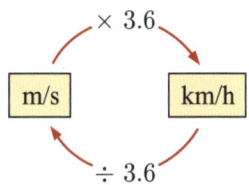

EXERCISE 19G.2

1 Using 10 m/s = 36 km/h, convert to km/h:
   a 30 m/s
   b 70 m/s
   c 5 m/s

2 Using 36 km/h = 10 m/s, convert to m/s:
   a 72 km/h
   b 144 km/h
   c 9 km/h

Example 11 ◀)) Self Tutor

a A sprinter runs at 11 m/s. Convert this speed to km/h.
b An aeroplane travels at 900 km/h. Convert this speed to m/s.

a   11 m/s
   $= (11 \times 3.6)$ km/h
   $= 39.6$ km/h

b   900 km/h
   $= (900 \div 3.6)$ m/s
   $= 250$ m/s

3 Convert to km/h:
- **a** 200 m/s
- **b** 45 m/s
- **c** 27 m/s
- **d** 800 m/s

4 Convert to m/s:
- **a** 50 km/h
- **b** 110 km/h
- **c** 21 km/h
- **d** 540 km/h

Example 12 ◀) Self Tutor

A 400 m sprinter finishes a race in 45 seconds. Find his speed in km/h.

$$\text{speed} = \frac{400 \text{ m}}{45 \text{ sec}}$$
$$= 8.888\,8\ldots\text{ m/s}$$
$$= (8.888\,8\ldots \times 3.6)\text{ km/h}$$
$$= 32\text{ km/h}$$

---

5 In 2009, Usain Bolt achieved a world record time of 19.19 seconds for the 200 metre sprint. Find his average speed, correct to 2 decimal places, in:
- **a** m/s
- **b** km/h.

6 Find the following speeds in km/h:
- **a** A sprinter runs 100 m in 9.7 seconds.
- **b** A greyhound races 500 m in 29 seconds.
- **c** A horse gallops 2000 m in 2 min 10 seconds.
- **d** A swimmer travels 1500 m in 15 minutes.

7 Sophie ran 300 m in 40 seconds.
- **a** Show that Sophie's speed was 27 km/h.
- **b** Does this mean that if Sophie ran for 1 hour, she would run 27 km? Explain your answer.

Review set 19A

1 A petrol pump delivers 42 litres of petrol into a car in 3 minutes. Write this rate in litres per minute.

2 Convert:
- **a** 150 beats per minute into beats per second
- **b** 54 m/s into km/h.

3 A freight train travels 770 km in 8 hours, while a truck on the highway travels 120 km in 85 minutes. Which mode of transport is faster?

4 The graph shows the progress of a car as it travels between cities.
- **a** How far does the car travel in 3 hours?
- **b** How long does it take for the car to travel 100 km?
- **c** Find the speed of the car.

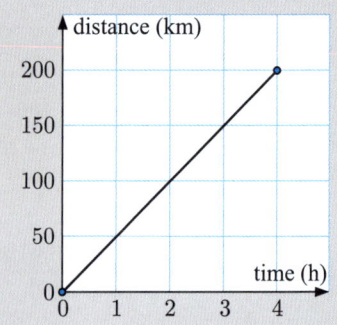

5 Water from a tap will fill a 9 litre watering can in 45 seconds. How long will it take to fill a 120 litre pond?

6 A runner travels 32.5 km in 2 hours and 30 minutes. Find his speed in:
   a km/h   **b** m/s.

7 Find the density of a 420 g paperweight with a volume of 150 cm^3.

8 Suppose the exchange rate between the British pound and the Singapore dollar is 1 pound = 1.78 Singapore dollars.
   a Convert 2000 pounds into Singapore dollars.
   b Convert 655 Singapore dollars into pounds.

9 Find the pressure when a force of 60 N is applied to an area of 25 m^2.

10 Find the density of a 5 cm by 30 cm by 60 cm piece of packing foam weighing 126 g.

11 Which of the chocolate bars is the better value for money?

12 The density of concrete is 2.4 g/cm^3. Find the volume of 2 kg of concrete.

Review set 19B

1 Margot receives £220 for an eight hour nursing shift. What is her hourly rate of pay?

2 Convert:
   a 80 km/h to m/s   **b** 15 cm per year to mm per month.

3 At a local market it costs £2.10 to buy 0.6 kg of rhubarb.
   a Find the price per kilogram of the rhubarb.
   b How much would it cost to buy 2.5 kg of rhubarb?

4 Trent rides his motorcycle for 3 hours. In this time he covers a distance of 198 km, and uses 11 litres of fuel.
   a Find:
      i Trent's average speed in km/h
      ii the petrol consumption of the motorcycle in km/litre.
   b Tomorrow Trent must travel 20 km to a friend's house. He can either ride his motorcycle, or catch the bus. Given that petrol costs 118 p/litre and the bus costs £1.90 per trip, which option is cheaper?

5 A pack of 12 fruit bars costs £4.92. Find the unit cost in pence per bar.

6 This solid cylinder weighs 200 g. If it was dropped into water, would it sink or float?

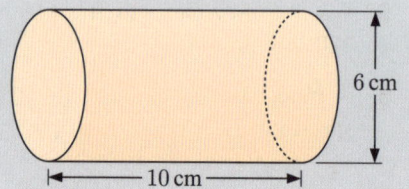

7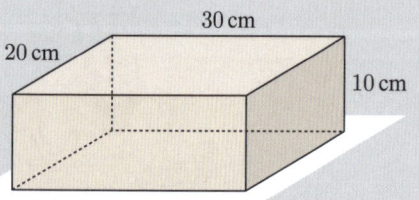

This box exerts 0.125 N/cm² of pressure on the floor. Find the force exerted by the box on the floor.

8 a Which of these packets of cereal is better value for money?

b The 750 g packet is put on special, and now costs only £3.49. Which packet is better value for money now?

£5.50   £3.99

9 The distance-time graph shows Sylvie's progress when driving her car from home to the beach.

a How far is the beach from Sylvie's home?
b How long did it take Sylvie to get to the beach?
c Find Sylvie's average speed for the whole journey.
d What was Sylvie's speed between C and D?
e At which points did the car change its speed?

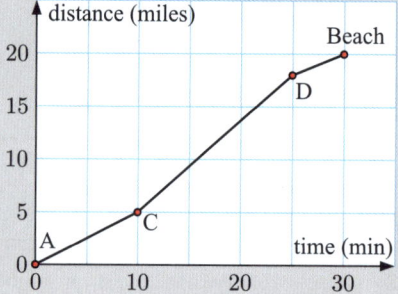

10 Lucy has travelled from the United States to Mexico. She wants to convert 500 US dollars into Mexican pesos. The current exchange rate is 1 US dollar = 13.1 Mexican pesos. How many pesos will Lucy receive?

11 The density of plastic is 1.8 g/cm³. Find the mass of this plastic cone.

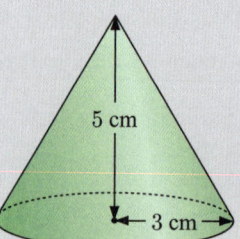

12 Alex drove 200 km in 4 hours.

a Find his average speed.
b Driving at this speed, how long would it take Alex to drive 325 km?
c Write Alex's average speed in metres per second.

20

Coordinate geometry

Contents:

- **A** The Cartesian plane
- **B** Linear relationships
- **C** Gradient
- **D** Parallel lines
- **E** Axes intercepts
- **F** The equation of a line
- **G** Graphing lines in the form $y = mx + c$
- **H** Graphing lines in the form $Ax + By = C$
- **I** Vertical and horizontal lines
- **J** Finding the equation of a line

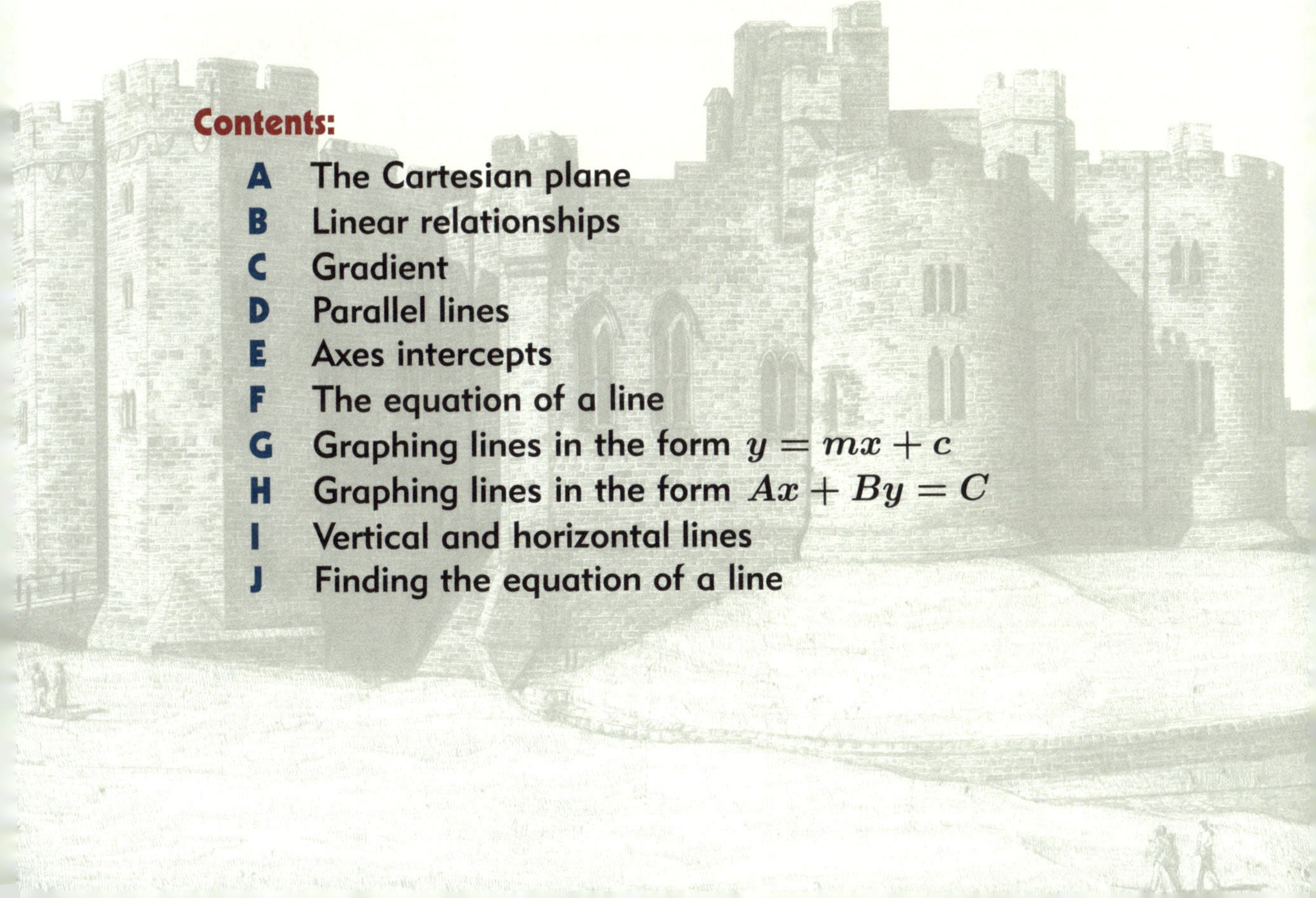

Opening problem

When Tiffany makes a mobile phone call, the *cost* of the call depends on the *time* that it lasts for. The relationship between the cost and the time of the call is shown on the graph alongside.

Things to think about:

a Is the relationship between *cost* and *time* linear?
b What is the *connection fee* or fixed cost charged for a very short phone call?
c By how much does the cost of the call increase for each minute the call lasts?
d Can you use this graph to determine:
   i the cost of a 3 minute call
   ii the length of a call which costs £2?

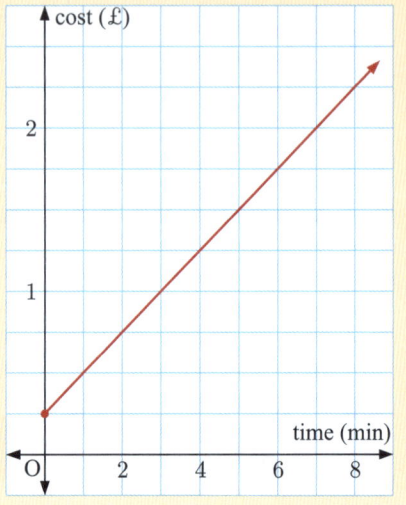

The graph in the **Opening Problem** is a straight line. We say there is a **linear relationship** between the variables *cost* and *time*.

Historical note

Sir Isaac Newton is recognised as one of the great mathematicians of all time. His achievements are remarkable considering mathematics, physics, and astronomy were enjoyable pastimes after his main interests of chemistry, theology, and alchemy.

Despite his obvious abilities, there is a story from Newton's childhood which indicates that even the greatest thinkers can find silly solutions for simple problems. He was asked to go out and cut a hole in the bottom of the barn door for the cats to go in and out. He decided to cut two holes: one for the cat and a smaller one for the kittens.

Sir Isaac Newton

After completing school, Newton was initially made to work on a farm, but when his uncle discovered his enthusiasm for mathematics it was decided that he should attend Cambridge University.

Newton's contribution to coordinate geometry included the introduction of negative values for coordinates. In his *Method of Fluxions*, Newton suggested eight new types of coordinate systems, one of which we know today as polar coordinates.

A THE CARTESIAN PLANE

The number grid alongside is a **Cartesian plane**, named after **René Descartes**. The numbers or **coordinates** on it allow us to locate the exact position of any point on the plane.

We start with a point of reference O called the **origin**. Through it we draw a horizontal line called the **x-axis**, and a vertical line called the **y-axis**.

The **x-axis** is an ordinary number line with positive numbers to the right of O and negative numbers to the left of O.

The **y-axis** has positive numbers above O and negative numbers below O.

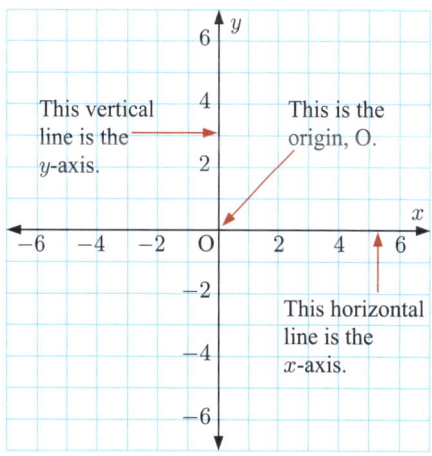

PLOTTING POINTS ON THE CARTESIAN PLANE

To specify the position of a point on the number plane, we use an **ordered pair** of coordinates in the form (x, y).

For example, on the grid alongside we see the point described by the ordered pair of coordinates $(3, 2)$. We say that 3 is the **x-coordinate** and 2 is the **y-coordinate**.

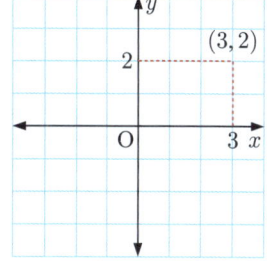

To help us identify particular points, we often refer to them using a capital letter. For example, consider the points A$(1, 3)$, B$(4, -3)$, and C$(-4, -2)$.

To plot the point A$(1, 3)$:
- start at the origin O
- move right along the x-axis 1 unit
- then move upwards 3 units.

To plot the point B$(4, -3)$:
- start at the origin O
- move right along the x-axis 4 units
- then move downwards 3 units.

To plot the point C$(-4, -2)$:
- start at the origin O
- move left along the x-axis 4 units
- then move downwards 2 units.

DEMO

The x-coordinate is always given first. It indicates horizontal movement away from the origin.

QUADRANTS

The x and y-axes divide the Cartesian plane into four regions referred to as **quadrants**. These quadrants are numbered in an **anti-clockwise direction** as shown:

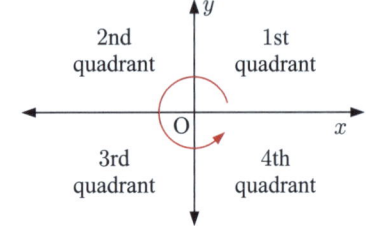

Example 1

Plot the points A(3, 5), B(−1, 4), C(0, −3), D(−3, −2), and E(4, −2) on the same set of axes.

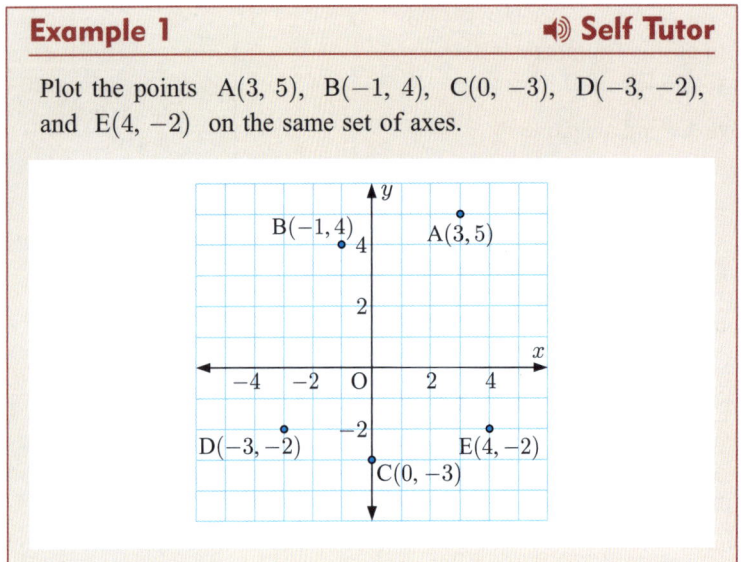

Start at O and move horizontally, then vertically.
→ is positive
← is negative
↑ is positive
↓ is negative.

EXERCISE 20A

1 State the coordinates of the points P, Q, R, S, and T:

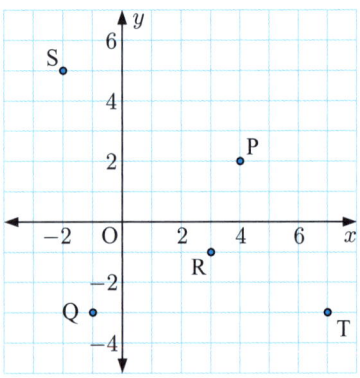

2 a On the same set of axes, plot the points:
A(2, 5), B(4, −2), C(−5, 0), D(−1, −4), E(−2, 3), F(0, 3), G(5, 1), and H(−5, −1).
b State the quadrant in which each point in **a** lies.

3 State the quadrants in which I would find points whose coordinates have:
   a the same sign   **b** different signs.

Example 2  Self Tutor

On a Cartesian plane, show all the points with positive x-coordinate and negative y-coordinate.

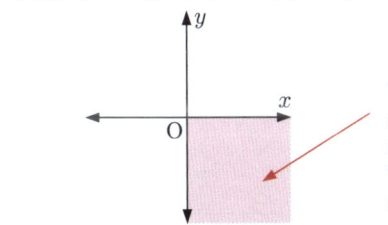

This shaded region contains all points where x is positive and y is negative.
The points on the axes are not included.

This region is the 4th quadrant.

4 On different sets of axes, show all the points with:
- **a** x-coordinate equal to -1
- **b** y-coordinate equal to 3
- **c** x-coordinate equal to 0
- **d** y-coordinate equal to 0
- **e** negative x-coordinate
- **f** positive y-coordinate
- **g** negative x and y-coordinates
- **h** negative x-coordinate and positive y-coordinate.

5
- **a** State the coordinates of A, B, and C.
- **b** Find the coordinates of point D such that ABCD is a rectangle.

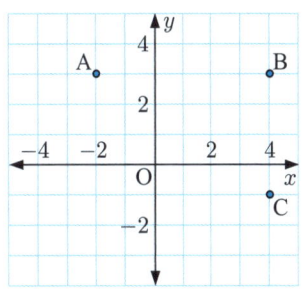

6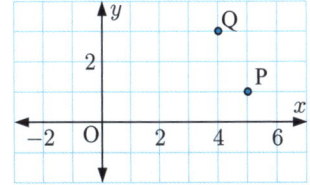

- **a** State the coordinates of P and Q.
- **b** Point R has the same y-coordinate as P, and lies on the y-axis. Find the coordinates of R.
- **c** Find the coordinates of S such that PQRS is a kite.

B LINEAR RELATIONSHIPS

Consider the pattern:

    •   • •   • • •
• •   • • • • • • •   . . . . .
 1st   2nd   3rd

We can construct a **table of values** which connects the diagram number n to the number of dots D. To go from one diagram to the next we need to add *two more* dots.

n	1	2	3	4
D	3	5	7	9

$+2$ $+2$ $+2$

The **equation** which connects n and D in this case is $D = 2n + 1$.

The number of dots D *depends* on the diagram number n. We say that:
- n is the **independent variable**, and place it on the horizontal axis
- D is the **dependent variable**, and place it on the vertical axis.

Since the points lie in a straight line, we say there is a **linear relationship** between n and D.

In this case the values in between the points are meaningless. For example, we cannot have a $2\frac{1}{2}$th diagram. We therefore do not connect the points with a straight line.

However, in other situations it may be sensible to connect the points with a straight line. We can use the line to answer questions involving values between the given points.

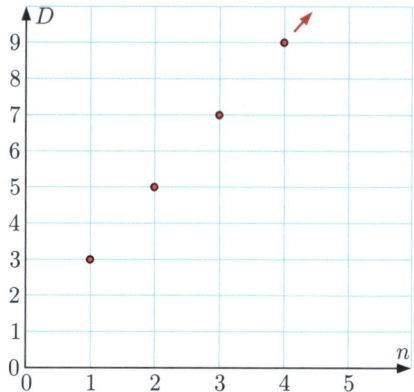

Example 3   ◀) Self Tutor

Max has 10 litres of fuel left in his car's petrol tank. When he fills it at the petrol station, petrol runs into the tank at 15 litres per minute. The petrol tank can hold 70 litres.

  a Identify the independent and dependent variables.
  b Make a table of values for the number of litres L of petrol in the tank after time t minutes, and plot the graph of L against t.
  c Is the relationship between L and t linear?
  d Is it sensible to join the points graphed with a straight line?
  e For every time increase of 1 minute, what is the change in L?
  f Find the number of litres of petrol in the tank after 1.5 minutes.
  g At what time will there be 50 litres of petrol in the tank?

a The *number of litres of petrol* in the tank depends on the *time* it has been filling.
∴ *time* is the independent variable and the *number of litres of petrol* is the dependent variable.

b Each minute the tank is filling adds another 15 litres of petrol.

t (min)	0	1	2	3	4
L (litres)	10	25	40	55	70

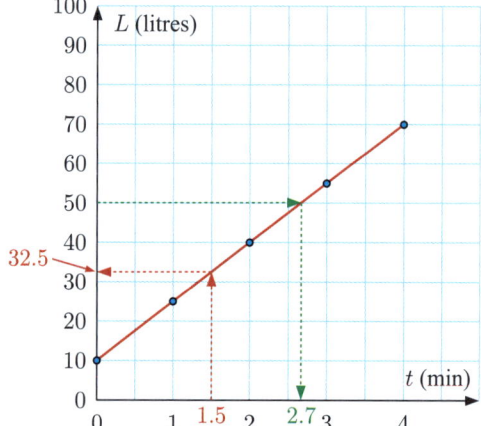

The independent variable is placed on the horizontal axis.

c The points lie in a straight line so the relationship is linear.
d Yes, as Max could add petrol for 2.5 minutes, say, or put 56 litres of petrol in the tank.
e For every time increase of 1 minute, L increases by 15 litres.
f After 1.5 minutes there are 32.5 litres of petrol in the tank.
g There are 50 litres of petrol in the tank after about 2.7 minutes.

EXERCISE 20B

1 Each week a department store employee receives a basic salary of £300. In addition, she is paid a bonus of £20 for each new member she signs up to the store's rewards club.

   Let I be the employee's income and m be the number of members she signs up to the rewards club.

   a Identify the independent and dependent variables.

   b Construct a table of values for I, for $m = 0, 1, 2, 3, ..., 10$.

   c Hence draw a graph of I against m.

   d Is the relationship linear?

   e Is it sensible to join the points with a straight line? Explain your answer.

   f For each new member signed, what is the employee's increase in income?

2 Simon is filling a large container with sports drink for his football team. It already contains 3 litres when he starts to fill it up using 2 litre bottles.

   a Make a table of values for the volume of sports drink S in the container after Simon has emptied n bottles of sports drink into it. Consider $n = 0, 1, 2, ..., 8$.

   b Identify the independent and dependent variables.

   c Plot the graph of S against n.

   d Is the relationship between S and n linear?

   e Is it sensible to join the points graphed with a straight line? Explain your answer.

   f For each full bottle of sports drink added, what is the change in S?

   g What volume of sports drink is in the container after Simon has emptied 2.5 bottles into it?

   h How many full bottles must be emptied into the container so it contains 15 litres in total?

3 Adrian's cookbook gives the cooking temperatures in degrees Fahrenheit (°F). However, Adrian is only familiar with degrees Celsius (°C).

   There is a linear relationship between °F and °C. The boiling point of water is 212°F or 100°C. The freezing point of water is 32°F or 0°C.

   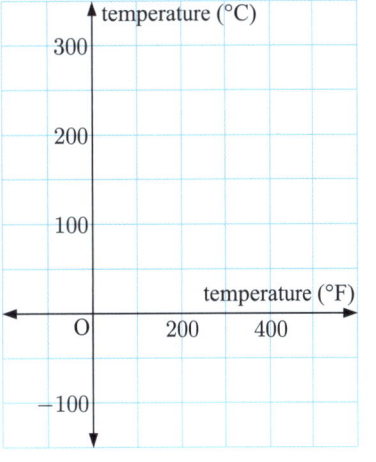

   a Draw a set of axes as shown. Use the boiling and freezing information to mark two points on your graph. Join the points with a straight line.

   b Find the point where the number of degrees Celsius equals the same number of degrees Fahrenheit. What is the temperature?

   c Help Adrian convert these temperatures into °C:
      i a cake must be cooked at 350°F
      ii a roast must reach an internal temperature of 145°F.

   d Use your graph to complete the following table:

Temperature in °F	0	180		
Temperature in °C			150	200

C GRADIENT

Discussion   Which line is steeper?

Look at the two lines alongside.

Justin thinks that *line 2* is steeper, because *line 2* rises 7 units vertically, whereas *line 1* only rises 6 units.

- Is Justin correct? Which line do you think is steepest?
- What does it mean to say that one line is *steeper* than another?
- How can we *measure* the steepness of a line?

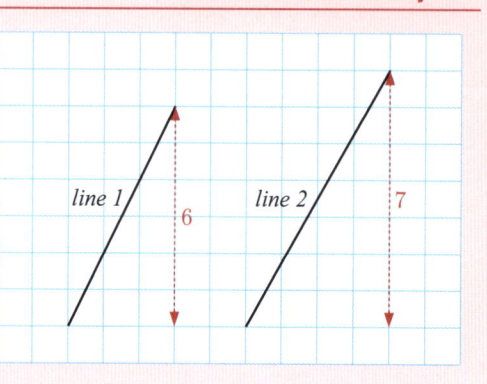

The **gradient** of a line is a measure of its steepness.

In the **Discussion** above, *line 2* rises higher than *line 1*, but *line 1* is *steeper* because it rises at a greater *rate* than *line 2*.

We can calculate the rate at which a line rises or falls by choosing two points on the line, then dividing the vertical step between them by the horizontal step between them.

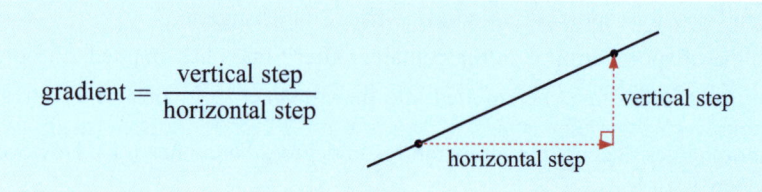

$$\text{gradient} = \frac{\text{vertical step}}{\text{horizontal step}}$$

For the lines in the **Discussion** above:

- The gradient of *line 1* is $\frac{6}{3} = 2$. For every 1 unit moved horizontally, the line moves 2 units upwards.
- The gradient of *line 2* is $\frac{7}{4} = 1.75$.
- *Line 1* has a higher gradient than *line 2*, so *line 1* is steeper than *line 2*.

For an upwards sloping line, if the horizontal step is positive then the vertical step is positive.

Upward sloping lines have a **positive gradient**.

For a downward sloping line, if the horizontal step is positive then the vertical step is negative.

Downward sloping lines have a **negative gradient**.

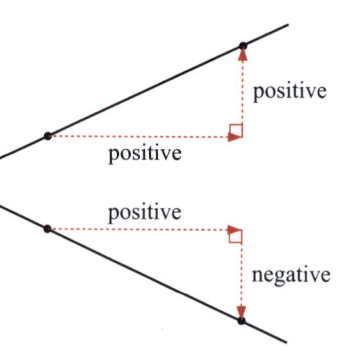

For a **horizontal line**, the vertical step is 0, so the **gradient of a horizontal line is 0**.

For a **vertical line**, the horizontal step is 0, so the **gradient of a vertical line is undefined**.

Example 4 ◀)) Self Tutor

Find the gradient of the line segment:
  a AB   b BC

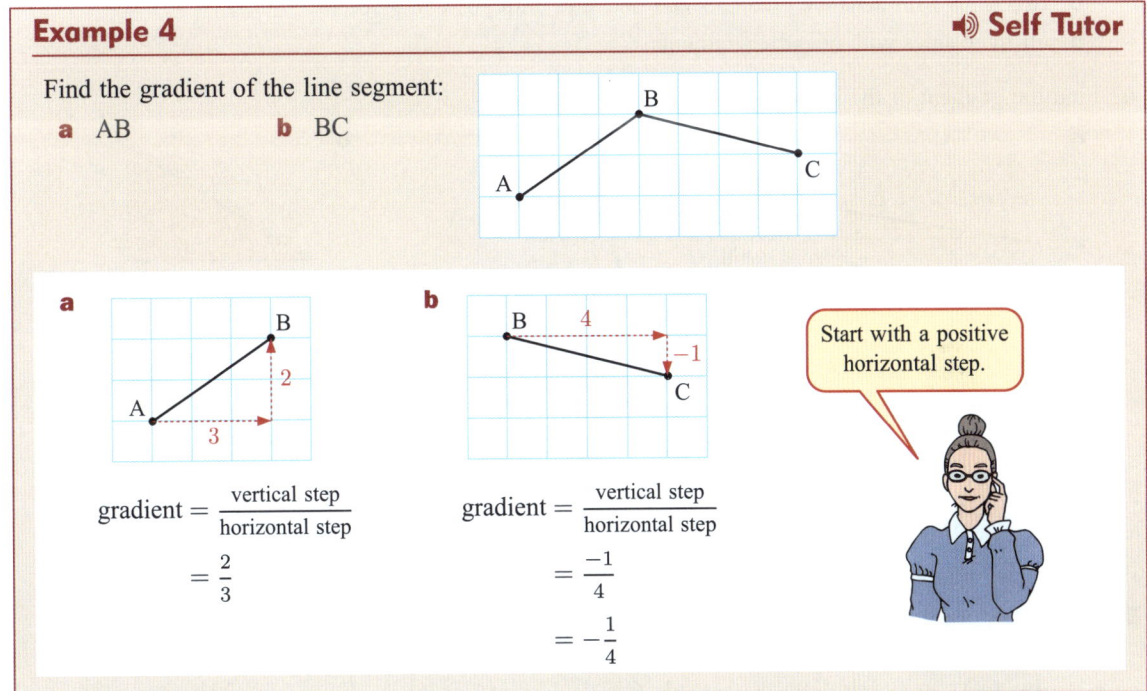

EXERCISE 20C.1

1 State the gradient of each line:

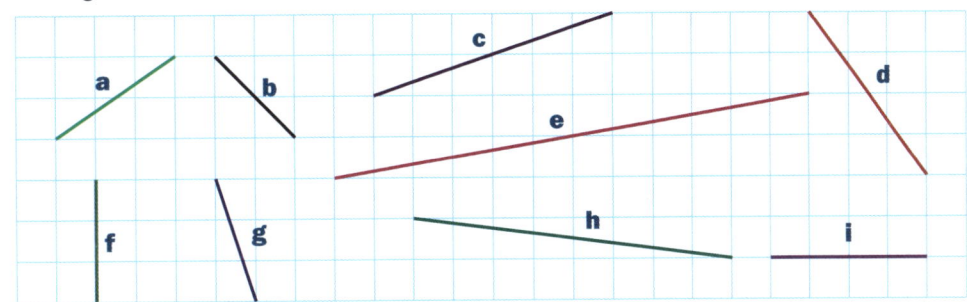

2 On grid paper, draw a line segment with gradient:

  a $\frac{1}{2}$ b $\frac{2}{3}$ c $-\frac{1}{4}$ d 2

  e 1   f -3 g $\frac{4}{3}$ h $-\frac{5}{2}$

  i 0   j $1\frac{1}{5}$ k 5   l undefined

We write 2 as $\frac{2}{1}$.

3 On grid paper, draw a triangle which has:
  a all three sides with positive gradients
  b two sides with positive gradients and one side with negative gradient
  c two sides with negative gradients and one side with positive gradient
  d all three sides with negative gradients.

4 Find the gradient of the following:

a

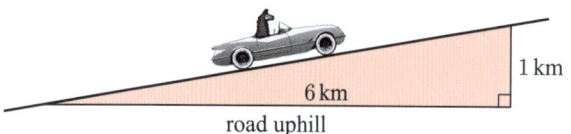
road uphill

b

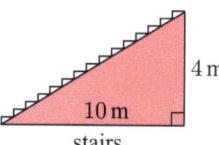
stairs

c

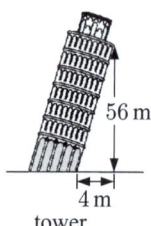
tower

d

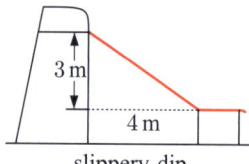
slippery-dip

e

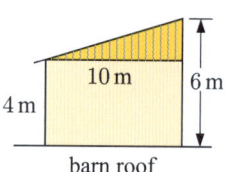
barn roof

5 a Determine the gradient of the line segment:
  i OA   **ii** OB   **iii** OC
  iv OD   **v** OE   **vi** OF
  vii OG **viii** OH **ix** OI

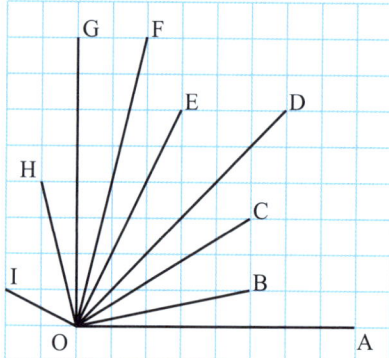

b Copy and complete the following statements:
  i The gradient of a horizontal line is ......
  ii The gradient of a vertical line is ......
  iii As line segments become steeper, their gradients ......

6

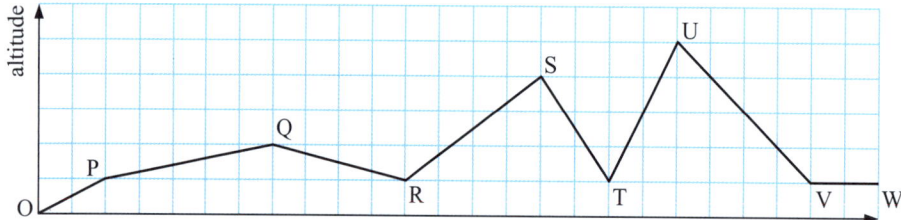

Imagine that you are walking across the countryside from O to W, from left to right. The vertical axis of the grid shows your altitude, or height above your starting point.

  a When are you going uphill?
  b When are you going downhill?
  c Where is the steepest positive gradient?
  d Where is the steepest negative gradient?
  e Where is the gradient 0?
  f Where is the least steep positive gradient?

Example 5

Draw a line through the point $(1, 1)$, with gradient $\frac{2}{3}$.

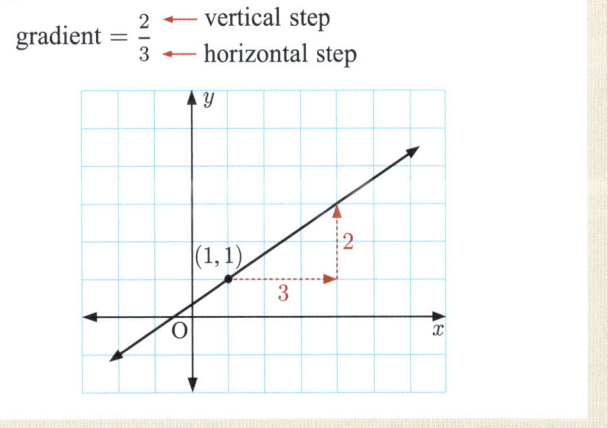

7 Draw a line through the point:

  a $(4, 1)$ with gradient $\frac{3}{2}$
  b $(3, -1)$ with gradient $\frac{1}{4}$
  c $(-2, 3)$ with gradient $-\frac{1}{5}$
  d $(0, 4)$ with gradient $-\frac{2}{3}$
  e $(0, -2)$ with gradient 3
  f $(-2, -1)$ with gradient -2
  g $(0, 6)$ with gradient $-\frac{5}{2}$
  h $(0, -3)$ with gradient $-\frac{3}{4}$.

Choose a positive horizontal step.

8 By plotting the points on graph paper, find the gradient of the line segment joining:

  a $O(0, 0)$ and $A(2, 6)$
  b $O(0, 0)$ and $B(-4, 2)$
  c $G(0, -1)$ and $H(2, 5)$
  d $K(1, 1)$ and $L(-2, -2)$
  e $M(3, 1)$ and $N(-1, 3)$
  f $P(-2, 4)$ and $Q(2, 0)$.

THE GRADIENT FORMULA

Although we can find gradients using steps on a diagram, it is often quicker to use a formula.

For points $A(x_1, y_1)$ and $B(x_2, y_2)$, the vertical step is $y_2 - y_1$, and the horizontal step is $x_2 - x_1$.

$\therefore$ the gradient is $\dfrac{y_2 - y_1}{x_2 - x_1}$.

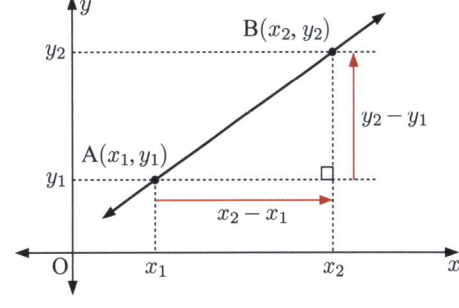

The **gradient** of the line through (x_1, y_1) and (x_2, y_2) is $\dfrac{y_2 - y_1}{x_2 - x_1}$.

Example 6

Find the gradient of PQ for P(1, 3) and Q(4, −2).

P(1, 3)   Q(4, −2)   gradient of PQ $= \dfrac{y_2 - y_1}{x_2 - x_1}$

x_1 y_1 x_2 y_2

$= \dfrac{-2 - 3}{4 - 1}$

$= \dfrac{-5}{3}$

$= -\dfrac{5}{3}$

By using a formula, we do not have to plot the points on a graph.

EXERCISE 20C.2

1 For each graph, find the gradient of the line using:

   i horizontal and vertical steps   **ii** the gradient formula.

a 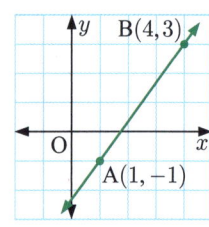   **b** 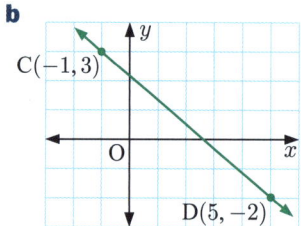   **c**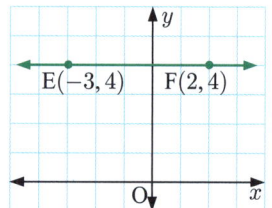

2 Use the gradient formula to find the gradient of the line segment joining:

   a O(0, 0) and A(2, 6)
   b O(0, 0) and B(−4, 2)
   c O(0, 0) and C(2, −12)
   d O(0, 0) and D(1, −5)
   e E(1, 0) and F(1, 5)
   f G(0, −1) and H(2, −1)
   g I(1, 1) and J(3, 3)
   h S(2, 3) and T(−2, −7)
   i J(4, 1) and K(−2, 3)
   j A(4, 6) and B(−8, −3)
   k P(1, −3) and Q(−5, −7)
   l M(6, 2) and N(−2, −16).

Example 7

Point A has coordinates $(a, 3)$ and point B has coordinates $(4, 6)$. The gradient of AB is $\dfrac{1}{2}$.

   a Show that $\dfrac{3}{4-a} = \dfrac{1}{2}$.
   b Hence solve for a.

a gradient of AB $= \dfrac{y_2 - y_1}{x_2 - x_1}$

$\therefore \dfrac{1}{2} = \dfrac{6 - 3}{4 - a}$

$\therefore \dfrac{3}{4 - a} = \dfrac{1}{2}$

b $\dfrac{3}{4-a} = \dfrac{1}{2}$

$\therefore \ \dfrac{3}{4-a} \times \dfrac{2}{2} = \dfrac{1}{2} \times \dfrac{(4-a)}{(4-a)}$ {creating a common denominator}

$\therefore \ 6 = 4 - a$ {equating numerators}

$\therefore \ a = -2$

3 Point A has coordinates $(2, 6)$ and point B has coordinates $(-8, a)$. The gradient of AB is 2.

  a Show that $\dfrac{a-6}{-10} = 2$.   **b** Hence solve for a.

4 Point P has coordinates $(4, 5)$ and point Q has coordinates $(a, 7)$. The gradient of PQ is $\dfrac{1}{4}$.

  a Show that $\dfrac{2}{a-4} = \dfrac{1}{4}$.   **b** Hence solve for a.

D PARALLEL LINES

In **Chapter 10**, we saw that **parallel** lines are a fixed distance apart and never meet.

Investigation 1 — Parallel lines

What to do:

1 In the graph alongside, AB is parallel to CD, and PQ is parallel to RS.

Copy and complete:

  a The gradient of AB = ...
  The gradient of CD = ...

  b The gradient of PQ = ...
  The gradient of RS = ...

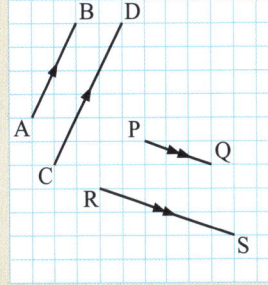

2 Draw two parallel lines of your choosing on a grid. Calculate the gradient of each line.

3 Copy and complete: Parallel lines have ......

From the **Investigation**, you should have found that:

> **parallel lines** have **equal gradient**.

Example 8

$A(2, 3)$, $B(-1, 5)$, $C(4, 7)$, and $D(13, 1)$ are four points on the Cartesian plane.
- **a** Show that AB is parallel to CD.
- **b** Is AD parallel to BC?

a gradient of AB $= \dfrac{5-3}{-1-2} = \dfrac{2}{-3} = -\dfrac{2}{3}$

gradient of CD $= \dfrac{1-7}{13-4} = \dfrac{-6}{9} = -\dfrac{2}{3}$

AB and CD have equal gradients, so they are parallel.

b gradient of AD $= \dfrac{1-3}{13-2} = \dfrac{-2}{11} = -\dfrac{2}{11}$

gradient of BC $= \dfrac{7-5}{4-(-1)} = \dfrac{2}{5}$

AD and BC have different gradients, so they are not parallel.

EXERCISE 20D

1 Consider the points $A(-2, 4)$, $B(4, 8)$, $C(7, 5)$, and $D(-11, -7)$.

  a Find the gradient of the line segment:
  - **i** AB
  - **ii** BC
  - **iii** CD
  - **iv** DA
  - **v** AC
  - **vi** BD

  b Which of the line segments are parallel?

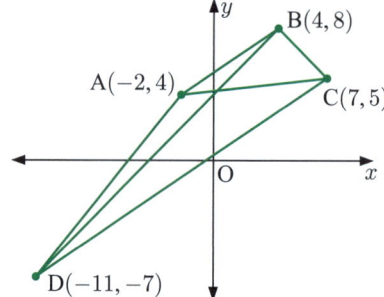

2 $P(-2, 6)$, $Q(5, 5)$, $R(3, -1)$, and $S(-5, -3)$ are four points on the Cartesian plane.
  - **a** Show that PS is parallel to QR.
  - **b** Is PQ parallel to RS?

3 Consider the points $A(7, 7)$, $B(16, 7)$, $C(1, -2)$, and $D(-8, -2)$.
  - **a** Show that:
    - **i** AB is parallel to CD
    - **ii** AD is parallel to BC.
  - **b** What kind of quadrilateral is ABCD?

4
  - **a** Plot the points $A(-6, -6)$, $B(-3, 2)$, $C(0, -3)$, $D(8, -5)$, $E(4, 2)$, and $F(2, 5)$ on the Cartesian plane.
  - **b** Find two parallel line segments involving these points.

5 Consider the points $P(-7, -3)$, $Q(5, 0)$, and $R(13, 2)$.
  - **a** Show that PQ is parallel to QR.
  - **b** What can you deduce about the points P, Q, and R?

6 Consider the points A(4, 8), B(2, 2), C(b, 10), and D(2, 1).

   a Suppose lines AB and CD are parallel. Find the value of b.

   b Suppose instead that lines AD and BC are parallel. Find the value of b.

E   AXES INTERCEPTS

When we are investigating lines on a number plane, it is useful to know where the line cuts the x and y-axes.

> The **x-intercept** of a line is the x-coordinate of the point where the line cuts the x-axis.
>
> The **y-intercept** of a line is the y-coordinate of the point where the line cuts the y-axis.

Example 9   ◀)) Self Tutor

Find the x and y-intercepts of the line given:

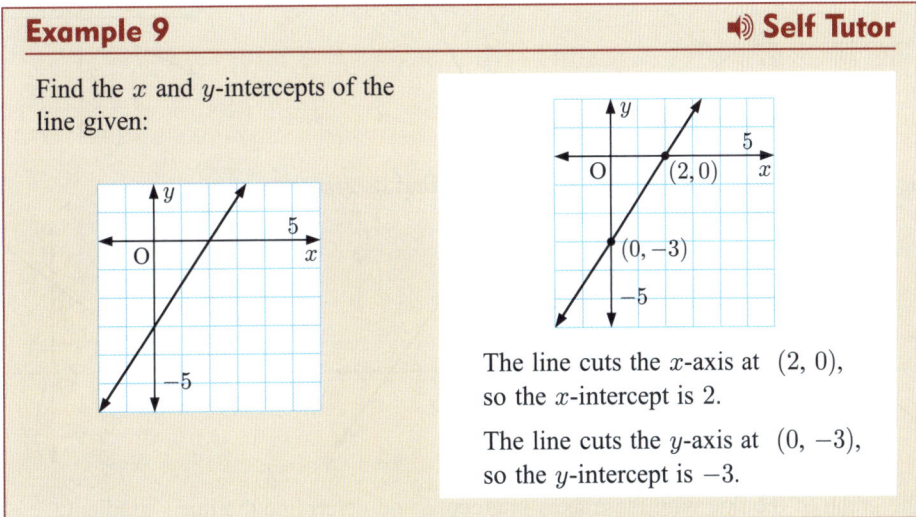

The line cuts the x-axis at $(2, 0)$, so the x-intercept is 2.

The line cuts the y-axis at $(0, -3)$, so the y-intercept is -3.

If the axes intercepts of a line are non-zero, we can quickly sketch the line and find its gradient.

Example 10   ◀)) Self Tutor

Draw the graph of the line with x-intercept -2 and y-intercept -3.
Find the gradient of the line.

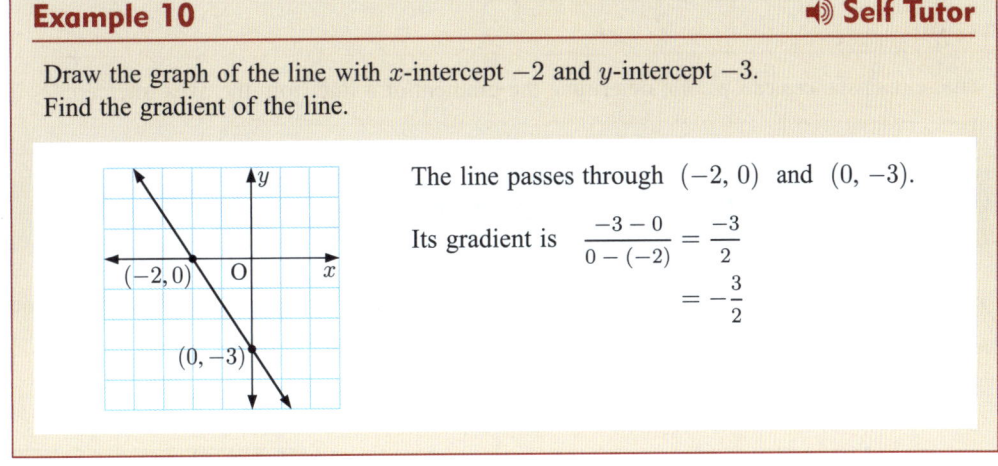

The line passes through $(-2, 0)$ and $(0, -3)$.

Its gradient is $\dfrac{-3 - 0}{0 - (-2)} = \dfrac{-3}{2}$

$= -\dfrac{3}{2}$

EXERCISE 20E

1 Find the x and y-intercepts of each line:

a 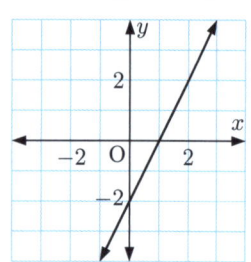   b 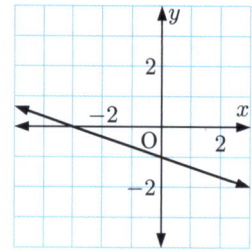   c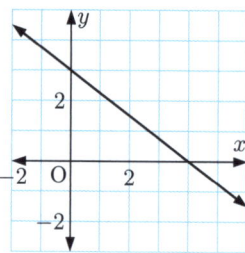

d 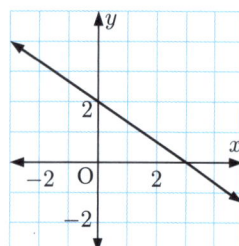   e 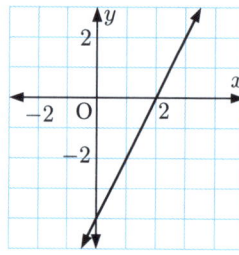   f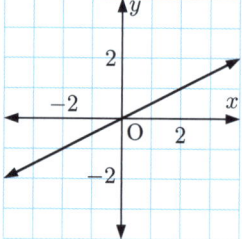

2 Find the x and y-intercepts of each line, and hence find its gradient:

a 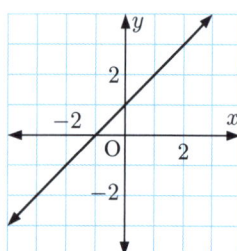   b 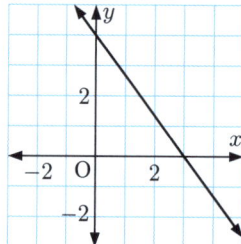   c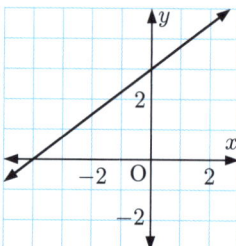

3 Draw the graph of the line with these axes intercepts, and determine the line's gradient:

a x-intercept -1 and y-intercept 2
b x-intercept 2 and y-intercept 6
c x-intercept 3 and y-intercept -5
d x-intercept -4 and y-intercept -2

Discussion

Under what circumstances can we *not* determine the gradient of a line from the axes intercepts?

F THE EQUATION OF A LINE

The **equation of a line** is a rule which connects the x and y-coordinates of **all** points on the line.

For any line that is not horizontal or vertical, the rule can be written in the form:
- $y = mx + c$ or
- $Ax + By = C$

For example, this table of values shows the x and y-coordinates of points on a line.

x	0	1	2	3	4
y	-3	-2	-1	0	1

The y-value is always 3 less that the x-value, so the equation of the line can be written as $y = x - 3$, or $x - y = 3$.

Example 11

◀)) **Self Tutor**

By inspection only, find the equation of the straight line passing through the following points:

a

x	1	2	3	4	5
y	−3	−6	−9	−12	−15

b

x	1	2	3	4	5
y	5	4	3	2	1

The line in **b** could be written as $y = -x + 6$.

a Each y-value is -3 times its corresponding x-value, so $y = -3x$.

b The sum of each x and y-value is always 6, so $x + y = 6$.

A point lies on a line if its coordinates satisfy the equation of the line.

Example 12

◀)) **Self Tutor**

Determine whether:
  a $(3, 7)$ lies on the line with equation $y = 2x + 1$
  b $(4, -2)$ lies on the line with equation $5x + 3y = 16$.

a When $x = 3$, we have
$$y = 2(3) + 1$$
$$= 6 + 1$$
$$= 7 \ \checkmark$$
So, $(3, 7)$ does lie on the line.

b Substituting $x = 4$ and $y = -2$ into the LHS gives
$$5(4) + 3(-2)$$
$$= 20 - 6$$
$$= 14 \ \times$$
Since LHS $\neq$ RHS, $(4, -2)$ does *not* lie on the line.

EXERCISE 20F

1 By inspection only, find the equation of the straight line passing through the following points:

a

x	1	2	3	4	5
y	4	8	12	16	20

b

x	2	3	4	5	6
y	4	5	6	7	8

c

x	0	1	2	3
y	0	−2	−4	−6

d

x	0	1	2	3
y	3	2	1	0

e

x	1	2	3	4	5
y	3	5	7	9	11

f

x	1	2	3	4	5
y	1	3	5	7	9

2 Determine whether the point:
 a (3, 4) lies on the line $y = 2x - 2$
 b (−1, 4) lies on the line $y = x + 6$
 c (−2, 10) lies on the line $y = -3x + 4$
 d $(\frac{1}{2}, -6)$ lies on the line $y = 2x - 8$
 e (2, −1) lies on the line $3x + y = 5$
 f (−3, 4) lies on the line $2x - 5y = -14$.

3 Consider the line $y = 2x - 3$.
 a Find the y-coordinate of the point on the line with x-coordinate 4.
 b Find the x-coordinate of the point on the line with y-coordinate −4.

4 Consider the line $y = 5 - 2x$.
 a Find the y-coordinate of the point on the line with x-coordinate −2.
 b Find the x-coordinate of the point on the line with y-coordinate −3.

5 Consider the line $3x - 5y = 30$.
 a Find the y-coordinate of the point on the line with x-coordinate 0.
 b Find the x-coordinate of the point on the line with y-coordinate 0.
 c Hence state the axes intercepts of the line.

G GRAPHING LINES IN THE FORM $y = mx + c$

If we are given the equation of a line in the form $y = mx + c$, we can draw its graph:
- by constructing a **table of values**
- using its **gradient** and **y-intercept**.

GRAPHING LINES BY CONSTRUCTING A TABLE OF VALUES

For any given value of x, we can use the equation of the line to find the corresponding value of y.

Consider the equation $y = \frac{1}{2}x - 1$.

When $x = -2$, $y = \frac{1}{2}(-2) - 1$
$= -1 - 1$
$= -2$

When $x = 2$, $y = \frac{1}{2}(2) - 1$
$= 1 - 1$
$= 0$

By continuing this process, we construct the table of values:

x	−3	−2	−1	0	1	2	3
y	$-2\frac{1}{2}$	−2	$-1\frac{1}{2}$	−1	$-\frac{1}{2}$	0	$\frac{1}{2}$

The graph of $y = \frac{1}{2}x - 1$ is the line which passes through these points.

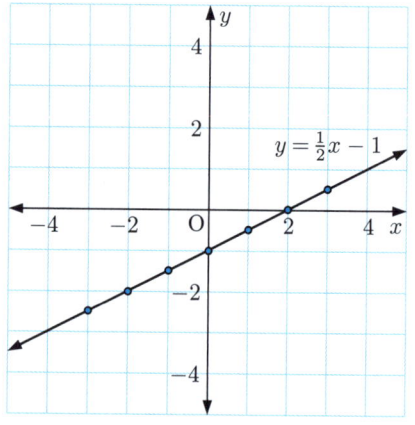

Example 13

Consider the equation $y = x - 2$.

a Construct a table of values using $x = -3, -2, -1, 0, 1, 2,$ and 3.
b Hence draw the graph of $y = x - 2$.

a

x	-3	-2	-1	0	1	2	3
y	-5	-4	-3	-2	-1	0	1

b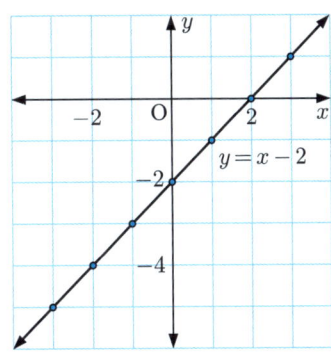

EXERCISE 20G.1

1 For each of the following equations:
  i Construct a table of values using $x = -3, -2, -1, 0, 1, 2,$ and 3.
  ii Draw the graph of the straight line.

  a $y = x$
  b $y = -x$
  c $y = 2x$
  d $y = -2x$
  e $y = 2x - 1$
  f $y = \frac{1}{4}x + 2$
  g $y = -\frac{1}{4}x + 2$
  h $y = 2 - x$
  i $y = -1.5x - 1$

  Check your answers using technology.

GRAPHING PACKAGE

2 For each of the following equations:
  i Construct a table of values using $x = -3, -2, -1, 0, 1, 2,$ and 3.
  ii Draw the graph of the straight line.
  iii State the gradient and y-intercept of the line.

  a $y = x + 3$
  b $y = -2x + 1$
  c $y = \frac{1}{3}x - 1$

Discussion

Discuss the graphs you drew in the previous Exercise.

Can we tell the gradient and y-intercept of the graph immediately from the equation of the line?

GRAPHING LINES FROM THE GRADIENT AND y-INTERCEPT

Investigation 2 — Graphing lines

In this Investigation, we discover how the equation of a line relates to its *gradient* and *y-intercept*.

What to do:

1 Draw the graph of each line given. Hence write down the gradient and y-intercept of each line.

To find the gradient of the line, choose any two points on the line and use the gradient formula.

	Equation of line	Gradient	y-intercept
a	$y = 2x + 1$	2	1
b	$y = 3x - 2$		
c	$y = -x + 3$		
d	$y = 4x$		
e	$y = -2x + 2$		
f	$y = \frac{1}{2}x - 1$		
g	$y = -\frac{1}{3}x + 1$		

2 Copy and complete: "For a line with equation $y = mx + c$, the gradient is ...... and the y-intercept is ...."

You should have discovered that:

> $y = mx + c$ is the equation of a straight line with gradient m and y-intercept c.

Example 14 ◀) Self Tutor

State the gradient and y-intercept of the line with equation:

a $y = 3x - 2$ b $y = 7 - 2x$ c $y = 0$

a $y = 3x - 2$ has $m = 3$ and $c = -2$
   $\therefore$ the gradient is 3 and the y-intercept is -2.
b $y = 7 - 2x$ can be written as $y = -2x + 7$, with $m = -2$ and $c = 7$
   $\therefore$ the gradient is -2 and the y-intercept is 7.
c $y = 0$ can be written as $y = 0x + 0$, with $m = 0$ and $c = 0$
   $\therefore$ the gradient is 0 and the y-intercept is 0.

To draw the graph of $y = mx + c$:

- Use the y-intercept c to plot the point $(0, c)$.
- Starting from $(0, c)$, use horizontal and vertical steps from the gradient m to locate another point on the line.
- Join the two points and extend the line in either direction.

Always let the horizontal step be positive.

Example 15

Draw the graph of: **a** $y = \frac{2}{3}x + 1$ **b** $y = -2x - 3$.

a For $y = \frac{2}{3}x + 1$:
- the y-intercept is $c = 1$
- the gradient is
$m = \dfrac{2 \leftarrow \text{vertical step}}{3 \leftarrow \text{horizontal step}}$

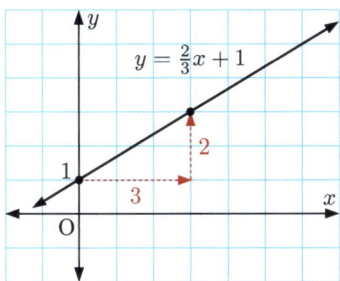

b For $y = -2x - 3$:
- the y-intercept is $c = -3$
- the gradient is
$m = \dfrac{-2 \leftarrow \text{vertical step}}{1 \leftarrow \text{horizontal step}}$

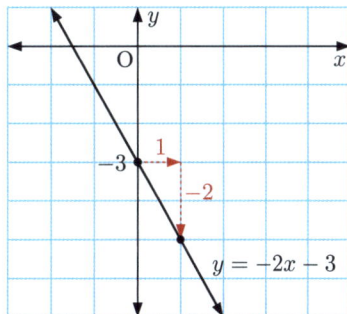

EXERCISE 20G.2

1 State the gradient and y-intercept of the line with equation:

a $y = 4x + 8$ **b** $y = -3x + 2$ **c** $y = 6 - x$

d $y = -2x + 3$ **e** $y = -2$ **f** $y = 11 - 3x$

g $y = \frac{1}{2}x - 5$ **h** $y = 3 - \frac{3}{2}x$ **i** $y = \frac{2}{5}x + \frac{4}{5}$

j $y = \dfrac{x+1}{2}$ **k** $y = \dfrac{2x-10}{5}$ **l** $y = \dfrac{11-3x}{2}$

> Equations in the form $y = mx + c$ are said to be in *gradient-intercept* form.

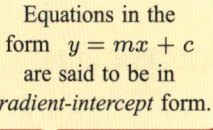

2 Draw the graph of:

a $y = x + 3$ **b** $y = -x + 4$ **c** $y = 2x + 2$

d $y = -3x - 2$ **e** $y = \frac{1}{2}x - 1$ **f** $y = \frac{2}{3}x + 4$

g $y = 3x$ **h** $y = -\frac{1}{2}x$ **i** $y = -2x + 1$

j $y = 3 - \frac{1}{3}x$ **k** $y = \frac{3}{4}x - 2$ **l** $y = -\frac{1}{4}x - 3$

3 Consider the line with equation $y = -\frac{2}{3}x + 2$.

a Find the:
  i gradient   **ii** y-intercept   **iii** x-intercept.

b Draw the graph of the line.

c Does the point $\left(4, -\frac{2}{3}\right)$ lie on the line?

H GRAPHING LINES IN THE FORM $Ax + By = C$

To draw the graph of a line in the form $Ax + By = C$, we:
- Find the y-intercept by letting $x = 0$.
- Find the x-intercept by letting $y = 0$.
- Join the points where the line cuts the axes and extend the line in either direction.

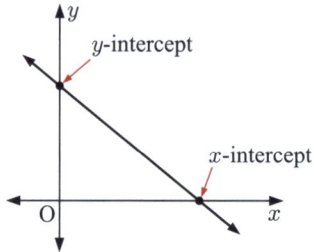

Example 16   ◀)) Self Tutor

Draw the graph of $4x - 3y = 12$.

When $x = 0$, $-3y = 12$
$\therefore\ y = -4$

So, the y-intercept is -4.

When $y = 0$, $4x = 12$
$\therefore\ x = 3$

So, the x-intercept is 3.

EXERCISE 20H

1 Draw the graph of:
- **a** $x + 3y = 6$
- **b** $3x - 2y = 12$
- **c** $2x + 5y = 10$
- **d** $4x + 3y = 6$
- **e** $x + y = 5$
- **f** $x - y = -3$
- **g** $3x - y = -6$
- **h** $7x + 2y = 14$
- **i** $4x + 9y = -18$

Equations in the form $Ax + By = C$ are said to be in *general* form.

2 Consider the line with equation $3x - 5y = 15$.
- **a** Find the:
  - **i** x-intercept
  - **ii** y-intercept.
- **b** Determine whether the following points lie on the line:
  - **i** $(-5, -6)$
  - **ii** $(1, -2)$
- **c** Draw the graph of the line, showing your results from **a** and **b**.

3 a Draw the graph of:
  - **i** $3x + 2y = 6$
  - **ii** $x + 4y = 8$
  - **iii** $2x - 5y = 10$
  - **iv** $5x - 4y = 40$

b Find the gradient of each line in **a**. What do you notice?

c By rearranging the equation into gradient-intercept form, show that the line with equation $Ax + By = C$ has gradient $-\dfrac{A}{B}$.

4 **a** Draw the graph of $3x + 4y = 24$.
  b Which of these lines is parallel to $3x + 4y = 24$?

   A $y = \frac{4}{3}x + 1$ **B** $y = -\frac{4}{3}x - 5$ **C** $y = \frac{3}{4}x + 6$ **D** $y = -\frac{3}{4}x + 7$

I VERTICAL AND HORIZONTAL LINES

VERTICAL LINES

For all points on a vertical line, the x-coordinate is constant regardless of the value of the y-coordinate.

The graph alongside shows the vertical lines $x = -1$ and $x = 3$.

> All **vertical** lines have equations of the form $x = a$.
> The gradient of a vertical line is **undefined**.

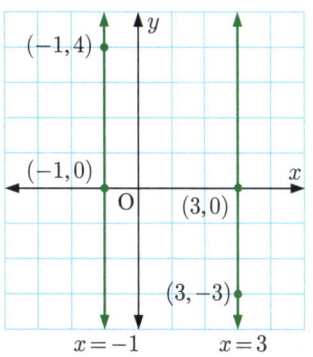

HORIZONTAL LINES

For all points on a horizontal line, the y-coordinate is constant regardless of the value of the x-coordinate.

The graph alongside shows the horizontal lines $y = 1$ and $y = -2$.

> All **horizontal** lines have equations of the form $y = c$.
> The gradient of a horizontal line is **zero**.

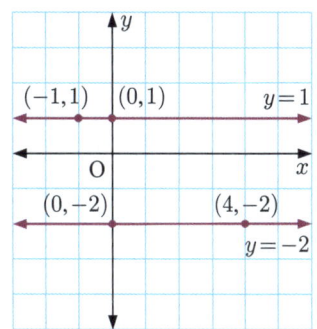

EXERCISE 20I

1 Draw the graph of each line and state its gradient:

  a $x = 2$ **b** $y = 3$ **c** $y = -4$ **d** $x = -3$
  e $y = 1.5$ **f** $x = -\frac{5}{2}$ **g** $y = 0$ **h** $y = -\frac{1}{4}$

2 Suppose l_1 is the line with equation $x = 5$, and l_2 is the line with equation $y = -3$.
  a Graph l_1 and l_2 on the same set of axes.
  b Suppose l_1 cuts the x-axis at A, and l_2 cuts the y-axis at B. Find:
    i the coordinates of A and B   **ii** the gradient of the line segment AB.

3 **a** On the same set of axes, graph the lines $x = -2$, $x = 6$, $y = 5$, and $y = -6$.
  b Consider the quadrilateral formed by the intersection points of these lines.
    i What type of quadrilateral is it?
    ii Find the area of the quadrilateral.
    iii Find the gradients of the diagonals of the quadrilateral.

J FINDING THE EQUATION OF A LINE

To determine the equation of a line, we need to know either:
- its gradient and at least one point which lies on the line, *or*
- two points which lie on the line.

If we know the gradient m and y-intercept c of a line, we can write down the equation of the line immediately as $y = mx + c$.

Example 17 ◀)) Self Tutor

Find the equation of the line with gradient 5 and y-intercept -3.

The line has gradient $m = 5$ and y-intercept $c = -3$.
So, the equation of the line is $y = 5x - 3$.

If we know the gradient and one point which lies on the line, we can substitute the coordinates of the point into the equation to find c.

Example 18 ◀)) Self Tutor

A line has gradient $\frac{2}{3}$ and passes through the point $(3, 1)$. Find the equation of the line.

The gradient $m = \frac{2}{3}$, so the equation has the form $y = \frac{2}{3}x + c$.

Since $(3, 1)$ lies on the line, we substitute $x = 3$, $y = 1$ into the equation.

$\therefore \ 1 = \frac{2}{3}(3) + c$

$\therefore \ 1 = 2 + c$

$\therefore \ c = -1$ So, the equation is $y = \frac{2}{3}x - 1$.

EXERCISE 20J

1 Write down the equation of the line with:
  a gradient 3 and y-intercept 4
  b gradient 7 and y-intercept -1
  c gradient -1 and y-intercept $\frac{1}{3}$
  d gradient $\frac{1}{2}$ and y-intercept 0
  e gradient 0 and y-intercept 5
  f gradient $-\frac{5}{2}$ and y-intercept $-\frac{2}{3}$.

2 Find the equation of the line which has gradient:
  a 2 and which passes through $(1, 8)$
  b -3 and which passes through $(4, -16)$
  c 5 and which passes through $(-3, -1)$
  d $\frac{3}{4}$ and which passes through $(8, 12)$
  e 0 and which passes through $(-2, 4)$
  f $-\frac{1}{2}$ and which passes through $(5, -9)$.

Example 19

Find the equation of the illustrated line:

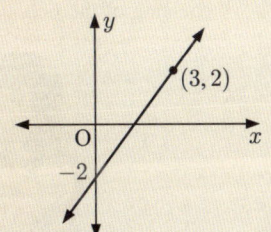

The line passes through $(0, -2)$ and $(3, 2)$.

∴ the gradient $m = \dfrac{2 - (-2)}{3 - 0} = \dfrac{4}{3}$

and the y-intercept $c = -2$

∴ the equation is $y = \dfrac{4}{3}x - 2$.

3 Find the equation of the illustrated line:

a 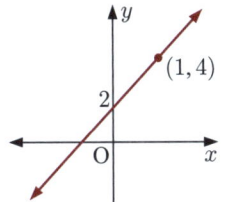 **b** 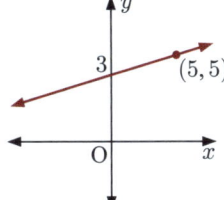 **c**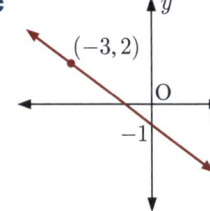

d 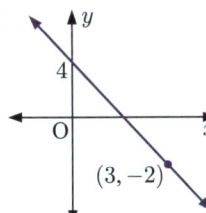 **e** 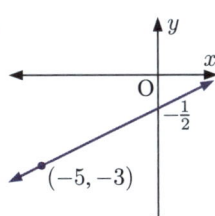 **f**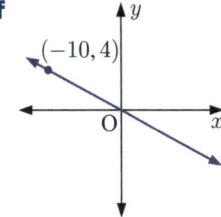

4 Find the equation of the line which has:
  a y-intercept -5, and which passes through $(3, 7)$
  b y-intercept 7, and which passes through $(5, -8)$
  c y-intercept -1, and which passes through $(-8, -4)$.

Example 20

Find the equation of the line which passes through $(-2, 1)$ and $(3, 16)$.

The line has gradient $m = \dfrac{16 - 1}{3 - (-2)} = \dfrac{15}{5} = 3$.

∴ the equation of the line has the form $y = 3x + c$.

Since $(-2, 1)$ lies on the line, we substitute $x = -2$, $y = 1$ into the equation.

∴ $1 = 3(-2) + c$
∴ $1 = -6 + c$
∴ $c = 7$

So, the equation of the line is $y = 3x + 7$.

> You can substitute the coordinates of *either* point into the equation.

5 Find the equation of the line which passes through:

   a $(2, 7)$ and $(5, 13)$ **b** $(3, 6)$ and $(6, 0)$

   c $(-3, -4)$ and $(6, 2)$ **d** $(-4, -3)$ and $(3, -3)$

   e $(-6, -14)$ and $(2, -2)$ **f** $(-8, 11)$ and $(-2, 3)$.

Check your answer by making sure *both* points satisfy the equation of the line.

6 Find the equation of the illustrated line:

a 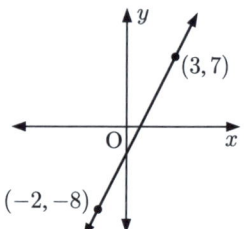   **b** 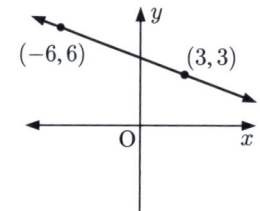   **c**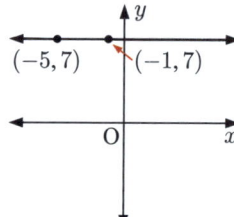

d 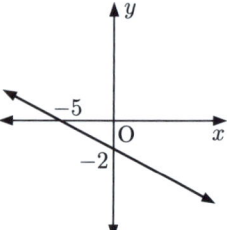   **e** 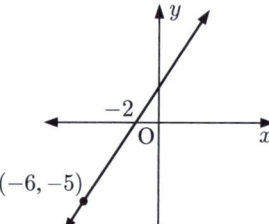   **f**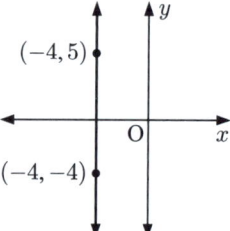

7 Find the rule connecting the variables:

a 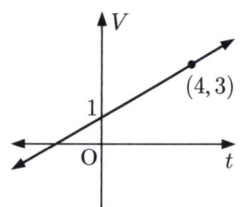   **b** 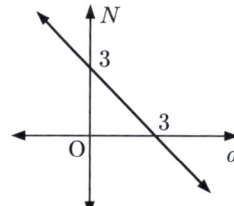   **c**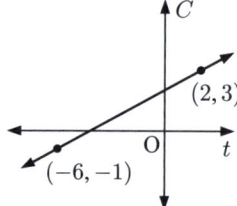

Example 21   ◀) Self Tutor

Find the equation of the line parallel to $y = -2x + 8$, and which passes through $(3, -1)$.

The line $y = -2x + 8$ has gradient -2.

$\therefore$ the required line also has gradient -2, and its equation has the form $y = -2x + c$.

Since $(3, -1)$ lies on the line, we substitute $x = 3$, $y = -1$ into the equation.

$\therefore\ -1 = -2(3) + c$

$\therefore\ c = 5$

So, the equation of the line is $y = -2x + 5$.

8 Find the equation of the line:

  a parallel to $y = 3x + 4$, and which passes through $(2, 4)$

  b parallel to $y = -\frac{1}{4}x - 2$, and which passes through $(8, -3)$

  c parallel to $2x - 5y = 6$, and which passes through $(-5, 1)$.

9 Lines 1 and 2 are parallel. Find the equation of:

  a line 1   **b** line 2.

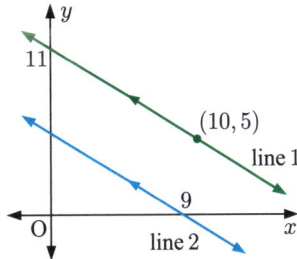

10

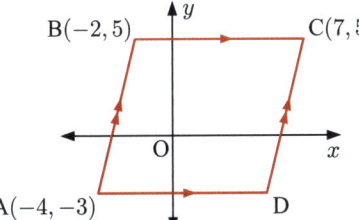

ABCD is a parallelogram. Find the equation of line:

  a AB   **b** BC   **c** CD   **d** AD

Review set 20A

1 State the coordinates of the points A, B, C, D, and E.

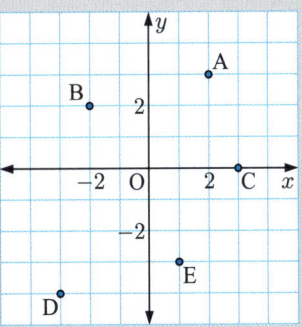

2 On a set of axes, illustrate all the points which have equal x and y-coordinates.

3 For the equation $y = 4 - 3x$:

  a Construct a table of values with $x = -3, -2, -1, 0, 1, 2,$ and 3.
  b Draw the graph of the straight line.
  c State the gradient and y-intercept of the line.

4 State the gradient of each line: **a**   **b**

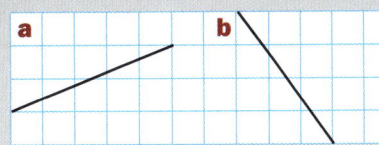

5 On the same set of axes, draw graphs of the lines with equations $x = -2\frac{1}{2}$ and $y = -4$.

6 An accountant charges a £70 consultation fee and then £120 per hour thereafter.
  a Identify the independent and dependent variables.
  b Make a table of values for the cost £C of an appointment with the accountant for t hours where $t = 0, 1, 2, 3, 4$.
  c Draw a graph of C against t.
  d Is the relationship between C and t linear?
  e Is it sensible to join the points graphed with a straight line? Explain your answer.
  f For every increase in t of 1 hour, what is the change in C?

7 State the gradient and y-intercept of the line with equation:
  a $y = 4x - 3$
  b $y = 2 - x$
  c $y = \dfrac{3x - 1}{2}$

8 **a** Plot the points P(-1, 3), Q(5, 5), R(8, 1), and S(-4, -3) on a set of axes.
  b Determine whether:
    i PQ is parallel to SR
    ii PS is parallel to QR.
  c What kind of quadrilateral is PQRS?

9 Find the x and y-intercepts of each line, and hence find its gradient:

  a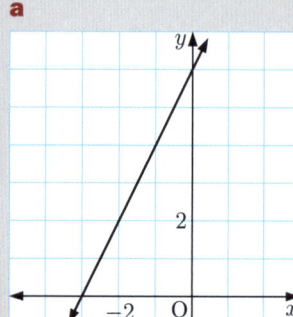
  b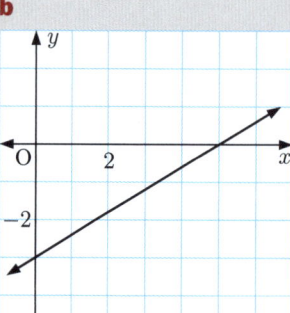
  c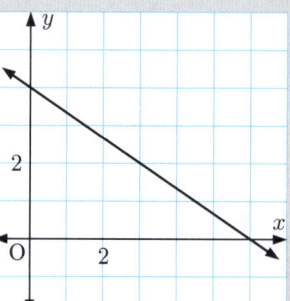

10 For each of the following lines:
  i State the gradient and y-intercept.
  ii Draw the graph of the line.
  a $y = 2x - 3$
  b $y = \dfrac{1}{3}x + 1$
  c $y = 8 - 3x$

11 Draw the graph of:
  a $y = \dfrac{3}{2}x - 2$
  b $2x - 3y = 30$
  c $y = -\dfrac{1}{4}x + 5$
  d $4x + 3y = -36$

12 Find the equation of each line:
  a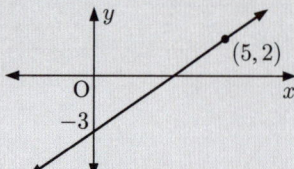
  b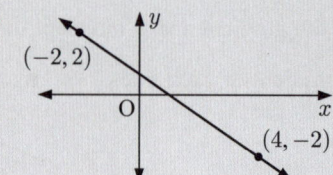

Review set 20B

1 Plot the points P(2, 5), Q(3, −3), R(−4, 0), and S(−2, −4) on a set of axes.

2 Find, by inspection, the equation of the straight line passing through these points:

x	0	1	2	3	4	5
y	3	2	1	0	−1	−2

3 A tank contains 400 litres of water. The tap is left on and 20 litres of water escape per minute. Suppose V is the volume of water remaining in the tank t minutes after the tap is turned on.
  - **a** Identify the independent and dependent variables.
  - **b** Make a table of values for V and t, for $t = 0, 2, 4, 6, 8, 10$ minutes.
  - **c** Draw the graph of V against t.
  - **d** Is the relationship between V and t linear?
  - **e** Is it sensible to join the points graphed with a straight line? Explain your answer.
  - **f** For every increase of 1 minute for t, what is the change in V?
  - **g** How much water is left in the tank after $2\frac{1}{2}$ minutes?
  - **h** How long will it take for the volume of water in the tank to fall to 170 litres?

4 On grid paper, draw a line with gradient:
  - **a** $\frac{2}{5}$
  - **b** $-\frac{4}{3}$
  - **c** 0

5 Find the gradient of the straight line through the points:
  - **a** (2, 1) and (8, 13)
  - **b** (−2, 3) and (1, −4)
  - **c** (−1, −5) and (9, 3)
  - **d** (2, −3) and (−7, 12).

6 Draw a line through the point:
  - **a** (3, 5) with gradient $\frac{1}{4}$
  - **b** (−2, 3) with gradient −2.

7 Determine whether the point:
  - **a** (−2, 5) lies on the line with equation $y = -2x + 1$
  - **b** (3, −2) lies on the line with equation $4x - 3y = 18$.

8 Draw the graph of:
  - **a** $y = -\frac{1}{3}x + 2$
  - **b** $y = \frac{3}{4}x - 3$
  - **c** $x - 4y = 12$

9 Find the equation of each line:
  - **a**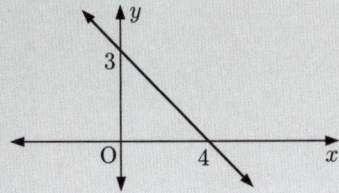
  - **b**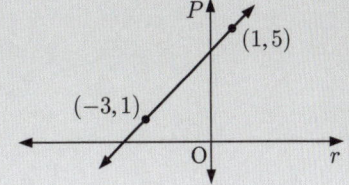

10 Which of these lines is parallel to $y = \frac{2}{3}x - 7$?
  - **A** $2x + 3y = 4$
  - **B** $2x - 3y = -1$
  - **C** $3x - 2y = 6$
  - **D** $3x + 2y = -2$

11 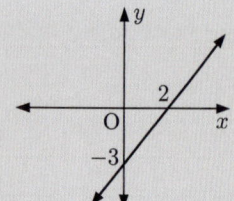 For the line given, find the:
  a x-intercept
  b y-intercept
  c equation of the line.

12 Find the equation of the line:
  a with gradient 2 and y-intercept -3
  b with gradient $-\frac{1}{2}$ and which passes through $(-6, 5)$
  c which passes through $(-4, 6)$ and $(5, -3)$
  d which is parallel to $y = \frac{3}{2}x - 1$, and which passes through $(-2, -8)$.

21

Simultaneous equations

Contents:
- **A** Trial and error solution
- **B** Graphical solution
- **C** Solution by equating values of y
- **D** Solution by substitution
- **E** Solution by elimination
- **F** Problem solving with simultaneous equations

Opening problem

At the summer sales, Cassandra buys a dress and a skirt. Together they cost £30. The dress cost £8 more than the skirt.

Suppose the dress cost £x, and the skirt cost £y.

Things to think about:

a Can you explain why:
   i $x + y = 30$ ii $x - y = 8$?

b How many solutions are there to the equation:
   i $x + y = 30$ ii $x - y = 8$?

c Which solution satisfies *both* equations at the same time?

We have previously studied equations involving one variable.

For example, the linear equation $3x + 5 = 14$ has exactly one solution, $x = 3$.

Some equations have more than one variable, and more than one solution.

Consider the equations in the **Opening Problem**:

- $x + y = 30$ has infinitely many solutions. They lie in a straight line on the Cartesian plane. $x = 1$, $y = 29$ is one solution, and $x = 2$, $y = 28$ is another.
- $x - y = 8$ also has infinitely many solutions which lie in a straight line. They include $x = 9$, $y = 1$, and $x = 10$, $y = 2$.

To find a solution which satisfies *both* equations at the same time, we need to solve the equations *simultaneously*. We say they are **simultaneous equations**.

> A solution to **simultaneous equations** satisfies *both* equations at the same time.

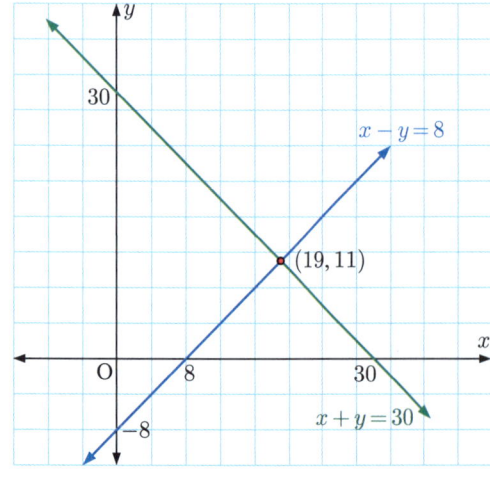

There is in fact only *one* solution to the simultaneous equations $\begin{cases} x + y = 30 \\ x - y = 8 \end{cases}$.

The solution occurs at the point where the two lines intersect.

The solution $x = 19$, $y = 11$ satisfies both equations at the same time, since $19 + 11 = 30$ and $19 - 11 = 8$.

So, Cassandra paid £19 for the dress and £11 for the skirt.

A TRIAL AND ERROR SOLUTION

One way to solve simultaneous equations is by **trial and error**. We list possible solutions to *one* of the equations, and look for the solution which also satisfies the *other* equation.

Example 1

Solve $\begin{cases} x + y = 8 \\ 2x - y = 7 \end{cases}$ simultaneously by trial and error.

We consider solutions to $x + y = 8$, and find the solution which also satisfies $2x - y = 7$.

The solution is $x = 5$, $y = 3$.

x	y	$x+y$	$2x-y$
0	8	8	-8
1	7	8	-5
2	6	8	-2
3	5	8	1
4	4	8	4
5	3	8	7 ✓

EXERCISE 21A

1 Determine whether the given values satisfy the pair of simultaneous equations:

   a $\begin{cases} x - y = 3 \\ 2x + y = 11 \end{cases}$ $(x = 5,\ y = 2)$ **b** $\begin{cases} x + y = 9 \\ 2x - y = 6 \end{cases}$ $(x = 5,\ y = 4)$

   c $\begin{cases} a + b = 2 \\ a - b = 8 \end{cases}$ $(a = 5,\ b = -3)$ **d** $\begin{cases} 2p + q = 7 \\ 3p + 2q = 10 \end{cases}$ $(p = 4,\ q = -1)$

2 Given that the solutions are integers, solve the following by trial and error:

   a $\begin{cases} x + y = 4 \\ 3x + 5y = 14 \end{cases}$ **b** $\begin{cases} x + y = 11 \\ 4x + 3y = 40 \end{cases}$ **c** $\begin{cases} y = x + 2 \\ 9x - 4y = 7 \end{cases}$

   d $\begin{cases} y = 6 + x \\ 8x - 3y = -3 \end{cases}$ **e** $\begin{cases} x + y = 4 \\ 2x - y = 5 \end{cases}$ **f** $\begin{cases} x + y = 6 \\ 2x + y = 10 \end{cases}$

   g $\begin{cases} a - b = 1 \\ 2a + 3b = 2 \end{cases}$ **h** $\begin{cases} p - q = 3 \\ 5p + 2q = 29 \end{cases}$ **i** $\begin{cases} 3x + 2y = 17 \\ -y + x = -21 \end{cases}$

Discussion

1 Attempt to solve the following simultaneous equations by trial and error:

   a $\begin{cases} 3x + 7y = 1 \\ 6x - 14y = -4 \end{cases}$ **b** $\begin{cases} 2x - y = 8 \\ -4x + 2y = 9 \end{cases}$ **c** $\begin{cases} x - y = 2 \\ 2y - 2x = -4 \end{cases}$

2 Discuss your answers to **1** with your class. In particular:

- For **a** and **b**, did anyone find a solution? *Is* there a solution?
- For **c**, did you all find the same solution? Are all of the solutions you found valid?
- Discuss the problems with using trial and error to solve simultaneous equations.

B GRAPHICAL SOLUTION

Suppose we are given two linear equations involving x and y. The solutions to each equation will form a straight line. If we graph the two equations on the same set of axes, any **point of intersection** corresponds to a **simultaneous solution** of the equations.

- If the lines are not parallel, the lines will meet in exactly one point. The simultaneous equations have *exactly one solution*.

- If the lines are parallel but not identical, the lines will never meet. The simultaneous equations have *no solutions*.

- If the lines are identical, there will be infinitely many points of intersection. The simultaneous equations have *infinitely many solutions*.

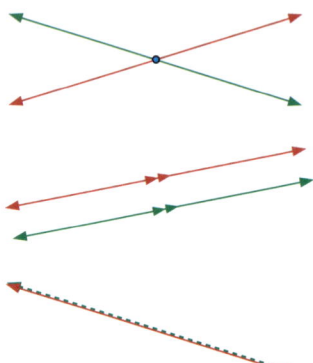

Example 2 ◀)) Self Tutor

Solve the following simultaneous equations graphically: $\begin{cases} y = 2x - 4 \\ 2x + 3y = 12 \end{cases}$

We draw the graphs of $y = 2x - 4$ and $2x + 3y = 12$ on the same set of axes.

The graphs meet at the point $(3, 2)$.

$\therefore$ the solution is $x = 3$, $y = 2$.

Check:

Substituting these values into:

- $y = 2x - 4$ gives $2 = 2(3) - 4$ ✓
- $2x + 3y = 12$ gives $2(3) + 3(2) = 12$ ✓

EXERCISE 21B

1 Solve the following simultaneous equations graphically:

   a $\begin{cases} y = 4x - 1 \\ y = 2x - 3 \end{cases}$ **b** $\begin{cases} y = 3x \\ y = -2x + 5 \end{cases}$ **c** $\begin{cases} y = x + 2 \\ y = -3x - 6 \end{cases}$

2 Solve the following simultaneous equations graphically:

   a $\begin{cases} y = 2x - 8 \\ 2x + 5y = 20 \end{cases}$ **b** $\begin{cases} 4x + y = 8 \\ 2x - 3y = 18 \end{cases}$ **c** $\begin{cases} 3x - y = -6 \\ 3x + 4y = -36 \end{cases}$

3 Try to solve the following simultaneous equations graphically. State the number of solutions in each case.

   a $\begin{cases} y = 4x + 1 \\ y = 4x - 3 \end{cases}$ **b** $\begin{cases} 4x - 2y = -8 \\ y = 2x + 4 \end{cases}$

4 The graphs of $y = \frac{1}{2}x + 1$ and $2x + 3y = 12$ are shown alongside.

   a Identify each line.

   b Estimate the solution which satisfies both equations. Round your solution to 1 decimal place.

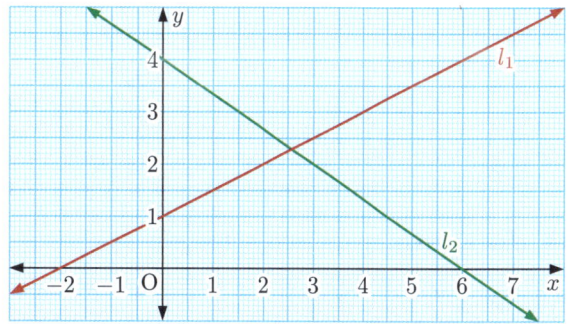

Discussion

Discuss the advantages and disadvantages of solving simultaneous equations by graphical methods.

C   SOLUTION BY EQUATING VALUES OF y

We will now consider some **algebraic** methods for solving linear simultaneous equations. Algebraic methods are often quicker than trial and error or graphing. They are also more accurate if the solutions are not integers.

If y is the subject of *both* equations, we equate the y values and solve for x.

Example 3   ◀)) Self Tutor

Solve the simultaneous equations: $\begin{cases} y = 2x - 1 \\ y = x + 3 \end{cases}$

If $y = 2x - 1$ and $y = x + 3$, then

$2x - 1 = x + 3$ {equating ys}

$\therefore \ 2x - 1 - x = x + 3 - x$ {subtracting x from both sides}

$\therefore \ x - 1 = 3$

$\therefore \ x = 4$ {adding 1 to both sides}

Now $y = x + 3$

$\therefore \ y = 4 + 3$

$\therefore \ y = 7$

The solution is $x = 4$, $y = 7$.

Check: In $y = 2x - 1$, $y = 2 \times 4 - 1 = 8 - 1 = 7$ ✓

Always check your solution in *both* equations.

EXERCISE 21C

1 Solve these simultaneous equations by equating values of y:

a $\begin{cases} y = x + 1 \\ y = 2x - 3 \end{cases}$
b $\begin{cases} y = x + 4 \\ y = -x + 2 \end{cases}$
c $\begin{cases} y = x + 2 \\ y = -2x + 5 \end{cases}$

2 Solve by equating values of y:

a $\begin{cases} y = -2x - 4 \\ y = x - 4 \end{cases}$
b $\begin{cases} y = -x + 4 \\ y = 2x - 8 \end{cases}$
c $\begin{cases} y = -2 - x \\ y = 2x + 13 \end{cases}$

d $\begin{cases} y = 3x + 7 \\ y = -x - 6 \end{cases}$
e $\begin{cases} y = 2x - 10 \\ y = 3x - 18 \end{cases}$
f $\begin{cases} y = 5 - 3x \\ y = 10 - 6x \end{cases}$

3 Try to solve by equating values of y:

a $\begin{cases} y = 2x + 3 \\ y = 2x - 2 \end{cases}$
b $\begin{cases} y = 2.3x - 3.6 \\ y = -1.2x + 5.7 \end{cases}$

Comment on your results.

D SOLUTION BY SUBSTITUTION

The method of **substitution** is used when a variable is given as the subject of *one* of the equations.

For example, in the equations $\begin{cases} y = 2x + 3 \\ 3x - 4y = 8 \end{cases}$ we see that y is the subject of the first equation.

We substitute the expression for y on the RHS into the second equation.

Example 4 ◀)) **Self Tutor**

Solve simultaneously by substitution: $\begin{cases} y = 2x + 3 \\ 3x - 4y = 8 \end{cases}$

$y = 2x + 3$ .... (1)
$3x - 4y = 8$ .... (2)

Substituting (1) into (2) gives $3x - 4(2x + 3) = 8$
$\therefore\ 3x - 8x - 12 = 8$
$\therefore\ -5x - 12 = 8$
$\therefore\ -5x = 20$
$\therefore\ x = -4$

Substituting $x = -4$ into (1) gives $y = 2(-4) + 3$
$\therefore\ y = -8 + 3$
$\therefore\ y = -5$

The solution is $x = -4,\ y = -5$.

Check: $3x - 4y = 3(-4) - 4(-5) = -12 + 20 = 8$ ✓

In equation (2), we replace y with $2x + 3$.

EXERCISE 21D

1 Use the method of substitution to solve:

a $\begin{cases} y = x + 2 \\ 3x + 2y = 19 \end{cases}$
b $\begin{cases} y = 2x + 1 \\ 5x - 4y = 2 \end{cases}$
c $\begin{cases} 4x + 3y = -14 \\ y = -3x - 3 \end{cases}$

d $\begin{cases} x = 3y + 2 \\ 2x - 5y = 1 \end{cases}$
e $\begin{cases} 7x - 2y = -6 \\ x = y - 3 \end{cases}$
f $\begin{cases} y = 3x + 2 \\ 4x + 2y = -1 \end{cases}$

2 Use the method of substitution to solve:

a $\begin{cases} 3x - 7y = 11 \\ x = 4y + 6 \end{cases}$
b $\begin{cases} y = 8 - 4x \\ 12x - y = 12 \end{cases}$
c $\begin{cases} 2x - 5y = 0 \\ y = 5 - 3x \end{cases}$

3 Find the point of intersection of the two lines:

a $y = 3 + x$ and $5x - 2y = 0$
b $x + 3y = 6$ and $y = x - 2$
c $x = 5 - y$ and $4x + y = 5$
d $2x + 3y = 12$ and $x = \dfrac{1}{3}(y - 4)$

4 Try to solve using substitution:

a $\begin{cases} y = 3x - 5 \\ 12x - 4y = 7 \end{cases}$
b $\begin{cases} 6x + 3y = 15 \\ y = 5 - 2x \end{cases}$

Comment on your results.

E SOLUTION BY ELIMINATION

Solution by **elimination** is used to solve simultaneous equations such as $\begin{cases} 3x + 4y = 10 \\ 5x - 4y = 6 \end{cases}$ where neither variable is given as the subject of an equation.

In this method, we make the coefficients of x (or y) the **same size** but **opposite in sign**.

We then **add** the equations, which has the effect of **eliminating** one of the variables.

We can add equations together without changing their solutions.

The method of elimination uses the fact that:

> If $a = b$ and $c = d$ then $a + c = b + d$.

This allows us to add the equations together without changing their solutions.

Example 5   ◀) Self Tutor

What equation results when $3x - y = 1$ and $-3x + 4y = 4$ are added vertically?

$$\begin{array}{r} 3x - y = 1 \\ -3x + 4y = 4 \\ \hline \text{Adding,} \quad 3y = 5 \end{array}$$

Add the LHSs together and the RHSs together.

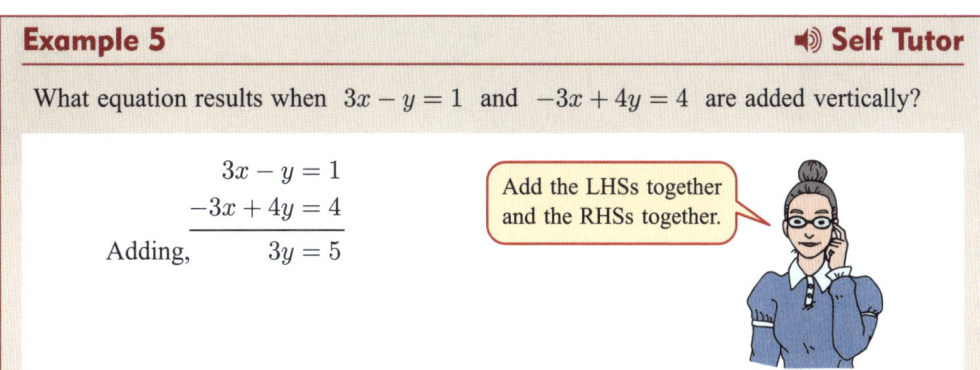

Example 6 — Self Tutor

Solve simultaneously, by elimination: $\begin{cases} 3x + 2y = 5 \\ x - 2y = 3 \end{cases}$

The coefficients of y are the same size but opposite in sign.

We **add** the LHSs and the RHSs to get an equation which contains x only.

$$3x + 2y = 5 \quad \ldots (1)$$
$$x - 2y = 3 \quad \ldots (2)$$

Adding, $\quad 4x \quad\quad = 8$

$\therefore\ x = 2$

Substituting $x = 2$ into (1) gives $\quad 3(2) + 2y = 5$

$\therefore\ 6 + 2y = 5$

$\therefore\ 2y = -1$

$\therefore\ y = -\dfrac{1}{2}$

The solution is $x = 2$, $y = -\dfrac{1}{2}$.

Check: In (2), $(2) - 2\left(-\dfrac{1}{2}\right) = 2 + 1 = 3$ ✓

EXERCISE 21E.1

1 What equation results when the following are added vertically?

a $\begin{cases} 5x + 3y = 12 \\ x - 3y = -6 \end{cases}$
b $\begin{cases} 2x + 5y = -4 \\ -2x - 6y = 12 \end{cases}$
c $\begin{cases} 4x - 6y = 9 \\ x + 6y = -2 \end{cases}$

d $\begin{cases} 12x + 15y = 33 \\ -18x - 15y = -63 \end{cases}$
e $\begin{cases} 5x + 6y = 12 \\ -5x + 2y = -8 \end{cases}$
f $\begin{cases} -7x + y = -5 \\ 7x - 3y = -11 \end{cases}$

2 Solve using the method of elimination:

a $\begin{cases} 2x + y = 3 \\ 3x - y = 7 \end{cases}$
b $\begin{cases} 4x + 3y = 7 \\ 6x - 3y = -27 \end{cases}$

c $\begin{cases} 2x + 5y = 16 \\ -2x - 7y = -20 \end{cases}$
d $\begin{cases} 3x + 5y = -11 \\ -3x - 2y = 8 \end{cases}$

e $\begin{cases} 4x - 7y = 41 \\ 3x + 7y = -6 \end{cases}$
f $\begin{cases} -4x + 3y = -25 \\ 4x - 5y = 31 \end{cases}$

You can choose to eliminate either x or y, depending on which is easier.

MULTIPLYING EQUATIONS BY A CONSTANT

In problems where the coefficients of x (or y) are *not* the same size and opposite in sign, we first need to **multiply** an equation by a constant. This will not change the solutions to the equations.

Example 7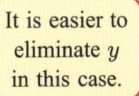

Solve $\begin{cases} 3x + 2y = -2 \\ 5x - y = 27 \end{cases}$ using the method of elimination.

$3x + 2y = -2$ .... (1)
$5x - y = 27$ .... (2)

We have $+2y$ in (1), so we obtain $-2y$ from (2) by multiplying both sides of (2) by 2.

$\ \ 3x + 2y = -2 \quad \{(1)\}$
$\ \ 10x - 2y = 54 \quad \{(2) \times 2\}$
Adding, $\ \ 13x \ = 52$
$\ \ \therefore\ x = 4$

Substituting $x = 4$ into (1), $\ 3(4) + 2y = -2$
$\ \therefore\ 12 + 2y = -2$
$\ \therefore\ 2y = -14$
$\ \therefore\ y = -7$

The solution is $x = 4$, $y = -7$.

Check: In (2), $5(4) - (-7) = 20 + 7 = 27$ ✓

It is easier to eliminate y in this case.

Example 8

Solve simultaneously, by elimination: $\begin{cases} 5x + 3y = 12 \\ 7x + 2y = 19 \end{cases}$

$5x + 3y = 12$ .... (1)
$7x + 2y = 19$ .... (2)

We can multiply (1) by 2 and (2) by -3:

$\ \ 10x + 6y = 24 \quad \{(1) \times 2\}$
$\ \ -21x - 6y = -57 \quad \{(2) \times -3\}$
Adding, $\ \ -11x \ = -33$
$\ \ \therefore\ x = 3$

Substituting $x = 3$ into (1), $\ 5(3) + 3y = 12$
$\ \therefore\ 15 + 3y = 12$
$\ \therefore\ 3y = -3$
$\ \therefore\ y = -1$

The solution is $x = 3$, $y = -1$.

Check: In (2), $7(3) + 2(-1) = 21 - 2 = 19$ ✓

Sometimes it helps to multiply *both* equations by constants.

EXERCISE 21E.2

1 Give the equation that results when both sides of the equation:

  a $3x + 4y = 2$ are multiplied by 3
  b $x - 4y = 7$ are multiplied by -2
  c $5x - y = -3$ are multiplied by 5
  d $7x + 3y = -4$ are multiplied by -3
  e $-2x - 5y = 1$ are multiplied by -4
  f $3x - y = -1$ are multiplied by -1.

2 Solve using the method of elimination:

  a $\begin{cases} 4x - 3y = 6 \\ -2x + 5y = 4 \end{cases}$
  b $\begin{cases} 2x - y = 9 \\ x + 4y = 36 \end{cases}$
  c $\begin{cases} 3x + 4y = 6 \\ x - 3y = -11 \end{cases}$

  d $\begin{cases} 4x + 3y = 17 \\ 5x - 9y = 34 \end{cases}$
  e $\begin{cases} 2x - 7y = -5 \\ 6x + 5y = -15 \end{cases}$
  f $\begin{cases} 5x + 8y = 8 \\ 9x + 2y = 33 \end{cases}$

3 Solve using the method of elimination:

  a $\begin{cases} x - 2y = -5 \\ 2x + y = -5 \end{cases}$
  b $\begin{cases} 2x + 3y = 7 \\ 3x - 2y = 4 \end{cases}$
  c $\begin{cases} 3x + 2y = 11 \\ 9x - 5y = 22 \end{cases}$

  d $\begin{cases} 4x - 3y = 6 \\ 6x + 7y = 32 \end{cases}$
  e $\begin{cases} 2x + 5y = 20 \\ 3x + 2y = 19 \end{cases}$
  f $\begin{cases} 3x - 2y = 10 \\ 4x + 3y = 19 \end{cases}$

  g $\begin{cases} -4x + 3y = -5 \\ 3x + 2y = -9 \end{cases}$
  h $\begin{cases} 7x - 3y = 29 \\ 3x + 4y = -14 \end{cases}$
  i $\begin{cases} 3x + 4y + 11 = 0 \\ 5x + 6y + 7 = 0 \end{cases}$

4 Consider the simultaneous equations $\begin{cases} 2x + 2y = 12 \\ 5x - 10y = 15 \end{cases}$.

  a Can you *divide* an equation by a non-zero constant without changing its solution? Explain your answer.
  b Divide *one* equation by a non-zero constant so you will be able to eliminate y.
  c Hence solve the simultaneous equations.

5 Use the method of elimination to attempt to solve:

  a $\begin{cases} 3x + y = 8 \\ 6x + 2y = 16 \end{cases}$
  b $\begin{cases} 2x + 5y = 8 \\ 4x + 10y = -1 \end{cases}$

Comment on your results.

F PROBLEM SOLVING WITH SIMULTANEOUS EQUATIONS

In this Section we deal with problems given in sentences. We need to interpret the information and use it to write two equations in two unknowns. We use the techniques we have learnt during the Chapter to solve the equations simultaneously, and hence answer the original problem.

Example 9

Two numbers have a difference of 7 and an average of 4. Find the numbers.

Let x be the larger number and y be the smaller number.

The difference between x and y is $\quad x - y = 7 \quad$ .... (1)

The average of x and y is $\quad \dfrac{x+y}{2} = 4 \quad$ .... (2)

$$\begin{aligned} x - y &= 7 \quad \{(1)\} \\ x + y &= 8 \quad \{(2) \times 2\} \end{aligned}$$

Adding, $\quad 2x = 15$

$\therefore \quad x = \dfrac{15}{2}$

Substituting $x = \dfrac{15}{2}$ into (1) gives $\dfrac{15}{2} - y = 7$

$$\therefore \quad y = \dfrac{15}{2} - 7$$

$$\therefore \quad y = \dfrac{1}{2}$$

The numbers are $\dfrac{1}{2}$ and $\dfrac{15}{2}$.

Check: (1) $\dfrac{15}{2} - \dfrac{1}{2} = 7 \; \checkmark \quad$ (2) $\dfrac{\frac{15}{2} + \frac{1}{2}}{2} = 4 \; \checkmark$

> When solving problems with simultaneous equations we must find two equations containing two unknowns.

Example 10

At a clearance sale, all CDs are sold for one price and all DVDs are sold for another price. Marisa bought 3 CDs and 2 DVDs for a total of £34.50. Nico bought 2 CDs and 5 DVDs for a total of £56. Find the cost of each item.

Let x pence be the cost of one CD, and y pence be the cost of one DVD.

3 CDs and 2 DVDs cost £34.50, so $\quad 3x + 2y = 3450 \quad$ .... (1)
2 CDs and 5 DVDs cost £56, so $\quad 2x + 5y = 5600 \quad$ .... (2)

We will eliminate x by multiplying equation (1) by 2 and equation (2) by -3.

$$\begin{aligned} \therefore \quad 6x + 4y &= 6900 \quad \{(1) \times 2\} \\ -6x - 15y &= -16\,800 \quad \{(2) \times -3\} \end{aligned}$$

Adding, $\quad -11y = -9900$

$\therefore \quad y = 900$

Substituting $y = 900$ into (1) gives $\quad 3x + 2(900) = 3450$

$\therefore \quad 3x + 1800 = 3450$

$\therefore \quad 3x = 1650$

$\therefore \quad x = 550$

The cost of one CD is £5.50, and the cost of one DVD is £9.

436 Simultaneous equations (Chapter 21)

EXERCISE 21F

1. The difference between two numbers is 84, and their sum is 278. What are the numbers?
2. Two numbers have a sum of 200 and a difference of 37. Find the numbers.
3. Find two numbers whose difference is 8 and whose average is 13.
4. Two hammers and a screwdriver cost a total of £34. A hammer and 3 screwdrivers cost a total of £32. Find the price of each type of tool.
5. Four adults and three children go to a theatre for £148. Two adults and five children are charged £116 for the same performance. Find the price of an adult's ticket and a child's ticket.

Example 11 ◀) Self Tutor

A carpenter makes cabinets and desks. These items use the same types of doors and drawers. Each cabinet has 2 doors and 3 drawers, and each desk has 1 door and 5 drawers. The carpenter has 41 doors and 100 drawers available. How many of each item should he make to use his entire supply of doors and drawers?

Suppose the carpenter makes x cabinets and y desks.

$\therefore 2x + y = 41$ .... (1) {total number of doors}
$3x + 5y = 100$ .... (2) {total number of drawers}

Item	Number	Doors	Drawers
Cabinet	x	$2x$	$3x$
Desk	y	y	$5y$
Total		41	100

We will eliminate y by multiplying equation (1) by -5.

$\quad\quad -10x - 5y = -205 \quad \{(1) \times -5\}$
$\quad\quad\quad 3x + 5y = 100 \quad \{(2)\}$

Adding, $-7x \quad\quad = -105$
$\therefore x = 15$

Substituting $x = 15$ into (1) gives $2(15) + y = 41$
$\therefore 30 + y = 41$
$\therefore y = 11$

The carpenter should make 15 cabinets and 11 desks.

6. A purse contains £3.75 in 5 pence and 20 pence coins. There are 33 coins altogether. How many of each type of coin are in the purse?
7. A yard contains rabbits and pheasants only. There are 35 heads and 98 feet in the yard. How many rabbits and pheasants does the yard contain?
8. Milk is sold in one litre and two litre cartons. A shop owner orders 120 litres of milk, and receives 97 cartons. How many of each type did she receive?

Example 12 ◀) Self Tutor

An equilateral triangle has sides of length $(3x - y)$ cm, $(x + 5)$ cm, and $(y + 3)$ cm. Find the length of each side.

The sides of an equilateral triangle are equal.
$\therefore y + 3 = x + 5$ .... (1) and
$\quad 3x - y = x + 5$ .... (2)

$(3x - y)$ cm $(x + 5)$ cm
$(y + 3)$ cm

Using (1), $y = x + 2$

Substituting into (2), $3x - (x + 2) = x + 5$
$\therefore \ 2x - 2 = x + 5$
$\therefore \ x = 7$

Substituting $x = 7$ into (1), $y + 3 = 7 + 5$
$\therefore \ y = 9$

Substituting $x = 7$ and $y = 9$ into the side lengths:

$3x - y = 21 - 9$ $x + 5 = 7 + 5$ $y + 3 = 9 + 3$
$\quad\quad\ = 12$ $\quad\quad = 12$ $\quad\quad = 12$

The sides of the triangle are all 12 cm long.

9 The figure alongside is a rectangle.
  a Find x and y.
  b Hence, find the area of the rectangle.

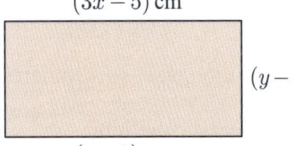

10 KLM is an equilateral triangle.
  a Find x and y.
  b Hence, find the perimeter of the triangle.

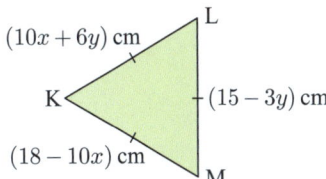

Review set 21A

1 By trial and error, find integers x and y which satisfy both $y = 2x + 1$ and $x + y = 7$.

2 Solve graphically: $\begin{cases} y = 2x - 1 \\ y = x - 2 \end{cases}$

3 Solve by substitution: $\begin{cases} y = 11 - 3x \\ 4x + 3y = -7 \end{cases}$

4 Solve by equating values of y:
  a $\begin{cases} y = 2x + 3 \\ y = x - 2 \end{cases}$
  b $\begin{cases} y = -5x + 1 \\ y = -5x - 1 \end{cases}$

5 Solve simultaneously:
  a $\begin{cases} 3x - 2y = 16 \\ y = 2x - 10 \end{cases}$
  b $\begin{cases} 3x - 5y = 11 \\ 4x + 3y = 5 \end{cases}$

6 Two pencils and a ruler cost 98 pence in total. One pencil and two rulers cost £1.24 in total. Find the cost of each item.

7
  a Rearrange $4x + y = 29$ to make y the subject of the formula.
  b Hence, use the method of substitution to solve simultaneously: $\begin{cases} 4x + y = 29 \\ 2x - 3y = 25 \end{cases}$

8 Solve by elimination: $\begin{cases} 2x - 3y = 18 \\ 4x + 5y = -8 \end{cases}$

9 The figure alongside is a rectangle.
  a Find x and y.
  b Hence, find the area of the rectangle.

Top: $(x + 2y)$ cm; Left: $(3y - 11)$ cm; Right: $(x - y)$ cm; Bottom: $(2x + 3)$ cm

Review set 21B

1 a By drawing their graphs on the same set of axes, find the point of intersection of $y = 3x - 2$ and $y = 2x + 1$.
  b Hence, solve the simultaneous equations: $\begin{cases} y = 3x - 2 \\ y = 2x + 1 \end{cases}$

2 Use the graph alongside to estimate the solution to the simultaneous equations $\begin{cases} y = 2 - 3x \\ 2x - 3y = -9 \end{cases}$.

Round your solution to 1 decimal place.

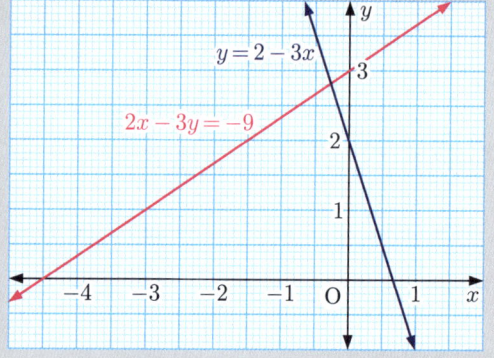

3 Solve by equating values of y: $\begin{cases} y = 16 - 3x \\ y = 2x - 4 \end{cases}$

4 Solve by substitution: $\begin{cases} y = 2x - 3 \\ 3x - 2y = 4 \end{cases}$

5 Solve simultaneously:
  a $\begin{cases} y = 2x - 5 \\ 3x - 2y = 11 \end{cases}$
  b $\begin{cases} 3x + 5y = 1 \\ 4x - 3y = 11 \end{cases}$

6 The difference between two numbers is 11, and their sum is 85. Find the numbers.

7 Sally has only 10 pence and 50 pence coins in her purse. She has 21 coins altogether with a total value of £5.30. How many of each coin type does she have?

8 Solve simultaneously:
  a $\begin{cases} 3x + y = -4 \\ y = -2x + 5 \end{cases}$
  b $\begin{cases} y = 3x + 2 \\ y = 3x - 5 \end{cases}$

9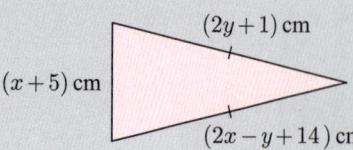

Sides: $(x+5)$ cm, $(2y+1)$ cm, $(2x - y + 14)$ cm

The perimeter of this triangle is 29 cm.
  a Find x and y.
  b Hence, find the length of the equal sides of the triangle.

22

Transformations

Contents:
- **A** Translations
- **B** Reflections and line symmetry
- **C** Rotations and rotational symmetry
- **D** Enlargements and reductions

Opening problem

How can we describe the transformation of the green semi-circle to produce the following images?

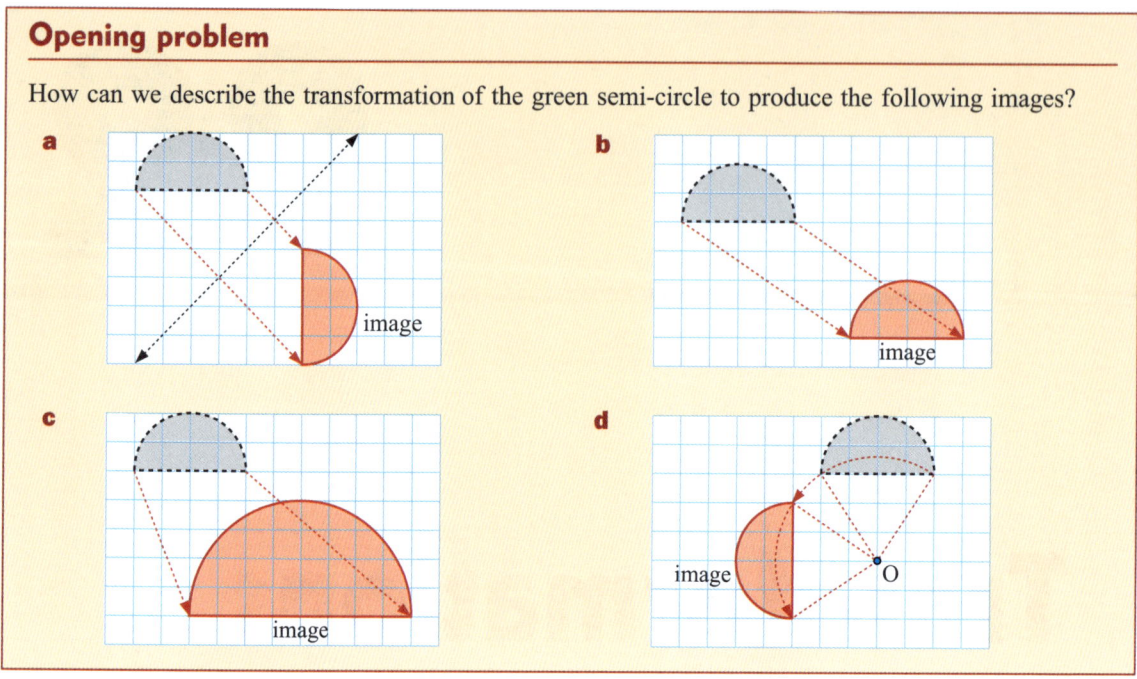

TRANSFORMATIONS

A change in the size, shape, orientation, or position of a figure is called a **transformation**. Reflections, rotations, translations, and enlargements are all examples of transformations.

When we perform a transformation, the original figure is called the **object**, and the new figure is called the **image**.

Historical note

Sir D'Arcy Wentworth Thompson (1860 - 1948) was a mathematical biologist from Scotland, most noted for his book *On Growth and Form*. He emphasised the similarities between some plants and animals which develop because a particular characteristic is physically beneficial to both organisms.

In the chapter *The Comparison of Related Forms*, Thompson described how differences in the forms of related animals could be described using mathematical transformations. For example, the pictures below show how a salmon can be transformed into a schnapper by using a curved grid.

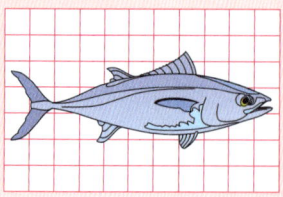
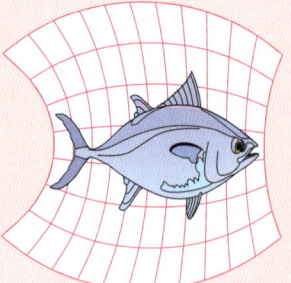

The transformations we will consider in this Chapter are:

- **translations**, where every point moves a fixed distance in a given direction

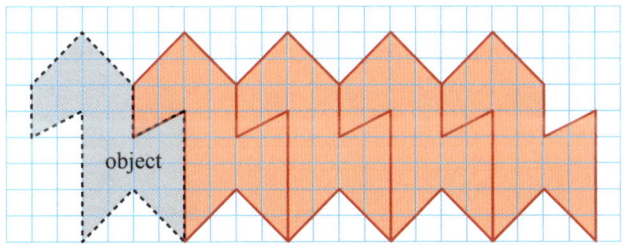

> When this shape is repeatedly translated it forms a **tessellation**. It can completely cover an area without gaps.

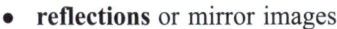

slide the original 4 units to the right to find each new image

- **reflections** or mirror images

- **rotations** about a point through a given angle

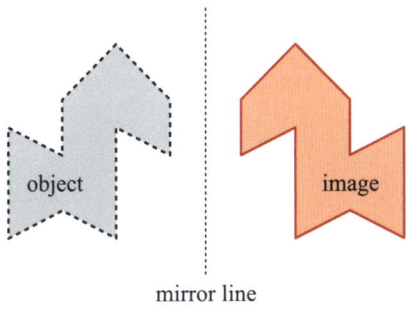

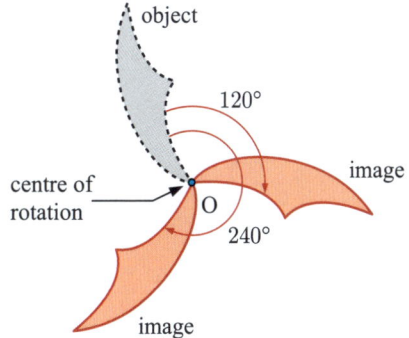

- **enlargements** and **reductions**, where objects are transformed into larger or smaller objects with the same shape.

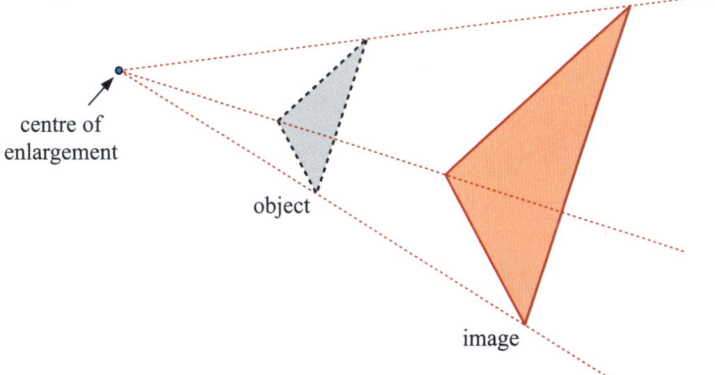

A TRANSLATIONS

A **translation** is a transformation in which every point on the figure moves a fixed distance in a given direction.

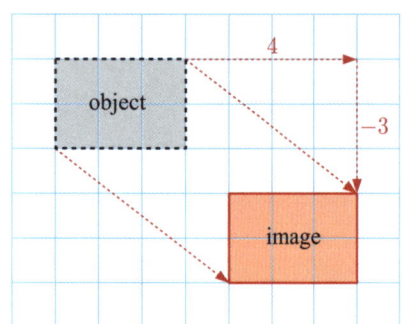

This object has been translated 4 units right and 3 units down.

We can use a **translation vector** to describe the horizontal and vertical movement of the translation.

In this case, translation vector is

$$\begin{pmatrix} 4 \\ -3 \end{pmatrix} \begin{matrix} \leftarrow \text{horizontal movement} \\ \leftarrow \text{vertical movement} \end{matrix}$$

Under a translation, the size and shape of an object does not change. The *position* of the object changes.

Example 1 ◀))) Self Tutor

a Translate this object 5 units left and 2 units upwards.
b Describe the translation using a translation vector.

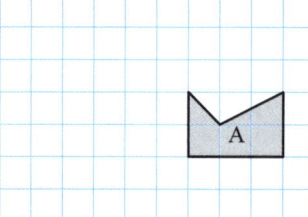

a
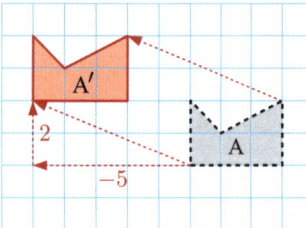

A′ is the image of the object A.

b The translation vector is $\begin{pmatrix} -5 \\ 2 \end{pmatrix}$.

EXERCISE 22A

1 Translate the given figures in the direction indicated. In each case, describe the translation using a translation vector.

PRINTABLE DIAGRAMS

a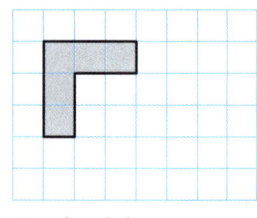
3 units right,
2 units down.

b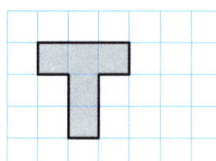
4 units right.

c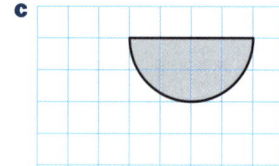
3 units down.

d	e	f
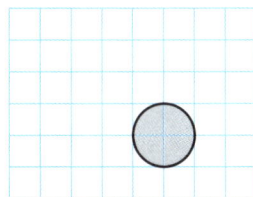	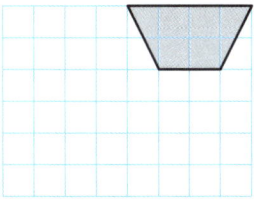	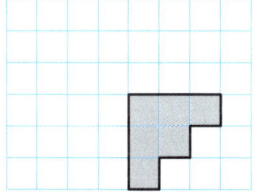
3 units left, 1 unit up.	1 unit left, 4 units down.	3 units left, 3 units up.

2 Describe the translation from A to A′ in words and using a translation vector.

a	b	c
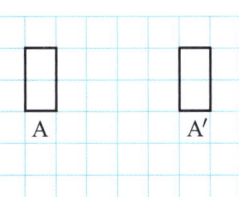	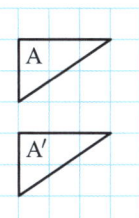	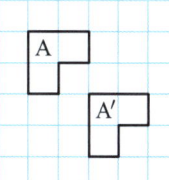

d	e	f
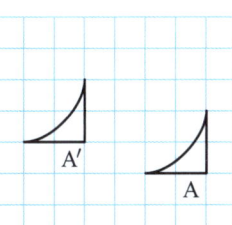	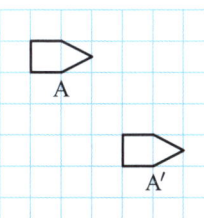	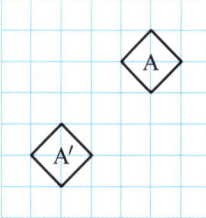

Example 2
◆) **Self Tutor**

a Describe what is meant by the translation vector $\begin{pmatrix} 2 \\ 4 \end{pmatrix}$.

b Translate the quadrilateral ABCD using the translation vector $\begin{pmatrix} 2 \\ 4 \end{pmatrix}$.

c State the coordinates of each vertex of the image.

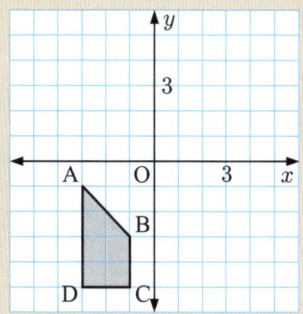

a The translation vector $\begin{pmatrix} 2 \\ 4 \end{pmatrix}$ means that the object moves 2 units to the right and 4 units upwards.

c The vertices of the image are A′(−1, 3), B′(1, 1), C′(1, −1), and D′(−1, −1).

b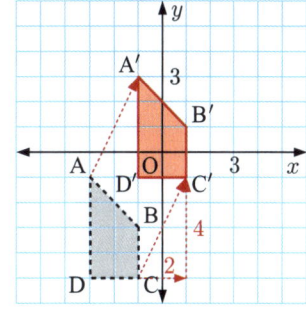

444   **Transformations** (Chapter 22)

3 Describe what is meant by the translation vector:

a $\begin{pmatrix} 1 \\ 4 \end{pmatrix}$ b $\begin{pmatrix} -2 \\ 3 \end{pmatrix}$ c $\begin{pmatrix} 3 \\ 0 \end{pmatrix}$ d $\begin{pmatrix} -1 \\ -3 \end{pmatrix}$ e $\begin{pmatrix} 0 \\ -5 \end{pmatrix}$

4 Translate each figure using the translation vector given, and state the coordinates of each vertex of the image:

a $\begin{pmatrix} -4 \\ 0 \end{pmatrix}$ 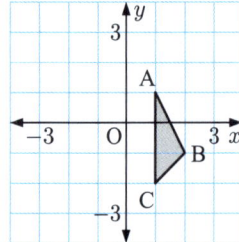   b $\begin{pmatrix} 2 \\ -1 \end{pmatrix}$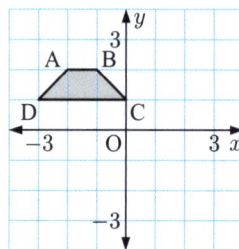

c $\begin{pmatrix} -3 \\ 4 \end{pmatrix}$ 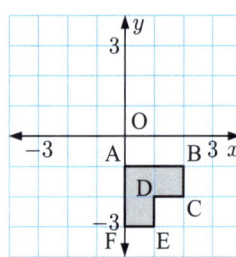   d $\begin{pmatrix} -2 \\ -2 \end{pmatrix}$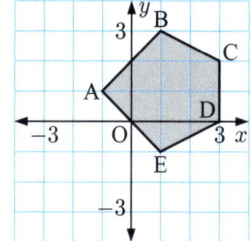

5 The figure A has been translated to B, then B has been translated to C.

a Give the translation vector from A to B.
b Give the translation vector from B to C.
c What translation vector would move A directly to C?

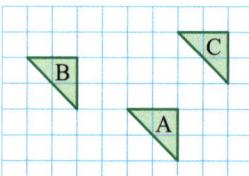

Example 3   ◀)) **Self Tutor**

Is this transformation a translation?

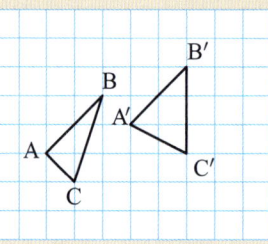

In a translation, every point on the figure moves the same distance in the same direction.

In this case, A has moved 3 units right and 1 unit up, while C has moved 4 units right and 1 unit up.

This transformation is **not** a translation.

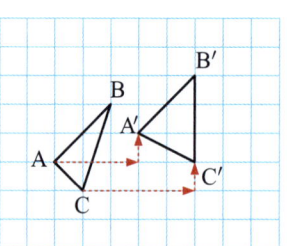

The triangles do not have the same shape, so the transformation is not a translation.

6 Are the following transformations translations? If so, describe the translation.

a 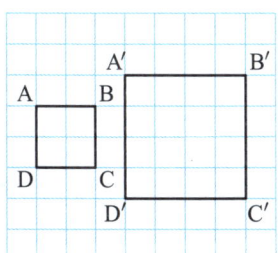   b 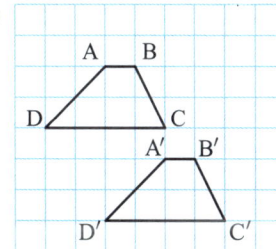   c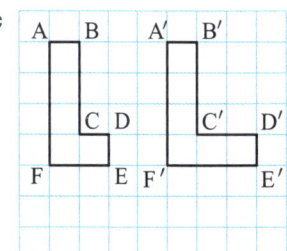

7 Consider the figures alongside.

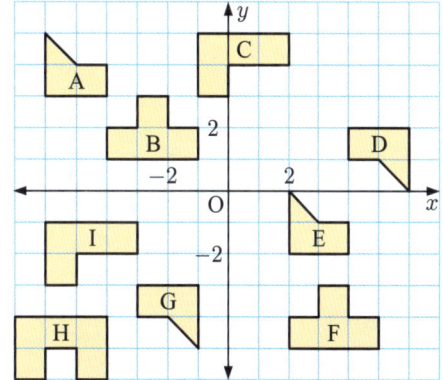

a Which figure is a translation of figure C? Describe the translation from figure C to this figure.

b Which figure is a translation of figure G? Describe the translation from figure G to this figure.

c Which of these figures cannot be translated to any other figure?

B REFLECTIONS AND LINE SYMMETRY

We encounter reflections every day as we look in a bathroom mirror, peer into a pond of water, or glance at the traffic in a car rear-view mirror.

Discussion

When you see a photo of yourself, is your image the same as when you see yourself in a mirror?

The object alongside has been **reflected** in the **mirror line** to form its image. In this case we call it the **mirror image**.

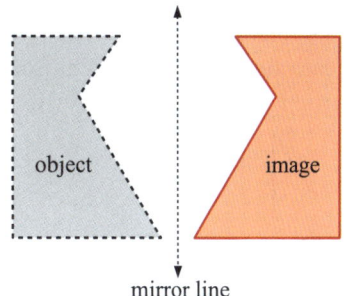

Activity 1   Reflections

You will need: A mirror, paper, pencil, ruler.

What to do:

1 Make two copies of the figures shown below:

   a   b

   c   d

2 Put the mirror along the mirror line m on one copy. Draw the reflection as accurately as you can on the second copy.

3 Cut out the second copy with its reflection and fold it along the mirror line. You should find that the two parts of the figure can be folded exactly onto one another along the mirror line.

In general, to reflect an object in a mirror line, we draw lines at right angles to the mirror line which pass through key points on the object. The image of each point will be the same distance from the mirror line as the point on the object, but on the opposite side of the mirror line.

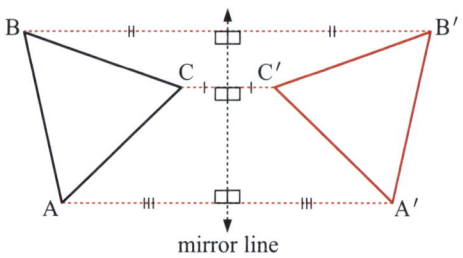

It is easier to reflect figures on grid paper because we can count squares and see right angles.

Example 4   ◀) Self Tutor

Reflect each figure in the given mirror line:

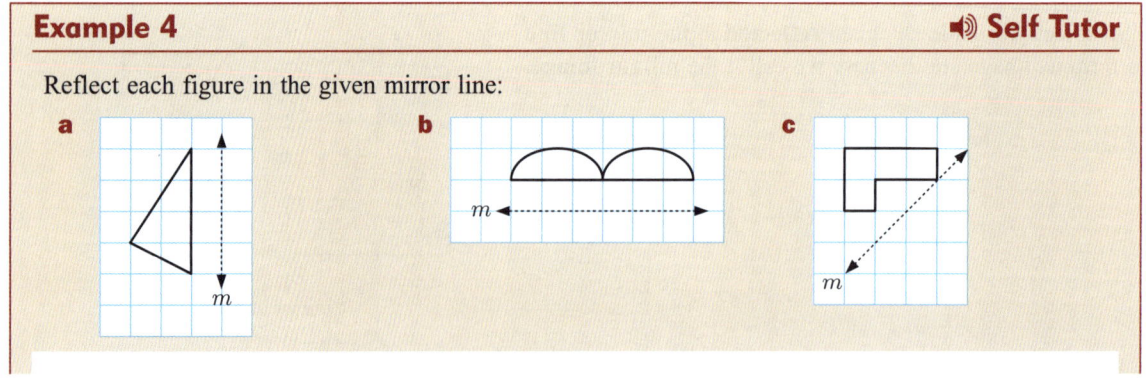

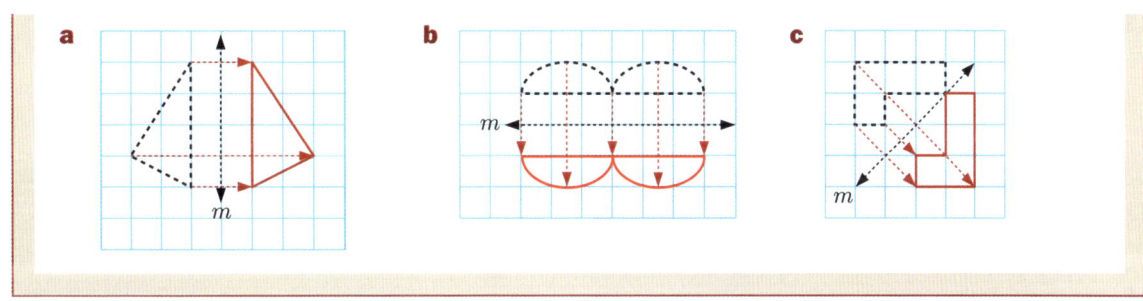

EXERCISE 22B.1

1 Copy each figure onto grid paper and reflect it in the given mirror line:

a 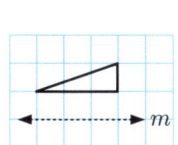 b 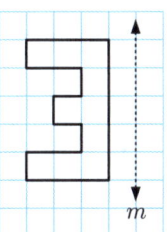 c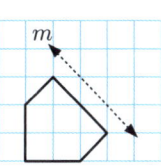

PRINTABLE FIGURES

d 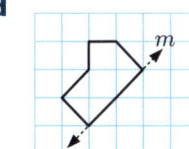 e 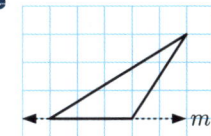 f

2 Which of the following transformations represent reflections?

a 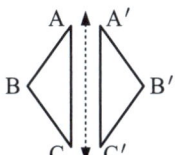 b 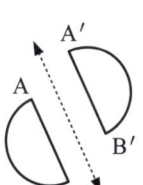 c 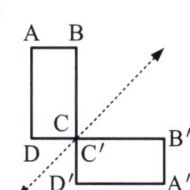 d

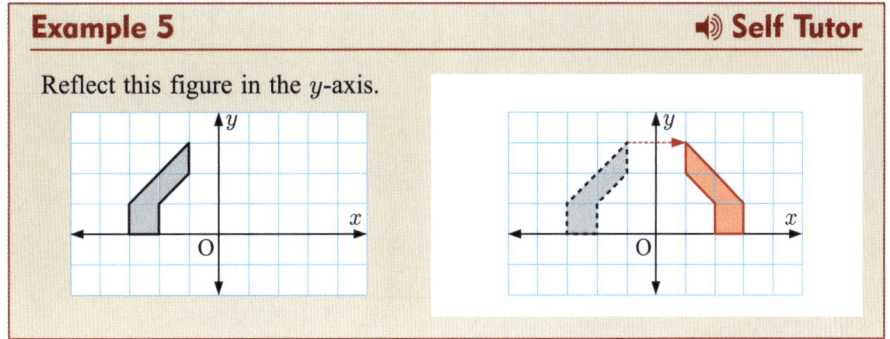

3 Reflect each figure in the y-axis:

a 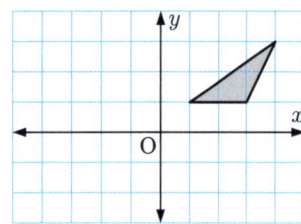 **b** 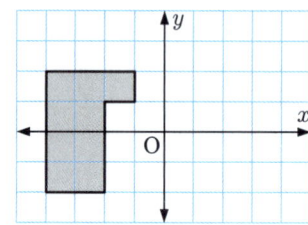 **c**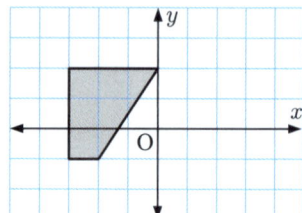

4 Reflect each figure in the x-axis:

a 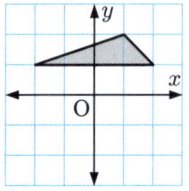 **b** 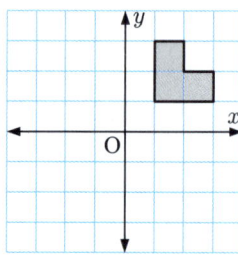 **c**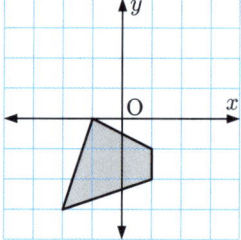

5 For each diagram, determine whether A′ is a reflection of A in the x-axis:

a 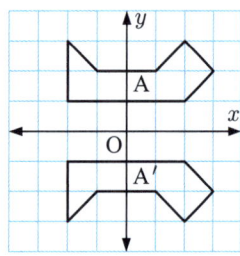 **b** 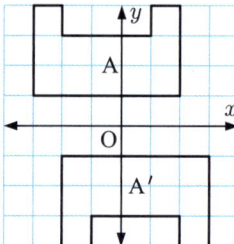 **c**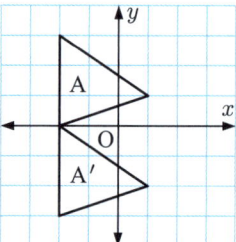

6 a Which two figures are reflections of each other?
   b Which of the axes is the mirror line for this reflection?
   c Which two figures are translations of each other?

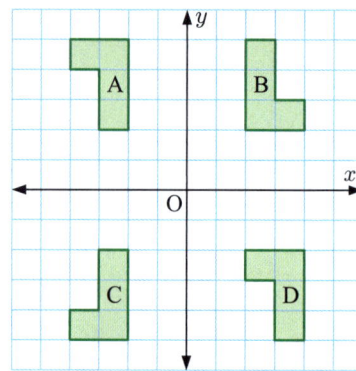

7 Reflect this figure in:
   a the x-axis
   b the y-axis
   c the line $x = 2$
   d the line $y = -1$
   e the line $y = x$.

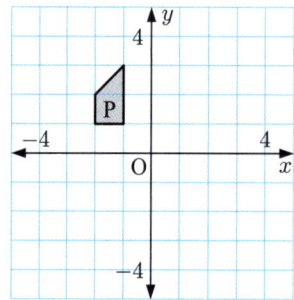

8 In each diagram, B is a reflection of A. Find the equation of the mirror line.

a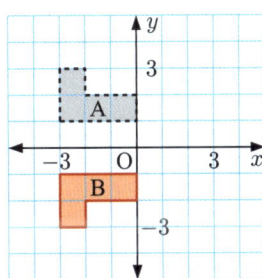

b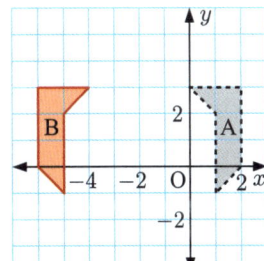

c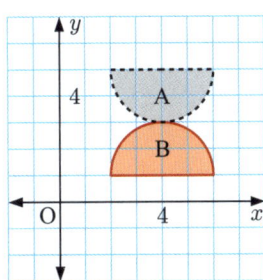

d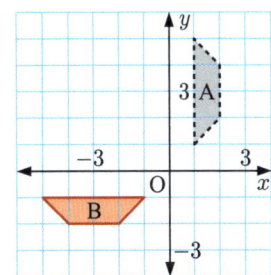

LINE SYMMETRY

A figure has a **line of symmetry** if it can be reflected in that line so that each half of the figure is reflected onto the other half of the figure.

A figure has a line of symmetry if it can be folded onto itself along that line.

For example, an isosceles triangle has one line of symmetry. It is the line from the apex to the midpoint of the base of the triangle.

Example 6   ◄» Self Tutor

For each figure, draw all lines of symmetry:

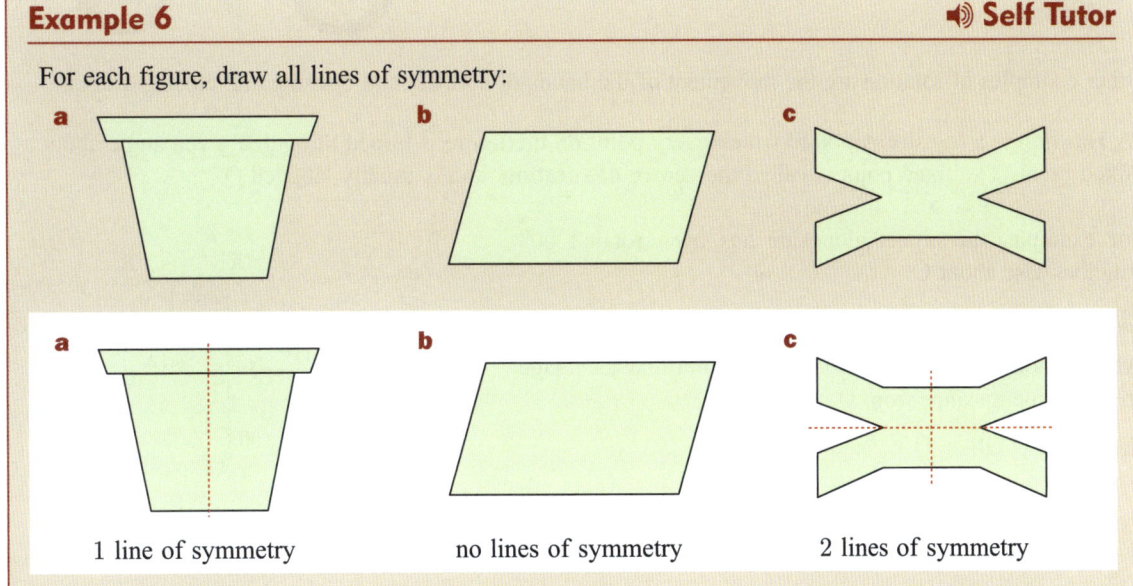

EXERCISE 22B.2

1 For each figure, draw all lines of symmetry:

a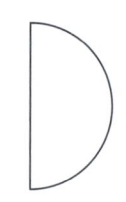

b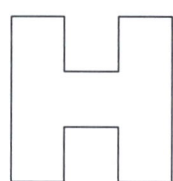

c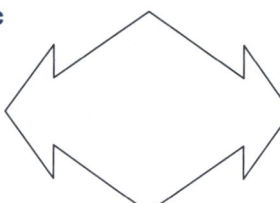

d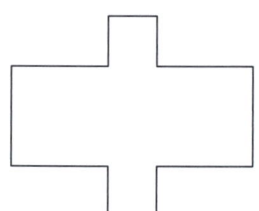

e

f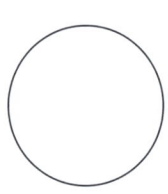

2 Draw each of the following figures, and draw all lines of symmetry. Record the number of lines of symmetry in each case.

- **a** a square
- **b** an equilateral triangle
- **c** a rectangle
- **d** a rhombus
- **e** a regular pentagon
- **f** a kite

C ROTATIONS AND ROTATIONAL SYMMETRY

When a wheel moves about its axle, we say that the wheel *rotates*.

The centre point on the axle is the **centre of rotation**.

The angle through which the wheel turns is the **angle of rotation**.

Other examples of rotation are the movement of the hands of a clock, and opening and closing a door.

A **rotation** is a transformation in which every point on the figure is turned through a given angle about a fixed point. The fixed point is called the **centre of rotation**, and is usually labelled O.

For example, the object alongside has been rotated 90° anticlockwise about O.

Click on the icon to view the rotation.

When a point is rotated about O, that point and its image are the same distance from O.

OA = OA′, OB = OB′, and so on.

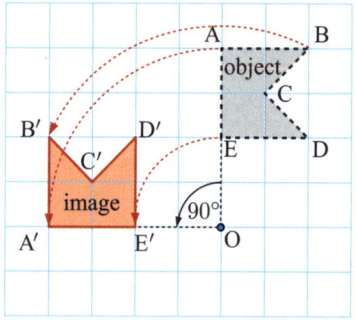

To completely describe a rotation we need to know:
- the **centre** of the rotation
- the **direction** of the rotation (clockwise or anticlockwise)
- the **angle** of the rotation.

We draw arcs of circles centred at O to make sure that a point and its image are the same distance from O.

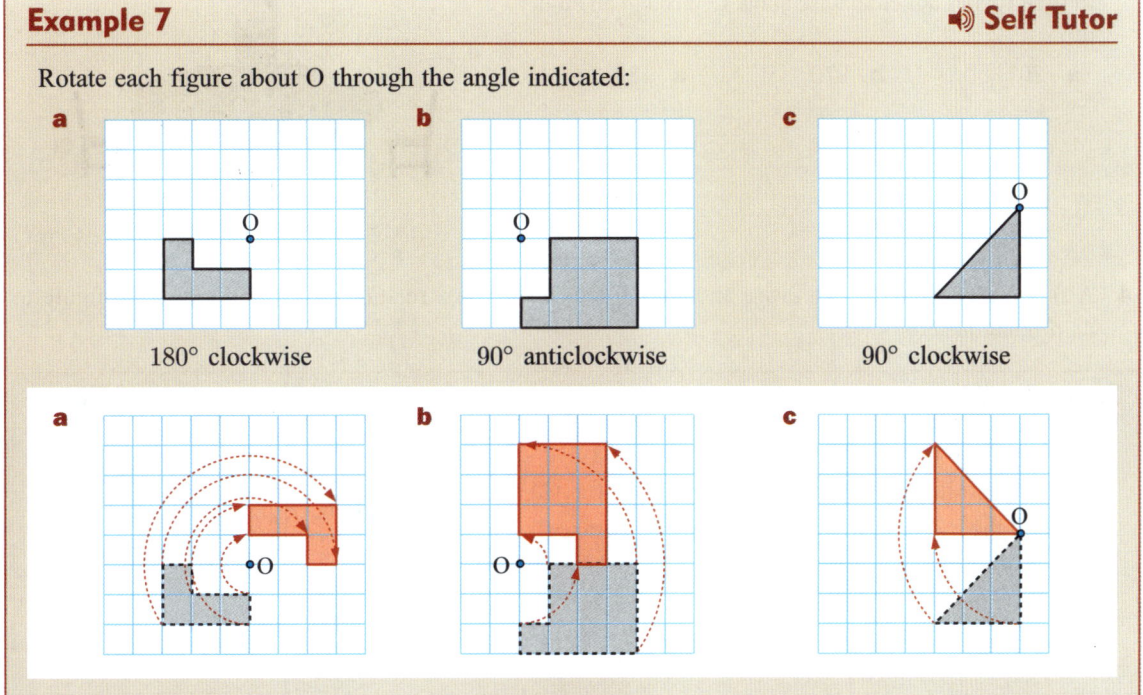

EXERCISE 22C.1

1 Rotate each figure about O through the angle given:

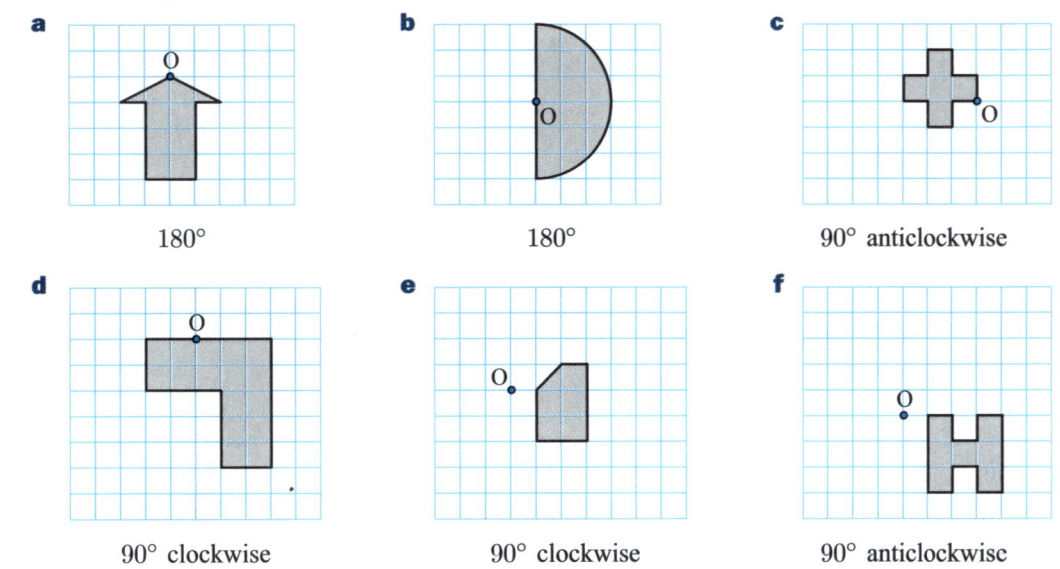

2 Which of the following letters of the alphabet either remain the same letter or become a different letter under a rotation less than 360°?

A B C D E F G H I J K L M
N O P Q R S T U V W X Y Z

3 Derek is sitting in chair A. Which chair will he move to if he rotates anticlockwise about O through an angle of:

  a 90° **b** 270° **c** 180°?

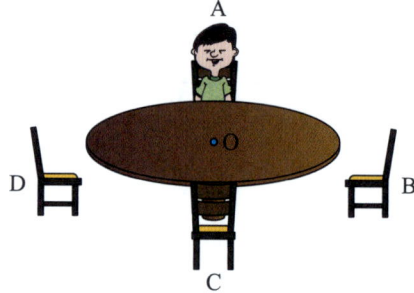

4 A figure is rotated 75° clockwise about a point O. Describe a rotation which will return the figure to its original position.

Example 8   ◀) Self Tutor

a State the vertex coordinates of triangle ABC.
b Rotate the triangle 90° clockwise about the origin O.
c State the coordinates of each vertex of the image.

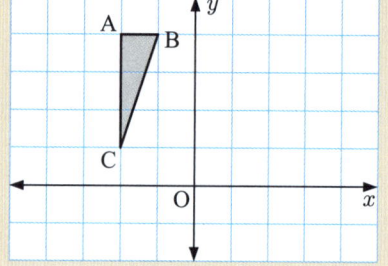

a The triangle has vertices A$(-2, 4)$, B$(-1, 4)$, and C$(-2, 1)$.

c The image triangle has vertices A′$(4, 2)$, B′$(4, 1)$, and C′$(1, 2)$.

b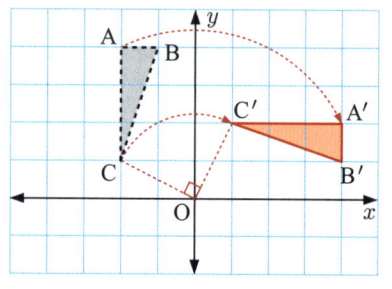

---

5 **a** State the vertex coordinates of triangle PQR.
  b Rotate the triangle 90° anticlockwise about the origin O.
  c State the coordinates of each vertex of the image.

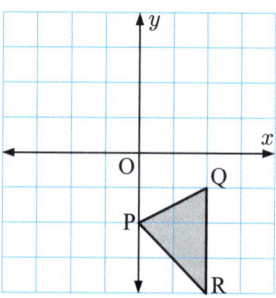

6 A quadrilateral has vertices A(−4, −3), B(−1, −3), C(−1, −4), and D(−4, −4).
   a Plot ABCD on a Cartesian plane.
   b Rotate ABCD 90° clockwise about the origin O.
   c State the coordinates of each vertex of the image.

7 a Which of the figures alongside is a rotation of A about the origin?
   b Determine the angle of rotation from figure A to this figure.

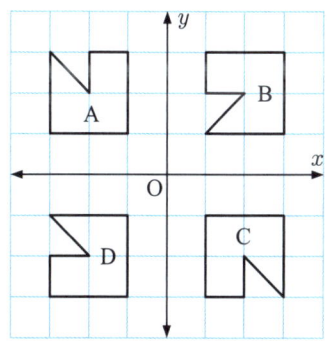

8 In each diagram, describe the centre, direction, and angle of rotation from P to Q.

a 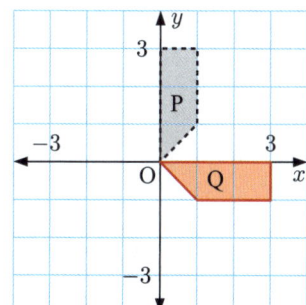 **b** 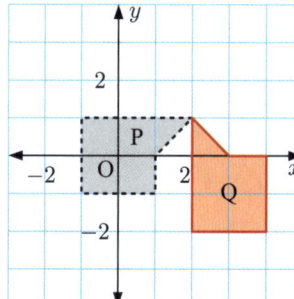 **c**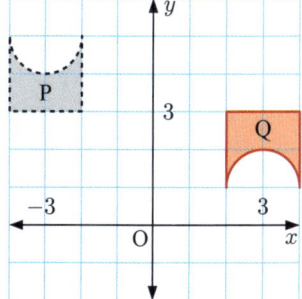

ROTATIONAL SYMMETRY

A shape has **rotational symmetry** if it can be rotated about a particular point through an angle **less than 360°** so that it maps onto itself.

The point through which the object rotates is called the **centre of rotational symmetry**.

For example, this propeller shape has rotational symmetry. If it is rotated about O through 180° then it will look identical to how it did at the start. O is the centre of rotational symmetry.

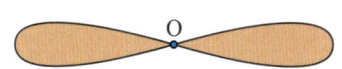

Note that *every* shape will map onto itself under a 360° rotation, but this is not rotational symmetry.

If a figure has more than one line of symmetry then it will also have rotational symmetry. The centre of rotational symmetry will be the point where the lines of symmetry meet.

However, a figure which has rotational symmetry does not necessarily have line symmetry. For example, consider the figure alongside.

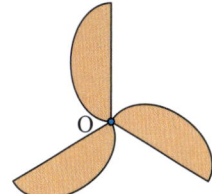

THE ORDER OF ROTATIONAL SYMMETRY

The **order of rotational symmetry** is the number of times a figure maps onto itself during one complete turn about the centre.

DEMO

For example,

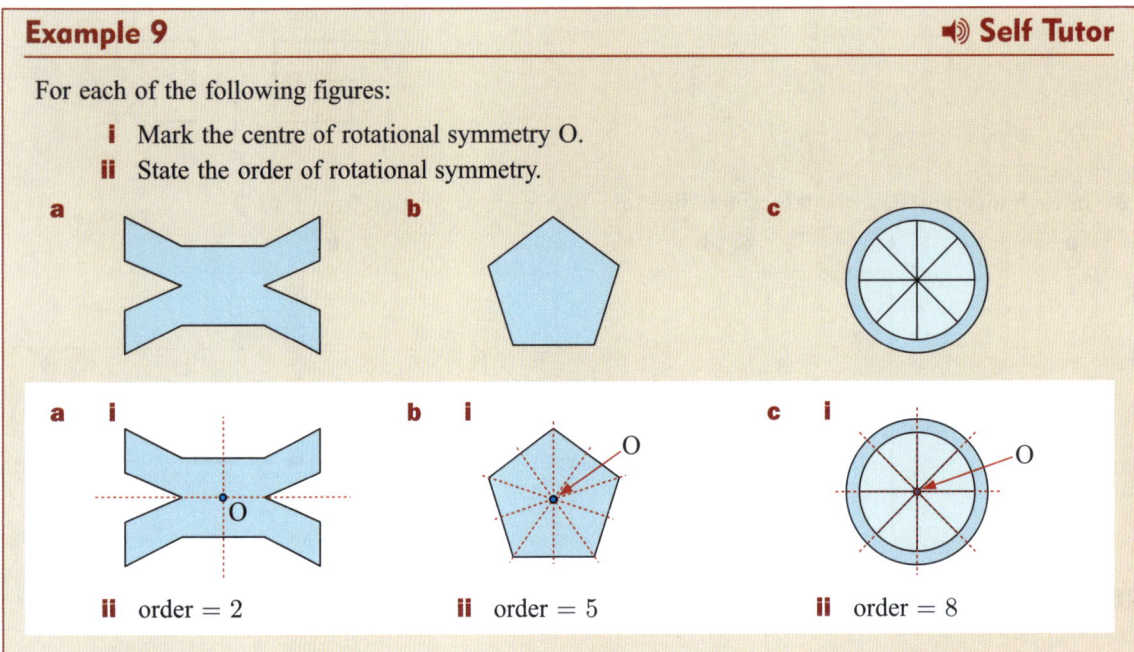

Example 9
◀)) **Self Tutor**

For each of the following figures:
  i Mark the centre of rotational symmetry O.
  ii State the order of rotational symmetry.

a, b, c figures shown above.

a ii order = 2
b ii order = 5
c ii order = 8

EXERCISE 22C.2

1 Which of the following shapes have rotational symmetry?

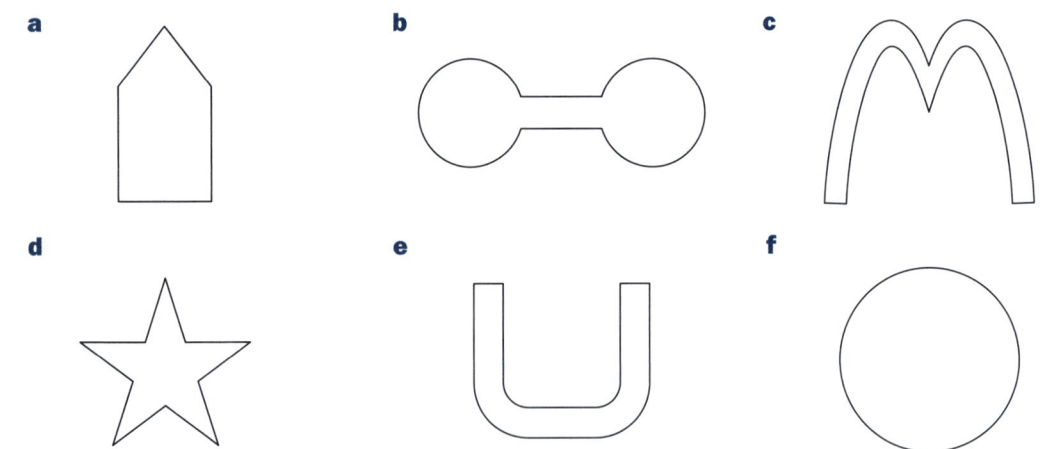

2 For each of the following figures:
  i Mark the centre of rotational symmetry O.   **ii** State the order of rotational symmetry.

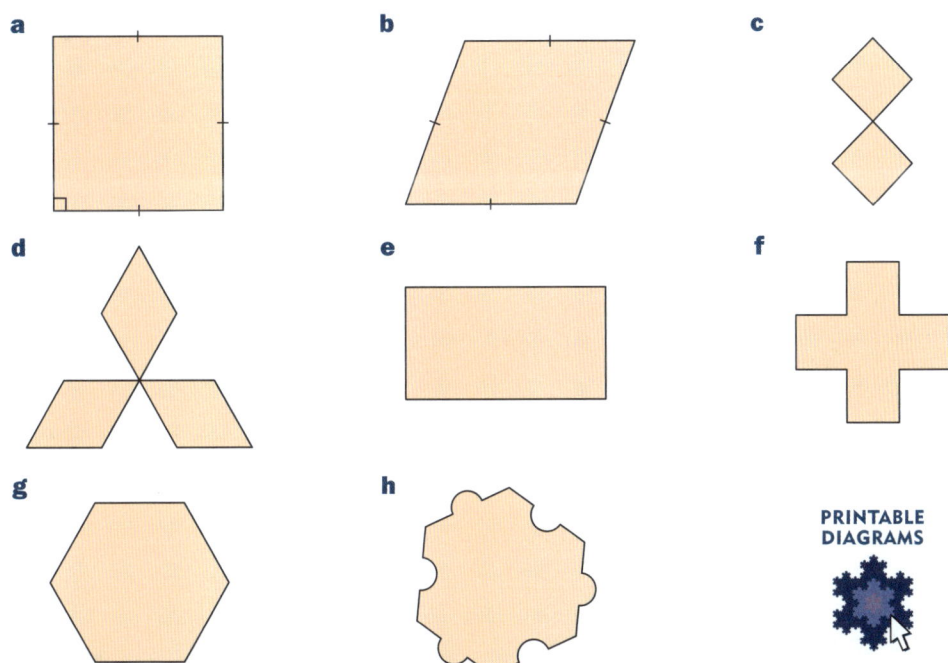

3 Draw a figure which has order of rotational symmetry: **a** 5   **b** 8.

D ENLARGEMENTS AND REDUCTIONS

We are all familiar with **enlargements** in the form of photographs, zoom tools in computer software, or looking through a microscope. For an enlargement the image is larger than the original.

Plans and maps are examples of **reductions**. In these cases the image is smaller than the original.

The **scale factor** of an enlargement (or reduction) is the ratio by which the side lengths of the object are enlarged (or reduced).

Consider the diagrams below. They are clearly not the same *size*, but they do have the same *shape*. Their side lengths have the same proportions.

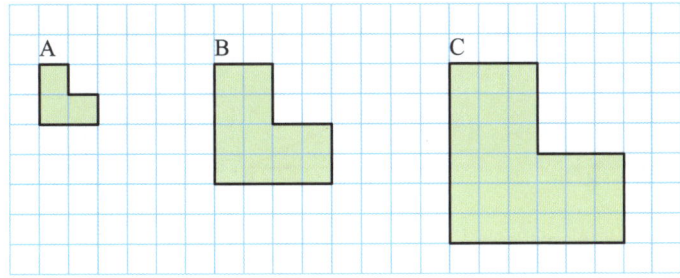

B is an enlargement of A, and is a reduction of C.

When A is enlarged to B, the side lengths are all doubled. The scale factor $k = 2$.

When C is reduced to B, the side lengths are all multiplied by $\frac{2}{3}$. The scale factor $k = \frac{2}{3}$.

If the scale factor is greater than 1, an **enlargement** occurs.

If the scale factor is less than 1, a **reduction** occurs.

A scale factor of 2 **does not** mean that the shape formed has twice the area of the original. In fact, it has $2 \times 2 = 4$ times the area. Count the squares inside A and B to confirm this.

EXERCISE 22D.1

1 In each diagram, A has been enlarged to A'. Find the scale factor in each case.

a 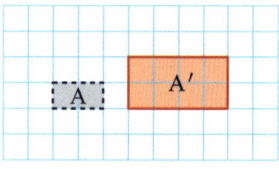 b 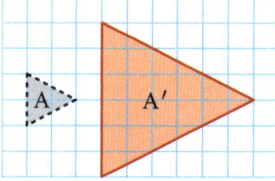 c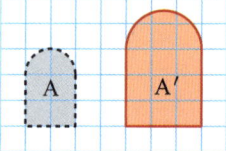

2 In each diagram, B' is a reduction of B. Find the scale factor in each case.

a 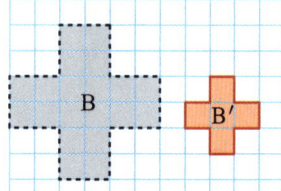 b 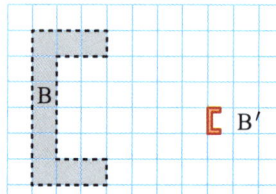 c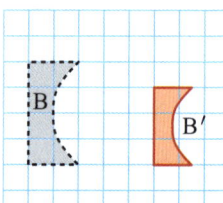

3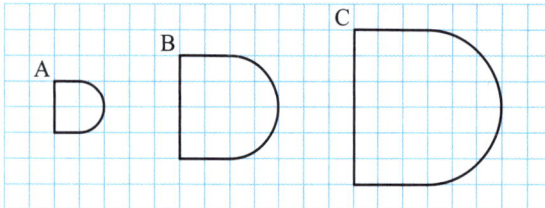

Describe the transformation from:

a A to B b B to A
c A to C d C to A
e B to C f C to B.

CENTRE OF ENLARGEMENT

In each diagram below, triangle PQR has been enlarged to triangle P'Q'R' using a scale factor of 3.

A 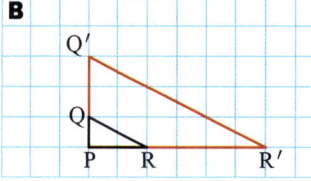 **B**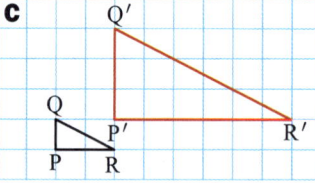

C

Discussion

1 What is the difference between the enlargements above?
2 What does "enlargement with scale factor 3" tell you?
3 What additional instructions do you need to be able to draw an enlargement in the correct position?

A 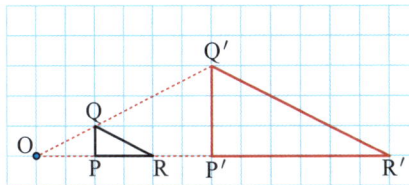 **C**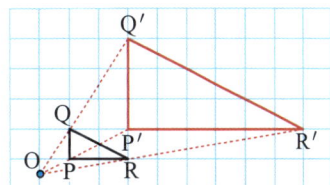

Consider again the diagrams **A** and **C** above. Suppose we draw a line through each vertex of the object and its corresponding point on the image. We see that the lines always meet at a point. We label this point O and call it the **centre of enlargement**.

Notice on each diagram that $OP' = 3 \times OP$, $OQ' = 3 \times OQ$, and $OR' = 3 \times OR$. This corresponds to the scale factor $k = 3$.

Discussion

What is the centre of enlargement for diagram **B** on page **456**?

Example 10   ◀)) Self Tutor

Find the image of each figure for the centre of enlargement O and scale factor given:

a scale factor 2   **b** scale factor 3   **c** scale factor $\frac{1}{2}$

 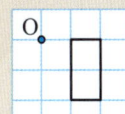 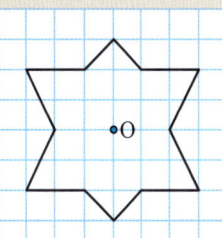

a 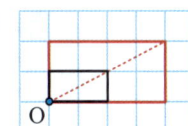   **b** 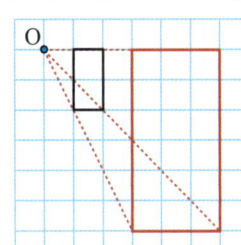   **c**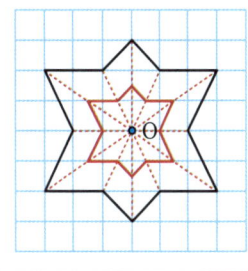

EXERCISE 22D.2

1 Find the image of each figure for the centre of enlargement O and scale factor given: **PRINTABLE DIAGRAMS**

a scale factor 2   **b** scale factor 3   **c** scale factor $\frac{1}{2}$

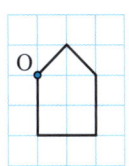 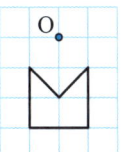 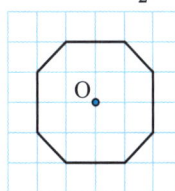 

d scale factor 2

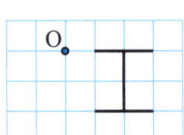

e scale factor $\frac{1}{3}$

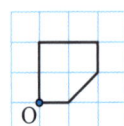

f scale factor 3

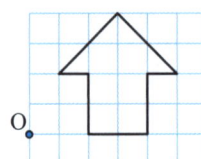

2 Copy the following diagrams. Locate the centre of enlargement in each case.

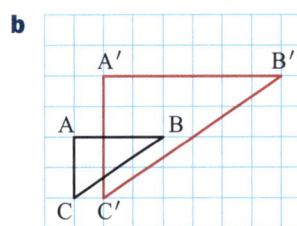

3 In this diagram, P has been enlarged to Q.
  a Locate the centre of the enlargement.
  b Find the scale factor for the enlargement.

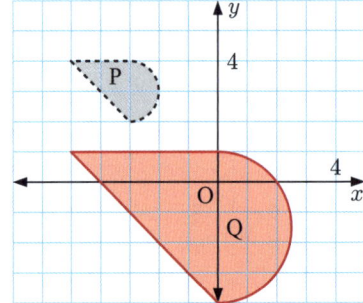

Activity 2 — Tessellations

Click on the icon to obtain this Activity.

TESSELLATIONS

Review set 22A

1 Translate the given figures in the direction indicated:

  a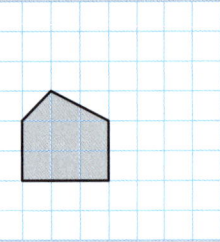
  3 units right

  b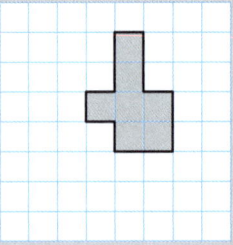
  2 units left, 2 units down

PRINTABLE DIAGRAMS

2 Figure A has been translated to B, then B has been translated to C.

    a Give the translation vector from A to B.

    b Give the translation vector from B to C.

    c What translation vector would move A directly to C?

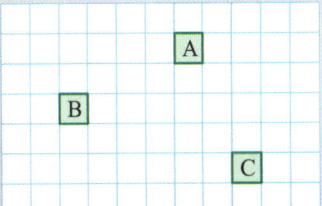

3 Copy the figure and reflect it in the mirror line shown.

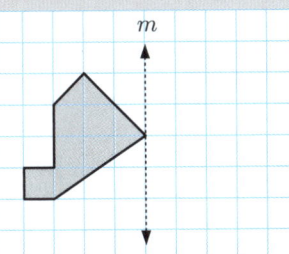

4 In this diagram, B is a reflection of A. Find the equation of the mirror line.

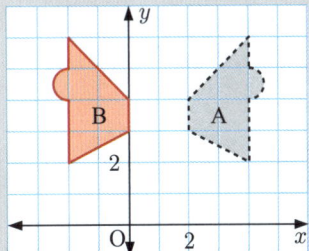

5 Draw a rectangle and all of its lines of symmetry.

6 Rotate each figure about O through the angle indicated:

    a

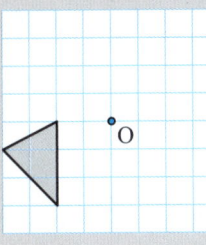

180°

    b

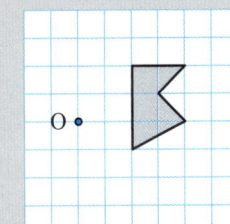

270° anticlockwise

7 Find the order of rotational symmetry for the following shapes:

    a

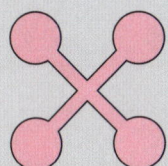

    b

8 **a** Which of the figures alongside is a rotation of A about the origin?

    b Determine the angle of rotation from figure A to this figure.

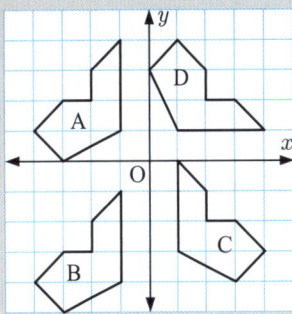

9 A triangle has vertices A(2, 1), B(4, 3), and C(3, 0).
  a Plot triangle ABC on the Cartesian plane.
  b Reflect △ABC in the x-axis, and state the coordinates of each vertex of the image.
  c Reflect △ABC in the y-axis, and state the coordinates of each vertex of the image.

10 Find the image of each figure for the centre of enlargement O and scale factor given:

  a scale factor 3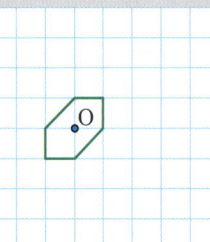

  b scale factor $\frac{1}{3}$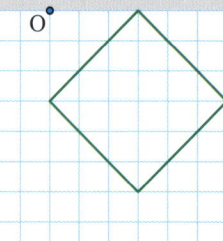

11 On separate diagrams, carry out the following transformations of the figure alongside: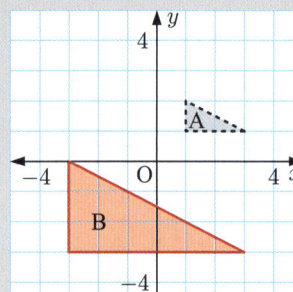

  a a translation of $\begin{pmatrix} 3 \\ -2 \end{pmatrix}$

  b a rotation about O, 90° clockwise

  c a reflection in the mirror line m

  d a reduction with scale factor $\frac{1}{2}$ and centre of enlargement O.

12 In this diagram, A has been enlarged to B.
  a Locate the centre of the enlargement.
  b Find the scale factor for the enlargement.

Review set 22B

1 Figure A has been translated to give the image B.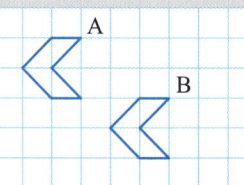
  a State the translation vector.
  b Describe the transformation needed to return B to A.

2 a Draw triangle ABC where A is (3, 3), B is (−3, −2), and C is (3, −2).

  b Translate the figure using the translation vector $\begin{pmatrix} 3 \\ 2 \end{pmatrix}$.

  c State the coordinates of each vertex of the image.

3 Reflect each figure in the axis indicated:

a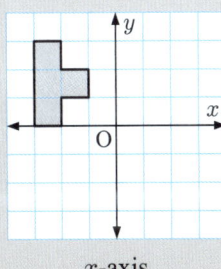

x-axis

b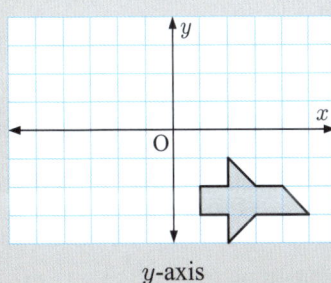

y-axis

4 Reflect this figure in:
   a the line $y = 1$
   b the line $y = x$.

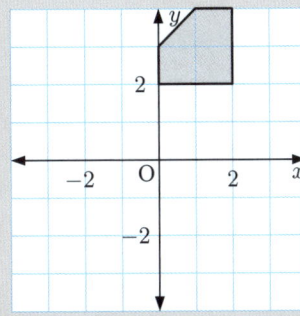

5 Copy the following figures and draw in all lines of symmetry:

a 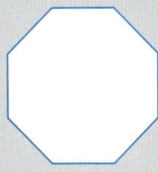   b    c N

6 a State the coordinates of each vertex of quadrilateral PQRS.
   b Rotate the quadrilateral 90° anticlockwise about the origin O.
   c State the coordinates of each vertex of the image.

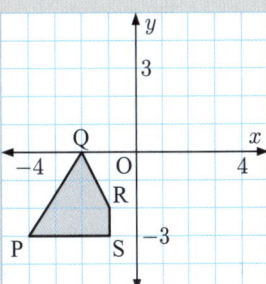

7 For the given figure:
   a Locate the centre of rotational symmetry.
   b Find the order of rotational symmetry.

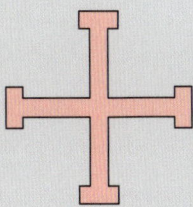

8 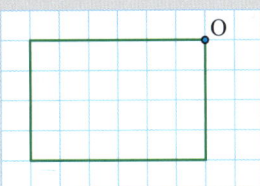 Reduce this figure with scale factor $\frac{1}{2}$ and centre O.

9 Copy the given diagram and locate the centre of enlargement.

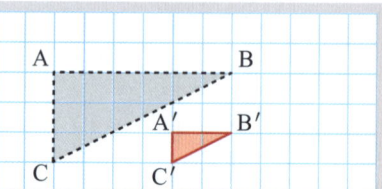

10 On separate diagrams, carry out the following transformations of the figure alongside:

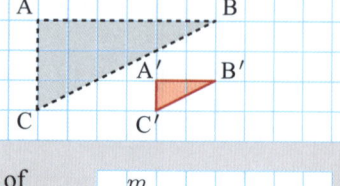

  a a translation of $\begin{pmatrix} -3 \\ 4 \end{pmatrix}$

  b a rotation about O, 180° clockwise

  c a reflection in the mirror line m

  d an enlargement with scale factor 2 and centre of enlargement O.

11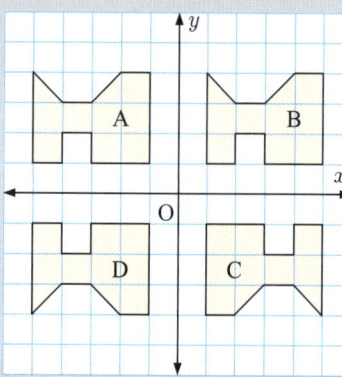

  a Which of the figures alongside is:
   i a *translation* of A
   ii a *reflection* of A
   iii a *rotation* of A?

  b Describe the transformation from C to D.

12 Describe the centre, direction, and angle of rotation from A to B.

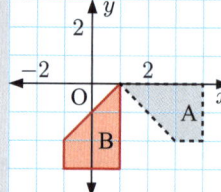

23

Similarity and congruence

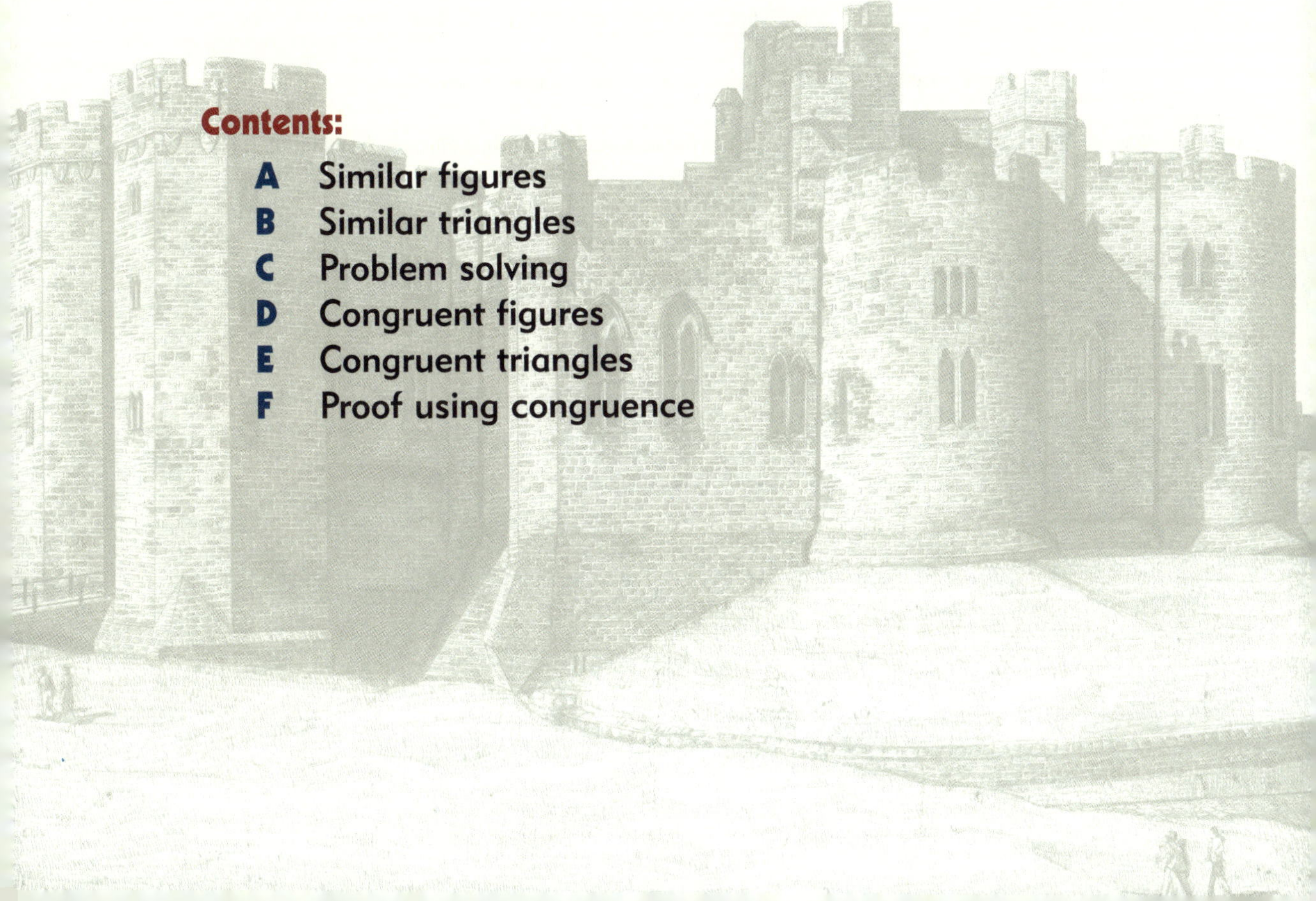

Contents:

- **A** Similar figures
- **B** Similar triangles
- **C** Problem solving
- **D** Congruent figures
- **E** Congruent triangles
- **F** Proof using congruence

Opening problem

Jane cut two triangular slices of cheesecake, and gave one to her brother Nathan.

"That's not fair", Nathan said, "your slice is bigger than mine".

Jane used a ruler to measure the sides of each slice. "See, my slice has sides 5 cm, 6 cm, and 7 cm, and so does yours. That means the slices are the same size."

"Not necessarily", said Nathan, "the slices might have the same sides, but the angles might be different".

Things to think about:

a Who do you think is correct?
b What mathematical argument can you use to justify your answer?

In this Chapter we will look at **similar figures**, which are figures with the same shape but not necessarily the same size. We will also study **congruent figures**, which are identical in both shape and size.

A SIMILAR FIGURES

The word *similar* suggests a comparison between objects which have some, but not all, properties in common. In mathematics, similar figures have the same **shape**, but not necessarily the same **size**.

> Two figures are **similar** if one is an enlargement of the other.

Common examples of similar figures include television images, photo enlargements, house plans, maps, and model cars.

A'B'C'D' is an enlargement of ABCD with scale factor 3. The two figures are therefore similar.

Notice that

$$\frac{A'B'}{AB} = \frac{B'C'}{BC} = \frac{C'D'}{CD} = \frac{D'A'}{DA} = 3,$$

so the corresponding side lengths are in the **same ratio**.

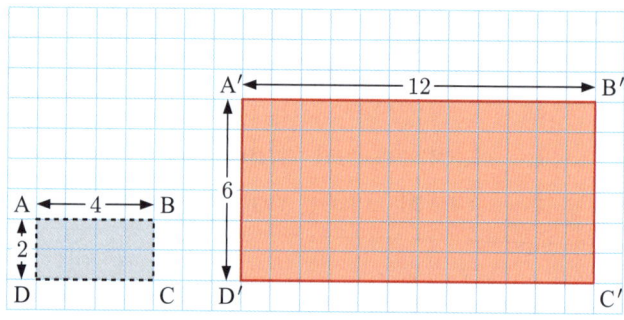

When a figure is enlarged or reduced, the sizes of its angles do not change. The figures are therefore **equiangular**.

> Two figures are **similar** if:
> - the figures are **equiangular** *and*
> - the corresponding side lengths are in the **same ratio**.

Example 1   ◀) Self Tutor

Determine whether the following pairs of figures are similar:

a 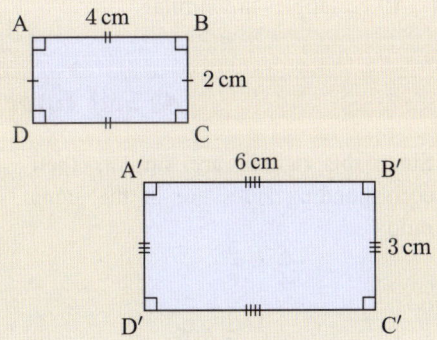   **b**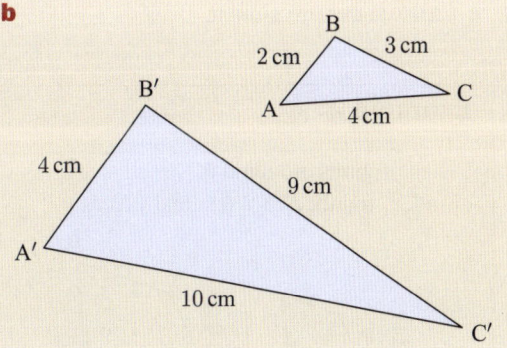

a $\dfrac{A'B'}{AB} = \dfrac{6}{4} = \dfrac{3}{2}$ and $\dfrac{B'C'}{BC} = \dfrac{3}{2}$

∴ the corresponding side lengths are in the same ratio.
The figures are also equiangular, so the figures are similar.

b $\dfrac{A'B'}{AB} = \dfrac{4}{2} = 2$ and $\dfrac{B'C'}{BC} = \dfrac{9}{3} = 3$

∴ the corresponding side lengths are *not* in the same ratio.
∴ the figures are not similar.

EXERCISE 23A

1 Determine whether the following pairs of figures are similar:

a 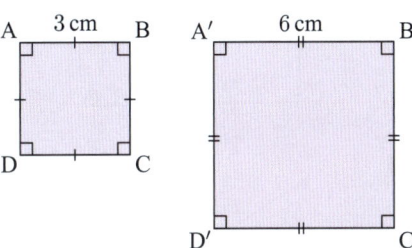   **b**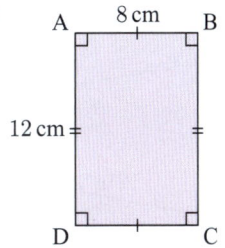

c 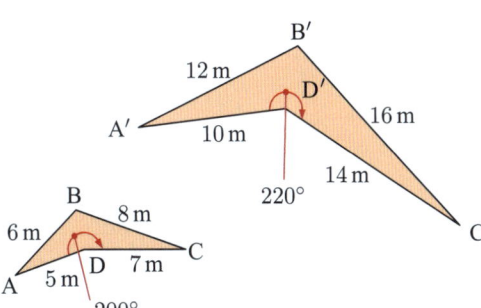   **d**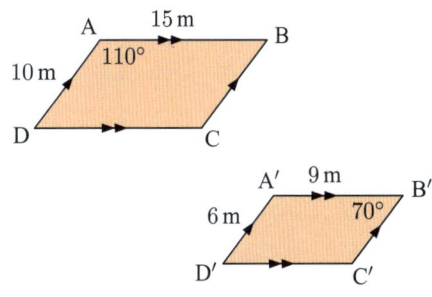

2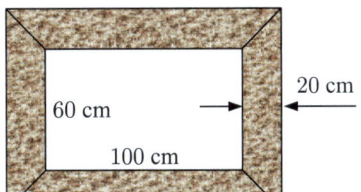

A 20 cm wide picture frame surrounds a painting which is 100 cm by 60 cm.
Are the two rectangles shown here similar?

3 Comment on the truth of the following statements. For any statement which is false, you should justify your answer with an illustration.
  a All circles are similar.
  b All parallelograms are similar.
  c All squares are similar.
  d All rectangles are similar.

Example 2   ◀)) Self Tutor

These figures are similar.
Find x, rounded to 2 decimal places.

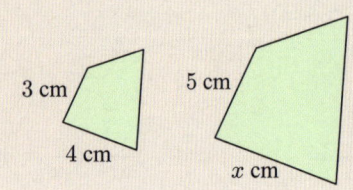

Since the figures are similar, their corresponding sides are in the same ratio.

$$\therefore \quad \frac{x}{4} = \frac{5}{3}$$

$$\therefore \quad x = \frac{5}{3} \times 4$$

$$\therefore \quad x = \frac{20}{3}$$

$$\therefore \quad x \approx 6.67$$

4 These figures are similar. Find x exactly:

a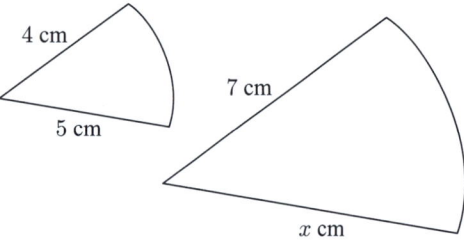

b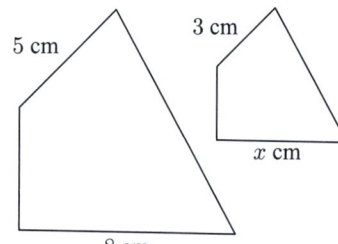

c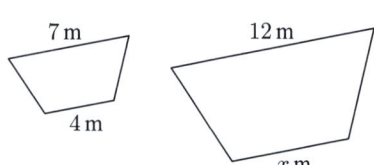

d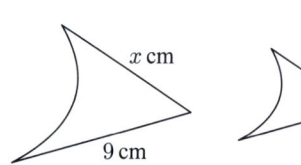

e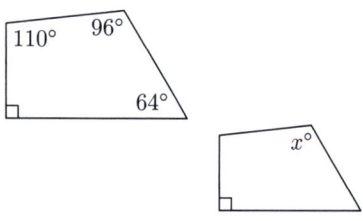

f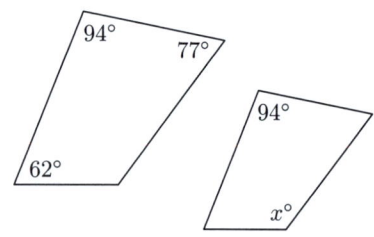

5 Narelle is drawing a scale diagram of her bedroom, which is a rectangle 4.2 m long by 3.6 m wide. On her diagram she draws her room 7 cm long.
  a How wide will her bedroom be on her diagram?
  b What is the scale factor for the diagram?

6 Find x given that triangle ABC is similar to triangle A'B'C':

a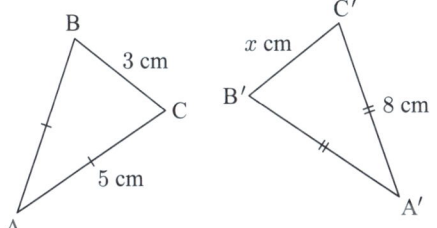

b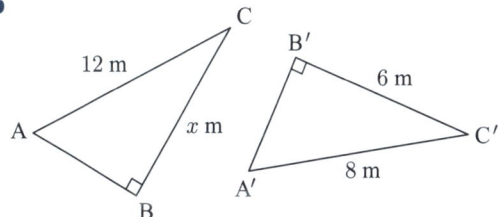

c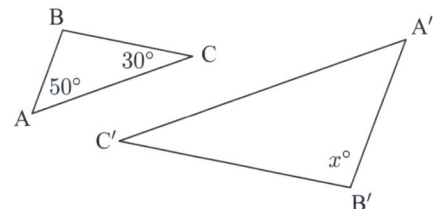

d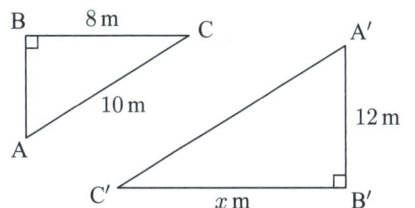

7 Sketch two quadrilaterals that:
  a are equiangular, but not similar
  b have sides in proportion, but are not similar.

8 Can you draw two triangles which are equiangular but not similar?

"In proportion" means "in the same ratio".

Discussion

Are these figures similar?

Justify your answer using the definition of similar figures.

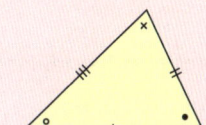 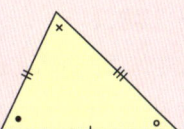

B SIMILAR TRIANGLES

In the previous Exercise we saw that quadrilaterals that are equiangular are not necessarily similar, and quadrilaterals that have sides in proportion are not necessarily similar.

However, if *triangles* are equiangular, then their corresponding sides *must* be in the same ratio, and vice versa. So, to show that two triangles are similar, we only need to show that **one** of these properties is true.

TESTS FOR TRIANGLE SIMILARITY

Two triangles are similar if either:
• they are equiangular   *or*   • their side lengths are in the same ratio.

Notice that:
• either of these properties is sufficient to prove that two triangles are similar
• since the angles of any triangle add up to 180°, if two angles of one triangle are equal to two angles of another triangle, then the remaining angles of the triangles must also be equal.

Example 3

Show that the following figures possess similar triangles:

a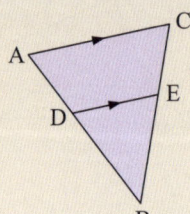

b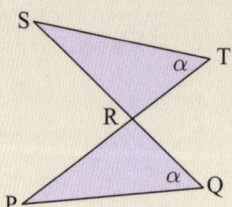

a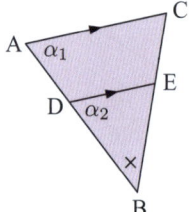

△s ABC and DBE are equiangular as:
- $\alpha_1 = \alpha_2$ {equal corresponding angles}
- The angle at B is common to both triangles.

∴ the triangles are similar.

b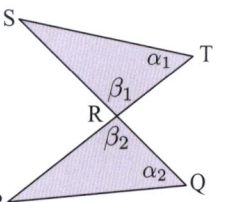

△s PQR and STR are equiangular as:
- $\alpha_1 = \alpha_2$ {given}
- $\beta_1 = \beta_2$ {vertically opposite angles}

∴ the triangles are similar.

> If two triangles are similar, we list corresponding vertices in the same order.

EXERCISE 23B.1

1 Show that the following figures possess similar triangles:

a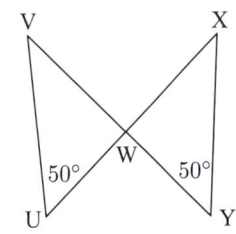

b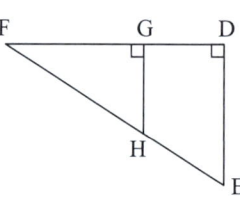

c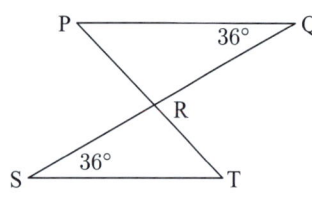

d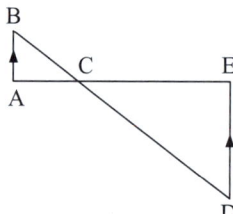

e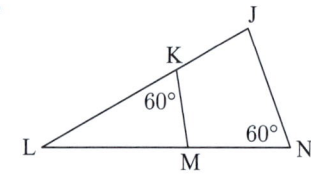

f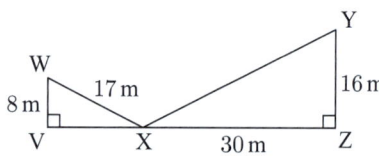

> If two angles of one triangle are equal in size to two angles of another triangle, then the remaining angles of the triangles must also be equal.

2 a Show that $\hat{ACB} = \alpha$.
  b Hence show that the three triangles in the given figure are all similar to each other.

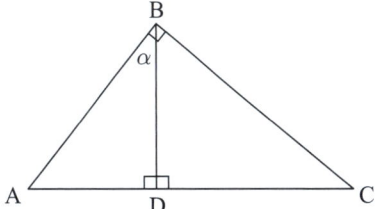

FINDING SIDE LENGTHS

Once we have established that two triangles are similar, we may use the fact that corresponding sides are in the same ratio to find unknown lengths.

Example 4

Establish that a pair of triangles is similar, and find x:

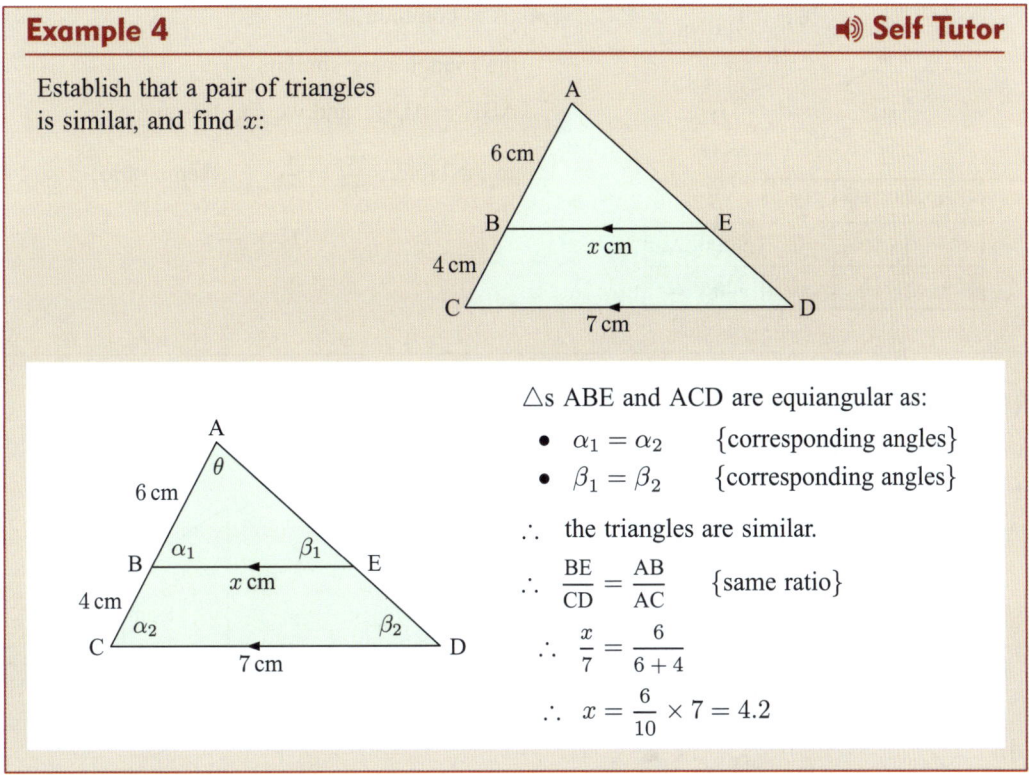

$\triangle$s ABE and ACD are equiangular as:
- $\alpha_1 = \alpha_2$ {corresponding angles}
- $\beta_1 = \beta_2$ {corresponding angles}

$\therefore$ the triangles are similar.

$\therefore \dfrac{BE}{CD} = \dfrac{AB}{AC}$ {same ratio}

$\therefore \dfrac{x}{7} = \dfrac{6}{6+4}$

$\therefore x = \dfrac{6}{10} \times 7 = 4.2$

When solving similar triangle problems, it may be useful to use a table. Consider the following method, written in the context of the Example above:

Step 1: Label equal angles.
Step 2: Show that the triangles are equiangular, and hence similar.
Step 3: Put the information in a table, showing the equal angles and the side lengths *opposite* these angles.
Step 4: Use the columns to write down the equation for the ratio of the corresponding sides.
Step 5: Solve the equation.

α	β	θ	
-	6	x	small $\triangle$
-	10	7	large $\triangle$

from which $\dfrac{6}{10} = \dfrac{x}{7}$

$\therefore x = 4.2$

Example 5

Establish that a pair of triangles is similar, and hence find x.

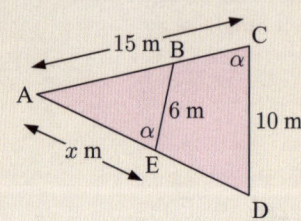

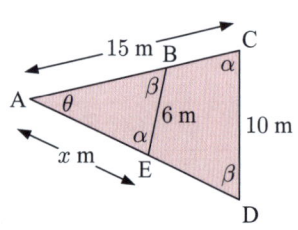

△s ABE and ADC are equiangular since:

- $\widehat{AEB} = \widehat{ACD}$ {given}
- $\widehat{A}$ is common

∴ the triangles are similar.

∴ $\widehat{ABE} = \widehat{ADC}$, and we call this angle β.

α	β	θ	
-	x	6	small △
-	15	10	large △

Using the table, $\dfrac{x}{15} = \dfrac{6}{10}$ {same ratio}

∴ $x = 15 \times \dfrac{6}{10}$

∴ $x = 9$

EXERCISE 23B.2

1 In each figure, establish that a pair of triangles is similar. Hence find x.

a 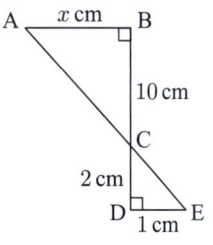 **b** 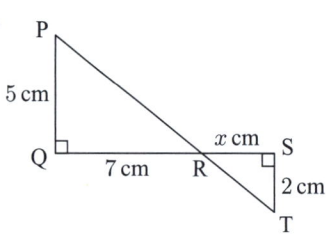 **c**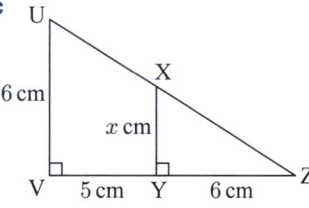

d 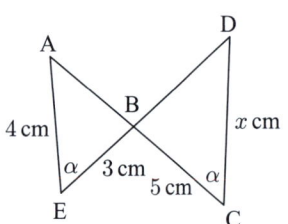 **e** 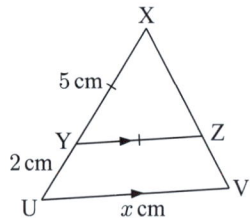 **f**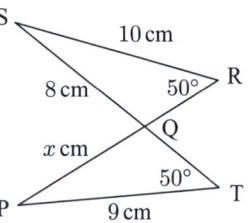

2 a Show that △ABC is similar to △EDC.
   b Find x.
   c Find the area of each triangle.

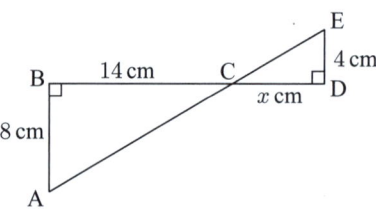

3 Given the figure alongside, find the length of BE.

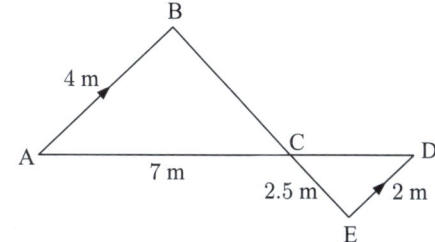

C PROBLEM SOLVING

The properties of similar triangles have been known since ancient times. However, even with the technologically advanced measuring instruments available today, similar triangles are still important for finding heights and distances which would otherwise be difficult to measure.

Step 1: Read the question carefully. Draw a diagram showing all of the given information.

Step 2: Introduce a variable such as x, for the unknown quantity to be found.

Step 3: Establish that a pair of triangles are similar, and hence write an equation involving the variable.

Step 4: Solve the equation.

Step 5: Answer the question in a sentence.

> Diagrams are very useful. Make sure your diagrams are neat and large enough.

Example 6 ◀) Self Tutor

When a 30 cm stick is stood vertically on the ground, it casts a 24 cm shadow. At the same time a man casts a shadow of length 152 cm. How tall is the man?

The sun shines at the same angle on both the stick and the man. We suppose this is angle $\alpha°$ to the horizontal.
Let the man be h cm tall.

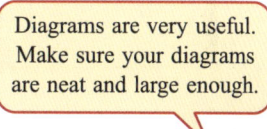

The triangles are equiangular and therefore similar.

$\therefore \ \dfrac{h}{30} = \dfrac{152}{24}$ {same ratio}

$\therefore \ h = \dfrac{152}{24} \times 30$

$\therefore \ h = 190$

The man is 190 cm tall.

$\alpha°$	$(90-\alpha)°$	$90°$	
h cm	152 cm	-	large △
30 cm	24 cm	-	small △

EXERCISE 23C

1 Find the height of the pine tree:

a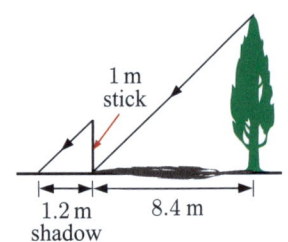

b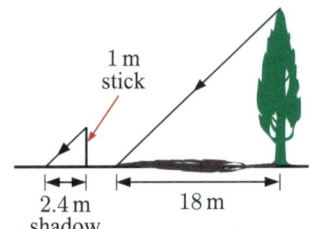

2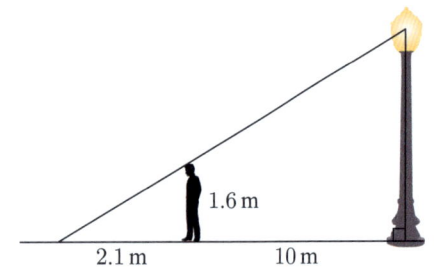

When a 1.6 m tall person stands 10 m from the base of an electric light pole, the shadow of the person is 2.1 m long.

Find the height of the globe above ground level.

3 A ramp is built to enable wheelchair access to a building that is 24 cm above ground level. The ramp has a constant slope of 2 in 15, which means that for every 15 cm horizontally it rises 2 cm. Calculate the length of the base of the ramp.

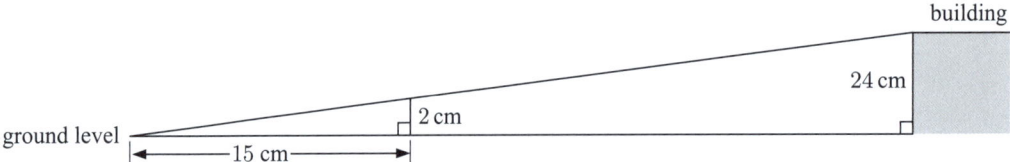

4 A piece of timber rests against both the top of a fence and the wall behind it, as shown.

a Find how far up the wall the timber reaches.
b Find the length of the timber.

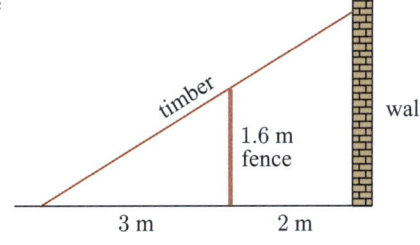

5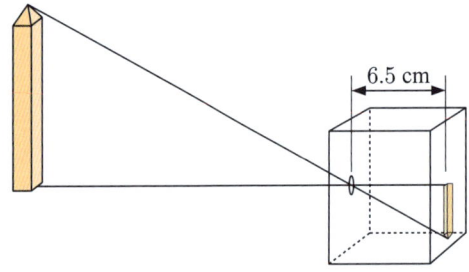

A pinhole camera displays an image on a screen as shown alongside. The monument shown is 21 m tall, and its image is 3.5 cm high. The distance from the pinhole to the image is 6.5 cm.

How far is the pinhole from the monument?

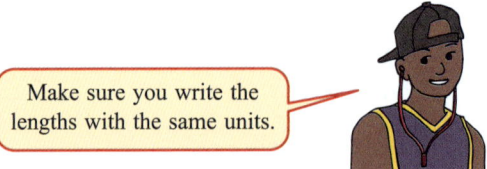

Make sure you write the lengths with the same units.

6 A, B, C, and D are pegs on the bank of a canal which has parallel straight sides. C and D are directly opposite each other. AB = 30 m and BC = 140 m.

When I walk from A directly away from the bank, I reach a point E, 25 m from A, where E, B, and D line up.

How wide is the canal?

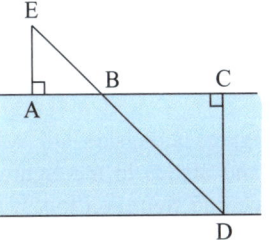

Activity 1 — Paper folding

Click on the icon to obtain this Activity.

D CONGRUENT FIGURES

> Two figures are **congruent** if they are identical in size and shape. They do not need to have the same orientation.

For example, the figures alongside are congruent even though one is a rotation of the other.

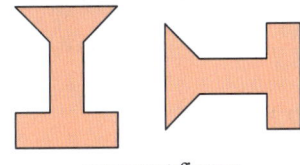

congruent figures

Activity 2 — Creating congruent figures

You will need: Two sheets of card, scissors.

What to do:

1 Draw a shape on one of the sheets of card.

2 Place the second sheet of card behind it, and hold them together tightly. Carefully cut out the shape, cutting through both sheets of card. This will give you two congruent figures.

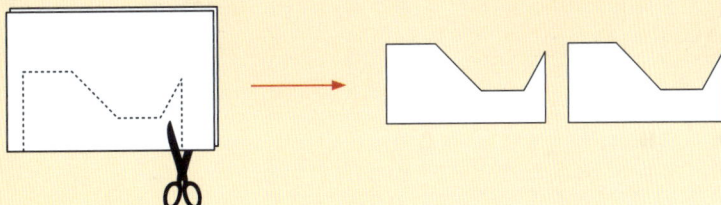

3 In a group or as a class, place both figures from each student in a box, and mix the figures up.

4 Try to pair up the congruent figures. How can you tell that two figures are congruent? What features are you looking for?

Two figures are **congruent** if:
- the figures are **equiangular**   *and* • the corresponding side lengths are equal.

The figures alongside are congruent. The corresponding sides and angles in the figures are identical. If we were to place one figure on top of the other, they would match each other perfectly.

Example 7   ◀)) **Self Tutor**

Are the following pairs of figures congruent?

a   **b**   **c**

a The figures do not have the same shape, so they are *not congruent*.
b The figures are identical in size and shape even though one is rotated. They are therefore *congruent*.
c The figures have the same shape, but they are not the same size. They are therefore *not congruent*.

EXERCISE 23D

1 Are the following pairs of figures congruent?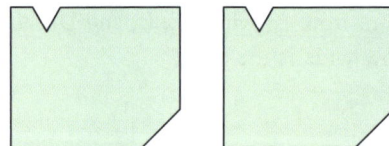

a

b

c

d

2 Which two of these figures are congruent?

A   B   C   D   E

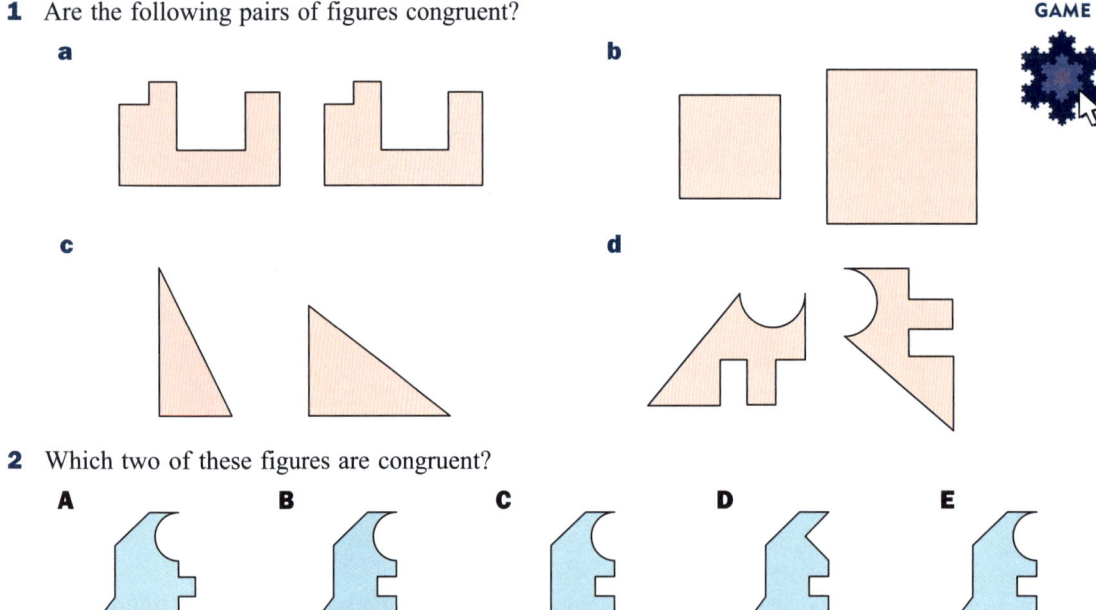

Similarity and congruence (Chapter 23) 475

3 Which of the figures below is congruent to the figure alongside?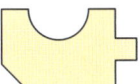

A 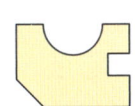   B    C    D

4 Quadrilaterals EFGH and ABCD are congruent.

Determine the:
- **a** length of side EF
- **b** size of angle $F\hat{G}H$
- **c** perimeter of EFGH.

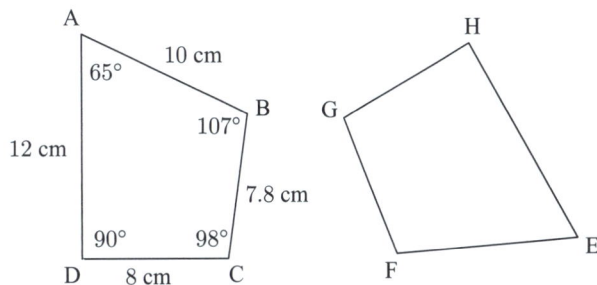

E   CONGRUENT TRIANGLES

The triangles alongside have identical side lengths and angles, so the triangles are **congruent**.

However, do we necessarily need *all* of the information given to conclude that the triangles are congruent?

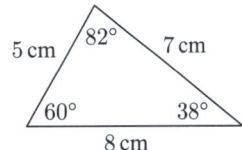 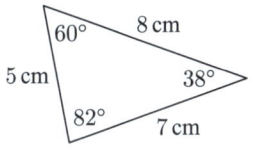

Investigation — Constructing triangles

In this Investigation we will discover conditions which allow us to conclude that two triangles are congruent.

You will need: Paper, ruler, protractor.

What to do:

1 **Three sides**
Draw a triangle with side lengths 5 cm, 8 cm, and 10 cm. How many different triangles can you construct?

2 **Two sides and an included angle**
Draw a triangle with two side lengths 8 cm and 12 cm, with an angle of 25° between these sides. How many different triangles can you construct?

3 **Two angles and a corresponding side**
Draw a triangle with two angles measuring 70° and 45°, with the side between these angles being 10 cm long. How many different triangles can you construct?

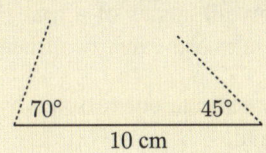

4 Right angle, hypotenuse, and a side

Draw a right angled triangle with hypotenuse 10 cm, and one other side 6 cm long. How many different triangles can you construct?

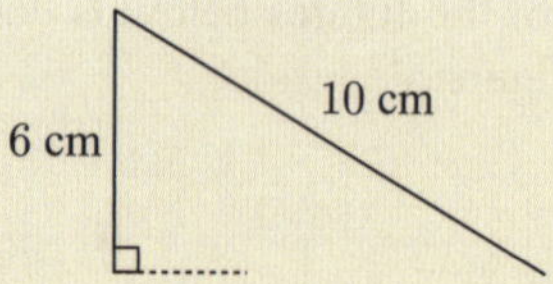

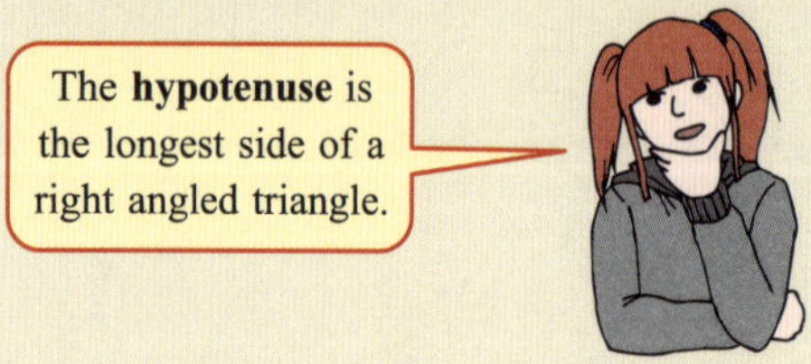

5 Two sides and a non-included angle

Draw a triangle with two side lengths 8 cm and 12 cm, with an angle of 25° between the 12 cm side and the third side as shown. How many different triangles can you construct?

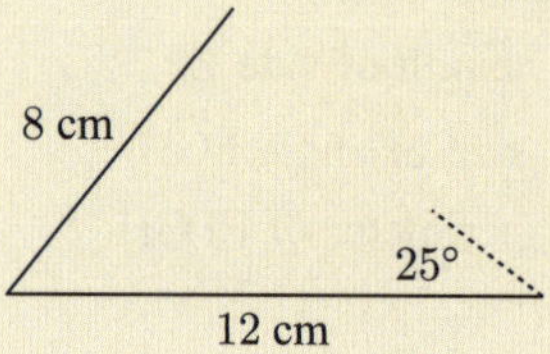

6 Three angles

Draw a triangle with angles 50°, 60°, and 70°. How many different triangles can you construct?

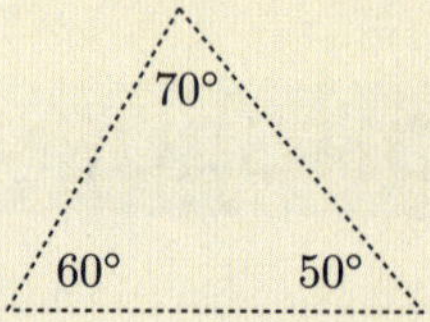

You should have made the following discoveries:

Two triangles are **congruent** if any one of the following is true:

- All corresponding sides are equal in length. **(SSS)**

- Two sides and the **included angle** are equal. **(SAS)**

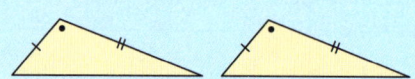

- Two angles and a pair of **corresponding sides** are equal. **(AAcorS)**

- For right angled triangles, the hypotenuses and one pair of sides are equal. **(RHS)**

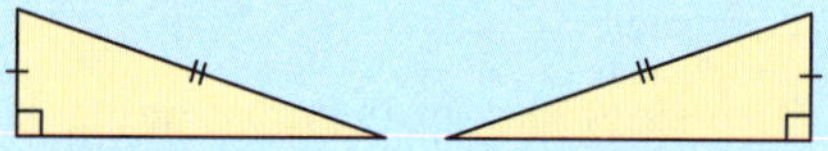

If we know two side lengths and a non-included angle, there may be two ways to construct the triangle. This is therefore *not* sufficient information to show that two triangles with these properties are congruent.

If we know all angles of a triangle, the triangle may still vary in size. This is therefore *not* sufficient information to show that two triangles with these angles are congruent.

We usually indicate our reason why two triangles are congruent by writing one of the abbreviations given above in bold.

Example 8

State whether these pairs of triangles are congruent, giving reasons for your answers.

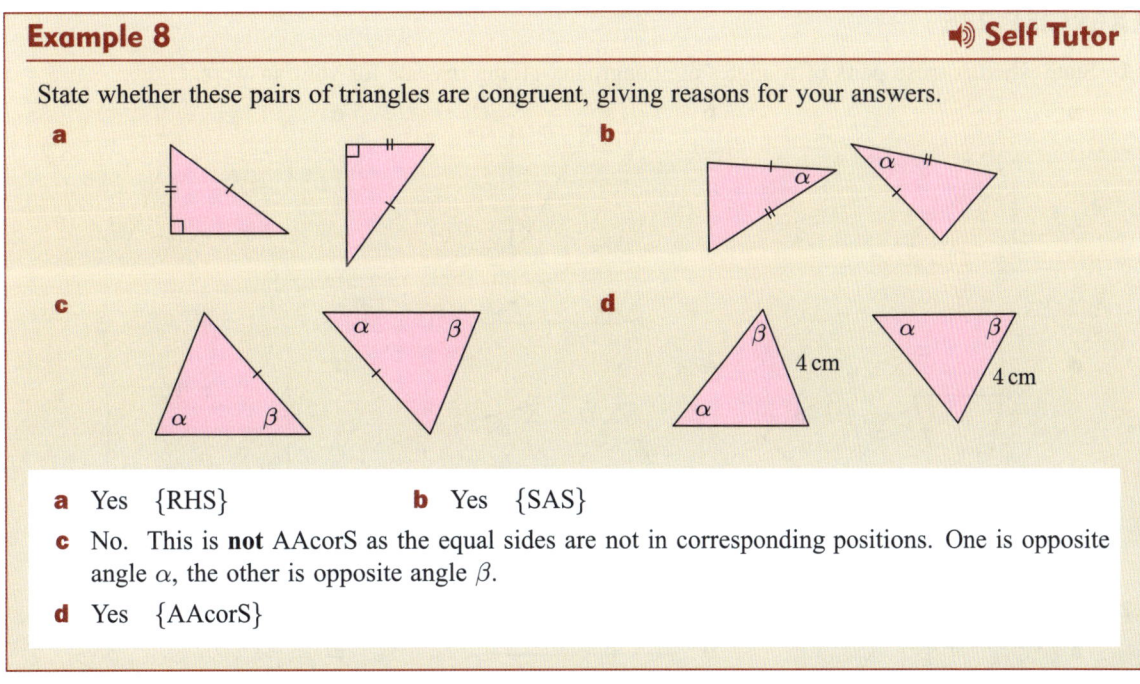

a Yes {RHS} **b** Yes {SAS}

c No. This is **not** AAcorS as the equal sides are not in corresponding positions. One is opposite angle α, the other is opposite angle β.

d Yes {AAcorS}

Once we have established that two triangles are congruent, we can deduce that the remaining corresponding sides and angles of the triangles are equal.

Example 9

a Show that these triangles are congruent.

b What can be deduced from this congruence?

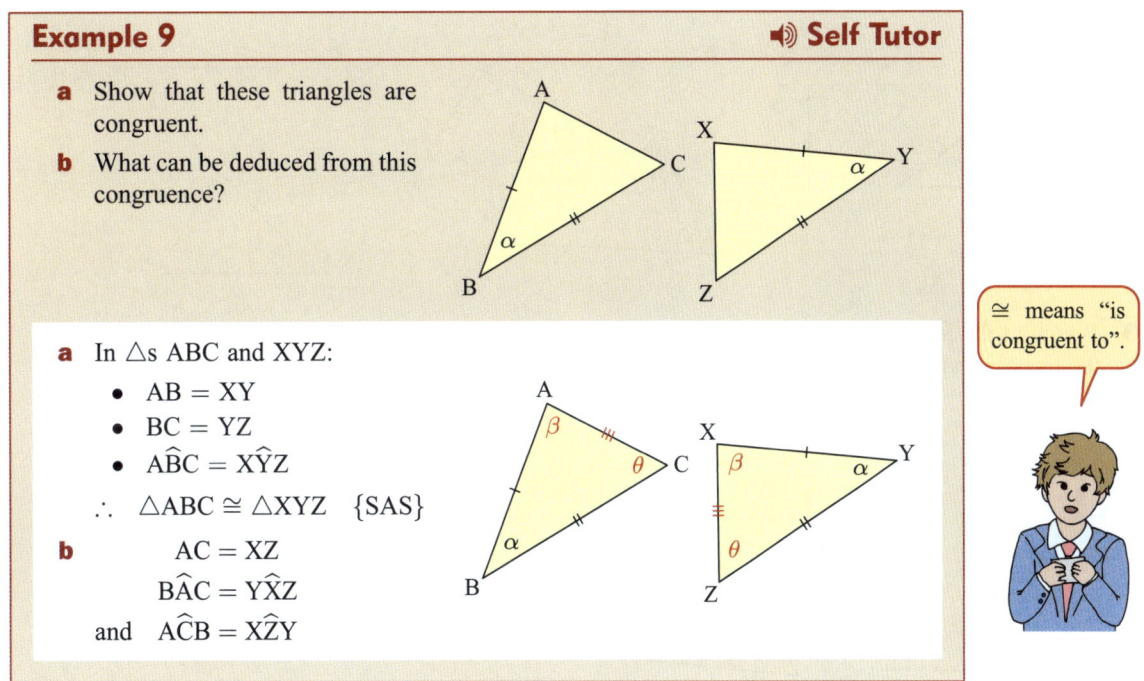

a In △s ABC and XYZ:
- AB = XY
- BC = YZ
- $A\hat{B}C = X\hat{Y}Z$

∴ △ABC ≅ △XYZ {SAS}

b AC = XZ
  $B\hat{A}C = Y\hat{X}Z$
 and $A\hat{C}B = X\hat{Z}Y$

≅ means "is congruent to".

When we describe congruent triangles, we label the vertices that are in corresponding positions in the same order. For instance, in the previous Example, we write △ABC ≅ △XYZ, not △ABC ≅ △YZX.

EXERCISE 23E

1 State whether these pairs of triangles are congruent, giving reasons for your answers:

a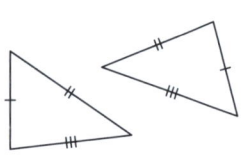
b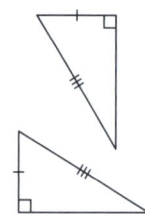
c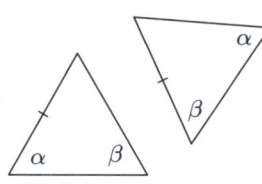

d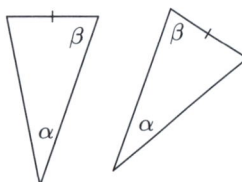
e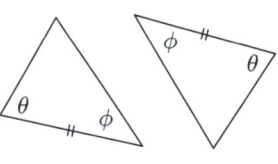
f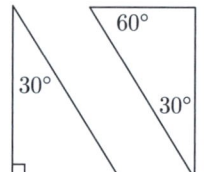

2 State whether these pairs of triangles are congruent, giving reasons for your answers:

a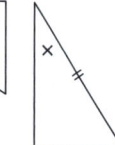
b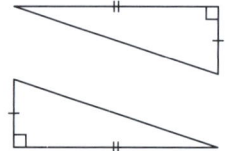
c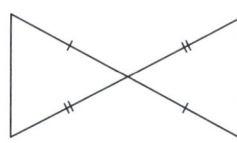

d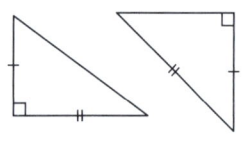
e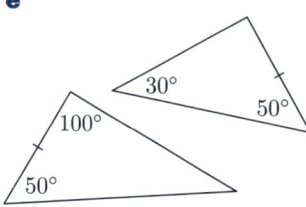
f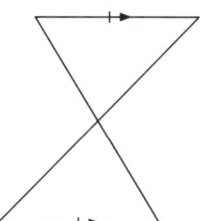

3 Which of the following triangles is congruent to the one alongside?

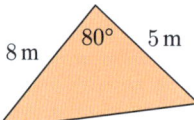

A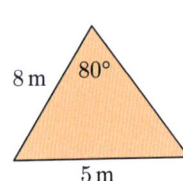
B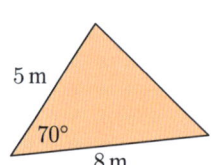
C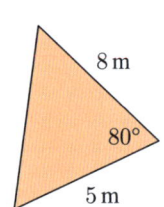
D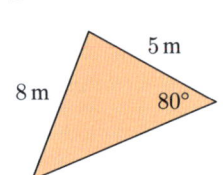

Similarity and congruence (Chapter 23) **479**

4 Which of these triangles are congruent to each other?

A 15 cm, 17 cm, α

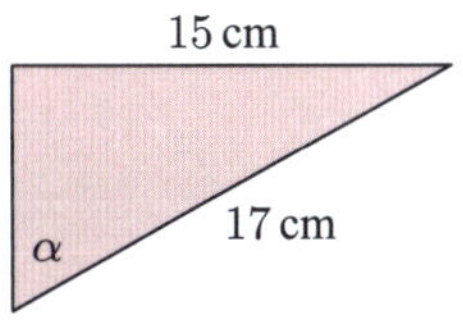

B 15 cm, 50°, right angle

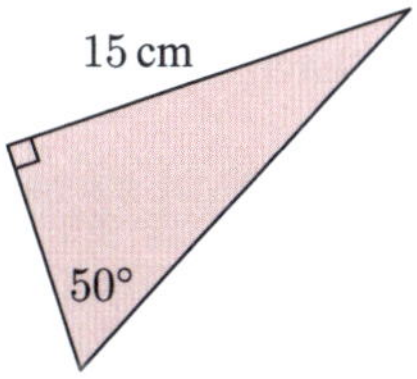

C 17 cm, 9 cm, right angle

D 15 cm, 40°, right angle

E 9 cm, 50°, 40°, right angle

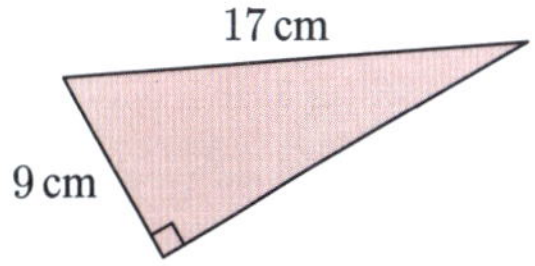

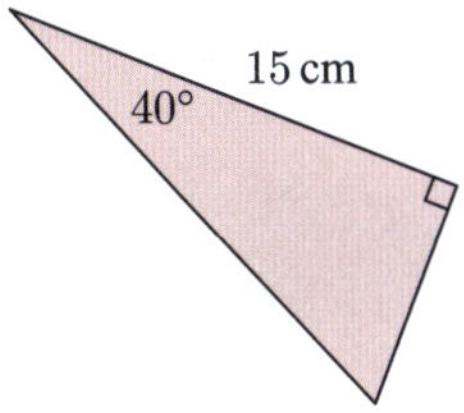

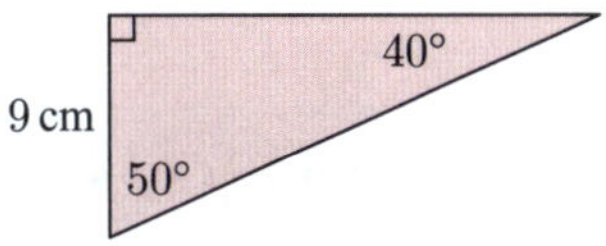

F 17 m, 15 m, 9 m, α

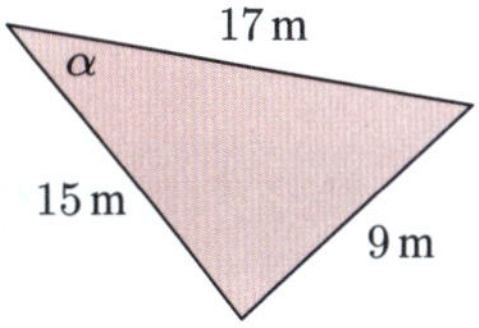

5 The following pairs of triangles are not drawn to scale, but the information on them is correct.
  i Determine whether the triangles are congruent.
  ii If the triangles are congruent, what else can we deduce about them?

a Triangle ABC with angles α at A, β at B, and tick mark on AC; Triangle PQR with β at Q, α at P, and tick mark on PR.

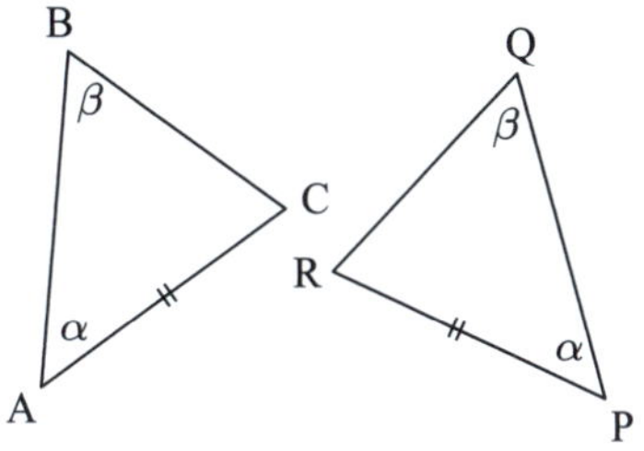

b Triangle JKL with markings; Triangle XYZ with markings.

c Triangle DEF with right angle at F; Triangle PQR with right angle at Q.

d Triangle RST with β at T; Triangle XYZ with β at Z.

e Triangles ABC and EDC sharing vertex C.

f Triangles PQT and RQS sharing vertex Q.

g Triangle ABC with α at B; Triangle DEF with α at F.

h Triangle DEF with α at F; Triangle XYW with α at Y.

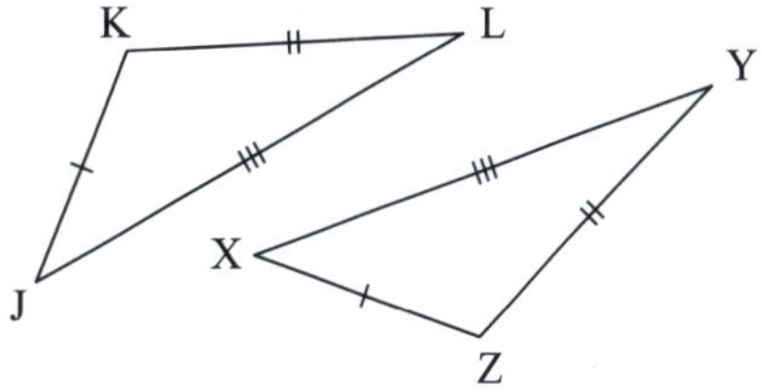

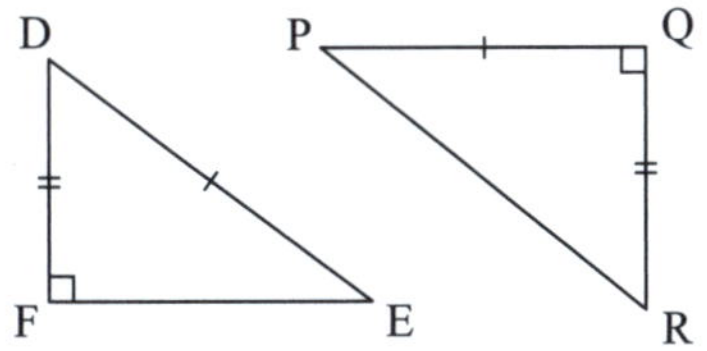

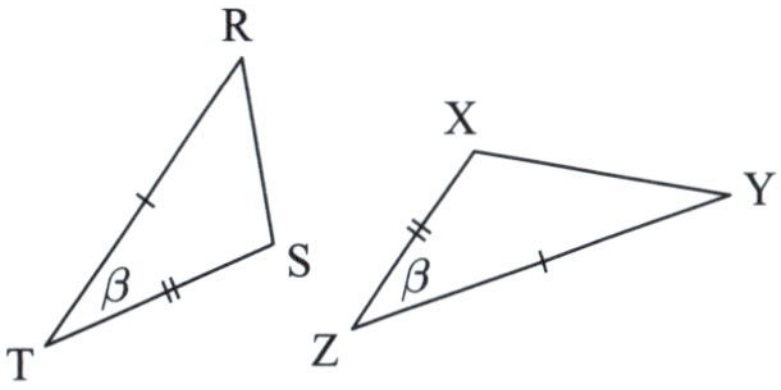

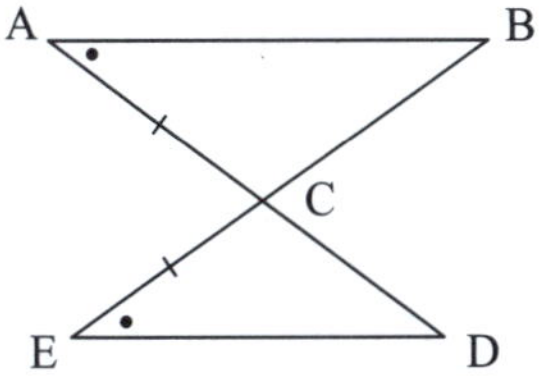

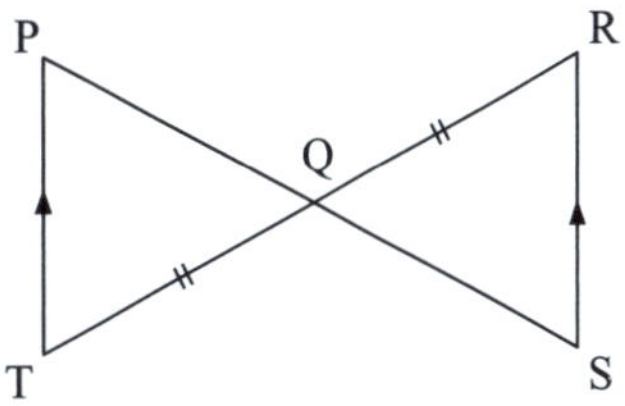

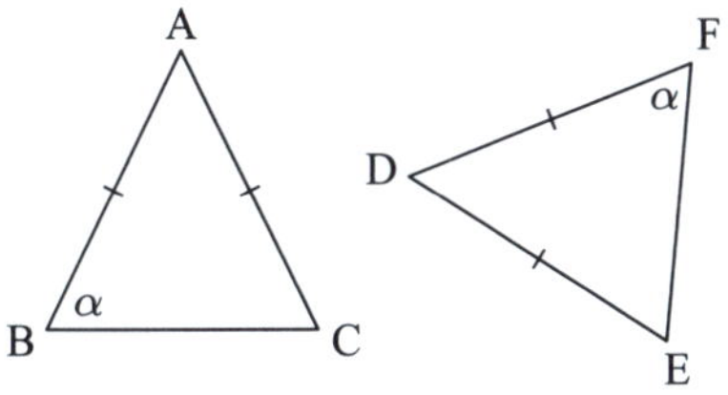

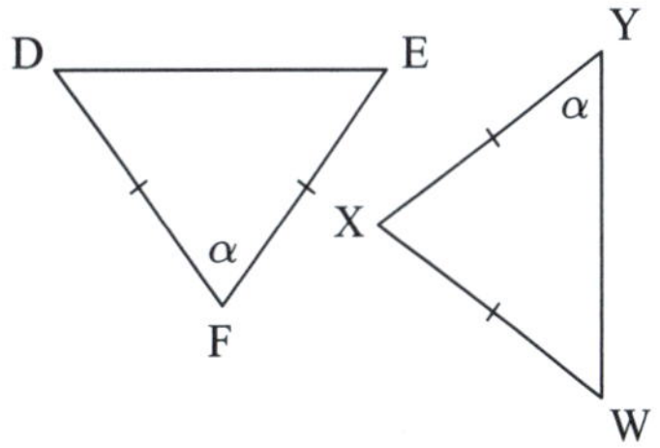

Discussion

We have seen that if two *triangles* have equal corresponding sides, then they are congruent.

Is the same true for *quadrilaterals*? Can we say that the quadrilaterals alongside are congruent?

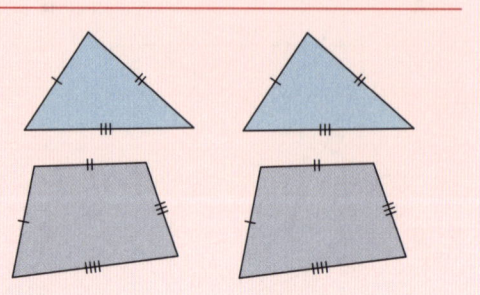

F PROOF USING CONGRUENCE

In **Chapter 11**, we studied the properties of isosceles triangles and special quadrilaterals. We can use congruence to prove many of these properties.

Example 10 ◀)) Self Tutor

Consider the isosceles triangle ABC.

M is the midpoint of BC.

a Use congruence to show that $B\hat{A}M = C\hat{A}M$.

b What property of isosceles triangles has been proven?

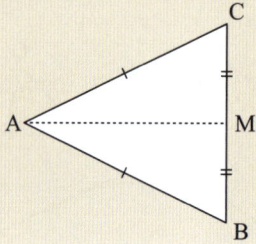

a In triangles ABM and ACM:
- AB = AC   {△ABC is isosceles}
- BM = CM   {M is the midpoint of BC}
- AM is common to both triangles.
- ∴ △ABM ≅ △ACM   {SSS}

Equating corresponding angles, $B\hat{A}M = C\hat{A}M$. (•)

b In any isosceles triangle, the line joining the apex to the midpoint of the base bisects the vertical angle.

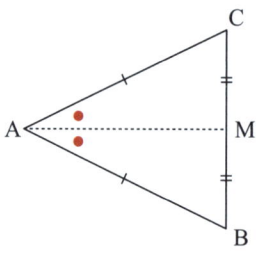

EXERCISE 23F

1 Consider the parallelogram ABCD.

a Copy and complete:

In triangles ABD and CDB:
- $A\hat{D}B$ = ...   {equal alternate angles}
- $A\hat{B}D$ = ...   {equal alternate angles}
- BD is common to both triangles
- ∴ △ABD ≅ △CDB   {......}

Equating corresponding angles, $D\hat{A}B$ = ...

b What property of parallelograms has been proven in **a**?

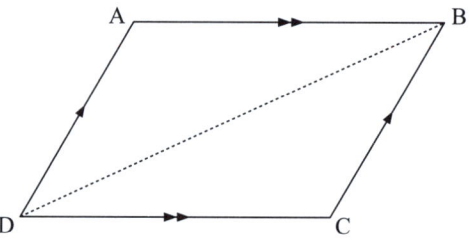

2 Consider the kite PQRS.
   a Show that △PQR ≅ △PSR.
   b Hence show that $Q\hat{P}R = S\hat{P}R$ and $Q\hat{R}P = S\hat{R}P$.
   c What property of kites has been proven?

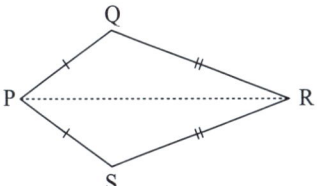

3 Consider the square ABCD.
   a Show that △ABC ≅ △DAB.
   b Hence, show that AC = DB.
   c What property of squares has been proven?

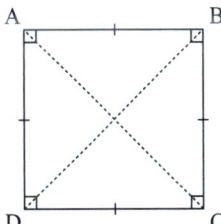

4 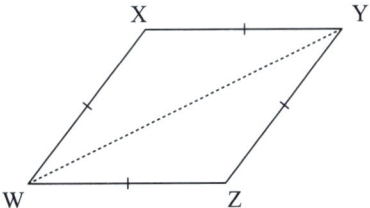 Consider the rhombus WXYZ.
   a Show that △WXY ≅ △YZW.
   b Hence, show that $X\hat{Y}W = Z\hat{W}Y$.
   c Hence, show that XY is parallel to WZ.
   d Likewise, show that XW is parallel to YZ.
   e What property of rhombuses has been proven?

Review set 23A

1 Determine whether these rectangles are similar.

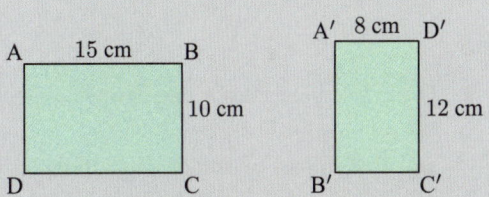

2 Are all rhombuses similar? Explain your answer.

3 Find x:

   a 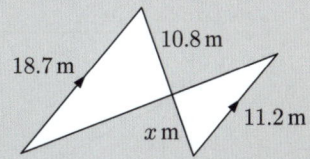 **b**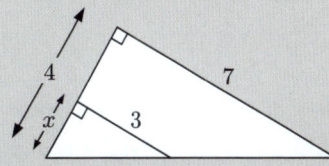

4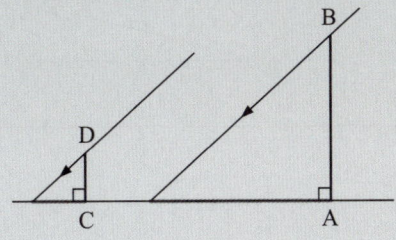

AB is a vertical flagpole of unknown height.
CD is a vertical stick 1.4 m long.
When the shadow of the flagpole is 12.3 m long, the shadow of the stick is 1.65 m long.
Find, rounded to 3 significant figures, the height of the flagpole.

5 **a** Explain why triangles ABE and ACD are similar.
   b Find the length of CD given that BE = 4 cm.

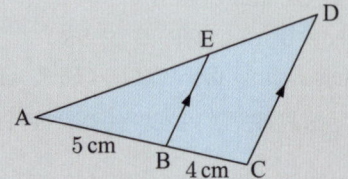

6 State whether each pair of figures is congruent:

   a    **b**    **c**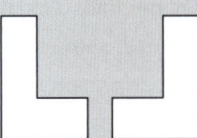

7 State whether each pair of triangles is congruent, giving reasons for your answers.

   a 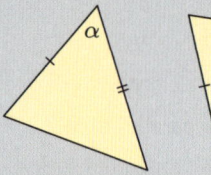   **b**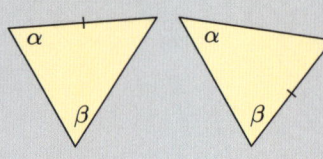

8 Consider the kite ABCD.
   a Show that △ABC ≅ △ADC.
   b Hence show that $A\hat{B}C = A\hat{D}C$.
   c What property of kites has been proven?

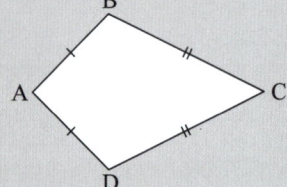

9 These triangles are not drawn to scale, but the information on them is correct.
   a Determine whether the triangles are congruent.
   b If the triangles are congruent, what can be deduced from the congruence?

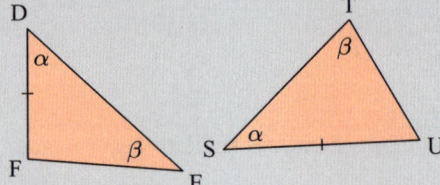

10 A square piece of paper is divided into four triangles as shown.
   a Show that triangles A and D are congruent.
   b Hence show that triangles B and C are congruent.
   c Find the area of each triangle.

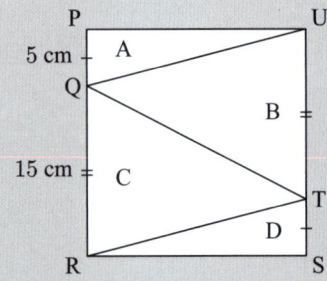

Review set 23B

1 Find x, given that the figures are similar:

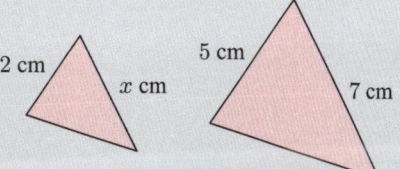

2 Which two of these figures are congruent?

   A   **B**   **C**   **D**

 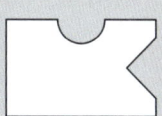 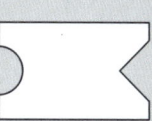 

3 Show that the following figures possess similar triangles:

  a   **b**   **c**

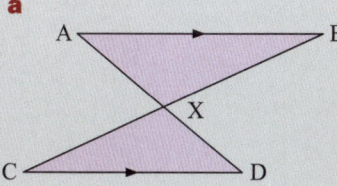 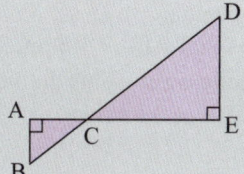 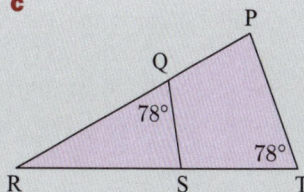

4

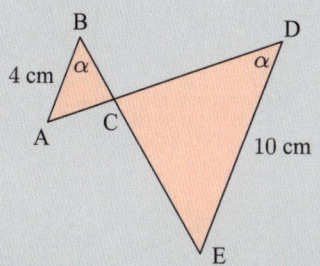

  a Show that $\triangle ABC$ is similar to $\triangle EDC$.
  b Given that CD = 8 cm, find the length of CB.

5 State whether these pairs of triangles are congruent, giving reasons for your answers:

  a   **b**   **c**

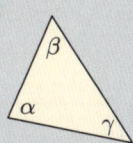 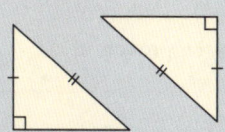 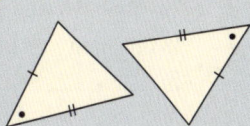

6 In the isosceles triangle PQR, PX is perpendicular to the base QR.

  a Show that $\triangle PQX \cong \triangle PRX$.
  b Hence show that $P\hat{Q}X = P\hat{R}X$.
  c What property of isosceles triangles has been proven?

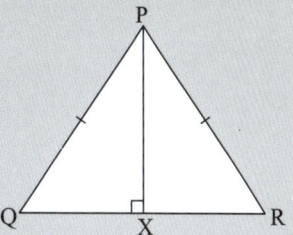

7 Establish that a pair of triangles is similar, and hence find x:

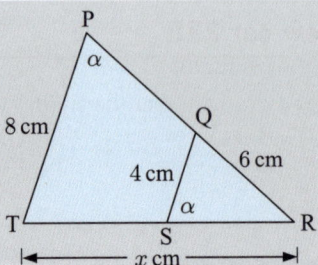

8

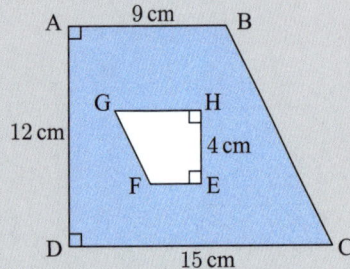

The trapeziums in the diagram alongside are similar.

a Find the length of:
    i EF   **ii** GH

b Find the blue shaded area.

9 P and Q are markers on the bank of a canal which has parallel sides. R and S are telegraph poles which are directly opposite each other. PQ = 30 m and QR = 100 m.
When I walk 20 m from P directly away from the bank, I reach the point T such that T, Q, and S line up.
How wide is the canal?

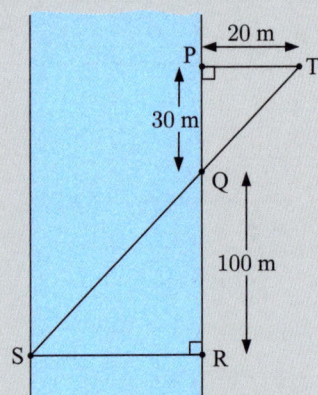

10 Consider the kite ABCD alongside.

  a Show that $\triangle ABC \cong \triangle ADC$.

  b Hence show that $B\hat{A}C = D\hat{A}C$.

  c Explain why $\triangle ABX \cong \triangle ADX$.

  d Show that BX and DX have the same length.

  e What property of kites has been proven in **d**?

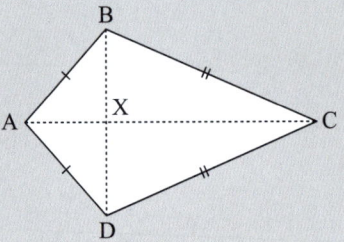

24

Trigonometry

Contents:

- **A** Scale diagrams in geometry
- **B** Labelling right angled triangles
- **C** The trigonometric ratios
- **D** Finding side lengths
- **E** Finding angles
- **F** Problem solving with trigonometry
- **G** The first quadrant of the unit circle

Opening problem

Sometimes it is difficult or even impossible to measure angles, heights, and distances directly.

Suppose you are in front of a building which has a flagpole on top.

You have no way of measuring the height of the flagpole directly, but what you *can* do is measure the angle from the horizontal to the base and to the top of the flagpole.

So, you make two measurements from point A, which is 40 m from the base of the building.

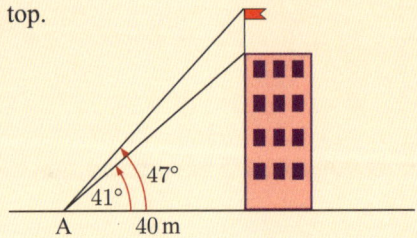

Things to think about:
- **a** How could you use these measurements to *estimate* the flagpole's height?
- **b** How accurate would you expect your estimation to be?
- **c** Is there a mathematical method for calculating the height of the flagpole to greater accuracy?

> **Trigonometry** is the study of the relationship between lengths and angles of a triangle.

We use trigonometry alongside algebra and geometry to find unknown lengths and angles of triangles.

A SCALE DIAGRAMS IN GEOMETRY

Scale diagrams can be used to **estimate** side lengths and angles in geometrical figures.

Example 1   ◉ Self Tutor

Jake has been contracted to install lights to shine on the front of a hotel. He wants to install the lights at ground-level at a point A on the near side of a drain. The drain is 36.5 m from the base of the hotel. From this point, the angle up to the top of the hotel is 50°. How high is the hotel?

We choose a suitable *scale*, in this case 1 mm ≡ 1 m.

We draw a horizontal line segment BA 36.5 mm long, and from point B draw a vertical line.

We then use a protractor to draw a 50° angle at A, and extend the line to meet the vertical. The point C, at the top of the hotel, is where the two lines meet.

Using a ruler, BC ≈ 43.5 mm

∴ the building is approximately 43.5 metres high.

EXERCISE 24A

1 **a** Convert this rough sketch into an accurate scale diagram. Use a scale of 1 cm ≡ 1 m.
   b Use your scale diagram to find the actual length of:
      i BC   **ii** AC

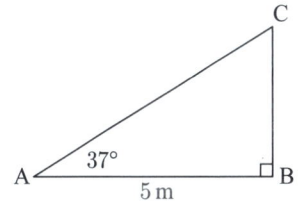

2 Use a scale diagram with scale 1 cm ≡ 10 m to find the height of the tree.

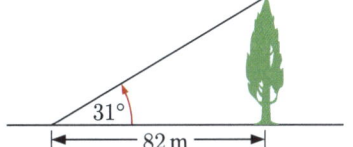

3 The triangular garden ABC has AB = 8 m, BC = 7.2 m, and AC = 5.9 m.
   a Draw a scale diagram of the garden with scale 1 cm ≡ 1 m.
   b Use your diagram to estimate the measures of the garden's angles.

You must use both your ruler and protractor.

4 Use a scale diagram to estimate the height of the flagpole in the **Opening Problem**.

Discussion

What are the most likely causes of error when using scale diagrams?

How accurate are the answers when using scale diagrams?

B LABELLING RIGHT ANGLED TRIANGLES

While scale diagrams allow us to estimate the side lengths and angles of triangles, our estimates may not be sufficiently accurate, especially if the scale diagram is much smaller than the actual situation. We can use **trigonometry** to calculate these triangle properties more accurately.

Before we can perform right angled triangle trigonometry, we need to label the sides of the triangle in a systematic way.

The **hypotenuse (HYP)** is the longest side of a right angled triangle, and is opposite the right angle.

For a given angle θ, the **opposite (OPP)** side is the side opposite the angle θ.

The remaining side is next to the angle θ, and so is called the **adjacent (ADJ)** side.

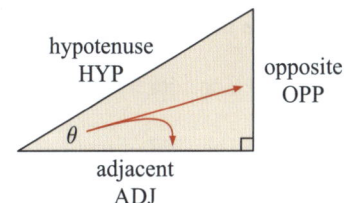

Example 2   🔊 Self Tutor

For the triangle given, name the:
 a hypotenuse
 b side opposite θ
 c side adjacent to θ.

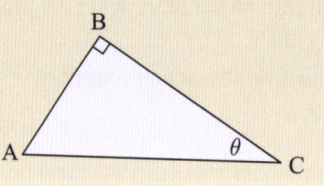

Locate the hypotenuse first. Then locate the opposite and adjacent sides for the angle you are working with.

a The hypotenuse is AC.
b The side opposite θ is AB.
c The side adjacent to θ is BC.

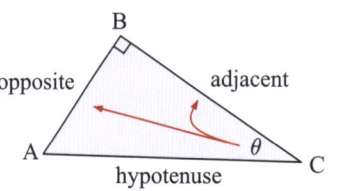

EXERCISE 24B

1 For each diagram below, name the:

   i hypotenuse   ii side opposite angle θ iii side adjacent to angle θ.

   a 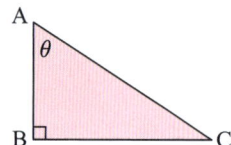   b 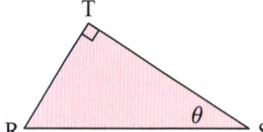   c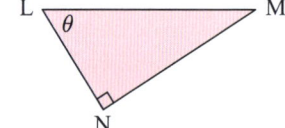

2 The hypotenuse of the right angled triangle shown has length a units. The other sides have lengths b units and c units. θ and ϕ are the two acute angles.
Find the length of the side:

 a opposite θ b adjacent to θ
 c opposite ϕ d adjacent to ϕ.

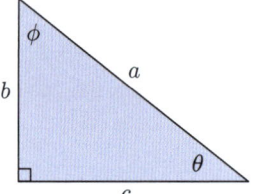

C THE TRIGONOMETRIC RATIOS

Having agreed on a consistent way to label right angled triangles, we can define three basic **trigonometric ratios** as the ratios of the side lengths.

Consider a right angled triangle with one angle θ.

The **sine** of angle θ is $\quad \sin\theta = \dfrac{\text{OPP}}{\text{HYP}}$.

The **cosine** of angle θ is $\quad \cos\theta = \dfrac{\text{ADJ}}{\text{HYP}}$.

The **tangent** of angle θ is $\quad \tan\theta = \dfrac{\text{OPP}}{\text{ADJ}}$.

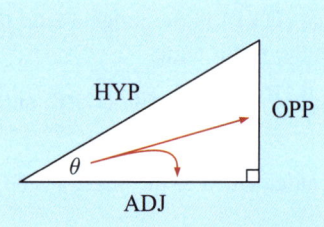

Investigation 1 — Trigonometric ratios

In this Investigation we explore the trigonometric ratios for triangles which are *similar*.

Consider the right angled triangle alongside.
It contains one angle of 39°.

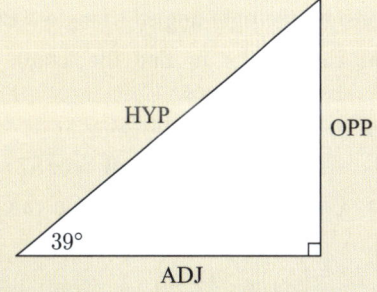

What to do:

1. Use a ruler to check that:
   - the hypotenuse (HYP) is 5.4 cm long
   - the side opposite the 39° angle (OPP) is 3.4 cm long
   - the side adjacent to the 39° angle (ADJ) is 4.2 cm long.

2. Copy and complete this table:

HYP	OPP	ADJ	$\dfrac{\text{OPP}}{\text{HYP}}$	$\dfrac{\text{ADJ}}{\text{HYP}}$	$\dfrac{\text{OPP}}{\text{ADJ}}$
5.4	3.4	4.2	$\dfrac{3.4}{5.4} \approx 0.63$		

3. **a** Use a ruler and protractor to construct a different right angled triangle with one angle of 39°.
   b Measure the sides of the triangle, and repeat step **2**. Comment on your results.
   c Compare your results with those of your classmates. Does the size of the triangle affect the trigonometric ratios?

4. Consider the triangles ABC and A'B'C'.

   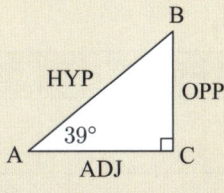 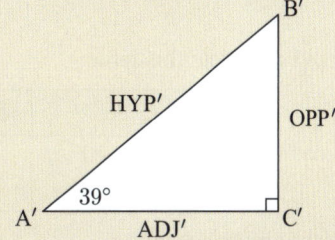

   a Show that the triangles are similar.
   b Hence, show that:

   i $\dfrac{\text{OPP}}{\text{HYP}} = \dfrac{\text{OPP}'}{\text{HYP}'}$ **ii** $\dfrac{\text{ADJ}}{\text{HYP}} = \dfrac{\text{ADJ}'}{\text{HYP}'}$ **iii** $\dfrac{\text{OPP}}{\text{ADJ}} = \dfrac{\text{OPP}'}{\text{ADJ}'}$

You should have discovered that for any right angled triangle with one angle 39°, the trigonometric ratios are constant. In particular, $\sin 39° \approx 0.63$, $\cos 39° \approx 0.78$, $\tan 39° \approx 0.81$.

Trigonometric ratios can be found using the software provided or a calculator. You should first check that your calculator is in DEGREE mode.

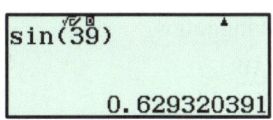  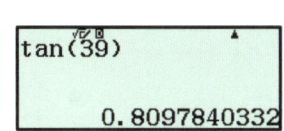

TRIGONOMETRIC RATIOS

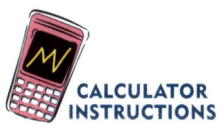

CALCULATOR INSTRUCTIONS

EXERCISE 24C

1 Use your calculator to evaluate, correct to 2 decimal places:
  a $\cos 18°$
  b $\sin 66°$
  c $\tan 23°$

2 Consider the right angled triangle ABC given.
  a Use a ruler to find the length of each side, rounded to 1 decimal place.
  b Hence estimate the value of:
   i $\sin 57°$
   ii $\cos 57°$
   iii $\tan 57°$
  c Check your answers using a calculator.

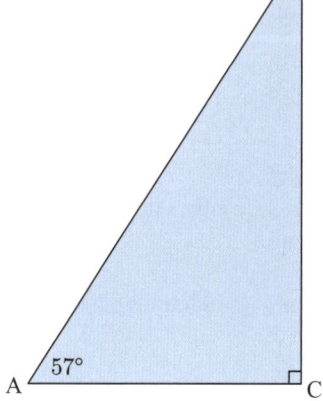

$\sin \theta = \dfrac{\text{OPP}}{\text{HYP}}$
$\cos \theta = \dfrac{\text{ADJ}}{\text{HYP}}$
$\tan \theta = \dfrac{\text{OPP}}{\text{ADJ}}$

3 a Use Pythagoras' theorem to find the length of the hypotenuse of this triangle. Round your answer to 2 decimal places.
  b Hence estimate the value of:
   i $\sin 28°$
   ii $\cos 28°$
   iii $\tan 28°$
  c Check your answers using a calculator.

4 a Copy and complete this table:

θ	$90° - \theta$	$\sin \theta$	$\cos \theta$	$\sin(90° - \theta)$	$\cos(90° - \theta)$
70°	20°				
35°	55°				

Comment on your results.

  b Use the diagram alongside to show that, for all $0° < \theta < 90°$:
   i $\sin(90° - \theta) = \cos \theta$
   ii $\cos(90° - \theta) = \sin \theta$.

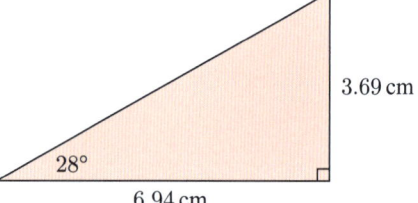

5 a Use your calculator to find:
   i $\sin 52°$
   ii $\cos 52°$
   iii $\dfrac{\sin 52°}{\cos 52°}$
   iv $\tan 52°$
  b Explain why $\tan \theta = \dfrac{\sin \theta}{\cos \theta}$.

6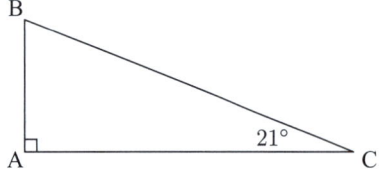

Use your calculator to find the following ratios:
  a $\dfrac{AB}{BC}$
  b $\dfrac{AC}{BC}$
  c $\dfrac{AB}{AC}$

Discussion

Suppose θ is one angle of a right angled triangle.

What range of possible values can $\sin \theta$ and $\cos \theta$ take?

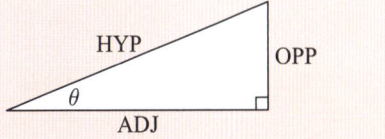

D FINDING SIDE LENGTHS

If we know the angles of a right angled triangle, the trigonometric ratios give us the ratios of its side lengths.

For example, in the triangle alongside, we know that $\dfrac{PQ}{PR} = \cos 36° \approx 0.809$.

So, if we also know one of the side lengths, we can use the trigonometric ratios to find the other side lengths.

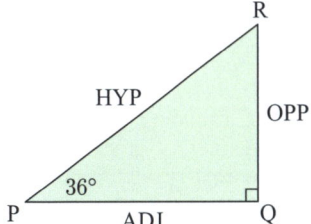

> *Step 1*: Redraw the figure and mark on it HYP, OPP, and ADJ relative to a given angle.
> *Step 2*: Choose an appropriate trigonometric ratio, and construct an equation.
> *Step 3*: Solve the equation to find the unknown side length.

Example 3   ◆) Self Tutor

Find x, rounding your answer to 2 decimal places:

a (triangle with 7 cm hypotenuse, 26° angle, x cm adjacent)

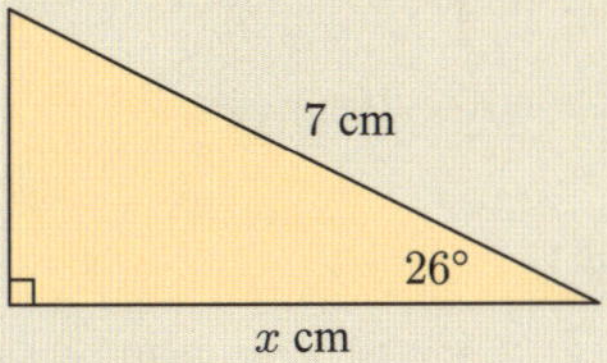

b (x m, 61° angle, 5 m opposite)

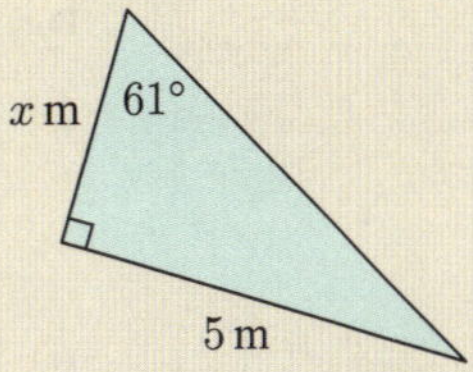

a 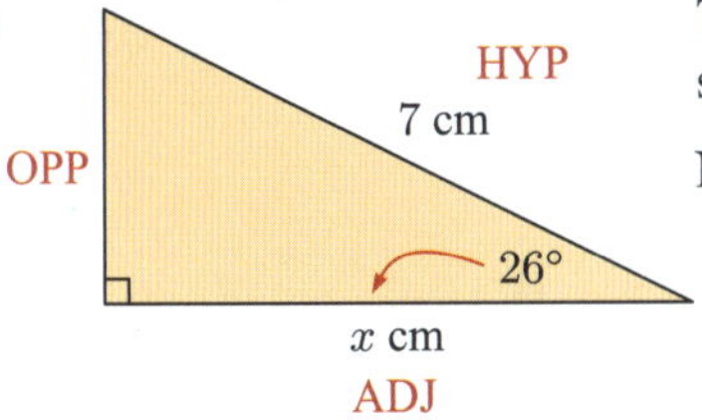 The relevant sides are ADJ and HYP, so we use the *cosine* ratio.

Now $\cos 26° = \dfrac{x}{7}$ $\{\cos \theta = \dfrac{\text{ADJ}}{\text{HYP}}\}$

$\therefore x = 7 \times \cos 26°$

$\therefore x \approx 6.29$

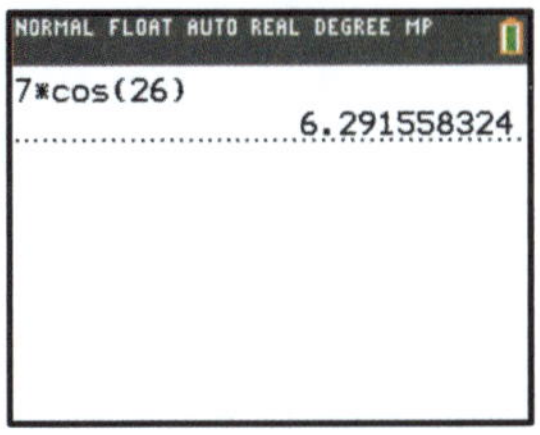

```
NORMAL FLOAT AUTO REAL DEGREE MP
7*cos(26)
                          6.291558324
```

b 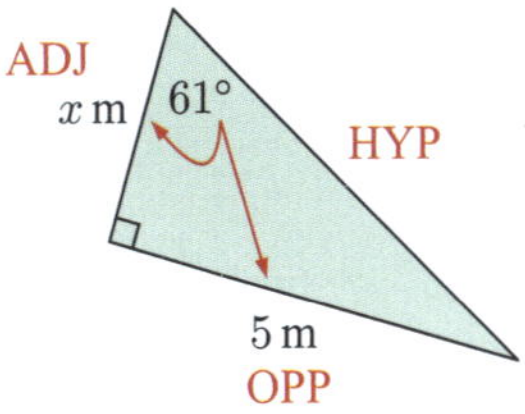 The relevant sides are OPP and ADJ, so we use the *tangent* ratio.

Now $\tan 61° = \dfrac{5}{x}$ $\{\tan \theta = \dfrac{\text{OPP}}{\text{ADJ}}\}$

$\therefore x \times \tan 61° = 5$

$\therefore x = \dfrac{5}{\tan 61°}$

$\therefore x \approx 2.77$

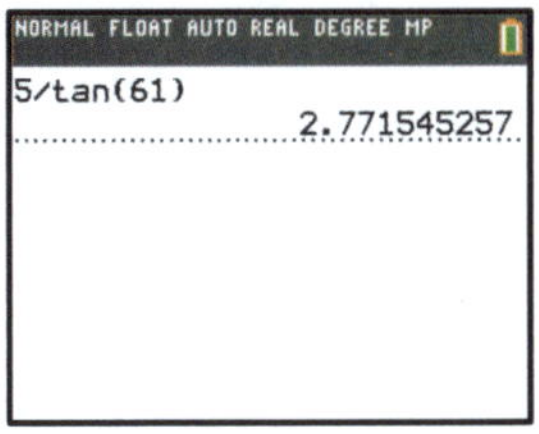

```
NORMAL FLOAT AUTO REAL DEGREE MP
5/tan(61)
                          2.771545257
```

EXERCISE 24D

1 Write a trigonometric equation connecting the angle and the sides given:

a 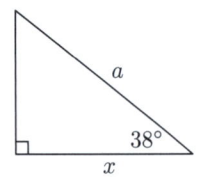   b 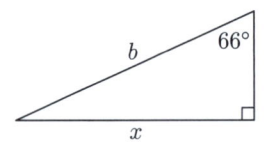   c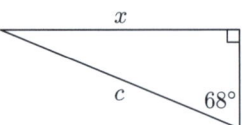

d 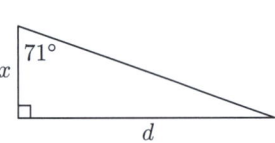   e 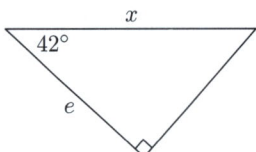   f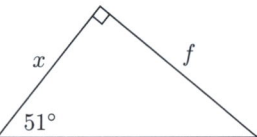

2 Find x, rounding your answer to 2 decimal places:

a 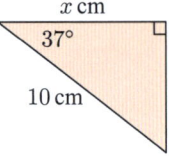   b 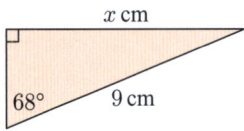   c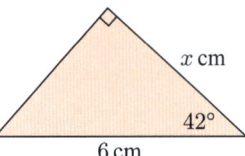

d 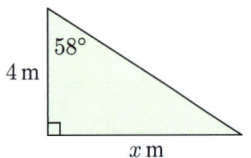   e 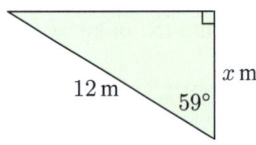   f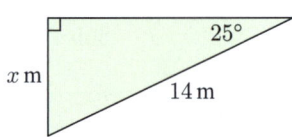

g 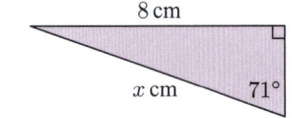   h 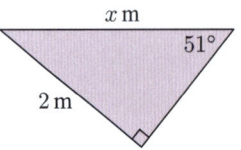   i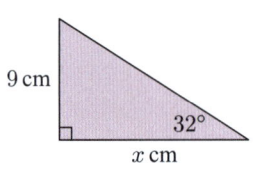

j 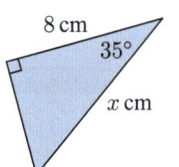   k 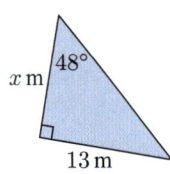   l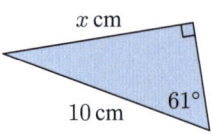

m 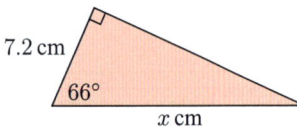   n 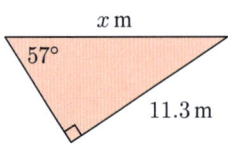   o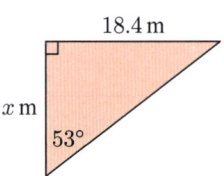

3 Find, to 1 decimal place, *all* the unknown angles and sides of:

a 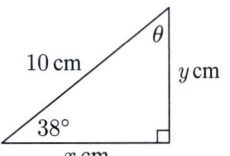   b 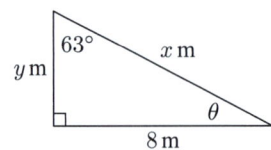   c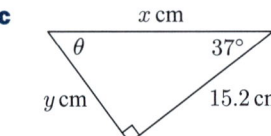

E FINDING ANGLES

If we are given two side lengths of a right angled triangle, we can find the other angles of the triangle.

In the triangle alongside, $\sin\theta = \frac{2}{7}$.

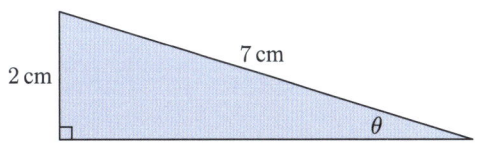

So, θ is the angle whose sine is $\frac{2}{7}$. We say θ is the **inverse sine** of $\frac{2}{7}$, and write $\theta = \sin^{-1}\left(\frac{2}{7}\right)$.

We can use a calculator to evaluate inverse sines. Click on the icon for instructions.

For the right angled triangle with hypotenuse 7 cm and opposite side 2 cm, $\theta \approx 16.6°$.

CALCULATOR INSTRUCTIONS

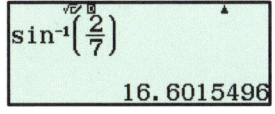

We define **inverse cosine** and **inverse tangent** in a similar way.

Example 4   ◀) Self Tutor

Find, to 1 decimal place, the measure of the angle marked θ.

a   9 m, 7 m, θ

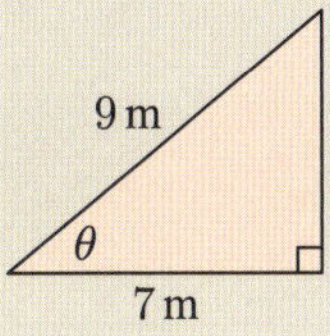

b   14 cm, 11 cm, θ

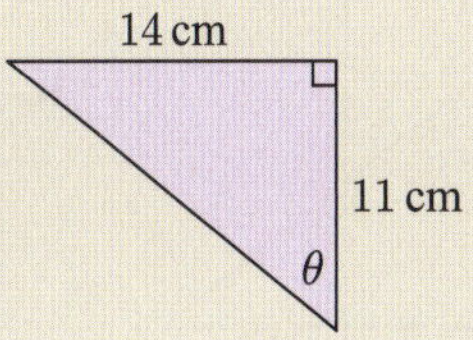

a   HYP 9 m, OPP, θ, 7 m ADJ

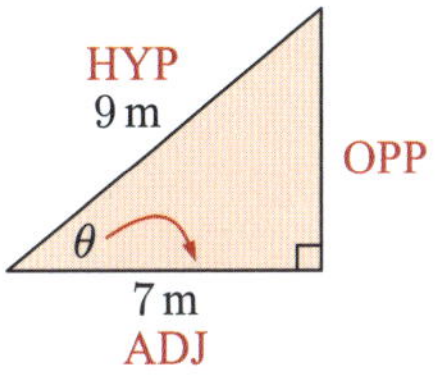

$\cos\theta = \frac{7}{9}$ $\{\cos\theta = \frac{\text{ADJ}}{\text{HYP}}\}$

$\therefore\ \theta = \cos^{-1}\left(\frac{7}{9}\right)$

$\therefore\ \theta \approx 38.9°$

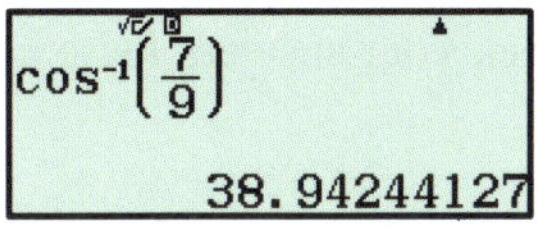

$\cos^{-1}\left(\frac{7}{9}\right)$
38.94244127

b   OPP 14 cm, ADJ 11 cm, HYP, θ

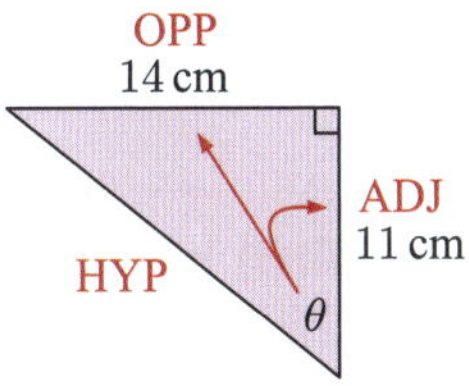

$\tan\theta = \frac{14}{11}$ $\{\tan\theta = \frac{\text{OPP}}{\text{ADJ}}\}$

$\therefore\ \theta = \tan^{-1}\left(\frac{14}{11}\right)$

$\therefore\ \theta \approx 51.8°$

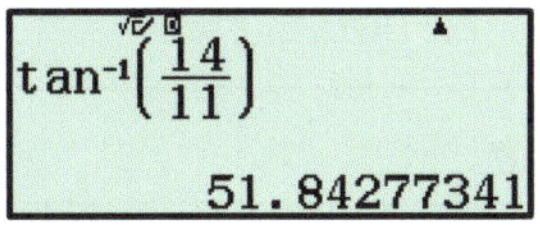

$\tan^{-1}\left(\frac{14}{11}\right)$
51.84277341

EXERCISE 24E

1 Consider the right angled triangle alongside.
   a Write $\sin\theta$ as a fraction.
   b Hence find θ.

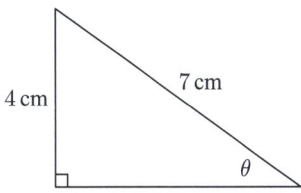

2 Find, to 1 decimal place, the measure of angle θ:

a 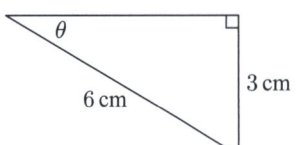 b 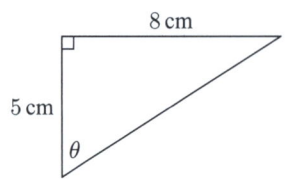 c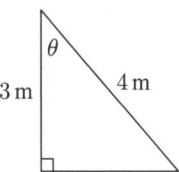

d 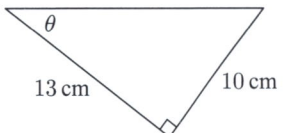 e 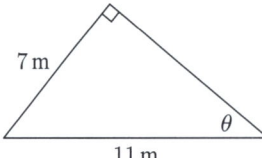 f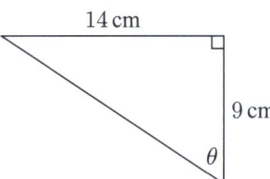

g 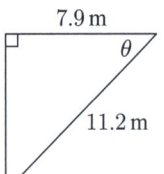 h 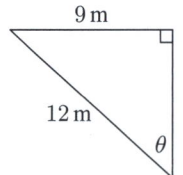 i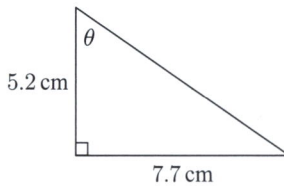

j 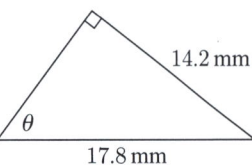 k 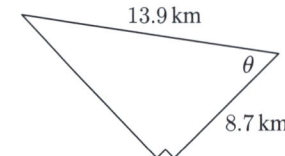 l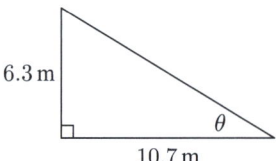

3 Use trigonometry to find, to 1 decimal place, all the unknown sides and angles in the following triangles. Check your answers for x using Pythagoras' theorem.

a 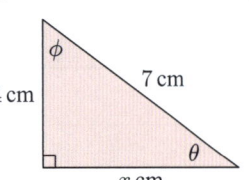 b 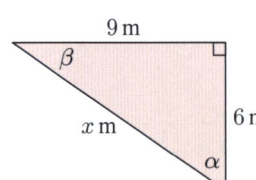 c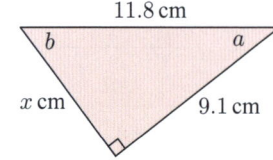

Puzzle

The spiral of Theodorus

The **spiral of Theodorus** consists of a series of right angled triangles.

The initial triangle ABC is isosceles with legs of length 1.

For each subsequent triangle, one of the legs is the hypotenuse of the previous triangle, and the other leg has length 1.

How many triangles can be drawn before they start to overlap?

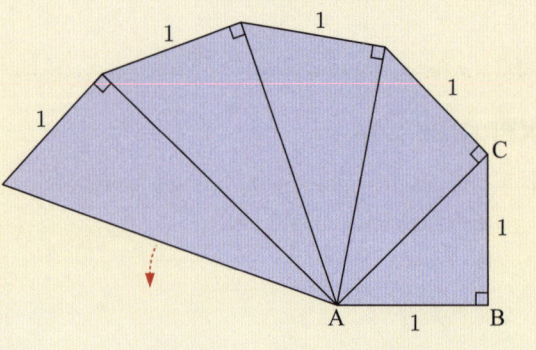

F PROBLEM SOLVING WITH TRIGONOMETRY

The trigonometric ratios can be used to solve problems involving right angled triangles.

When solving these problems, you should follow the steps below:

Step 1: Draw a **diagram** to illustrate the situation.

Step 2: Mark on the diagram the **unknown** angle or side that needs to be calculated. We often use x for a length and θ for an angle.

Step 3: Locate a **right angled triangle** in your diagram.

Step 4: Write an **equation** using one of the trigonometric ratios.

Step 5: **Solve** the equation to find the unknown.

Step 6: **Write** your answer in sentence form.

Example 5   ◀)) Self Tutor

A ladder leaning against a vertical wall reaches 3.5 m up the wall, and makes an angle of 55° with the ground. Find the length of the ladder.

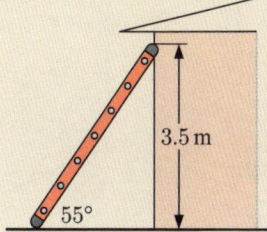

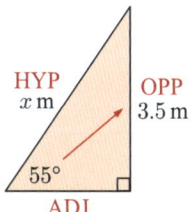

Let the ladder be x m long.

$\sin 55° = \dfrac{3.5}{x}$ $\{\sin \theta = \dfrac{\text{OPP}}{\text{HYP}}\}$

$\therefore\ x \times \sin 55° = 3.5$

$\therefore\ x = \dfrac{3.5}{\sin 55°}$

$\therefore\ x \approx 4.27$

$\therefore\ $ the ladder is about 4.27 m long.

EXERCISE 24F.1

1 From a point 25 metres from the base of a flagpole, the angle to the top of the pole is 35°. Find the height of the flagpole.

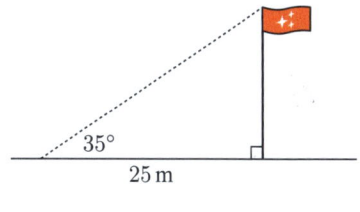

2 Lucas starts at the base of a hill. He walks up a steep path at an angle of 22° for 100 metres. Find his height above ground level.

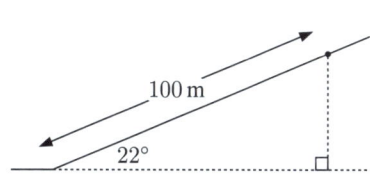

3 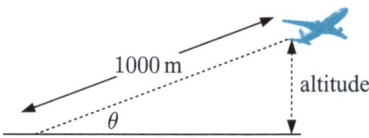 An aeroplane takes off at a constant angle to the ground. At the time when it has flown 1000 m, its altitude is 320 m. Find the angle θ at which the aeroplane takes off.

4 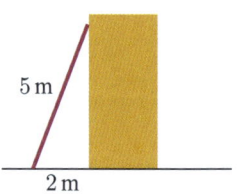 The feet of a 5 m long ladder are placed 2 m from a wall. Find the angle that the ladder makes with the ground.

Example 6 ◀) Self Tutor

Determine the length of the roofing beam required to support the roof shown alongside:

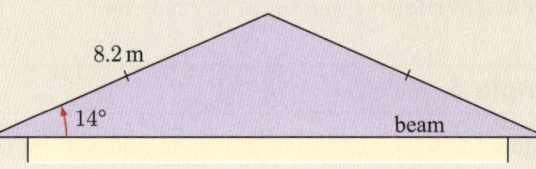

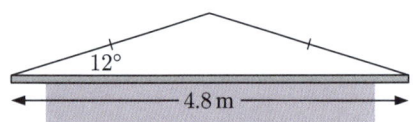

Suppose the beam is $2x$ m long.

$\cos 14° = \dfrac{x}{8.2}$ $\{\cos \theta = \dfrac{\text{ADJ}}{\text{HYP}}\}$

$\therefore\ x = 8.2 \times \cos 14°$

$\therefore\ 2x = 2 \times 8.2 \times \cos 14°$

≈ 15.9

$\therefore\ $ the beam is about 15.9 m long.

5

A beam of length 4.8 metres supports a garage roof. The pitch of the roof is 12°. Find the length of the sloping sides.

We can use the properties of isosceles triangles to locate a right angle.

6 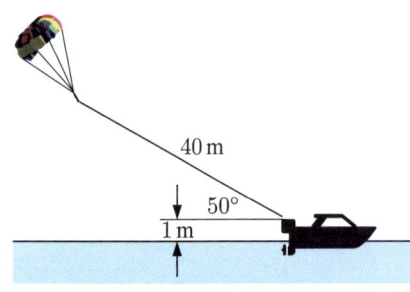 A parasailer is towed behind a boat. The towing cable is 40 metres long, and makes an angle of 50° with the deck of the boat. How high is the parasailer above the water?

7 Find the perimeter of this rectangle.

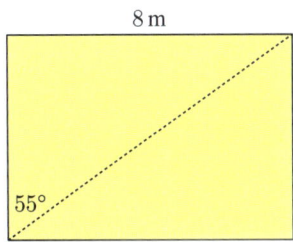

8 An isosceles triangle has sides 7 cm, 7 cm, and 8 cm in length. Find the measure of the base angles.

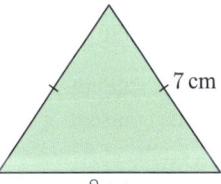

9 An isosceles triangle has equal sides of length 13 cm, and base angles of 40°. Find the length of the base of the triangle.

10 A rhombus has sides of length 15 cm, and one diagonal of length 20 cm. Find the measure of the angles of the rhombus.

11 A 7 m long ladder leaning against a vertical wall makes an angle of 50° to the horizontal. The foot of the ladder is pushed towards the wall until an angle of 65° is obtained. How much further up the wall does the ladder now reach?

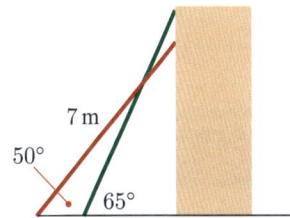

12 Find the height of the flagpole in the **Opening Problem** on page **486**.

PROBLEMS INVOLVING BEARINGS

In **Chapter 10** we saw that **bearings** are used to describe directions relative to due north.

We can use trigonometry to solve problems involving bearings.

Example 7 ◀⁾ Self Tutor

A cyclist rides to a point 21.3 km west and 13.8 km north of their starting point. Find, to the nearest degree, the bearing of the finish from the starting point.

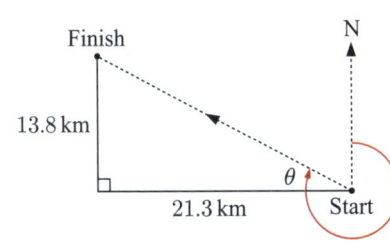

$\tan \theta = \dfrac{13.8}{21.3}$ $\{\tan \theta = \dfrac{\text{OPP}}{\text{ADJ}}\}$

$\therefore \theta = \tan^{-1}\left(\dfrac{13.8}{21.3}\right) \approx 32.9°$

$\therefore 270° + \theta \approx 303°$

So, the bearing of the finish from the starting point is about 303°.

EXERCISE 24F.2

1 A ship sails to a point 15 km north and 11 km east of its starting point. Find, to the nearest degree, the bearing of the finish from the starting point.

2 A small plane flies between two country airfields. Its destination is 18.4 km south and 27.3 km west of its origin. Find, to the nearest degree, the bearing on which the plane flies.

3 Tamara swims from a jetty to a reef. The reef is 400 m east and 150 m south of the jetty.
  a How far did Tamara swim?
  b On what bearing did Tamara swim?
  c Find the bearing on which Tamara must swim to return to the jetty.

4 An orienteer is studying the map for a competition. Each grid unit represents 1 km. Find the distance and true bearing from:
  a the start to A
  b A to B
  c B to C
  d C to D
  e D to the finish.

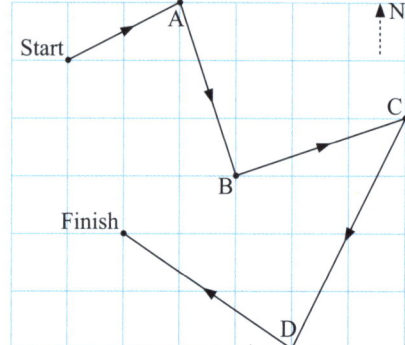

G   THE FIRST QUADRANT OF THE UNIT CIRCLE

The **unit circle** is the circle with centre O(0, 0) and radius 1 unit.

Consider point P(a, b) which lies on the unit circle in the first quadrant.

OP makes an angle θ with the x-axis.

Notice that $\sin \theta = \dfrac{b}{1} = b$ and $\cos \theta = \dfrac{a}{1} = a$

So,   P has coordinates ($\cos \theta$, $\sin \theta$).

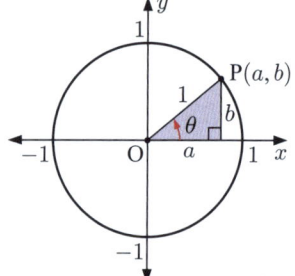

Example 8 — Self Tutor

  a State exactly the coordinates of point P.
  b Find the coordinates of P correct to 3 decimal places.

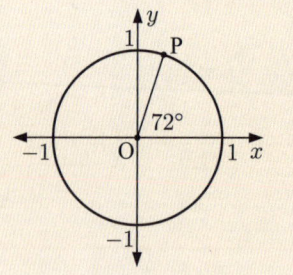

  a P is ($\cos 72°$, $\sin 72°$)
  b P is ≈ (0.309, 0.951)

EXERCISE 24G.1

1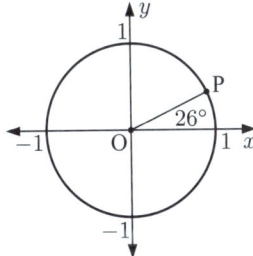
    a State exactly the coordinates of P.
    b Find the coordinates of P correct to 3 decimal places.

2 Point A has coordinates $(0.2588, 0.9659)$.
  a State the value of:
    i $\cos 75°$ **ii** $\sin 75°$
  b Use $\tan \theta = \dfrac{\sin \theta}{\cos \theta}$ and your answers to **a** to find $\tan 75°$.
  c Check your answer to **b** by finding $\tan 75°$ on your calculator.

3 Suppose P is the point on the unit circle at angle $45°$ to the x-axis.
  a Explain why triangle ONP is isosceles.
  b Use Pythagoras' theorem to show that $ON = NP = \sqrt{\dfrac{1}{2}}$.
  c Use your calculator to verify that $\sqrt{\dfrac{1}{2}} = \dfrac{1}{\sqrt{2}}$.
  d Hence state the exact coordinates of P.
  e Find the exact values of:
    i $\cos 45°$ **ii** $\sin 45°$ **iii** $\tan 45°$

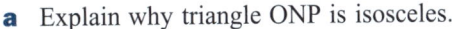

4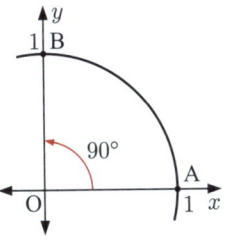
  a Find the coordinates of points A and B.
  b Find the value of:
    i $\cos 90°$ **ii** $\sin 90°$ **iii** $\tan 90°$
  c Find the value of:
    i $\cos 0°$ **ii** $\sin 0°$ **iii** $\tan 0°$

5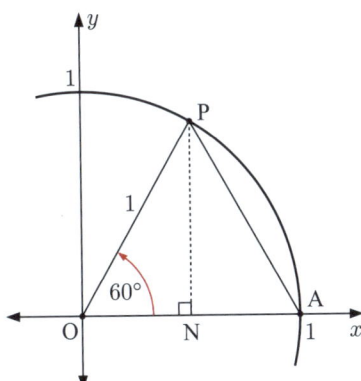

P is the point on the unit circle such that angle AOP is $60°$. PN is drawn perpendicular to the x-axis.
  a Explain why triangle AOP is equilateral.
  b State the exact length of ON.
  c Use Pythagoras' theorem to show that $PN = \sqrt{\dfrac{3}{4}}$.
  d Use your calculator to verify that $\sqrt{\dfrac{3}{4}} = \dfrac{\sqrt{3}}{2}$.
  e Hence state the exact coordinates of P.
  f Find the exact value of:
    i $\cos 60°$ **ii** $\sin 60°$ **iii** $\tan 60°$
  g Find the size of $O\hat{P}N$.
  h Find the exact value of:
    i $\cos 30°$ **ii** $\sin 30°$ **iii** $\tan 30°$

IMPORTANT ANGLES

From the previous Exercise you should have discovered trigonometric ratios of some important angles:

θ	$\cos\theta$	$\sin\theta$	$\tan\theta$
$0°$	1	0	0
$30°$	$\frac{\sqrt{3}}{2}$	$\frac{1}{2}$	$\frac{1}{\sqrt{3}}$
$45°$	$\frac{1}{\sqrt{2}}$	$\frac{1}{\sqrt{2}}$	1
$60°$	$\frac{1}{2}$	$\frac{\sqrt{3}}{2}$	$\sqrt{3}$
$90°$	0	1	undefined

You should memorise these results or be able to quickly deduce them from diagrams.

EXERCISE 24G.2

1 State the value of:
- **a** $\sin 30°$
- **b** $\cos 90°$
- **c** $\tan 60°$
- **d** $\cos 45°$
- **e** $\tan 0°$
- **f** $\sin 90°$

2 Using the table above, show that:
- **a** $\sin 30° + \cos 60° = 1$
- **b** $\sin^2 30° + \cos^2 30° = 1$
- **c** $\cos^2 45° + \sin^2 45° = 1$
- **d** $\sin 30° \cos 60° + \sin 60° \cos 30° = 1$
- **e** $\sin^2 30° + \sin^2 45° + \sin^2 60° = \frac{3}{2}$

We use the notation $\sin^2\theta$ to mean $\sin\theta \times \sin\theta$.

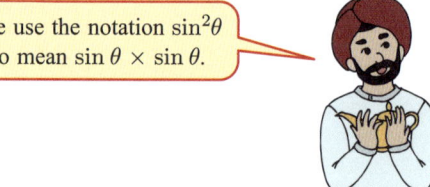

3 Without using a calculator, find the value of:
- **a** $\sin^2 60°$
- **b** $\dfrac{\sin 30°}{\cos 30°}$
- **c** $\tan^2 60°$
- **d** $\cos 0° + \sin 90°$
- **e** $\cos^2 30°$
- **f** $1 - \tan 0°$
- **g** $\dfrac{\sin 60°}{\cos 60°}$
- **h** $1 - \cos 60°$
- **i** $2 + \sin 30°$

4 Find the exact value of the unknown:

a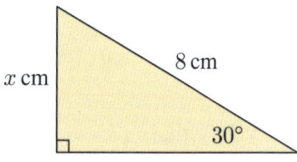

b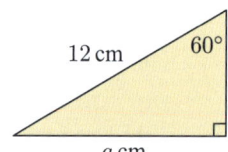

c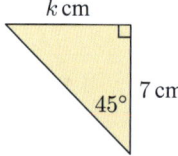

d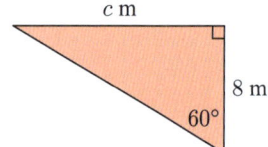

e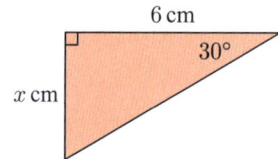

f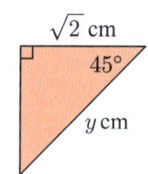

Trigonometry (Chapter 24) 501

Historical note

The word *trigonometry* is derived from the Greek word *trigonometria*, meaning "triangle measuring".

This subject was first studied in ancient Egypt and Babylon in the 2nd millennium BC. The Egyptians used trigonometry in the construction of the pyramids, and the scribe **Ahmes** included a problem of trigonometry in the *Rhind Mathematical Papyrus*. The Babylonian astronomers, meanwhile, studied the relationships between angles and distances in their astronomy.

The Greek mathematician **Hipparchus of Nicaea** (180 - 125 BC) compiled the first trigonometric table, including the corresponding measurements of arc and chord for a series of angles. His work was extended by **Menelaus of Alexandria** (70 - 140 AD), who recorded Menelaus' theorem in his book *Sphaerica*, and **Claudius Ptolemy**, who expanded Hipparchus' table of values in his book *Almagest*.

Hipparchus

Review set 24A

1 a Convert this sketch into an accurate scale diagram using the scale 1 cm ≡ 1 m.
  b Use your scale diagram to estimate the length of PQ.
  c Use trigonometry to find the length of PQ, rounded to 2 decimal places.

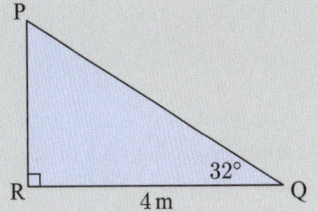

2 For this right angled triangle, name the:
  a hypotenuse
  b side opposite angle θ
  c side adjacent to angle θ.

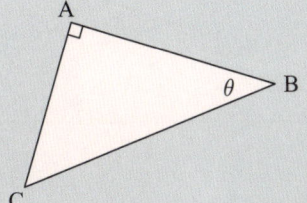

3 Consider the right angled triangle PQR shown.
  a Use a ruler to find the length in cm of each side, rounded to 1 decimal place.
  b Hence estimate the value of:
    i $\sin 26°$ **ii** $\cos 26°$ **iii** $\tan 26°$
  Check your answers using a calculator.

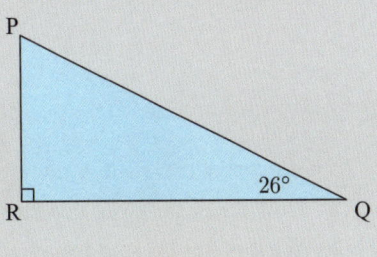

4 Write a trigonometric equation connecting the variables.

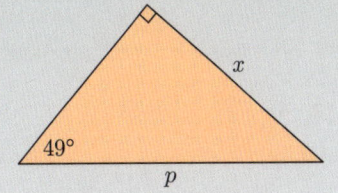

5 Find x, rounding your answers to 1 decimal place:

a 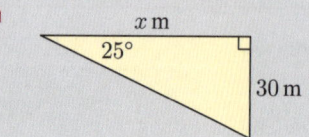 b 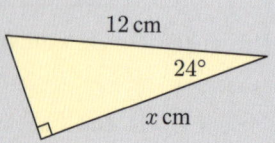 c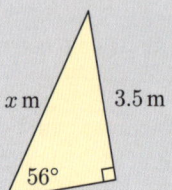

6 Find, to 1 decimal place, the measure of angle θ:

a 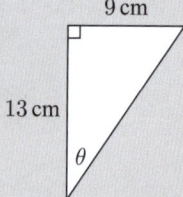 b 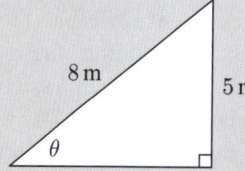 c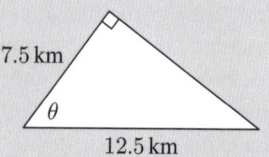

7 Find, to 1 decimal place, all the unknown sides and angles in this triangle.

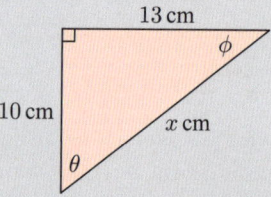

8 The shadow of a tree is 17.5 m long. The angle from the end of the shadow to the tree top is 38°. Find the height of the tree.

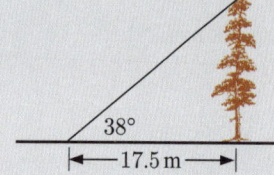

9 A 95 cm long ramp is needed to climb a step of height 30 cm. Find the angle of incline θ of the ramp.

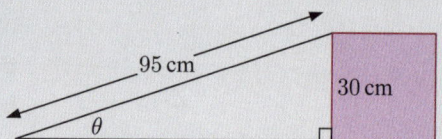

10 Find the perimeter of this triangle.

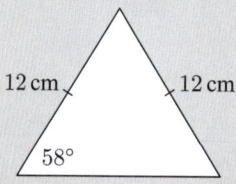

11 A motorcyclist travels to a point 87 km west and 63 km north of her starting point.
  a How far is she from her starting point?
  b What is her bearing from her starting point?

12 Without using a calculator, find the value of:

  a $\dfrac{\sin^2 30°}{\cos^2 30°} + \sin 90°$ **b** $\sin^2 60° + \tan 45° - \cos 0°$

Review set 24B

1 Find the length of the side:
   a opposite θ
   b adjacent to θ
   c opposite ϕ
   d adjacent to ϕ.

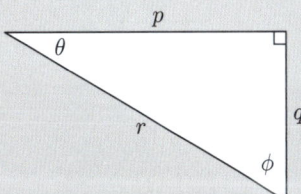

2 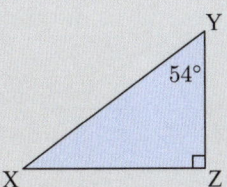 Use your calculator to find the following ratios:
   a $\dfrac{YZ}{XY}$ **b** $\dfrac{XZ}{XY}$ **c** $\dfrac{XZ}{YZ}$

3 Find x, rounding your answer to 2 decimal places:

   a 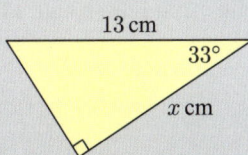   **b** 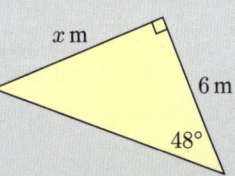   **c**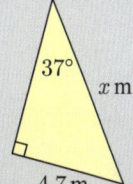

4 Find, to 1 decimal place, the measure of the angle marked θ:

   a 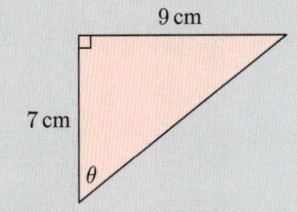   **b** 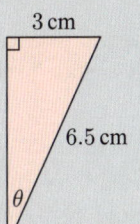   **c**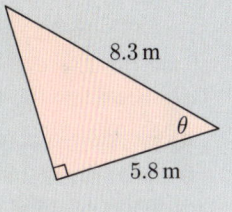

5 To find x, Ella performed these steps:
$$\tan 13° = \frac{5}{x}$$
$$\therefore\ x \times \tan 13° = 5$$
$$\therefore\ x = \frac{5}{\tan 13°}$$
$$\therefore\ x \approx 21.7$$

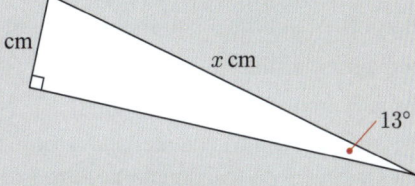

   a Explain the error Ella has made.   **b** Find the correct value of x.

6 A boundary fence is reinforced with a series of metal poles as shown. Each pole makes an angle of 65° with the ground, and it enters the ground 1.2 m from the fence. How long is each pole?

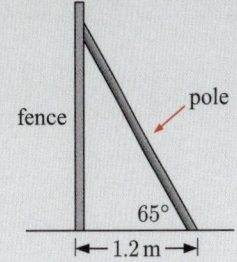

7 Find the value of y, rounded to 1 decimal place:

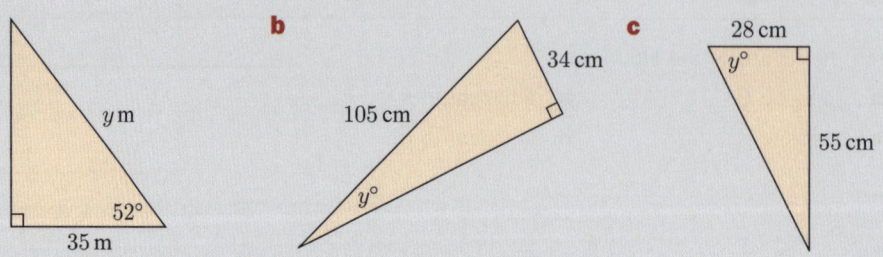

8 Find, to 1 decimal place, all the unknown angles and sides of this triangle.

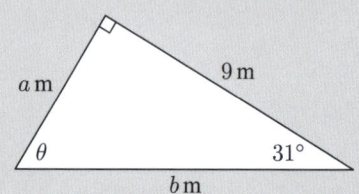

9

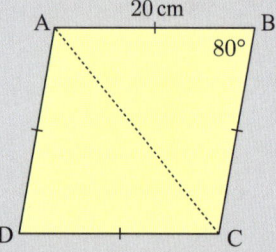

ABCD is a rhombus. Find the length of the diagonal AC.

10 Two office buildings are located 10 m apart. A laser beam is shone from A to B as shown. The angle made between the laser beam and the taller building is 11°.

  a How far is it from A to B?

  b The larger building is twice as tall as the shorter building. Find the height of each building, to the nearest metre.

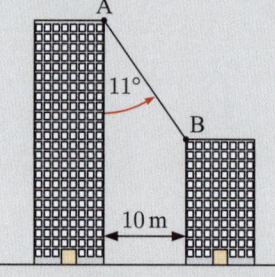

11 A ship leaves port A and travels to the island port B which is 60 km south and 40 km west of A.

  a Draw a diagram of the situation.

  b How far is B from A?

  c The ship wishes to sail back directly to A from B. On what bearing does the captain need to sail?

12 Find the exact value of x:

  a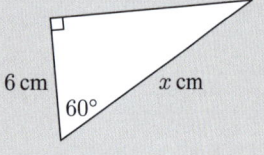

  b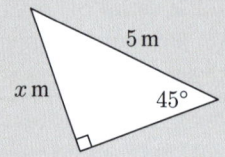

  c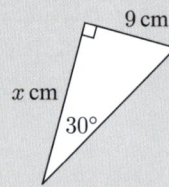

25

Quadratic equations and functions

Contents:

- **A** Quadratic equations
- **B** The Null Factor law
- **C** Solving quadratic equations
- **D** Problem solving with quadratic equations
- **E** Quadratic functions
- **F** Graphs of quadratic functions
- **G** Axes intercepts
- **H** Axis of symmetry
- **I** Vertex

Opening problem

Matheus is a baker specialising in wedding cakes. The guests at a wedding should each receive a slice of cake with the rectangular cross-section shown. The icing should be the same thickness x cm on the top and sides.

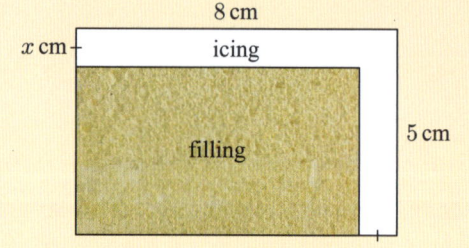

Things to think about:

a What possible values could x take?

b Can you explain why the area of the icing is given by $(-x^2 + 13x)$ cm^2?

c Matheus wants the guests to experience 30% icing and the rest filling. Can you hence explain why $x^2 - 13x + 12 = 0$?

d How many solutions does the equation $x^2 - 13x + 12 = 0$ have?

e How thick should the icing be?

A QUADRATIC EQUATIONS

A **quadratic equation** is an equation which can be written in the form $ax^2 + bx + c = 0$, where a, b, and c are constants, $a \neq 0$.

For example:
- $2x^2 - 3x + 5 = 0$ and $x^2 + 3x = 0$ are quadratic equations
- $x^2 + 5 + \dfrac{1}{x} = 0$ is not a quadratic equation.

Quadratic equations may have two, one, or zero solutions, as demonstrated in the table below:

Equation	$ax^2 + bx + c = 0$ form	Solutions	
$x^2 - 4 = 0$	$x^2 + 0x - 4 = 0$	$x = 2$ or $x = -2$	two solutions
$(x - 2)^2 = 0$	$x^2 - 4x + 4 = 0$	$x = 2$	one solution
$x^2 + 4 = 0$	$x^2 + 0x + 4 = 0$	none, as x^2 is always $\geqslant 0$	zero solutions

Example 1   ◀)) Self Tutor

Show that $x = 2$ and $x = -5$ are solutions to the quadratic equation $x^2 + 3x - 10 = 0$.

If $x = 2$, $\quad x^2 + 3x - 10$
$\quad = 2^2 + 3 \times 2 - 10$
$\quad = 4 + 6 - 10$
$\quad = 0$

If $x = -5$, $\quad x^2 + 3x - 10$
$\quad = (-5)^2 + 3 \times (-5) - 10$
$\quad = 25 - 15 - 10$
$\quad = 0$

$\therefore \;\; x = 2$ and $x = -5$ are both solutions to the quadratic equation $x^2 + 3x - 10 = 0$.

EXERCISE 25A

1 State whether each of the following is a quadratic equation:

a $x^2 - x + 6 = 0$ **b** $2x^3 + 5x - 4 = 0$ **c** $4x^2 - 7 = 8x$

d $-6x^2 - 11 = 0$ **e** $3x^2 + x - \dfrac{2}{x} = 0$ **f** $\dfrac{2}{3}x^2 - \dfrac{1}{2}x = 0$

2 For each of the following quadratic equations, determine which of the numbers in brackets are solutions.

a $x^2 - 7x + 6 = 0$ $\{1, 2, 4, 6, 7\}$ **b** $x^2 + 2x - 8 = 0$ $\{-5, -4, -1, 1, 2\}$

c $x^2 - 5x = 0$ $\{-2, 0, 3, 5, 7\}$ **d** $x^2 - 6x + 9 = 0$ $\{-3, 1, 3, 4, 6\}$

e $2x^2 + 3x - 2 = 0$ $\{-2, -1, -\dfrac{1}{2}, \dfrac{1}{2}, 1\}$

3 Explain why the quadratic equation $x^2 + 5 = 0$ has no solutions.

B THE NULL FACTOR LAW

In the previous Section we used trial and error to find solutions to quadratic equations. To find solutions in a more systematic way, we can try to write the quadratic equation in a factored form. We can then use the **Null Factor law**.

Investigation   The Null Factor law

What to do:

1 Complete the following products:

a $3 \times 5 =$ **b** $0 \times 5 =$ **c** $1 \times 2 =$ **d** $1 \times 0 =$

e $2 \times -4 =$ **f** $-6 \times -7 =$ **g** $0 \times -9 =$ **h** $-5 \times 8 =$

i $-10 \times 0 =$ **j** $7 \times -11 =$ **k** $2 \times 4 \times 5 =$ **l** $3 \times -6 \times 4 =$

m $7 \times 5 \times 0 =$ **n** $4 \times 4 \times 5 =$ **o** $-8 \times 0 \times 7 =$ **p** $0 \times 13 \times 0 =$

2 Circle the products which have a value of zero.

3 Look at the numbers in the products which you have circled. What do you notice?

WORKSHEET

In the previous **Investigation**, you should have discovered the **Null Factor law**:

> When the product of two or more numbers is zero, at least one of them must be zero.
> So, if $ab = 0$ then $a = 0$ or $b = 0$.

For example:
- If $2xy = 0$ then $x = 0$ or $y = 0$.
- If $x(x - 2) = 0$ then $x = 0$ or $x - 2 = 0$.
- If $xyz = 0$ then $x = 0$, $y = 0$, or $z = 0$.

If we are given an equation where the LHS is factorised and the RHS is zero, we can use the Null Factor law to find solutions.

Example 2

Solve for x:

a $5x(x+2) = 0$ **b** $(x+4)(x-1) = 0$

a
$$5x(x+2) = 0$$
$\therefore\ 5x = 0$ or $x + 2 = 0$ {Null Factor law}
$\therefore\ x = 0$ or $x = -2$ {solving linear equations}
So, $x = 0$ or -2

b
$$(x+4)(x-1) = 0$$
$\therefore\ x + 4 = 0$ or $x - 1 = 0$ {Null Factor law}
$\therefore\ x = -4$ or $x = 1$ {solving linear equations}
So, $x = -4$ or 1

EXERCISE 25B

1 Explain what can be deduced from:
  a $ac = 0$
  b $bd = 0$
  c $abc = 0$
  d $3x = 0$
  e $x(x-3) = 0$
  f $x^2 = 0$
  g $(x-5)y = 0$
  h $x^2y = 0$

2 Solve for x:
  a $2x(x-1) = 0$
  b $x(x+5) = 0$
  c $3x(x+2) = 0$
  d $(x-1)^2 = 0$
  e $-x(x-4) = 0$
  f $-2x(x+3) = 0$
  g $x(2x+1) = 0$
  h $3x(4x-3) = 0$
  i $-x(3x+5) = 0$

3 Solve for x:
  a $(x-1)(x-5) = 0$
  b $(x+2)(x-4) = 0$
  c $(x+3)(x+7) = 0$
  d $(x+7)(x-11) = 0$
  e $2x(x-8) = 0$
  f $(x+12)(x-5) = 0$
  g $-3x(x+7) = 0$
  h $(2x+1)(x-3) = 0$
  i $(x+6)(3x-1) = 0$
  j $(2x+1)(x+6) = 0$
  k $4(x-3)^2 = 0$
  l $(x-31)(x+11) = 0$
  m $(x+4)(4x-1) = 0$
  n $-3x(7x+3) = 0$
  o $(2-x)(3x+4) = 0$

C SOLVING QUADRATIC EQUATIONS

For quadratic equations written in other forms, we cannot apply the Null Factor law directly. Instead, we first need to factorise the quadratic using the techniques studied in **Chapter 13**.

Step 1: If necessary, rearrange the equation so the RHS is **zero**.
Step 2: **Fully factorise** the LHS.
Step 3: Use the **Null Factor law**: if $ab = 0$ then $a = 0$ or $b = 0$.
Step 4: **Solve** the resulting linear equations.

Quadratic equations and functions (Chapter 25)

EQUATIONS IN THE FORM $ax^2 + bx = 0$

To solve quadratic equations of the form $ax^2 + bx = 0$ where $a \neq 0$, we first take out x as a common factor. We can then use the Null Factor law.

Example 3

Solve for x: $\quad x^2 = 4x$

$$x^2 = 4x$$
$$\therefore \; x^2 - 4x = 0 \quad \text{\{subtracting } 4x \text{ from both sides to make RHS} = 0\}$$
$$\therefore \; x(x-4) = 0 \quad \text{\{factorising the LHS\}}$$
$$\therefore \; x = 0 \quad \text{or} \quad x - 4 = 0 \quad \text{\{Null Factor law\}}$$
$$\therefore \; x = 0 \quad \text{or} \quad x = 4$$

WARNING ON INCORRECT CANCELLING

Given the equation $x^2 = 4x$ in the Example above, you may be tempted to divide both sides by x, giving $x = 4$.

While $x = 4$ is a solution to the equation, it is not the only solution. By dividing both sides by x, we have lost the solution $x = 0$.

From this example we conclude that:

> We should never cancel a common factor involving a variable unless we are sure that this factor is non-zero.

EXERCISE 25C.1

1 Solve for x:

    a $x^2 - x = 0$ **b** $x^2 - 13x = 0$ **c** $x^2 + 8x = 0$

    d $x^2 + 3x = 0$ **e** $2x + x^2 = 0$ **f** $5x - x^2 = 0$

    g $12x - x^2 = 0$ **h** $x^2 + 7x = 0$ **i** $x^2 - 4x = 0$

    j $2x^2 - 7x = 0$ **k** $3x^2 - 15x = 0$ **l** $2x^2 + 8x = 0$

2 Solve:

    a $x^2 = 3x$ **b** $x^2 = 10x$ **c** $a^2 = a$

    d $x^2 = -6x$ **e** $2x^2 = -7x$ **f** $5x = x^2$

    g $8x = x^2$ **h** $3y^2 = 18y$ **i** $5x^2 = 6x$

    j $7x + x^2 = x$ **k** $x^2 - 4 = 2x - 4$ **l** $2x^2 + x = 3x$

SOLVING EQUATIONS USING THE DIFFERENCE BETWEEN TWO SQUARES

In equations such as $x^2 - 9 = 0$ and $4x^2 - 1 = 0$, the LHS can be factorised as the **difference between two squares**:

$$a^2 - b^2 = (a+b)(a-b)$$

Example 4

Solve for x:

a $x^2 - 4 = 0$

b $4x^2 - 25 = 0$

a
$$x^2 - 4 = 0$$
$$\therefore (x+2)(x-2) = 0$$
$$\therefore x+2 = 0 \quad \text{or} \quad x-2 = 0$$
$$\therefore x = -2 \quad \text{or} \quad x = 2$$
$$\therefore x = \pm 2$$

b
$$4x^2 - 25 = 0$$
$$\therefore (2x)^2 - 5^2 = 0$$
$$\therefore (2x+5)(2x-5) = 0$$
$$\therefore 2x+5 = 0 \quad \text{or} \quad 2x-5 = 0$$
$$\therefore x = -\frac{5}{2} \quad \text{or} \quad x = \frac{5}{2}$$
$$\therefore x = \pm\frac{5}{2}$$

EXERCISE 25C.2

1 Solve for x:
  - **a** $x^2 - 16 = 0$
  - **b** $x^2 - 9 = 0$
  - **c** $x^2 - 49 = 0$
  - **d** $x^2 - 64 = 0$
  - **e** $x^2 - 144 = 0$
  - **f** $x^2 - 5 = 0$

2 Explain why $x^2 + 4 = 0$ has no real solutions.

3 Solve for x:
  - **a** $2x^2 - 8 = 0$
  - **b** $2x^2 - 72 = 0$
  - **c** $3x^2 - 27 = 0$
  - **d** $5x^2 - 20 = 0$
  - **e** $-3x^2 + 12 = 0$
  - **f** $-2x^2 + 8 = 0$

4 Solve for x:
  - **a** $9x^2 - 4 = 0$
  - **b** $1 - 9x^2 = 0$
  - **c** $4 - 25x^2 = 0$
  - **d** $4x^2 - 1 = 0$
  - **e** $4x^2 - 9 = 0$
  - **f** $2 - 8x^2 = 0$

EQUATIONS OF THE FORM $x^2 + bx + c = 0$

For quadratic equations in the general form $x^2 + bx + c = 0$, we look for **perfect squares** or else use **sum and product** factorisation.

Example 5

Solve for x:

a $x^2 - 3x + 2 = 0$

b $x^2 = x + 12$

c $x^2 + 4 = 4x$

a
$$x^2 - 3x + 2 = 0$$
$$\therefore (x-1)(x-2) = 0 \quad \text{\{the numbers } -1 \text{ and } -2 \text{ have sum } -3 \text{ and product } 2\text{\}}$$
$$\therefore x - 1 = 0 \quad \text{or} \quad x - 2 = 0$$
$$\therefore x = 1 \text{ or } 2$$

b
$$x^2 = x + 12$$
$$\therefore x^2 - x - 12 = 0 \quad \text{\{subtracting } x+12 \text{ from both sides to make RHS} = 0\text{\}}$$
$$\therefore (x-4)(x+3) = 0 \quad \text{\{the numbers } -4 \text{ and } 3 \text{ have sum } -1 \text{ and product } -12\text{\}}$$
$$\therefore x - 4 = 0 \quad \text{or} \quad x + 3 = 0$$
$$\therefore x = 4 \text{ or } -3$$

c
$$x^2 + 4 = 4x$$
$$\therefore \ x^2 - 4x + 4 = 0 \quad \{\text{subtracting } 4x \text{ from both sides to make RHS} = 0\}$$
$$\therefore \ x^2 - 2 \times x \times 2 + 2^2 = 0$$
$$\therefore \ (x-2)^2 = 0 \quad \{\text{perfect square factorisation}\}$$
$$\therefore \ x - 2 = 0$$
$$\therefore \ x = 2$$

Sometimes each term in an equation contains a **constant common factor**. To solve these equations we first remove the common factor.

Example 6 ◀) Self Tutor

Solve for x: $\quad 3x^2 + 21x + 30 = 0$

$$3x^2 + 21x + 30 = 0 \quad \{\text{common factor} = 3\}$$
$$\therefore \ 3(x^2 + 7x + 10) = 0$$
$$\therefore \ 3(x+2)(x+5) = 0 \quad \{\text{the numbers 2 and 5 have sum 7 and product 10}\}$$
$$\therefore \ x + 2 = 0 \quad \text{or} \quad x + 5 = 0$$
$$\therefore \ x = -2 \text{ or } -5$$

EXERCISE 25C.3

1 Solve for x:
- **a** $x^2 - 7x + 10 = 0$
- **b** $x^2 + 6x + 8 = 0$
- **c** $x^2 + 11x + 10 = 0$
- **d** $x^2 - 8x + 12 = 0$
- **e** $x^2 - 5x + 4 = 0$
- **f** $x^2 - 11x + 24 = 0$
- **g** $x^2 + 10x + 25 = 0$
- **h** $x^2 - 3x - 18 = 0$
- **i** $x^2 + 7x - 18 = 0$
- **j** $x^2 - 22x + 121 = 0$
- **k** $x^2 - 6x + 9 = 0$
- **l** $x^2 - 5x - 6 = 0$
- **m** $x^2 + 11x - 60 = 0$
- **n** $x^2 + 18x - 63 = 0$
- **o** $x^2 - 12x - 64 = 0$
- **p** $x^2 - 19x + 70 = 0$

Look for factorisations which are perfect squares.

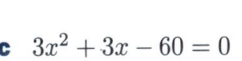

2 Solve for x:
- **a** $2x^2 + 4x - 30 = 0$
- **b** $-x^2 + 12x - 36 = 0$
- **c** $3x^2 + 3x - 60 = 0$
- **d** $-3x^2 + 21x - 36 = 0$
- **e** $5x^2 - 5x - 210 = 0$
- **f** $-4x^2 - 32x - 48 = 0$

3 Solve:
- **a** $x^2 + 4x = 12$
- **b** $x^2 - 14x = 15$
- **c** $x^2 + 2 = 3x$
- **d** $d^2 = 3d + 28$
- **e** $x^2 = 20 + x$
- **f** $8 = x^2 + 7x$
- **g** $x^2 = 5x + 24$
- **h** $2x^2 + 2x = 24$
- **i** $k^2 = 4k + 45$
- **j** $3x^2 = 30x - 48$
- **k** $x^2 + 1 = 2x$
- **l** $y^2 = 19y + 20$

4 Solve for x:
- **a** $x(x+2) = 15$
- **b** $x(x-3) = 40$
- **c** $(x-4)(x+5) = -8$
- **d** $x^2 - 4 = x + 2$
- **e** $2(x+5) = x^2 + 11$
- **f** $5 - x^2 = 2x - 3$

LEARNING ALGEBRA

D PROBLEM SOLVING WITH QUADRATIC EQUATIONS

Many problems expressed in words, and geometric problems involving right angled triangles, can be described in algebra by a **quadratic equation**.

PROBLEM SOLVING METHOD

- Carefully **read the question** until you understand it. A **sketch** may be useful.
- Decide on the **unknown** quantity. Label it with a variable such as x.
- Use the given information to construct an **equation**.
- **Solve** the equation using **factorisation** and the **Null Factor law**.
- **Check** that any solutions satisfy the original problem.
- Where appropriate, write your answer to the question in **sentence form**.

Example 7 ◀) Self Tutor

The sum of a number and its square is 30. Find the number.

Let the number be x.

So, $x + x^2 = 30$ {the number plus its square is 30}
$\therefore\ x^2 + x = 30$ {rearranging}
$\therefore\ x^2 + x - 30 = 0$ {making RHS $= 0$}
$\therefore\ (x+6)(x-5) = 0$ {factorising}
$\therefore\ x + 6 = 0$ or $x - 5 = 0$ {Null Factor law}
$\therefore\ x = -6$ or $x = 5$

$\therefore\ $ the numbers are -6 and 5.

Check: If $x = -6$, we have $-6 + (-6)^2 = -6 + 36 = 30$ ✓
     If $x = 5$, we have $5 + 5^2 = 5 + 25 = 30$ ✓

EXERCISE 25D

1 The sum of a number and its square is 42. Find the number.

2 When a number is squared, the result is five times the original number. Find the number.

3 When a number is subtracted from its square, the result is 56. Find the number.

4 The product of two consecutive integers is 56. Find the integers.

5 The sum of the squares of two consecutive odd numbers is 202. Find the numbers.

6 Two numbers differ by 4. The product of the two numbers is 77. What are the numbers?

Example 8

A rectangle has length 3 cm greater than its width, and the area of the rectangle is 28 cm². Find the dimensions of the rectangle.

If the rectangle has width x cm, then its length is $(x+3)$ cm.

$$\therefore\ x(x+3) = 28 \qquad \{\text{width} \times \text{length} = \text{area}\}$$
$$\therefore\ x^2 + 3x = 28 \qquad \{\text{expanding}\}$$
$$\therefore\ x^2 + 3x - 28 = 0 \qquad \{\text{making RHS} = 0\}$$
$$\therefore\ (x+7)(x-4) = 0 \qquad \{\text{factorising}\}$$
$$\therefore\ x+7 = 0 \ \text{ or } \ x-4 = 0 \qquad \{\text{Null Factor law}\}$$
$$\therefore\ x = -7 \text{ or } 4$$
$$\therefore\ x = 4 \qquad \{\text{lengths must be positive}\}$$

$\therefore$ the rectangle is 4 cm × 7 cm.

7 The length of a rectangle is 4 cm more than its width, and the area of the rectangle is 96 cm². Find the width of the rectangle.

8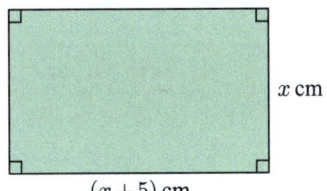

The area of this rectangle is 50 cm². Find:
a the value of x
b the perimeter of the rectangle.

9 A triangle has altitude 4 cm less than its base. If the area of the triangle is $38\frac{1}{2}$ cm², find the length of its base.

10 A rectangle has sides which differ in length by 3 cm. If the area of the rectangle is 154 cm², find its perimeter.

11
a Show that the total shaded area of the rectangle is $(x^2 + 8x + 15)$ m².
b Given that the total shaded area is 48 m², find x.

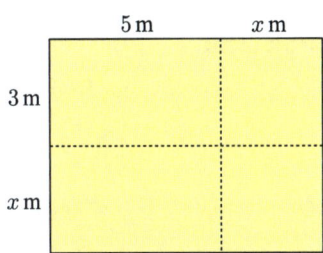

12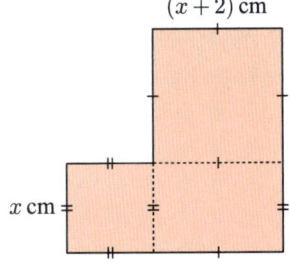

a Show that the total area of this shape is $(3x^2 + 6x + 4)$ cm².
b Given that the total area of the shape is 76 cm², find:
  i the value of x
  ii the perimeter of the shape.

13 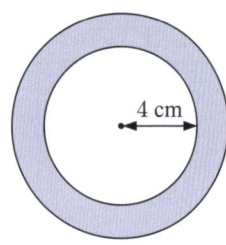 The shaded region has area 33π cm². Find the radius of the outer circle.

14 Answer the **Opening Problem** on page 506.

E QUADRATIC FUNCTIONS

A **quadratic function** is a relationship between two variables which can be written in the form $y = ax^2 + bx + c$ where x and y are the variables, and a, b, and c are constants, $a \neq 0$.

FINDING y GIVEN x

For any value of x, the corresponding value of y can be found by substitution into the function.

Example 9 ◀)) **Self Tutor**

Suppose $y = 2x^2 + 4x - 5$. Find the value of y when:
a $x = 0$ **b** $x = 3$

a When $x = 0$,
$y = 2(0)^2 + 4(0) - 5$
$= 0 + 0 - 5$
$= -5$

b When $x = 3$,
$y = 2(3)^2 + 4(3) - 5$
$= 18 + 12 - 5$
$= 25$

EXERCISE 25E.1

1 Which of the following are quadratic functions?
   a $y = 15x - 8$ **b** $y = \frac{1}{3}x^2 + 6$ **c** $3y + 2x^2 - 7 = 0$ **d** $y = 15x^3 + 2x - 16$

2 Suppose $y = 3x^2 + 7$. Find the value of y when:
   a $x = 2$ **b** $x = 5$ **c** $x = -3$

3 For each function, find the value of y for the given value of x:
   a $y = x^2 + 5x - 14$ when $x = 2$ **b** $y = 2x^2 + 9x$ when $x = -5$
   c $y = -2x^2 + 3x - 6$ when $x = 3$ **d** $y = 4x^2 + 7x + 10$ when $x = -2$

FINDING x GIVEN y

When we substitute a value for y, we are left with a quadratic equation which we need to solve for x. Since the equation is quadratic, there may be 0, 1, or 2 possible values for x.

Example 10

Suppose $y = x^2 - 6x + 8$. Find the value(s) of x for which:
a $y = 15$
b $y = -1$

a When $y = 15$, $x^2 - 6x + 8 = 15$
$\therefore x^2 - 6x - 7 = 0$
$\therefore (x+1)(x-7) = 0$
$\therefore x = -1$ or $x = 7$

b When $y = -1$, $x^2 - 6x + 8 = -1$
$\therefore x^2 - 6x + 9 = 0$
$\therefore (x-3)^2 = 0$
$\therefore x = 3$

EXERCISE 25E.2

1 Suppose $y = x^2 - 2x - 3$. Find the value(s) of x for which:
a $y = 0$
b $y = -4$
c $y = 5$

2 For each quadratic function, find the value(s) of x for the given value of y:
a $y = x^2 + 6x + 10$ when $y = 1$
b $y = x^2 + 5x + 8$ when $y = 2$
c $y = x^2 - 5x + 1$ when $y = -3$
d $y = 3x^2$ when $y = -3$

F GRAPHS OF QUADRATIC FUNCTIONS

The graph of a quadratic function is called a **parabola**. The parabola is one of the **conic sections**.

Historical note — Conic sections

Conic sections are curves which can be obtained by cutting a cone with a plane.

The name parabola comes from the Greek word for **thrown** because when an object is thrown, its path makes a parabolic arc.

There are many other examples of parabolas in everyday life. For example, parabolic mirrors are used in car headlights, heaters, satellite dishes, and radio telescopes, because of their special geometric properties.

You may like to explore the conic sections for yourself by cutting an ice cream cone. Cutting parallel to the side produces a parabola, as shown in the diagram.

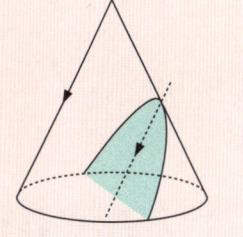

The simplest quadratic function is $y = x^2$. Its graph can be drawn from a table of values.

x	-3	-2	-1	0	1	2	3
y	9	4	1	0	1	4	9

The graph has a minimum turning point at $(0, 0)$. We call this the **vertex** of the parabola.

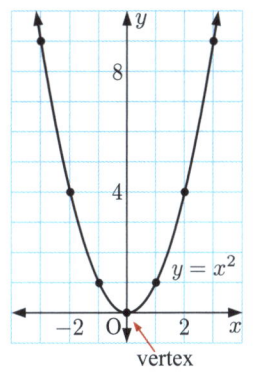

Example 11

Draw the graph of $y = x^2 + 2x - 3$ using a table of values from $x = -3$ to $x = 3$.

Consider $y = x^2 + 2x - 3$
When $x = -3$, $y = (-3)^2 + 2(-3) - 3$
$= 9 - 6 - 3$
$= 0$

We can do the same for the other values of x:

x	-3	-2	-1	0	1	2	3
y	0	-3	-4	-3	0	5	12

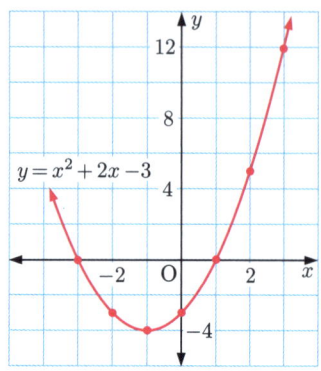

EXERCISE 25F

1 Using a table of values from $x = -3$ to $x = 3$, draw the graph of:

a $y = x^2 - 4$
b $y = x^2 + 2x$
c $y = -x^2 + 1$
d $y = x^2 - 2x + 8$
e $y = -x^2 + 2x + 1$
f $y = 2x^2 + 3x$
g $y = -2x^2 + 4$
h $y = x^2 + x + 4$
i $y = -x^2 + 4x - 9$

Check your graphs using technology.

GRAPHING PACKAGE

CALCULATOR INSTRUCTIONS

2 a Which of the graphs in **1** have the shape:

i  ii  ?

b Use 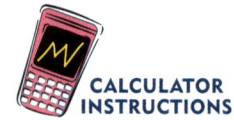 to copy and complete:

"The graph of $y = ax^2 + bx + c$ has shape ...... if $a > 0$, and shape ...... if $a < 0$".

G AXES INTERCEPTS

- An **x-intercept** of a function is a value of x where its graph meets the x-axis.
- A **y-intercept** of a function is a value of y where its graph meets the y-axis.

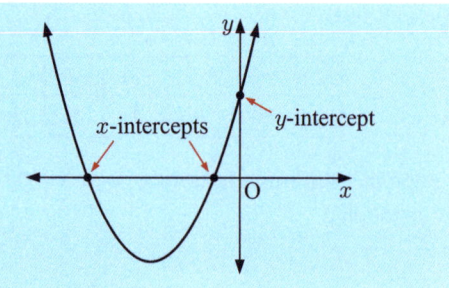

The y-intercept is found by letting $x = 0$ in the equation of the function.

Example 12

Find the y-intercept of:

a $y = 2x^2 - 3x + 5$

b $y = (x+3)(x-6)$

a When $x = 0$, $y = 2(0)^2 - 3(0) + 5$
$= 5$
$\therefore$ the y-intercept is 5.

b When $x = 0$, $y = (0+3)(0-6)$
$= 3 \times -6$
$= -18$
$\therefore$ the y-intercept is -18.

The x-intercept is found by letting $y = 0$ in the equation of the function. So, to find the x-intercepts of $y = ax^2 + bx + c$, we solve the equation $ax^2 + bx + c = 0$.

This quadratic equation may have two, one, or no solutions, so the function may have two x-intercepts, one x-intercept, or no x-intercepts.

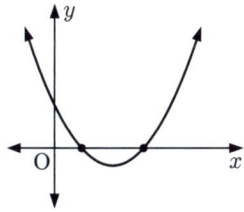

two x-intercepts

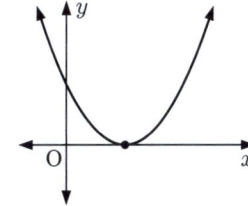

one x-intercept

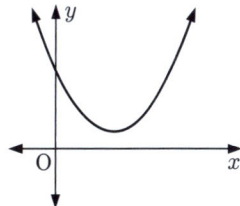

no x-intercepts

Example 13

Find the x-intercept(s) of:

a $y = (x+2)(x-7)$

b $y = x^2 + 6x + 9$

a When $y = 0$, $(x+2)(x-7) = 0$
$\therefore x = -2$ or 7
$\therefore$ the x-intercepts are -2 and 7.

b When $y = 0$, $x^2 + 6x + 9 = 0$
$\therefore (x+3)^2 = 0$
$\therefore x = -3$
$\therefore$ the x-intercept is -3.

EXERCISE 25G.1

1 State the x and y-intercepts of each graph:

a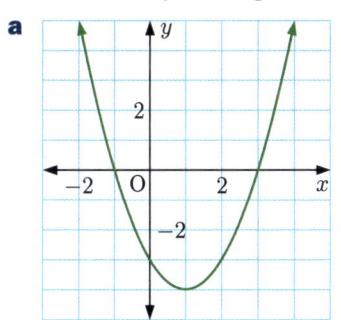

b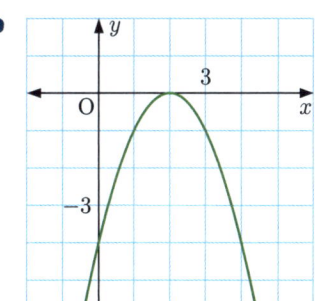

c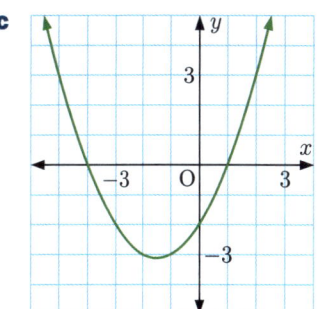

2 Find the y-intercept of:
- **a** $y = x^2 - 3x + 6$
- **b** $y = 3x^2 - 2x - 4$
- **c** $y = -2x^2 + 6x - 7$
- **d** $y = (x+2)(x+6)$
- **e** $y = (x-3)(x+5)$
- **f** $y = 2(x-4)(x+4)$

3 Find the x-intercept(s) of:
- **a** $y = (x+2)(x-3)$
- **b** $y = x(x-4)$
- **c** $y = (x+3)(x-7)$
- **d** $y = (x-1)^2$
- **e** $y = 2(x-3)(x+4)$
- **f** $y = -(x+5)^2$

It is easier to find the x-intercept(s) if the quadratic is in factorised form.

4 Find the x-intercept(s) of:
- **a** $y = x^2 - 5x$
- **b** $y = x^2 - 9$
- **c** $y = x^2 + 4x + 4$
- **d** $y = x^2 - 5x - 14$
- **e** $y = -x^2 + 12x - 36$
- **f** $y = 3x^2 - 3x - 6$

5 Find the axes intercepts of:
- **a** $y = (x+5)(x-4)$
- **b** $y = x^2 - 8x + 16$
- **c** $y = x^2 - 5x - 24$

6 The graph of $y = x^2 + x - 6$ is shown alongside.
- **a** Use the graph to find the solutions to $x^2 + x - 6 = 0$.
- **b** Use factorisation to check your answer.

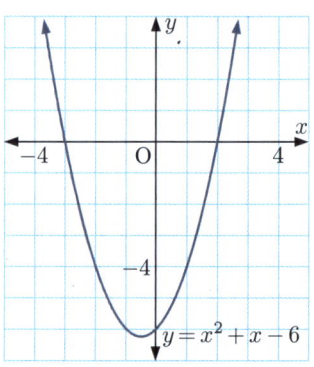

Example 14 ◀) Self Tutor

Sketch the graph of $y = -2(x+1)(x-2)$ by considering:
- **a** the value of a
- **b** the y-intercept
- **c** the x-intercepts.

$y = -2(x+1)(x-2)$

a $a = -2$ which is < 0, so the parabola opens downwards.

b When $x = 0$,
$y = -2(0+1)(0-2)$
$= -2 \times 1 \times -2$
$= 4$
$\therefore$ the y-intercept is 4.

c When $y = 0$,
$-2(x+1)(x-2) = 0$
$\therefore x = -1$ or $x = 2$
$\therefore$ the x-intercepts are -1 and 2.

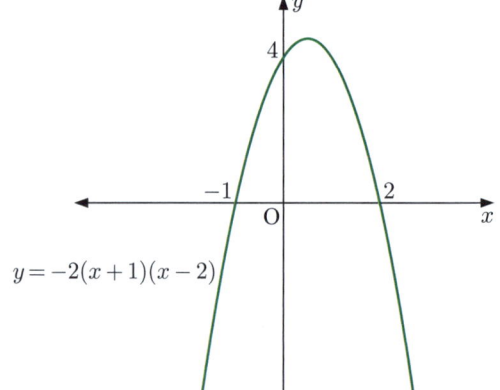

7 Sketch the graph of the quadratic function which has:

   a x-intercepts -1 and 1, and y-intercept -1

   b x-intercepts -3 and 1, and y-intercept 2

   c x-intercepts 2 and 5, and y-intercept -4

   d x-intercept 2 and y-intercept 4.

> If a quadratic function has only one x-intercept, its graph *touches* the x-axis.

8 Sketch the graph of each quadratic function by considering:

    i the value of a **ii** the y-intercept   **iii** the x-intercepts.

   a $y = (x+2)(x-2)$ **b** $y = (x-1)(x+3)$ **c** $y = (x+2)^2$

   d $y = -(x-2)(x+1)$ **e** $y = 3(x+1)^2$ **f** $y = -3(x-4)(x-1)$

   g $y = x^2 - 2x - 3$ **h** $y = x^2 + 2x - 8$ **i** $y = -x^2 - 5x + 6$

Activity

Click on the icon to practise matching a quadratic function with its graph.

QUADRATIC FUNCTIONS

USING GRAPHS TO ESTIMATE SOLUTIONS TO QUADRATIC EQUATIONS

If we are given the graph of $y = ax^2 + bx + c$, we can estimate the solutions to $ax^2 + bx + c = 0$ by observing the x-intercepts of the graph. This is especially useful if the equation cannot be solved by factorisation.

Example 15   ◀)) Self Tutor

Use the graph alongside to estimate the solutions to $x^2 - 3x + 1 = 0$.

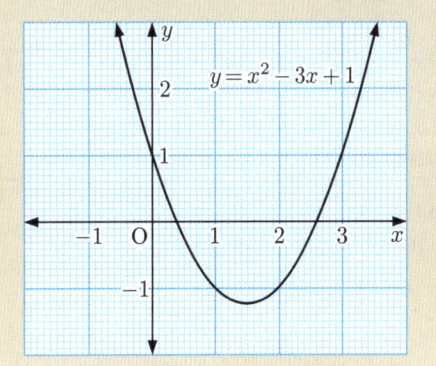

The x-intercepts of $y = x^2 - 3x + 1$ are ≈ 0.4 and ≈ 2.6.

$\therefore$ the solutions to $x^2 - 3x + 1 = 0$ are $x \approx 0.4$ or 2.6.

EXERCISE 25G.2

1 Use the graph given to estimate the solutions to:

a $x^2 - 4x + 2 = 0$

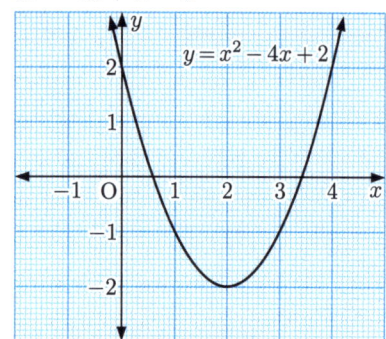

b $-x^2 - x + 3 = 0$.

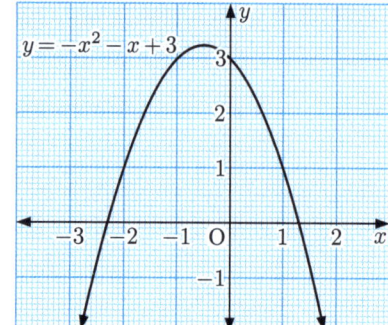

2 Use the graph alongside to estimate the solutions to:

a $x^2 + 6x + 7 = 0$
b $x^2 + 6x + 7 = 1$
c $x^2 + 6x + 7 = -2$.

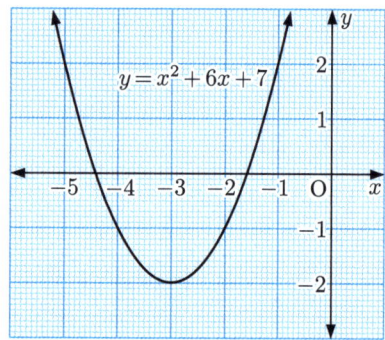

H AXIS OF SYMMETRY

The graphs of all quadratic functions are symmetrical about a vertical line passing through the vertex. This line is called the **axis of symmetry**.

If the graph has two x-intercepts, then the axis of symmetry is midway between them.

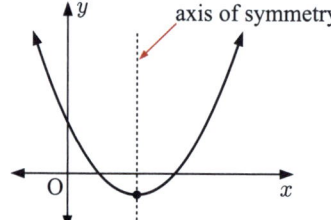

The equation of a vertical line has the form $x = k$.

Example 16   ◀)) Self Tutor

Find the equation of the axis of symmetry for the quadratic graph below.

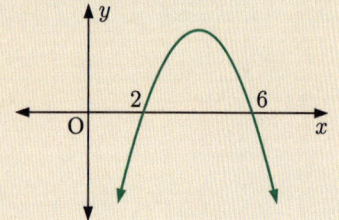

The x-intercepts are 2 and 6, and 4 is midway between 2 and 6.

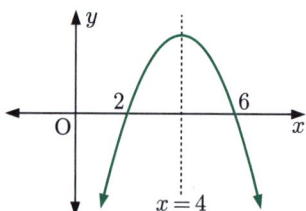

∴ the axis of symmetry is $x = 4$.

EXERCISE 25H

1 For each graph, find the equation of the axis of symmetry:

a 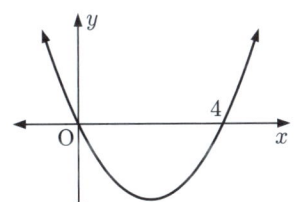   b 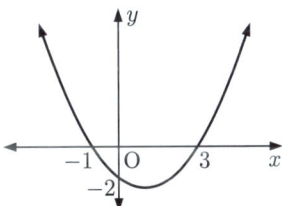   c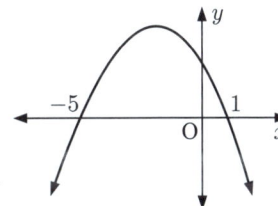

d 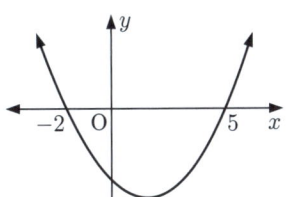   e 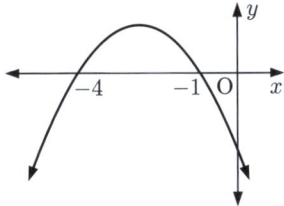   f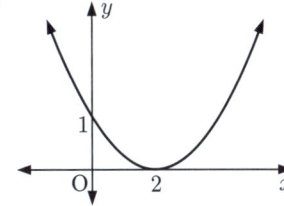

2 For each of the following quadratic functions:

    i Find the x-intercept(s).   **ii** Find the equation of the axis of symmetry.

  a $y = (x-2)(x-4)$ **b** $y = -(x+1)(x-5)$ **c** $y = 2(x+3)(x-3)$

  d $y = x(x+5)$ **e** $y = -3(x+4)^2$ **f** $y = 4(x+6)(x-9)$

3 The quadratic function alongside has x-intercept -5 and axis of symmetry $x = -1$. Find the other x-intercept.

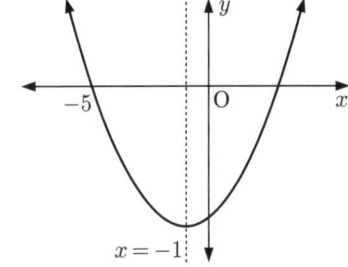

I   VERTEX

The **vertex** is the **turning point** of a quadratic function. For the function $y = ax^2 + bx + c$, the vertex is the point at which the function has:

- a **maximum value** for $a < 0$

- a **minimum value** for $a > 0$.

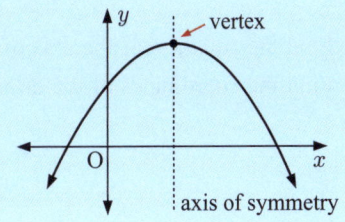

The axis of symmetry gives us the x-coordinate of the vertex. We substitute this value into the function to find the y-coordinate of the vertex.

Example 17

Consider the quadratic function $y = -x^2 + 2x + 3$.
- **a** Find the axes intercepts.
- **b** Find the equation of the axis of symmetry.
- **c** Find the coordinates of the vertex.
- **d** Sketch the function, showing the features you have found.

a When $x = 0$, $y = 3$
$\therefore$ the y-intercept is 3.

When $y = 0$, $-x^2 + 2x + 3 = 0$
$\therefore \; -(x^2 - 2x - 3) = 0$
$\therefore \; -(x + 1)(x - 3) = 0$
$\therefore \; x = -1$ or 3
$\therefore$ the x-intercepts are -1 and 3.

b 1 is midway between -1 and 3.
$\therefore$ the axis of symmetry is $x = 1$.

c When $x = 1$,
$y = -(1)^2 + 2(1) + 3$
$= -1 + 2 + 3$
$= 4$
$\therefore$ the vertex is $(1, 4)$.

d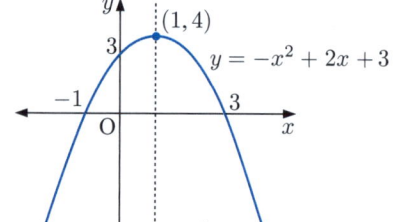

EXERCISE 25I

1 For each quadratic function, write down the coordinates of the vertex.

a 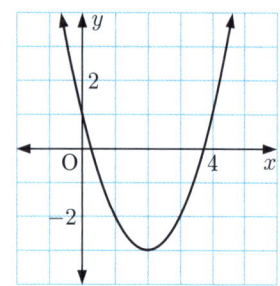 **b** 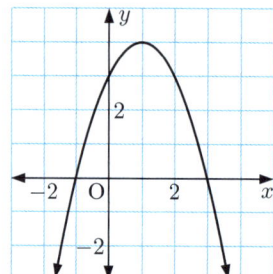 **c**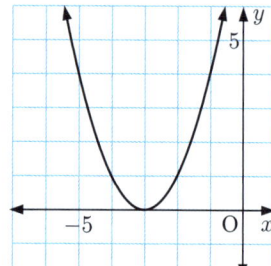

2 The quadratic function $y = x^2 - 4x - 5$ has x-intercepts -1 and 5.
- **a** Find the equation of the axis of symmetry.
- **b** Find the coordinates of the turning point.

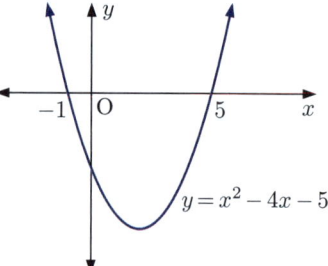

3 Consider the quadratic function $y = (x + 2)(x - 6)$.
- **a** Find the x-intercepts.
- **b** Find the equation of the axis of symmetry.
- **c** Find the coordinates of the turning point.

4 For each of the following quadratic functions:
  i Find the axes intercepts.
  ii Find the equation of the axis of symmetry.
  iii Find the coordinates of the vertex.
  iv Hence, sketch the graph of the function.

  a $y = x(x-2)$
  b $y = -(x-1)(x+3)$
  c $y = 2(x-3)^2$
  d $y = x^2 - 2x - 8$
  e $y = 4x - x^2$
  f $y = -x^2 - x + 6$

Review set 25A

1 Determine whether the following are quadratic equations:
  a $4x^2 - 8x = 0$
  b $x^2 = x + \dfrac{1}{x}$
  c $2x + 7 = 0$

2 Solve for x:
  a $x(x-3) = 0$
  b $(x+5)(x-2) = 0$
  c $-(x+7)^2 = 0$

3 Solve for x:
  a $x^2 + 6x = 0$
  b $3x^2 - 75 = 0$
  c $x^2 + 7x - 30 = 0$
  d $x^2 - 5x - 24 = 0$
  e $2x^2 - 12x = 14$
  f $x(x+2) = 35$

4 When the square of a number is subtracted from the original number, the result is -110. Find the number.

5 The area of the square is twice the area of the rectangle. Find the dimensions of each figure.

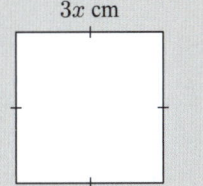
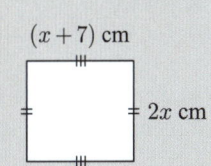

6 For the quadratic function $y = x^2 - 3x - 15$, find:
  a the value of y when $x = 4$
  b the values of x when $y = 3$.

7 Use a table of values from $x = -3$ to $x = 3$ to draw the graph of $y = x^2 + x - 3$.

8 Find the x-intercepts of:
  a $y = 5x(x+4)$
  b $y = 2x^2 + 6x - 56$

9 Use this graph to estimate the solutions to $x^2 - 5x + 1 = 0$.

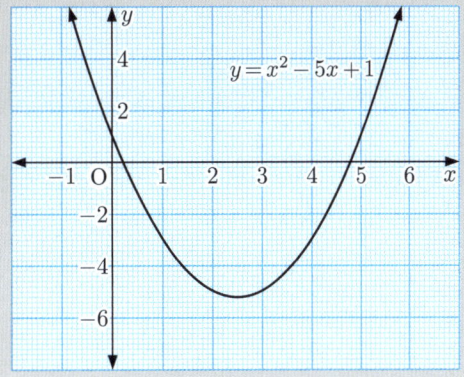

10 Find the equation of the axis of symmetry for each graph:

a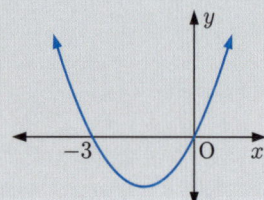

b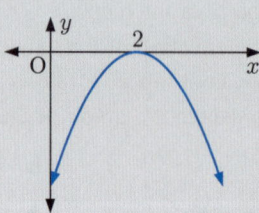

11 Consider the quadratic function $y = x^2 + 2x - 24$.
   a Find the axes intercepts.
   b Find the equation of the axis of symmetry.
   c Find the coordinates of the vertex.
   d Hence, sketch the graph of the function.

Review set 25B

1 Explain why the quadratic equation $x^2 + 9 = 0$ has no solutions.

2 What can be deduced from $2pqr = 0$?

3 Solve for x:
   a $x^2 = -4x$
   b $4 - 9x^2 = 0$
   c $5x^2 - 80 = 0$

4 Solve for x:
   a $x^2 - 6x - 27 = 0$
   b $x^2 - 20 = 8x$
   c $x^2 + 11 = 4(1 - 2x)$

5 A rectangular plot has length 5 m more than its width. If the area of the plot is 84 m², find its dimensions.

6 The sum of the squares of three consecutive even numbers is 308. What are the numbers?

7 Suppose $y = x^2 + x - 12$. Find the values of x for which $y = 30$.

8 Use a table of values from $x = -3$ to $x = 3$ to draw the graph of $y = -x^2 + 4x + 10$.

9 Find the x and y-intercepts of:
   a $y = (x + 7)(x - 3)$
   b $y = x^2 - 13x + 30$

10 Sketch the graph of $y = x^2 - 2x - 15$ by considering:
   a the value of a
   b the y-intercept
   c the x-intercepts.

11 A quadratic function has x-intercept 2 and axis of symmetry $x = -3$. Find the other x-intercept.

12 The graph of $y = x^2 + x - 2$ is shown alongside.
   a Find the coordinates of A and B.
   b Find the equation of the axis of symmetry.
   c Find the coordinates of the turning point.

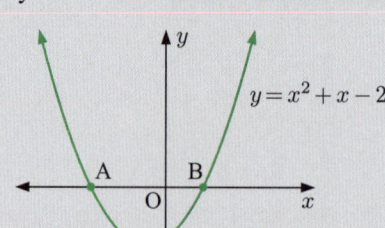

26

Proportion

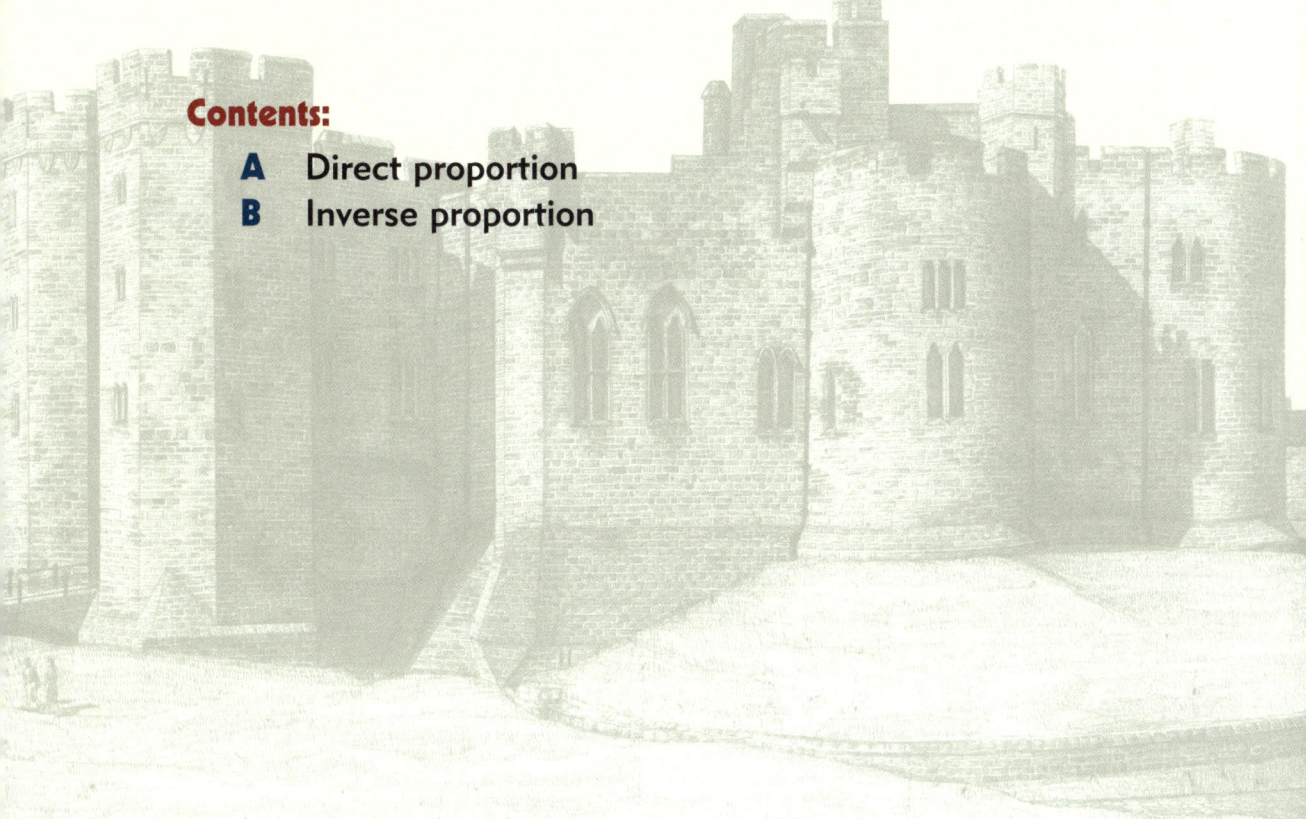

Contents:
- **A** Direct proportion
- **B** Inverse proportion

Opening problem

A fairground stall sells bottles of soft drink for £2 each. Suppose we buy n bottles and the total cost is £C.

To study the relationship between the *number of bottles* and the *total cost*, we can use a table of values or a graph.

n	0	1	2	3	4	5
C	0	2	4	6	8	10

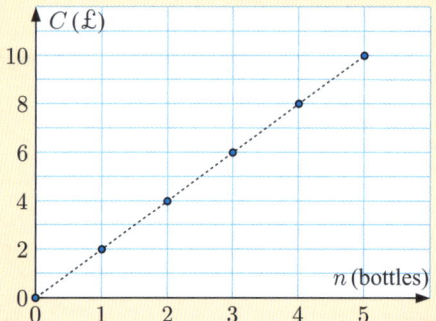

The total cost *depends* on the number of bottles bought.

The number of bottles is the independent variable, so it is placed on the x-axis.
The total cost is the dependent variable, so it is placed on the y-axis.

Since we can only buy a whole number of bottles, the graph of C against n consists of discrete points. However, an imagined line passing through these points would also pass through the origin.

Things to think about:

a Which of the following are true?
  i doubling the number of bottles doubles the total cost
  ii halving the number of bottles halves the total cost
  iii increasing the number of bottles by 30% increases the total cost by 30%.

b How can we describe the relationship between n and C?

A DIRECT PROPORTION

The rectangles alongside are **similar**. Their sides are in the same **ratio**.

The ratios of the side lengths can be written as the *proportion* $1 : 3 = 2 : 6$

A rectangle with sides x cm and y cm will be similar to both of the other two if $x : y = 1 : 3$. This occurs when $y = 3x$, for any x.

Since the sides are in proportion, we say that y is **directly proportional** to x.

Two variables are **directly proportional** if multiplying one of them by a number results in the other one being multiplied by the same number.

In the **Opening Problem**, C is directly proportional to n.

- When n is tripled from 1 to 3, C is tripled from 2 to 6.
- If we plot C against n, the points lie in a straight line with gradient 2, and which passes through the origin. The variables are connected by the formula $C = 2n$.

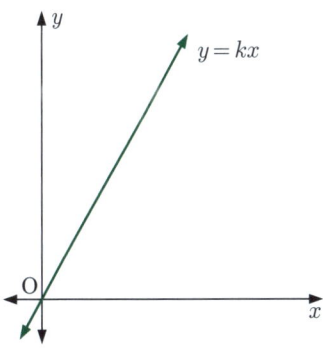

If two quantities x and y are **directly proportional**, we write $y \propto x$.

If $y \propto x$ then $y = kx$ where k is a constant called the **proportionality constant**.

When y is graphed against x, the graph is a straight line with **gradient** k, which passes through the **origin**.

$\propto$ reads "*is directly proportional to*".

Example 1 ◀) Self Tutor

At the village bakery, fruit buns are sold for 60 pence each.

a Use a graph to show that £y, the total price paid, is directly proportional to x, the number of fruit buns bought.

b Find a formula connecting x and y.

a

x	0	1	2	3	4	5
y	0	0.60	1.20	1.80	2.40	3.00

The points lie in a straight line passing through $(0, 0)$, so $y \propto x$.

b The gradient of the line $= \dfrac{0.60 - 0}{1 - 0} = 0.6$

$\therefore$ the proportionality constant k is 0.6.

$\therefore$ the formula connecting x and y is $y = 0.6x$.

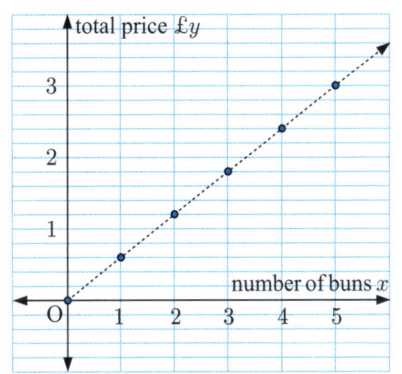

EXERCISE 26A.1

1 Which graph indicates that y is directly proportional to x?

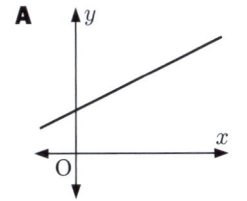 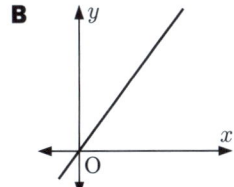 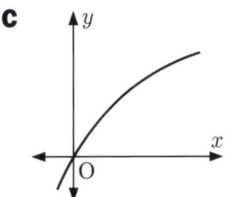 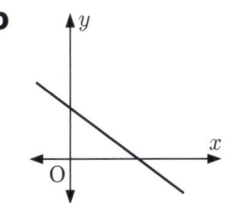

2 For each table of values:

  i Plot the graph of y against x.
  ii Determine whether y is directly proportional to x. If it is, find the proportionality constant.

a

x	0	1	2	3	4
y	0	3	6	9	12

b

x	0	1	2	3	4
y	3	5	7	9	11

c

x	0	1	3	5	6
y	0	4	8	12	16

d

x	2	3	5	6	8
y	5	7.5	12.5	15	20

3 The table alongside shows the total wages £W earned for working h hours.

h	0	1	2	3	4	5
W	0	12.50	25.00	37.50	50.00	62.50

  a Draw a graph of W against h.
  b Explain why W and h are directly proportional.
  c Find a formula connecting W and h.

4 A dripping tap leaks water at a rate of 10 ml per minute.

  a Copy and complete the table of values alongside:

Time (t minutes)	0	2	4	6	8	10
Volume of water leaked (V ml)						

  b Plot the graph of V against t.
  c Are V and t directly proportional? If so, find the proportionality constant.

5 A dry cleaner charges £3.20 for cleaning each of the first three items, then £2.20 for each extra item.

  a Copy and complete the table of values alongside:

Number of items (n)	0	1	2	3	4	5
Cost (£C)						

  b Plot C against n.
  c Are C and n directly proportional? If so, find the proportionality constant.

6 For each of the following equations:

  i Draw the graph of y against x.
  ii Determine whether y is directly proportional to x.

  a $y = 4x$ **b** $y = x + 2$ **c** $y = \dfrac{7x}{3}$ **d** $y = 6 - x$

7 Suppose y is directly proportional to x. State what happens to:

  a y if x is doubled
  b y if x is trebled
  c x if y is doubled
  d x if y is halved
  e y if x is increased by 20%
  f x if y is decreased by 30%
  g y if 2 is added to x
  h y if 3 is subtracted from x.

8 The law connecting the circumference C and radius r of a circle is $C = 2\pi r$.

  a Explain why $C \propto r$.
  b Find the proportionality constant.
  c State what happens to:
    i C if r is doubled
    ii r if C is increased by 50%.

PROBLEM SOLVING USING DIRECT PROPORTION

If two variables are directly proportional, we can use either a formula or a multiplier to find the value of one variable given the value of the other. In general, using a multiplier is the quicker method.

Example 2

Suppose $y \propto n$ and that $y = 40$ when $n = 3$. Find n when $y = 137$.

Method 1:

Since $y \propto n$, $y = kn$ where k is the proportionality constant.

When $n = 3$, $y = 40$

$\therefore \ 40 = k \times 3$

$\therefore \ \frac{40}{3} = k$

$\therefore \ y = \frac{40}{3}n$

So, when $y = 137$,

$137 = \frac{40}{3}n$

$\therefore \ 137 \times \frac{3}{40} = n$

$\therefore \ n \approx 10.3$

Method 2:

n	3	?
y	40	137

$\times \frac{137}{40}$

To change y from 40 to 137, we multiply by $\frac{137}{40}$.

Since $y \propto n$, we must also multiply n by $\frac{137}{40}$

$\therefore \ n = 3 \times \frac{137}{40} \approx 10.3$

EXERCISE 26A.2

1. Suppose $y \propto x$ and that $y = 35$ when $x = 7$. Find:
   a y when $x = 28$
   b x when $y = 5$.

2. Suppose $y \propto x$ and that $y = 20$ when $x = 6$. Find:
   a y when $x = 18$
   b x when $y = 70$.

3. Complete the table alongside given that $y \propto x$:

x	5	15	
y	20		120

4. The amount of petrol Jill's car uses each day is directly proportional to the distance she travels. Yesterday she travelled 81 km, and used 6 litres of petrol.
   a If Jill travelled 54 km in one day, how much petrol would she use?
   b Jill used 9 litres of petrol in one day. How far did she travel?

5. The resistance R ohms to the flow of electricity in a wire varies in direct proportion to the length l cm of the wire. A 10 cm length of wire has resistance 0.06 ohms.
   a Find:
      i the law connecting R and l
      ii the resistance when the wire is 50 cm long
      iii the length of wire which has a resistance of 3 ohms.
   b Plot the graph of R against l.

530 Proportion (Chapter 26)

6 A 4 litre can of paint will cover 18 m² of wall. Find:
   a the area of wall which can be covered using 10 litres of paint
   b the amount of paint required to paint a room with wall area 40.5 m².

7 Rachael eats l lollies every 5 days. Find, in terms of l:
   a the number of lollies Rachael eats in 20 days
   b the number of days it will take Rachael to eat 30 lollies.

8 The speed of a falling object is directly proportional to the time for which it falls. The speed of an object which has fallen for 5 seconds is 49 m/s.
   a Find the speed of an object which falls for 8 seconds.
   b How long will it take a falling object to reach a speed of 100 m/s?

9 Alongside is a conversion graph between British pounds (£x) and euros (€y).
   a Is y directly proportional to x? Explain your answer.
   b Convert:
      i £200 to euros
      ii €180 to pounds.

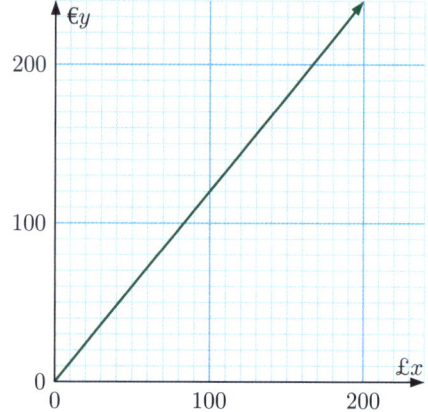

10 The dimensions of a model car are directly proportional to the dimensions of the actual car it was designed from. The actual car is 4 m long, and the model car is 30 cm long.
   a The actual car is 1.4 m wide. Find the width of the model car.
   b The model car is 9 cm high. Find the height of the actual car.

B INVERSE PROPORTION

Discussion

If two painters can paint a house in three days, how long would it take four painters to paint the house, each working at the same rate?

The two variables in this example are the *number of painters* and the *time taken*.

The four painters will be able to do twice as much work in the same amount of time. So it will therefore take them a half of the time, or $1\frac{1}{2}$ days, to complete the job.

In a case like this where doubling one variable halves the other, we have an **inverse proportion**.

> Two variables are **inversely proportional** or **vary inversely** if, when one is *multiplied* by a constant, the other is *divided* by the same constant.

Dividing by k is the same as multiplying by $\frac{1}{k}$, so if one of the variables is multiplied by 2, the other must be multiplied by $\frac{1}{2}$.

Consider again the example of two painters completing a job in three days.

Suppose x is the number of painters and y is the number of days to complete the job.

The table below shows some possible combinations of x and y.

x	1	2	3	4	6
y	6	3	2	$1\frac{1}{2}$	1

When these points are plotted, they form part of a **hyperbola**.

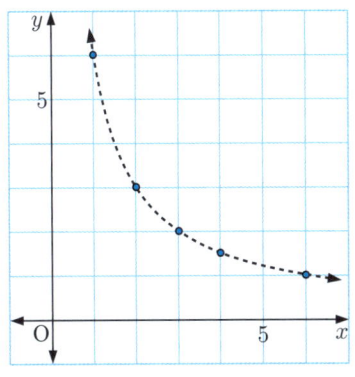

Suppose we include in the table a row for the values of $\frac{1}{x}$.

x	1	2	3	4	6
$\frac{1}{x}$	1	$\frac{1}{2}$	$\frac{1}{3}$	$\frac{1}{4}$	$\frac{1}{6}$
y	6	3	2	$1\frac{1}{2}$	1

The points on a graph of y against $\frac{1}{x}$ form a straight line which passes through the origin, so y is directly proportional to $\frac{1}{x}$.

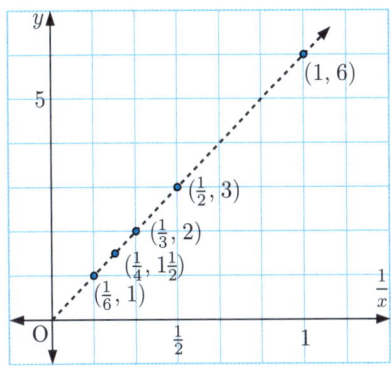

> If y is **inversely proportional** to x, then y is **directly proportional** to $\frac{1}{x}$.
>
> Consequently, $\quad y = \dfrac{k}{x} \quad$ or $\quad xy = k$.

In the painters example, notice that $xy = 6$ for all points in the table. So, in this case $k = 6$.

Example 3

Suppose M is inversely proportional to t, and that $M = 200$ when $t = 3$. Find:

a M when $t = 5$
b t when $M = 746$

Since M is inversely proportional to t, $M \propto \dfrac{1}{t}$.

a

t	3	5
M	200	

$\times \frac{5}{3}$

t is multiplied by $\dfrac{5}{3}$

$\therefore$ M is multiplied by $\dfrac{3}{5}$ {as $M \propto \dfrac{1}{t}$}

$\therefore$ $M = 200 \times \dfrac{3}{5} = 120$

b

t	3	
M	200	746

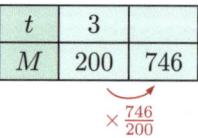

M is multiplied by $\dfrac{746}{200}$

$\therefore$ t is multiplied by $\dfrac{200}{746}$ {as $M \propto \dfrac{1}{t}$}

$\therefore$ $t = 3 \times \dfrac{200}{746} \approx 0.80$

EXERCISE 26B

1 a Copy and complete the table of values:

x	1	2	4	8
$\dfrac{1}{x}$				
y	40	20	10	5

b Plot the graph of y against $\dfrac{1}{x}$.

c Explain why y is inversely proportional to x.

d Calculate xy for each point in the table.

2 For each of the following tables, calculate the value of xy for each point. Hence determine whether x and y are inversely proportional. If an inverse proportionality exists, determine the law connecting the variables, and draw the graph of y against x.

a

x	1	2	3	6
y	12	6	4	2

b

x	2	3	6	9
y	12	8	4	3

c

x	7	6	12	1
y	12	14	7	84

3 Which graph(s) could indicate that y is inversely proportional to x?

A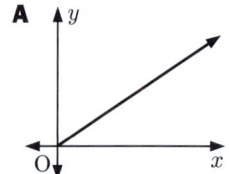

B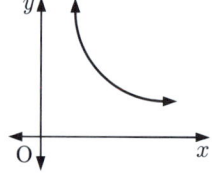

C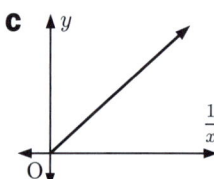

D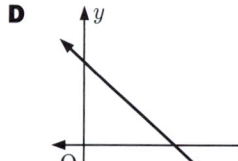

If y is inversely proportional to x, then y is directly proportional to $\dfrac{1}{x}$.

4 The formula connecting average speed s, distance travelled d, and time taken t, is $s = \dfrac{d}{t}$.
Complete the following statements by inserting "directly" or "inversely":

   a If d is fixed, s is ...... proportional to t.   **b** If t is fixed, s is ...... proportional to d.
   c If s is fixed, d is ...... proportional to t.

5 Suppose y is inversely proportional to x. Describe what happens to:

   a y if x is multiplied by 4 **b** y if x is multiplied by $\dfrac{7}{9}$ **c** x if y is multiplied by $\dfrac{8}{3}$.

6 Suppose y is inversely proportional to x, and that $y = 20$ when $x = 2$. Find:

   a the value of y when $x = 5$ **b** the value of x when $y = 100$.

7 Suppose p is inversely proportional to n, and that $p = 24$ when $n = 6$. Find:

   a the value of p when $n = 8$ **b** the value of n when $p = 12$.

8 Copy and complete this table given that $y \propto \dfrac{1}{x}$:

x	6	10	
y	25		3

Example 4   ◀)) Self Tutor

The velocity V of a body travelling a fixed distance is inversely proportional to the time taken t to complete the journey. When the velocity is 40 cm/s, the time taken is 280 seconds. Find the time taken when the velocity is 50 cm/s.

$V \propto \dfrac{1}{t}$

t	280	
V	40	50

$\times \dfrac{50}{40}$

V is multiplied by $\dfrac{50}{40} = \dfrac{5}{4}$

$\therefore\ t$ is multiplied by $\dfrac{4}{5}$ {as $V \propto \dfrac{1}{t}$}

$\therefore\ t = 280 \times \dfrac{4}{5} = 224$

$\therefore$ the time taken is 224 seconds.

---

9 The time taken to complete a certain job is inversely proportional to the number of workers doing the task. If 20 workers could do the job in 6 days, find how long it would take 15 workers to do the job.

10 Four people are hired to set up 5000 seats for a concert. Working together, they could complete the job in 12 hours. However, one person is unwell, and is unable to help set up the chairs.

   a How many chairs will the remaining three people be able to set up in 12 hours?
   b How long will it take them to complete the job?

11 A rectangle with dimensions x m by y m has area 60 m².

   a Explain why y is inversely proportional to x.
   b Copy and complete this table:

x	5		15
y		10	

   c Plot the graph of y against x.

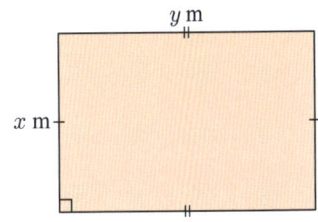

12 The chamber of an airtight cylinder is filled with the yellow gas fluorine. For a given mass of gas kept at a constant temperature, the volume is inversely proportional to the pressure. When $V = 10$ cm^3, the pressure $P = 40$ units.

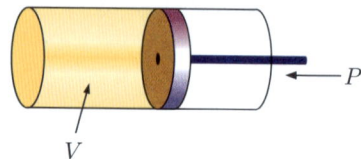

  a Find the pressure when the volume is 20 cm^3.
  b Find the volume when the pressure is 100 units.
  c Plot the graph of P against V.

13 The average court time for each player in a netball team is inversely proportional to the number of players the team has. If the team has 10 players, the average court time for each player is 28 minutes.

  a If the team has 12 players, what is the average court time for each player?
  b If the average court time is 35 minutes, how many players does the team have?
  c Given that 7 players are allowed on court at one time, how long does each game last?

Review set 26A

1 Which of the following graphs indicates that y is directly proportional to x?

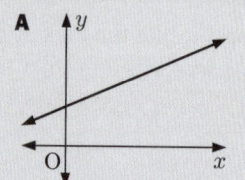 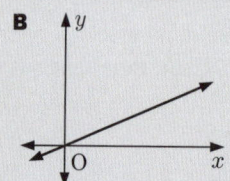 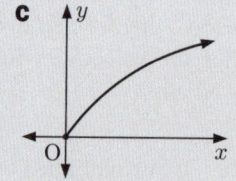 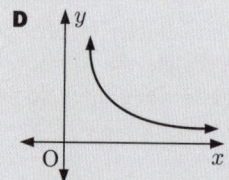

2 Consider the table of values alongside.

x	2	3	6	11
y	6	9	18	33

  a Plot the graph of y against x.
  b Explain why y is directly proportional to x.
  c Find a formula connecting x and y.

3 Copy and complete this table given that $y \propto x$:

x	12	60	
y	3		20

4 Suppose x is inversely proportional to y, and that $y = 65$ when $x = 10$. Find:
  a the law connecting x and y **b** y when $x = 12$ **c** x when $y = 150$.

5 The variable M is directly proportional to t. What happens to M if t is trebled?

6 A school's quiz night raises money in direct proportion to the number of people attending. Last year 85 people attended the quiz night, and the event raised £1275. This year the entry fees were the same, but the school only raised £1080. How many people attended this year?

7 Darren thinks that y is inversely proportional to x. Explain why Darren is incorrect.

x	4	10	50	100
y	50	20	5	2

8 The variables x and y in the table are inversely proportional. Find a and b.

x	3	6	a
y	8	b	60

9 If 3 workers can build a house in 60 days, how long would it take 5 workers to build the house?

10 The Dentists' Association has found that the average number of fillings required at a patient's checkup is inversely proportional to the time the patient spends cleaning their teeth each day.

Over the last 10 years, Thomas has spent 40 seconds brushing his teeth each day, and has needed 3 fillings. If Thomas increases his brushing time to 2 minutes each day for the next 10 years, how many fillings will he need in this period?

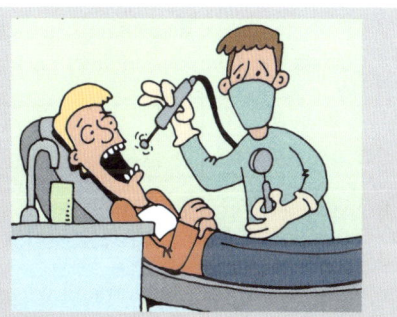

Review set 26B

1 The variables p and q in the table are directly proportional. Find a and b.

p	4	7	a
q	12	b	42

2 If 7 litres of petrol are needed to drive 100 km, how far could you travel on 16 litres of petrol?

3 The table alongside shows the total price £P of buying n books at £36 each.

Number of books (n)	1	2	3	4
Total price (£P)	36	72	108	144

  a Show that P and n are directly proportional.
  b Find the law connecting P and n.
  c Plot the graph of P against n.

4 Which equation indicates that:

  a y is directly proportional to x
  b y is inversely proportional to x?

  A $y = 3x + 1$ **B** $y = \dfrac{3}{x}$ **C** $y = 3x^2$ **D** $y = \dfrac{1}{3}x$

5 Suppose $y \propto x$ and that $y = 77.9$ when $x = 4.1$.

  a Find the proportionality constant.
  b Find the value of y when $x = 6.3$.

6 Alongside is a conversion graph of the relationship between inches (x in) and centimetres (y cm).

  a Explain why y is directly proportional to x.
  b Convert 3 inches to centimetres. Give your answer correct to 1 decimal place.

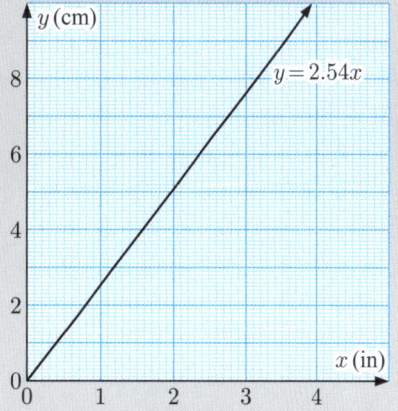

7 For each table, determine whether x and y are inversely proportional. If an inverse proportionality exists, find the law connecting x and y.

a
x	2	4	8
y	10	5	2

b
x	3	6	9
y	12	6	4

8 Suppose y is inversely proportional to x, and that $y = 40$ when $x = 3$. Find x when $y = 15$.

9 Jamela thinks that if y is inversely proportional to x, then a 25% increase in y will produce a 25% decrease in x. Explain why Jamela is incorrect.

10 6 pumps can empty a lake in 45 minutes. How long would 10 pumps take to empty the lake?

27

Further functions

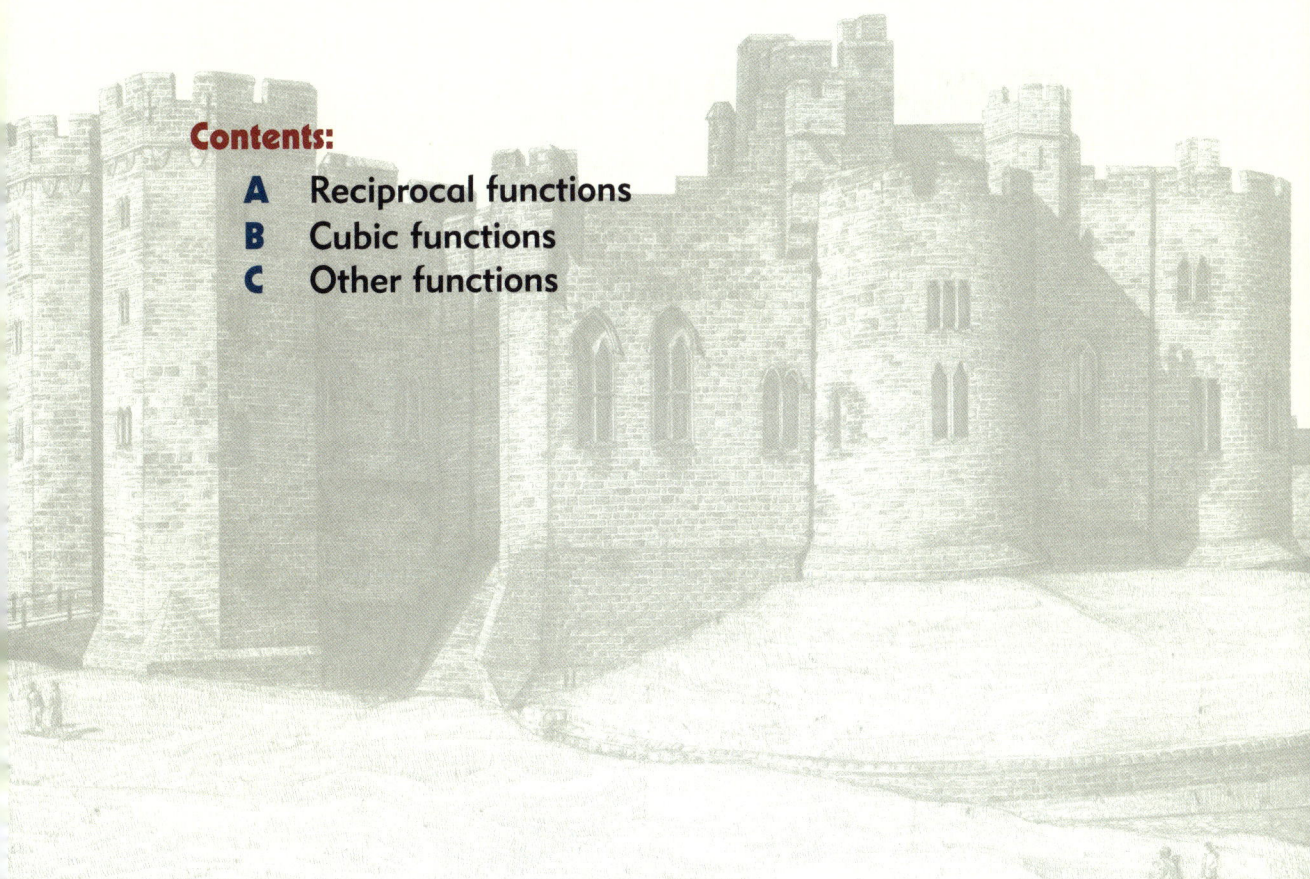

Contents:
- **A** Reciprocal functions
- **B** Cubic functions
- **C** Other functions

538 Further functions (Chapter 27)

Opening problem

A theme park water slide is constructed according to the function $H = -0.0125x^3 + 0.31x^2 - 2.35x + 12$, where H is the height in metres of the slide above the ground x metres from the starting platform.

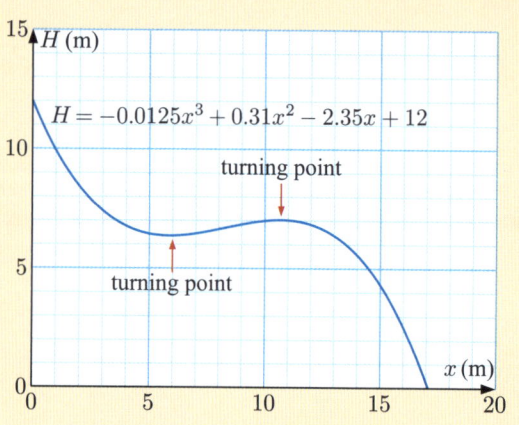

Things to think about:

a What type of function is this?

b How high is the starting platform above the ground?

c Supporting structures hold up the water slide at the turning points. Can you use the graph to estimate the height of each structure?

In this Chapter we complete our study of functions by considering reciprocal functions, cubic functions, and properties of other functions which do not fit into any of the categories we have studied.

A RECIPROCAL FUNCTIONS

In the graph alongside, y is inversely proportional to x. The variables are connected by the equation $y = \dfrac{20}{x}$.

However, this is not the complete graph of $y = \dfrac{20}{x}$, because we have only considered positive values of x and y.

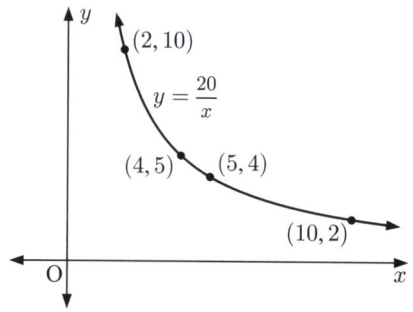

By extending the graph to include negative values of x and y, we can draw the complete graph of $y = \dfrac{20}{x}$.

$y = \dfrac{20}{x}$ is an example of a **reciprocal function**.

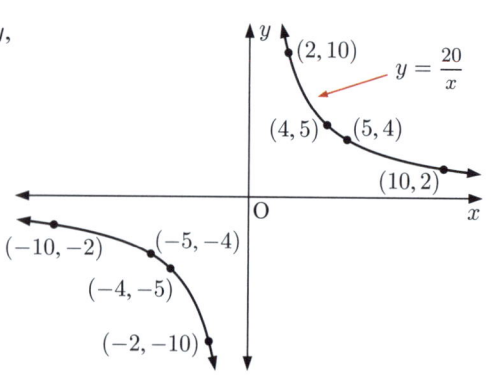

Further functions (Chapter 27) 539

A **reciprocal function** is a function of the form $y = \dfrac{k}{x}$, where $k \neq 0$ is a constant.

The graph of a reciprocal function is called a **rectangular hyperbola**.

The simplest example of a reciprocal function is $y = \dfrac{1}{x}$.

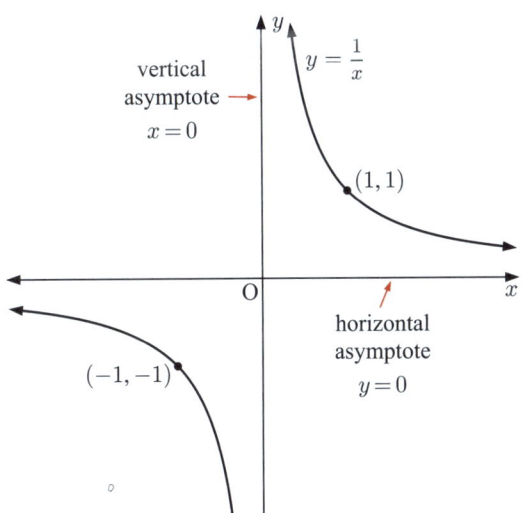

Notice that:

- The graph of $y = \dfrac{1}{x}$ has two branches.
- $y = \dfrac{1}{x}$ is undefined when $x = 0$.
- The graph of $y = \dfrac{1}{x}$ exists in the first and third quadrants only.
- The graph gets closer and closer to the horizontal line $y = 0$, but never reaches it. We say that $y = 0$ is a **horizontal asymptote**.
- The graph gets closer and closer to the vertical line $x = 0$, but never reaches it. We say that $x = 0$ is a **vertical asymptote**.

Example 1 ◀) Self Tutor

Complete the table of values for the reciprocal function $y = \dfrac{8}{x}$.

x	-8	-4	-2	-1	1	2	4	8
y								

Hence draw a graph of the function.

When $x = -8$, $y = \dfrac{8}{-8} = -1$.

When $x = -4$, $y = \dfrac{8}{-4} = -2$.

Using this process for the other values of x, we complete the table of values:

x	-8	-4	-2	-1	1	2	4	8
y	-1	-2	-4	-8	8	4	2	1

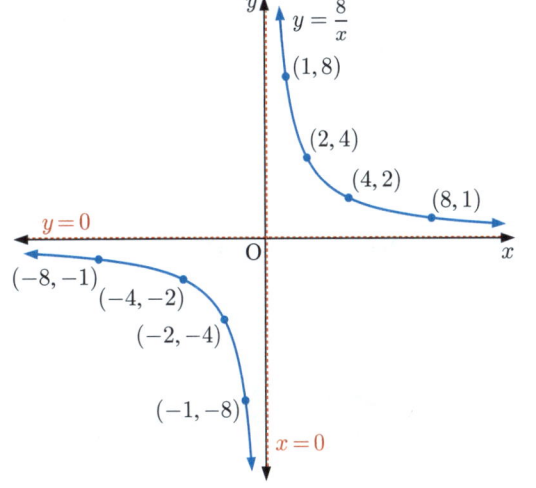

EXERCISE 27A

1 Copy and complete the table of values for the given function. Hence sketch the graph of the function.

  a $y = \dfrac{6}{x}$

x	-6	-4	-2	-1	1	2	4	6
y								

  b $y = -\dfrac{6}{x}$

x	-6	-4	-2	-1	1	2	4	6
y								

2 a Sketch the graphs of $y = \dfrac{1}{x}$, $y = \dfrac{2}{x}$, and $y = \dfrac{4}{x}$ on the same set of axes.

  For each function, complete this table of values to help you:

x	-4	-2	-1	1	2	4
y						

  DEMO

  b Describe the effect of varying k on the graph of $y = \dfrac{k}{x}$.

3 a Sketch the graphs of $y = -\dfrac{1}{x}$, $y = -\dfrac{2}{x}$, and $y = -\dfrac{4}{x}$ on the same set of axes.

  For each function, complete this table of values to help you:

x	-4	-2	-1	1	2	4
y						

  b Comment on the shape of the graph of $y = \dfrac{k}{x}$ when $k < 0$.

4 Determine the equation of each reciprocal function:

  a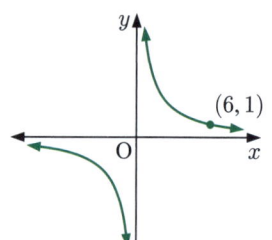
  b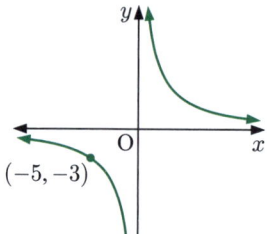
  c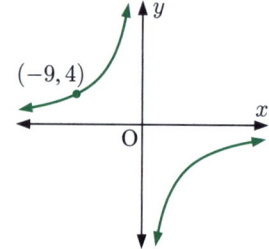

Discussion

- Can you explain why the graph of $y = \dfrac{k}{x}$ never cuts:
  ▸ the y-axis
  ▸ the x-axis?

B CUBIC FUNCTIONS

A **cubic function** is a function of the form $y = ax^3 + bx^2 + cx + d$, where a, b, c, and d are constants, $a \neq 0$.

The simplest cubic function is $y = x^3$. Its graph can be drawn from a table of values:

When $x = -2$, $y = (-2)^3 = -8$
When $x = -1$, $y = (-1)^3 = -1$
When $x = 0$, $y = 0^3 = 0$
When $x = 1$, $y = 1^3 = 1$
When $x = 2$, $y = 2^3 = 8$

x	-2	-1	0	1	2
y	-8	-1	0	1	8

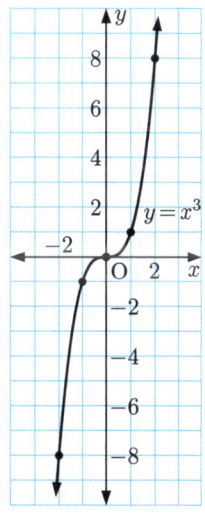

Example 2 ◀)) Self Tutor

Sketch the graph of $y = x^3 - 2x$ using a table of values from $x = -2$ to $x = 2$.

When $x = -2$, $y = (-2)^3 - 2(-2)$
$ = -8 + 4$
$ = -4$

Using this process for the other values of x, we complete the table of values:

x	-2	-1	0	1	2
y	-4	1	0	-1	4

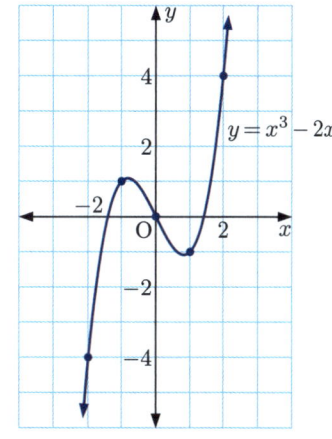

Example 3 ◀)) Self Tutor

Find the x-intercepts of each cubic function:

a $y = x(x - 3)(x - 4)$ **b** $y = (x + 1)^2(x - 5)$

a When $y = 0$,
$x(x - 3)(x - 4) = 0$
$\therefore x = 0, 3,$ or 4
$\therefore$ the x-intercepts are $0, 3,$ and 4.

b When $y = 0$,
$(x + 1)^2(x - 5) = 0$
$\therefore x = -1$ or 5
$\therefore$ the x-intercepts are -1 and 5.

The x-intercepts are found by letting $y = 0$.

EXERCISE 27B

1 Determine whether each function is a cubic function:

   a $y = 2x^3 + 3x^2 - x + 5$ **b** $y = x^3 - x^2 + \dfrac{1}{x}$ **c** $y = -x^3 - x + 7$

   d $y = \dfrac{1}{2}x^3 + \dfrac{3}{4}x^2 - 1$ **e** $y = x^4 + 2x^3 - x + 3$ **f** $y = 4 - 5x + 3x^3$

2 Use a table of values from $x = -2$ to $x = 2$ to sketch the graph of each cubic function:

   a $y = x^3 + 2$ **b** $y = x^3 - 3x$ **c** $y = -x^3 + 4x$

   d $y = x^3 - x^2$ **e** $y = x^3 - 2x - 4$ **f** $y = -x^3 + 3x^2 - 3$

3 Find the x-intercept(s) of each cubic function:

   a $y = x(x-2)(x-6)$ **b** $y = (x+5)(x+2)(x-1)$ **c** $y = -x(x+4)^2$

   d $y = 5x^2(x-2)$ **e** $y = \dfrac{1}{2}(x-1)^2(x+3)$ **f** $y = -\dfrac{1}{3}(x+5)^3$

4 Use the x-intercepts to match each cubic function with its graph:

   a $y = x^2(x-3)$ **b** $y = -x(x+2)(x-3)$ **c** $y = -(x+3)^3$

   d $y = (x+2)^2(x-1)$ **e** $y = (x-2)^3$ **f** $y = (x+3)(x+1)(x-2)$

A 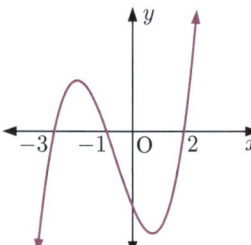   **B** 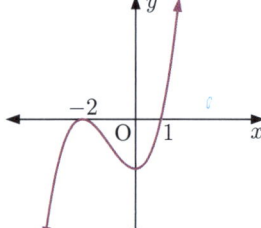   **C**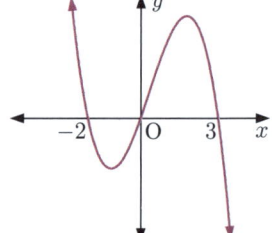

D 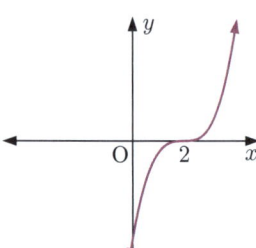   **E** 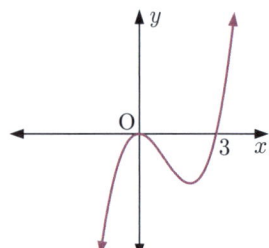   **F**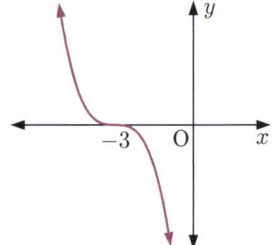

5 The graph of $y = x^3 - 5x + 1$ is shown below. Use the graph to estimate the solutions to the equation $x^3 - 5x + 1 = 0$.

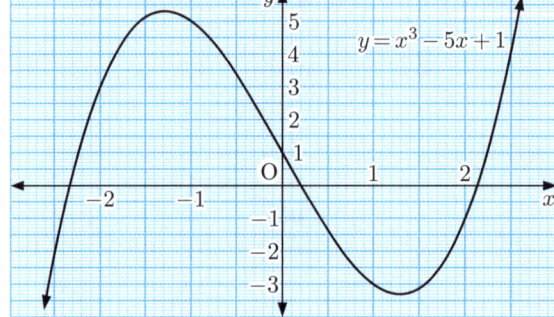

Discussion

Is it possible for a cubic function to have:
- three x-intercepts
- two x-intercepts
- one x-intercept
- no x-intercepts?

Where possible, sketch an example of such a curve.

C OTHER FUNCTIONS

Real world situations are not always modelled by simple linear or quadratic functions that we are familiar with. However, even if we are unfamiliar with the function, we can often still use the graph of the function to answer questions about the situation.

Example 4 ◆) Self Tutor

When a pain killing injection is administered, the effectiveness of the injection after t minutes is $E = 4t \times (0.78)^t$ units.

a Find the effectiveness of the injection after 1 minute.

b Find the maximum effectiveness of the injection, and the time at which it occurred.

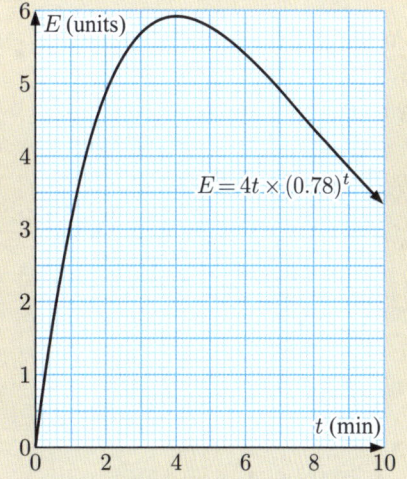

a When $t = 1$, $E \approx 3.1$
So, the effectiveness of the injection after 1 minute is ≈ 3.1 units.

b The maximum effectiveness of the injection is ≈ 5.9 units, and it occurs after ≈ 4 minutes.

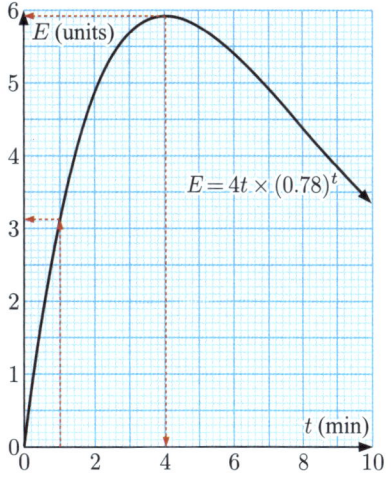

EXERCISE 27C

1 At the start of her ride, a cyclist accelerates from rest until she reaches her maximum speed. Her speed is given by $s = -\frac{1}{48}x^3 + \frac{1}{4}x^2$ m/s, where x is the time in seconds, $0 \leqslant x \leqslant 8$.

   a Find the speed of the cyclist after 6 seconds.

   b At what time did the cyclist have speed 2 m/s?

Check your answers using technology.

GRAPHING PACKAGE

CALCULATOR INSTRUCTIONS

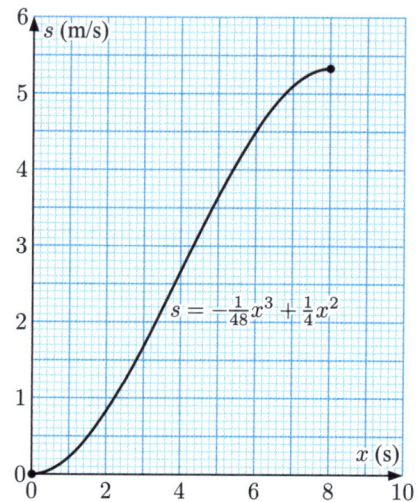

2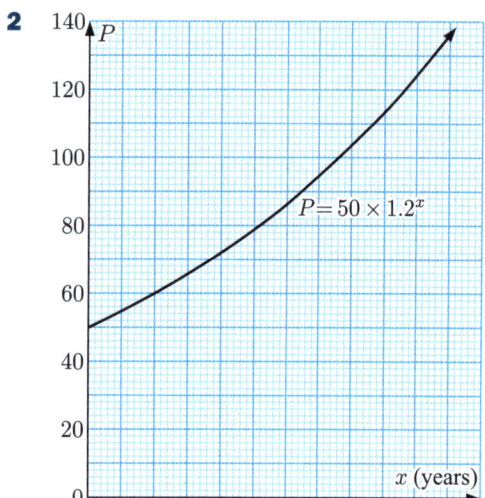

The population of puffins nesting in a series of cliffs is given by $P = 50 \times 1.2^x$, where x is the number of years after the initial observation.

   a Find the population of puffins:

      i initially   **ii** after 2 years.

   b How long does it take for the population to reach 100?

Check your answers using technology.

3 A crate with a square base x m long and an open top is to be constructed with volume 6 m³.

   a Explain why:

      i the height of the crate is $\frac{6}{x^2}$ m

      ii the outer surface area of the crate is $\left(x^2 + \frac{24}{x}\right)$ m².

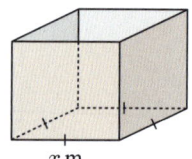

   b Use the graph alongside to estimate the base side length which minimises the outer surface area of the crate.

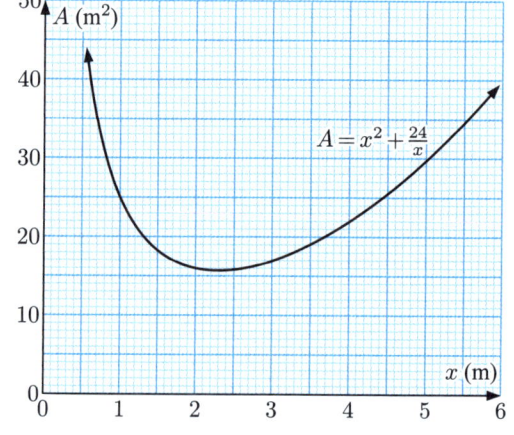

4 When a tablet is swallowed, the quantity of medicine in the bloodstream after t hours is given by $M = 2.2t(0.6)^t$ mg.

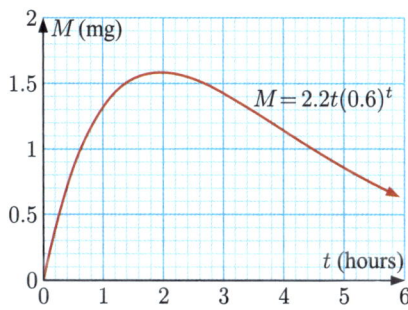

  a Find the quantity of medicine in the bloodstream after:
    i 1 hour   **ii** 5 hours.
  b Find the maximum quantity of medicine in the bloodstream and the time when this occurs.

Review set 27A

1 Copy and complete this table of values for $y = \dfrac{9}{x}$:

x	-9	-3	-1	1	3	9
y						

Hence sketch a graph of the function.

2 Nick has constructed the table of values alongside for a reciprocal function, but he has made an eror.

x	-8	-4	-2	2	4	8
y	4	8	16	-16	8	-4

  a What is the reciprocal function he is using?
  b What was his error?

3 Using a table of values from $x = -2$ to $x = 2$, sketch the graph of $y = x^3 + x^2 - 2x$.

4 Using x-intercepts, match each cubic equation with its graph:
  a $y = x(x-2)(x+4)$ **b** $y = (x+2)(x+4)^2$ **c** $y = -x^2(x-2)$

A 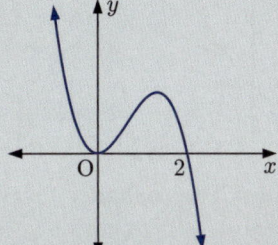   **B** 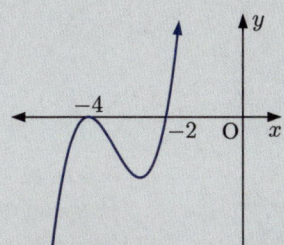   **C**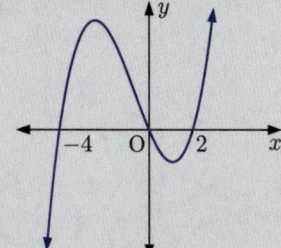

5 The temperature on Mt Snowden during one day can be modelled by
$T = -0.01x^3 + 0.285x^2 - 1.8x + 10.8$ °C, where x is the number of hours after midnight, $0 \leqslant x \leqslant 24$.

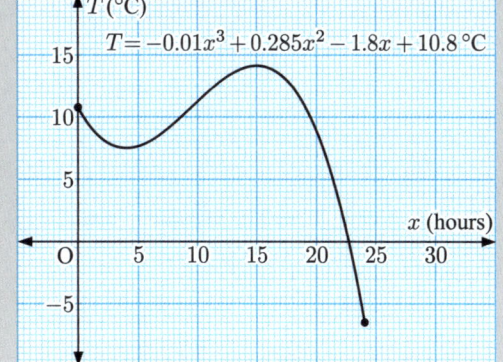

  a Find the temperature at noon.
  b Find the maximum temperature, and the time when this occurred.
  c Find the minimum temperature, and the time when this occurred.
  d For how many hours was the temperature above 10°C?

Review set 27B

1 Copy and complete this table of values for $y = -\dfrac{12}{x}$:

x	-12	-6	-3	-1	1	3	6	12
y								

Hence sketch the graph of the function.

2 Determine the equation of each reciprocal graph:

  a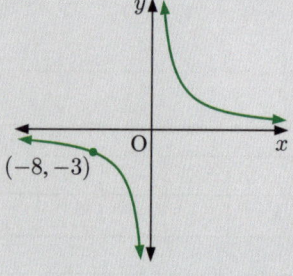

  b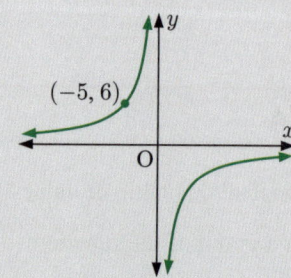

3 Find the x-intercept(s) of each cubic function:

  a $y = -x(x+1)(x-5)$ **b** $y = x^2(x-3)$ **c** $y = -(x+2)^3$

4 The graph of $y = -x^3 + 2x^2 + 2x - 2$ is shown alongside. Use the graph to estimate:

  a the solutions to the equation $-x^3 + 2x^2 + 2x - 2 = 0$

  b the maximum value of $-x^3 + 2x^2 + 2x - 2$ for $x > 0$.

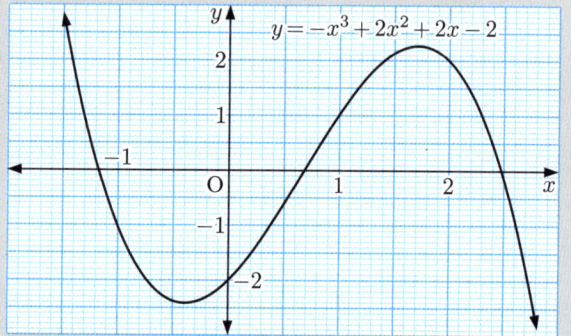

5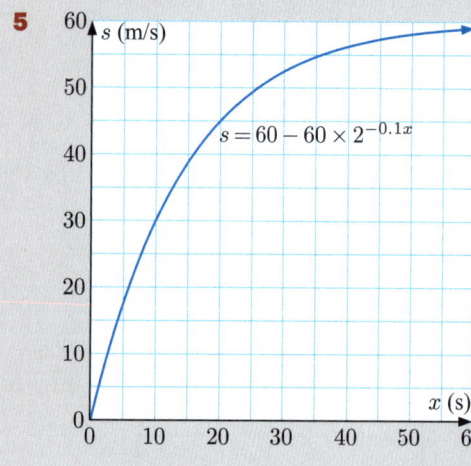

The speed of a skydiver x seconds after jumping from a plane is $s = 60 - 60 \times 2^{-0.1x}$ m/s.

  a Find the speed of the skydiver after 5 seconds.

  b Find the time taken for the speed of the skydiver to reach 45 m/s.

28

Number sequences

Contents:
- **A** Number sequences
- **B** Arithmetic sequences
- **C** Geometric sequences
- **D** Fibonacci-type sequences

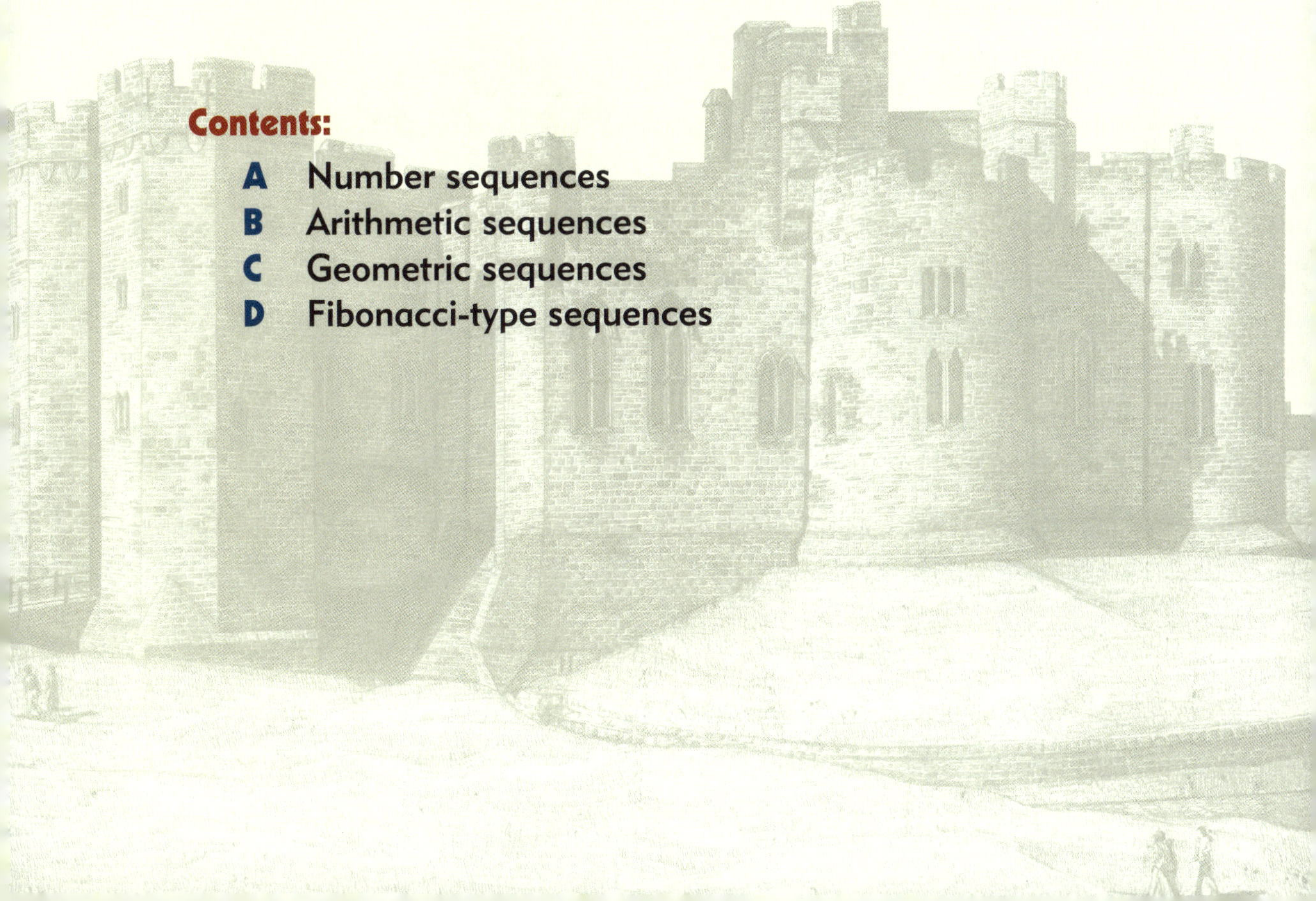

Opening problem

The list of numbers alongside is a *sequence*. The numbers are written in a particular order. The string of dots indicates that the sequence continues forever.

5, 9, 13, 17, 21, ...

Things to think about:

a Can you explain how the number sequence relates to the matchstick pattern below?

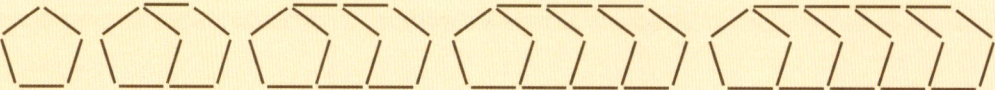

b Can you see a pattern which shows how one number relates to the next?
c What is the next term of the sequence?
d Can you find the 100th term of the sequence, without having to find all of the previous terms?

In this Chapter, we will study **number sequences**. We will see that, for certain types of sequences, there are rules which allow us to find any member of the sequence.

A NUMBER SEQUENCES

A **number sequence** is an ordered list of numbers defined by a rule.

Consider the illustrated pattern of balls:

The first layer has just one blue ball.
The second layer has three pink balls.
The third layer has five black balls.
The fourth layer has seven green balls.

The number of balls in each layer forms a **sequence** of numbers which could be continued forever:
1, 3, 5, 7, 9, 11, 13, 15, 17, ...

The numbers in a sequence are called the **terms** or **members** of the sequence.

We will look at two ways of describing a number sequence.

(1) A **term-to-term rule**, which describes how to obtain each term from one or more of the preceding terms.
The term-to-term rule for the pattern of balls is "start at 1, then add 2 each time".

(2) A **position-to-term rule**, which gives the nth term of the sequence in terms of n.
The position-to-term rule for the pattern of balls is $2n - 1$.
We can verify this by checking that:
    the 1st term of the sequence is $2(1) - 1 = 1$ ✓
    the 2nd term of the sequence is $2(2) - 1 = 3$ ✓
    the 3rd term of the sequence is $2(3) - 1 = 5$ ✓

We can use this rule to find, for example, the 10th term of the sequence. The 10th term is $2(10) - 1 = 19$.

Example 1 ◀)) Self Tutor

Find the first four terms of the sequence:
- **a** with term-to-term rule: start at 19, then subtract 3 each time
- **b** with nth term $4n + 7$.

a The first four terms of the sequence are 19, 16, 13, 10.
-3 -3 -3

b The 1st term of the sequence is $4(1) + 7 = 11$
The 2nd term of the sequence is $4(2) + 7 = 15$
The 3rd term of the sequence is $4(3) + 7 = 19$
The 4th term of the sequence is $4(4) + 7 = 23$
So, the first four terms of the sequence are 11, 15, 19, 23.

EXERCISE 28A

1 Write down the first four terms of the sequence with term-to-term rule:
- **a** start at 7, then add 6 each time
- **b** start at 40, then subtract 5 each time
- **c** start at 5, then multiply by 2 each time
- **d** start at 250, then divide by 5 each time.

2 Find the first four terms of the sequence with nth term:
- **a** $3n + 1$
- **b** $5n - 2$
- **c** $22 - 4n$
- **d** $n^2 + 1$
- **e** $n^2 - 2n + 4$
- **f** 2^n

3 Consider the sequence with nth term $8n + 5$.
- **a** Find the first four terms.
- **b** Find the 10th term.
- **c** Write a term-to-term rule for the sequence.

4 A sequence has nth term $n^2 + n + c$, and the third term is 20. Find the value of c.

5 Consider the sequence with nth term $50 - n^2$. Find:
- **a** the 6th term
- **b** the sum of the first 3 terms
- **c** the first term of the sequence which is negative.

6 The nth terms of two sequences are given by the position-to-term rules $4n + 1$ and $45 - 3n$. Write down the numbers between 10 and 30 which are terms of *both* sequences.

7 A sequence is generated by the rule "multiply the previous term by 2, then add 5". The first term of the sequence is 10. Find the next three terms of the sequence.

8 **a** Write a position-to-term rule for the sequence of square numbers 1, 4, 9, 16, 25, ...
  b Write down the 12th term of the sequence.

9 The number of dots in each figure of the pattern forms a sequence.
- **a** Find the first 5 terms of the sequence.
- **b** Write a term-to-term rule for the sequence.

10 The **triangular numbers** are numbers which can be represented by a triangular arrangement of dots.

   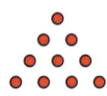 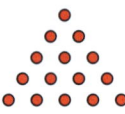

1 dot   3 dots   6 dots

  a Find the first five numbers in the sequence of triangular numbers.
  b The position-to-term rule for the sequence of triangular numbers is $\frac{1}{2}n^2 + \frac{1}{2}n$.
    i Verify this rule for the first five triangular numbers.
    ii Use the rule to find the 10th triangular number.

Example 2   ◀) Self Tutor

A sequence has nth term $5n - 3$.
  a Show that 42 is a term of the sequence. **b** Is 63 a term of the sequence?

  a Suppose the nth term is 42.
    $\therefore 5n - 3 = 42$
    $\therefore 5n = 45$
    $\therefore n = 9$
  So, 42 is the 9th term of the sequence.

  b Suppose the nth term is 63.
    $\therefore 5n - 3 = 63$
    $\therefore 5n = 66$
    $\therefore n = 13\frac{1}{5}$
  But n must be an integer, so 63 is *not* a member of the sequence.

11 A sequence has nth member $88 - 7n$.
  a Show that 39 is a member of the sequence. **b** Is 8 a member of the sequence?
  c Is -52 a member of the sequence?

12 A sequence has nth member $n^2 + 5$.
  a Find the 4th member of the sequence. **b** Is 41 a member of the sequence?
  c Is 80 a member of the sequence?

Example 3   ◀) Self Tutor

Find the next two terms of each sequence:
  a 3, 10, 17, 24, 31, ...   **b** 2, 5, 9, 14, 20, ...

  a The terms increase by 7 each time.
    Continuing this pattern, the next two terms are 38 and 45.
  b The difference between the terms increases by 1 each time.
    Continuing this pattern, the next two terms are 27 and 35.

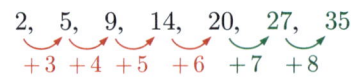

13 Find the next two terms of each sequence:

    a 4, 9, 14, 19, 24, ...
    b 60, 52, 44, 36, 28, ...
    c $-3, -2\frac{1}{2}, -2, -1\frac{1}{2}, -1, ...$
    d 2, 4, 8, 16, 32, ...
    e 5, 7, 10, 14, 19, ...
    f 9, 12, 16, 21, 27, ...
    g 8, 12, 18, 26, 36, ...
    h 15, 20, 27, 36, 47, ...

14 Consider the sequence 2, 3, 5, ...

    a Gemma thinks the 4th and 5th terms are 7 and 11.

        i Explain the rule that Gemma is using.
        ii Using Gemma's rule, find the 6th and 7th terms.

    b Harry is using a different rule. He thinks the 4th and 5th terms are 8 and 12. Using Harry's rule, find the 6th and 7th terms.

B ARITHMETIC SEQUENCES

An **arithmetic sequence** or **arithmetic progression** is a sequence in which each term differs from the previous one by the same fixed number. We call this number the **common difference**.

For example:

- 1, 5, 9, 13, 17, ... is an arithmetic sequence with common difference 4
  +4 +4 +4 +4

- 42, 37, 32, 27, 22, ... is an arithmetic sequence with common difference -5.
  -5 -5 -5 -5

Example 4   Self Tutor

Determine whether the following sequences are arithmetic:

    a 6, 10, 16, 22, 28, ...
    b 23, 20, 17, 14, 11, ...

a 6, 10, 16, 22, 28, ...
    +4 +6 +6 +6

The difference is *not* the same throughout the sequence, so the sequence is *not* arithmetic.

b 23, 20, 17, 14, 11, ...
    -3 -3 -3 -3

The difference is the same throughout the sequence, so the sequence is arithmetic with common difference -3.

EXERCISE 28B.1

1 Determine whether the following sequences are arithmetic:

    a 2, 5, 8, 11, 14, ...
    b 5, 9, 13, 18, 22, ...
    c 29, 23, 16, 10, 4, ...
    d 11, 4, -3, -10, -17, ...

2 Write down the common difference for these arithmetic sequences:

    a 3, 7, 11, 15, ...
    b 5, 13, 21, 29, ...
    c 14, 9, 4, -1, ...
    d $-4, -1\frac{1}{2}, 1, 3\frac{1}{2}, ...$

3 Find the unknowns, given that the following sequences are arithmetic:
   a 4, 10, □, 22, 28, ...
   b 13, 20, 27, □, 41, ...
   c 19, □, 11, 7, △, ...
   d 22, □, 4, △, −14, ...

4 Determine whether the number of dots in the figures forms an arithmetic sequence. Explain your answer.

   a

   b

5 Write down the common difference for the arithmetic sequence of numbers given by:
   a the green arrow
   b the blue arrow
   c the purple arrow
   d the red arrow.

1	2	3	4	5	6	7	8	9	10
11	12	13	14	15	16	17	18	19	20
21	22	23	24	25	26	27	28	29	30
31	32	33	34	35	36	37	38	39	40
41	42	43	44	45	46	47	48	49	50
51	52	53	54	55	56	57	58	59	60
61	62	63	64	65	66	67	68	69	70
71	72	73	74	75	76	77	78	79	80
81	82	83	84	85	86	87	88	89	90
91	92	93	94	95	96	97	98	99	100

THE nTH TERM OF AN ARITHMETIC SEQUENCE

If we are given the first few terms of an arithmetic sequence, we can find an expression for the nth term of the sequence.

Investigation — The nth term of an arithmetic sequence

What to do:

1 Consider the sequence with nth term $5n + 3$.
   a Write down the first four terms of the sequence.
   b Explain why the sequence is arithmetic, and state the common difference.

2 Repeat **1 a** for the sequence with nth term:
   a $2n - 1$
   b $7n + 5$
   c $n^2 + 2$
   d $-3n + 20$
   e $\frac{1}{2}n + 1$
   f $n^3 - 4$
   g $-6n + 50$
   h $-\frac{1}{4}n + 2$

3 What type of rule for the nth term will produce an arithmetic sequence?

4 How can we identify the common difference of the arithmetic sequence from the expression for the nth term?

From the **Investigation**, you should have found that:

- For any arithmetic sequence, the nth term can be written as a **linear** expression.
- The coefficient of n in the expression is equal to the common difference of the arithmetic sequence.

Example 5   ◀⁾ Self Tutor

Write an expression for the nth term of:
  a 9, 13, 17, 21, ...
  b 23, 18, 13, 8, ...

a The sequence is arithmetic with common difference 4.
  $\therefore$ the expression for the nth term has the form $4n + c$.
  The 1st term is 9, so $4(1) + c = 9$
  $\therefore\ 4 + c = 9$
  $\therefore\ c = 5$
  So, the nth term of the sequence is $4n + 5$.

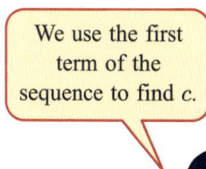

We use the first term of the sequence to find c.

b The sequence is arithmetic with common difference -5.
  $\therefore$ the expression for the nth term has the form $-5n + c$.
  The 1st term is 23, so $-5(1) + c = 23$
  $\therefore\ -5 + c = 23$
  $\therefore\ c = 28$
  So, the nth term of the sequence is $-5n + 28$.

EXERCISE 28B.2

1 Consider the arithmetic sequence 10, 17, 24, 31, 38, ...
   a State the common difference of the sequence.
   b Find an expression for the nth term of the sequence.

2 Write an expression for the nth term of each arithmetic sequence:
   a 8, 11, 14, 17, ...
   b 14, 19, 24, 29, ...
   c 35, 29, 23, 17, ...
   d 5, 13, 21, 29, ...
   e $-7, -3, 1, 5,$ ....
   f 22, 13, 4, -5, ...

3 **a** Find an expression for the nth term of the sequence 5, 11, 17, 23, ...
   b Hence find the 20th term of the sequence.

4 **a** Find an expression for the nth term of the sequence 31, 28, 25, 22, ...
   b Hence find the 10th term of the sequence.
   c Is -20 a term of the sequence?

5 Rick is using matchsticks to represent houses.
   a Draw a diagram to represent 4 houses.
   b How many matchsticks does Rick need to make:
      i 1 house
      ii 2 houses
      iii 3 houses
      iv 4 houses?

1 house

2 houses

3 houses

   c Write an expression for the number of matchsticks needed to represent n houses.
   d How many matchsticks does Rick need to represent 15 houses?

C GEOMETRIC SEQUENCES

A **geometric sequence** or **geometric progression** is a sequence in which each term can be obtained from the previous one by multiplying by the same non-zero constant. This constant is called the **common ratio**.

For example:

- 2, 6, 18, 54, ... is a geometric sequence with common ratio 3
  $\times 3 \quad \times 3 \quad \times 3$

- 80, 40, 20, 10, ... is a geometric sequence with common ratio $\frac{1}{2}$.
  $\times \frac{1}{2} \quad \times \frac{1}{2} \quad \times \frac{1}{2}$

Example 6   ◆) Self Tutor

Determine whether the following sequences are geometric:
  a 1, 4, 16, 64, ...
  b 60, 30, 10, 5, ...

  a 1, 4, 16, 64, ...
      $\times 4 \quad \times 4 \quad \times 4$

The ratio is the same throughout the sequence, so the sequence is geometric with common ratio 4.

  b 60, 30, 10, 5, ...
      $\times \frac{1}{2} \quad \times \frac{1}{3} \quad \times \frac{1}{2}$

The ratio is not the same throughout the sequence, so the sequence is not geometric.

EXERCISE 28C

1 Determine whether the following sequences are geometric:
  a 5, 10, 20, 40, ...
  b 4, 12, 36, 72, ...
  c 45, 15, 5, $\frac{5}{3}$, ...
  d 1, 5, 20, 100, ...

2 Which of the following is a geometric progression?
  A 1, 2, 8, 16, ...
  B $\frac{1}{3}, \frac{2}{9}, \frac{2}{27}, \frac{4}{81}$, ...
  C 2, 4, 6, 8, ...
  D 2, $\frac{2}{5}, \frac{2}{25}, \frac{2}{125}$, ...

3 Write down the common ratio for these geometric sequences:
  a 2, 10, 50, 250, ...
  b 1, 7, 49, 343, ...
  c 60, 30, 15, 7.5, ...
  d 3, 6, 12, 24, ...
  e 4000, 400, 40, 4, ...
  f 8, 12, 18, 27, ...

4 For the geometric sequence with first two terms given, find b and c:
  a 4, 8, b, c, ...
  b 8, 2, b, c, ...
  c 15, 5, b, c, ...

5 A sequence has nth term 3^n.
  a Write down the first four terms of the sequence.
  b Show that the sequence is geometric, and state the common ratio.

6 A sequence has nth term $\left(\frac{1}{4}\right)^n$.
  a Write down the first four terms of the sequence.
  b Show that the sequence is geometric, and state the common ratio.

7 The first two terms of a geometric sequence are 16 and 40.
  a Find the next term of the sequence.
  b Find the largest whole number term in the sequence.

D FIBONACCI-TYPE SEQUENCES

Historical note — The Fibonacci sequence

In his 1202 book *Liber Abaci*, **Leonardo Pisano Bigollo** (more commonly known today as **Fibonacci**) posed a problem involving a population of rabbits.

He supposed that in the first month a pair of baby rabbits were put together, and that each month after that:

- every pair of baby rabbits matured into a pair of adult rabbits
- every pair of adult rabbits produced a new pair of baby rabbits.

Fibonacci posed the question: How many pairs of rabbits will there be after one year?

Leonardo Fibonacci

In Fibonacci's rabbit puzzle, the number of rabbit pairs at the end of each month forms this sequence:
  1, 1, 2, 3, 5, 8, 13, 21, ...

This sequence is known as the **Fibonacci sequence**.

Notice that the sequence starts 1, 1, and each term after that is the sum of the previous two terms:
  $2 = 1 + 1$
  $3 = 2 + 1$
  $5 = 3 + 2$
  $8 = 5 + 3$, and so on.

> In a **Fibonacci-type sequence**, the first two terms are given, and each term after that is the sum of the previous two terms.

For example, the sequence below is a Fibonacci-type sequence with starting values 1 and 4:
  1, 4, 5, 9, 14, 23, 37, ...
       ↑   ↑   ↑   ↑   ↑
      4+1 5+4 9+5 14+9 23+14

Example 7 ◀)) Self Tutor

Write down the next three terms of each Fibonacci-type sequence:
  a 2, 5, 7, ...   **b** 3, 8, 11, ...

  a 2, 5, 7, 12, 19, 31
              ↑   ↑   ↑
             7+5 12+7 19+12

  b 3, 8, 11, 19, 30, 49
                  ↑   ↑   ↑
                11+8 19+11 30+19

EXERCISE 28D

1 Write down the next three terms of each Fibonacci-type sequence:
  a 1, 3, 4, ...
  b 4, 6, 10, ...
  c 2, 7, 9, ...
  d 5, 6, 11, ...
  e 4, 11, 15, ...
  f 3, 10, 13, ...

2 Which of these is a Fibonacci-type sequence?
  A 3, 7, 10, 17, ...
  B 5, 12, 17, 22, ...
  C 4, 12, 20, 32, ...
  D 6, 13, 19, 32, ...

3 Find the missing values in these Fibonacci-type sequences:
  a 1, 8, □, 17, ...
  b 3, □, 13, 23, ...
  c □, 6, 8, ▽, ...
  d 6, □, 13, ▽, ...
  e □, 15, ▽, 36, ...
  f □, ▽, 17, 30, ...

4 Find the 8th term of the Fibonacci-type sequence starting 2, 7, 9, ...

5 Consider a Fibonacci-type sequence with starting values 4 and 4.
  a Find the next three terms of the sequence.
  b Find the smallest term of the sequence which is greater than 100.

6 A Fibonacci-type sequence has starting values 2 and 8. Find the smallest number which occurs in both this sequence *and* the arithmetic sequence with nth term $5n + 6$.

7 A Fibonacci-type sequence has first term 5 and second term b.
  a Write an expression, in terms of b, for the third term of the sequence.
  b Show that the fifth term of the sequence is $10 + 3b$.
  c Given that the fifth term is 43, find the value of b.

8 The Fibonacci-type sequence alongside is incomplete.
Let the first term of the sequence be a.   ...., 9, ..., ..., 41
  a Write an expression, in terms of a, for:
    i the third term of the sequence
    ii the fourth term of the sequence.
  b Find a, and hence complete the sequence.

9 The first two terms of a Fibonacci-type sequence are x and y.
  a Find, in terms of x and y, an expression for the:
    i third term of the sequence
    ii sixth term of the sequence.
  b Given that the third term is 12 and the sixth term is 56, find the values of x and y.

Discussion

Consider Fibonacci's rabbit puzzle in the **Historical Note** on page **555**.

Can you explain why the number of rabbit pairs in a given month is equal to the sum of the number of rabbit pairs in the previous two months?

Review set 28A

1 Write down the first four terms of the sequence with term-to-term rule:
   a start at 8, then add 5 each time
   b start at 19, then subtract 7 each time.

2 A sequence has nth term $n^2 + 2n$. Find:
   a the first 4 terms of the sequence
   b the 10th term of the sequence.

3 Find the next two terms of this sequence: 3, 7, 12, 18, 25, ...

4 Find the unknowns given that the following sequences are arithmetic:
   a 9, 17, □, 33, 41, ...
   b 27, □, 15, △, 3, ...

5 Find an expression for the nth term of each arithmetic sequence:
   a 13, 16, 19, 22, ...
   b 60, 51, 42, 33, ...

6 **a** Find an expression for the nth term of the sequence 12, 19, 26, 33, ...
   b Hence find the 8th term of the sequence.
   c Is 100 a term of the sequence?

7 Write down the common ratio for these geometric sequences:
   a 3, 12, 48, 192, ...
   b 45, 30, 20, $\frac{40}{3}$, ...

8 For the geometric sequence with first two terms given, find b and c:
   a 4, 20, b, c, ...
   b 1, $\frac{3}{4}$, b, c, ...

9 Write down the next three terms of these Fibonacci-type sequences:
   a 4, 7, 11, ...
   b 6, 8, 14, ...

10 Suppose the first two terms of a Fibonacci-type sequence are a and b.
   a Find, in terms of a and b, an expression for the third term of the sequence.
   b Show that the seventh term of the sequence is $5a + 8b$.
   c Given that the third term is 7 and the seventh term is 53, find the values of a and b.

11 The number of dots in each figure of the pattern forms a sequence.

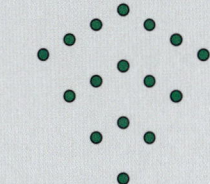

   a Find the first 5 terms of the sequence.
   b Explain why the position-to-term rule for the sequence is n^2.
   c Use this rule to find the 15th term in the sequence.

Review set 28B

1 Find the first four terms of the sequence with nth term:
   a $2n + 7$
   b $3n^2 + n$

2 **a** Write a position-to-term rule for the sequence of numbers 25, 32, 39, 46, ...
   b Find the 10th term of the sequence.

3 A sequence is generated by the rule "multiply the previous term by 3, then subtract 2". The first term of the sequence is 2.
Find the next three terms of the sequence.

4 Write down the common difference for these arithmetic sequences:
   a 5, 17, 29, 41, ...
   b 6, $5\frac{1}{2}$, 5, $4\frac{1}{2}$, ...

5 Consider this sequence of matchstick figures:

   a Draw the next matchstick figure.
   b How many matchsticks are needed to make the:
      i 1st figure   **ii** 2nd figure   **iii** 3rd figure   **iv** 4th figure?
   c Write an expression for the number of matchsticks needed to make the nth figure.
   d How many matchsticks are needed to make the 20th figure?

6 Determine whether the following sequences are geometric:
   a 1, 5, 25, 125, ...
   b 120, 60, 40, 20, ...

7 A sequence has nth term $\left(\frac{3}{2}\right)^n$.
   a Write down the first four terms of the sequence.
   b Show that the sequence is geometric, and state the common ratio.

8 Find the missing values in these Fibonacci-type sequences:
   a 5, □, 12, ▽, ...
   b □, ▽, 13, 22, ...

9 Determine whether the number of dots in the figures forms an arithmetic sequence. Explain your answer.
   a (dot patterns)   **b** (dot patterns)

10 A Fibonacci-type sequence has starting values 4 and 11.
   a Find the next three terms of the sequence.
   b Find the smallest term of the sequence which is greater than 100.

11 Consider the sequence 3, 6, 9, ...
   a Anthony thinks the 4th and 5th terms are 12 and 15.
      i Explain the rule Anthony is using.
      ii Using Anthony's rule, find the 6th and 7th terms.
   b Chelsea is using a different rule. She thinks the 4th and 5th terms are 15 and 24.
      i Explain the rule Chelsea is using.
      ii Using Chelsea's rule, find the 6th and 7th terms.

29

Probability

Contents:

- **A** Probability
- **B** Experimental probability
- **C** Sample space
- **D** Theoretical probability
- **E** Venn diagrams
- **F** The addition law of probability
- **G** Tables and grids
- **H** Independent events
- **I** Dependent events
- **J** Tree diagrams
- **K** Expectation

Opening problem

Players A, B, C, and D are the final four left in a knock-out tennis tournament. In the semi-finals, A will play B, and C will play D. The winners of these matches will play in the grand final.

The table alongside shows the probabilities of each player beating the others. For example, the yellow shaded square indicates that A has a 40% chance of beating B.

		\multicolumn{4}{c}{loser}			
		A	B	C	D
winner	A		40%	55%	47%
	B	60%		42%	54%
	C	45%			35%
	D	53%	46%	65%	

Things to think about:

a What is the probability that B will beat A?

b What is the sum of the probability that C will beat D and the probability that D will beat C? Can you explain why this must be true?

c Can you complete the table with the probability that C will beat B?

d How can we illustrate the possible outcomes of the finals series?

e Which player is most likely to win the grand final?

Probability deals with the **chance** or likelihood of an event occurring.

We can determine probabilities based on:

- the results of an experiment
- what we theoretically expect to happen.

The study of chance has important applications in physical and biological sciences, economics, politics, sport, life insurance, quality control, production planning, and many other areas.

Probability theory can be applied to card and dice games to try to increase our chances of success. It may therefore appear that an understanding of probability encourages gambling. However, a better knowledge of probability theory actually helps us to understand why the majority of habitual gamblers lose in the long term.

Historical note

Chevalier de Méré (1607 - 1684) was a French aristocrat and gambler in the 17th century. He wanted to know the answer to this question:

"Should I bet even money on the occurrence of at least one 'double six' when rolling a pair of dice 25 times?"

De Méré's experience of playing dice games convinced him that the answer was yes, but he did not know how to prove it. He therefore asked his friend, the French mathematician **Blaise Pascal** (1623 - 1662), for help.

In a series of letters between Pascal and fellow mathematician **Pierre de Fermat** (1607 - 1665), the problem was solved. In the process, they became interested in solving other questions of this kind, and together they laid the foundations of a new branch of mathematics called **theoretical probability**.

Pierre de Fermat

A PROBABILITY

The **probability** of an event is a measure of the chance that it will occur.

Probabilities can be given as percentages from 0% to 100%, as proper fractions, or decimal numbers between 0 and 1.

An **impossible** event has 0% chance of happening, and is assigned the probability 0.

A **certain** event has 100% chance of happening, and is assigned the probability 1.

All other events are assigned a probability between 0 and 1.

The number line below shows how we could interpret different probabilities:

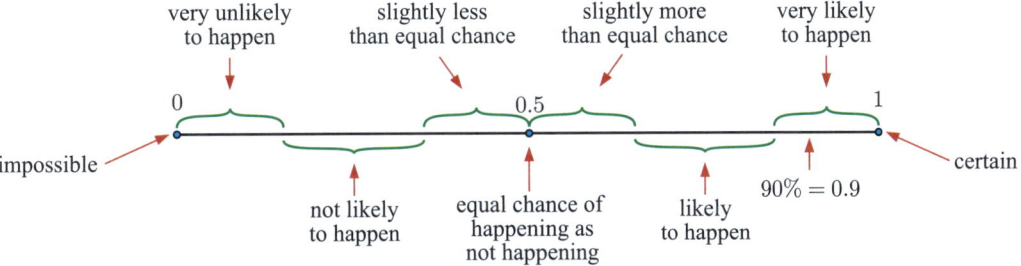

For example, suppose that the weather forecast says there is a 90% chance of rain tomorrow. We would say it is *very likely* that it will rain tomorrow.

Example 1 ◀)) Self Tutor

Use phrases to describe each probability:
a There is a 75% chance that Ted will be on time for school tomorrow.
b There is a 40% chance that Claire will burn her toast.

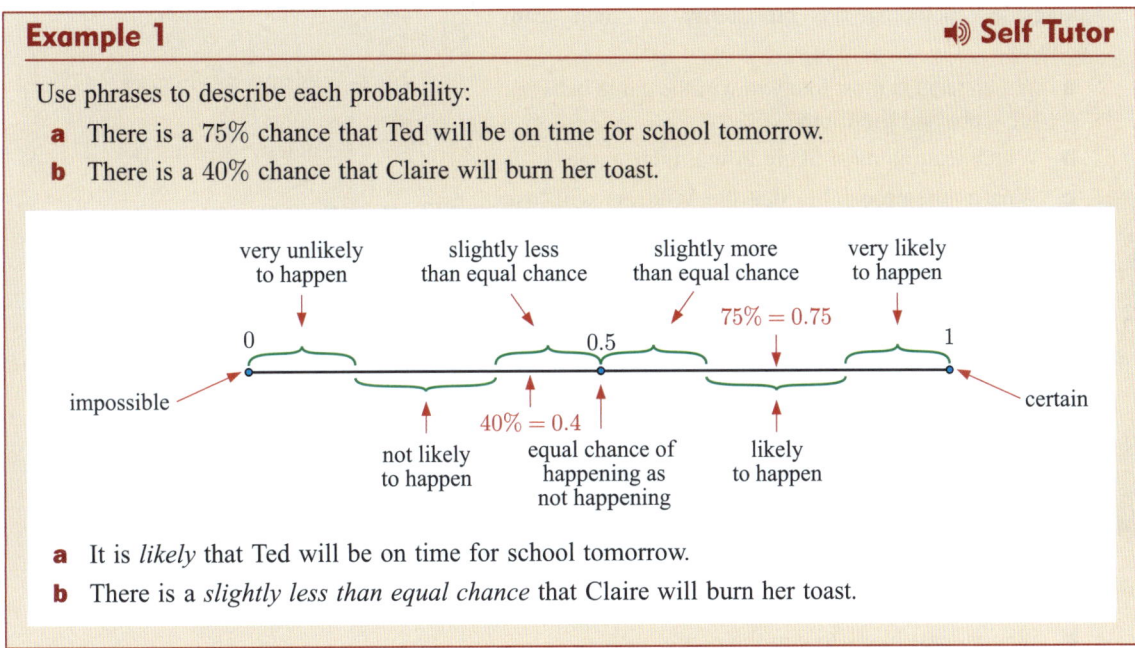

a It is *likely* that Ted will be on time for school tomorrow.
b There is a *slightly less than equal chance* that Claire will burn her toast.

In probability we can use capital letters such as E to represent events. The probability that event E occurs is written $P(E)$.

In **Example 1**, we could let E be the event that Ted will be on time for school tomorrow. In this case $P(E) = 0.75$.

COMPLEMENTARY EVENTS

Two events are **complementary** if exactly one of them *must* occur.

The probabilities of complementary events sum to 1.

The **complement** of event E is denoted E'. It is the event that E *does not* occur.

> For any event E with **complementary** event E',
> $P(E) + P(E') = 1$ or $P(E') = 1 - P(E)$.

If E is the event that Ted will be on time for school tomorrow, then E' is the event that Ted will *not* be on time for school tomorrow, and $P(E') = 1 - 0.75 = 0.25$.

EXERCISE 29A

1 Use phrases to describe each probability:

  a There is a 25% chance that Ella will score a goal in her next football match.

  b There is a 60% chance that the restaurant will be booked out on Saturday night.

  c There is a 5% chance that William will forget to take his lunch to school tomorrow.

2 The next four games in a baseball season are listed alongside, including the probability of each team winning.

  a Which team is most likely to win the game between the Cubs and the Lions?

  b Which team is most likely to win their game?

  c What is the probability that the Wildcats will lose their game?

  d Is the following statement true or false? "It is *likely* that the Angels will beat the Eagles."

32%	Cubs vs Lions	68%
59%	Wildcats vs Flames	41%
73%	Eagles vs Angels	27%
20%	Pumas vs Strikers	80%

3 Suppose a bag is filled with balls, and one ball is chosen at random. Use a phrase to describe the probability of choosing a *red* ball, if the bag contains:

  a 1 red ball and 1 blue ball
  b 5 red balls
  c 2 blue balls and 3 green balls
  d 1 red ball and 10 blue balls.

4 Five students are competing in a long distance race. Each student's probability of winning is given alongside.

  a Who is most likely to win the race?
  b Who is least likely to win the race?
  c Find the sum of the probabilities given. Explain your result.
  d Find the probability that either Julie or Tran will win the race.

Julie	20%
Edward	22%
Rob	7%
Tran	15%
Patricia	36%

e Describe in words, the probability that Rob will win the race.

f Let E be the event that Edward will win the race.
  i Find $P(E)$.
  ii State the complementary event E'.
  iii Find $P(E')$.

5 Suppose S is the event that it will snow tomorrow, and $P(S) = 0.03$.

  a State the complementary event S', in words.   **b** Find $P(S')$.
  c Use a phrase to describe the probability of:
    i S occurring   **ii** S' occurring.

B EXPERIMENTAL PROBABILITY

In experiments involving chance, we use the following terms to describe what we are doing and the results we are obtaining.

- The **number of trials** is the total number of times the experiment is repeated.
- The **outcomes** are the different results possible for one trial of the experiment.
- The **frequency** of a particular outcome is the number of times that this outcome is observed.
- The **relative frequency** of an outcome is the frequency of that outcome expressed as a fraction or percentage of the total number of trials.

$$\text{relative frequency} = \frac{\text{frequency}}{\text{number of trials}}$$

For example, suppose a small plastic cone was tossed into the air 300 times. It fell on its *side* 203 times and on its *base* 97 times. We say that:

- the number of trials is 300
- the possible outcomes are *side* and *base*
- the frequency of *side* is 203, and the frequency of *base* is 97
- the relative frequency of *side* $= \frac{203}{300} \approx 0.677$

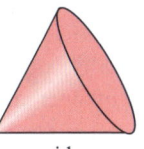 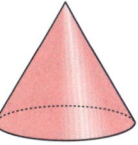
side   base

- the relative frequency of *base* $= \frac{97}{300} \approx 0.323$.

In the absence of any further data, the relative frequency of each outcome is our best estimate of the probability of it occurring.

The **experimental probability** is the **relative frequency** of the outcome.

Experimental probabilities are usually written as decimals.

We write $P(side) \approx 0.677$, $P(base) \approx 0.323$.

Example 2

George has kept a record of the number of apples he eats each day.

Estimate the probability that on a randomly selected day, George eats:

a exactly 1 apple   **b** at least 3 apples.

Number of apples	Frequency
0	10
1	17
2	12
3	3
4	1

George has recorded data for $10 + 17 + 12 + 3 + 1 = 43$ days.

a $P(1 \text{ apple}) \approx \dfrac{\text{frequency}}{\text{number of trials}} \approx \dfrac{17}{43} \approx 0.395$

b $P(\text{at least 3 apples}) \approx \dfrac{\text{frequency}}{\text{number of trials}}$

$\approx \dfrac{3+1}{43}$ {on 3 days he ate 3 apples, on 1 day he ate 4 apples}

≈ 0.093

EXERCISE 29B.1

1 When a coin is tossed 100 times, it falls heads 47 times. Find the experimental probability that the coin falls:

a heads   **b** tails.

2 Pat works in London, so it is very convenient to travel by public transport. At a minimum she will travel to work on the Tube, and she will often also use the Tube to return home and to go to meetings during the day. The table alongside summarises the *trips per day* she makes over 190 working days.

Estimate the probability that Pat will make:

a exactly four trips per day   **b** at least four trips per day.

Trips per day	Frequency
1	37
2	81
3	48
4	17
5	6
6	1

3 José recorded the lengths of internet commercials in seconds. His results are summarised in the table.

Estimate the probability that a randomly chosen internet commercial will last:

a between 20 and 40 seconds   **b** at least a minute

c between 20 seconds and a minute.

Length	Frequency
$0 \leqslant t < 20$	17
$20 \leqslant t < 40$	38
$40 \leqslant t < 60$	19
$t \geqslant 60$	4

4

Hours slept	Frequency
$5 \leqslant h < 6$	7
$6 \leqslant h < 7$	29
$7 \leqslant h < 8$	
$8 \leqslant h < 9$	39

Nathan recorded how long he slept each night for 121 nights. A table of his results is shown alongside.

a Find the number of nights in which Nathan slept between 7 and 8 hours.

b Estimate the probability that tonight he will sleep:

  i between 6 and 7 hours   **ii** at least 7 hours
  iii between 5 and 8 hours.

5 A group of randomly selected people are asked what type of animal they own as a pet. The relative frequency of "none" is $\frac{1}{5}$.

Animal	Frequency
dog	$2x + 8$
cat	$x + 18$
bird	$32 - x$
none	24
other	$24 - x$

  a Explain why the experimental probability that a randomly selected person *does* have a pet $= 0.8$.

  b Show that $\dfrac{24}{x + 106} = \dfrac{1}{5}$ and hence solve for x.

  c Estimate the probability that a randomly selected person owns a pet bird.

6 Lauren teaches a class of 29 students. She recorded the number of students present in her class each day for 40 days. On exactly half of the days, there was no more than one student absent.

  a Find the values of x and y.

  b Estimate the probability that tomorrow Lauren's class will have between 25 and 27 students inclusive.

Number of students	Frequency
24	1
25	2
26	x
27	10
28	y
29	12

Historical note

In the late 17th century, English mathematicians compiled and analysed mortality tables which showed the number of people who died at different ages. From these tables they could estimate the probability that a person would be alive at a future date. This led to the establishment of the first life insurance company in 1699.

Life insurance companies use statistics on **life expectancy** and **death rates** to calculate the premiums to charge people who insure with them.

NUMBER OF TRIALS

Investigation 1 — Dice rolling experiment

You will need:

At least one six-sided die with numbers 1 to 6 on its faces. Several dice would be useful to speed up the experiment.

What to do:

1 List the possible outcomes for the uppermost face when the die is rolled.

2 Suppose a die was rolled many times. Discuss what you would expect the relative frequency of rolling a 2 to be.

3 Roll a die 20 times, and count the number of times a 2 is rolled. Hence, calculate the relative frequency of rolling a 2.

4 Pool your results with another student, so in total you have data for 40 rolls. Calculate the relative frequency of rolling a 2 for 40 rolls.

5 Use the simulation to roll a die 60, 80, 100, 150, 200, 250, 500, and 1000 times. In each case, record the relative frequency of rolling a 2.

6 Plot a graph of relative frequency against the number of rolls. What do you notice?

7 What do you think will happen to the relative frequency of rolling a 2 as the number of rolls increases further?

SIMULATION

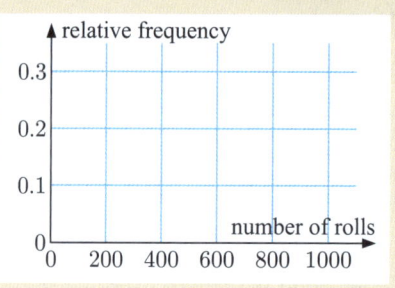

From the **Investigation**, you should have found that as the number of rolls increased, the relative frequency of rolling a 2 approached a particular value. Our estimate of the probability of rolling a 2 becomes more *accurate* as the number of trials increases.

In general, the larger the number of trials, the more accurate the estimate of the probability will be.

EXERCISE 29B.2

1 Connor visits the same bakery each day for lunch. On 13 out of his last 40 visits, the bakery had sold out of pies.

   a Estimate the probability that the bakery will have sold out of pies on his next visit.

   b How can Connor improve the accuracy of this estimate?

2 Ricky and Melia have 70 and 145 paper clips respectively. They both dropped their paper clips onto 6 cm by 6 cm squared paper. They then counted how many paper clips fell completely inside squares, and how many landed on a grid line:

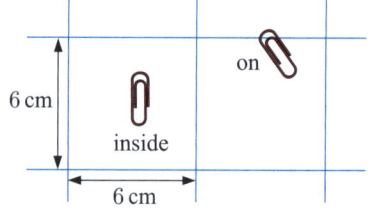

	Inside squares	On grid line
Ricky	54	16
Melia	113	32

   a Estimate the probability of a clip falling inside a square using:
      i Ricky's results
      ii Melia's results.

   b Whose estimate is likely to be more *accurate*? Explain your answer.

3 During the first day at a carnival, 247 people played the coconut shy game. 175 people did not win a prize, 64 people won a minor prize, and the remainder won a major prize.

   a Estimate the probability that the next player will win:
      i a major prize
      ii any prize.

b On the second day of the carnival, 350 people played the game. 237 people did not win a prize, 85 people won a minor prize, and the remainder won a major prize.

  i Combine the results from the first and second day.
  ii Use the combined results to estimate the probability that the next player will win any prize.

c Compare the estimates in **a ii** and **b ii**. Which estimate is likely to be more accurate? Explain your answer.

FREQUENCY TREES

In some surveys, more than one piece of information is gathered.

For example, suppose 50 people are asked whether they like eating eggs, and whether they like eating bacon. The results of the survey were:

- 41 people said that they like eggs
- of the people who like eggs, 26 also like bacon
- of the people who *dislike* eggs, 6 like bacon.

To represent this information, we can draw a diagram known as a **frequency tree**. We write down the frequencies of outcomes in circles, and connect the circles with arrows to show how the frequencies are related.

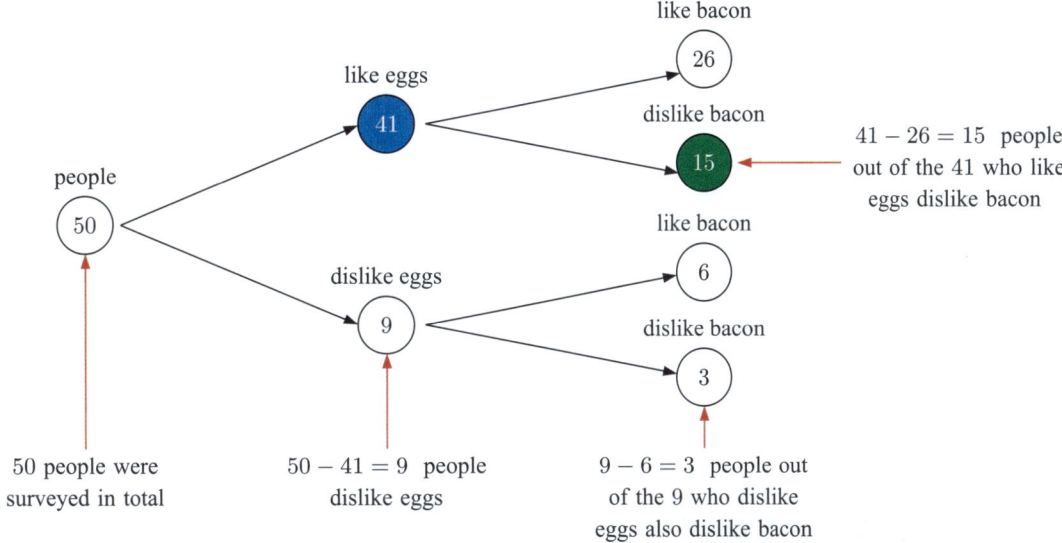

We can use the frequency tree to help us calculate probabilities involving both pieces of information.

For example, suppose a person who likes eggs is randomly selected.

P(a person known to like eggs, dislikes bacon) = $\dfrac{15}{41}$ ← number of people who like eggs and dislike bacon
← total number of people who like eggs

EXERCISE 29B.3

1 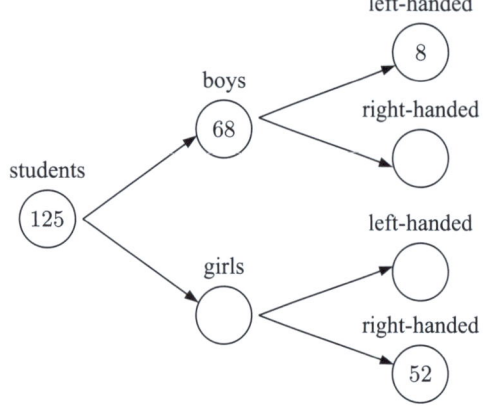 The frequency tree alongside shows the number of boys and girls studying GCSE at a particular school, and whether they are left-handed or right-handed.

    a Copy and complete the frequency tree.

    b A girl is randomly selected from the year level. Calculate the probability that she will be:

        i left-handed   **ii** right-handed.

2 Peter likes to count the number of cars that pass his apartment building. Over the past week, he counted:

- 277 cars in total
- 99 white cars of which 57 were small
- 69 large cars that were not white.

  a Copy and complete this frequency tree.

  b Estimate the probability that the next white car to pass the building will be large.

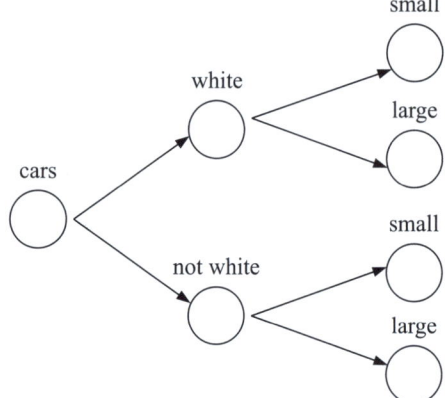

3 A popular restaurant requires patrons to make reservations if they want to have lunch or dinner on a weekend. Over the past month the restaurant received:

- 198 weekend reservations in total
- 102 reservations for Saturday of which 35 were for lunch
- 63 reservations for dinner on Sunday.

  a Draw a frequency tree to display this information.

  b Estimate the probability that the next reservation on a Sunday will be for lunch.

C   SAMPLE SPACE

> The **sample space** of an experiment is the set of *all* its possible outcomes.
>
> The sample space is denoted by the symbol "$\mathcal{E}$".

The simplest way to represent a sample space is to **list** the possible outcomes using **set notation**.

> An **event** is a set of outcomes in the sample space that have a particular property.

For example:
- The sample space for spinning this spinner is $\mathcal{E} = \{1, 2, 3, 4, 5, 6, 7, 8\}$.
- If A is the event that an even number is spun, then $A = \{2, 4, 6, 8\}$.

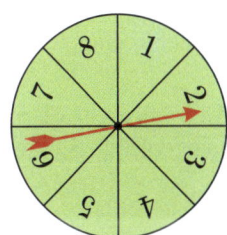

Notice how every outcome in A is also in the sample space $\mathcal{E}$. We say that A is a **subset** of $\mathcal{E}$.

Example 3 ◆) Self Tutor

Suppose the octahedral die alongside is rolled once. Using set notation, list:
a the sample space $\mathcal{E}$
b the outcomes in event A, that an odd number is rolled.

a $\mathcal{E} = \{1, 2, 3, 4, 5, 6, 7, 8\}$ **b** $A = \{1, 3, 5, 7\}$

EXERCISE 29C

1 Using set notation, list the sample spaces for the following experiments:
  a tossing a coin
  b spinning the spinner alongside.

2 A normal six-sided die is rolled once. Using set notation, list:
  a the sample space $\mathcal{E}$
  b the outcomes in event A, that a prime number is rolled.

3 Lucy randomly selects one letter from the alphabet.
  a Write down the sample space $\mathcal{E}$.
  b List the outcomes in:
    i A, the event that Lucy selects a consonant
    ii B, the event that Lucy selects a letter in the word EVENT.

D THEORETICAL PROBABILITY

The sample space when rolling a single die is $\{1, 2, 3, 4, 5, 6\}$.

Since the die is a cube and therefore symmetrical, we expect that each of the six outcomes will be **equally likely** to occur. We say that the **theoretical probability** of any given outcome occurring is 1 in 6, or $\frac{1}{6}$.

For example, $P(\text{rolling a 2}) = \frac{1}{6}$.

If a sample space has n outcomes which are **equally likely** to occur when the experiment is performed once, then each outcome has probability $\frac{1}{n}$ of occurring.

Consider the event of *rolling a prime number* with an ordinary die. Of the 6 possible outcomes, the three outcomes 2, 3, and 5 all correspond to this event. So, the probability of rolling a prime number is 3 in 6, or $\frac{3}{6}$.

When the outcomes of an experiment are equally likely, the probability that an event A occurs is:

$$P(A) = \frac{\text{number of outcomes corresponding to } A}{\text{number of outcomes in the sample space}}$$

Example 4   ◀)) Self Tutor

A small child has a collection of shapes.
  a She chooses one of the shapes at random. Find the probability that it is a triangle.
  b The child collects all the circles, then chooses one of them at random to give to her mother. Find the probability that the circle she chooses is green.

a P(triangle) = $\frac{5}{17}$ ◀—— There are 5 triangles.
                                  ◀—— There are 17 shapes to choose from.

b P(the circle is green) = $\frac{2}{7}$ ◀—— 2 of the circles are green.
                                            ◀—— There are 7 circles to choose from.

EXERCISE 29D

1 A spinner with the numbers 1 to 5 written on equal sectors is spun once. Find the probability of spinning:

  a a 4   **b** a 1 or a 2   **c** an odd number.

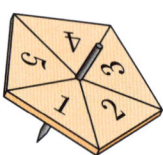

2 A symmetrical octahedral die has the numbers 1 to 8 marked on its faces. If it is rolled once, determine the probability of getting:

  a a 4   **b** a number less than 5   **c** a number greater than 8.

3 A bag contains 4 red and 3 green buttons. One button is randomly selected from the bag. Find the probability that the button is:

  a red   **b** green   **c** red or green.

4  A 52 card pack is well shuffled, then one card is dealt from the top of the pack. Find the probability that the card is:

  a a Jack   **b** a black card
  c a diamond   **d** a diamond or an ace.

5 Find the probability that a person randomly selected in the street has his or her birthday in:

   a May   **b** November   **c** February.

6 Candice has baked 6 blueberry muffins and 12 raspberry muffins. She chooses one to eat at random. Let E be the event that the muffin is blueberry.

   a Explain the meaning of the event E'.

   b Find $P(E)$.

   c Find $P(E')$.

E' is the **complementary event** of E.

7 37 people applied for a job at a toy store.

11 were selected for an interview, and from these people, 3 were offered a position at the store.

   a Find the probability that a randomly selected applicant was:

      i offered a position

      ii interviewed, but not offered a position

      iii not selected for an interview.

   b Find the sum of the probabilities in **a**. Explain your answer.

   c One of the applicants who was interviewed is selected at random. Find the probability that he or she was offered a position.

8 Amy is invited to select a treat from a bag containing 6 strawberry lollies, 3 mints, and 5 chocolates.

   a If Amy selects a treat at random, find the probability that she selects:

      i a mint   **ii** a mint or a chocolate.

   b Amy selects a chocolate, eats it, then passes the bag to Beth. If Beth selects a treat at random, find the probability that she selects:

      i a mint   **ii** a chocolate.

9 There are 5 different pairs of socks in Vanessa's sock drawer. The light in her room is not working, so she takes 2 socks from her drawer at random. Find the probability that Vanessa has taken a matching pair.

E   VENN DIAGRAMS

In a **Venn diagram**, we draw a rectangle to represent the sample space and circles within it to represent the events. We then write the individual outcomes in the regions they correspond to.

Consider the spinner alongside.

Let A be the event that an even number is spun.

$\therefore \ A = \{2, 4, 6, 8\}$ and
$A' = \{1, 3, 5, 7\}$

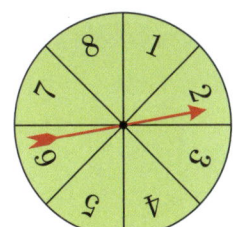

The Venn diagram alongside shows event A. The outcomes in A are written in the circle and the other outcomes in the sample space are written outside the circle.

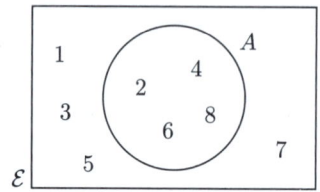

REPRESENTING MORE THAN ONE EVENT

If we have more than one event to consider, we draw *overlapping* circles.

For the spinner above, suppose A is an even number and B is a multiple of 3.

$\therefore$ $A = \{2, 4, 6, 8\}$ and $B = \{3, 6\}$.

To draw the Venn diagram for these events, we notice that the outcome 6 is in both A and B.

We write $A \cap B = \{6\}$ and place 6 in the region where the circles for A and B overlap.

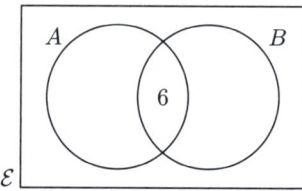

The remaining outcomes for A are placed in the circle for A but *not* the circle for B.

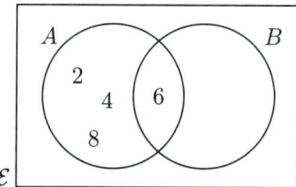

The remaining outcome for B is placed in the circle for B but *not* the circle for A.

The outcomes in neither event are placed outside both circles.

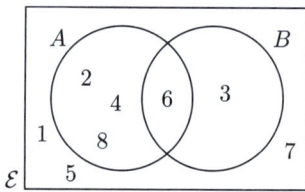

"AND" AND "OR"

We have seen that if both events A **and** B occur, we write $A \cap B$.

$A \cap B$ corresponds to the region where the circles for A and B overlap.

In probability, if we talk about A **or** B occurring, we consider this to mean A or B or *both*.

We write A **or** B as $A \cup B$.

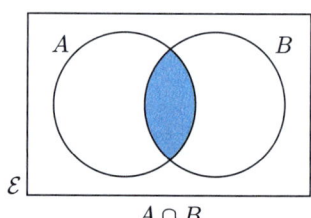

$A \cap B$

For the events A and B with the spinner above, $A \cup B = \{2, 3, 4, 6, 8\}$.

$A \cup B$ corresponds to the region including either or both circles.

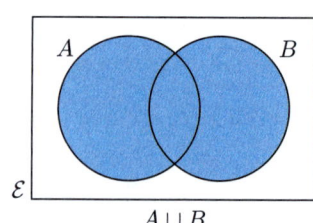

$A \cup B$

Example 5

A wheel with sectors numbered 1 to 20 is spun once. Let A be the event of spinning an odd number, and B be the event of spinning a number greater than 13.

a Represent A and B on a Venn diagram.

b Find the probability that the number spun is:

  i odd *and* greater than 13
  ii odd *or* greater than 13.

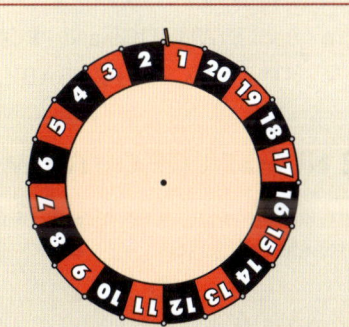

a $A = \{1, 3, 5, 7, 9, 11, 13, 15, 17, 19\}$
$B = \{14, 15, 16, 17, 18, 19, 20\}$

b **i** 3 of the 20 elements are in the region $A \cap B$.

$\therefore$ P(odd *and* greater than 13) $= \dfrac{3}{20}$

ii 14 of the 20 outcomes are in the region $A \cup B$.

$\therefore$ P(odd *or* greater than 13) $= \dfrac{14}{20} = \dfrac{7}{10}$

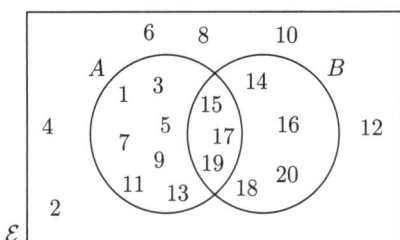

EXERCISE 29E.1

1 A regular six-sided die is rolled once. Let A be the event that a composite number is rolled.

  a Represent A on a Venn diagram.
  b Find the probability that A occurs.

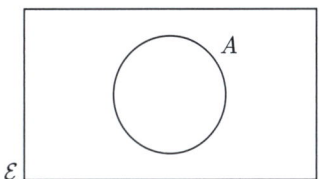

2 The octagonal spinner alongside is spun once. Let A be the event that the spinner lands on a red sector, and B be the event that an even number is spun.

  a Copy and complete this Venn diagram.

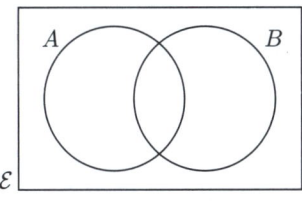
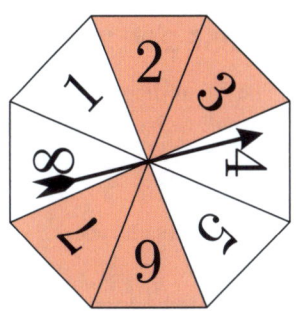

  b Find the probability that the spinner lands on:

  i a red even number
  ii a red sector *or* an even number.

3 The numbers from 1 to 25 are placed in a bag, and one number is selected at random. Let A represent selecting a multiple of 3, and B represent selecting a multiple of 4.

  a Represent A and B on a Venn diagram.
  b Find the probability of selecting:

  i a multiple of 3
  ii a multiple of 3 *and* 4
  iii a multiple of 4, but not a multiple of 3
  iv a multiple of 3 or 4, but not both.

4 A letter of the English alphabet is chosen at random. Use a Venn diagram to find the probability that the letter is in:
  a both STATISTICS and PROBABILITY
  b STATISTICS, but not PROBABILITY
  c STATISTICS or PROBABILITY
  d neither word.

THE NUMBER OF OUTCOMES IN EVENTS

If there are too many outcomes to fit on the Venn diagram, we can instead write the *number of outcomes* in each region in brackets.

Example 6

This Venn diagram shows the number of students in a music class who can play piano (P) and guitar (G).

Determine the probability that a randomly selected student can play:
  a piano, but not guitar
  b either instrument.

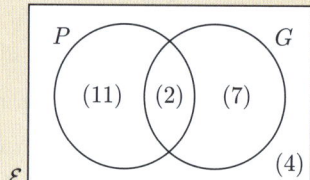

Total number of students $= 11 + 2 + 7 + 4 = 24$

a P(piano but not guitar) $= P(P \cap G')$
$$= \frac{11}{24}$$

b P(either instrument) $= P(P \text{ or } G \text{ or both})$
$$= P(P \cup G)$$
$$= \frac{11 + 2 + 7}{24}$$
$$= \frac{20}{24}$$
$$= \frac{5}{6}$$

EXERCISE 29E.2

1 The Venn diagram alongside shows the number of people in a small office who can touch type (T).
Find the probability that a randomly selected person in the office:
  a can touch type
  b cannot touch type.

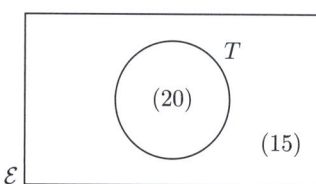

2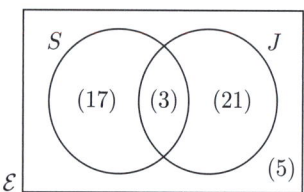

A group of university students were asked whether they study full time (S) and if they have a job (J).
The results are shown in the Venn diagram.
  a How many students participated in the survey?
  b Find the probability that a randomly selected student:
    i studies full time *and* has a job
    ii studies full time
    iii has a job, but does not study full time.

Example 7

In a class of 24 boys, 16 play football, 11 play rugby, and 5 play both of these sports.
  a Display this information on a Venn diagram.
  b Find the probability that a randomly selected boy plays exactly one of the sports.

a Let F represent boys who play football and R represent boys who play rugby.

Now $a + b = 16$ {16 play football}
$b + c = 11$ {11 play rugby}
$b = 5$ {5 play both}
$a + b + c + d = 24$ {there are 24 in the class}
$\therefore \ b = 5, \ a = 11, \ c = 6, \ d = 2.$

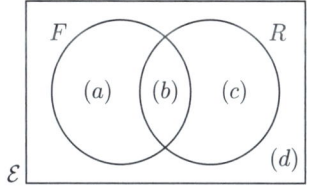

b P(exactly one of the sports) $= \dfrac{11 + 6}{24}$

$= \dfrac{17}{24}$

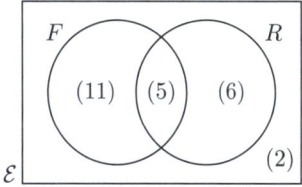

3 In a building with 58 apartments, 45 households have children, 19 have pets, and 5 have neither children nor pets.
  a Draw a Venn diagram to display this information.
  b Find the probability that a randomly selected household:
   i does not have children
   ii has children or pets or both
   iii has children or pets, but not both
   iv has pets, but not children
   v has children, but not pets.

4 On a particular day, 500 people visited a carnival. 300 people rode the Ferris wheel and 350 people rode the roller coaster. Each person rode at least one of these attractions.
  a Display this information on a Venn diagram.
  b Find the probability that a randomly selected person who went to the carnival rode:
   i both attractions
   ii the Ferris wheel but not the roller coaster.

F THE ADDITION LAW OF PROBABILITY

Investigation 2 — The addition law of probability

Click on the icon to obtain this Investigation.

ADDITION LAW

From the **Investigation** you should have discovered the **addition law of probability**:

> For two events A and B, $\quad P(A \cup B) = P(A) + P(B) - P(A \cap B)$
> which means: $\quad P(\text{either } A \text{ or } B \text{ or both}) = P(A) + P(B) - P(\text{both } A \text{ and } B)$.

MUTUALLY EXCLUSIVE EVENTS

In a probability experiment, not all events we consider will have common outcomes.

For example, suppose we randomly select an integer from 1 to 6.

Let A be the event that a prime number is selected, so $A = \{2, 3, 5\}$.

Let B be the event that a composite number is selected, so $B = \{4, 6\}$.

The events A and B have no common outcomes, so we can draw a Venn diagram so their circles do not overlap.

$$P(A \cap B) = 0$$

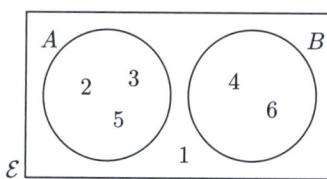

> Two events are **mutually exclusive** or **disjoint** if they have no common outcomes.
> If A and B are mutually exclusive events then $P(A \cap B) = 0$.
> For mutually exclusive events, $P(A \cup B) = P(A) + P(B)$.

Example 8 ◀) Self Tutor

Suppose $P(A) = 0.2$, $P(B) = 0.6$, and $P(A \cap B) = 0.1$.
a Are A and B mutually exclusive events? Explain your answer.
b Find $P(A \cup B)$.

a A and B are *not* mutually exclusive, since $P(A \cap B) \neq 0$.
b $P(A \cup B) = P(A) + P(B) - P(A \cap B)$
$\qquad\qquad = 0.2 + 0.6 - 0.1$
$\qquad\qquad = 0.7$

EXERCISE 29F

1 An ordinary die with faces 1, 2, 3, 4, 5, and 6 is rolled once. Consider these events:

  A: rolling a 1 $\qquad\qquad\qquad$ B: rolling a 3
  C: rolling an odd number $\qquad$ D: rolling an even number
  E: rolling a prime number $\qquad$ F: rolling a result greater than 3.

List the pairs of events which are mutually exclusive.

2 Suppose $P(A) = 0.7$, $P(B) = 0.2$, and $P(A \cap B) = 0.15$.
  a Are A and B mutually exclusive events? Explain your answer.
  b Find $P(A \cup B)$.

3 Suppose $P(X) = 0.3$, $P(X \cup Y) = 0.7$, and X and Y are mutually exclusive. Find $P(Y)$.

4 Suppose $P(C) = 0.6$ and $P(D) = 0.7$. Explain why C and D are not mutually exclusive.

Probability (Chapter 29) 577

G TABLES AND GRIDS

So far we have considered experiments with only *one* operation, such as a single spin of a spinner.

TABLES

Provided the sample space is small, we can use a **table** to systematically list the outcomes of an experiment with more than one operation.

Each *column* of the table corresponds to an operation.

Each *row* of the table corresponds to an outcome of the experiment.

Example 9   ◀) Self Tutor

Melissa draws a ball from each of the bags alongside.

Construct a table showing the possible outcomes.

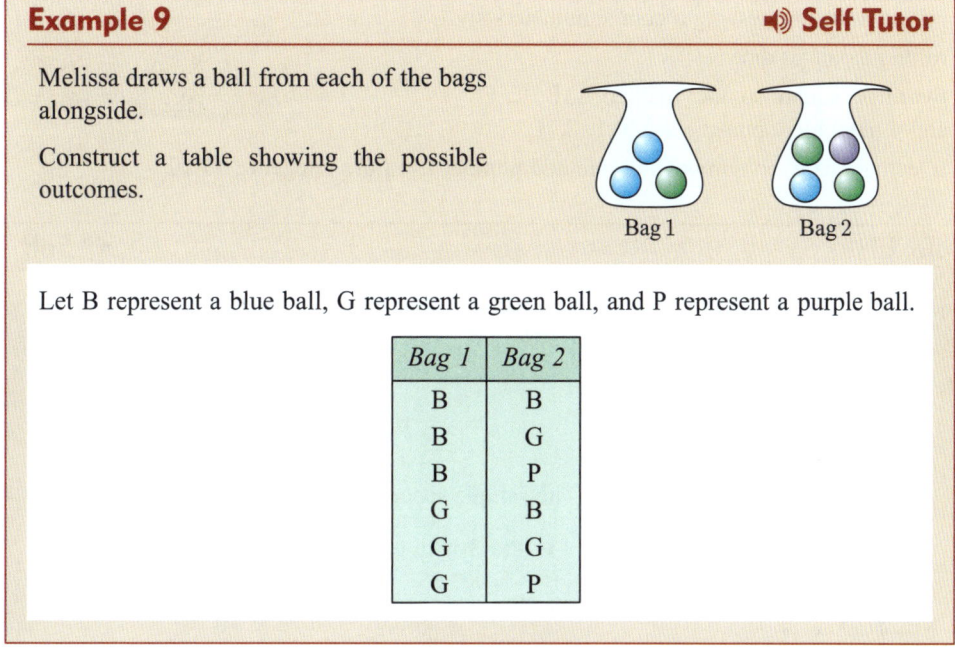

Bag 1   Bag 2

Let B represent a blue ball, G represent a green ball, and P represent a purple ball.

Bag 1	Bag 2
B	B
B	G
B	P
G	B
G	G
G	P

GRIDS

A **grid** is an efficient way to represent the sample space of an experiment involving two operations. Each point on the grid represents a possible outcome.

Example 10   ◀) Self Tutor

Use a 2-dimensional grid to illustrate the possible outcomes when tossing a coin and spinning the spinner shown.

Let H represent a "head" and T represent a "tail".

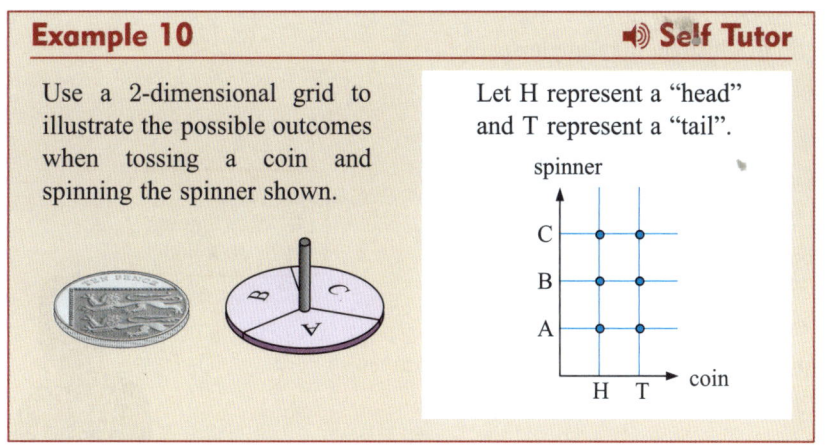

Each point on the grid represents one of the possible outcomes: {HA, HB, HC, TA, TB, TC}

578   Probability   (Chapter 29)

EXERCISE 29G

1 Construct a table of outcomes for:
   a the genders of a 2-child family
   b the order in which 3 men can be lined up
   c tossing 4 coins simultaneously
   d the order in which 4 different rowing teams A, B, C, and D could finish a race.

2 Draw a 2-dimensional grid to illustrate the sample space for:
   a rolling a die and tossing a coin simultaneously
   b rolling a pair of dice
   c twirling a square spinner marked A, B, C, D and a triangular spinner marked 1, 2, 3
   d selecting one even number and one odd number from the integers 1 to 10.

Example 11   ◀)) Self Tutor

Two square spinners, each with 1, 2, 3, and 4 on their edges, are twirled simultaneously.
   a Draw a 2-dimensional grid of the possible outcomes.
   b Hence, find the probability of getting:
      i   a 3 with each spinner   ii a 3 and a 1   iii an even result with each spinner.

a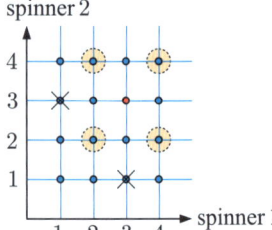

b The sample space has 16 members.
   i $P(\text{a 3 with each spinner}) = \frac{1}{16}$ {•}
   ii $P(\text{a 3 and a 1}) = \frac{2}{16} = \frac{1}{8}$ {crossed points}
   iii $P(\text{an even result with each spinner})$
       $= \frac{4}{16} = \frac{1}{4}$ {circled points}

3 Two coins are tossed simultaneously.
   a Use a 2-dimensional grid to illustrate the sample space.
   b Hence, find the probability of getting:
      i   two tails   ii a head and a tail   iii at least one tail.

4 A coin and an ordinary die are tossed and rolled simultaneously.
   a Draw a 2-dimensional grid to illustrate the sample space.
   b Hence, find the probability of getting:
      i   a tail and a 6   ii a tail or a 6
      iii neither a 2 nor a 6   iv neither a tail nor a 5
      v   a head and an odd number   vi a head or an odd number.

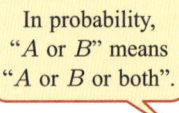

In probability, "A or B" means "A or B or both".

5 A triangular spinner labelled A, B, and C is spun, and an ordinary die is rolled simultaneously.

   a Draw a 2-dimensional grid to illustrate the sample space.

   b Hence, find the probability of getting:

      i B and 5   **ii** A and a prime number   **iii** A or C, and a multiple of 3.

6 A 4-sided die and a 6-sided die are rolled simultaneously. The possible results are shown on the 2-dimensional grid. Determine the probability of getting:

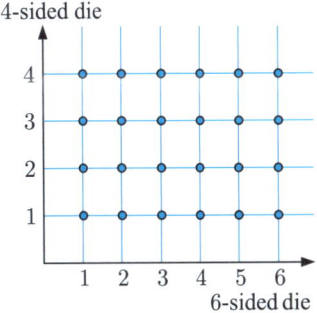

   a two 3s   **b** a 1 and a 2

   c a 1 or a 2   **d** at least one 4

   e exactly one 4   **f** no 4s

   g a sum of 5   **h** a sum greater than 5

   i an even sum   **j** an odd sum.

7 Two students each select a day of the week at random. Find the probability that:

   a Tuesday and Friday are selected

   b at least one student selects a day from the weekend

   c the selected days start with the same letter

   d between them, the selected days contain at least 15 letters.

8 In a role-playing game, Melanie and Neil duel by each rolling an eight-sided die. Neil's character has a higher level, so to win the duel Melanie must roll a number at least 2 greater than Neil's roll. Find the probability that Melanie will win the duel.

H   INDEPENDENT EVENTS

We will now look more closely at **compound events** involving two operations.

> Two events are **independent** if the occurrence of each event does not affect the occurrence of the other.

Consider tossing a coin and rolling a die simultaneously.

The two events "getting a head" and "rolling a 5" are independent events, since the outcome of the coin toss has no effect on the outcome from the dice roll.

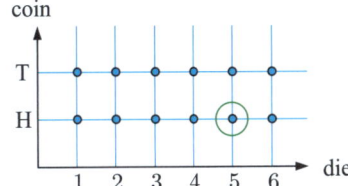

From the grid, we see there are 12 possible outcomes, only one of which is a head *and* a five.

∴   P(a head and a 5) = $\frac{1}{12}$

Also,   P(a head) = $\frac{1}{2}$ and   P(a 5) = $\frac{1}{6}$.

∴   P(a head) × P(a 5) = $\frac{1}{2} \times \frac{1}{6} = \frac{1}{12}$.

Notice that   P(a head **and** a 5) = P(a head) × P(a 5).

> If two events A and B are **independent** then
> $P(A \cap B) = P(A) \times P(B)$.

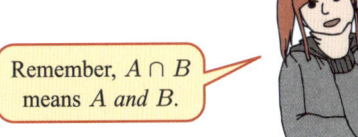

Remember, $A \cap B$ means *A and B*.

Example 12

A coin is tossed and a square spinner labelled A, B, C, and D is spun simultaneously. Determine the probability of getting a tail and a B.

The events are independent, since the outcome from the coin does not affect the outcome from the spinner, and vice versa.

$\therefore$ P(a tail and a B) = P(a tail) $\times$ P(a B)
$= \dfrac{1}{2} \times \dfrac{1}{4}$
$= \dfrac{1}{8}$

EXERCISE 29H

1 The spinners alongside are each spun once. Find the probability that both spinners will land on green.

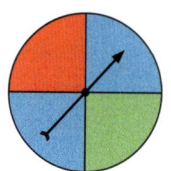
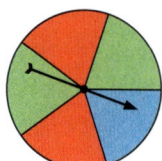

2 A coin and a 6-sided die are tossed simultaneously. Find the probability of getting a tail and a number larger than 2.

3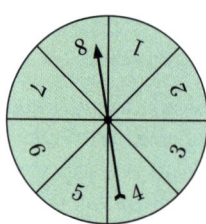
Tony and Ben each spin the spinner alongside. Find the probability that Tony spins a prime number, and Ben spins a multiple of 3.

Example 13

Sunil has probability $\dfrac{4}{5}$ of hitting a target and Monika has probability $\dfrac{5}{6}$.

If they both fire simultaneously at the target, determine the probability that:
a they both hit
b they both miss.

Let S be the event of Sunil hitting the target and M be the event of Monika hitting it.

a P(both hit) = P(S and M)
$= $ P(S) $\times$ P(M)
$= \dfrac{4}{5} \times \dfrac{5}{6}$
$= \dfrac{2}{3}$

b P(both miss) = P(S' and M')
$= $ P(S') $\times$ P(M')
$= \dfrac{1}{5} \times \dfrac{1}{6}$
$= \dfrac{1}{30}$

4 Golfers George and Brett each have putts on a green. George has probability 0.6 of holing his putt, and Brett has probability 0.3 of holing his putt. Find the probability that:
a they both hole their putts
b they both miss their putts.

5 On any given day at a car dealership, Bronwyn has probability 0.2 of selling a car. At another dealership, Dwayne has probability 0.15 of selling a car. Find the probability that, on a given day:
   a they both sell a car
   b neither of them sell a car
   c Bronwyn sells a car but Dwayne does not.

6 Each time Morris the dog gets thrown a frisbee, he has probability 0.7 of catching it. If a frisbee is thrown to Morris three times, find the probability that Morris:
   a catches it all three times
   b misses the first throw, but catches the next two
   c catches the first throw, but misses the next two.

I   DEPENDENT EVENTS

Suppose a hat contains 5 red and 3 blue tickets. One ticket is randomly chosen, its colour is noted, and it is then thrown away. A second ticket is randomly selected.

If the first ticket was red,   P(second ticket is red) $= \frac{4}{7}$ ← 4 reds remaining
← 7 to choose from

If the first ticket was blue, P(second ticket is red) $= \frac{5}{7}$ ← 5 reds remaining
← 7 to choose from

So, the probability of the second ticket being red *depends* on what colour the first ticket was.

> Two events are **dependent** if they are **not independent**.
>
> The occurrence of one of the events *does* affect the occurrence of the other event.

The rule for finding compound event probabilities for dependent events is different from the rule for independent events.

> If A and B are dependent events, then $P(A \cap B) = P(A) \times P(B$ given that A has occurred$)$.

Example 14   ◆)) Self Tutor

A bag contains 3 orange and 4 green balls. Two balls are randomly selected one after the other from the bag, the first being removed before the second ball is chosen. Find the probability that:
a both balls are orange
b the first ball is orange and the second ball is green.

a   P(both balls are orange)
   = P(first ball is orange and the second ball is orange)
   = P(first ball is orange) × P(second ball is orange given that the first ball is orange)
   $= \frac{3}{7} \times \frac{2}{6}$ ← 2 orange balls remaining
← 6 balls to choose from

$= \dfrac{6}{42}$

$= \dfrac{1}{7}$

b P(first ball is orange and the second ball is green)

$=$ P(first ball is orange) $\times$ P(second ball is green given that the first ball is orange)

$= \dfrac{3}{7} \times \dfrac{4}{6}$ ⟵ 4 green balls remaining
   ⟵ 6 balls to choose from

$= \dfrac{12}{42}$

$= \dfrac{2}{7}$

EXERCISE 29I

1 A box contains 5 red and 2 purple balls. Two balls are randomly selected from the box. The first is *not* replaced in the box before the second is selected. Determine the probability that:

  a both balls are red   **b** the first ball is purple and the second is red.

2 Amelie has a bag containing 4 red apples and 6 green ones. She selects one apple at random, eats it, and then takes another, also at random.

  a Determine the probability that:

   i both apples were red

   ii both apples were green

   iii the first was red and the second was green

   iv the first was green and the second was red.

  b Find the sum of the probabilities in **a**. Explain your answer.

3 Marjut has a carton containing 10 cans of soup. 4 cans are tomato and the rest are pumpkin. She selects two cans at random without looking at the labels.

  a Find the probability that both cans are:

   i tomato soup   **ii** pumpkin soup.

  b Hence find the probability that Marjut selects one can of each flavour.

4 A raffle has 150 tickets which are placed in a barrel. Two tickets are drawn at random from the barrel to decide the prizes. John has 3 tickets in the raffle. Determine his probability of winning:

  a first and second prize   **b** second prize but not first prize.

  c none of the prizes.

5 When Jamie plays football, he kicks with his right foot 80% of the time, and his left foot 20% of the time. His right foot kicks are accurate 70% of the time, and his left foot kicks are accurate 40% of the time. Find the probability that Jamie's next kick will be:

  a with his right foot and accurate

  b with his left foot and inaccurate.

Example 15 ◀) Self Tutor

A hat contains 15 tickets with the numbers 1 to 15 printed on them.
If three tickets are drawn from the hat without replacement, determine the probability that they are all multiples of 3.

The multiples of 3 are $\{3, 6, 9, 12, 15\}$.

$\therefore$ P(three multiples of 3)
= P(1st is a multiple *and* 2nd is a multiple *and* 3rd is a multiple)
$= \dfrac{5}{15}$ ◀—— 5 multiples out of 15 numbers
$\quad \times \dfrac{4}{14}$ ◀—— 4 multiples out of 14 numbers after a successful first draw
$\quad \quad \times \dfrac{3}{13}$ ◀— 3 multiples out of 13 numbers after two successful draws
≈ 0.0220

6 Reyn has a carton containing a dozen eggs, 5 of which have a double-yolk. He randomly selects 3 eggs from the carton to bake a cake. Find the probability that:

  a they all have a double-yolk   **b** none of them have a double-yolk.

7 A bag contains three white and five red marbles. Three marbles are selected from the bag without replacement. Determine the probability that:

  a they are all red   **b** they are all white   **c** the first is red and the others are white.

8 A box contains 15 tickets with the numbers 1 to 15 printed on them. If three tickets are drawn from the box without replacement, find the probability that:

  a they are all odd   **b** they are all even   **c** the first two are odd and the third is even.

J TREE DIAGRAMS

Another way to display the outcomes of an experiment with more than one operation is a **tree diagram**.

The tree diagram alongside shows the sample space when a coin is tossed and a triangular spinner marked A, B, C is spun simultaneously.

The first set of branches show the possible outcomes when tossing a coin. For each of these outcomes, there are three possible outcomes for spinning the spinner.

The path shown in red represents the outcome "TB".

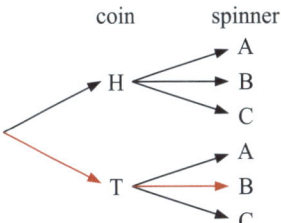

EXERCISE 29J.1

1 Use a tree diagram to illustrate the sample space for:

  a tossing a 10 pence coin and a 20 pence coin simultaneously

  b rolling a 6-sided die and tossing a coin simultaneously

  c the genders of a 3-child family.

2 Bag A contains red and white marbles and bag B contains blue and yellow marbles. A bag is selected and one marble is taken from it. Draw a tree diagram to illustrate the sample space.

3 Hats A, B, and C each contain pink and white tickets. A hat is selected and then two tickets are taken from it. Draw a tree diagram to illustrate the sample space.

USING TREE DIAGRAMS TO CALCULATE PROBABILITIES

Consider the archers Sunil and Monika in **Example 13**.

We can include the probabilities for each archer hitting or missing the target on the branches.

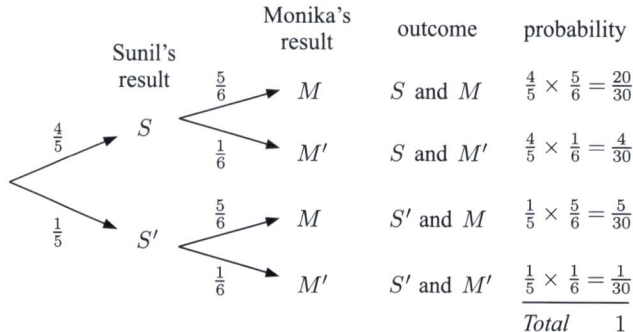

Notice that:
- There are *four* alternative paths, and each path shows a particular outcome.
- All outcomes are represented.
- The probability of each outcome is obtained by **multiplying** the probabilities along its path.

Example 16

Stephano is having computer problems. His desktop computer will only boot up 90% of the time, and his laptop will only boot up 70% of the time. Stephano attempts to boot both machines.

a Draw a tree diagram to illustrate this situation.
b Use the tree diagram to determine the chance that:
  i both will boot up
  ii only the desktop computer boots up.

a D = desktop computer boots up
L = laptop boots up

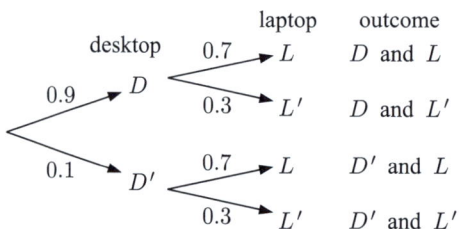

b **i** P(both boot up)
$= \text{P}(D \text{ and } L)$
$= \text{P}(D) \times \text{P}(L)$
$= 0.9 \times 0.7$
$= 0.63$

ii P(desktop boots up but laptop does not)
$= \text{P}(D \text{ and } L')$
$= \text{P}(D) \times \text{P}(L')$
$= 0.9 \times 0.3$
$= 0.27$

If two or more branches meet the description of the event, the probability for each branch is found and the results are **added**.

Probability (Chapter 29) 585

Example 17

Bag A contains 4 red jelly beans and 1 yellow jelly bean. Bag B contains 2 red and 3 yellow jelly beans. A bag is randomly selected by tossing a coin, and one jelly bean is removed from it. Determine the probability that it is yellow.

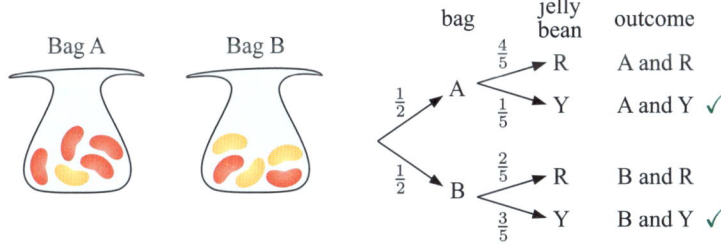

To get a yellow we need either Bag A and yellow, **or**, Bag B and yellow. We **add** the probabilities for these outcomes.

$P(\text{yellow}) = P(A \text{ and } Y) + P(B \text{ and } Y)$
$= \frac{1}{2} \times \frac{1}{5} + \frac{1}{2} \times \frac{3}{5}$ {branches marked ✓}
$= \frac{4}{10}$
$= \frac{2}{5}$

EXERCISE 29J.2

1 Suppose this spinner is spun twice.
  a Draw a tree diagram to illustrate the sample space.
  b Determine the probability that:
    i blue appears on both spins
    ii green appears on both spins
    iii different colours appear on the two spins
    iv blue appears on *either* spin.

2 The probability of the race track being muddy next week is estimated to be $\frac{1}{4}$. If it is muddy, the horse Rising Tide will start favourite with probability $\frac{2}{5}$ of winning. If it is dry, Rising Tide has a $\frac{1}{20}$ chance of winning.
  a Display the sample space of possible results on a tree diagram.
  b Determine the probability that Rising Tide will win next week.

3 Tennis star Boris gets his first serve in 72% of the time. If he gets his first serve in, he wins the point 85% of the time. If not, he only wins the point 50% of the time.
Find the probability that Boris will win the next point he serves.

4 Machine A cans 60% of the fruit at a factory. Machine B cans the rest. Machine A spoils 3% of its product, while Machine B spoils 4%. Determine the probability that the next can inspected at this factory will be spoiled.

5 Box A contains 2 black and 3 white blocks. Box B contains 5 black and 1 white block. A box is chosen by the flip of a coin, and one block is taken at random from that box. Determine the probability that the block is white.

6 Balls numbered 1 to 10 are placed in a bag. Two of the balls are drawn out at random.
Find the probability that the numbers on the balls are consecutive.

7 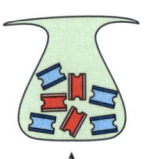 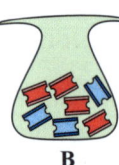 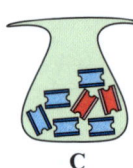 Three bags contain different numbers of blue and red tickets. A bag is selected using a die which has three A, two B, and one C face.

One ticket is selected at random from the chosen bag. Determine the probability that it is blue.

8 Consider the **Opening Problem** on page 560.

  a Copy and complete the tree diagram alongside.
  b Find the probability that player A will win the grand final.
  c Who is most likely to win the grand final?
  d Suppose that instead, A played C and B played D in the semi-finals. Who would now be most likely to win the grand final?

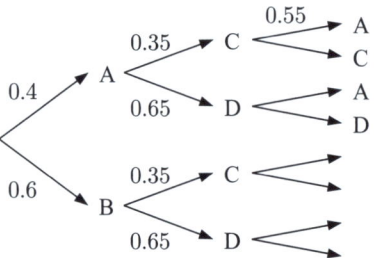

Activity   The finals series

Click on the icon to obtain an Activity on finals series which are the best of 3 or 5 matches.

FINALS SERIES

K   EXPECTATION

When a chance experiment is performed many times, we cannot predict *exactly* what will happen. However, we can use probability to calculate how many times we *expect* a particular event to occur.

For example, when a fair coin is tossed 100 times, we would *expect* half of the results to be heads. We therefore expect $100 \times \frac{1}{2} = 50$ heads.

Similarly, if a normal six-sided die is rolled 300 times, we would expect a "4" to be rolled $\frac{1}{6}$ of the time. We therefore expect $300 \times \frac{1}{6} = 50$ results to be a "4".

> If there are n trials of an experiment, and the probability of an event occurring in each trial is p, then the **expectation** of the occurrence of that event is np.

Example 18   ◀)) Self Tutor

When an archer fires at a target, there is probability $\frac{2}{5}$ that he hits the bullseye.

In a competition, the archer is required to fire 40 arrows. How many times would you expect him to hit the bullseye?

$p = P(\text{bullseye}) = \frac{2}{5}$ and $n = 40$

∴ the expected number of bullseyes is $np = 40 \times \frac{2}{5} = 16$

EXERCISE 29K

1. A goalkeeper has probability $\frac{3}{10}$ of saving a penalty attempt. How many goals would he expect to save out of 90 penalty shots?

2. During the snow season there is a $\frac{3}{7}$ probability of snow falling on any particular day. If Dan skis for five weeks, on how many days could he expect to see snow falling?

3. A hat contains three yellow discs and four green discs. A disc is drawn from the hat and its colour recorded. The disc is then returned to the hat and the procedure is repeated 350 times. On how many occasions would you expect a green disc to be drawn?

4. Two dice are rolled simultaneously 180 times. On how many occasions would you expect to get a double?

5. In a random survey of a district, people are asked whether they will vote for politician A, B, or C. The results are shown alongside.

A	B	C
165	87	48

    a How many people took part in the survey?
    b Estimate the probability that a randomly chosen voter in the district will vote for:
       i A   ii B   iii C.
    c There are 7500 people in the district. How many of these would you expect to vote for:
       i A   ii B   iii C?

6. The pair of spinners alongside are spun 600 times. How many times would you expect the sum of the spins to equal 5?

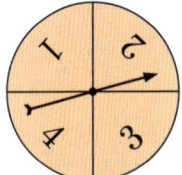 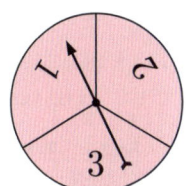

Review set 29A

1. Write a phrase to describe the probability of these events:
   a There is a 95% chance that there will be a test next week.
   b There is a 45% chance that it will snow tomorrow.

2. A survey of forty five 18 year olds was conducted. It was found that 19 enjoyed camping.
   a Estimate the probability that a randomly selected 18 year old likes camping.
   b How can the accuracy of this estimate be improved?

3. A letter of the alphabet is chosen at random. Find the probability that the letter is:
   a M or N
   b contained in the word CHOCOLATE
   c after the letter S in the alphabet.

4. A coin is tossed, and a spinner with equal sectors numbered 1, 2, 3, 4, and 5 is spun.
   a Use a 2-dimensional grid to illustrate the sample space.
   b Find the probability of getting:
      i a head and a 5   ii a head or a 5.

5 A marketing company was commissioned to investigate the main reason teenagers used the internet. The results of the survey are shown alongside.

Reason	Frequency	Relative frequency
Homework	29	
Social media	43	
Playing games	15	
Streaming music	69	
Total		

   a How many teenagers were surveyed?

   b Copy and complete the table.

   c Estimate the probability that a randomly selected teenager mainly uses the internet for:

      i homework   **ii** something other than streaming music.

6 Alec and Joe sit for an examination in Chemistry. Alec has a 95% chance of passing and Joe has a 25% chance of passing. Determine the probability that:

   a both pass   **b** both fail

   c Joe passes *and* Alec fails

   d Alec passes *and* Joe fails.

7 A bag of mixed lollies contains 10 mints and 6 chocolate caramels. A second bag contains 8 mints and 8 chocolate caramels. A bag is randomly chosen by tossing a coin, and a lolly is then taken from it.

   a Construct a tree diagram to show the sample space.

   b Hence find the probability that the lolly selected will be:

      i a chocolate caramel   **ii** a mint.

8 From past experience, a surfer has probability 0.83 of catching a wave. In one week she tries to catch 75 waves. How many waves do you expect her to catch?

9 A hat contains 12 tickets with the numbers 1 to 12 printed on them. If two tickets are drawn from the hat without replacement, find the probability that they are both prime numbers.

10 Each morning when Harold has a shower, there is a 90% chance that the hot water is working. The probability that Harold has a long shower is 80% if the hot water is working, and 10% if the hot water is not working.

   a Find the probability that, on any given day, Harold will have a long shower.

   b During a 365 day year, how many long showers would you expect Harold to have?

11 The probability of a delayed flight on a foggy day is $\frac{9}{10}$. When it is not foggy, the probability of a delayed flight is $\frac{1}{12}$. The probability of a foggy day is $\frac{1}{20}$.

   a Construct a tree diagram to show this information.

   b Hence, find the probability of:

      i a foggy day and a delayed flight   **ii** a delayed flight

      iii a flight which is not delayed.

   c Comment on your answers to **b ii** and **iii**.

12 A class has 25 students. 15 have blue eyes, 9 have fair hair, and 3 have both blue eyes and fair hair.
  a Represent this information on a Venn diagram.
  b Hence, find the probability that a randomly selected student from the class:
   i has neither blue eyes nor fair hair
   ii has blue eyes, but not fair hair.

13 Let $P(A) = 0.2$ and $P(B) = 0.7$. Find $P(A \cap B)$ given that A and B are:
  a mutually exclusive
  b independent.

Review set 29B

1 What is meant by saying that two events are *independent*?

2 When a box of drawing pins was dropped onto the floor, 49 pins landed on their backs and 32 landed on their sides. Estimate, to 2 decimal places, the probability of a drawing pin landing:
  a on its back
  b on its side.

back   side

3 Donna kept records of the number of clients she interviewed over 38 consecutive days.
  a Find the value of x, and hence write down the unknown values in the table.
  b Estimate the probability that tomorrow Donna will interview:
   i no clients
   ii four or more clients
   iii less than three clients.

Number of clients	Frequency
0	$4 - x$
1	6
2	12
3	$3x - 1$
4	6
5	x
6	0
7	2

4 300 Year 9 and 10 students at a school are asked if they like the current school uniform. Of these students:
  - 150 are in Year 10
  - 69 Year 10 students dislike the uniform
  - 108 Year 9 students like the uniform.

  a Copy and complete this frequency tree.
  b A Year 9 student is randomly selected. Estimate the probability that they dislike the uniform.

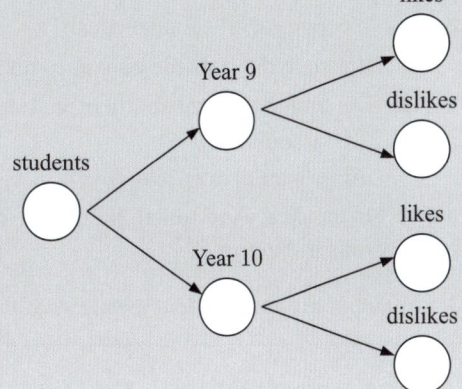

5 Use a tree diagram to illustrate the sample spaces for the following:
  a Bags A, B, and C contain green and yellow tickets. A bag is selected and then a ticket is taken from it.
  b Martin and Justin play tennis. The first to win three sets wins the match.

6 The digits 1, 6, and 9 are placed in random order to create a 3-digit number.
  a Use a table to help you list all the possible outcomes.
  b Find the probability that the number will be a perfect square.

7 Bag X contains three white and two red marbles. Bag Y contains one white and three red marbles. A bag is randomly chosen and two marbles are drawn from it.
  a Illustrate the information on a tree diagram.
  b Find the probability that the two marbles drawn are different colours.

8 In an office of 15 employees, 10 drive a car to work, 6 use public transport, and 3 use both.
  a Represent this information on a Venn diagram.
  b Find the probability that a randomly selected employee uses neither a car nor public transport to get to work.

9 A water polo player has probability $\frac{1}{4}$ of scoring a goal each time she shoots. If she has 24 shots at goal, how many goals would you expect her to score?

10 Shelley draws 3 cards without replacement from the container alongside. She will win a prize if all 3 cards are the same colour.

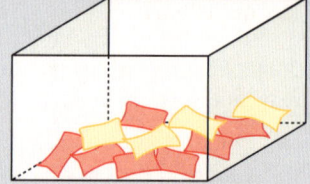

  a Find the probability that Shelley will win a prize.
  b Suppose the rules change so that Shelley now draws her cards *with* replacement.
    i Do you think this increases or decreases her probability of winning? Explain your answer.
    ii Find the probability of Shelley winning under the new rules.

11 A coin is tossed and a die is rolled simultaneously.
  a Illustrate the sample space on a grid.
  b Find the probability of getting:
    i a head and an even number
    ii a head and a non-3
    iii a 5 or a 6
    iv a head or an even number.

12 Two dice are rolled simultaneously.
  a Illustrate the possible outcomes on a 2-dimensional grid.
  b Determine the probability of getting:
    i a double 5
    ii at least one 4
    iii a sum greater than 9
    iv a sum of 5 or 6.
  c If the dice were rolled 900 times, on how many occasions would you expect the sum of the rolls to be prime?

13 X and Y are independent events such that $P(X) = 0.6$ and $P(X \cap Y) = 0.3$. Find $P(X \cup Y)$.

30

Statistics

Contents:

- **A** Populations and samples
- **B** Discrete data
- **C** Continuous data
- **D** Measuring the centre
- **E** Measuring the spread

Statistics (Chapter 30)

Opening problem

Roland owns two hotels, one in Leeds and one in Manchester. He wants to find out whether there is a difference in the number of nights guests stay at the hotels.

He therefore inspects the last 40 reservations placed for each hotel, and records the number of nights the guests stayed.

Leeds								
2	3	1	2	4	2	6	3	4	5
8	3	1	3	4	2	1	2	4	5
3	6	2	3	2	1	3	6	2	4
8	1	5	7	2	1	8	5	3	2

Manchester								
2	4	4	5	3	6	2	3	1	7
2	3	4	3	5	6	5	2	4	7
3	2	8	1	7	3	1	2	5	6
4	5	6	4	5	4	8	1	3	7

Things to think about:

a What is the best way to organise this data?
b How can the data be displayed?
c What is the *most common* length of stay at each hotel?
d How can Roland best measure:
  i the *average* length of stay for each hotel   **ii** the *spread* of each data set?
e Can a reliable conclusion be drawn from the data? What factors could affect the reliability of the conclusion?
f How could Roland improve the accuracy of his investigation?

In **Chapter 14** we saw how to interpret tables, graphs, and charts. These are common ways to display **data** that has been collected from the real world.

Statistics is the study of data collection and analysis. In this Chapter, we will see how statistics can help us draw meaningful conclusions about the world around us.

Historical note

After **William the Conqueror** (1028 - 1087) invaded and conquered England in 1066, his followers overtook estates previously occupied by Saxons. Confusion reigned over who owned what.

In 1086 William ordered that a census be conducted to record population, wealth, and land ownership. A person's wealth was recorded in terms of land, animals, farm implements, and the number of peasants on the estate. All this information was collated in what is now called the **Domesday Book**. Regarded as the greatest public record of Medieval Europe, the Domesday Book is displayed in the National Archives in Kew.

William the Conqueror

A   POPULATIONS AND SAMPLES

> A **population** is a collection of individuals or objects about which we want to draw conclusions.

The population might be the people living in a certain suburb, the trees in a forest, or the bottles from a production line.

> A **census** is the process of collecting data from the whole population.

A census is the most accurate way to investigate a population of interest. It involves collecting data from *every* member of the population.

Activity — Censuses in the United Kingdom

Since 1801, a census has been conducted every 10 years in the United Kingdom. It involves collecting information from every household or individual on a particular night.

A different organisation is responsible for the census in each country. More information can be found at:

- www.ons.gov.uk/census for England and Wales
- www.scotlandcensus.gov.uk for Scotland
- www.nisra.gov.uk/statistics/census for Northern Ireland.

What to do:

1. Research the type of information that is collected in the census.
2. What is the purpose of performing the census?
3. The 2011 census cost approximately £480 million, more than twice the amount spent on the 2001 census.
    a. Is it always *practical* to obtain data from every member of a population?
    b. How can we collect data about a population without the time and cost of a census?

In most situations, it is impractical or impossible to obtain data from the entire population. Instead, we can collect data from *some* members of the population.

> A **sample** is a group of individuals from the population.
> A **survey** is the process of collecting data from a sample.

Conclusions based on a sample will never be as accurate as conclusions based on the entire population. However, if the sample is chosen carefully, we can use it to draw reliable conclusions about the whole population.

EXERCISE 30A

1. A quality inspector wants to determine whether a machine is filling boxes with the correct volume of cereal.
    a. What is the population of interest?
    b. Explain why conducting a census would not be practical in this case.

2 Decide whether conducting a census or collecting a sample would be most appropriate for investigating:
  a the eating preferences of students in your class
  b the shoe sizes of teenagers in Bristol
  c opinions on changes to the communal garden of a block of flats
  d the lengths of trout in a river
  e the diameter of apples picked from trees in an orchard.

3 40 students from a school with 820 students are selected to complete a questionnaire on their school uniform.
  a What is the population size?   **b** What is the size of the sample?

4 Hugh's hotel contains 15 floors, each with 25 rooms. He wants to assess how well the rooms are being cleaned. He randomly selects one room from each floor to inspect after it has been cleaned.
  a Give a reason why Hugh did not perform a census.
  b What is the population size?
  c What is the size of the sample?
  d Why do you think Hugh might have chosen one room from each floor rather than all the rooms on one floor? Explain your answer.

Discussion

"Jedi" are the main protagonists in the *Star Wars* films. In the 2001 census, approximately 390 000 people in England and Wales recorded "Jedi" as their religion, making it the fourth largest "religion" in the region.

- In a census or survey with a questionnaire, will all participants necessarily answer questions:
  ▸ truthfully   ▸ correctly?
- Is a questionnaire necessarily the most *accurate* way to collect data?
- What is meant by the term "response order bias"? What effect does this have on surveys and elections?

B DISCRETE DATA

A **discrete variable** takes exact numerical values. It is often a result of **counting**.

For example, in the **Opening Problem**, the *number of nights stayed* is a discrete variable. It can only take an exact numerical value such as 1, 2, 3, 4, 5, ...

ORGANISING AND DISPLAYING DISCRETE DATA

A **frequency table** can be used to organise numerical data.

The data can then be displayed using a **bar chart**.

For the Leeds hotel data, we have:

Frequency table

Number of nights	Tally	Frequency
1							6
2										10
3									8
4						5
5						4
6					3
7			1
8					3

Bar chart

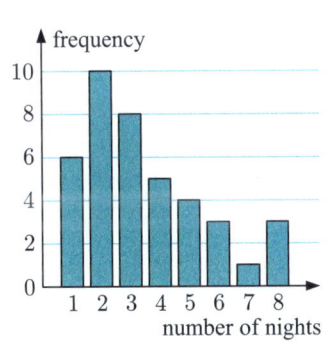

We leave gaps between the bars to show that the variable is discrete.

Alternatively, we can use vertical lines to show the frequency of each data value. The graph alongside is called a **vertical line chart**.

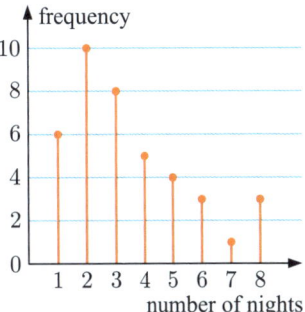

DESCRIBING THE DISTRIBUTION OF A DATA SET

Many data sets show **symmetry** about the **mode**, which is the most frequently occurring value.

If we place a curve over the bar chart alongside, we see that this curve shows symmetry. We call this a **symmetrical distribution**.

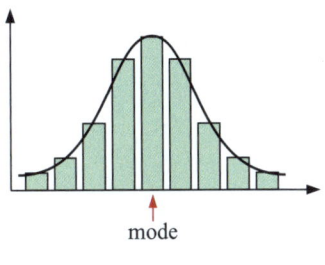

The distribution for the Leeds hotel data is shown alongside. We say it is **positively skewed** because, compared with the symmetrical distribution, it has been "stretched" on the right or positive side of the mode.

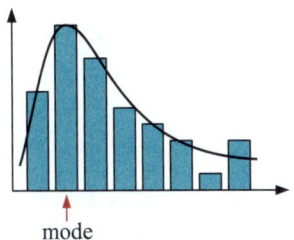

So, we have:

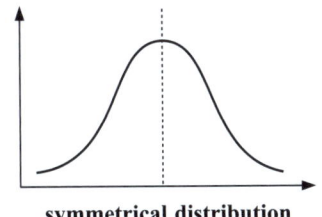

symmetrical distribution

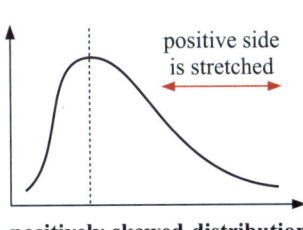

positively skewed distribution

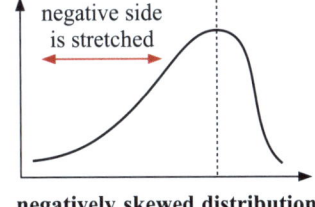

negatively skewed distribution

Outliers are data values that are either much larger or much smaller than the general body of data.

Outliers appear separated from the body of data on a bar chart.

If an outlier is a genuine piece of data, it should be retained for analysis. However, if it is found to be the result of an error in the data collection process, it should be removed from the data.

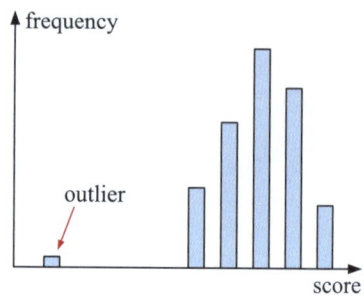

EXERCISE 30B

1 A randomly selected sample of shoppers were asked how many times they shopped at the supermarket in the past week. A bar chart was constructed for the results.

   a How many shoppers were in the sample?

   b How many of the shoppers shopped once or twice?

   c What percentage of the shoppers shopped more than four times?

   d Describe the distribution of the data.

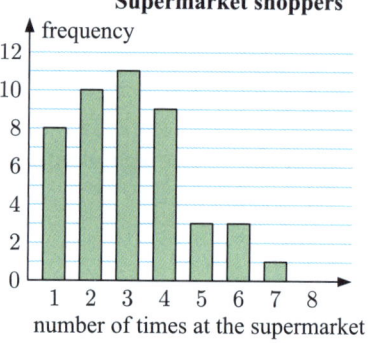

2 Employees of a company were asked how many times they left the office on business appointments during one week. The following vertical line chart was constructed from the data:

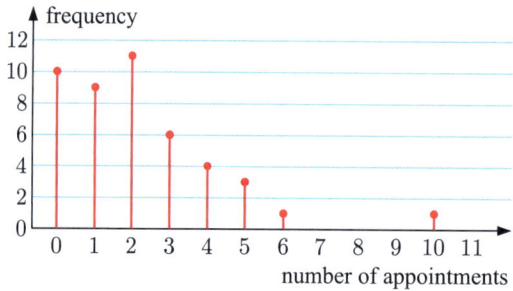

   a How many employees did not leave the office on a business appointment?

   b State the mode of the data.

   c What percentage of the employees left the office more than 5 times?

   d Describe the distribution of the data.

   e How would you describe the data value "10"?

3 Rhiannon is a ten-pin bowler. She recorded the number of pins she knocked down in each frame during the last 6 games she played in the table alongside.

   a Find the number of frames in which Rhiannon knocked down 7 pins.

   b Draw a bar chart to display the data.

   c Describe the distribution of the data. Are there any outliers?

Number of pins	Frequency
3	1
4	3
5	5
6	12
7	
8	13
9	8
Total	60

4 20 students were asked how many TV sets they have in their household. The following data was collected:

2 1 0 3 1 2 1 3 4 0 0 2 2 0 1 1 0 1 0 1

 a Organise the data in a frequency table.
 b Draw a bar chart to display the data.
 c How would you describe the distribution of the data? Are there any outliers?
 d How many households had no TV sets?
 e What percentage of the households had three or more TV sets?

5 The number of toothpicks in a box is stated as 50, but the actual number of toothpicks has been found to vary. To investigate this, the number of toothpicks in a box was counted for a sample of 60 boxes. The results were:

50 52 51 50 50 51 52 49 50 48 51 50 47 50 52
48 50 49 51 50 49 50 52 51 50 50 52 50 53 48
50 51 50 50 49 48 51 49 52 50 49 49 50 52 50
51 49 52 52 50 49 50 49 51 50 50 51 50 53 48

 a Use a frequency table to organise this data.
 b Display the data using a vertical line chart.
 c Describe the distribution of the data.
 d What percentage of the boxes contained exactly 50 toothpicks?

Example 1   ◀)) Self Tutor

Consider the hotel data in the **Opening Problem** on page **592**.

 a Organise the Manchester hotel data using a frequency table.
 b Draw a multiple bar chart to display the data for both hotels.
 c Compare the distribution of the Manchester hotel data with that of the Leeds hotel. At which hotel do guests generally stay longer?

a   Manchester hotel

Number of nights	Tally	Frequency
1	||||	4
2	卌 |	6
3	卌 ||	7
4	卌 ||	7
5	卌 |	6
6	||||	4
7	||||	4
8	||	2

b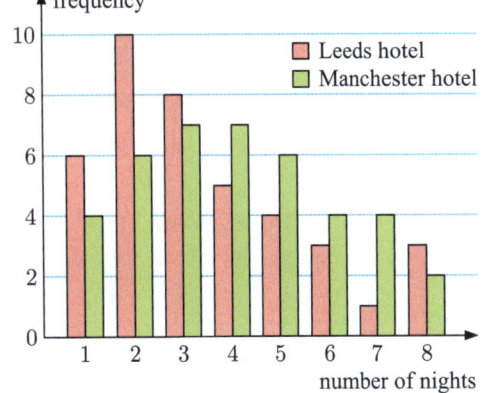

 c From the multiple bar chart, both sets of data are positively skewed.
 However, the mode for the Manchester hotel is higher, and its distribution is less skewed than that for the Leeds hotel.
 Guests generally stay longer at the Manchester hotel.

6

Number of phone calls	Frequency	
	13 year olds	18 year olds
0	5	1
1	8	2
2	13	3
3	8	4
4	6	4
5	3	6
6	3	8
7	2	7
8	1	5
9	0	4
10	0	3
11	1	3

Fifty 13 year olds and fifty 18 year olds were asked how many phone calls they made during the previous day.

a Use the table to construct a multiple bar chart.

b Compare the distributions of the data sets. Which age group generally made more phone calls?

In a *multiple* bar chart, the bars for two data sets are drawn next to one another.

7 Emilio has 6 hens: 3 Rhode Island Reds and 3 Buff Orpingtons. He recorded the number of eggs collected from each breed per day for three weeks.

Rhode Island Red
4 5 1 5 3 3 3
5 3 3 0 3 4 2
2 3 3 4 4 2 1

Buff Orpington
2 1 3 3 0 3 0
1 0 2 1 1 0 0
1 1 0 4 0 0 1

a Organise the data into a frequency table including a column for each breed.

b Draw a multiple bar chart to display the data.

c Describe the distribution of each data set.

d Which breed generally laid fewer eggs over the three week period? Explain your answer.

8 The composite bar chart alongside shows the number of people living in each flat in two apartment buildings.

a Which building has:

  i unoccupied flats

  ii more flats with 3 occupants?

b Construct a frequency table for:

  i building A   **ii** building B.

c Draw a multiple bar chart to display the data.

d Describe the distribution of each data set.

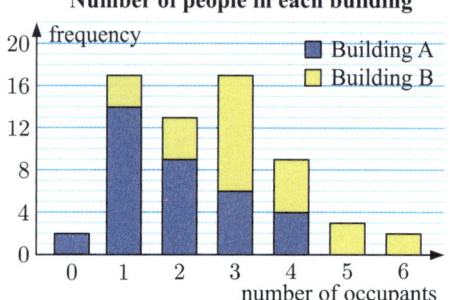

In a *composite* bar chart, the bars for two data sets are drawn one above the other.

Investigation 1 — Grouped discrete data

In situations where there are lots of different numerical values recorded, it may not be practical to use an ordinary frequency table, or to display the data using a bar chart or vertical line chart. Instead, we *group* the data into **class intervals**.

What to do:

1 A local hardware store is studying the number of people visiting the store at lunch time. Over 30 consecutive weekdays they recorded the data:

```
37  30  17  13  46  23  40  28  38  24
23  22  18  29  16  35  24  18  24  44
32  54  31  39  32  38  41  38  24  32
```

a Construct a frequency table for the *ungrouped* data, and draw a bar chart.

b Explain why the bar chart you drew in **a** is not very useful.

c Copy and complete the table alongside by counting the number of data values between 10 and 19 inclusive, between 20 and 29 inclusive, and so on.

Number of people	Tally	Frequency
10 - 19		
20 - 29		
30 - 39		
40 - 49		
50 - 59		
Total		30

d Hence draw a bar chart for the data "grouped" into the intervals "10 - 19", "20 - 29", ...

e In which of your bar charts can you more easily observe the distribution of the data?

2 A traffic surveyor recorded the number of cars passing through an intersection each day for several weeks. Their results are summarised in the bar chart below.

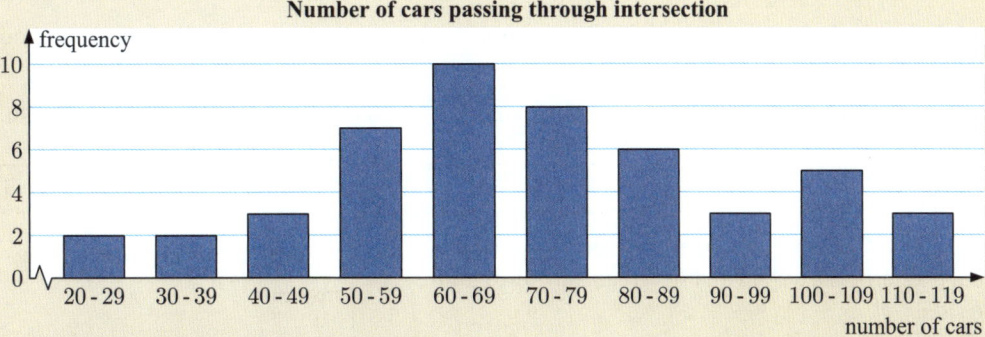

Number of cars passing through intersection

a Describe the distribution of the data.

b On how many days did between 80 and 89 (inclusive) cars pass through the intersection?

c How many days were included in the survey?

d From the bar chart, can we determine:

  i the number of days in which *exactly* 40 cars passed through the intersection

  ii the mode of the distribution?

Explain your answers.

Discussion

In what ways are discrete data and categorical data:
- different
- similar?

C CONTINUOUS DATA

A **continuous variable** takes values within a certain continuous range. It is usually a result of **measuring**.

When data is recorded for a continuous variable, there will be many different values. The data is therefore organised using **class intervals**.

We commonly use a **frequency histogram** to display continuous data. A frequency histogram is similar to a bar chart, but because the data is continuous, the bars are joined together.

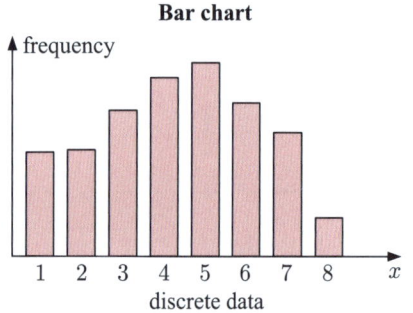
Bar chart
discrete data

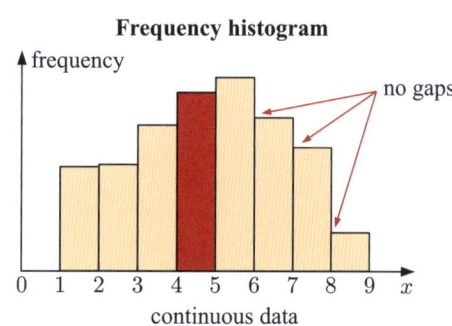
Frequency histogram
continuous data

The *sides* of the bars are labelled on the horizontal axis to show the end points of each class interval. For example in the frequency histogram above, the red shaded bar corresponds to data values in the class interval $4 \leqslant x < 5$.

The **modal class** is the class of values that appears most often. On a frequency histogram, the modal class has the highest bar.

In this course you are not required to draw histograms, but you should know how to interpret them.

Example 2   ◀)) Self Tutor

The weights of parcels (w kg), sent on a given day from a post office are shown in the frequency histogram alongside.

a Describe the distribution of the data.
b State the modal class.
c How many parcels were sent in total on this particular day?
d Over the next month, 564 parcels are sent from the post office. Estimate the number which weigh more than 4 kg.

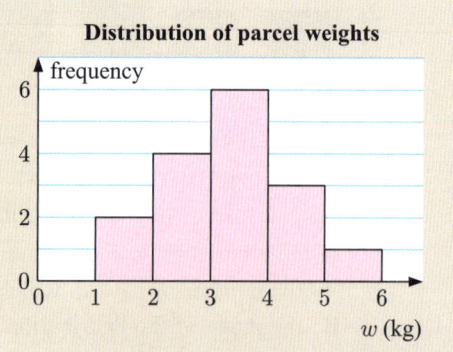
Distribution of parcel weights

a The distribution is approximately symmetrical.

b The modal class is $3 \leqslant w < 4$ kg.

c $2 + 4 + 6 + 3 + 1 = 16$ parcels were sent in total.

d Of the 16 parcels sent on the one day, $\frac{4}{16} = \frac{1}{4}$ of them weighed more than 4 kg.

∴ for the next month we expect $\frac{1}{4} \times 564 = 141$ parcels to weigh more than 4 kg.

EXERCISE 30C

1

Shot put distances thrown

During a training session, Daniel performed 20 throws of the shot put. The results are shown in the frequency histogram alongside.

a Describe the distribution of the data.

b State the modal class.

c Calculate the percentage of times Daniel threw the shot put further than 16 m.

2 A frequency histogram of the weights w of players in a volleyball squad is given alongside.

a Explain why *weight* is a continuous variable.

b Describe the distribution of the data.

c State and interpret the modal class.

d How many players are in this squad?

e Calculate the proportion of players who weighed more than 95 kg.

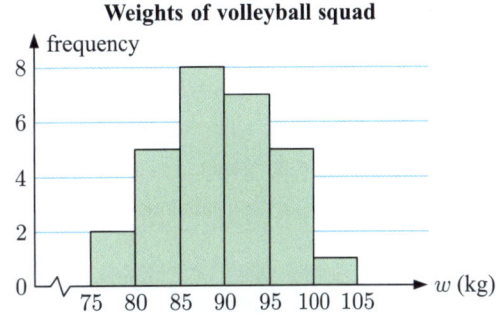

Weights of volleyball squad

3 A botanist takes a random sample of seedlings from a nursery, and measures their height h in millimetres. He drew the frequency histogram alongside to display his results.

a Describe the distribution of the data.

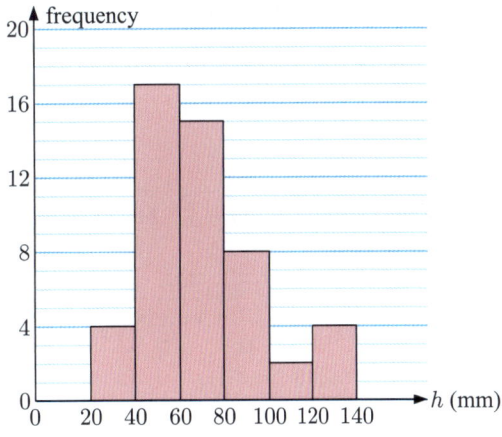

Heights of seedlings in a nursery

b **i** Write down the class intervals the botanist used to organise the data.
  ii Hence copy and complete the table.
c Use your table to answer the following:
  i How many of the seedlings are at least 100 mm tall?
  ii What percentage of the seedlings are between 60 and 80 mm tall?

Height (h mm)	Frequency
$20 \leqslant h < 40$	

d There are 857 seedlings in the nursery. Estimate the number of seedlings which measure:
  i less than 100 mm   **ii** between 40 and 100 mm.

4 Lewis is a physical education teacher. He recorded the times each of his students took to sprint 100 m. He drew two histograms to display his results, one for each gender.

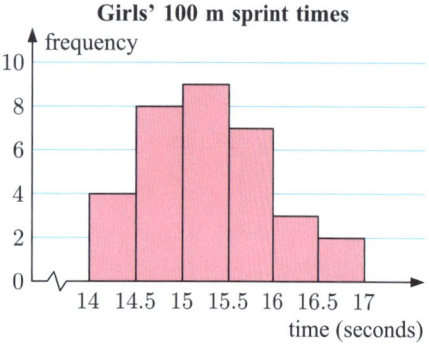
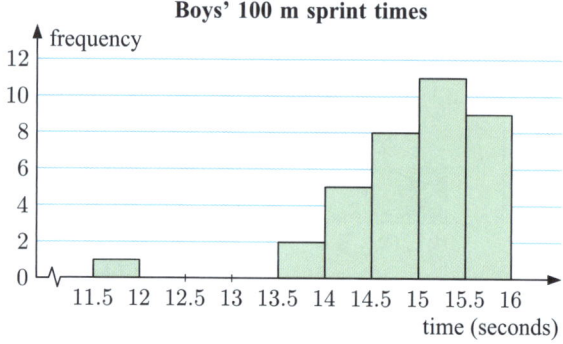

a For each data set:
  i describe its distribution   **ii** state the modal class.
b Which gender had more students with times in the interval $15 \leqslant t < 16$ seconds?
c Which gender generally had faster times for the 100 m sprint?
d Which of Lewis' data sets contains an outlier? Explain your answer.

D MEASURING THE CENTRE

We can get a better understanding of a data set if we can locate the **middle** or **centre** of the data, and get an indication of its **spread**. Knowing one of these without the other is often of little use.

There are three statistics that are used to measure the **centre** of a data set. They are the **mean**, the **median**, and the **mode**.

THE MEAN

The **mean** $\bar{x}$ of a data set is the statistical name for its arithmetic average.

$$\text{mean} = \frac{\text{the sum of the data values}}{\text{the number of data values}}$$

or $\bar{x} = \dfrac{\sum x}{n}$

where $\sum x$ is the sum of the data.

The mean is not necessarily a member of the data set.

$\bar{x}$ is read "x bar".
$\sum$ is read "the sum of all".

THE MEDIAN

> The **median** is the *middle value* of an ordered data set.

The median splits an ordered data set into halves. Half of the data are less than or equal to the median, and half are greater than or equal to it.

We *order* a data set by listing it from smallest to largest.

If there are n data, the median is the $\left(\dfrac{n+1}{2}\right)$th ordered data value.

For example:

If $n = 13$, $\dfrac{n+1}{2} = 7$ so the median is the 7th ordered data value.

If $n = 14$, $\dfrac{n+1}{2} = 7.5$ so the median is the average of 7th and 8th ordered data values.

If there is an even number of data values, the median is not necessarily a member of the data set.

THE MODE

> The **mode** is the most frequently occurring value in the data set.

If there are two values which occur most frequently, we say the data set is **bimodal**.

If a data set has more than two modes, we do not use the mode as a measure of the centre of the data set.

Example 3 ◀◉ Self Tutor

The number of small aeroplanes flying into a remote airstrip over a 15-day period is given below:

$$5 \quad 7 \quad 0 \quad 3 \quad 4 \quad 6 \quad 4 \quad 0 \quad 5 \quad 3 \quad 6 \quad 9 \quad 4 \quad 2 \quad 8$$

For this data set, find the: **a** mean **b** median **c** mode.

a mean $= \dfrac{5+7+0+3+4+6+4+0+5+3+6+9+4+2+8}{15} \quad \longleftarrow \dfrac{\sum x}{n}$

$= \dfrac{66}{15}$

$= 4.4$ aeroplanes

b The ordered data set is: $\;\; 0 \;\; 0 \;\; 2 \;\; 3 \;\; 3 \;\; 4 \;\; 4 \;\; \boxed{4} \;\; 5 \;\; 5 \;\; 6 \;\; 6 \;\; 7 \;\; 8 \;\; 9$

Since $n = 15$, $\dfrac{n+1}{2} = 8$

∴ the median is the 8th data value

∴ the median $= 4$ aeroplanes.

c 4 is the value that occurs the most often

∴ the mode $= 4$ aeroplanes.

Equal or approximately equal values of the mean, mode, and median *may* indicate a *symmetrical distribution* of data. However, you should always consider a graph of the distribution before calling the data set symmetric.

EXERCISE 30D.1

1 For each of the following data sets, find the:

    i mean   **ii** median   **iii** mode.

    Use technology to check your answers.

    a 12, 17, 20, 24, 30, 30, 42
    b 8, 8, 8, 10, 11, 11, 12, 12, 16, 20, 20, 24
    c 7.9, 8.5, 9.1, 9.2, 9.9, 10.0, 11.1, 11.2, 11.2, 12.6, 12.9
    d 427, 423, 415, 405, 445, 433, 442, 415, 435, 448, 429, 427, 403, 430, 446, 440, 425, 424, 419, 428, 441

2 Consider the following two data sets:

    Data set A: 5, 6, 6, 7, 7, 7, 8, 8, 9, 10, 12
    Data set B: 5, 6, 6, 7, 7, 7, 8, 8, 9, 10, 20

    a Find the mean for both *data set A* and *data set B*.
    b Find the median for both *data set A* and *data set B*.
    c Explain why the mean of *data set A* is less than the mean of *data set B*.
    d Explain why the median of *data set A* is the same as the median of *data set B*.

3 The selling price of ten houses in a particular suburb are:

    £234 000,   £240 000,   £240 000,   £255 000,   £258 000,
    £267 000,   £291 000,   £304 000,   £365 000,   £505 000

    a Find the mean, median, and modal selling prices.
    b Explain why the mode is an unsatisfactory measure of the middle in this case.
    c A young couple wants to buy a house in this suburb. They want to determine the price of a typical house in the suburb based on this data set. Which measure of centre do you recommend that they use? Explain your answer.

4 The following raw data is the daily rainfall (to the nearest millimetre) of a city for the month of February 2014:

    0, 4, 1, 0, 0, 0, 2, 9, 3, 0, 0, 0, 8, 27, 5, 0, 0, 0, 0, 8, 1, 3, 0, 0, 15, 1, 0, 0

    a Find the mean, median, and mode for the data.
    b   **i** Identify any outliers in the data set.
         ii Should the outliers be removed before finding the measures of centre? Explain your answer.

5 A basketball team scored 38, 52, 43, 54, 41, and 36 points in their first six matches.

    a Find the mean number of points scored for the first six matches.
    b What score does the team need to shoot in their next match to maintain the same mean score?
    c The team scores only 20 points in their seventh match. Find the mean number of points scored for the seven matches.
    d The team scores 42 points in their eighth and final match.
        i Will their previous mean score increase or decrease? Explain your answer.
        ii Find the mean score for all eight matches.

Example 4

Each student in a class of 20 is assigned a number between 1 and 10 to indicate his or her fitness. The results are: 7, 9, 8, 8, 10, 9, 8, 7, 8, 6, 9, 5, 6, 8, 9, 7, 7, 8, 10, 8

Organise the data into a frequency table, and hence calculate the:

a mean **b** median **c** mode.

a

Score	Tally	Number of students	Product
5	|	1	$1 \times 5 = 5$
6	||	2	$2 \times 6 = 12$
7	||||	4	$4 \times 7 = 28$
8	⊁||	7	$7 \times 8 = 56$
9	||||	4	$4 \times 9 = 36$
10	||	2	$2 \times 10 = 20$
Total		20	157

The mean score

$$= \frac{\text{total of scores}}{\text{number of scores}}$$

$$= \frac{157}{20}$$

$$= 7.85$$

b There are 20 scores, so the median is the average of the 10th and 11th ordered scores.

Score	Number of students	
5	1	← 1st student
6	2	← 2nd and 3rd student
7	4	← 4th, 5th, 6th, and 7th student
8	7	← 8th, 9th, **10th, 11th**, 12th, 13th, 14th student
9	4	
10	2	

The 10th and 11th students both scored 8, so the median $= 8$.

c Looking down the *number of students* column, the highest frequency is 7. This corresponds to a score of 8, so the mode $= 8$.

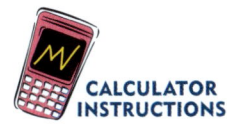
CALCULATOR INSTRUCTIONS

STATISTICS PACKAGE

---

6 3 coins were tossed simultaneously 40 times, and the number of heads for each toss was recorded.

Calculate the:
  a mode **b** median
  c mean.

Number of heads	Frequency
0	6
1	16
2	14
3	4
Total	40

7 The bar chart alongside gives the value of donations for an overseas aid organisation, collected in a particular street.

  a Construct a frequency table from the bar chart.
  b Determine the total number of donations.
  c Find the:
    i mean **ii** median **iii** mode.
  d Which of the measures of centre can be found easily from the bar chart only?

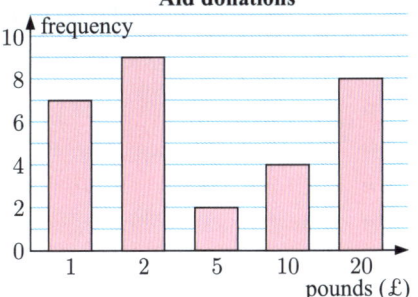

8 Hui breeds ducks. The number of ducklings surviving for each pair after one month is recorded in the table.

Number of survivors	Frequency
0	1
1	2
2	5
3	9
4	20
5	30
6	9
Total	76

  a Calculate the:
   i mean   **ii** mode   **iii** median.
  b Is the data skewed?
  c How does the skewness of the data affect the measures of the centre of the distribution?

9 Consider the **Opening Problem** on page **592**.
  a For each data set, find the: **i** mean   **ii** mode   **iii** median.
  b At which hotel do guests generally stay longer?

10 Chiara planted 19 seedlings, which she divided into groups A and B. Plants in group A were fertilised and plants in group B were not.
The heights of Chiara's plants in centimetres after one month were:
  Group A: 18.8, 21.2, 17.6, 22.8, 20.0, 15.7, 19.7, 24.1, 21.0, 18.2
  Group B: 11.8, 14.3, 17.6, 20.6, 15.4, 18.5, 27.9, 18.4, 16.2
  a Calculate the mean and median for each data set.
  b Do you think the fertiliser is effective? Explain your answer.

11 Two electronics stores recorded the total value of sales made each day for a week.
  Store A: £5200, £4750, £6230, £9410, £6780, £11 250, £6220
  Store B: £8250, £7460, £12 980, £15 260, £9480, £11 150, £10 270
  a Find the mean and median sales for each store.
  b Which store generally made more sales?
  c Which measure of centre should the stores use when predicting their total sales for the coming year? Explain your answer.

12 The vertical line charts below show the review scores (out of 10) for two Italian restaurants.

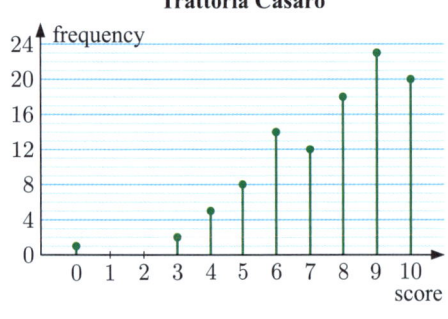
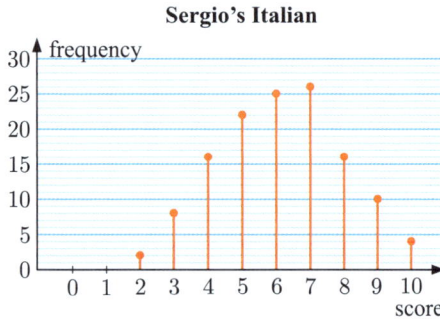

  a Based on the vertical line charts, which restaurant generally had better review scores?
  b Construct a frequency table for each data set.
  c Calculate the mean, median, and mode for each data set.
  d How do the numerical measures of centre compare with your initial observations?

Example 5

Linda has taken four Mathematics tests so far this year. Each test has been out of 20 marks, and her average mark has been 15.

What mark does Linda need in the 5th test to raise her average to 16?

Average mark = $\dfrac{\text{sum of marks}}{4} = 15$

$\therefore$ sum of marks = $15 \times 4 = 60$

Let Linda's mark for the 5th test be x.

$\therefore$ we require $\dfrac{60 + x}{5} = 16$

$\therefore \ 60 + x = 80$

$\therefore \ x = 20$

So, Linda needs a mark of 20 in the 5th test.

13 Jackie has played 8 games of netball this season, scoring an average of 17 goals per game. How many goals does she need to score in the next game to increase her average to 18?

14 A sample of 12 measurements has a mean of 8.5, and a sample of 20 measurements has a mean of 7.5. Find the mean of all 32 measurements.

15 On Saturday, Derek picked pears from 32 trees. He picked an average of 17 pears per tree. On Sunday he picked some more pears, averaging 12 pears per tree. Over the whole weekend he picked an average of 14 pears per tree. How many trees did Derek pick from on Sunday?

Discussion

Develop at least two examples to show how the measures of centre are affected by outliers.

Which of the measures of centre is most affected by the presence of an outlier?

Which of the measures of centre are unaffected by the presence of an outlier?

ESTIMATING THE MEAN OF GROUPED DATA

When data is presented in **class intervals**, the actual data values are not known. This makes it impossible to calculate the exact mean of the data set.

To *estimate* the mean of the data, we use the **midpoint** of an interval to represent all of the scores within the interval.

For example, if we have distances in the interval $50 \text{ km} \leqslant d < 100 \text{ km}$, we *estimate* that all of the data in that interval corresponds to the distance 75 km.

Example 6

The Department for Transport collected data regarding the distance each of its trams travelled in one day. This is shown in the table.

Estimate the mean distance travelled by the trams.

Distance (d miles)	Frequency
$100 \leqslant d < 200$	10
$200 \leqslant d < 300$	15
$300 \leqslant d < 400$	16
$400 \leqslant d < 500$	9

Distance (d miles)	Frequency	Interval midpoint	Product
$100 \leqslant d < 200$	10	150	1500
$200 \leqslant d < 300$	15	250	3750
$300 \leqslant d < 400$	16	350	5600
$400 \leqslant d < 500$	9	450	4050
Total	50		14 900

$\therefore$ mean

$= \dfrac{\text{sum of data values}}{\text{the number of data values}}$

$\approx \dfrac{14\,900}{50}$

≈ 298 miles

EXERCISE 30D.2

1 The daily maximum temperatures for Manila over a one year period are given below.

Maximum temperature (t °C)	Frequency
$24 \leqslant t < 26$	1
$26 \leqslant t < 28$	8
$28 \leqslant t < 30$	32
$30 \leqslant t < 32$	107
$32 \leqslant t < 34$	174
$34 \leqslant t < 36$	43

Estimate the mean maximum temperature.

2 Nick served a tennis ball 200 times. The speeds of the serves are summarised in the table alongside.

  a Find the modal class of the data.
  b If possible, find the:
   i number of serves faster than 170 km h^{-1}
   ii number of serves slower than 162 km h^{-1}
   iii percentage of serves between 155 km h^{-1} and 175 km h^{-1}.
  c Estimate the mean speed of the serves.

Speed (s km h^{-1})	Frequency
$150 \leqslant s < 155$	18
$155 \leqslant s < 160$	28
$160 \leqslant s < 165$	35
$165 \leqslant s < 170$	43
$170 \leqslant s < 175$	41
$175 \leqslant s < 180$	35

3 The table alongside shows the number of runs scored by Clive during his team's cricket season.

  a How many times did Clive bat?
  b How many times did Clive score at least 20 runs?
  c Estimate the mean number of runs scored by Clive.

Number of runs	Frequency
0 - 9	3
10 - 19	4
20 - 29	9
30 - 39	5
40 - 49	2

Investigation 2 — Measuring the centre of a population from a sample

At a particular school, there are 200 Year 11 students. The data below shows the grade each student received for mathematics in the previous year, converted to a numerical value (A ≡ 5, B ≡ 4, ...).

```
2 2 4 3 4 3 4 3 3 2 3 5 5 3 3 3 2 3 2 4 3 3 4 4 4
3 2 2 3 3 4 2 4 4 1 3 4 3 2 3 3 4 4 2 3 1 4 3 3 4
3 3 3 3 4 3 3 3 3 3 2 3 3 1 2 3 3 4 4 2 3 2 4 2 4
4 3 2 2 3 2 3 3 3 4 5 3 2 3 2 3 4 3 3 4 3 4 5 2 3
2 4 2 4 3 2 3 3 4 4 3 2 4 4 4 3 4 3 4 2 4 3 2 5 2
4 3 3 3 4 2 2 4 2 4 2 1 3 3 3 2 3 4 2 3 3 3 3 3 4
5 4 1 4 1 3 3 4 4 3 2 4 3 3 2 3 2 3 4 2 2 4 3 5 3
3 5 3 2 3 3 4 3 2 3 3 2 2 3 4 2 3 4 4 4 2 3 4 3 4
```

What to do:

1. Click on the icon to access the data in the **statistics package**.

    STATISTICS PACKAGE

    a Look at the bar chart for the population data and describe its distribution.

    b Write down the mean, median, and mode for the *population*.

2. Marianne is a Year 10 student. She wants to investigate the mathematics grades of her peers.

    a Explain why Marianne would want to collect data from a *sample* of students, rather than the entire population.

    b Click on the icon to run a simulation which randomly selects 10 students from the population. Generate 20 samples and record the mean, median, and mode for each sample in a table like the one below.

    SIMULATION

Sample number	Mean	Median	Mode
1			
2			
⋮			

    c Explain why the measures of centre of a sample are *estimates* of the population measures.

3. Marianne is unsure if collecting data from a sample of 10 students will give accurate results. Click on the icon to investigate the effect of changing the sample size on the mean, median, and mode.

    DEMO

    a Copy and complete the table alongside.

    b Comment on how each sample measure of centre compares to the corresponding population measure as the sample size increases.

    c Explain why a sample of size 200 is not sensible in this scenario.

    d Is a sample of size 150 necessarily practical in this case?

    e How large do you think Marianne's sample should be? Give reasons for your answer.

Sample size	Mean	Median	Mode
10			
20			
50			
75			
100			
150			
200			

E MEASURING THE SPREAD

Knowing the middle of a data set can be quite useful, but for a more complete picture of the data set we also need to know its **spread** or **variation**.

The most simple measure of the spread of a data set is the **range**.

$$\text{range} = \text{maximum data value} - \text{minimum data value}$$

Example 7   ◆) Self Tutor

Find the range of the data set: 5, 3, 8, 4, 9, 7, 5, 6, 2, 3, 6, 8, 4.

range = maximum data value − minimum data value = 9 − 2 = 7

The higher the range of a data set is, the more "spread out" its data values are.

EXERCISE 30E

1 Find the range of each data set:
   a 5, 6, 6, 6, 7, 7, 7, 8, 8, 8, 8, 9, 9, 9, 9, 9, 10, 10, 11, 11, 11, 12, 12
   b 11, 13, 16, 13, 25, 19, 20, 19, 19, 16, 17, 21, 22, 18, 19, 17, 23, 15
   c 23.8, 24.4, 25.5, 25.5, 26.6, 26.9, 27, 27.3, 28.1, 28.4, 31.5

2 While Jarrod was eating a bag of mandarins, he counted the number of seeds in each. The numbers he counted were:

   3   4   7   11   2   6   3   14   10   6   9

   Calculate the range of the data.

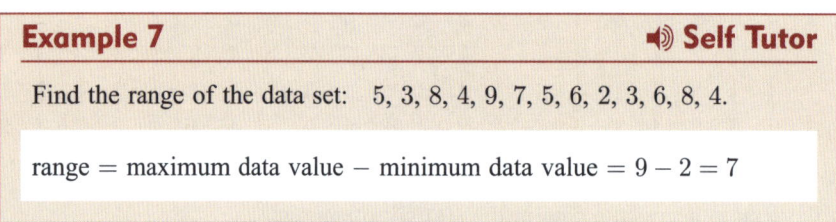

3 The table below shows the number of tows performed by a tow truck driver each day over a 45 day period.

Number of tows	3	4	5	6	7	8	9
Frequency	2	5	6	9	12	8	3

Find the:
   a mean   **b** median   **c** range.

4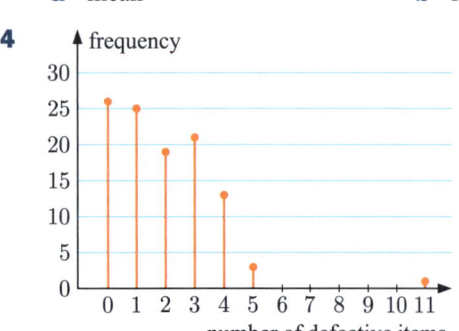

The vertical line chart alongside shows the number of defective items found by a quality inspector per day.

   a Calculate the range of the data.
   b Identify the outlier.
   c Do you think the range you calculated provides a good measure of the data's spread? Explain your answer.

5 Kylie and Chris were asked to listen to 20 songs, and give each a rating out of 20. The results are shown below.

Kylie					Chris				
14	11	16	8	10	15	11	9	12	16
7	10	5	20	13	14	10	14	9	17
12	3	19	6	11	18	13	12	12	16
4	15	10	19	16	11	12	18	14	10

a For each data set, find the:
  i median   **ii** range.
b In general, who gave the higher ratings?
c Who had greater variation in their ratings?

6 The Year 6 and Year 10 students at a school were asked how many times they visited their grandparents in the last month. The results are shown in the graphs below.

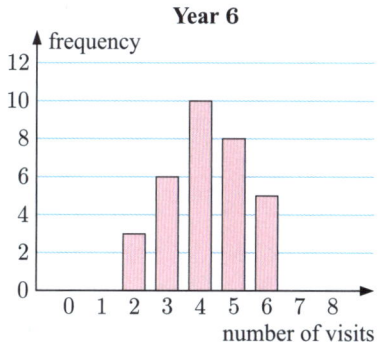
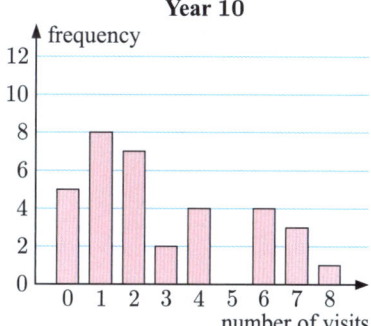

a For each data set, find the:
  i median   **ii** range.
b Which class:
  i generally visited their grandparents more often
  ii had greater variation in their number of visits?

Discussion

Suppose we have data that has been grouped into class intervals.

1 Explain why we cannot calculate the range of the data *exactly*.

2 How can we *estimate* the range of the data?

Depth (d km)	Frequency
$50 \leqslant d < 100$	5
$100 \leqslant d < 150$	12
$150 \leqslant d < 200$	9
$200 \leqslant d < 250$	17
$250 \leqslant d < 300$	20
$300 \leqslant d < 350$	15

Review set 30A

1. Decide whether conducting a census or collecting a sample would be most appropriate for investigating:
   a the weights of strawberries in a punnet
   b the blood pressure levels of Scottish men.

2. Describe the data distribution shown:

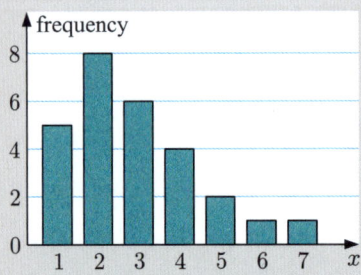

3. A class of 20 students was asked "How many bedrooms are there in your house?" The following data was collected:

   3 2 3 2 2 4 3 4 2 3 2 1 2 2 3 2 4 2 3 2

   a Is the data discrete or continuous?
   b Are there any outliers in the data?
   c Organise the data in a frequency table.
   d Construct a vertical line chart to display the data.

4. Consider the set of data: 17 14 9 12 23 14 12 18 9 15 6 14 21 13 10
   Find the:
   a mode
   b mean
   c median
   d range.

5. A class consists of 30 students. The mean height of students is 172 cm. The mean height of the males is 176 cm, and the mean height of the females is 166 cm. How many males are in the class?

6. The masses of eggs in a carton marked "50 g eggs" are recorded alongside.

   a Describe the distribution of the data.
   b What is the modal class? Explain what this means.
   c Copy and complete this table:

Mass (m g)	Frequency
$48 \leqslant m < 49$	
$49 \leqslant m < 50$	
$50 \leqslant m < 51$	
$51 \leqslant m < 52$	
$52 \leqslant m < 53$	

   Masses of eggs in a carton

   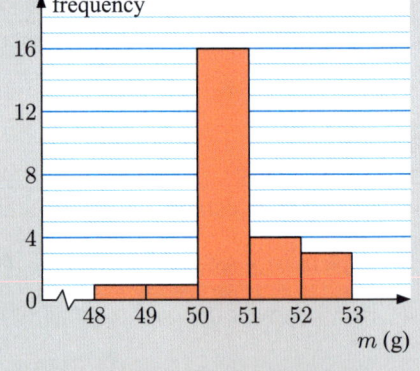

   d How many eggs were in the carton?
   e Estimate the mean mass of an egg in the carton.

7. Consider the ordered data set: 1, 2, 2, a, b, 4, 4, 6, 6, 8
   The mean and median of the data set are both equal to 4. Find a and b.

8 The Davis and Douglas families kept a record of the amount they spent at the supermarket each week for 10 weeks:

    Davis: £102.50   £115.95   £107.60   £122.15   £131.05
                  £111.15   £120.50   £127.55   £100.95   £113.40

    Douglas: £109.80   £86.75   £94.50   £129.75   £72.05
                      £133.05   £121.05   £97.60   £73.80   £105.35

  a Find the mean and range of each data set.
  b Which family generally spent more at the supermarket each week?
  c Which family had the greater variation in the amount they spent each week?

9 Consider the following data sets:

    Data set A: 5.3, 5.1, 6.5, 8.7, 7.6, 6.6, 10.0, 6.1, 6.1, 1.8, 7.6, 5.7, 3.8, 3.8, 9.5
    Data set B: 10.6, 10.4, 8.4, 10.3, 8.2, 11.4, 9.5, 9.5, 8.6, 9.6, 10.0, 10.9, 10.3, 10.4, 12.9
    Data set C: 4.5, 4.1, 4.7, 5.3, 5.2, 5.4, 5.1, 4.4, 5.1, 6.0, 5.1, 4.6, 4.6, 4.1, 5.3

  List the data sets in ascending order of:

  a median   **b** range.

Review set 30B

1 A shop manager wants to know what customers think of the services provided by the shop. The manager questions the first 10 customers who enter the shop on Monday morning.

In this situation, what is:

  a the population   **b** the sample?

2 For the data set alongside, find the:

                  13 16 15 17 14 13 13 15 16 14
  a mean   **b** median   **c** mode.   16 14 15 15 15 13 17 14 12 14

3 A sample of 15 measurements has a mean of 14.2, and a sample of 10 measurements has a mean of 12.6. Find the mean of the combined sample of 25 measurements.

4 The numbers of people at a judo class each week were:

    10 8 10 9 7 11 9 11 10
    10 9 8 9 9 11 8 10 9
    10 11 10 7 9 11 10 8

  a Organise the data in a frequency table.
  b Draw a vertical line chart to display the data.
  c Describe the distribution of the data.

5 As punishment for misbehaving, a class of students had to pick up litter during break. The table shows the number of litter pieces picked up by the students.

  a How many students were in the class?
  b For this data, find the:

    i mean   **ii** mode   **iii** median.

  c Draw a bar chart to display the data.
  d Describe the distribution of the data.

Pieces of litter	Frequency
6	1
7	3
8	2
9	6
10	7
11	8
12	5

6 The bar charts below show the number of stars reviewers gave for two mobile apps:

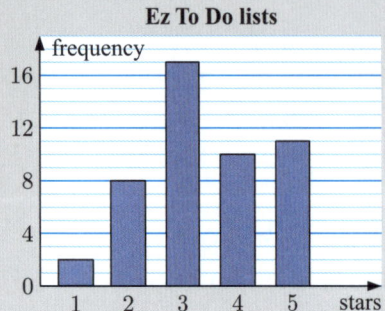
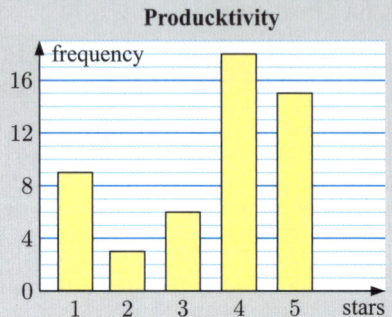

  a Construct a stars frequency table for each app.
  b Calculate the mean, median, and mode for each data set.
  c Which app generally received more stars per review?
  d Explain why we cannot compare the spread of the data sets using the range.

7 The table alongside shows how many people used the swimming pool at a gym each day over 40 days.
Estimate the mean number of people who used the swimming pool each day.

Number of people	Frequency
20 - 29	5
30 - 39	11
40 - 49	11
50 - 59	9
60 - 69	4
	40

8 The lengths of newborn babies at a hospital were recorded over a one month period. The results are shown in the frequency histogram.

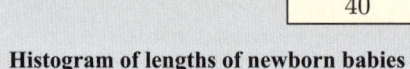

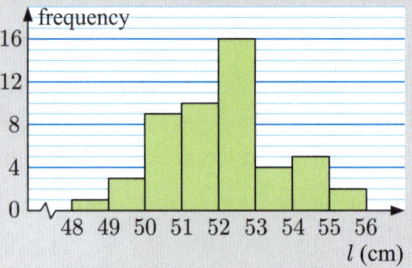

  a Describe the distribution of the data.
  b How many babies were 52 cm or more?
  c What percentage of babies had lengths in the interval $50 \text{ cm} \leqslant l < 53 \text{ cm}$?
  d In the following month, 67 babies were born at the same hospital.
Estimate the number of babies that measured less than 51 cm in length.

9 Nia and Rex asked the students in their respective classes to shoot a basketball 10 times each. The number of goals scored by each student in each class were recorded:

  Nia's class: 2 6 7 6 6 7 7 5 3 5 5 6 5
  3 4 8 1 6 3 3 6 8 7 2 4
  Rex's class: 0 4 2 1 3 2 1 2 3 1 2 1 1
  2 2 4 4 0 1 4 2 2 2

  a Draw vertical line charts for each class.
  b Describe the distribution of each data set.
  c Based on the vertical line charts, predict which class has:
    i the higher median   **ii** the lower range.
  d Calculate the median and range for each class to check your answers to **b**.

31

Bivariate statistics

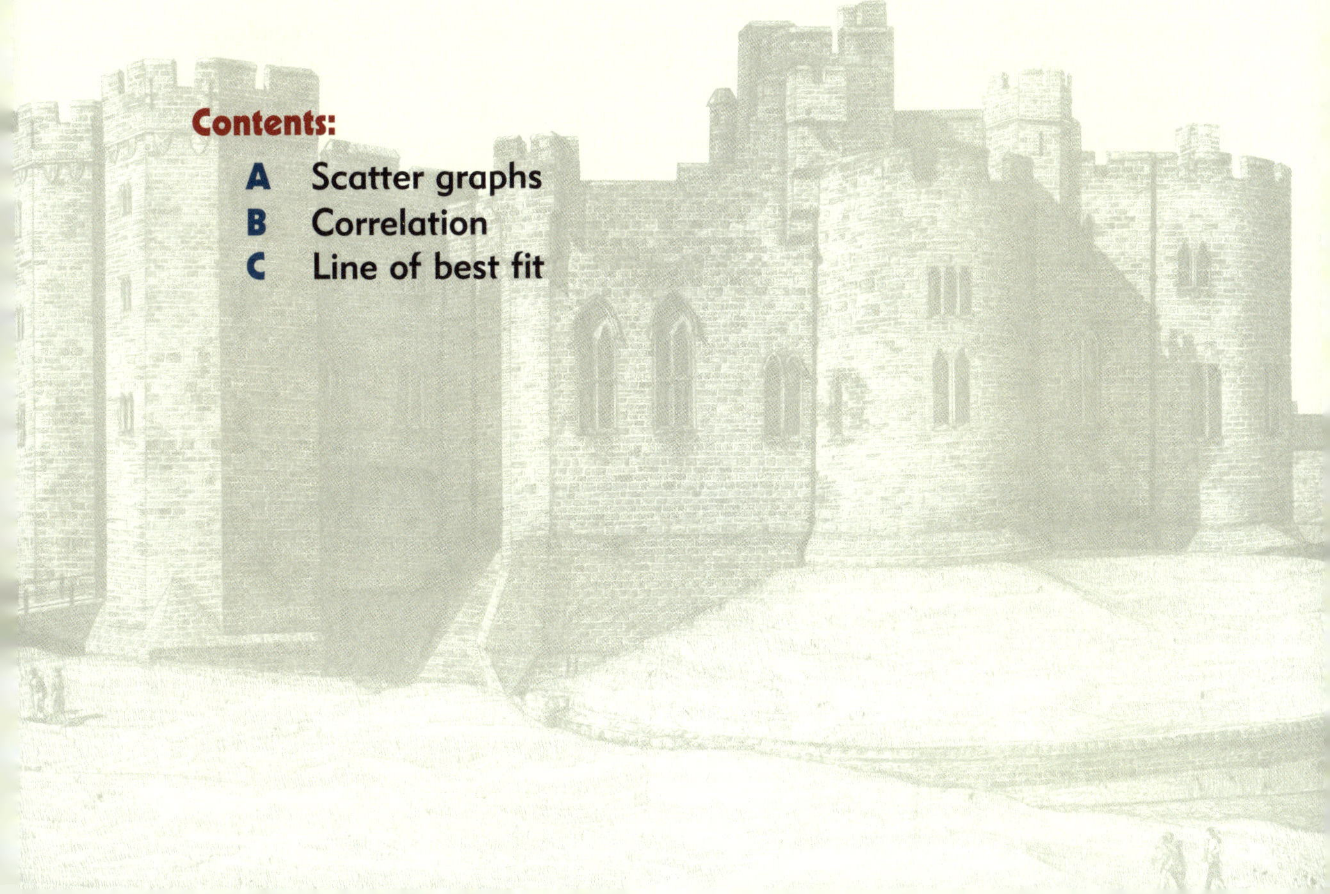

Contents:
- **A** Scatter graphs
- **B** Correlation
- **C** Line of best fit

Opening problem

The relationship between the *height* and *weight* of members of a football team is to be investigated. Data for each player is given below.

Player	1	2	3	4	5	6	7	8	9
Height (cm)	203	189	193	187	186	197	180	186	188
Weight (kg)	106	93	95	86	85	92	78	84	93

Player	10	11	12	13	14	15	16	17	18
Height (cm)	181	179	191	178	178	186	190	189	193
Weight (kg)	84	86	92	80	77	90	86	95	89

Things to think about:

a Which is the *dependent* variable?
b How could we display this data?
c Does an increase in the *height* of a player generally correspond to an increase or a decrease in their *weight*?
d How can we use this data to estimate the weight of a player who is 200 cm tall? How reliable will this estimate be?

We often want to know how two variables are **related**. We want to know whether an increase in one variable results in an increase or a decrease in the other. We call this **bivariate statistics** because we are dealing with *two* variables.

To analyse the relationship between two variables, we first need to decide which is the **dependent** variable and which is the **independent** variable. The value of the dependent variable *depends* on the value of the independent variable.

For example:

- the height of a girl *depends* on her age
- Jonathon's pay rate *depends* on the number of years he has worked for the company.

Having made this decision, we can then draw a **scatter graph** or **scatter diagram** to display the data. The independent variable is placed on the horizontal axis, and the dependent variable is placed on the vertical axis.

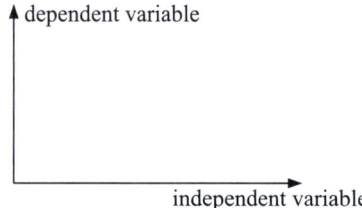

A SCATTER GRAPHS

In the **Opening Problem**, the *height* and *weight* of each football player was measured.

We suspect that the weight of a footballer *depends* on his height, so we place height on the horizontal axis and weight on the vertical axis.

The data from each individual footballer is then displayed as a point on the scatter graph.

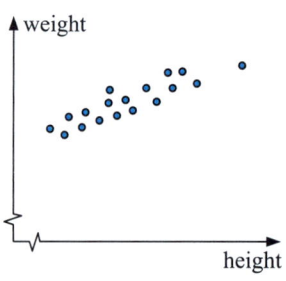

Example 1 ◀)) Self Tutor

This scatter graph shows the ages and work done by 5 employees.

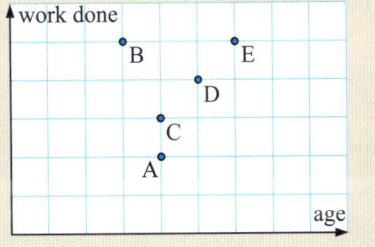

a Which person is the oldest?
b Who has done the most work?
c Who has done the least work?

*A scatter graph is sometimes called a **scatter diagram**.*

a E has the largest value on the *age* axis.
∴ E is the oldest.
b B and E have the highest values on the *work done* axis.
∴ B and E have done the most work.
c A has the lowest value on the *work done* axis.
∴ A has done the least work.

EXERCISE 31A

1 Liesl measured the height and weight of eight objects in her room. The results are shown in the scatter graph alongside.

a Which object is the lightest?
b Which object is the tallest?
c Which two objects are the same height?
d Which two objects have the same weight?

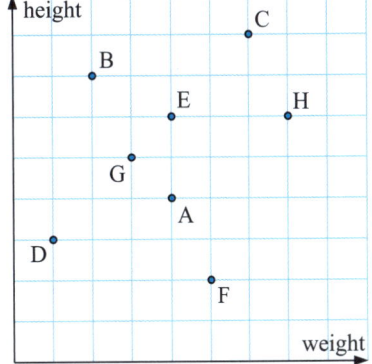

2 This scatter graph shows the productivity of five employees, and the number of hours of training they have received.

a Who has received the most hours of training?
b Who is the least productive?
c Are there employees with equal productivity?
d Do you think that the training given to the employees has been worthwhile? Explain your answer.

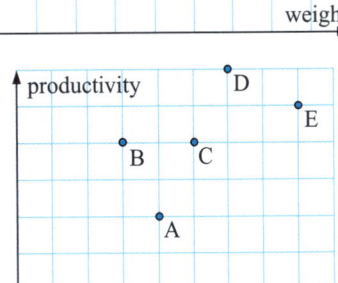

3 Peter's football team played seven games during an end-of-season carnival. The results of the games are displayed in the scatter diagram.

a In which game did the team score the most goals?
b In which game did the team concede the least goals?
c How many of the games ended in a draw?
d How many games did Peter's team win?

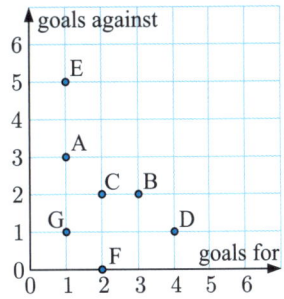

4 Six students were asked how far they lived from school, and how long it took them to travel to school. The results are shown in the table.

Student	P	Q	R	S	T	U
Distance to school (km)	8	5	2	12	6	4
Travel time (minutes)	10	5	20	15	15	7

    a Draw a scatter graph to display the data, with *distance* on the horizontal axis.

    b Which student lives furthest from the school?

    c Which student took the least time to travel to school?

    d One of the students walks to school. Which student do you think it is? Explain your answer.

> In this case time is the *dependent* variable because the *travel time* depends on the *distance to school*.

B   CORRELATION

Correlation is a measure of the strength of the relationship or association between two variables.

We can use a scatter graph to describe the correlation between two variables.

Step 1: Look at the scatter graph for any **pattern**.

For a generally *upward* shape, we say that the correlation is **positive**.

As the independent variable increases, the dependent variable generally increases.

For a generally *downward* shape, we say that the correlation is **negative**.

As the independent variable increases, the dependent variable generally decreases.

For *randomly scattered* points with no upward or downward trend, we say there is **no correlation**.

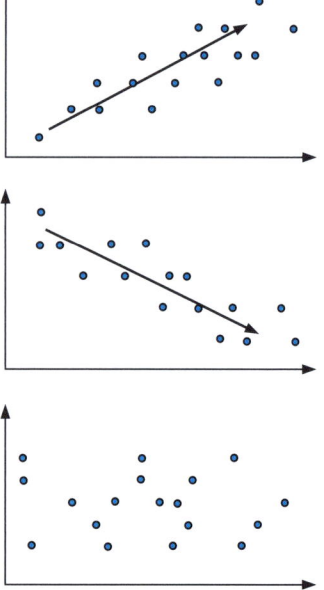

Step 2: Look at the pattern of points to see if the relationship is **linear**.

    The relationship is approximately linear.   The relationship is not linear.

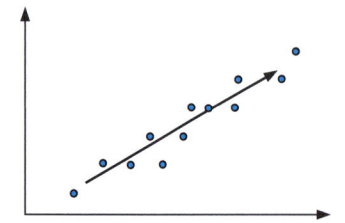 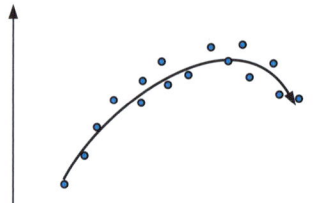

Step 3: Look at the spread of points to judge the **strength** of the correlation.

These scatter graphs show strength classifications for positive relationships:

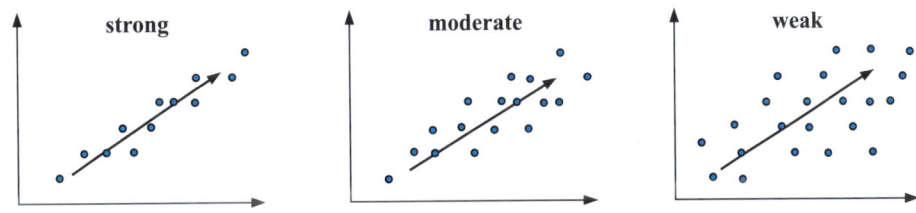

These scatter graphs show strength classifications for negative relationships:

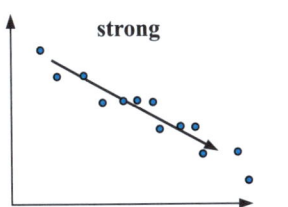
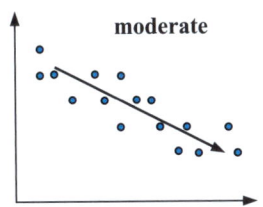
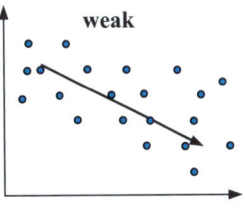

Step 4: Look for any **outliers**. These are isolated points which do not follow the trend formed by the main body of data.

If an outlier is the result of a recording or graphing error, it should be corrected or discarded. However, if the outlier is a genuine piece of data, it should be kept.

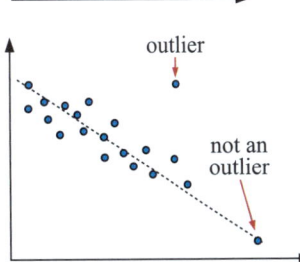

Example 2 ◀)) Self Tutor

Alexander researched the elevation above sea level and mean annual temperature of 12 cities around the world. The results are given in this table.

Elevation (m)	600	850	150	300	100	200	500	450	750	30	300	50
Mean annual temperature (°C)	15	10	16	15	25	20	21	19	9	27	22	28

a Draw a scatter graph of the data.

b Describe the relationship between *elevation* and *mean temperature*.

a
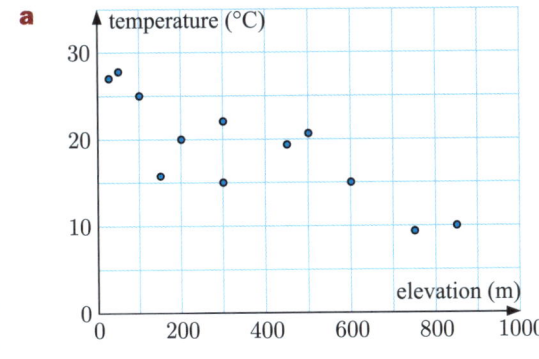

b There is a moderate negative linear correlation between *elevation* and *mean temperature*.

What factors other than elevation affect the mean annual temperature of a city?

EXERCISE 31B.1

1 For each scatter graph:
  i State whether there is positive, negative, or no association between the variables.
  ii Decide whether the relationship between the variables is linear.
  iii Describe the strength of the association (zero, weak, moderate, or strong).
  iv Determine whether there are any outliers.

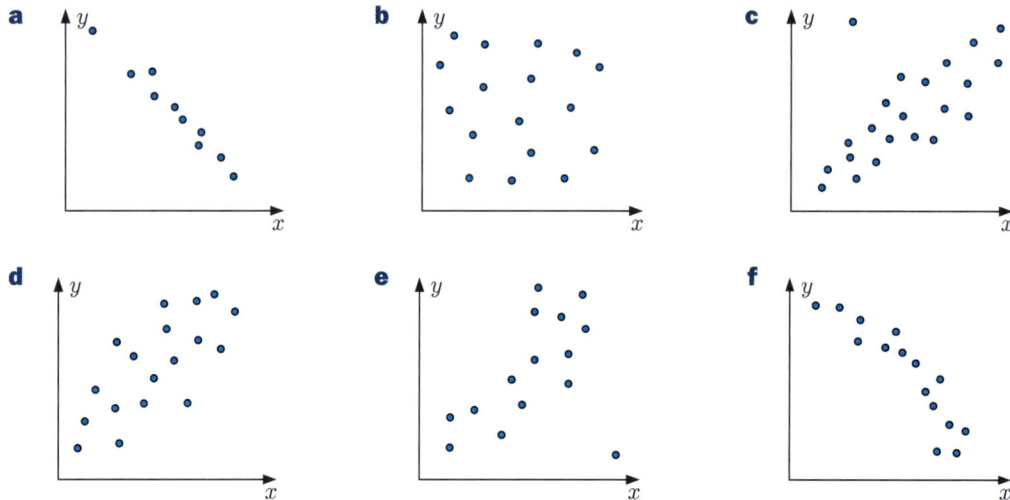

2 a There is a positive association between x and y. What happens to y as x increases?
  b There is a negative correlation between T and d. What happens to d as T increases?
  c Copy and complete:
    If there is no association between two variables then the points on the scatter graph are ....

3 A class of 15 students was asked how many text messages they had sent and received in the last week. The results are shown below:

Student	A	B	C	D	E	F	G	H	I	J	K	L	M	N	O
Messages sent	5	0	12	9	17	15	10	4	8	18	25	17	0	6	13
Messages received	8	0	15	7	19	11	8	7	12	15	21	16	4	6	16

  a Draw a scatter diagram to display the data.
  b Describe the relationship between *messages sent* and *messages received*.

4 a 10 students were asked for their exam marks in Physics and Mathematics. Their percentages are given in the table below.

Student	A	B	C	D	E	F	G	H	I	J
Physics	75	83	45	90	70	78	88	50	55	95
Mathematics	68	70	50	65	60	72	75	40	45	80

  i Draw a scatter graph of the data, with the Physics marks on the horizontal axis.
  ii Comment on the relationship between the Physics and Mathematics marks.

b The same students were asked for their Art exam results. Their percentages were:

Student	A	B	C	D	E	F	G	H	I	J
Art	75	70	80	85	82	70	60	75	78	65

Draw a scatter graph to see if there is any relationship between the Physics marks and the Art marks of each student.

5 The following table shows the sales of hot drinks in a popular café each month, along with the average daily temperature for the month.

Month	Jan	Feb	Mar	Apr	May	Jun	Jul	Aug	Sep	Oct	Nov	Dec
Temperature (°C)	5	12	13	18	20	21	22	17	14	10	7	6
Sales (£ × 1000)	22	25	20	15	16	12	12	8	10	15	16	18

a Draw a scatter diagram of the data, with the independent variable *temperature* along the horizontal axis.

b Comment on the relationship between the sales and the temperature.

CAUSALITY

Correlation between two variables does not necessarily mean that one variable *causes* the other to change.

For example:

- Households in a suburb were asked how many bedrooms their house has and how many cars they own. A strong positive correlation was found between the variables.

  This does *not* mean that having more bedrooms will allow you to buy more cars, nor does it mean that buying more cars will create more bedrooms in your house.

  Rather, the strong positive correlation occurs because both the *number of bedrooms* and the *number of cars* are closely related to a third variable, *income*. Households with higher incomes are more likely to have houses with more bedrooms, and to buy more cars.

- The heights of 20 trees in a forest and the marks attained by 20 students on a Chemistry test were found to have a strong positive correlation.

  Obviously, the marks a student obtained had no influence on the tree they were paired with. It was merely a coincidence that the students that achieved higher marks were paired with taller trees.

> If a change in one variable *causes* a change in the other variable then we say that a **causal relationship** exists between them.
>
> In cases where a causal relationship is not apparent, we cannot conclude that a causal relationship exists based on high correlation alone.

EXERCISE 31B.2

1 When the following pairs of variables were measured, a strong, positive correlation was observed. Discuss whether a causal relationship exists between the variables. If not, suggest a third variable to which they may both be related:

  a The amount of fertiliser applied to a pepper plant and the average length of its fruit.

  b The number of people visiting a park and the number of people visiting the beach.

  c The average distance travelled to work each day and the amount of money spent on petrol each week.

  d The number of car crashes and the number of break-ins in capital cities.

2 Tori and Stuart spent 8 weeks training for an athletics competition. Tori's event is the 200 m sprint and Stuart's event is the 110 m hurdles. They each recorded the best times they achieved during training at the end of each week:

Week number	1	2	3	4	5	6	7	8
Tori's time (s)	27.5	27.2	27.6	27.0	26.4	26.1	26.3	25.9
Stuart's time (s)	20.3	19.9	19.4	19.7	19.2	19.1	19.0	18.8

  a Draw a scatter graph of:

  i Tori's times against Stuart's times

  ii Tori's times against the week number

  iii Stuart's times against the week number.

  b Comment on the relationship between each pair of variables in **a**.

  c Is it reasonable to conclude that there is a causal relationship between Tori's times and Stuart's times? Explain your answer.

Activity   **Global warming**

Click on the icon to access this Activity about global warming.

GLOBAL WARMING

C LINE OF BEST FIT

If there is a strong linear correlation between two variables x and y, then it is reasonable to draw a **line of best fit** through the data.

The line of best fit can be used to estimate the value of y for any value of x.

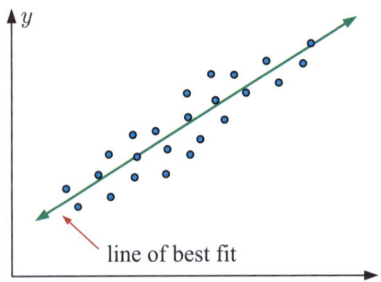

line of best fit

Bivariate statistics (Chapter 31) 623

For a bivariate data set involving x and y, we use these steps to find the **line of best fit by eye**:

Step 1: Find the means $\overline{x}$ and $\overline{y}$ of the x and y values respectively.

Step 2: Plot the **mean point** $(\overline{x}, \overline{y})$ on a scatter plot of the data.

Step 3: Draw a line through the mean point which fits the trend of the data and which has about as many points above the line as below it.

Example 3   ◆) Self Tutor

On a hot day, six cars were left in the sun in a car park. The length of time each car was left in the sun was recorded, as well as the temperature inside the car at the end of the period.

Car	A	B	C	D	E	F
Time (x minutes)	50	5	25	40	15	45
Temperature (y °C)	47	28	36	42	34	41

a Calculate $\overline{x}$ and $\overline{y}$.

b Draw a scatter graph of the data. Plot the mean point $(\overline{x}, \overline{y})$ on the scatter graph, and draw a line of best fit through this point.

c Predict the temperature of a car which has been left in the sun for:

  i 35 minutes   **ii** 75 minutes.

a $\overline{x} = \dfrac{50 + 5 + 25 + 40 + 15 + 45}{6} = 30, \quad \overline{y} = \dfrac{47 + 28 + 36 + 42 + 34 + 41}{6} = 38$

b
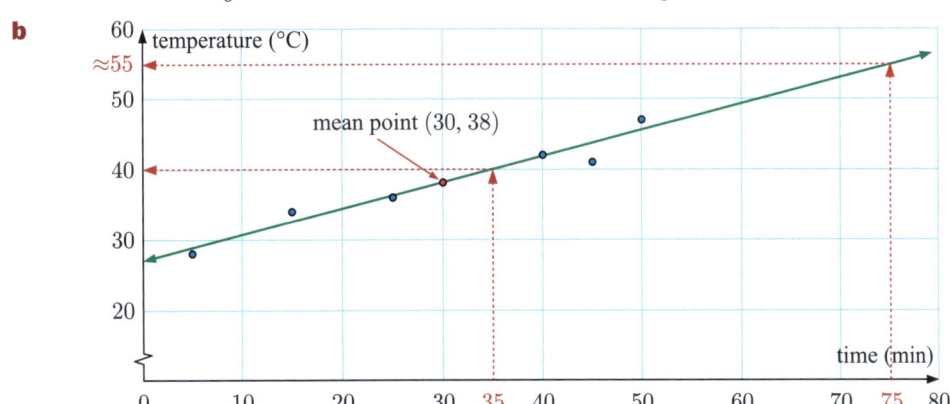

c   **i** When $x = 35$, $y \approx 40$.
The temperature of a car left in the sun for 35 minutes will be approximately 40°C.

  ii When $x = 75$, $y \approx 55$.
The temperature of a car left in the sun for 75 minutes will be approximately 55°C.

INTERPOLATION AND EXTRAPOLATION

Given a bivariate data set, the data values with the lowest and highest values of x are called the **poles**.

If we use values of x **in between** the poles to estimate y, we say we are **interpolating** between the poles.

If we use values of x **outside** the poles to estimate y, we say we are **extrapolating** outside the poles.

As a general rule, it is reasonable to interpolate between the poles, but unreliable to extrapolate outside them.

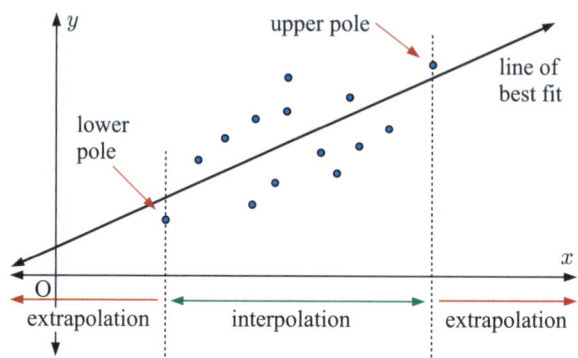

In **Example 3** on the previous page:

- The estimate in **c i** is an interpolation, so we would expect this estimate to be reliable.
- The estimate in **c ii** is an extrapolation, and therefore may not be reliable. We cannot assume that the linear trend observed in the data will continue up to a time of 75 minutes.

EXERCISE 31C

1 Consider the bivariate data alongside.

x	5	10	2	13	6
y	11	3	18	5	13

   a Find $\overline{x}$ and $\overline{y}$.

   b Draw a scatter graph of the data.

   c Does the data appear to be positively correlated or negatively correlated?

   d Plot the mean point $(\overline{x}, \overline{y})$ on the scatter graph, and draw a line of best fit through this point.

   e Estimate the value of y when $x = 8$.

2 The table alongside shows the *percentage of unemployed adults* and the *number of major thefts per day* in eight large cities.

City	Unemployment (%)	Thefts
A	7	113
B	6	67
C	10	117
D	8	88
E	9	120
F	6	38
G	3	61
H	7	76

   a Find the mean unemployment percentage $\overline{x}$ and the mean number of thefts $\overline{y}$.

   b Draw a scatter graph of the data.

   c Plot $(\overline{x}, \overline{y})$ on the scatter graph.

   d Draw the line of best fit on the scatter graph.

   e Another city has 15% unemployment.

      i Estimate the number of major thefts per day for that city.

      ii Comment on the reliability of your estimate.

3 Each month, an opinion poll shows the approval rating of the Prime Minister and the Opposition leader. The approval ratings for the last 10 polls are shown below:

Prime Minister ($x\%$)	55	59	68	61	46	42	38	45	42	44
Opposition ($y\%$)	37	35	31	35	43	40	42	37	41	39

   a Calculate $\overline{x}$ and $\overline{y}$.

b Draw a scatter diagram of the data. Plot the mean point $(\bar{x}, \bar{y})$ on the scatter diagram, and draw a line of best fit through this point.

c In a new opinion poll, the Prime Minister's approval rating is 47%. Estimate the approval rating of the Opposition leader.

4 A café manager believes that during April the *number of people wanting dinner* is related to the *temperature at noon*. Over a 13 day period, the number of diners and the noon temperature were recorded.

Temperature (x °C)	8	10	13	15	15	12	10	13	17	16	18	14	12
Number of diners (y)	63	70	74	81	77	65	75	87	91	75	96	82	88

a Find the mean point $(\bar{x}, \bar{y})$.

b Draw a scatter graph of the data. Plot $(\bar{x}, \bar{y})$ on the scatter graph, and draw a line of best fit through this point.

c Estimate the number of diners at the café when the temperature is:
  i 11°C
  ii 4°C.

d Comment on the reliability of your estimates in **c**.

5 Consider the data in the table below:

x	2	8	4	3	9	6	1	5	10	7
y	21.0	9.1	18.1	50.0	8.3	12.3	21.9	14.3	5.3	11.2

a Draw a scatter graph for the data and circle the outlier.

b Find the mean point $(\bar{x}, \bar{y})$ and plot it on your scatter graph. Draw a line of best fit through the mean point.

c Recalculate $\bar{x}$ and $\bar{y}$ without the outlier and repeat **b**.

d Compare your two lines of best fit. Which line of best fit better describes the trend seen in the *main body* of data?

Review set 31A

1 The scatter graph shows the number of defective items made by each employee of a factory, plotted against the employee's number of weeks of experience.
Describe the correlation between the variables.

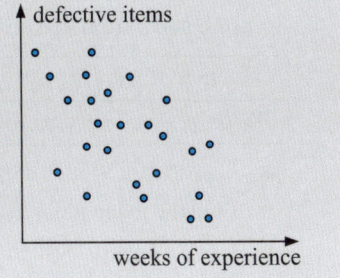

2 Five hockey clubs were surveyed on their players' average hours of general fitness training, and the number of injuries to players during matches.

    a Which club had the fewest injuries?

    b Which club's players had the fewest hours of fitness training?

    c Write a sentence describing the general trend of the graph.

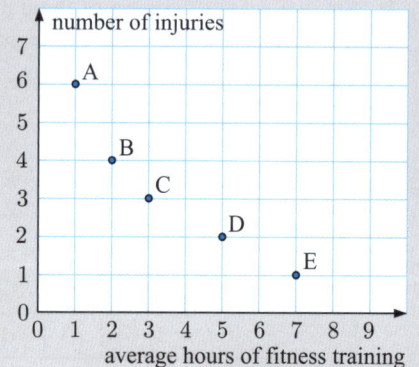

3 Sam and Richard competed in a series of 8 chess games. The table below shows how many pieces remained for each player at the end of each game.

Sam	4	2	6	2	4	1	5	7
Richard	2	1	3	7	4	1	3	5

    a Construct a scatter graph of the data.

    b Describe the correlation between the variables. Are there any outliers?

    c Is it reasonable to conclude that an increase in the number of pieces Sam has left *causes* an increase in the number of pieces Richard has left? Explain your answer.

4 Following an outbreak of a deadly virus, medical authorities begin taking records of the number of cases. Their records are shown below.

Days after outbreak (n)	2	3	4	5	6	7	8	9	10	11
Diagnosed cases (d)	8	14	33	47	80	97	118	123	139	153

    a Produce a scatter graph of d against n.

    b Plot the point $(\overline{n}, \overline{d})$ on the scatter graph, and draw the line of best fit by eye.

    c Estimate the number of diagnosed cases on day 14. Comment on the reliability of your estimate.

5 The whorls on a cone shell get wider as you go from the top of the shell towards the bottom. Measurements from a shell are summarised in the following table:

Position of whorl (p)	1	2	3	4	5	6	7	8
Width of whorl (w cm)	0.7	1.2	1.4	2.0	2.0	2.7	2.9	3.5

    a Construct a scatter diagram for the data.

    b Find the mean point $(\overline{p}, \overline{w})$ and plot it on your scatter diagram.

    c Draw a line of best fit by eye.

    d   **i** If a cone shell has 14 whorls, what width do you expect the 14th whorl to have?

         ii How reliable is this prediction?

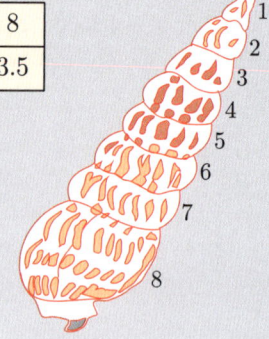

6 The table below shows the number of games won, and the final position on the Premier League ladder, for the Liverpool Football Club from 2005-06 to 2016-17:

Games won (x)	25	20	21	25	18	17	14	16	26	18	16	22
Position (y)	3	3	4	2	7	6	8	7	2	6	8	4

a Would you expect x and y to be positively or negatively correlated? Explain your answer.

b Draw a scatter graph of the data.

c Find the mean point $(\overline{x}, \overline{y})$ and plot it on your scatter graph. Draw a line of best fit on your scatter graph.

d Suppose Liverpool wins 22 games next season. Predict their position on the ladder.

Review set 31B

1 This scatter graph displays the land area and population of 10 countries.

a Which country has the smallest population?

b Which country has the largest area?

c Which two countries have the same population?

d Which country is the most densely populated?

2 The scatter graph alongside shows the number of pages and chapters in the novels on Rashida's bookshelf.

a Which book has the:

    i most pages   **ii** least chapters?

b Copy and complete: As the number of pages increases, the number of chapters generally ......

3 Consider the bivariate data alongside.

x	11	7	13	3	12	8
y	17	14	20	5	26	8

a Find $\overline{x}$ and $\overline{y}$.

b Draw a scatter diagram of the data.

c Does the data appear to be positively correlated or negatively correlated?

d Plot the mean point $(\overline{x}, \overline{y})$ on the scatter diagram, and draw a line of best fit through this point.

e Estimate the value of y when $x = 5$. How reliable is this estimate?

4 The table below shows the average daily maximum temperatures in Bath for each month. $t = 1$ corresponds to January, $t = 2$ corresponds to February, and so on.

Month (t)	1	2	3	4	5	6	7	8	9	10	11	12
Temperature (T °C)	7.6	7.9	10.5	13.3	16.7	19.7	21.7	21.5	18.8	14.6	10.7	8.0

  a Would you expect the relationship between T and t to be linear? Explain your answer.
  b Draw a scatter graph of T against t to confirm your answer to **a**.

5 In a Cardiff shopping mall, David asked 10 people how many coins they had in their wallet or purse, and the total value of those coins.

Number of coins	5	8	11	7	5	10	2	10	0	12
Value of coins	£1.10	£3.82	£4.56	£2.90	£3.51	£4.54	£2.50	£1.02	£0	£6.23

Let n be the number of coins, and v be the value of the coins.
  a Draw a scatter graph of the data.
  b Describe the correlation between the variables. Do you think a causal relationship exists between them?
  c Find the mean point and plot it on your scatter graph. Draw a line of best fit by eye through the mean point.
  d Terese has 20 coins in her purse.
    i Estimate the total value of these coins.
    ii How reliable is your prediction?

6 Consider the relationship between a *number* and the *number of factors* it has.
  a Would you expect the correlation between these variables to be:
    i positive or negative
    ii strong, moderate, or weak?
  Explain your answers.
  b Copy and complete this table:

Number (x)	1	2	3	4	5	....	16	17	18	19	20
Number of factors (y)	1	2	2	3		....				2	6

  c Draw a scatter graph of the data.
  d Describe the correlation between the variables.

7 The following table gives peptic ulcer rates per 1000 people for differing family incomes.

Income (I £1000s)	20	25	30	35	40	50	60	80	100
Peptic ulcer rate (R)	8.3	7.7	6.9	9.1	5.9	4.7	3.6	2.6	1.2

  a Draw a scatter graph of the data.
  b Identify the outlier.
  c Draw a line of best fit by eye on your scatter graph.
  d Estimate the peptic ulcer rate in families with an income of £55 000.
  e Explain why the model is inadequate for families with an income in excess of £120 000.

32

Vectors

Contents:

- **A** Directed line segment representation
- **B** Vector equality
- **C** Vector addition
- **D** Vector subtraction
- **E** Vectors in component form
- **F** The vector between two points
- **G** Scalar multiplication

630 Vectors (Chapter 32)

Opening problem

Flight 172 is travelling at 500 miles per hour in a northerly direction.

Flight 347 is also travelling north, but at half the speed of flight 172.

Flight 066 is travelling south at the same speed as flight 172.

Things to think about:

a How can we illustrate the motion of flight 172 on a scale diagram?

b How can we illustrate the motions of the other two flights so we can clearly distinguish between them?

VECTORS AND SCALARS

To fully describe the motion of the planes in the **Opening Problem**, we need to consider not only their *speeds*, but also their *directions*.

> Quantities which have only magnitude are called **scalars**.
>
> Quantities which have both magnitude and direction are called **vectors**.

Velocity is a vector quantity which includes both **speed** *and* **direction**.

Other examples of vector quantities include acceleration, force, displacement, and momentum.

For example, when using force we consider how hard we push an object, and what direction we push it in.

A DIRECTED LINE SEGMENT REPRESENTATION

Consider a bus which is travelling at 100 km h^{-1} to the south-east. We can represent the motion of the bus using an arrow on a scale diagram.

The **length of the arrow** represents the size or magnitude of the velocity. The **arrowhead** shows the direction of travel.

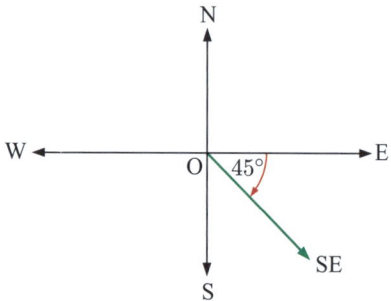

Scale: 1 cm represents 50 km h^{-1}

NOTATION

- The vector from O to A can be written as
$\overrightarrow{OA}$ or   **a**   or $\vec{a}$.

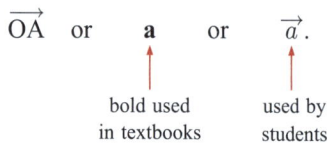

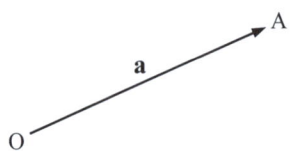

- The **magnitude** or **length** of the vector $\overrightarrow{OA}$ can be written as $|\overrightarrow{OA}|$ or OA or $|\mathbf{a}|$ or $|\vec{a}|$.
- $\overrightarrow{AB}$ is the vector which **emanates** from A and **terminates** at B. $\overrightarrow{AB}$ is the **position vector** of B relative to A.

Example 1 ◁)) Self Tutor

On a scale diagram, draw a vector representing a velocity of 40 m s^{-1} on the bearing $075°$.

Scale: 1 cm $\equiv 10$ m s^{-1}

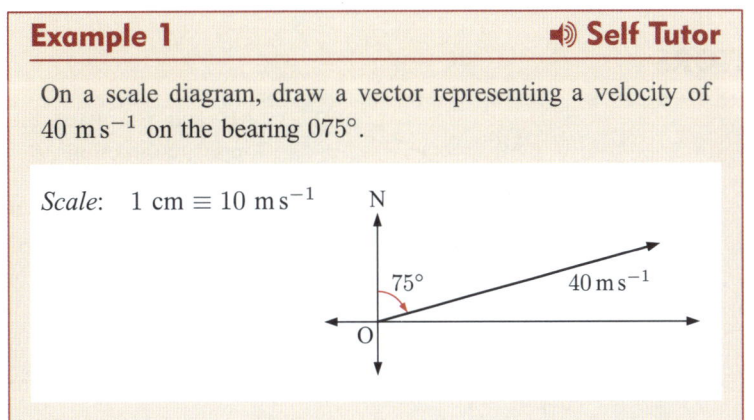

EXERCISE 32A

1 Using a scale of 1 cm represents 10 units, draw a vector which represents:
  a a velocity of 40 km h^{-1} to the south-west
  b a velocity of 35 m s^{-1} to the north
  c a displacement of 25 m in the direction $120°$
  d an aeroplane taking off at an angle of $12°$ to the runway, with speed 60 m s^{-1}.

2 If ⟶ represents a force of 45 Newtons due east, draw a directed line segment to represent a force of:
  a 75 N due west
  b 60 N south-west.

3 On a scale diagram, draw a vector representing:
  a a velocity of 60 km h^{-1} in a north-easterly direction
  b a displacement of 25 km on the bearing $055°$
  c an aeroplane taking off at an angle of $10°$ to the runway, with speed 90 km h^{-1}.

B VECTOR EQUALITY

EQUAL VECTORS

Two vectors are **equal** if they have the same magnitude *and* direction.

Equal vectors are **parallel** and in the same direction, and are **equal in length**.

This means that arrows representing equal vectors are translations of one another.

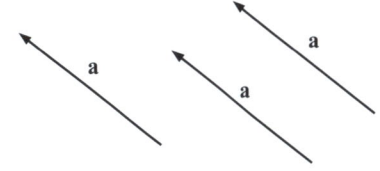

THE ZERO VECTOR

> The **zero vector**, **0**, is a vector of length 0.
> It is the only vector with no direction.

When we write the zero vector by hand, we usually write $\vec{0}$.

NEGATIVE VECTORS

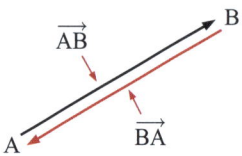

$\vec{AB}$ and $\vec{BA}$ have the same length but opposite directions.

We say that $\vec{BA}$ is the **negative** of $\vec{AB}$ and write $\vec{BA} = -\vec{AB}$.

Given the vector **a** shown, we can draw the vector $-\mathbf{a}$.

a and $-\mathbf{a}$ are parallel and equal in length, but are opposite in direction.

Example 2 ◀)) Self Tutor

ABCD is a parallelogram in which $\vec{AB} = \mathbf{a}$ and $\vec{BC} = \mathbf{b}$.

Find vector expressions for:

a $\vec{BA}$ **b** $\vec{CB}$ **c** $\vec{AD}$ **d** $\vec{CD}$

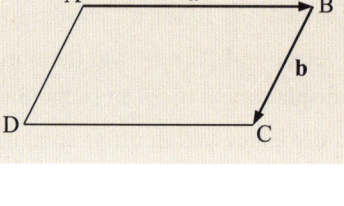

a $\vec{BA} = -\mathbf{a}$ {the negative vector of $\vec{AB}$}
b $\vec{CB} = -\mathbf{b}$ {the negative vector of $\vec{BC}$}
c $\vec{AD} = \mathbf{b}$ {parallel to and the same length as $\vec{BC}$}
d $\vec{CD} = -\mathbf{a}$ {parallel to and the same length as $\vec{BA}$}

EXERCISE 32B

1 State the vectors which are:

a equal in magnitude
b parallel
c in the same direction
d equal
e negatives of one another.

2 PQRS is a parallelogram in which $\vec{PQ} = \mathbf{a}$ and $\vec{PS} = \mathbf{b}$.
Find vector expressions for:

a $\vec{QR}$ **b** $\vec{SR}$ **c** $\vec{SP}$ **d** $\vec{RS}$

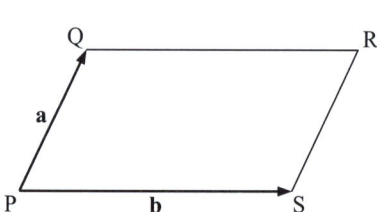

3 ABCD is a rhombus. Let $\overrightarrow{AB} = \mathbf{a}$ and $\overrightarrow{BC} = \mathbf{b}$.
Which of the following statements are true?

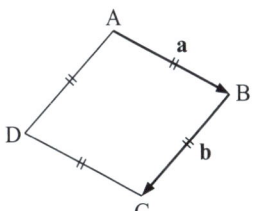

- **a** $\overrightarrow{AD} = \mathbf{b}$
- **b** $\overrightarrow{CD} = \mathbf{a}$
- **c** $\overrightarrow{DC} = \mathbf{b}$
- **d** $|\mathbf{a}| = |\mathbf{b}|$
- **e** $|\overrightarrow{DA}| = -|\mathbf{b}|$
- **f** $|-\mathbf{a}| = |-\mathbf{b}|$

C VECTOR ADDITION

Suppose we have three towns A, B, and C.

A trip from A to B, followed by a trip from B to C, is equivalent to a trip from A to C.

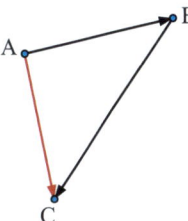

This can be expressed in vector form as the sum $\overrightarrow{AB} + \overrightarrow{BC} = \overrightarrow{AC}$, where the $+$ sign could mean "*followed by*".

After considering diagrams like the one above, we can define vector addition geometrically:

> To add **a** and **b** : *Step 1:* Draw **a**.
> *Step 2:* At the arrowhead end of **a**, draw **b**.
> *Step 3:* Join the beginning of **a** to the arrowhead end of **b**.
> This is vector **a** + **b**.

So, given  we have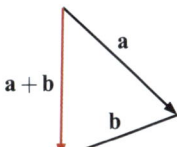

 **DEMO**

Example 3 ◀) Self Tutor

Find a single vector which is equal to:

- **a** $\overrightarrow{AB} + \overrightarrow{BE}$
- **b** $\overrightarrow{DC} + \overrightarrow{CA} + \overrightarrow{AE}$
- **c** $\overrightarrow{CB} + \overrightarrow{BD} + \overrightarrow{DC}$

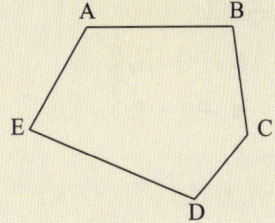

a $\overrightarrow{AB} + \overrightarrow{BE} = \overrightarrow{AE}$ **b** $\overrightarrow{DC} + \overrightarrow{CA} + \overrightarrow{AE} = \overrightarrow{DE}$ **c** $\overrightarrow{CB} + \overrightarrow{BD} + \overrightarrow{DC} = \overrightarrow{CC} = \mathbf{0}$
{zero vector}

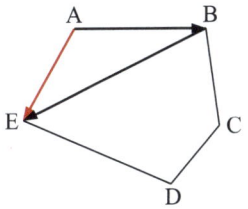 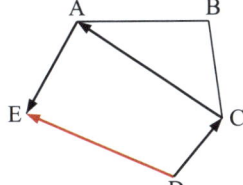 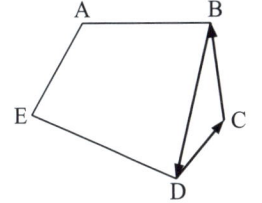

THE ZERO VECTOR

Having defined vector addition, we are now able to state that:

> The **zero vector**, **0**, is a vector of length 0.
>
> For any vector **a**: $a + 0 = 0 + a = a$
> $a + (-a) = (-a) + a = 0$

EXERCISE 32C

1 Copy the given vectors **p** and **q** and show how to construct $p + q$:

a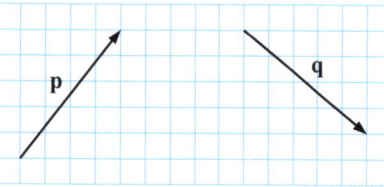

b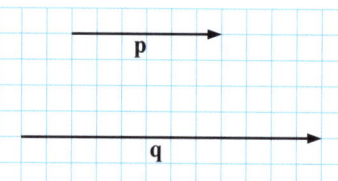

c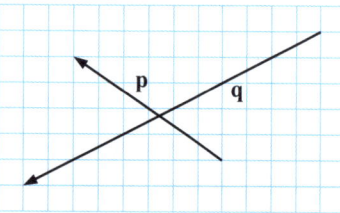

d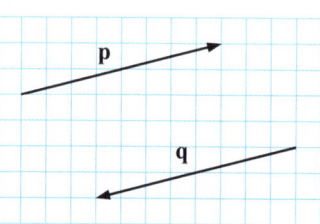

e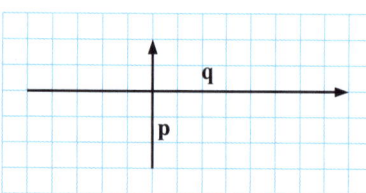

f

2 Find a single vector which is equal to:

a $\vec{QR} + \vec{RS}$ **b** $\vec{PQ} + \vec{QR}$

c $\vec{PS} + \vec{SR} + \vec{RQ}$ **d** $\vec{PR} + \vec{RQ} + \vec{QS}$

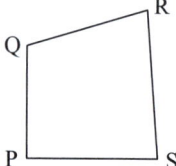

3 a Consider the vectors:

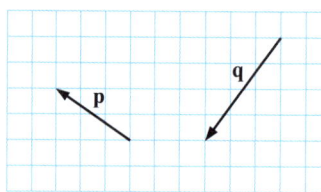

Use vector diagrams to find: **i** $p + q$ **ii** $q + p$

b For any two vectors **p** and **q**, is $p + q = q + p$? Explain your answer.

4 Write an expression in terms of **a**, **b**, **c**, and **d**, for:

a $\overrightarrow{PS}$
b $\overrightarrow{PR}$
c $\overrightarrow{QT}$
d $\overrightarrow{PT}$

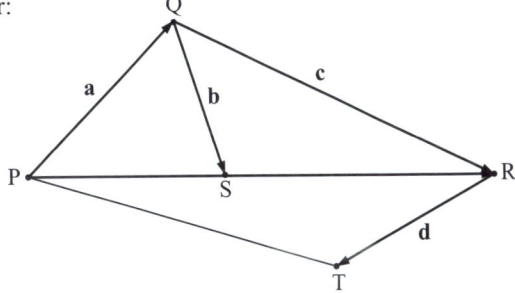

5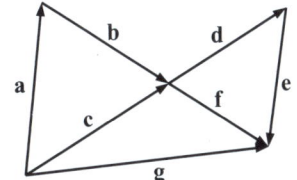

Write as a single vector:

a $\mathbf{a} + \mathbf{b}$
b $\mathbf{d} + \mathbf{e}$
c $\mathbf{c} + \mathbf{f}$
d $\mathbf{a} + \mathbf{b} + \mathbf{f}$
e $\mathbf{c} + \mathbf{d} + \mathbf{e}$

D VECTOR SUBTRACTION

To subtract one vector from another, we simply **add its negative**.

$$\mathbf{a} - \mathbf{b} = \mathbf{a} + (-\mathbf{b})$$

For example, given 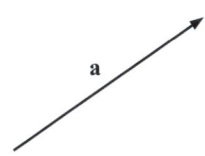 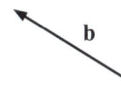 we have 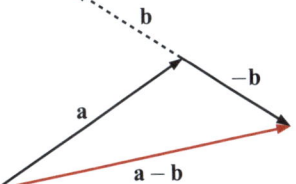  DEMO

Example 4   ◀)) Self Tutor

For the given vectors **s** and **t**, find $\mathbf{s} - \mathbf{t}$.

EXERCISE 32D

1 For the following vectors **p** and **q**, show how to construct $\mathbf{p} - \mathbf{q}$:

a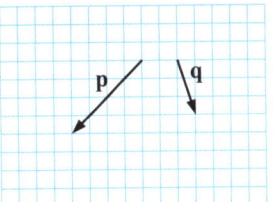
b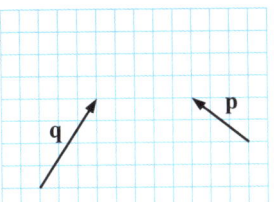
c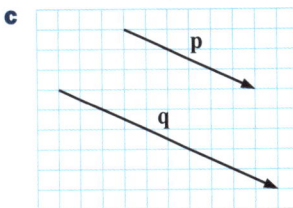

d 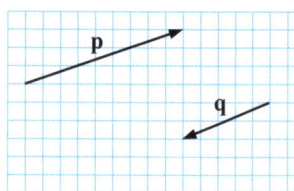 **e** 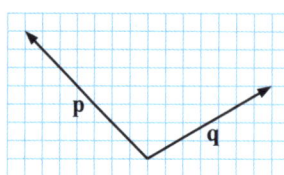 **f**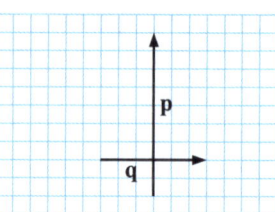

Example 5 ◄)) Self Tutor

For points P, Q, R, and S, simplify the following vector expressions:

a $\vec{QR} - \vec{SR}$ **b** $\vec{QR} - \vec{SR} - \vec{PS}$

a $\vec{QR} - \vec{SR}$
$= \vec{QR} + \vec{RS}$ {as $\vec{RS} = -\vec{SR}$}
$= \vec{QS}$

b $\vec{QR} - \vec{SR} - \vec{PS}$
$= \vec{QR} + \vec{RS} + \vec{SP}$
$= \vec{QP}$

2 For points P, Q, R, and S, simplify the following vector expressions:

a $\vec{QR} + \vec{RS}$ **b** $\vec{PS} - \vec{RS}$ **c** $\vec{RS} + \vec{SR}$

d $\vec{RS} + \vec{SP} + \vec{PQ}$ **e** $\vec{QP} - \vec{RP} + \vec{RS}$ **f** $\vec{RS} - \vec{PS} - \vec{QP}$

3 a Explain why $\vec{AB} = \mathbf{b} - \mathbf{a}$.
  b Write vector expressions for:
  i $\vec{BC}$ **ii** $\vec{CA}$

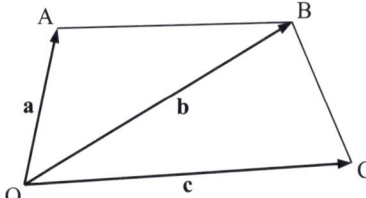

4 Use the diagram to simplify:

a a + c **b** h + f **c** j − i
d d − c **e** e − b **f** −f − h

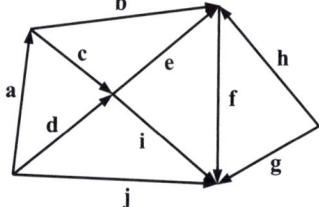

E VECTORS IN COMPONENT FORM

So far we have examined vectors in their geometric representation.

We have used arrows where:
- the **length** of the arrow represents size or magnitude
- the **arrowhead** indicates direction.

However, we can also represent vectors by describing the horizontal and vertical steps required to go from the starting point to the finishing point.

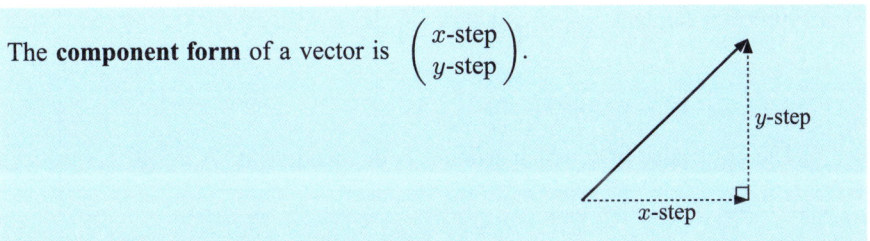

The **component form** of a vector is $\begin{pmatrix} x\text{-step} \\ y\text{-step} \end{pmatrix}$.

The x-step is positive if we move to the right, and negative if we move to the left.

The y-step is positive if we move upwards, and negative if we move downwards.

For example:

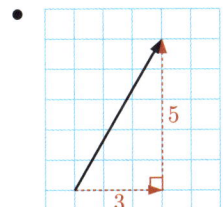

We move 3 units to the right and 5 units upwards,

so the vector is $\begin{pmatrix} 3 \\ 5 \end{pmatrix}$.

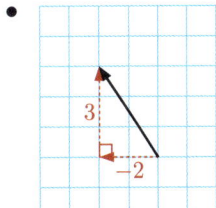

We move 2 units to the left and 3 units upwards,

so the vector is $\begin{pmatrix} -2 \\ 3 \end{pmatrix}$.

The **position vector of point** $A(a_1, a_2)$ relative to the origin $O(0, 0)$ is $\overrightarrow{OA} = \begin{pmatrix} x\text{-step} \\ y\text{-step} \end{pmatrix} = \begin{pmatrix} a_1 \\ a_2 \end{pmatrix}$.

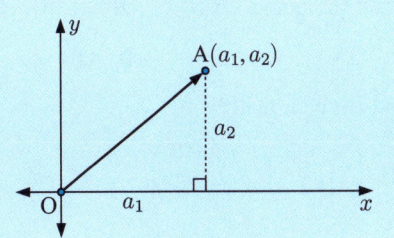

VECTOR EQUALITY

Two vectors are equal if and only if their x-components are equal *and* their y-components are equal.

$\begin{pmatrix} p \\ q \end{pmatrix} = \begin{pmatrix} r \\ s \end{pmatrix}$ if and only if $p = r$ and $q = s$.

Example 6

Find k given that $\begin{pmatrix} k^2 \\ k-1 \end{pmatrix} = \begin{pmatrix} 9 \\ -4 \end{pmatrix}$.

We have $k^2 = 9$ and $k - 1 = -4$.

The only value of k which satisfies both equations is $k = -3$.

EXERCISE 32E.1

1 Draw arrow diagrams to represent the vectors:

    a $\begin{pmatrix} 4 \\ 2 \end{pmatrix}$ **b** $\begin{pmatrix} 0 \\ 3 \end{pmatrix}$ **c** $\begin{pmatrix} -2 \\ 5 \end{pmatrix}$ **d** $\begin{pmatrix} 3 \\ 4 \end{pmatrix}$

2 Write the illustrated vectors in component form:

a 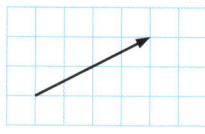   **b** 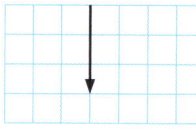   **c**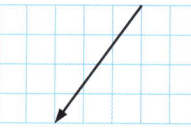

d 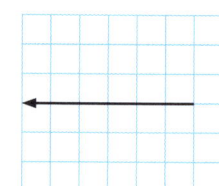   **e** 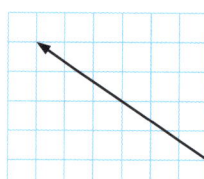   **f**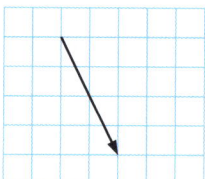

3 Given the points A(3, 4), B(-1, 2), C(2, -1), and D(0, -7), find:

    a $\overrightarrow{OA}$ **b** $\overrightarrow{OB}$ **c** $\overrightarrow{OC}$ **d** $\overrightarrow{OD}$

4 Find k given that:

    a $\begin{pmatrix} k+2 \\ k^2 \end{pmatrix} = \begin{pmatrix} -2 \\ 16 \end{pmatrix}$ **b** $\begin{pmatrix} 3 \\ k+4 \end{pmatrix} = \begin{pmatrix} k^2+2 \\ 3 \end{pmatrix}$ **c** $\begin{pmatrix} 4 \\ k^2+8 \end{pmatrix} = \begin{pmatrix} k^2-3k \\ 6k \end{pmatrix}$

VECTOR ADDITION

Consider the addition of vectors $\mathbf{a} = \begin{pmatrix} a_1 \\ a_2 \end{pmatrix}$ and $\mathbf{b} = \begin{pmatrix} b_1 \\ b_2 \end{pmatrix}$.

The horizontal step for $\mathbf{a} + \mathbf{b}$ is $a_1 + b_1$,

and the vertical step for $\mathbf{a} + \mathbf{b}$ is $a_2 + b_2$.

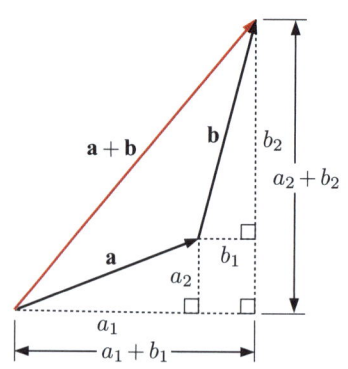

If $\mathbf{a} = \begin{pmatrix} a_1 \\ a_2 \end{pmatrix}$ and $\mathbf{b} = \begin{pmatrix} b_1 \\ b_2 \end{pmatrix}$ then $\mathbf{a} + \mathbf{b} = \begin{pmatrix} a_1 + b_1 \\ a_2 + b_2 \end{pmatrix}$.

Example 7

Given $\mathbf{a} = \begin{pmatrix} 2 \\ 5 \end{pmatrix}$ and $\mathbf{b} = \begin{pmatrix} 1 \\ -3 \end{pmatrix}$, find $\mathbf{a} + \mathbf{b}$.

Check your answer graphically.

$$\mathbf{a} + \mathbf{b} = \begin{pmatrix} 2 \\ 5 \end{pmatrix} + \begin{pmatrix} 1 \\ -3 \end{pmatrix}$$
$$= \begin{pmatrix} 2+1 \\ 5+-3 \end{pmatrix}$$
$$= \begin{pmatrix} 3 \\ 2 \end{pmatrix}$$

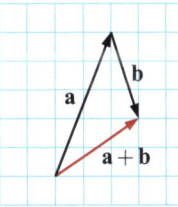

NEGATIVE VECTORS

For the vector $\mathbf{a} = \begin{pmatrix} 5 \\ 2 \end{pmatrix}$, we notice that $-\mathbf{a} = \begin{pmatrix} -5 \\ -2 \end{pmatrix}$.

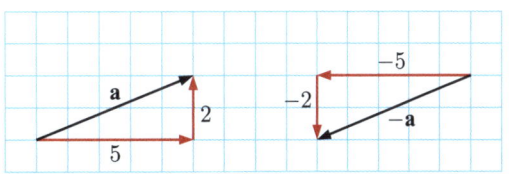

If $\mathbf{a} = \begin{pmatrix} a_1 \\ a_2 \end{pmatrix}$ then $-\mathbf{a} = \begin{pmatrix} -a_1 \\ -a_2 \end{pmatrix}$.

ZERO VECTOR

The zero vector is $\mathbf{0} = \begin{pmatrix} 0 \\ 0 \end{pmatrix}$.

For any vector $\mathbf{a}$:

$\mathbf{a} + \mathbf{0} = \mathbf{0} + \mathbf{a} = \mathbf{a}$

$\mathbf{a} + (-\mathbf{a}) = (-\mathbf{a}) + \mathbf{a} = \mathbf{0}$

The zero vector is the only vector which has no direction.

VECTOR SUBTRACTION

To subtract one vector from another, we **add its negative**.

If $\mathbf{a} = \begin{pmatrix} a_1 \\ a_2 \end{pmatrix}$ and $\mathbf{b} = \begin{pmatrix} b_1 \\ b_2 \end{pmatrix}$ then $\mathbf{a} - \mathbf{b} = \mathbf{a} + (-\mathbf{b})$
$$= \begin{pmatrix} a_1 \\ a_2 \end{pmatrix} + \begin{pmatrix} -b_1 \\ -b_2 \end{pmatrix}$$
$$= \begin{pmatrix} a_1 - b_1 \\ a_2 - b_2 \end{pmatrix}$$

If $\mathbf{a} = \begin{pmatrix} a_1 \\ a_2 \end{pmatrix}$ and $\mathbf{b} = \begin{pmatrix} b_1 \\ b_2 \end{pmatrix}$ then $\mathbf{a} - \mathbf{b} = \begin{pmatrix} a_1 - b_1 \\ a_2 - b_2 \end{pmatrix}$.

Example 8

Given $\mathbf{p} = \begin{pmatrix} 3 \\ -2 \end{pmatrix}$ and $\mathbf{q} = \begin{pmatrix} 1 \\ 4 \end{pmatrix}$, find:

a $\mathbf{p} - \mathbf{q}$

b $\mathbf{q} - \mathbf{p}$

a $\mathbf{p} - \mathbf{q} = \begin{pmatrix} 3 \\ -2 \end{pmatrix} - \begin{pmatrix} 1 \\ 4 \end{pmatrix}$

$= \begin{pmatrix} 3 - 1 \\ -2 - 4 \end{pmatrix}$

$= \begin{pmatrix} 2 \\ -6 \end{pmatrix}$

b $\mathbf{q} - \mathbf{p} = \begin{pmatrix} 1 \\ 4 \end{pmatrix} - \begin{pmatrix} 3 \\ -2 \end{pmatrix}$

$= \begin{pmatrix} 1 - 3 \\ 4 - -2 \end{pmatrix}$

$= \begin{pmatrix} -2 \\ 6 \end{pmatrix}$

THE MAGNITUDE OF A VECTOR

By the theorem of Pythagoras,

If $\mathbf{a} = \begin{pmatrix} a_1 \\ a_2 \end{pmatrix}$, the **magnitude** or **length** of $\mathbf{a}$ is $|\mathbf{a}| = \sqrt{a_1^2 + a_2^2}$.

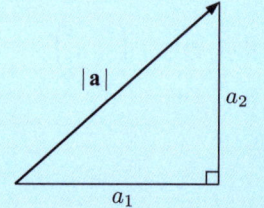

Example 9

Find the length of: **a** $\begin{pmatrix} 5 \\ 2 \end{pmatrix}$ **b** $\begin{pmatrix} -4 \\ 3 \end{pmatrix}$

a The length of $\begin{pmatrix} 5 \\ 2 \end{pmatrix} = \sqrt{5^2 + 2^2}$

$= \sqrt{25 + 4}$

$= \sqrt{29}$ units

b The length of $\begin{pmatrix} -4 \\ 3 \end{pmatrix} = \sqrt{(-4)^2 + 3^2}$

$= \sqrt{16 + 9}$

$= 5$ units

EXERCISE 32E.2

1 Find:

a $\begin{pmatrix} 2 \\ 3 \end{pmatrix} + \begin{pmatrix} 1 \\ 5 \end{pmatrix}$

b $\begin{pmatrix} 5 \\ 4 \end{pmatrix} + \begin{pmatrix} -1 \\ 6 \end{pmatrix}$

c $\begin{pmatrix} 0 \\ 3 \end{pmatrix} + \begin{pmatrix} 7 \\ -4 \end{pmatrix}$

d $\begin{pmatrix} -4 \\ -1 \end{pmatrix} + \begin{pmatrix} 2 \\ 6 \end{pmatrix}$

e $\begin{pmatrix} -5 \\ 0 \end{pmatrix} + \begin{pmatrix} 1 \\ -6 \end{pmatrix}$

f $\begin{pmatrix} -5 \\ -9 \end{pmatrix} + \begin{pmatrix} -3 \\ 7 \end{pmatrix}$

2 Given $\mathbf{a} = \begin{pmatrix} 2 \\ -3 \end{pmatrix}$, $\mathbf{b} = \begin{pmatrix} 3 \\ -1 \end{pmatrix}$, and $\mathbf{c} = \begin{pmatrix} -2 \\ -3 \end{pmatrix}$, find:

a $\mathbf{a} + \mathbf{b}$ **b** $\mathbf{b} + \mathbf{a}$ **c** $\mathbf{b} + \mathbf{c}$ **d** $\mathbf{c} + \mathbf{b}$

e $\mathbf{a} + \mathbf{c}$ **f** $\mathbf{c} + \mathbf{a}$ **g** $\mathbf{a} + \mathbf{a}$ **h** $\mathbf{b} + \mathbf{a} + \mathbf{c}$

3 Given $\mathbf{m} = \begin{pmatrix} 3 \\ 4 \end{pmatrix}$ and $\mathbf{n} = \begin{pmatrix} 1 \\ -2 \end{pmatrix}$, find:

   a $\mathbf{m} + \mathbf{n}$ **b** $-\mathbf{m}$ **c** $-\mathbf{n}$

4 Find:

   a $\begin{pmatrix} 7 \\ 4 \end{pmatrix} - \begin{pmatrix} 1 \\ 2 \end{pmatrix}$ **b** $\begin{pmatrix} 8 \\ 8 \end{pmatrix} - \begin{pmatrix} 3 \\ 6 \end{pmatrix}$ **c** $\begin{pmatrix} 2 \\ 5 \end{pmatrix} - \begin{pmatrix} 6 \\ 0 \end{pmatrix}$

   d $\begin{pmatrix} -2 \\ 1 \end{pmatrix} - \begin{pmatrix} 4 \\ 5 \end{pmatrix}$ **e** $\begin{pmatrix} 5 \\ -2 \end{pmatrix} - \begin{pmatrix} -3 \\ 1 \end{pmatrix}$ **f** $\begin{pmatrix} 6 \\ -4 \end{pmatrix} - \begin{pmatrix} -5 \\ -1 \end{pmatrix}$

5 Given $\mathbf{p} = \begin{pmatrix} -1 \\ 3 \end{pmatrix}$, $\mathbf{q} = \begin{pmatrix} -2 \\ -3 \end{pmatrix}$, and $\mathbf{r} = \begin{pmatrix} 3 \\ -4 \end{pmatrix}$, find:

   a $\mathbf{p} - \mathbf{q}$ **b** $\mathbf{q} - \mathbf{r}$ **c** $\mathbf{p} + \mathbf{q} - \mathbf{r}$

   d $\mathbf{p} - \mathbf{q} - \mathbf{r}$ **e** $\mathbf{q} - \mathbf{r} - \mathbf{p}$ **f** $\mathbf{r} + \mathbf{q} - \mathbf{p}$

6 Find the length of each vector:

   a $\begin{pmatrix} 1 \\ 4 \end{pmatrix}$ **b** $\begin{pmatrix} 6 \\ 0 \end{pmatrix}$ **c** $\begin{pmatrix} 3 \\ -2 \end{pmatrix}$ **d** $\begin{pmatrix} -1 \\ -5 \end{pmatrix}$ **e** $\begin{pmatrix} -4 \\ 2 \end{pmatrix}$ **f** $\begin{pmatrix} -6 \\ -1 \end{pmatrix}$

F   THE VECTOR BETWEEN TWO POINTS

Consider two points $A(a_1, a_2)$ and $B(b_1, b_2)$ with position vectors **a** and **b** respectively.

To move from A to B, we can travel $-\mathbf{a}$ back to the origin, then **b** to get to B.

$\therefore \overrightarrow{AB} = -\mathbf{a} + \mathbf{b}$

$\phantom{\therefore \overrightarrow{AB}} = -\begin{pmatrix} a_1 \\ a_2 \end{pmatrix} + \begin{pmatrix} b_1 \\ b_2 \end{pmatrix}$

$\phantom{\therefore \overrightarrow{AB}} = \begin{pmatrix} b_1 - a_1 \\ b_2 - a_2 \end{pmatrix}$

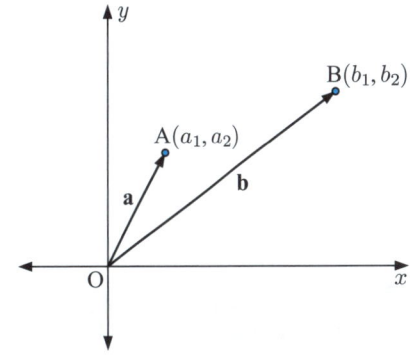

> The **position vector** of $B(b_1, b_2)$ relative to $A(a_1, a_2)$ is $\overrightarrow{AB} = \begin{pmatrix} x\text{-step} \\ y\text{-step} \end{pmatrix} = \begin{pmatrix} b_1 - a_1 \\ b_2 - a_2 \end{pmatrix}$.

The distance between A and B is $|\overrightarrow{AB}| = \sqrt{(b_1 - a_1)^2 + (b_2 - a_2)^2}$ which is the result we used in coordinate geometry.

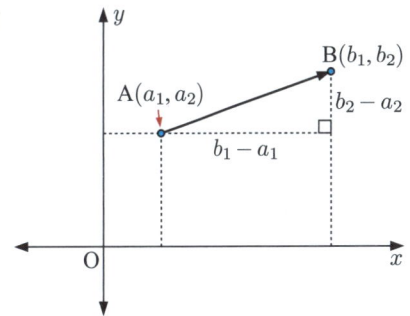

Example 10

Given the points A(2, −3) and B(4, 2), find:

a $\vec{AB}$

b $\vec{BA}$.

a $\vec{AB} = \begin{pmatrix} 4-2 \\ 2--3 \end{pmatrix} = \begin{pmatrix} 2 \\ 5 \end{pmatrix}$

b $\vec{BA} = \begin{pmatrix} 2-4 \\ -3-2 \end{pmatrix} = \begin{pmatrix} -2 \\ -5 \end{pmatrix}$

From this Example, we notice that $\vec{BA} = -\vec{AB}$.

EXERCISE 32F

1 Given points A(3, 4), B(6, 1), C(−5, 3), and D(−1, −2), find:

   a $\vec{AB}$ **b** $\vec{AC}$ **c** $\vec{AD}$

   d $\vec{BD}$ **e** $\vec{CB}$ **f** $\vec{CA}$

2 For each of the following pairs of points, find: **i** $\vec{AB}$ **ii** the distance AB.

   a A(3, 5) and B(1, 2)   **b** A(−2, 1) and B(3, −1)

   c A(3, 4) and B(0, 0)   **d** A(11, −5) and B(−1, 0)

3 Suppose $\vec{AB} = \begin{pmatrix} 2 \\ 5 \end{pmatrix}$ and $\vec{BC} = \begin{pmatrix} -1 \\ 3 \end{pmatrix}$.

   a Explain why $\vec{AC} = \vec{AB} + \vec{BC}$.   **b** Hence find $\vec{AC}$.

4 Given $\vec{AB} = \begin{pmatrix} 1 \\ 4 \end{pmatrix}$ and $\vec{AC} = \begin{pmatrix} -2 \\ 1 \end{pmatrix}$, find $\vec{BC}$.

5 Given $\vec{AB} = \begin{pmatrix} -3 \\ 2 \end{pmatrix}$, $\vec{BD} = \begin{pmatrix} 0 \\ 4 \end{pmatrix}$, and $\vec{CD} = \begin{pmatrix} 1 \\ -3 \end{pmatrix}$, find $\vec{AC}$.

G SCALAR MULTIPLICATION

Numbers such as 1 and −2 are called *scalars* because they have size but no direction.

2**a** and −2**a** are examples of multiplying a vector by a scalar.

2**a** is a short way to write **a** + **a**, and similarly −2**a** = (−**a**) + (−**a**).

For we have 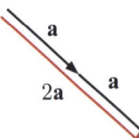 and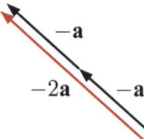

So, 2**a** has the same direction as **a** and is twice as long as **a**, and

−2**a** is in the opposite direction to **a** and is twice as long as **a**.

Example 11

Given $\mathbf{r} = \begin{pmatrix} 3 \\ 2 \end{pmatrix}$ and $\mathbf{s} = \begin{pmatrix} 2 \\ -2 \end{pmatrix}$, find geometrically:

a $2\mathbf{r} + \mathbf{s}$ **b** $\mathbf{r} - 2\mathbf{s}$

a
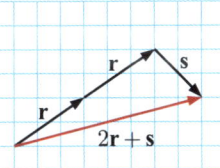

$2\mathbf{r} + \mathbf{s} = \begin{pmatrix} 8 \\ 2 \end{pmatrix}$

b
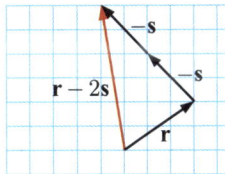

$\mathbf{r} - 2\mathbf{s} = \begin{pmatrix} -1 \\ 6 \end{pmatrix}$

In component form:

If k is a scalar then $k \begin{pmatrix} a \\ b \end{pmatrix} = \begin{pmatrix} ka \\ kb \end{pmatrix}$.

Each component is multiplied by k.

We can now check the results of **Example 11** using the component form:

In **a**, $2\mathbf{r} + \mathbf{s} = 2\begin{pmatrix} 3 \\ 2 \end{pmatrix} + \begin{pmatrix} 2 \\ -2 \end{pmatrix}$
$= \begin{pmatrix} 6 \\ 4 \end{pmatrix} + \begin{pmatrix} 2 \\ -2 \end{pmatrix}$
$= \begin{pmatrix} 8 \\ 2 \end{pmatrix}$

In **b**, $\mathbf{r} - 2\mathbf{s} = \begin{pmatrix} 3 \\ 2 \end{pmatrix} - 2\begin{pmatrix} 2 \\ -2 \end{pmatrix}$
$= \begin{pmatrix} 3 \\ 2 \end{pmatrix} - \begin{pmatrix} 4 \\ -4 \end{pmatrix}$
$= \begin{pmatrix} -1 \\ 6 \end{pmatrix}$

Example 12

Sketch any two vectors $\mathbf{p}$ and $\mathbf{q}$ such that:

a $\mathbf{p} = 2\mathbf{q}$ **b** $\mathbf{p} = -\frac{1}{2}\mathbf{q}$.

Let **q** be

a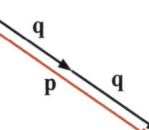

b

644 Vectors (Chapter 32)

EXERCISE 32G

1 Find:

  a $3\begin{pmatrix} 2 \\ 5 \end{pmatrix}$
  b $5\begin{pmatrix} -1 \\ 4 \end{pmatrix}$
  c $-2\begin{pmatrix} 6 \\ 0 \end{pmatrix}$
  d $\frac{1}{2}\begin{pmatrix} 4 \\ 14 \end{pmatrix}$

  e $-4\begin{pmatrix} -2 \\ 7 \end{pmatrix}$
  f $\frac{1}{3}\begin{pmatrix} -9 \\ 2 \end{pmatrix}$
  g $2\begin{pmatrix} 3 \\ 8 \end{pmatrix} + \begin{pmatrix} 1 \\ -5 \end{pmatrix}$
  h $\begin{pmatrix} 10 \\ 3 \end{pmatrix} - 4\begin{pmatrix} 0 \\ 4 \end{pmatrix}$

2 Suppose $r = \begin{pmatrix} 2 \\ 3 \end{pmatrix}$ and $s = \begin{pmatrix} 4 \\ -2 \end{pmatrix}$.

  Find each vector sum using   **i** geometry   **ii** component form:

  a $2r$
  b $-3s$
  c $\frac{1}{2}r$
  d $r - 2s$

  e $3r + s$
  f $2r - 3s$
  g $\frac{1}{2}s + r$
  h $\frac{1}{2}(2r + s)$

3 Sketch any two vectors **p** and **q** such that:

  a $p = q$
  b $p = -q$
  c $p = 3q$
  d $p = \frac{3}{4}q$
  e $p = -\frac{3}{2}q$

4 Write vector expressions for **c**, **d**, **e**, and **f** in terms of **a** and **b**:

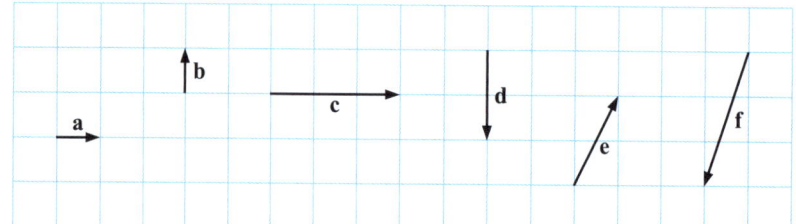

5 ABCD is a rhombus. Suppose $\overrightarrow{OA} = a$ and $\overrightarrow{OB} = b$.
Write, in terms of **a** and **b**, a vector expression for:

  a $\overrightarrow{CA}$
  b $\overrightarrow{BD}$
  c $\overrightarrow{CD}$

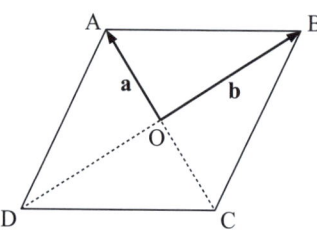

Review set 32A

1 Using a scale of 1 cm represents 10 units, sketch a vector to represent:

  a an aeroplane taking off at an angle of 8° to the runway with a speed of 60 m s^{-1}
  b a displacement of 45 m in the direction 060°.

2 For the given vectors **p**, **q**, and **r**, show how to construct:

  a $p + r$
  b $r - q - p$

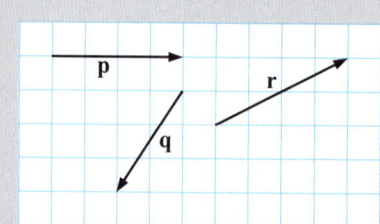

3 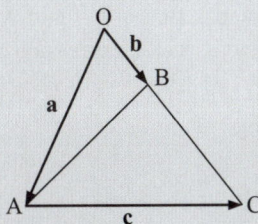 In the figure alongside, $\overrightarrow{OA} = \mathbf{a}$, $\overrightarrow{OB} = \mathbf{b}$, and $\overrightarrow{AC} = \mathbf{c}$.
Find, in terms of $\mathbf{a}$, $\mathbf{b}$, and $\mathbf{c}$:

  a $\overrightarrow{CA}$ **b** $\overrightarrow{AB}$ **c** $\overrightarrow{OC}$ **d** $\overrightarrow{BC}$

4 For points A, B, C, and D, simplify the following vector expressions:

  a $\overrightarrow{AB} + \overrightarrow{BD}$ **b** $\overrightarrow{BC} - \overrightarrow{DC}$ **c** $\overrightarrow{AB} - \overrightarrow{CB} + \overrightarrow{CD} - \overrightarrow{AD}$

5 Draw arrow diagrams to represent the vectors:

  a $\begin{pmatrix} 1 \\ 4 \end{pmatrix}$ **b** $\begin{pmatrix} -2 \\ 0 \end{pmatrix}$ **c** $\begin{pmatrix} 3 \\ -1 \end{pmatrix}$ **d** $\begin{pmatrix} -3 \\ -3 \end{pmatrix}$

6 Find:

  a $\begin{pmatrix} 5 \\ 1 \end{pmatrix} + \begin{pmatrix} -2 \\ 6 \end{pmatrix}$ **b** $\begin{pmatrix} -1 \\ 4 \end{pmatrix} - \begin{pmatrix} -3 \\ -2 \end{pmatrix}$ **c** $6 \begin{pmatrix} -2 \\ 9 \end{pmatrix}$

7 P, Q, and R are three points on a grid. The vector from P to Q is $\begin{pmatrix} 3 \\ 5 \end{pmatrix}$, and the vector from Q to R is $\begin{pmatrix} -1 \\ 4 \end{pmatrix}$. Calculate the vector from P to R.

Review set 32B

1 What can be said about vectors $\mathbf{p}$ and $\mathbf{q}$ if:

  a $|\mathbf{p}| = |\mathbf{q}|$ **b** $\mathbf{p} = 2\mathbf{q}$?

2 How are $\overrightarrow{AB}$ and $\overrightarrow{BA}$ related?

3 Write the illustrated vectors in component form:

  a 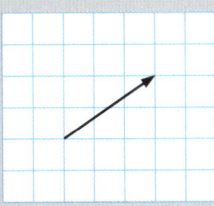 **b** 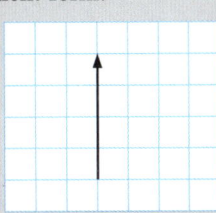 **c**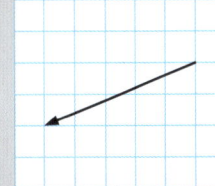

4 Find:

  a $\begin{pmatrix} -1 \\ 2 \end{pmatrix} + \begin{pmatrix} 7 \\ -8 \end{pmatrix}$ **b** $\begin{pmatrix} 4 \\ -10 \end{pmatrix} - \begin{pmatrix} -3 \\ 6 \end{pmatrix}$ **c** $\begin{pmatrix} 3 \\ 0 \end{pmatrix} - \frac{1}{2} \begin{pmatrix} 6 \\ -4 \end{pmatrix}$

5 For $\mathbf{m} = \begin{pmatrix} 3 \\ -1 \end{pmatrix}$ and $\mathbf{n} = \begin{pmatrix} 2 \\ 4 \end{pmatrix}$, find:

  a $\mathbf{m} - 2\mathbf{n}$ **b** $|\mathbf{m} + \mathbf{n}|$.

6 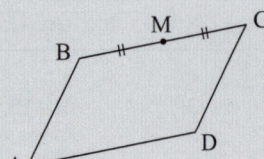 In parallelogram ABCD, $\overrightarrow{AB} = \mathbf{p}$, $\overrightarrow{BC} = \mathbf{q}$, and M is the midpoint of line segment BC. Find a vector expression for:

    a $\overrightarrow{CD}$ **b** $\overrightarrow{BM}$ **c** $\overrightarrow{MD}$ **d** $\overrightarrow{AD}$

7 Suppose $\mathbf{p} = \begin{pmatrix} 4 \\ 3 \end{pmatrix}$, $\mathbf{q} = \begin{pmatrix} 3 \\ -5 \end{pmatrix}$, and $\mathbf{r} = \begin{pmatrix} 0 \\ -4 \end{pmatrix}$. Find:

    a $2\mathbf{p} + \mathbf{q}$ **b** $\mathbf{p} - \mathbf{q} - \mathbf{r}$ **c** the length of $\mathbf{q}$.

8 For $A(-1, 1)$ and $B(2, -3)$, find:

    a $\overrightarrow{AB}$ **b** the distance AB.

ANSWERS

648 Answers

EXERCISE 1A
1. **a** 3 **b** −15 **c** 10 **d** 9 **e** −38 **f** 6
   g −7 **h** 0

2.
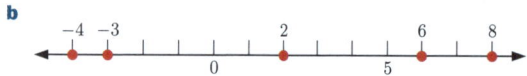

   Since −1 is to the right of −6, −1 is greater than −6. We could also say −6 is less than −1.

3. 2 is greater than negative 5. This statement is true.

4. **a** true **b** true **c** false **d** true **e** false **f** true

5. **a** $8 > 6$ **b** $18 > 7$ **c** $-9 < -4$
   d $-3 < 15$ **e** $20 > -15$ **f** $-6 < -2$

6. **a**

   The greatest number is 5. The least number is −9.

   b
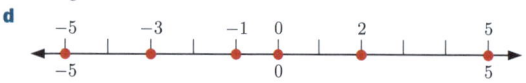

   The greatest number is 8. The least number is −4.

   c

   The greatest number is 9. The least number is −9.

   d

   The greatest number is 5. The least number is −5.

7. −2, −1, 1, 4

8. −4, −3, −1, 0, 2, 5, 6

9. 8, 6, 0, −2, −5, −7
10. −10, −7, −2, 0, 7, 8
11. Moscow −6°C, Oslo −4°C, Tokyo 1°C, Ulaanbaatar 3°C, Melbourne 19°C, Singapore 33°C

12. **a** $2 + 3 = 5$ **b** $4 - 2 = 2$ **c** $2 + 2 = 4$
    $1 + 3 = 4$ $3 - 2 = 1$ $2 + 1 = 3$
    $0 + 3 = 3$ $2 - 2 = 0$ $2 + 0 = 2$
    $-1 + 3 = 2$ $1 - 2 = -1$ $2 - 1 = 1$
    $-2 + 3 = 1$ $0 - 2 = -2$ $2 - 2 = 0$
    $-3 + 3 = 0$ $-1 - 2 = -3$ $2 - 3 = -1$

13. **a** −4 **b** −1 **c** −11 **d** −1 **e** 0 **f** −7
14. **a** 10 **b** 2 **c** −4 **d** −4 **e** −6 **f** −4
    g −8 **h** −12

EXERCISE 1B
1. **a** 1 **b** 7 **c** −7 **d** −1 **e** 1 **f** 7
   g −7 **h** −1
2. **a** −4 **b** 8 **c** −8 **d** 4 **e** −4 **f** 8
   g −8 **h** 4
3. **a** −1 **b** −8 **c** −2 **d** 4 **e** −10 **f** 9
   g 3 **h** −17
4. 2nd floor above ground level
5. **a** −9 **b** −2 **c** −9 **d** −21 **e** −12 **f** 3
   g −10 **h** −1 **i** 6

6. **a** 1 **b** 6 **c** 3 **d** 2 **e** −7 **f** 17
   g −17 **h** −27 **i** 16
7. **a** 9 **b** 3 **c** 9 **d** 2 **e** 16 **f** 15
8. −1°C

EXERCISE 1C
1. **a** 24 **b** −24 **c** −24 **d** 24 **e** −24 **f** −24
   g 24 **h** 24
2. **a** 36 **b** −36 **c** −36 **d** 36 **e** 33 **f** −33
   g −33 **h** 33 **i** 56 **j** −56 **k** 56 **l** −56
3. **a** −6 **b** −30 **c** 14 **d** −50 **e** −48 **f** −45
   g −88 **h** −33 **i** −81 **j** 24 **k** −55 **l** 42
4. **a** $\square = 1$ **b** $\square = -2$ **c** $\square = -11$ **d** $\square = -4$
   e $\square = -6$ **f** $\square = -2$ **g** $\square = 4$ **h** $\square = -1$
   i $\square = 6$ **j** $\square = -3$ **k** $\square = -3$ **l** $\square = -10$
5. **a** 280 m **b** £120
6. **a** −200 **b** 63 **c** 20 **d** 200 **e** 90 **f** 48

EXERCISE 1D
1. **a** 5 **b** −5 **c** −5 **d** 5 **e** 5 **f** 5
   g −5 **h** −5 **i** 1 **j** −1 **k** −1 **l** 1
   m 11 **n** −11 **o** 11 **p** −11
2. **a** $\square = -4$ **b** $\square = -6$ **c** $\square = -4$ **d** $\square = -25$
   e $\square = -9$ **f** $\square = -12$ **g** $\square = -8$ **h** $\square = -5$
   i $\square = -40$ **j** $\square = -9$ **k** $\square = 3$ **l** $\square = -28$
   m $\square = -8$ **n** $\square = -120$ **o** $\square = 144$ **p** $\square = 12$
3. **a** debt share is £50 000 **b** −9°C (drops 9°C per hour)
4. **a** $\square = -10$ **b** $\square = -1$ **c** $\square = -16$ **d** $\square = -5$
   e $\square = -5$ **f** $\square = 10$ **g** $\square = -40$ **h** $\square = -2$
   i $\square = 18$ **j** $\square = 72$ **k** $\square = 10$ **l** $\square = -9$

EXERCISE 1E
1. **a** D **b** C **c** A **d** E **e** B
2. **a** 2×3^2 **b** $2^2 \times 3 \times 5$ **c** 2×5^3 **d** $3^2 \times 5^3$
   e $2^3 \times 5 \times 7$ **f** $3^3 \times 7^2$ **g** $3^4 \times 5^2$ **h** $7^5 \times 11^3$
3. **a** 8 **b** 16 **c** 27 **d** 50 **e** 100 **f** 135
   g 315 **h** 1188
4. **a** 11 664 **b** 42 017 500 **c** 104 544
   d 178 200 **e** 5 282 739 **f** 54 925 000
5. **a** −9 **b** 9 **c** −1 **d** −1 **e** 25 **f** −25
   g 16 **h** −16
6. **a** 1 **b** 1 **c** −1 **d** −1 **e** 1
   For an even power of −1, the answer is positive.
   For an odd power of −1, the answer is negative.
7. **a** 15 625 **b** 729 **c** 161 051 **d** 6561
   e −16 384 **f** 2401 **g** 64 **h** 529
8. **a** 2^1 **b** 2^2 **c** 2^4 **d** 2^6
9. **a** 3^1 **b** 3^3 **c** 3^4 **d** 3^6
10. **a** 10^2 **b** 10^3 **c** 10^5 **d** 10^6
11. **a** 2^3 **b** 5^2 **c** 6^2 **d** 5^3

EXERCISE 1F
1. **a** 36 **b** 81 **c** 121 **d** 144
2. $31^2 = 961$ 3. **a** .... even **b** .... odd
4. **a** 7 **b** 8 **c** 11 **d** 12
5. **a** 17 **b** 24 **c** 39 **d** 48

Answers 649

6 a $1 \times 3 + 1 = 4 = 2^2$ $\quad 5 \times 7 + 1 = 36 = 6^2$
$2 \times 4 + 1 = 9 = 3^2$ $\quad 6 \times 8 + 1 = 49 = 7^2$
$3 \times 5 + 1 = 16 = 4^2$ $\quad 7 \times 9 + 1 = 64 = 8^2$
$4 \times 6 + 1 = 25 = 5^2$
b i 400 **ii** 899

EXERCISE 1G
1 a 8 **b** 0 **c** -10 **d** 6 **e** 6 **f** 24
2 a 35 **b** 25 **c** 6 **d** 15 **e** 4 **f** 18
g 21 **h** 19 **i** -4 **j** 23 **k** 11 **l** 13
3 a 1 **b** -7 **c** 1 **d** 2 **e** 4 **f** -15
g 16 **h** 8 **i** 4
4 No, $-3^2 = -9$ and $(-3)^2 = 9$. **5** £480 000 profit
6 -13 **7 a** £918 profit **b** £153 average profit
8 a 21 **b** 21 **c** 0 **d** 3 **e** 0 **f** 6
g 33 **h** 9 **i** 5 **j** 25 **k** 1 **l** 4
m 2 **n** 60 **o** 24 **p** -7 **q** -5 **r** -9
9 a 241 **b** 369 **c** 23 **d** 8 **e** -258 **f** 8
10 a -2 **b** 24 **c** -2 **d** 21 **e** 9 **f** 12
11 a -11 **b** 1 **c** 18 **d** -41 **e** -245 **f** -23
12 a $7 + 3 - 4 = 6$ **b** $4 \times 6 - 3 = 21$ **c** $12 \div 4 \times 3 = 9$
13 a $(9 - 7) \times 4 = 8$ **b** $80 \div (8 \times 2) = 5$
c $80 \div 8 \times 2 = 20$ **d** $4 \times 8 - (7 - 1) = 26$
e $4 \times (8 - 7) - 1 = 3$ **f** $4 \times (8 - 7 - 1) = 0$
g $(5 + 2) \times 6 - 3 = 39$ **h** $5 + 2 \times (6 - 3) = 11$
i $(5 + 2) \times (6 - 3) = 21$

EXERCISE 1H
1 a yes **b** no **c** no **d** yes **e** no **f** yes
2 a 1, 2, 3, 6 **b** 1, 2, 4, 8 **c** 1, 2, 5, 10
d 1, 3, 5, 15 **e** 1, 2, 3, 4, 6, 12
f 1, 13 **g** 1, 2, 3, 6, 9, 18
h 1, 2, 4, 5, 10, 20 **i** 1, 2, 3, 4, 6, 8, 12, 24
j 1, 5, 25 **k** 1, 2, 3, 6, 7, 14, 21, 42
l 1, 2, 3, 6, 9, 18, 27, 54 **m** 1, 3, 7, 9, 21, 63
n 1, 23 **o** 1, 2, 3, 4, 6, 8, 9, 12, 18, 24, 36, 72
3 a 1×6, 2×3 **b** 1×8, 2×4
c 1×9, 3×3 **d** 1×12, 2×6, 3×4
e 1×14, 2×7 **f** 1×28, 2×14, 4×7
g 1×45, 3×15, 5×9
h 1×48, 2×24, 3×16, 4×12, 6×8
i 1×72, 2×36, 3×24, 4×18, 6×12, 8×9
j 1×100, 2×50, 4×25, 5×20, 10×10
4 1, 2, 3, 4, 6, 12
5 a 3, 6, 9, 12, 15 **b** 5, 10, 15, 20, 25
c 7, 14, 21, 28, 35 **d** 9, 18, 27, 36, 45
e 12, 24, 36, 48, 60
6 a 63 **b** 78
7 a 2 **b** 4 **c** 2 **d** 5 **e** 6

EXERCISE 1I
1 {2, 3, 5, 7, 11, 13, 17, 19, 23, 29} **2** 71, 73 **3** 28
4 a 2^3 **b** 3^3 **c** 5^3 **d** 2^7 **e** 7^3 **f** 3^6
g 19^2 **h** 11^3
5 a $54 = 2 \times 3^3$ **b** $108 = 2^2 \times 3^3$
c $360 = 2^3 \times 3^2 \times 5$ **d** $228 = 2^2 \times 3 \times 19$

e $196 = 2^2 \times 7^2$ **f** $756 = 2^2 \times 3^3 \times 7$
g $936 = 2^3 \times 3^2 \times 13$ **h** $1225 = 5^2 \times 7^2$
i $588 = 2^2 \times 3 \times 7^2$ **j** $945 = 3^3 \times 5 \times 7$
k $910 = 2 \times 5 \times 7 \times 13$ **l** $1274 = 2 \times 7^2 \times 13$

EXERCISE 1J
1 a 4 **b** 3 **c** 14 **d** 8 **e** 12 **f** 5
g 11 **h** 6 **i** 14 **j** 16 **k** 32 **l** 13
m 9 **n** 14 **o** 28
2 15 m × 15 m **3** 45 cm **4** 6 bouquets

EXERCISE 1K
1 a 40 **b** 12 **c** 40 **d** 90 **e** 60 **f** 140
g 108 **h** 630
2 a 12 **b** 70 **c** 24 **d** 360 **e** 120 **f** 180
g 126 **h** 300
3 75 **4** 90 m
5 a 45, 90, 135, 180, 225, ... **b** 180 minutes
60, 120, 180, 240, 300, ...
90, 180, 270, 360, 450, ...
6 60 minutes **7 a** 24 days **b** 120 days

REVIEW SET 1A
1 a -5 **b** 2 **c** -3 **d** 11
2 A: 3, B: -1, C: 7, D: 0, E: -4
3 a $-6, -4, -3, 0, 2, 3, 7$ **b** 13
4 a -3 **b** 3 **c** -6 **d** 12
5 $-1°C$ **6** 21 pianos
7 a 24 **b** 4 **c** -8 **d** 1 **e** 64 **f** -25
8 a 64 **b** 200 **c** 350
9 a 3 **b** 9 **c** 10
10 a 16 **b** -13 **c** 30 **d** 6 **e** 11
11 a $3 \times 5 - 4 = 11$ **b** $8 - 6 \div 3 = 6$
12 a 1, 3, 7, 21 **b** 1, 2, 4, 8, 16, 32 **c** 1, 37
13 1×66, 2×33, 3×22, 6×11 **14** 41, 43, 47
15 a $450 = 2 \times 3^2 \times 5^2$ **b** $212 = 2^2 \times 53$
16 a 30 **b** 44 **c** 40
17 a 2 **b** 8 **c** 9
18 a Amy scored 48 points, Sean scored -8 points
b 56 points
19 104 minutes

REVIEW SET 1B
1 a $3 > -1$ **b** $-7 < -4$ **c** $-2 < 5$ **d** $0 > -10$
2 a $-4, -2, -1, 1, 3$
b $-5, -3, 1, 2, 8$
3 a -7 **b** -21
4 a negative × negative = positive **b** 77
5 5, 3, 1, 0, $-1, -4, -6$ **b** 11
6 a -3 **b** -72 **c** -9 **d** -1 **e** 10 **f** -24
7 a Ying **b** Cathy
c i 19 minutes **ii** 21 minutes **iii** 8 minutes
8 49, 64, and 81 **9 a** -32 **b** 280 **c** 36
10 729 and 784 **11 a** 51 **b** 9

12 a $12 \div (6-2) = 3$ **b** $(6+4) \div (2+3) = 2$
   c $18 \div (1 + 2 \times 4) = 2$
13 8   **14** 8, 16, 24, 32, 40   **15** $54 = 2 \times 3^3$
16 91, 93, 95   **17 a** 6   **b** 90
18 18 nectarines   **19** 60 lollies

EXERCISE 2A.1

1 a proper fraction   **b** improper fraction
   c proper fraction   **d** mixed number
   e improper fraction   **f** mixed number
2 a $\frac{7}{4}$ **b** $\frac{7}{2}$ **c** $\frac{21}{5}$ **d** $\frac{19}{8}$ **e** $-\frac{4}{3}$ **f** $-\frac{16}{5}$
3 a number line with $-1\frac{1}{3}$, $-\frac{2}{3}$, $\frac{1}{3}$, $\frac{5}{3}$
   b number line with $-1\frac{1}{4}$, $-\frac{1}{4}$, $\frac{3}{4}$, $2\frac{1}{4}$
   c number line with $-\frac{10}{7}$, $-\frac{2}{7}$, $\frac{3}{7}$, $\frac{6}{7}$, $1\frac{4}{7}$
4 a -2 **b** -2 **c** -2

EXERCISE 2A.2

1 a $\frac{3}{12}$ **b** $\frac{8}{12}$ **c** $\frac{10}{12}$ **d** $\frac{5}{12}$
2 a $\frac{14}{20}$ **b** $\frac{15}{20}$ **c** $\frac{13}{20}$ **d** $\frac{3}{20}$
3 a $\frac{21}{30}, \frac{18}{30}, \frac{20}{30}$ **b** $\frac{3}{5}, \frac{2}{5}, \frac{7}{10}$
4 a $-\frac{1}{6}, \frac{1}{4}, \frac{3}{8}, \frac{2}{3}, \frac{5}{6}$ **b** $-\frac{1}{3}, -\frac{5}{18}, -\frac{2}{9}, \frac{1}{6}, \frac{2}{9}$
5 a $\frac{2}{3}$ **b** $\frac{5}{6}$ **c** $\frac{5}{3}$ **d** $\frac{3}{5}$ **e** $-\frac{1}{6}$ **f** $-\frac{1}{3}$
   g $\frac{3}{4}$ **h** $-\frac{2}{5}$ **i** $\frac{11}{16}$ **j** $-\frac{3}{2}$
6 a $\frac{3}{8}$ **b** $\frac{5}{8}$
7 a 60 players   **b i** $\frac{5}{12}$ **ii** $\frac{2}{15}$ **iii** $\frac{11}{20}$
8 a 9   **b** 3   **c** 4   **d** 4   **e** 4   **f** 2
   g 6   **h** 1
9 a -5 **b** 3   **c** -6 **d** 6   **e** -6 **f** 2

EXERCISE 2B.1

1 a $\frac{7}{9}$ **b** $\frac{3}{7}$ **c** $\frac{9}{8}$ **d** $\frac{5}{11}$
2 a $\frac{5}{4}$ **b** $\frac{1}{2}$ **c** $\frac{7}{10}$ **d** $\frac{1}{12}$
   e $\frac{11}{8}$ **f** $\frac{1}{12}$ **g** $\frac{48}{35}$ **h** $-\frac{1}{18}$
3 a $\frac{11}{9}$ or $1\frac{2}{9}$ **b** $\frac{11}{8}$ or $1\frac{3}{8}$ **c** $\frac{23}{30}$ **d** $\frac{7}{60}$
4 a $\frac{4}{7}$ **b** $\frac{1}{4}$ **c** $\frac{15}{8}$ or $1\frac{7}{8}$ **d** $-\frac{8}{21}$
   e $\frac{7}{5}$ or $1\frac{2}{5}$ **f** $\frac{11}{24}$ **g** $\frac{43}{45}$ **h** $\frac{9}{20}$
5 a $\frac{23}{40}$ **b** $\frac{17}{40}$ **6 a** $2\frac{1}{6}$ cups   **b** $\frac{5}{6}$ of a cup
7 a $\frac{107}{210}$ **b** $-\frac{17}{120}$

EXERCISE 2B.2

1 a $\frac{2}{15}$ **b** $\frac{3}{8}$ **c** $\frac{4}{15}$ **d** 1   **e** $\frac{8}{15}$ **f** $\frac{15}{28}$
   g $\frac{12}{5}$ or $2\frac{2}{5}$ **h** $\frac{5}{4}$ or $1\frac{1}{4}$ **i** $\frac{3}{7}$ **j** $\frac{39}{80}$
   k $\frac{1}{4}$ **l** $\frac{15}{56}$ **m** $\frac{121}{25}$ or $4\frac{21}{25}$ **n** $\frac{2}{3}$
   o $\frac{13}{8}$ or $1\frac{5}{8}$ **p** 4   **q** $\frac{8}{27}$ **r** $\frac{30}{7}$ **s** $\frac{21}{10}$
   t 1

2 a 3   **b** $\frac{3}{8}$ **c** 2   **d** $\frac{8}{5}$ **e** $\frac{21}{20}$ **f** $\frac{8}{3}$
   g $\frac{11}{8}$ **h** $\frac{5}{18}$ **i** 10   **j** $\frac{16}{9}$ **k** $\frac{2}{5}$ **l** $\frac{20}{33}$
3 a $-\frac{8}{15}$ **b** $-\frac{3}{4}$ **c** $-\frac{18}{35}$ **d** $\frac{35}{12}$ or $2\frac{11}{12}$
4 a $\frac{24}{5}$ or $4\frac{4}{5}$ **b** $\frac{81}{256}$ **c** $\frac{5}{2}$ or $2\frac{1}{2}$ **d** $\frac{2}{5}$
   e 12   **f** $\frac{12}{11}$ or $1\frac{1}{11}$ **g** $\frac{14}{3}$ or $4\frac{2}{3}$ **h** $\frac{1}{2}$
   i $\frac{7}{6}$ or $1\frac{1}{6}$ **j** $\frac{3}{5}$ **k** $\frac{13}{5}$ or $2\frac{3}{5}$ **l** $\frac{1}{10}$
   m 11   **n** $\frac{74}{33}$ or $2\frac{8}{33}$ **o** $\frac{22}{5}$ or $4\frac{2}{5}$
5 a 20   **b** 9   **c** 25   **d** £15   **e** 25 m   **f** 56 kg
6 490 km   **7** $\frac{3}{10}$ **8** $\frac{7}{25}$
9 a $\frac{4}{27}$ **b** $\frac{20}{27}$ **10** 6 loaves   **11** $\frac{39}{80}$ of a bottle

EXERCISE 2C

1 a $1 + \frac{2}{10}$ **b** $1 + \frac{2}{100}$ **c** $1 + \frac{2}{100} + \frac{3}{1000} + \frac{4}{10\,000}$
   d $9 + \frac{9}{100} + \frac{9}{10\,000}$ **e** $\frac{3}{100} + \frac{8}{1000} + \frac{2}{10\,000}$
2 a 4.5   **b** 2.69   **c** 0.502   **d** 0.038
   e 2.005   **f** 4.0104
3 a 7000   **b** 70   **c** $\frac{7}{10}$ **d** $\frac{7}{100}$ **e** $\frac{7}{1\,000\,000}$
4 a $\frac{1}{5}$ **b** $\frac{17}{100}$ **c** $\frac{37}{50}$ **d** $\frac{1}{25}$ **e** $\frac{1}{40}$ **f** $\frac{1}{125}$
   g $\frac{5}{8}$ **h** $-\frac{4}{5}$
5 a number line with 0.4, 0.7, 1.1, 1.5
   b number line with -1.4, -0.6, -0.3, 0.5, 1.2
   c number line with 0.1, 0.35, 0.6, 0.95
6 a A: 0.7, B: 2.4   **b** A: -0.18, B: -0.02
7 a 0.9   **b** 0.85   **c** 0.4   **d** 0.12   **e** 0.62   **f** 0.375
8 a 0.023, 0.032, 0.203, 0.302, 0.32
   b 1.035, 1.35, 1.503, 1.53, 3.15
   c $\frac{2}{5}$, 0.45, $\frac{1}{2}$, 0.54, 0.6
9 a 0.64, 0.604, 0.406, 0.4006, 0.064
   b $\frac{7}{20}$, 0.261, 0.2, 0.16, $\frac{3}{25}$

EXERCISE 2D.1

1 a 6.88   **b** 11.2   **c** 22.84   **d** 24.976   **e** 13.1
   f 29.98   **g** 7.996   **h** 31.3232   **i** 39.966
2 a 4.62   **b** 4.44   **c** 7.88   **d** 12.15   **e** 4.773
   f 7.902   **g** 43.138   **h** 1.763   **i** 0.026 21
3 18.35 kg   **4** 3.86 m   **5 a** £18.25   **b** £1.75

EXERCISE 2D.2

1 a 57   **b** 4.6   **c** 600   **d** 0.01   **e** 0.07
   f 0.0032   **g** 0.0022   **h** 0.000 009 1   **i** 0.000 200 5
2 a 4.8   **b** 6.3   **c** 0.36   **d** 0.84   **e** 0.09
   f 0.0088   **g** -6 **h** -0.12 **i** 0.72
3 a 8   **b** 5   **c** 8   **d** 40   **e** 0.15
   f 0.2   **g** 0.016   **h** -17 **i** 6
4 17.6 litres   **5** 4 times   **6** 0.2 g   **7** 300 cups
8 £8.85
9 a 153.75   **b** 98.38   **c** $4.\dot{3}$ **d** 32
   e ≈ 0.691 **f** ≈ 2.05 **g** ≈ 23.2 **h** ≈ 0.254

Answers

EXERCISE 2E.1

1 **a** $6\% = \frac{6}{100}$ **b** $51\% = \frac{51}{100}$ **c** $27\% = \frac{27}{100}$ **d** $86\% = \frac{86}{100}$

2 **a** $\frac{1}{4}$ **b** $1\frac{3}{10}$ **c** $\frac{13}{20}$ **d** $\frac{2}{5}$ **e** $2\frac{1}{10}$ **f** 1 **g** $\frac{3}{25}$ **h** $\frac{1}{50}$ **i** $\frac{9}{40}$ **j** $\frac{1}{40}$ **k** $\frac{31}{40}$ **l** $\frac{249}{400}$

3 **a** 0.66 **b** 0.29 **c** 0.5 **d** 0.75 **e** 1.8 **f** 2.05 **g** 3 **h** 1.28 **i** 0.0001 **j** 0.003 **k** 0.105 **l** 0.5625

4 **a** 17% **b** 55% **c** 9% **d** 80% **e** 4% **f** 200% **g** 40% **h** 350% **i** 205% **j** 364% **k** 8.8% **l** 140.9%

5 **a** 25% **b** 30% **c** 35% **d** 44% **e** 160% **f** 54% **g** 47.5% **h** 30%

6 **a** 50% **b** 56% **c** 90% **d** 70% **e** 65% **f** $\approx 94.3\%$

7 $\approx 86.4\%$ **8** $\approx 123\%$ **9** 17%

10 **a** 4 **b** 15 **c** 9 **d** 9 **e** 35 **f** 150 **g** 9 **h** 30

11 **a** £4.80 **b** £16 **c** 4.8 litres **d** 52.5 kg

12 24 marks **13** 84 marks **14** 76 h 48 min

15 £2128.50 **16** **a** £4.05 **b** £58.05

17 110 matches **18** 12 weeks

EXERCISE 2E.2

1 **a** £160 **b** 16 litres **c** 1300 ml **d** 3200 kg **e** £1300 **f** 420 km **g** 350 litres **h** 700 kg **i** £2100

2 **a** £520 **b** 1152 kg **c** £78.75 **d** 104 ml **e** 210 kg **f** £24.80

3 1700 cars **4** £75 500 **5** 300 students

REVIEW SET 2A

1 number line with $-\frac{7}{4}$, $-\frac{1}{4}$, $\frac{3}{4}$, $2\frac{1}{4}$ marked between -2 and 3

2 **a** $\frac{1}{6}$ **b** $\frac{2}{5}$ **c** $\frac{7}{4}$ or $1\frac{3}{4}$ **d** $\frac{21}{23}$ **e** $-\frac{2}{9}$

3 **a** $-\frac{1}{12}$ **b** 4.64 **c** $\frac{25}{36}$ **d** 0.0072 **e** $\frac{27}{7}$ or $3\frac{6}{7}$ **f** 0.007

4 **a** 5 **b** $\frac{5}{100}$ **c** $\frac{5}{10000}$ **d** $\frac{5}{10}$

5 $\frac{7}{32}$ **6** **a** 30 quarters **b** $12\frac{3}{4}$ watermelons

7 **a** 12 m **b** £30.40 **c** $11\frac{2}{3}$ kg

8 **a** $5\frac{1}{2}$ **b** -12 **c** $1\frac{1}{3}$ **d** $3\frac{1}{3}$

9 **a** $\frac{31}{100}$ **b** $\frac{14}{25}$ **c** $\frac{3}{8}$ **d** $-\frac{9}{20}$

10 **a** 0.8 **b** 0.95 **c** 0.44 **d** 0.54

11 **a** 0.83 **b** 0.274 **c** 1.52 **d** 0.004

12 **a** 60% **b** 75% **c** 8% **d** 200%

13 $46.\dot{6}\%$ **14** 20% **15** 3 minutes **16** 150 staff

REVIEW SET 2B

1 number line with -1.3, -0.4, 0.2, 0.6, 1.1 marked between -2 and 1

2 **a** $5\frac{1}{15}$ **b** 0.185 **c** 1.984

3 **a** 3 **b** 10 **c** -33 **4** £1.89

5 **a** 0.6 **b** 6 **c** 60 **d** 0.6

6 **a** $\frac{3}{4}$ **b** $\frac{23}{12}$ **c** $\frac{3}{2}$ **d** $\frac{3}{10}$

7 $2\frac{5}{6}$ cans **8** **a** $\frac{23}{30}$ **b** $\frac{5}{18}$

9 £8.20 **10** **a** $0.68\dot{6}\dot{3}$ **b** ≈ 1.56

11 **a** $\frac{12}{25}$ **b** $\frac{3}{20}$ **c** $\frac{11}{200}$ **d** $\frac{1}{1000}$

12 72% **13** **a** 150 kg **b** 250 kg

14 **a** 93.75% **b** 62.5%

15 1800 students **16** **a** 41 marks **b** 50 marks

EXERCISE 3A.1

1 **a** cm **b** mm **c** m **d** mm **e** cm

2 **a** 52 000 m **b** 1150 mm **c** 165 cm **d** 6300 mm **e** 62 500 cm **f** 8 100 000 mm

3 **a** 4.8 m **b** 5.4 cm **c** 5.28 km **d** 2 m **e** 5.8 km **f** 7 km

4 **a** 42 100 m **b** 2.1 m **c** 7.5 cm **d** 1.5 km **e** 185 cm **f** 425 mm **g** 280 000 cm **h** 16.5 m **i** 250 000 mm

5 5.3 km **6** 12.375 km **7** 256 666 lengths

8 **a** 24% **b** 7.5% **c** $\approx 11.43\%$ **d** $\approx 4.38\%$

EXERCISE 3A.2

1 **a** 25.4 cm **b** 60.96 cm **c** 3.048 m **d** ≈ 12.872 km **e** 36.576 m **f** ≈ 122.284 km

2 15.24 cm **3** ≈ 11.887 m **4** ≈ 60.24 miles

EXERCISE 3B

1 **a** cm^2 **b** cm^2 **c** ha **d** m^2 **e** km^2 **f** mm^2

2 **a** 0.23 cm^2 **b** $36\,000$ m^2 **c** 0.0726 m^2 **d** $7\,600\,000$ mm^2 **e** 0.853 ha **f** $35\,400\,000$ cm^2 **g** 1354 mm^2 **h** $4\,320\,000$ cm^2 **i** 4820 mm^2

3 **a** $3\,000\,000$ m^2 **b** 70 ha **c** $6\,600\,000$ m^2 **d** 6.6 km^2 **e** 500 cm^2 **f** 0.0025 m^2 **g** 0.052 cm^2 **h** $720\,000\,000\,000$ mm^2

4 **a** ≈ 84.987 ha **b** ≈ 1.265 acres **c** ≈ 88.955 acres **d** $\approx 33\,185.4$ m^2

5 **a** 12% **b** 15% **c** 56.25% **d** $\approx 65.38\%$

6 100 rectangles **7** 2100 chickens

8 **a** ≈ 2.381 ha **b** ≈ 5.883 acres

EXERCISE 3C

1 **a** 250 mm^3 **b** $83\,000$ cm^3 **c** 0.598 cm^3 **d** $0.009\,81$ m^3 **e** 0.6351 m^3 **f** $81\,500$ mm^3

2 **a** 21.25% **b** $\approx 4.29\%$

3 ≈ 2608.7 cm^3 **4** 120 cm^3

EXERCISE 3D

1 **a** Ml **b** ml **c** kl **d** litres **e** ml

2 **a** 3760 ml **b** 47.32 kl **c** 3500 litres **d** 423 ml **e** 54 000 ml **f** 0.058 34 kl

3 **a** 18.75% **b** $\approx 6.67\%$

4 68% **5** 13 750 bottles **6** 9 tanks

7 **a** 25 ml **b** 3200 m^3 **c** 7320 litres **8** 1.8 litres

EXERCISE 3E

1 a kg b mg c g d t e mg
2 a 7000 mg b 0.007 kg c 0.58 kg
  d 580 000 g e 56 000 mg f 450 mg
  g 3.2 kg h 1870 kg i 0.047 835 kg
  j 4.653 g k 2 830 000 g l 63 200 g
  m 0.074 682 t n 1.7×10^9 mg o 91.275 kg
3 16%   **4** $\approx 16.67\%$ **5** 5 g   **6** 1.56 tonnes
7 150 000 lollies   **8** a 5.136 t b £2311.20

EXERCISE 3F.1

1 a 2700 s b 2280 s c 4200 s d 238 s
  e 7528 s f 19 187 s
2 a 210 min b 24 min c 313 min d 4428 min
3 a $\approx 8.67\%$ b 30%   **4** 36 minutes
5 a 52 days b 16 days c 1461 days d 0.25 days
6 a 744 h b 44 640 min c 86 400 seconds
7 a 90 years b 700 years c 2000 years
  d 550 years e 25 years f 3750 years
8 a 4 centuries b 572 centuries
9 a 51.15 centuries b 5.115 millennia
10 a 5 h 8 min b 7 h 11 min 10 s
  c 4 h 5 min d 2 h 35 min
11 a 2 h 25 min b 7 h 44 min c 5 h 48 min
  d 6 h 35 min e 4 h 50 min f 3 h 37 min
  g 6 h 14 min h 14 h 26 min
12 a 53 min b 9:03 am
13 a 10:15 am b 4:56 am c 4:20 pm
  d 8:05 pm the previous day e 12:32 pm
  f 8:30 pm the previous day
14 6:50 am   **15** 3 h 17 min
16 a 11:17 pm on January 13th b 2 days 11 h 14 min

EXERCISE 3F.2

1 a 0957 b 1106 c 1600 d 1425 e 0800
  f 0106 g 2058 h 1200 i 0002
2 a 11:40 am b 3:46 am c 4:34 pm d 7:00 pm
  e 8:00 am f 11:30 pm g 12:23 pm h 8:40 pm
3 a 2 h 40 min b 9 h 25 min c 2 h 50 min
  d 6 h 25 min e 11 h 25 min
4 13 h 41 min
5 So there is no confusion between am and pm.

REVIEW SET 3A

1 a 3280 m b 75.5 cm c 0.32 m
2 a 19.5 cm^2 b 64 000 cm^2 c 2.5 km^2
3 a 2.6 cm^3 b 8 m^3 c 1 200 000 cm^3
  d 56 ml e 4 litres f 2700 cm^3
4 a 0.056 g b 0.45 kg c 250 kg
5 40 000 staples   **6** 3 trips
7 a 2500 ml b 4 000 000 litres c 0.038 kl
8 a height in metres, area of base in m^2, thickness of concrete base in cm, weight in tonnes, capacity in kl
  b The width of concrete is a much smaller measurement than the height of the fountain, so we use smaller units for width.
  c 3600 kg d 400 litres

9 a 1800 s b 287 s c 8220 s   **10** 744 h
11 a 4 h 42 min b 8 h 25 min
12 a 48.44 centuries b 4.844 millennia   **13** 4:20 pm
14 a 0609 b 1319 c 2246   **15** 8 h 47 min

REVIEW SET 3B

1 a 1.56 km b 265 mm c 180 cm
2 7.5 km   **3** $\approx 11.95\%$ **4** 50 ml
5 a 0.35 g b 0.25 t c 16 800 g
6 a 540 mm^2 b 5.6 m^2 c 800 000 m^2
  d 600 mm^3 e 18 000 cm^3 f 0.025 m^3
7 9000 cartons   **8** 1.2 m^3
9 a ≈ 50.5875 ha b $\approx 505\,875$ m^2 **10** 7.5 t
11 a 6970 kg b 4.572 m c 50.8 cm
12 a 15 days b 22 days c 2922 days
13 a 5:20 pm b 10:37 am   **14** 11 h 4 min
15 a 9:26 am b 3:40 pm c 9:46 pm   **16** 2 h 36 min

EXERCISE 4A.1

1 a 20 b 70 c 70 d 100
  e 350 f 560 g 410 h 600
2 a 100 b 700 c 600 d 900
  e 300 f 1000 g 13 400 h 10 100
3 a 3000 b 3000 c 9000 d 12 000
  e 10 000 f 31 000 g 50 000 h 92 000
4 a 27 500 b 6820 c 704 000 d 21 100 000
5 a 440 b 2100 c 264 000 d 30 000
  e 49 000 f 10 000 g 43 200 h 2 141 000
  i 110 000 j 370 000 k 7 207 000 l 4 010 000
6 a £190 b £19 000 c 380 km
  d 800 ft e 29 000 people f 10 m
  g 9 kl h £270 000 i 500 000 sheep
7 a 190 000 000 swallows b 187 000 000 swallows
8 a 5 295 000 people b 5 300 000 people

EXERCISE 4A.2

1 a 1 b 7 c 8 d 12 e 128
2 a 2.4 b 3.6 c 4.9 d 6.4 e 4.3
3 a 4.24 b 2.73 c 5.63 d 4.38 e 6.52
4 a 0.2 b 0.18 c 0.184 d 0.1838
5 a 42 b 6.24 c 0.046 d 0.25
  e 10.3 f 1.0 g 0.037 64 h 0.0076
  i 70 j 0.0059 k 3.00 l 0.000 35
6 a 4.3 b 9.13 c 0.8 d 0.05 e 0.73
  f 0.002 g 0.5 h 6.17 i 2.429
7 a 2.64 b 6.47 c 101.00 d 396.89 e 1.19
  f 2767.70 g 202.33 h 50.15 i 1.99
8 7.45 cm $\approx$ 7.5 cm is correct, however Julie should have used the original value of 7.45 cm to round to the nearest integer, not 7.5. So 7.45 cm $\approx$ 7 cm.

EXERCISE 4B

1 a $\pm\frac{1}{2}$ cm b $\pm\frac{1}{2}$ ml c ± 50 ml d ± 250 g
2 between 67.5 kg and 68.5 kg
3 a $7.25 \leqslant x < 7.35$ b $11.455 \leqslant x < 11.465$
  c $275 \leqslant x < 285$
4 between 36.35°C and 36.45°C

5 7.845 km **6** 25 499 people
7 **a** 6.4 m **b** 6.05 m **c** 10 cm
8 between 520 m and 600 m **9** $19.4 \text{ kg} \leqslant w < 19.6 \text{ kg}$
10 $1.5 \text{ s} \leqslant d < 1.7 \text{ s}$ **11** 15 buckets
12 $246 \text{ cm} \leqslant p < 250 \text{ cm}$ **13** $788 \text{ cm} < l < 792 \text{ cm}$

EXERCISE 4C

1 **a** 320 **b** 400 **c** 420 **d** 540 **e** 900
 f 4200 **g** 3600 **h** 3000 **i** 14 000
2 **a** 2400 **b** 4200 **c** 9000 **d** 15 000 **e** 28 000
 f 150 000 **g** 90 000 **h** 360 000 **i** 720 000
3 **a** 12 **b** 45 **c** 70 **d** 12 **e** 18 **f** 300
4 **a** 100 **b** 1000 **c** 10 000 **d** 300 **e** 2000
 f 200 **g** 75 **h** 250 **i** 2000
5 **a** 4 **b** 3 **c** 5
6 **a** 9291 **b** 62 382 **c** 347 723 **d** 36
7 **a** 15.875 286 2
 b $\dfrac{8^2 - 4 \times 4}{3} = 16$, yes the answer in part **a** is reasonable.
8 12 000 words **9** 16 000 bricks **10** 80 minutes
11 10 000 vines **12** 50 km h^{-1}
13 **a** £200 **b** £600 **c** £160 **d** £30
 e £50 **f** £120 000
14 **a** £1000
 b **a** is an over-estimate as both numbers were rounded up.
15 £170 **16** £850 **17** £3000

REVIEW SET 4A

1 **a** 3580 **b** 3600 **c** 4000
2 **a** 390 km **b** 4 kl **c** 70 000 students
3 **a** £13.70 **b** £14
4 **a** 29 **b** 28.9 **c** 28.91
5 **a** 3.1 **b** 3.14
6 **a** 42.54 **b** 2.02 **c** 0.07
7 **a** $10.25 \leqslant x \leqslant 10.35$ **b** $2.085 \leqslant x < 2.095$
8 $48 \text{ cm} \leqslant p < 52 \text{ cm}$
9 **a** accurate within ± 0.5 cm
 b between 35.5 cm and 36.5 cm
10 **a** £160 **b** £280
11 **a** 600 **b** 24 000 **c** 50
12 2800 tiles **13** £10 000

REVIEW SET 4B

1 **a** 4610 **b** 4600 **c** 5000
2 **a** 700 km **b** 20 000 tonnes **c** 160 000 people
3 **a** £69.75 **b** £173
4 **a** 55 **b** 55.0 **c** 55.04
5 **a** 0.727 **b** 0.17
6 **a** 0.79 **b** 4.60 **c** 2286.76
7 $6.815 \text{ kg} \leqslant w < 6.825 \text{ kg}$ **8** $85.8 \text{ cm} \leqslant p < 86.2 \text{ cm}$
9 **a** £36 **b** £60
10 **a** 540 **b** 800 **c** 50
11 2000 minutes ($\approx$ 33 h 20 min) **12** $\approx$ 7 h 17 min

EXERCISE 5A

1 **a** $5x$ **b** $2c$ **c** $7q$ **d** $4fg$ **e** $6pq$ **f** $9rs$
 g $6ab$ **h** $4mn$ **i** $5ab$ **j** $2pq$ **k** jkl **l** dhp

2 **a** $pq + r$ **b** $4x + 5y$ **c** $2a - b$ **d** $ab - c$
 e $b - ac$ **f** $f - 7g$ **g** $ac + ad$ **h** $12 - 6rs$
 i $3(x + y)$ **j** $5(d - 1)$ **k** $8(w - x)$ **l** $pq(r - 2)$
3 **a** $2b$ **b** $3q$ **c** $2x + 4y$ **d** $3c + e$
 e $3 + 2y + z$ **f** $4a + 7$ **g** $4g + 2$ **h** $3 - 3d$
 i s **j** $s - 2t$ **k** $5 + 3r$ **l** $2 + 2a + 2b$
4 **a** $a \times a \times a \times a$ **b** $f \times f$
 c $4 \times p \times p \times p$ **d** $3 \times t \times t \times t \times t \times t$
 e $5 \times x \times x \times y$ **f** $7 \times f \times f \times g \times g \times g$
 g $5 \times a \times 5 \times a$ **h** $5 \times a \times a$
 i $p \times p + 2 \times q$ **j** $p \times p \times p - 3 \times q \times q$
5 **a** $3k^2$ **b** $4a^3$ **c** $2d^4$ **d** $4pq^2$
 e $3f^2 g^2$ **f** $w^2 xy^3$ **g** $m + m^2$ **h** $n^3 + n$
 i $y^2 - z^4$ **j** $a^2 + 7a$ **k** $8b - b^3$ **l** $2pq^2 + 6r^2 s$
 m $2h^2 - hj$ **n** $3x + 5x^3$ **o** $a^2 + 2b^3 - ab^2$
6 **a** $6y$ **b** $12x^2$ **c** $12a^2 c$ **d** $9d^2$
 e $6s^2 t^2$ **f** $2a^4$ **g** $16y^3$ **h** $12g^2$
 i $12a^3$ **j** $36b^5$ **k** $-3x^2$ **l** $-2x^3$

EXERCISE 5B

1 **a** 3 **b** -8 **c** 1 **d** -1 **e** 4 **f** -5
 g 3 **h** -2
2 **a** 5 **b** -5 **c** 14 **d** -7 **e** -1 **f** 6
 g 3 **h** -2
3 **a** expression **b** 6
 c **i** 2 **ii** 5 **iii** -7 **iv** -2 **d** 1
4 **a** 3 **b** 4 **c** 5
5 **a** expression **b** equation **c** equation
 d equation **e** expression **f** expression
6 **a** $3x$ and $5x$ **b** $2y$ and $-3y$ **c** $2x$ and $-x$
 d $-2y$ and $3y$ **e** $-5x$ and x **f** x^2 and $-7x^2$
 g $2x$ and $-5x$ **h** $5x$ and $\dfrac{x}{2}$

EXERCISE 5C

1 **a** $x + 6$ **b** $q + 11$ **c** $3b + 3$ **d** $2a + 7$ **e** $2d$
 f $2q + 5$ **g** 2 **h** $-2k + 3$ **i** $-2 - p$
2 **a** $2y$ **b** $3z$ **c** $2g^2$ **d** cannot be simplified **e** w^2
 f cannot be simplified **g** $2x$ **h** $9ab$ **i** $4m$
3 **a** 0 **b** $7p$ **c** cannot be simplified **d** $6pq$
 e $4ab$ **f** $2q^2$ **g** $12w$ **h** $13xy$ **i** $3z$ **j** 0
 k $9d - 9$ **l** $7g - 7g^2$ **m** $4s + 4s^2$
 n cannot be simplified **o** cannot be simplified
4 **a** $10a$ **b** $-2a$ **c** $2a$ **d** $-10a$
 e $8x$ **f** $6x$ **g** $-6x$ **h** $-8x$
 i cannot be simplified **j** $-13d$ **k** $-3d$
 l $3d$ **m** $2 - 2b$ **n** $-2t$ **o** $-m - 7$
5 **a** $9x$ **b** x **c** $-9x$
 d cannot be simplified **e** $5k - 5$ **f** $-10n$
 g $-15m$ **h** $4 - 5j$ **i** $7y$ **j** $7y$
 k $4y$ **l** $8y$
6 **a** $-2x - 1$ **b** $5t + 3$ **c** $-2x - y$ **d** $6pq - 4$
 e $10cd$ **f** $-a$ **g** $5x^2 - 2$ **h** $-3n - 3$
 i $-5v - 5w$ **j** $-3x^2$ **k** $5a - 7b$ **l** $-5z - 7$
 m $p - 2pq$ **n** $-7mn - 2m$

EXERCISE 5D
1. **a** $9+2$ **b** $5+a$ **c** $m+3n$ **d** $d+e+f$
2. **a** 8×6 **b** $6p$ **c** $4mn$ **d** bde
3. **a** $\frac{6}{5}$ **b** $\frac{d}{3}$ **c** $\frac{m}{5n}$ **d** $\frac{p+q}{x}$
4. **a** $\frac{6+10}{2}$ **b** $\frac{9+d}{2}$ **c** $\frac{k+4v}{2}$ **d** $\frac{d+e+f}{3}$
5. **a** $8-5$ **b** $s-6$ **c** $8-p$
6. In each case we let the number be x.
   a $x+3$ **b** $x-5$ **c** $\frac{x}{2}$ **d** $3x$
   e $\frac{x}{4}$ **f** $12-x$ **g** $2x+1$ **h** $5x-6$
7. **a** $p+8$ **b** $g-3$ **c** $n+2$ **d** $c+4$
   e $x-3$ **f** $4f$ **g** $\frac{h}{3}$ **h** $2a+4$
   i $2p+14$ **j** xy^2 **k** $4^2 c \ (=16c)$
   l $(ab)^2$ **m** p^2+q^2
8. **a** ...... $4-s$ **b** ...... $(27-b)$ girls
   c ...... $y+1$ **d** ...... $x+1,\ x+2$
   e ...... $d+2$ **f** ...... $a-1,\ a-2$
   g ...... $m-1$ and $m+1$ **h** ...... $s+3$

EXERCISE 5E
1. **a** £100 **b** £$20a$ **c** £ad
2. **a** 8 years old **b** $(14-x)$ years old
3. **a** £55 **b** £$(100-15h)$ **c** £$(100-hp)$
4. $(w-6)$ kg 5. $(20+m-n)$ people
6. £$(0.6a+0.9p)$ 7. **a** $0.8x$ m **b** $(600-0.8x)$ m
8. **a** 45 km **b** st km
9. **a** 12 cupcakes **b** $\frac{c}{n}$ cupcakes 10. $(bm+nr)$ pounds

REVIEW SET 5A
1. **a** $7a$ **b** $3b$ **c** $2ab$ **d** $5gh$
2. **a** $a+3b$ **b** a^2+2a **c** b^2+2b **d** $8a$
   e $3a-a^2$ **f** $4x^2-x$
3. **a** $7 \times t \times t \times t \times t$ **b** $3 \times b \times 3 \times b \times 3 \times b$
   c $2 \times b \times b - 3 \times c \times c \times c$
4. **a** 3 **b** -2 **c** $\frac{1}{2}$
5. **a** equation **b** expression **c** expression **d** equation
6. **a** 5 terms **b** $12gh$ and $3gh$ **c** -2
   d 4 **e** $5g+15gh-2h+4$
7. **a** $7a-3$ **b** $6x$ **c** $5bc$
   d $2a+3b$ **e** $13k-6$ **f** $17f-5g$
8. **a** $\frac{3+6}{2}$ **b** $\frac{5+a}{2}$ **c** $\frac{p+q}{2}$ **d** $\frac{a^2}{2}$
9. **a** $h+3$ **b** $2g-5$ **c** $2m-4$ **d** $\frac{a^2}{2}$
10. **a** 12 apples **b** $(18-3a)$ apples **c** $(18-ad)$ apples
11. **a** $x+(x+1)$ **b** $(50x+20(x+4))$ pence
12. $(9-4x)$ m 13. $(60p+95b)$ pence

REVIEW SET 5B
1. **a** mn **b** rs **c** mnp **d** abc
2. **a** $2x^2-3$ **b** b^4 **c** a^3-a^2 **d** $12y^3$
   e $5b^2-2b$ **f** s^2-st
3. **a** $8x^3$ **b** $6a^3$ **c** $-18x^3$
4. **a** $4x$ and $-7x$ **b** 3 terms
5. **a** expression **b** 5 terms **c** 4
   d $3x^2$ and $-x^2$, $-2x$ and x
6. **a** 3 and -1 **b** x and $2x$
7. **a** cannot be simplified **b** $2p^2$ **c** $6b-3$
   d n^2+7n **e** $2-5z$ **f** $12u-5t$
8. **a** $3+x+z$ **b** $-6xy$ **c** $x-2$
9. **a** ...... $3+a$ **b** ...... $x+2,\ x+4$ **c** ...... $g+\frac{g}{2}$
10. **a** £14 **b** £$(20-3p)$ **c** £$(20-cp)$
11. $(20-2n+m)$ muffins 12. $(16+b)$ years old
13. $(6+k+n+8)$ km

EXERCISE 6A
1. D 2. **a** $9:7$ **b** $11:13$
3. **a** $8:9$ **b** $8:5$ **c** $5:9$ **d** $9:5$ **e** $8:5:9$
4. **a** $8:3$ **b** $3:7$ **c** $35:45$ **d** $300:50$
   e $500:3000$ **f** $400:2500$ **g** $9000:150$
   h $12:8000$ **i** $240:40$
5. **a** Darryl : Toby $= 2:3$
   b working : exercising $= 60:10$
   c flour : sugar : sultanas $= 4:3:1$
6. **a** $25:35:50$ **b** $72:77:68$ **c** $35:47.5:52$
   d $400:1200:850$

EXERCISE 6B
1. **a** $\frac{7}{9}$ **b** $\frac{2}{9}$ 2. **a** $\frac{8}{11}$ **b** $\frac{3}{11}$
3. **a** **i** $3:1$ **ii** $\frac{3}{4}$ **iii** 75%
   b **i** $3:7$ **ii** $\frac{3}{10}$ **iii** 30%
   c **i** $1:4$ **ii** $\frac{1}{5}$ **iii** 20%
4. **a** $\frac{4}{14}$ **b** 50% **c** 36 kg
5. **a** $\frac{8}{20}$ **b** 35% **c** £15
6. **a** **i** $\frac{2}{5}$ **ii** $\frac{3}{5}$ **b** 42%

EXERCISE 6C
1. **a** $3:4$ **b** $2:1$ **c** $1:3$ **d** $3:5$
   e $1:2$ **f** $7:4$ **g** $1:2$ **h** $3:4$
   i $5:4$ **j** $1:2:3$ **k** $20:1$ **l** $1:2:4$
2. **a** $12:7$ **b** $3:2:5$ **c** $2:1:4$ **d** $6:4:9$
3. **a** $7:10$ **b** $2:7$ **c** $7:2$ **d** $1:16$
   e $3:2$ **f** $5:9$ **g** $3:50$ **h** $2:17$
   i $3:8$ **j** $241:140$ **k** $7:80$ **l** $50:3$
4. **a** $1:4$ **b** $8:5$ **c** $8:1$ **d** $1:5$ **e** $20:3$
5. **a** $8:9:10$ **b** $1:2:3$ **c** $9:12:16$
6. **a** $1:3:2$ **b** $3:6:2$
7. **a** equal **b** not equal **c** equal
   d not equal **e** not equal **f** equal
8. **a** Annette $1:8$, Bert $1:10$, Claire $1:9$, Derek $1:8$
   b Annette and Derek

EXERCISE 6D
1. **a** $x=12$ **b** $x=3$ **c** $x=10$ **d** $x=28$ **e** $x=35$
   f $x=42$ **g** $x=96$ **h** $x=56$ **i** $x=36$
2. 16 litres 3. 1 kg 4. 1.25 litres

Answers 655

5 a 12 cm   **b** 24 cm   **6** 9 more child care workers
7 a 2 : 7   **b** 2700 terns   **8** 40 cola drinks
9 30 litres   **10 a** 16 lilies   **b** 6 sunflowers
11 a $\frac{9}{28}$ **b i** 3.2 kg   **ii** 5.6 kg

EXERCISE 6E

1 a 10 litres   **b** 5 litres
2 a i $\frac{4}{7}$ **ii** $\frac{3}{7}$ **b i** £20 000   **ii** £15 000
3 a 2500   **b** 1500
4 a £15   **b** 3 : 2
  c Pranay gets 12 kg, Samar gets 8 kg.
5 a 21 minutes   **b** 27 minutes
6 a For every 3 parts mashed avocado, the guacamole contains 2 parts yoghurt.
  b mashed avocado   **c i** $\frac{3}{5}$ **ii** $\frac{2}{5}$ **d** 100 g
  e 360 g avocado, 240 g yoghurt
7 £760   **8 a** 56 marbles   **b** Katrina: 24, Lee: 32
9 Barry: £150 000, Robin: £100 000, Maurice: £350 000
10 a 64 kg   **b** 80 kg   **c** 16 kg

EXERCISE 6F

1 a 200   **b** 1000   **c** 25 000   **d** 500 000
2 a 2.5 m   **b** 10 m   **c** 50 m   **d** 87.5 m
3 a 5 cm   **b** 17.5 cm   **c** 12 cm   **d** 23.3 cm
4 a 10 m   **b** 8 m
5 a AB = 180 m, BC = 96 m, CD = 220 m, DA = 104 m
  b 19 200 m²

6

Scale 1 : 200

7

Scale 1 : 2000

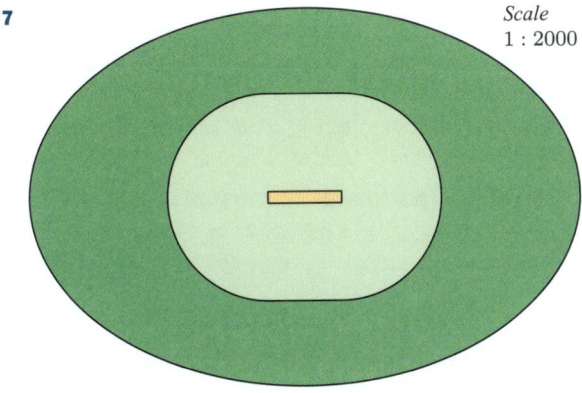

8

Scale 1 : 500

9 a

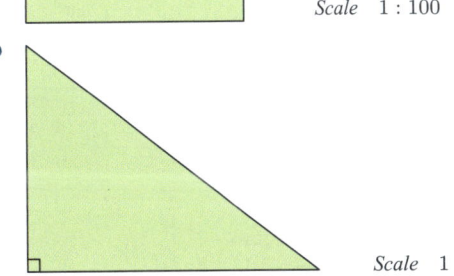

Scale 1 : 100

  b

Scale 1 : 1000

10 a 250
  b i 18 m   **ii** 6.25 m × 2.5 m   **iii** 11.25 m²
11 a i 80 km   **ii** 100 km   **b i** £12   **ii** £18
12 a 16 000 000   **b** 128 km
  c i 528 km   **ii** 496 km
13 a 1 000 000
  b i A → B → E → G, 125 km
   ii A → C → F → G, 140 km
   iii A → C → D → E → G, 140 km
14 a 250   **b** 2.5 m   **c** 79.5 m

REVIEW SET 6A

1 5 : 9   **2 a** $\frac{4}{11}$ **b** $\frac{7}{11}$
3 a 5 : 6   **b** 8 : 3   **c** 5 : 2   **4** 8 : 5   **5** 43%
6 a $x = 15$ **b** $x = 40$ **c** $x = 63$ **7** 24 lions
8 a James 200 km, Lucy 150 km
  b James 360 km, Lucy 270 km
9 £100 000, £80 000   **10 a** 48 min   **b** 150 min
11 a 20 m   **b** 15 m   **12 a** 100   **b** 75 cm

REVIEW SET 6B

1 a water : soft drink = 3 : 1
  b unplugged : recharging = 60 : 7
2 a 13 : 12   **b** $\frac{13}{25}$ **c** 52%
3 a 4 : 15   **b** 7 : 20   **4** 30 ml
5 a equal   **b** not equal   **6** £150, £350
7 a 15 safety personnel   **b** 160 competitors
8 84 p per litre   **9 a** 5000   **b** 1 000 000
10 25 t gravel, 15 t sand, 5 t cement   **11** 40 chocolates
12 a 38.8 m × 20 m   **b** 26 m   **c i** 2 m   **ii** 4 m
  d 5.6 m   **e** 5.2 m   **f** 33.44 m²   **g** 20.8 m

EXERCISE 7A

1 a $3^5 = 243$ **b** $2^4 = 16$ **c** $5^6 = 15\,625$
  d $3^7 = 2187$ **e** $7^5 = 16\,807$ **f** $(-2)^3 = -8$
  g x^2 **h** y^3 **i** a^5 **j** n^3 **k** x^9 **l** y^8
2 a $2^2 = 4$ **b** $3^2 = 9$ **c** $7^3 = 343$
  d $10^1 = 10$ **e** x^4 **f** y^4 **g** c^2 **h** b^3
3 a $2^6 = 64$ **b** $10^8 = 100\,000\,000$ **c** $3^6 = 729$
  d $2^{12} = 4096$ **e** x^{10} **f** p^9 **g** t^{12} **h** z^{12}
4 a $3^4 = 81$ **b** $5^5 = 3125$ **c** $7^2 = 49$
  d $2^8 = 256$ **e** $2^6 = 64$ **f** 11
  g $3^6 = 729$ **h** $5^6 = 15\,625$

5 **a** c^6 **b** b^5 **c** y^{15} **d** y^{10} **e** q^7 **f** z^{30}
g t^3 **h** a^{2+n} **i** g^3 **j** n^{10} **k** k^7 **l** p^6
6 **a** $(7^2)^3 = 7^6$ **b** $2^4 \div 2 = 2^3$ **c** $2^2 \times 2^7 = 2^9$
d $(x^3)^4 = x^{12}$ **e** $a^5 \times a^5 = a^{10}$ **f** $5^9 \div 5^3 = 5^6$
7 **a** $5a^2$ **b** $15q^3$ **c** $16x^3y^4$ **d** $7t$

EXERCISE 7B
1 **a** p^2q^2 **b** x^4y^4 **c** a^6b^6 **d** $a^3b^3c^3$
e $8a^3$ **f** $243d^5$ **g** $32k^5$ **h** $25g^2h^2$
2 **a** $\dfrac{a^2}{b^2}$ **b** $\dfrac{b^3}{8}$ **c** $\dfrac{j^4}{k^4}$ **d** $\dfrac{16}{z^4}$
e $\dfrac{16}{x^2}$ **f** $\dfrac{32}{b^5}$ **g** $\dfrac{q^4}{16}$ **h** $\dfrac{27}{b^3}$
3 **a** $\dfrac{4}{25}$ **b** $\dfrac{27}{64}$ **c** $\dfrac{16}{81}$ **d** $\dfrac{1}{32}$
4 **a** $4a^4$ **b** $9b^6$ **c** $16c^8$ **d** $32d^{10}$
e j^2k^6 **f** x^3y^6 **g** $18g^3$ **h** $25r^4s^2$
5 **a** $\dfrac{j^2k^2}{4}$ **b** $\dfrac{4}{c^2d^2}$ **c** $\dfrac{27p^3}{q^3}$ **d** $\dfrac{z^4}{25}$

EXERCISE 7C
1 **a** $\dfrac{3^4}{3^4} = \dfrac{81}{81} = 1$ **b** $\dfrac{3^4}{3^4} = 3^{4-4} = 3^0$
c Comparing the results of **a** and **b**, $3^0 = 1$.
2 **a** 1 **b** 1 **c** 1 $(x \neq 0)$ **d** 5 **e** 9 **f** 5
g 11 **h** p^6 $(p \neq 0)$ **i** 1 **j** 1 **k** 7 **l** 1
3 **a** 1 $(n \neq 0)$ **b** 1 $(k \neq 0)$ **c** x $(y \neq 0)$
d a^3 $(b \neq 0)$

EXERCISE 7D
1 **a** 3^{-2} **b** 5^{-1} **c** 2^{-2} **d** 3^{-2} **e** 7^{-1}
2 **a** $\tfrac{1}{5}$ **b** $\tfrac{1}{4}$ **c** $\tfrac{1}{8}$ **d** $\tfrac{1}{10}$ **e** $\tfrac{1}{9}$
f $\tfrac{1}{4}$ **g** $\tfrac{1}{121}$ **h** $\tfrac{1}{49}$ **i** $\tfrac{1}{27}$ **j** $\tfrac{1}{128}$
3 **a** $2\tfrac{2}{3}$ **b** $1\tfrac{1}{3}$ **c** $\tfrac{6}{7}$ **d** $5\tfrac{4}{5}$
4 **a** $\dfrac{1}{x}$ **b** $\dfrac{1}{k}$ **c** $\dfrac{1}{t^3}$ **d** $\dfrac{1}{r^5}$
5 **a** 3 **b** 5 **c** $\tfrac{7}{3}$ **d** $\tfrac{2}{5}$ **e** 10
f $\tfrac{36}{25}$ **g** $\tfrac{16}{81}$ **h** $\tfrac{27}{8}$
6 **a** $\dfrac{2}{x}$ **b** $\dfrac{1}{2x}$ **c** $\dfrac{1}{3q}$ **d** $\dfrac{3}{q}$ **e** $\dfrac{7}{a^2}$ **f** $\dfrac{1}{49a^2}$
g $\dfrac{1}{25z^2}$ **h** $\dfrac{5}{z^2}$ **i** $\dfrac{s}{t}$ **j** $\dfrac{1}{st}$ **k** $\dfrac{g}{h^3}$ **l** $\dfrac{1}{g^3h^3}$
m $\dfrac{4c}{d^2}$ **n** $\dfrac{1}{16c^2d^2}$ **o** $\dfrac{4}{c^2d^2}$ **p** $\dfrac{m^3}{7}$
7 **a** 2^2 **b** 2^{-2} **c** 3^2 **d** 3^{-2} **e** 2^4
f 2^{-4} **g** 5^2 **h** 5^{-2} **i** 3^4 **j** 3^{-4}
k 7^2 **l** 7^{-2} **m** 2^{-10} **n** 5^{-3}
o 2^0 or 3^0 or 5^0 or 7^0 **p** 2^{-a}

EXERCISE 7E.1
1 **a** 10^2 **b** 10^3 **c** 10^1 **d** 10^5
e 10^{-1} **f** 10^{-2} **g** 10^{-4} **h** 10^8
2 C
3 **a** $376 = 3.76 \times 10^2$ **b** $8000 = 8 \times 10^3$
c $0.04 = 4 \times 10^{-2}$ **d** $0.00507 = 5.07 \times 10^{-3}$
e $9\,040\,000 = 9.04 \times 10^6$ **f** $0.000\,000\,23 = 2.3 \times 10^{-7}$

4 **a** 4.25×10^2 **b** 4.25×10^5 **c** 4.25×10^0
d 4.25×10^{-1} **e** 2.01×10^1 **f** 2.01×10^4
g 2.01×10^{-3} **h** 2.01×10^6 **i** 3.87×10^3
j 3.87×10^{-2} **k** 3.87×10^7 **l** 3.87×10^{-4}
5 **a** 4.0075×10^4 km **b** 4×10^{-4} mm
c 4×10^7 bacteria **d** 1.4162×10^{-7}
e 1×10^{-2} mm **f** 1.5×10^7 °C
6 **a** 50 000 **b** 3000 **c** 18 000 000 **d** 810
e 650 000 **f** 11 **g** 275 000 000 **h** 8 000 000
7 **a** 0.03 **b** 0.00009 **c** 0.007
d 0.00041 **e** 0.0000082 **f** 0.761
g 0.000 000 325 **h** 0.000 000 02
8 **a** 0.000 000 475 m **b** 7 530 000 000 people
c 0.000 012 m **d** 40 100 km **e** 0.000 001 5 kg
9 **a** 1.817×10^7 **b** 9.34×10^{10} **c** 4.1×10^{-4}
10 **a** 602 000 000 000 000 000 000 000
b 6.02×10^{29} atoms **c** ≈ 1.66 tonnes
11 **a** 2.2×10^8 **b** 3.5×10^{-5} **c** 4.9×10^3
d 6.41×10^8 **e** 1.01×10^{-6} **f** 9×10^{-8}
12 **a** 9.11×10^{-28} g **b** 1.67×10^{-24} g **c** proton
13 **a** 6×10^7 **b** 2.4×10^7 **c** 4×10^{12}
d 8.1×10^{19} **e** 9×10^{-5} **f** 4.9×10^{-5}
g 4×10^{-1} **h** 2×10^{11} **i** 2.5×10^3
14 **a** 9×10^8 **b** 3.2×10^4 **c** 1.4×10^8
d 5×10^4 **e** 1.13×10^{-2} **f** 2.7×10^5
g 4.6×10^9 **h** 3.77×10^{-4} **i** 5.67×10^4
j 2.37×10^8 **k** 3.625×10^{-7} **l** 2.3×10^{-6}
15 **a** France **b** 2.5×10^{10} kg **c** 3.41×10^7 tonnes
d 2.81×10^6 tonnes **e** 400 times more

EXERCISE 7E.2
1 **a** 6.00×10^{-10} **b** 2.39×10^{10} **c** 2.19×10^{10}
d 4.01×10^{10} **e** 6.24×10^{-10} **f** 1.03×10^{-7}
2 **a** 9.55×10^9 **b** 9.62×10^{-7} **c** 4.90×10^{-14}
d 6.50×10^{-7} **e** 4.00×10^{-7} **f** 4.37×10^{14}
3 **a** 4.80×10^8 mm **b** 3.15×10^7 seconds
c 3.15×10^{10} seconds **d** 5×10^{-7} kg
4 **a** 8.64×10^4 km **b** 6.05×10^5 km **c** 6.31×10^7 km
5 **a** 1.8×10^{10} m **b** 2.59×10^{13} m **c** 9.46×10^{15} m

REVIEW SET 7A
1 **a** k^9 **b** b^{12} **c** p^8 **2** **a** 5^7 **b** 5^6 **c** 5^0
3 **a** $3b^2$ **b** $-3m^3$ **c** $6x^3y^3$
4 **a** $25x^2$ **b** $27m^3n^3$ **c** $\dfrac{p^4}{16q^4}$
5 **a** 1 **b** 13 **c** 6 **6** **a** $\tfrac{1}{9}$ **b** $\tfrac{1}{36}$ **c** $\tfrac{1}{1000}$
7 **a** 5^{-1} **b** b^9 **c** x^{-8}
8 **a** $\dfrac{1}{5c}$ **b** $\dfrac{7}{k^2}$ **c** $\dfrac{1}{64d^6}$
9 **a** 9×10^0 **b** 3.49×10^4 **c** 7.5×10^{-3}
10 **a** 2 810 000 **b** 2.81 **c** 0.00281
11 **a** 4.26×10^8 **b** 6×10^3 **c** 2.95×10^6
d 6.15×10^{-2}
12 **a** 2.57×10^6 km **b** 1.80×10^7 km **c** 9.37×10^8 km

REVIEW SET 7B

1 **a** x^6 **b** c^5 **c** d^{33}
2 **a** $3^8 = 6561$ **b** $2^0 = 1$ **c** y^{-3}
3 **a** $3^5 \div 3^2 = 3^3$ **b** $(x^3)^2 = x^6$ **c** $(4a)^3 \times a^3 = 64a^6$
4 **a** $-8x^3$ **b** $9m^4$ **c** $\dfrac{m^3}{64n^3}$
5 **a** -2 **b** 1 $(y \neq 0)$ **c** $\dfrac{27}{125}$
6 **a** $\dfrac{c^2}{d^2}$ **b** $\dfrac{q^3}{64}$ **c** $\dfrac{a^2b^2}{64}$
7 **a** $\dfrac{5}{4}$ **b** $\dfrac{49}{4}$ **c** $\dfrac{9}{100}$
8 **a** $\dfrac{b}{a^2}$ **b** $\dfrac{25}{a^4}$ **c** $\dfrac{p^3}{5}$
9 **a** 2.6357×10^2 **b** 5.11×10^{-4} **c** 8.634×10^8
10 **a** 2.78 **b** $39\,900\,000$ **c** $0.002\,081$
11 **a** 6.4×10^7 **b** 6×10^6 **c** 1.383×10^5
   d 5.72×10^{-3}
12 2.1×10^{-7} km

EXERCISE 8A

1 **a** $2x + 14$ **b** $3x - 6$ **c** $4a + 12$ **d** $5a + 5c$
   e $6b - 18$ **f** $7m + 28$ **g** $2n - 2p$ **h** $4p - 4q$
   i $15 + 3x$ **j** $5y - 5x$ **k** $8t - 64$ **l** $6d + 6e$
   m $40 - 4j$ **n** $7y + 7n$ **o** $2n - 24$ **p** $88 - 8d$
2 **a** $18x + 9$ **b** $3 - 9x$ **c** $10a + 15$ **d** $11 - 22n$
   e $18x + 6y$ **f** $5x - 10y$ **g** $12b + 4c$ **h** $2a - 4b$
   i $7a - 35b$ **j** $24 + 36d$ **k** $24 - 32y$ **l** $30b + 18a$
   m $22x - 11y$ **n** $7c - 63d$ **o** $6m + 42n$ **p** $64a - 8c$
3 **a** $x^2 + 2x$ **b** $5x - x^2$ **c** $2a^2 + 4a$ **d** $5b - 3b^2$
   e $ab + 2ac$ **f** $a^3 + a$ **g** $3x - 4x^2$ **h** $18x - 3x^2$
   i $5x^2 - 20x$ **j** $4a - 4a^2$ **k** $7b^2 + 14b$ **l** $a^2 + ab$
   m $3b - 8b^2$ **n** $m^2 - 3mn$ **o** $c^2 - 4ac$ **p** $24p - 42p^2$
4 **a** $-2x - 4$ **b** $-3x - 12$ **c** $-4x + 8$ **d** $-25 + 5x$
   e $-a - 2$ **f** $-x + 3$ **g** $-5 + x$ **h** $-2x - 1$
   i $-12 + 3x$ **j** $-20x + 8$ **k** $-15 + 20c$ **l** $-14 + 10x$
5 **a** $-a^2 - a$ **b** $-b^2 - 4b$ **c** $-5c + c^2$
   d $-2x^2 - 4x$ **e** $-2x + 2x^2$ **f** $-3y^2 - 6y$
   g $-20a + 4a^2$ **h** $-18b + 12b^2$ **i** $-2xy^2 + x^2y$
6 **a** $3x + 11$ **b** $7x + 2$ **c** $25 - 12x$ **d** $10x - 2$
   e $11 - 10x$ **f** $8 + 2x$ **g** $9 + 21x$ **h** $4x + 18$
   i $4x + 7$ **j** $6 + 2x$ **k** $-5x - 5$ **l** $-47 + 15x$
   m $7x + x^2$ **n** $7x + x^2$ **o** $4x - x^2$ **p** $2x^2 - 2x$
   q $7x + x^2$ **r** $x^2 + 4x$
7 **a** $5x + 6$ **b** $a + 3b$ **c** $-a + 5b$
   d $-2x + 26$ **e** -15 **f** $17n - 15$
   g $x - y$ **h** $a^2 + 7a - 15$ **i** $x^2 + 2x + 12$
   j $2a^2 + 3a$ **k** $-2a^2 + a$ **l** $x^2 - y^2$
   m $-2x + 16$ **n** $9x - 9$ **o** $5x^2 - 16x$

EXERCISE 8B

1 **a** $x^2 + 3x + 2$ **b** $x^2 + 6x + 8$ **c** $x^2 + 4x - 5$
   d $x^2 + 4x + 3$ **e** $y^2 + 5y + 6$ **f** $a^2 + 10a + 21$
   g $x^2 - 4$ **h** $x^2 - 2x - 8$ **i** $x^2 + 4x - 21$
   j $x^2 - 7x - 18$ **k** $x^2 - x - 12$ **l** $x^2 + 4x - 12$
   m $x^2 - 7x + 12$ **n** $x^2 - 13x + 40$ **o** $x^2 - 15x + 44$
   p $2x^2 + x - 3$ **q** $3x^2 - 10x - 8$ **r** $4x^2 - 2x - 12$
   s $12x^2 + 11x + 2$ **t** $-6x^2 - x + 1$
   u $-2x^2 + 7x + 30$ **v** $12x^2 - 5x - 3$
   w $-5x^2 + 16x + 16$ **x** $-2x^2 + 11x + 40$
2 **a** $x^2 + 2x + 1$ **b** $x^2 + 8x + 16$ **c** $x^2 - 4x + 4$
   d $x^2 - 10x + 25$ **e** $9 + 6y + y^2$ **f** $9 - 6y + y^2$
   g $4x^2 + 4x + 1$ **h** $4x^2 - 4x + 1$ **i** $1 + 8a + 16a^2$
   j $1 - 8a + 16a^2$ **k** $a^2 + 2ab + b^2$ **l** $a^2 - 2ab + b^2$
3 **a** $x^2 - 4$ **b** $y^2 - 25$ **c** $a^2 - 49$ **d** $b^2 - 16$
   e $9 - x^2$ **f** $36 - y^2$ **g** $1 - a^2$ **h** $64 - b^2$
   i $4x^2 - 1$ **j** $9a^2 - 4$ **k** $9 - 25b^2$ **l** $25 - 16y^2$

EXERCISE 8C

1 **a** $x^2 + 4x + 4$ **b** $a^2 + 6a + 9$ **c** $x^2 + 10x + 25$
   d $x^2 + 22x + 121$ **e** $4x^2 + 4x + 1$ **f** $4x^2 + 16x + 16$
   g $9x^2 + 12x + 4$ **h** $16x^2 + 24x + 9$ **i** $49x^2 + 14x + 1$
   j $4x^2 + 4xy + y^2$ **k** $36 + 12x + x^2$ **l** $9 + 30x + 25x^2$
2 **a** $(5+3)^2 = 8^2 = 64$, $5^2 + 3^2 = 25 + 9 = 34$
   b $5^2 + 2 \times 5 \times 3 + 3^2 = 25 + 30 + 9 = 64$
3 **a** $x^2 - 8x + 16$ **b** $x^2 - 2x + 1$ **c** $x^2 - 12x + 36$
   d $d^2 - 6d + 9$ **e** $16 - 8a + a^2$ **f** $49 - 14x + x^2$
   g $9x^2 - 6x + 1$ **h** $36 - 12d + d^2$ **i** $4x^2 - 20x + 25$
   j $9 - 24a + 16a^2$ **k** $9a^2 - 12ab + 4b^2$ **l** $9 - 30x + 25x^2$
4 **a** $4x^2 + 12x + 9$ **b** $16a^2 - 8a + 1$ **c** $9y^2 + 30y + 25$
   d $9a^2 - 24a + 16$ **e** $4x^2 - 28x + 49$
   f $64 + 48x + 9a^2$ **g** $4 + 20b + 25b^2$
   h $36 - 60x + 25x^2$ **i** $16 - 40y + 25y^2$

EXERCISE 8D

1 **a** $x^2 - 9$ **b** $x^2 - 9$ **c** $9 - x^2$ **d** $9 - x^2$
   e $x^2 - 4$ **f** $4 - x^2$ **g** $x^2 - 36$ **h** $a^2 - 16$
   i $b^2 - 1$ **j** $p^2 - q^2$ **k** $25 - n^2$ **l** $49 - y^2$
2 **a** $4x^2 - 1$ **b** $25x^2 - 4$ **c** $16a^2 - 9$
   d $9b^2 - 25$ **e** $16x^2 - 1$ **f** $1 - 16x^2$
   g $49 - 4y^2$ **h** $9 - 4x^2$ **i** $9x^2 - 4$
3 $31 \times 29 = (30+1)(30-1)$
   $= 30^2 - 1^2$
   $= 900 - 1$
   $= 899$
4 **a** $2x^2 + 2x - 15$ **b** $-2x - 5$ **c** $-2x^2 - 4x$
   d $5x^2 - 2x$

REVIEW SET 8A

1 **a** $2x + 22$ **b** $35x - 5x^2$ **c** $-12 + 8x$
2 **a** $x^2 - 2x - 24$ **b** $x^2 - 13x + 36$ **c** $2x^2 + x - 10$
3 **a** $x^2 + 16x + 64$ **b** $x^2 - 8x + 16$ **c** $49 - 28x + 4x^2$
4 **a** $x^2 - 49$ **b** $1 - a^2$ **c** $16x^2 - 25$
5 **a** $3x - 12$ **b** $-10x + 15$ **c** $x - x^2$
6 **a** $4x - 1$ **b** $-2x - 5$ **c** $-9x + 11$
7 **a** $x^2 + 11x + 28$ **b** $x^2 + 6x - 16$ **c** $-5x^2 - x + 6$
   d $2a^2 - 21a + 27$
8 **a** $x^2 + 8x - 6$ **b** $n - 9$
9 **a** $x^2 + 8x + 16$ **b** $x^2 - 20x + 100$
   c $4x^2 + 20x + 25$ **d** $9 - 24x + 16x^2$
10 **a** $x^2 - 81$ **b** $9x^2 - 4$ **c** $16 - 9x^2$

REVIEW SET 8B

1 **a** $3x - 24$ **b** $-12 + 6x$ **c** $2x^3 + 2x^2$
2 **a** $x^2 + 6x - 7$ **b** $y^2 - 5y - 24$ **c** $3x^2 + 8x - 3$
3 **a** $14 + 5x$ **b** $25 - 3x$ **c** $-3x + x^2$
4 **a** $x^2 + 12x + 36$ **b** $4x^2 - 20x + 25$ **c** $9x^2 - 49$
5 **a** $-3x - 12$ **b** $xy + 3x$ **c** $2a^2 - 10a$
6 **a** $x^2 + 5x - 36$ **b** $2x^2 + 5x - 42$ **c** $2a^2 - 11a + 12$
7 **a** $y^2 + 6y + 9$ **b** $9x^2 + 12x + 4$ **c** $16a^2 - 8ab + b^2$
8 **a** $25 - x^2$ **b** $9y^2 - 16$ **c** $36a^2 - 25$
9 **a** $-x^2 - 6x - 9$ **b** $y^2 + 5y - 40$
10 **a** $8x + 20$ **b** $3x^2 + 6x - 18$

EXERCISE 9A

1 **a** $x = 1$ **b** $x = 9$ **c** $x = 5$ **d** $x = 5$
   e $x = -2$ **f** $x = 4$ **g** $x = -4$ **h** $x = 13$
   i $x = 40$ **j** $x = -175$ **k** $x = -5$ **l** $x = -4$
2 **a** $x = 2$ **b** $x = 3$ **c** $x = -2$ **d** $x = 5$
   e $k = -4$ **f** $m = 1$

EXERCISE 9B

1 **a** $x = 5$ **b** $x = 9$ **c** $5x = 15$ **d** $7x = x + 6$
2 **a** $x = 4$ **b** $2x = 4$ **c** $3x = -3$ **d** $4x = 3x + 2$
3 **a** $x = 16$ **b** $x - 1 = 5$ **c** $3x = 14$ **d** $3x - 4 = -40$
4 **a** $x = -10$ **b** $x = -9$ **c** $2 - x = 5$ **d** $2x - 1 = 11$

EXERCISE 9C

1 **a** -3 **b** $+8$ **c** $\div 2$ **d** $\times 5$ **e** -12 **f** $\times 6$
   g $+5$ **h** $\div 9$ **i** $-\frac{2}{3}$ **j** $\times 13$ **k** $\div 15$ **l** $+\frac{4}{5}$
2 **a** x **b** x **c** x **d** x **e** p **f** q
   g $8r$ **h** s
3 **a** $x = 6$ **b** $x = 10$ **c** $x = 7$ **d** $x = 6$
   e $x = 6$ **f** $x = -2$ **g** $x = -35$ **h** $x = -11$
   i $x = -32$ **j** $x = -3$ **k** $x = -7$ **l** $x = -9$
   m $x = -9$ **n** $x = 4$ **o** $x = 9$ **p** $x = -18$
4 **a** $x = -4$ **b** $x = -27$ **c** $x = -4$ **d** $x = -90$
   e $x = 36$ **f** $x = 0$ **g** $x = 44$ **h** $x = 14$
5 **a** $a = -10$ **b** $b = 1\frac{1}{3}$ **c** $c = -9$ **d** $d = -3$
   e $e = 10$ **f** $f = 60$ **g** $z = -2$ **h** $w = -72$

EXERCISE 9D

1 **a** BU: $x \xrightarrow{\times 7} 7x \xrightarrow{+3} 7x + 3$
   UD: $7x + 3 \xrightarrow{-3} 7x \xrightarrow{\div 7} x$

b BU: $x \xrightarrow{+3} x + 3 \xrightarrow{\times 7} 7(x + 3)$
   UD: $7(x + 3) \xrightarrow{\div 7} x + 3 \xrightarrow{-3} x$

c BU: $x \xrightarrow{-2} x - 2 \xrightarrow{\times 5} 5(x - 2)$
   UD: $5(x - 2) \xrightarrow{\div 5} x - 2 \xrightarrow{+2} x$

d BU: $x \xrightarrow{\times 5} 5x \xrightarrow{-2} 5x - 2$
   UD: $5x - 2 \xrightarrow{+2} 5x \xrightarrow{\div 5} x$

e BU: $x \xrightarrow{\div 3} \frac{x}{3} \xrightarrow{+1} \frac{x}{3} + 1$
   UD: $\frac{x}{3} + 1 \xrightarrow{-1} \frac{x}{3} \xrightarrow{\times 3} x$

f BU: $x \xrightarrow{+1} x + 1 \xrightarrow{\div 3} \frac{x + 1}{3}$
   UD: $\frac{x + 1}{3} \xrightarrow{\times 3} x + 1 \xrightarrow{-1} x$

g BU: $x \xrightarrow{\div 8} \frac{x}{8} \xrightarrow{-5} \frac{x}{8} - 5$
   UD: $\frac{x}{8} - 5 \xrightarrow{+5} \frac{x}{8} \xrightarrow{\times 8} x$

h BU: $x \xrightarrow{-5} x - 5 \xrightarrow{\div 8} \frac{x - 5}{8}$
   UD: $\frac{x - 5}{8} \xrightarrow{\times 8} x - 5 \xrightarrow{+5} x$

i BU: $x \xrightarrow{\times 2} 2x \xrightarrow{-6} 2x - 6$
   UD: $2x - 6 \xrightarrow{+6} 2x \xrightarrow{\div 2} x$

j BU: $x \xrightarrow{\div -3} \frac{x}{-3} \xrightarrow{+10} \frac{x}{-3} + 10$
   UD: $\frac{x}{-3} + 10 \xrightarrow{-10} \frac{x}{-3} \xrightarrow{\times -3} x$

k BU: $x \xrightarrow{-7} x - 7 \xrightarrow{\times 8} 8(x - 7)$
   UD: $8(x - 7) \xrightarrow{\div 8} x - 7 \xrightarrow{+7} x$

l BU: $x \xrightarrow{-3} x - 3 \xrightarrow{\div 4} \frac{x - 3}{4}$
   UD: $\frac{x - 3}{4} \xrightarrow{\times 4} x - 3 \xrightarrow{+3} x$

2 **a** BU: $x \xrightarrow{\times 3} 3x \xrightarrow{+2} 3x + 2 \xrightarrow{\div 5} \frac{3x + 2}{5}$
   UD: $\frac{3x + 2}{5} \xrightarrow{\times 5} 3x + 2 \xrightarrow{-2} 3x \xrightarrow{\div 3} x$

b BU: $x \xrightarrow{\times 3} 3x \xrightarrow{\div 5} \frac{3x}{5} \xrightarrow{+2} \frac{3x}{5} + 2$
   UD: $\frac{3x}{5} + 2 \xrightarrow{-2} \frac{3x}{5} \xrightarrow{\times 5} 3x \xrightarrow{\div 3} x$

c BU: $x \xrightarrow{+2} x + 2 \xrightarrow{\times 3} 3(x + 2) \xrightarrow{\div 5} \frac{3(x + 2)}{5}$
   UD: $\frac{3(x + 2)}{5} \xrightarrow{\times 5} 3(x + 2) \xrightarrow{\div 3} x + 2 \xrightarrow{-2} x$

Answers   **659**

d BU: $\boxed{x} \xrightarrow{\times 7} \boxed{7x} \xrightarrow{-1} \boxed{7x-1} \xrightarrow{\div 6} \boxed{\dfrac{7x-1}{6}}$

UD: $\boxed{\dfrac{7x-1}{6}} \xrightarrow{\times 6} \boxed{7x-1} \xrightarrow{+1} \boxed{7x} \xrightarrow{\div 7} \boxed{x}$

e BU: $\boxed{x} \xrightarrow{\times 7} \boxed{7x} \xrightarrow{\div 6} \boxed{\dfrac{7x}{6}} \xrightarrow{-1} \boxed{\dfrac{7x}{6}-1}$

UD: $\boxed{\dfrac{7x}{6}-1} \xrightarrow{+1} \boxed{\dfrac{7x}{6}} \xrightarrow{\times 6} \boxed{7x} \xrightarrow{\div 7} \boxed{x}$

f BU: $\boxed{x} \xrightarrow{-1} \boxed{x-1} \xrightarrow{\times 7} \boxed{7(x-1)} \xrightarrow{\div 6} \boxed{\dfrac{7(x-1)}{6}}$

UD: $\boxed{\dfrac{7(x-1)}{6}} \xrightarrow{\times 6} \boxed{7(x-1)} \xrightarrow{\div 7} \boxed{x-1} \xrightarrow{+1} \boxed{x}$

g BU: $\boxed{x} \xrightarrow{\times 5} \boxed{5x} \xrightarrow{\div 6} \boxed{\dfrac{5x}{6}} \xrightarrow{-3} \boxed{\dfrac{5x}{6}-3}$

UD: $\boxed{\dfrac{5x}{6}-3} \xrightarrow{+3} \boxed{\dfrac{5x}{6}} \xrightarrow{\times 6} \boxed{5x} \xrightarrow{\div 5} \boxed{x}$

h BU: $\boxed{x} \xrightarrow{-3} \boxed{x-3} \xrightarrow{\times 5} \boxed{5(x-3)} \xrightarrow{\div 6} \boxed{\dfrac{5(x-3)}{6}}$

UD: $\boxed{\dfrac{5(x-3)}{6}} \xrightarrow{\times 6} \boxed{5(x-3)} \xrightarrow{\div 5} \boxed{x-3} \xrightarrow{+3} \boxed{x}$

i BU: $\boxed{x} \xrightarrow{\times 5} \boxed{5x} \xrightarrow{-3} \boxed{5x-3} \xrightarrow{\div 6} \boxed{\dfrac{5x-3}{6}}$

UD: $\boxed{\dfrac{5x-3}{6}} \xrightarrow{\times 6} \boxed{5x-3} \xrightarrow{+3} \boxed{5x} \xrightarrow{\div 5} \boxed{x}$

j BU: $\boxed{x} \xrightarrow{\times -2} \boxed{-2x} \xrightarrow{\div 3} \boxed{\dfrac{-2x}{3}} \xrightarrow{+1} \boxed{1-\dfrac{2x}{3}}$

UD: $\boxed{1-\dfrac{2x}{3}} \xrightarrow{-1} \boxed{\dfrac{-2x}{3}} \xrightarrow{\times 3} \boxed{-2x} \xrightarrow{\div -2} \boxed{x}$

k BU: $\boxed{x} \xrightarrow{\times -2} \boxed{-2x} \xrightarrow{+1} \boxed{1-2x} \xrightarrow{\div 3} \boxed{\dfrac{1-2x}{3}}$

UD: $\boxed{\dfrac{1-2x}{3}} \xrightarrow{\times 3} \boxed{1-2x} \xrightarrow{-1} \boxed{-2x} \xrightarrow{\div -2} \boxed{x}$

l BU:
$\boxed{x} \xrightarrow{\times -1} \boxed{-x} \xrightarrow{+1} \boxed{1-x} \xrightarrow{\times 2} \boxed{2(1-x)} \xrightarrow{\div 3} \boxed{\dfrac{2(1-x)}{3}}$

UD:
$\boxed{\dfrac{2(1-x)}{3}} \xrightarrow{\times 3} \boxed{2(1-x)} \xrightarrow{\div 2} \boxed{1-x} \xrightarrow{-1} \boxed{-x} \xrightarrow{\div -1} \boxed{x}$

EXERCISE 9E

1 **a** $x=2$ **b** $x=5$ **c** $x=4$ **d** $x=-8$
   e $x=4$ **f** $x=\frac{3}{8}$ **g** $x=1\frac{1}{2}$ **h** $x=17$
   i $x=-2$ **j** $x=2$ **k** $x=2$ **l** $x=-4\frac{1}{4}$

2 **a** $x=4$ **b** $x=22$ **c** $x=16$ **d** $x=9$
   e $x=-21$ **f** $x=-24$ **g** $x=63$ **h** $x=-56$
   i $x=22$

3 **a** $x=4$ **b** $x=-2$ **c** $x=4$ **d** $x=5$
   e $x=2$ **f** $x=7$ **g** $x=15$ **h** $x=7$

   i $x=4$

4 **a** $x=11$ **b** $x=9$ **c** $x=3\frac{1}{2}$ **d** $x=-7$
   e $x=-4$ **f** $x=-28$ **g** $x=-5$ **h** $x=2$
   i $x=-11$

5 **a** $x=10$ **b** $x=2$ **c** $x=4\frac{1}{2}$ **d** $x=-\frac{2}{3}$
   e $x=-3$ **f** $x=0$ **g** $x=1\frac{1}{3}$ **h** $x=\frac{1}{2}$
   i $x=0$

6 **a** $a=3$ **b** $x=448$ **c** $x=5$ **d** $x=1$
   e $n=4$ **f** $a=-8$ **g** $x=2\frac{1}{2}$ **h** $x=75$
   i $x=3$ **j** $n=-3$ **k** $k=7$ **l** $z=-\frac{4}{5}$

EXERCISE 9F

1 **a** $x=2$ **b** $x=-5$ **c** $x=15$ **d** $y=0$
   e $x=3$ **f** $m=\frac{5}{2}=2\frac{1}{2}$ **g** $x=2\frac{8}{9}$ **h** $p=15$
   i $t=6$ **j** $d=-10$ **k** $x=-6$ **l** $x=12$

2 **a** $x=6$ **b** $x=2$ **c** $x=-4$ **d** $x=3$
   e $x=4$ **f** $x=-2\frac{1}{2}$

3 **a** $x=1$ **b** $x=1$ **c** $x=\frac{1}{10}$ **d** $x=4$
   e $x=1$ **f** $x=-1\frac{1}{3}$

4 **a** $x=3$ **b** $t=5$ **c** $x=1$ **d** $y=\frac{1}{2}$
   e $a=3$ **f** $p=-3$

5 **a** $7(a+3)=21+7a$ reduces to $0=0$.
   b The equation is true for all real values of a. So, infinitely many values of a satisfy this equation.

6 **a** We get $3=4$ which is absurd. So, this equation has no solution.
   b No values of a satisfy the equation.

7 **a** $x=\frac{1}{3}$ **b** $x=2\frac{2}{9}$ **c** $x=3$ **d** $p=-\frac{1}{4}$

8 **a** $x=-8$ **b** $x=1\frac{3}{5}$ **c** $x=-1$ **d** $x=-2$
   e $x=5$ **f** $x=-1$ **g** $x=1$ **h** $x=-2$

EXERCISE 9G

1 **a** $x=\frac{10}{9}$ **b** $x=2$ **c** $x=-3$ **d** $x=\frac{13}{11}$
   e $x=\frac{24}{19}$ **f** $x=\frac{7}{8}$

2 **a** $x=\frac{1}{3}$ **b** $x=-\frac{1}{4}$ **c** $x=\frac{3}{2}$ **d** $x=-\frac{1}{6}$

3 **a** $x=\frac{21}{2}$ **b** $x=\frac{36}{5}$ **c** $x=\frac{10}{3}$ **d** $x=-\frac{12}{5}$
   e $x=\frac{7}{6}$ **f** $x=\frac{6}{5}$ **g** $x=-28$ **h** $x=-18$

4 **a** $x=-2$ **b** $x=\frac{8}{3}$ **c** $x=-\frac{1}{2}$

5 **a** $x=-1$ **b** $x=-\frac{7}{2}$ **c** $x=\frac{1}{4}$

EXERCISE 9H

1 The number is 9.   **2** The number is 5.
3 The number is 8.   **4** The numbers are 86 and 87.
5 The smallest integer is 35.   **6** The number is 91.
7 The number is $4\frac{1}{4}$.   **8** 7 packs   **9** 6 shirts
10 There are 7 passengers in each row.
11 Paige ate 5 sweets.   **12** Carol drank 580 ml of water.
13 Emma scored 37 runs, Alex scored 42 runs, and Toni scored 14 runs.
14 Isaac buys 6 shirts.   **15** eight 5-cent coins
16 6 apples   **17** Ellie's son is 10 years old now
18 Adrian is 7 years old now

Answers — p.660

EXERCISE 9I

1 a $S \leqslant 40$ **b** $A \geqslant 18$ **c** $a > 3$ **d** $b \leqslant -3$
  e $d < 5$ **f** $-20 \geqslant x$ **g** $4 < y$ **h** $z \geqslant 0$

2 a $x > 2$ **b** $b < 5$ **c** $c \geqslant 2\frac{1}{2}$ **d** $d \leqslant -7$
  e $a < -19$ **f** $p > -3$

3 a–**h** (number line graphs)

4 a true **b** false **c** false **d** false **e** true **f** false

5 a (number line graph) **b** $x = 3$

6 a 16 **b** 55

EXERCISE 9J

1 a $a > 2$ **b** $b \leqslant -3$
  c $s < 6$ **d** $c < 10$
  e $x \geqslant 5$ **f** $b < -4$
  g $t > -5$ **h** $k < -6$
  i $m \leqslant -60$

2 a $x > 3$ **b** $m \leqslant 4$
  c $a \geqslant \frac{10}{3}$ **d** $a < -2$
  e $b > 5$ **f** $s > -2$
  g $a \leqslant 2$ **h** $b < 3$

  i $b \geqslant \frac{1}{3}$ **j** $n < -2$
  k $x < 13$ **l** $b \geqslant \frac{7}{4}$

3 a $x > 9$ **b** $b \leqslant -5$
  c $c \geqslant 16$ **d** $x < 9$
  e $x \geqslant -\frac{14}{3}$ **f** $x > -6$
  g $x \leqslant 20$ **h** $x > \frac{5}{2}$
  i $x < \frac{28}{3}$

4 a $a < 15$ **b** $b \geqslant -1$
  c $c \leqslant -2$ **d** $a < -6$
  e $x \geqslant \frac{17}{3}$ **f** $x \leqslant 3$

5 a $x < 1$ **b** $c \geqslant -2$ **c** $b < -2$ **d** $a \geqslant 21$
  e $d > 14$ **f** $p \leqslant 0$

6 a $x > 2$ **b** $x \leqslant -3$
  c $x < -1$ **d** $x < \frac{4}{5}$
  e $x \leqslant -3$ **f** $x \geqslant -2$

7 a i $x = 10$ **ii** $x = -7$ **iii** $x = 4$
  b i $x = 9$ **ii** $x = 29$ **iii** $x = -4$

REVIEW SET 9A

1 $\times 6$ **2 a** $3x = -9$ **b** $4x = -10$
3 $x = -3$ **4** $x = -6$
5 a $x = 15$ **b** $x = 7$ **c** $a = -13$ **d** $t = -21$

Answers 661

6 a $\boxed{x} \xrightarrow{\div 6} \boxed{\dfrac{x}{6}} \xrightarrow{+1} \boxed{\dfrac{x}{6}+1}$

b $\boxed{x} \xrightarrow{\times 3} \boxed{3x} \xrightarrow{-4} \boxed{3x-4} \xrightarrow{\times 4} \boxed{4(3x-4)}$

c $\boxed{x} \xrightarrow{\times -4} \boxed{-4x} \xrightarrow{+2} \boxed{2-4x} \xrightarrow{\div 3} \boxed{\dfrac{2-4x}{3}}$

7 a $\boxed{\dfrac{x}{7}-3} \xrightarrow{+3} \boxed{\dfrac{x}{7}} \xrightarrow{\times 7} \boxed{x}$

b $\boxed{4(x+1)} \xrightarrow{\div 4} \boxed{x+1} \xrightarrow{-1} \boxed{x}$

c $\boxed{1+3x} \xrightarrow{-1} \boxed{3x} \xrightarrow{\div 3} \boxed{x}$

8 a $x=2$ **b** $x=6$ **c** $x=-3$ **d** $x=-2$
  e $x=1$ **f** $x=-2$

9 a $x=5$ **b** $x=\tfrac{1}{2}$ **c** $x=-9$ **d** $x=\tfrac{3}{5}$
  e $x=1$ **f** $x=\tfrac{6}{7}$

10 a $x=\tfrac{26}{3}$ **b** $x=\tfrac{14}{15}$ **c** $x=-\tfrac{5}{2}$

11 The number is 13. **12** The smallest integer is 20.

13 The number is 7. **14** twelve 5 p coins

15 a  **b** (number line: 2 to 4, open circle)

16 a $x \leqslant \tfrac{7}{2}$ **b** $x < -\tfrac{9}{2}$
  (number lines)

c $x \leqslant -\tfrac{4}{7}$
  (number line)

17 a $x > -10$ **b** $x \leqslant \tfrac{1}{3}$
  (number lines)

c $-10 < x \leqslant \tfrac{1}{3}$

REVIEW SET 9B

1 a dividing by 5 **b** adding 7

2 a $a=7$ **b** $b=-9$ **c** $c=-10$ **d** $d=-96$

3 a $\boxed{x} \xrightarrow{+3} \boxed{x+3} \xrightarrow{\div 2} \boxed{\dfrac{x+3}{2}}$

b $\boxed{x} \xrightarrow{\times 5} \boxed{5x} \xrightarrow{-9} \boxed{5x-9}$

c $\boxed{x} \xrightarrow{\div 7} \boxed{\dfrac{x}{7}} \xrightarrow{-2} \boxed{\dfrac{x}{7}-2}$

4 a $\boxed{3x-1} \xrightarrow{+1} \boxed{3x} \xrightarrow{\div 3} \boxed{x}$

b $\boxed{\dfrac{8x+10}{3}} \xrightarrow{\times 3} \boxed{8x+10} \xrightarrow{-10} \boxed{8x} \xrightarrow{\div 8} \boxed{x}$

c $\boxed{5-\dfrac{2x}{3}} \xrightarrow{-5} \boxed{\dfrac{-2x}{3}} \xrightarrow{\times 3} \boxed{-2x} \xrightarrow{\div -2} \boxed{x}$

5 a $x=4$ **b** $x=-48$ **c** $x=-1$
6 a $x=1$ **b** $x=-\tfrac{1}{2}$ **c** $x=-2$
7 a $x=7$ **b** $x=-\tfrac{3}{4}$ **c** $x=4\tfrac{1}{4}$
8 a $x=5$ **b** $x=2$ **c** $x=1\tfrac{2}{3}$ **d** $x=\tfrac{2}{3}$
9 a i $x+6=21$ **ii** $x=15$
  b i $\tfrac{x}{3}=5$ **ii** $x=15$ **c** yes
10 a $x=-5$ **b** $x=-\tfrac{24}{35}$ **c** $x=-2\tfrac{2}{3}$
11 The number is 5. **12** The larger integer is 19.
13 The number is 7. **14** 6 writing pads **15** 13 chogokin
16 a (number line) **b** (number line)

c (number line)

17 a $x \leqslant \tfrac{3}{2}$ **b** $x < 10$ **c** $x \geqslant 0$

EXERCISE 10A

1 a 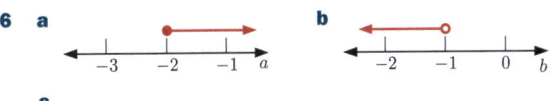 (A—B segment) **b** (crossing lines at A) **c** (parallel arrows)

d (line through A B C) **e** (multiple lines through A)

2 a AB or BA **b** XY, YX, XZ, ZX, YZ, or ZY
3 a PQ, PR, QR **b** QP and PR
4 a B **b** C **c** C **d** B
5 a (P X Q collinear) **b** (lines through M to E, H, G, F)

c (S T U V collinear)

d (lines through J K M N) **e** (lines through G with A F C E B D)

6 a Any three of AC, AE, BA, BC, BE, CA, CB, CE, EA, EB, EC.
  b 3 lines
  c i intersect at F **ii** are collinear **iii** are parallel

EXERCISE 10B

1 a C **b** A **c** D **d** B
2 a (angle PQR) **b** (angle RQP) **c** (reflex angle at F, EFG) **d** (angle CAB)

3 a i $25°$ **ii** $90°$ **iii** $45°$ **iv** $115°$
  b i acute **ii** right **iii** acute **iv** obtuse

4 a 38° **b** 89° **c** 120°

5 a i b **ii** g **iii** d
  b i reflex **ii** obtuse **iii** acute

6 a X, Y, Z **b** B **c** S, Y, P **d** C, Z, R **e** J, K, L **f** E, F, G

7 a i 68° **ii** 117° **iii** 112°
  b i 70° **ii** 80° **iii** 65°

8 ≈ 28.5°   **9** ≈ 110°

EXERCISE 10C

1 a supplementary **b** neither **c** complementary
  d neither **e** neither **f** supplementary

2 a 75° **b** 3° **c** 47°

3 a 51° **b** 123° **c** 90°

4 a supplementary **b** neither **c** complementary
  d neither

5 a The size of the angle complementary to $x°$ is $(90-x)°$.
  b The size of the angle supplementary to $y°$ is $(180-y)°$.
  c Two lines are perpendicular if they meet at 90°.

6 a $p = 125$ **b** $q = 38$ **c** $k = 94$ **d** $b = 85$
  e $q = 26$ **f** $t = 45$ **g** $s = 21$ **h** $a = 90$
  i $g = 30$

7 a $r = 266$ **b** $z = 120$ **c** $m = 236$

8 a $s = 50$ **b** $b = 115$ **c** $m = 31$ **d** $s = 75$
  e $j = 161$

EXERCISE 10D

1 a a and c, b and d **b** p and q, r and s **2** **B** and **C**

3 a c **b** d **c** d **d** b **4** **A**, **C**, and **D**

5 a s **b** s **c** q **d** q **6** **B** and **D**

7 a z **b** z **c** x **d** y

8 a corresponding **b** alternate **c** co-interior
  d corresponding **e** corresponding
  f vertically opposite **g** vertically opposite
  h co-interior **i** alternate

EXERCISE 10E

1 a $x = 124$ {equal corresponding angles}
  b $b = 82$ {supplementary co-interior angles}
  c $q = 42$ {equal alternate angles}
  d $y = 57$ {equal corresponding angles}
  e $k = 62$ {equal alternate angles}
  f $a = 135$ {equal corresponding angles}
  g $x = 147$ {equal alternate angles}
  h $y = 73$ {supplementary co-interior angles}
  i $d = 15$ {equal corresponding angles}

2 a $a = 76$ {vertically opposite angles}
    $b = 104$ {supplementary co-interior angles}
  b $a = 117$ {equal corresponding angles}
    $b = 117$ {vertically opposite angles}
  c $a = 38$ {vertically opposite angles}
    $b = 38$ {equal alternate angles}
  d $a = 145$ {angles at a point}
    $b = 35$ {supplementary co-interior angles}
  e $m = 96$ {supplementary co-interior angles}
    $n = 84$ {supplementary co-interior angles}
  f $a = 36$ {equal corresponding angles}
    $b = 36$ {equal alternate angles}

3 a $x = y$ {equal alternate angles}
  b $a + b = 180$ {supplementary co-interior angles}
  c $p = q$ {equal corresponding angles}
  d $a + b = c$ {equal alternate angles}

4 a parallel {equal alternate angles}
  b not parallel {co-interior angles do not sum to 180°}
  c not parallel {alternate angles are not equal}
  d parallel {equal corresponding angles}
  e parallel {angles on a straight line, equal corresponding angles}
  f parallel {angles on a straight line, equal corresponding angles}

5 The two lines are not parallel so the corresponding angles rule does not apply.

6 a The figure contains a pair of parallel lines.
    {co-interior angles sum to 180°}
    ∴ $a = 120$ {supplementary co-interior angles}
  b The figure contains a pair of parallel lines.
    {angles on a straight line, equal corresponding angles}
    ∴ $a = 115$ {equal corresponding angles, vertically opposite angles}

EXERCISE 10F

1 a 55° **b** 140° **c** 330° **d** 255°

2 a 304° **b** 248° **c** 064° **d** 141°

3 a i 040° **ii** 130° **iii** 310° **iv** 090° **v** 220°
    vi 270°
  b i 130° **ii** 048° **iii** 228° **iv** 068° **v** 310°
    vi 248°

4 a i 045° **ii** 225° **iii** 105° **iv** 285° **v** 135°
    vi 315°
  b i 120° **ii** 300° **iii** 070° **iv** 250° **v** 035°
    vi 215°

Answers 663

5 a east, 090° b southeast, 135° c south, 180°
  d west, 270° e southwest, 225° f northwest, 315°
  g north, 0° h northeast, 045°

6 a i 8 km, 070° ii 11 km, 155°
  b i

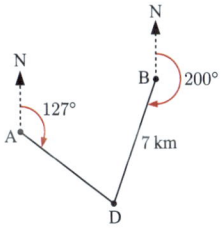

Scale: 1 cm = 4 km

  ii ≈ 6.4 km on bearing 127°

EXERCISE 10G

1 a, b

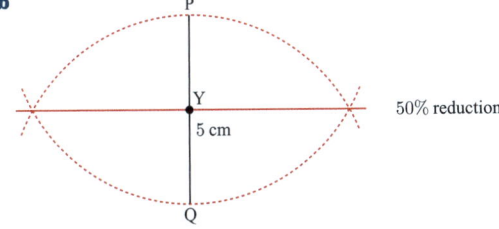

50% reduction

  c PY = QY = 2.5 cm

2 a, b

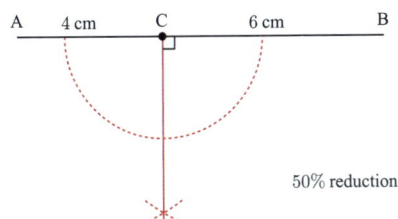

50% reduction

3 a, b

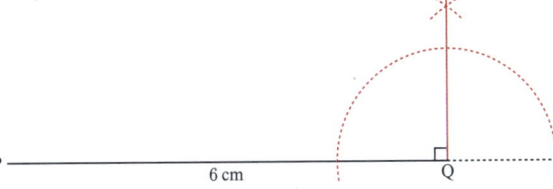

4 a, b

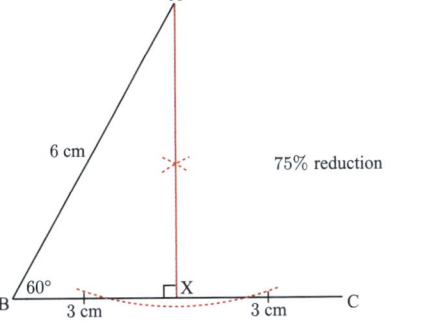

75% reduction

  c BX = 3 cm

5 a, b

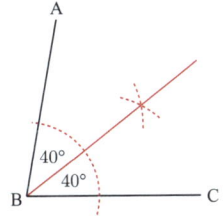

6 a, b

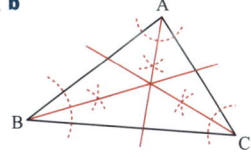 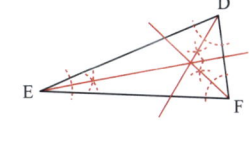

  d The three angle bisectors of a triangle are concurrent.

7 a

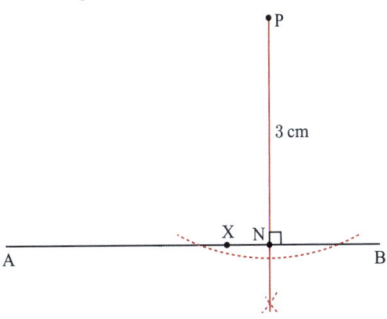

  b Let the perpendicular from P meet AB at N. If X lies on AB not at N, triangle PNX is right angled at N, and PN is always ⩽ PX.
  ∴ PN is the shortest distance.

8 a, b, c, d

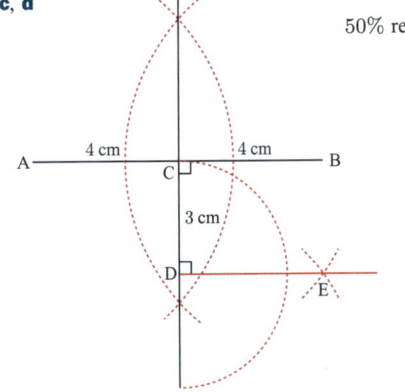

50% reduction

  e $A\widehat{C}D = C\widehat{D}E = 90°$
  ∴ CB ∥ DE {co-interior angles are supplementary}

9 a

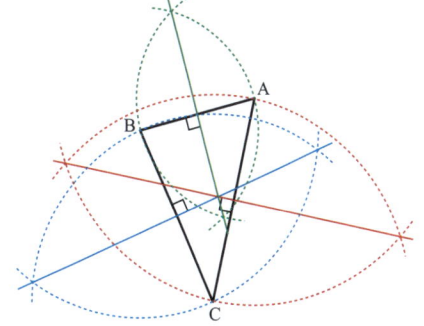

c "The three perpendicular bisectors of the sides of a triangle are concurrent (meet at the same point)."

10 a, b

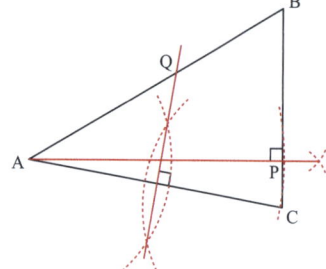

11 a, b

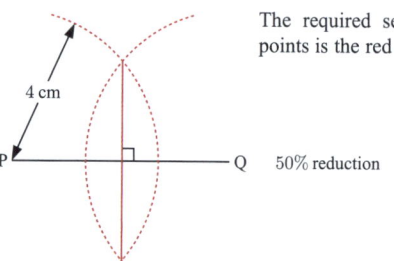

The required set of points is the red line.

12

B C
40 m • coin

A D

Scale: 1 cm = 20 m

The required set of points is the red line.

REVIEW SET 10A

1 a 37°   **b** 50°
2 a $a = 16$ **b** $b = 27$ **c** $c = 120$ **3** two points
4

C B
A P D

5 a $x = 62$ {vertically opposite angles, supplementary co-interior angles}
  b $x = 61$ {equal alternate angles}
  c $x = 88$ {angles on a line (twice)}
6 a $m = 116$ {equal alternate angles}
  b $m = 81$ {equal corresponding angles}
  c $m = 141$ {supplementary co-interior angles}
7 a $x = y$ {vertically opposite angles, equal corresponding angles}
  b $a + b = 180$ {supplementary co-interior angles}

8 a 060°   **b** 240°   **c** 140°   **d** 320°   **e** 290°
  f 110°
9 a parallel {vertically opposite angles, equal corresponding angles}
  b parallel {angles at a point, supplementary co-interior angles}

10 a, b   **c** PX = QX = 3 cm

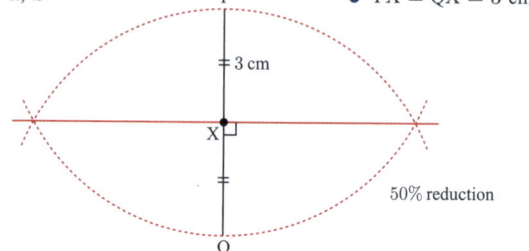

11 a

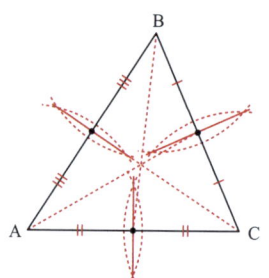

b All 3 altitudes of a triangle meet at one point.

REVIEW SET 10B

1 a Any two of AC, BA, BC, CA, CB
  b i they are collinear   **ii** they intersect at D
2 a 25°   **b** 92°
3 a $x = 110$ {equal corresponding angles}
  b $c = 126$ {angles at a point}
4 a $a = 35$ {complementary angles}
  b $b = 45$ {angles on a line}
5

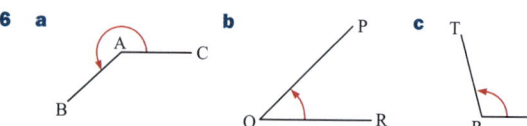

6 a   **b**   **c**
7 a i f **ii** d **iii** b
  b i reflex   **ii** obtuse   **iii** acute
8 a 338°   **b** 22° {angles at a point}
9 a 110°   **b** 290°   **c** 205°   **d** 025°   **e** 250°
  f 070°

10 a supplementary **b** neither
c complementary **d** neither

11 parallel {vertically opposite angles, supplementary co-interior angles, equal alternate angles}

12

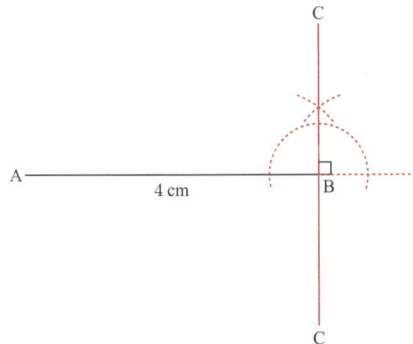

EXERCISE 11A

1 a polygon
b not a polygon as a side is curved, crossed, and not closed
c polygon **d** polygon
e not a polygon as sides are not straight **f** polygon
g not a polygon as sides cross over **h** polygon
i not a polygon as sides are not straight **j** polygon
k polygon **l** not a polygon as figure is not closed

2 a triangle **b** quadrilateral **c** hexagon
d heptagon **e** octagon **f** nonagon

3 a quadrilateral **b** triangle **c** decagon
d pentagon **e** quadrilateral **f** decagon
g heptagon **h** nonagon

4 a all angles not equal **b** all angles not equal
c all sides not equal in length **d** all sides not equal in length

5 a 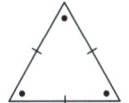 **b** 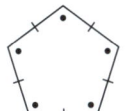 **c**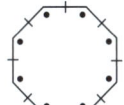

6 a  **b** 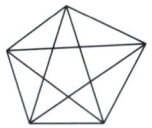 **c**

EXERCISE 11B

1 a isosceles **b** scalene **c** equilateral
2 a acute angled **b** right angled **c** obtuse angled
3 a $a = 130$ **b** $b = 90$ **c** $c = 43$ **d** $d = 129$
e $e = 53$ **f** $x = 131$
4 $50°$ **5 a** true **b** false **c** false **d** false
6 a AB **b** AC **c** BC **d** AB **e** BC
f AB **g** AB and BC **h** AC **i** AC
7 a $a = 55$ **b** $b = 55$ **c** $x = 20$ **d** $a = 55$, $b = 77$
e $x = 45$, $y = 70$ **f** $a = 45$, $b = 85$, $c = 130$, $d = 50$

8 a $\widehat{CBD} = 180° - \widehat{ABD}$ {angles on a line}
$= 180° - 90°$
$= 90°$

b $\widehat{BDC} = 55°$ **c** $\widehat{ADB} = 55°$

EXERCISE 11C

1 a $x = 74$ **b** $x = 64$ **c** $x = 69$ **d** $x = 24$
e $x = 45$ **f** $x = 22.5$ **g** $x = 140$ **h** $x = 103$
i $x = 33$

2 a $x = 12$ {equal sides of isosceles triangle}
b $x = 8$ {isosceles triangles, QR = QS = PS}
c $x = 90$ {line from apex to midpoint of base in an isosceles triangle is perpendicular to base}
d $x = 50$ {line from apex to midpoint of base in an isosceles triangle is perpendicular to base}
e $x = 5$ {isosceles triangle, $\widehat{QPR} = 57°$}
f $x = 90$ {line bisecting angle at apex of an isosceles triangle is perpendicular to base}
g $x = 140$ {if line from apex perpendicular to base bisects the base, it is an isosceles triangle}
h $x = 4$ {equilateral triangle, all angles $60°$}

3 both $67°$ **4 a** $x = 66$ **b** isosceles, BC = AC
5 a $\widehat{ABD} = 48°$ **b** $\triangle ABD$ is isosceles **c** $x = 3$
6 $\widehat{AEB} = 50°$

7 a The interior angles of an equilateral triangle are equal and sum to $180°$. Each angle is $\dfrac{180°}{3} = 60°$.

b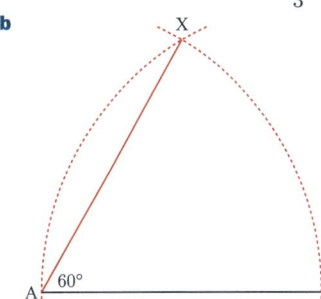

Step 1:
With centre A and radius AB, draw an arc above AB.
Step 2:
With centre B, and the *same* radius, draw an arc above AB to cut the first one at X.
Step 3:
Join A and X.

EXERCISE 11D.1

1 a $x = 65$ **b** $x = 93$ **c** $x = 35$ **d** $x = 40$
e $x = 50$ **f** $x = 90$
2 a $a = 65$, $b = 91$ **b** $m = 60$, $n = 65$ **c** $a = 70$

EXERCISE 11D.2

1 a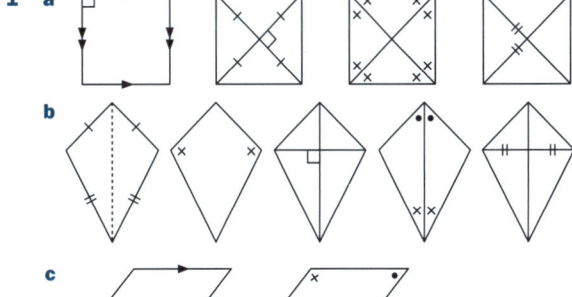

b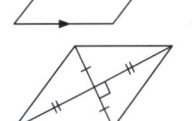

c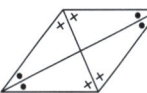

2 **a** $x = 90$, $y = 5$ **b** $x = 50$ **c** $x = 20$
  d $x = 60$, $y = 90$ **e** $a = 6$, $b = 8$ **f** $x = 14$
3 **a** true **b** false **c** false **d** true
4 **a** a square, a rhombus, or a kite {diagonals intersect at right angles}
  b a rhombus {diagonals bisect each other at right angles, and diagonals are not equal in length}
  c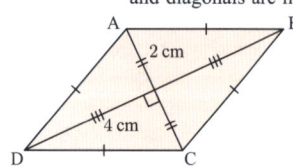

5 **a** $a = 50$, $b = 120$ {co-interior angles}
  b $a = 42$ {diagonals of rhombus bisect angles}
    $b = 96$ {angles of a triangle}
  c $x = 2$ {opposite sides of parallelogram are equal}
    $a = 40$ {equal alternate angles, parallel lines}
6 **a** $x = 180 - x$ {opposite angles of a parallelogram}
    $\therefore\ x = 90$
    $\therefore\ $ ABCD is a rectangle.
  b The triangle with marked angles is isosceles. Using angle sum of a triangle, $x = 45$ and $2x = 90$. The figure is a square.
7 true

EXERCISE 11E
1 **a** $540°$ **b** $900°$ **c** $720°$ **d** $1800°$ **e** $2340°$
2 **a** $x = 108$ {angle sum of a pentagon}
  b $x = 150$ {angle sum of a hexagon}
  c $x = 60$ {angle sum of a pentagon}
  d $x = 120$ {angle sum of a hexagon}
  e $x = 95$ {angle sum of a pentagon}
  f $x = 125$ {angle sum of a heptagon}
  g $x = 135$ {angle sum of an octagon}
  h $x = 70$ {angle sum of an octagon}
  i $x = 54$ {angle sum of a nonagon}
3 $135°$ **4** **a** $x = 60$ **b** $150°$ **5** 13 sides
6 **a** 3 reflex angles **b**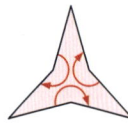

7 **a**

Regular polygon	Number of sides	Sum of angles	Size of each angle
triangle	3	180°	60°
quadrilateral	4	360°	90°
pentagon	5	540°	108°
hexagon	6	720°	120°
octagon	8	1080°	135°
decagon	10	1440°	144°

  b **i** The sum of the angles of an n-sided polygon is $(n-2) \times 180°$.
    ii The size of each angle of a regular n-sided polygon is $\dfrac{(n-2) \times 180°}{n}$.
  c $150°$
8 **a** $120°$ **b** $30°$ **c** $90°$ **d** $60°$ **e** $60°$ **f** $30°$

EXERCISE 11F
1 **a** F **b** D **c** A **d** I **e** L **f** G
  g C **h** E **i** K **j** J **k** B **l** H
2 **a** diameter
3 **a** A diameter is made up of two radii, one on each side of the centre.
  b **i** 8 cm **ii** 6 cm
4 **a** 3 cm **b** 5 cm **c** 6 cm **d** 10 cm **e** 2 cm
  f 8 cm
5 **a**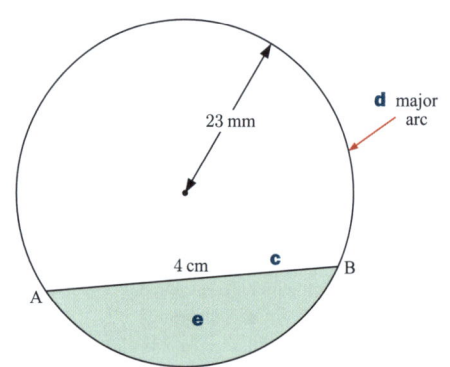

  b 46 mm

REVIEW SET 11A
1 **a** quadrilateral **b** hexagon **c** octagon
2 **a** right angled **b** obtuse angled
3 **a** scalene, right angled **b** isosceles, obtuse angled
  c equilateral, acute angled
4 **a** $x = 90$, $y = 60$ {supplementary co-interior angles}
  b $y = 70$ {equal corresponding angles}
    $x = 40$ {angle sum of isosceles triangle}
  c $x = 4$ {isosceles triangle, line from apex bisecting vertical angle bisects the base}
5 **a** square **b** rhombus **c** rectangle
6 **a** $x = 12$ **b** $x = 90$, $y = 42$ **c** $x = 70$
7 $\widehat{ABC} = 48°$ **8** $a = 90$, $b = 50$, $c = 70$, $d = 120$
9 $1620°$ **10** $x = 130$
11 **a** An **arc** is a part of a circle. It joins any two different points on the circle.

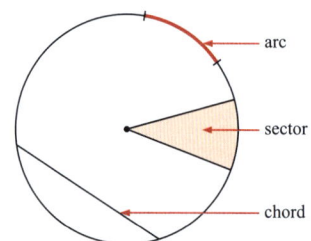

  b A **sector** of a circle is the region between two radii and the circle.
  c A **chord** of a circle is a line which joins any two points of the circle.
12 **a** True, the minor arc is always shorter than the major arc, so it must also be shorter than the semi-circle.
  b False, a chord may be shorter than the radius if the endpoints of the chord are close to each other.

REVIEW SET 11B

1 a 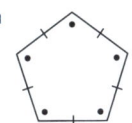 **b**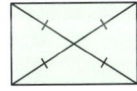

2 a isosceles **b** scalene
3 a isosceles **b** obtuse angled
4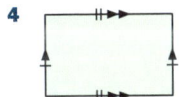

5 a $x = 138$ {base angles of isosceles triangles, angles on a straight line}
  b $x = 20$ {angle sum of a triangle}
  c $x = 82$ {base angles of isosceles triangle, angles on a line}
    $y = 16$ {angle sum of a triangle}
6 a true **b** true
7 a $x = 55$ {opposite angles in a parallelogram are equal}
  b $x = 45$ {angles on a line, base angles of isosceles triangle}
    $y = 4$ {isosceles triangles}
8 a $x = 90$, a rectangle **b** a rhombus
  c equal alternate angles, a trapezium
9 $135°$ **10 a** 2 reflex angles **b**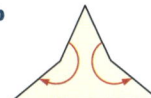

11 a A **semi-circle** is a half of a circle.
  b A **major arc** is the longer distance around the circle between 2 points.
  c A **minor segment** is the smaller region between a chord and the circle.
  d A **tangent** to a circle is a line which touches the circle but does not enter it.

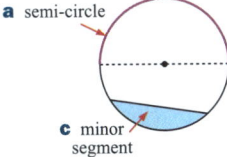
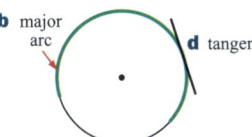

12 a, b, c, e

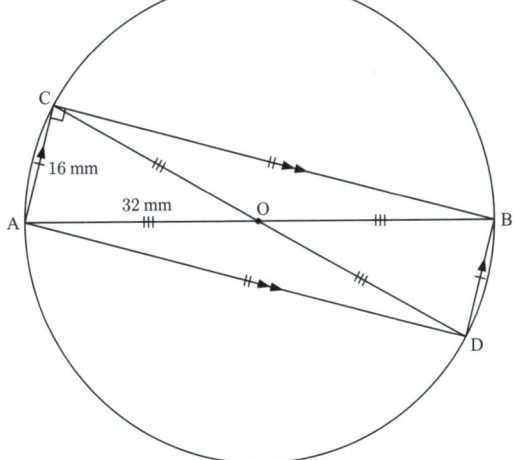

d $\widehat{ACB} = 90°$ **f** a rectangle

EXERCISE 12A

1 a cylinder **b** triangular-based pyramid (tetrahedron)
  c pentagonal prism

2 a 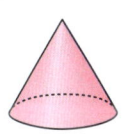 **b** 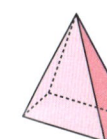 **c**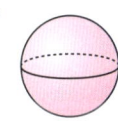
   d 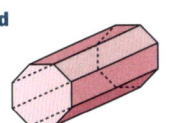 **e**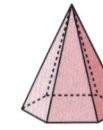

3 a a cylinder **b** a sphere **c** a rectangular prism
  d a cone **e** a triangular-based pyramid **f** a cylinder
4 a 6 vertices, 9 edges, 5 faces
  b 5 vertices, 8 edges, 5 faces
  c 12 vertices, 18 edges, 8 faces
5 a rectangle **b** triangle
6 a 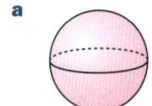 **b** 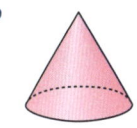 **c**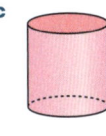
   d 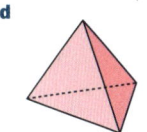 **e** 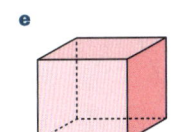 **f**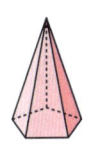

EXERCISE 12B

1 a 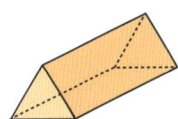 **b**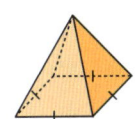

a triangular prism   a square-based pyramid

c   **d**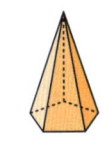

a cylinder   a pentagonal-based pyramid

e 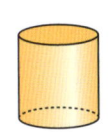 **f**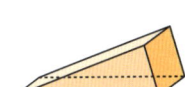

a triangular-based pyramid (tetrahedron)   a triangular prism

2 Note: Other answers are possible.

a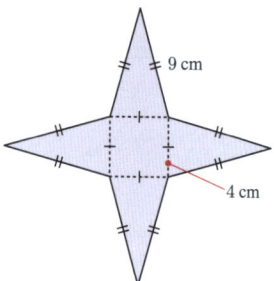

b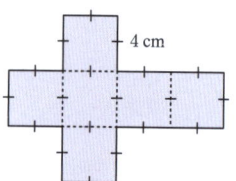

c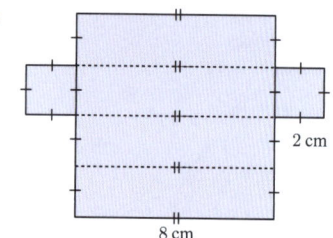

d, e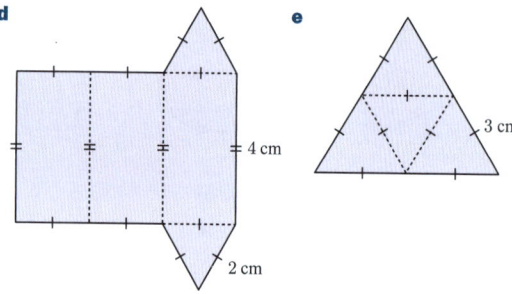

f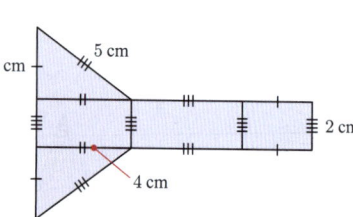

3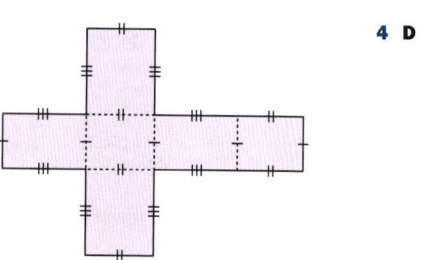

4 D

5 Other answers are possible.

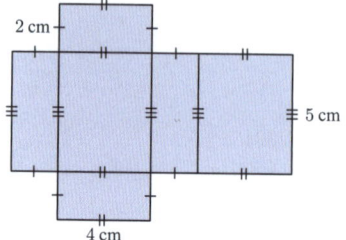

EXERCISE 12C

1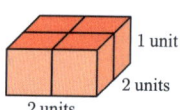

2 a 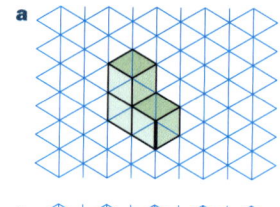   b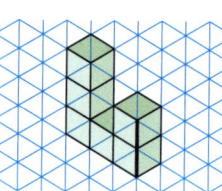

c 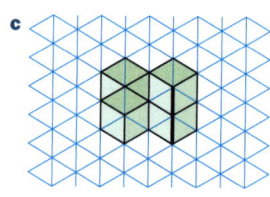   d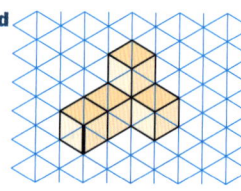

e 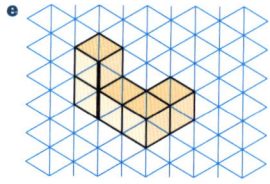   f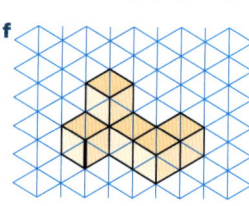

3 a 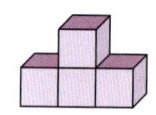   b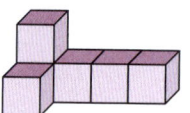

c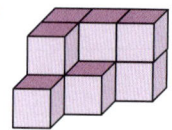

EXERCISE 12D.1

1 a 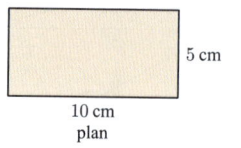   b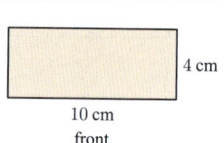
   10 cm plan   10 cm front

c    d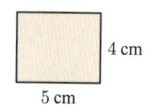
   10 cm back   5 cm left

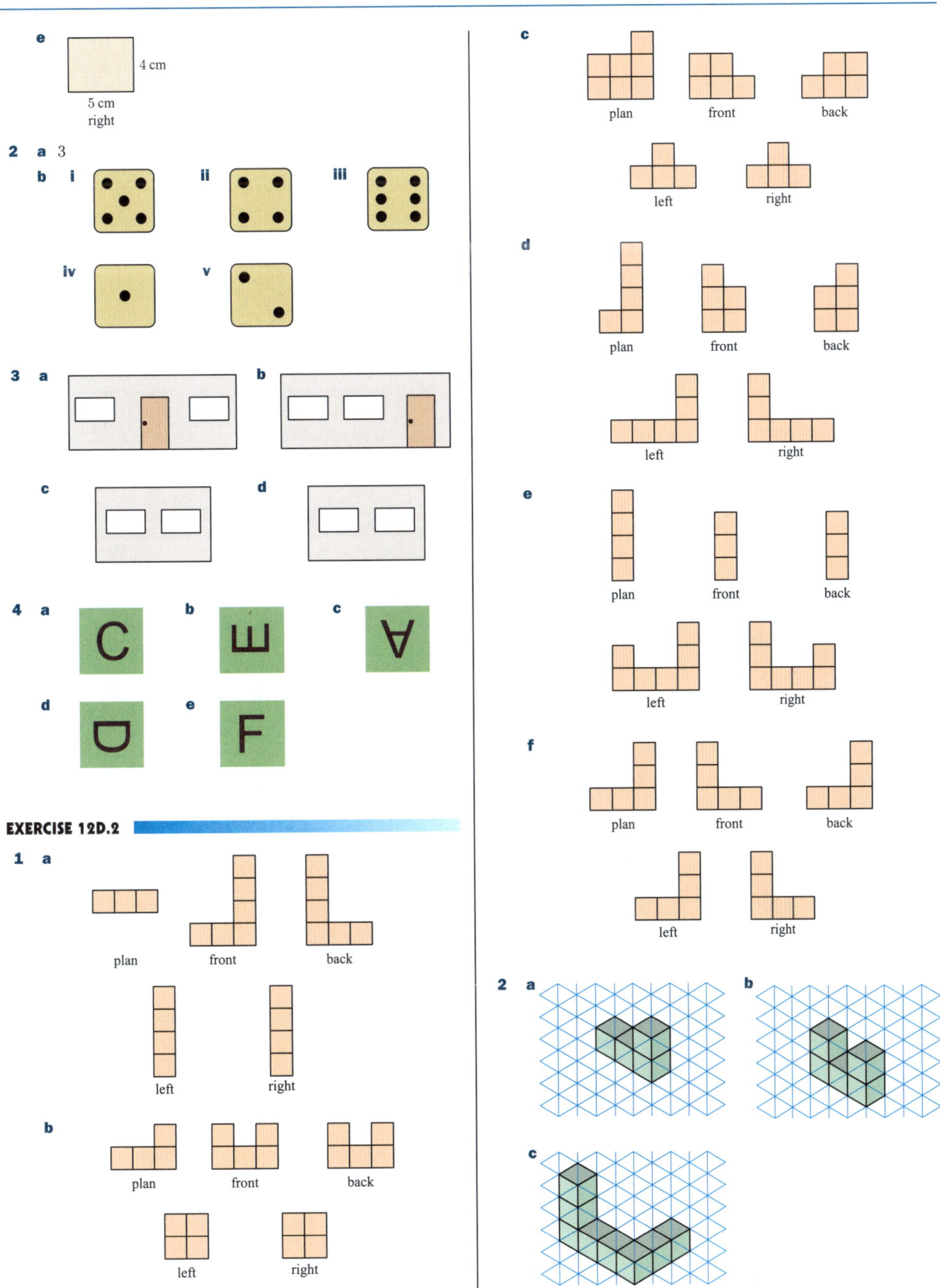

3 The left and right elevations should be reflections of each other in a vertical line. It is impossible for a block solid to have these views.

4 **a** C   **b** A   **c** D   **d** B

5

REVIEW SET 12A

1 **a** triangular prism   **b** pentagonal-based pyramid

2 Note: Other answers are possible.

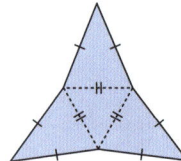

3 **a** rectangular prism   **b** 8 vertices, 12 edges, 6 faces

4

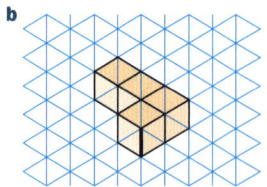

5 a **b**

6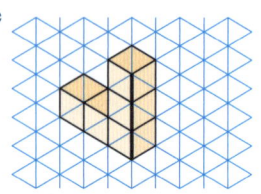

7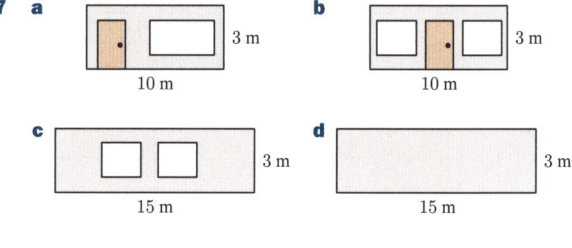

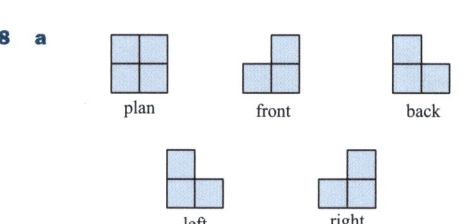

8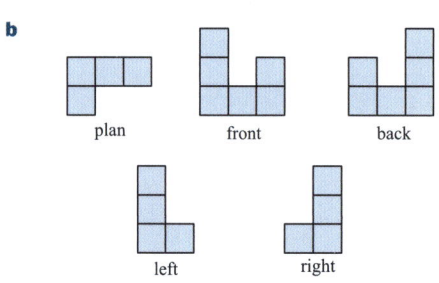

REVIEW SET 12B

1 a **b**

2 Note: Other answers are possible.   **3** a cube

Answers 671

4 a, b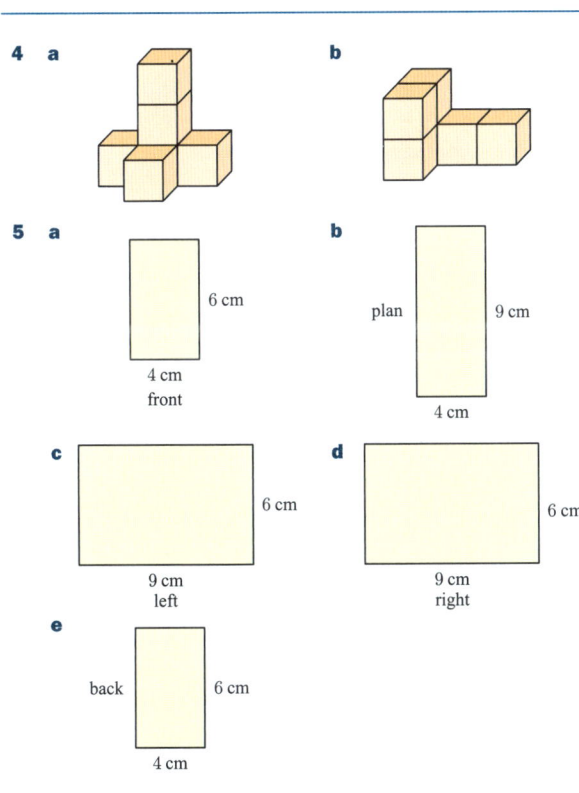

5 a 6 cm, 4 cm front
b plan 9 cm, 4 cm
c 6 cm, 9 cm left
d 6 cm, 9 cm right
e back 6 cm, 4 cm

6 Note: Other answers are possible.
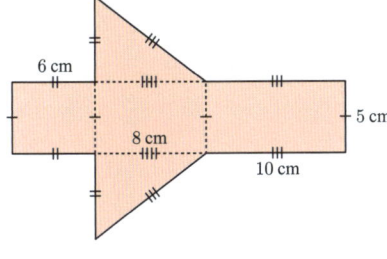

7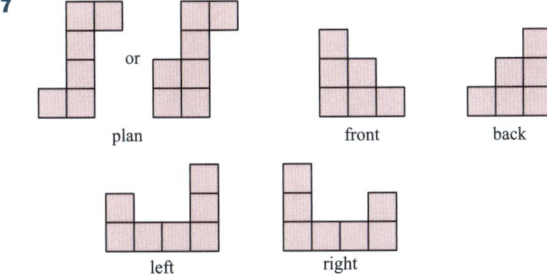

8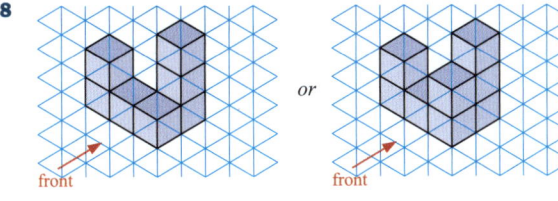

EXERCISE 13A

1 a 3 **b** b **c** 7 **d** q **e** $9c$ **f** $4d$
 g 1 **h** $4z$ **i** $14x$

2 a ab **b** st **c** $6x$ **d** f **e** c **f** r^2
 g $2d$ **h** $5a$ **i** $7b^2$ **j** $13gh$ **k** $6jk$ **l** $2xy$
 m 10 **n** $5b$ **o** $2pq$

3 a $(b+6)$ **b** $7(1+b)$ **c** $y(y-3)$
 d $8(x-4)$ **e** $3(x-8)$ **f** $5d(d+1)$

4 a 1 **b** 1 **c** $(x-1)$

EXERCISE 13B

1 a $3(x-2)$ **b** $8(d+5)$ **c** $10(x-3)$ **d** $2(2+x)$
 e $5(p+q)$ **f** $a(b+d)$ **g** $7(6-5x)$ **h** $h(g+1)$
 i $y(1+z)$ **j** $p(1-q)$ **k** $e(3+f)$ **l** $x(9-y)$
 m $k(j-3h)$ **n** $s(r-t)$ **o** $y(3x-2z)$ **p** $m(4n+l)$

2 a $x(x+7)$ **b** $3x(x+3)$ **c** $4x(2-x)$
 d $5x(1-5x)$ **e** $3x(2x+5)$ **f** $x^2(x+2)$
 g $x(x^2+3)$ **h** $x^2(8x-1)$ **i** $xy(x-y)$
 j $2xy(y+4x)$ **k** $4(x^2-2x+3)$ **l** $9x(x-5-3x^2)$

3 a $x(x+2)$ units^2 **b** (x^2+2x) units^2

4 a $5(2-a)$ **b** $4(3b-5)$ **c** $8(2d-c)$ **d** $a(3-b)$
 e $p(q-1)$ **f** $y(1-y)$ **g** $7y(y-2)$ **h** $d(2d-c)$
 i $z(1-z)$

5 a $-6(x+4)$ **b** $-8(1+x)$ **c** $-2(x+y)$
 d $-5(2p+q)$ **e** $-a(b+1)$ **f** $-7a(1+3a)$
 g $-8g(2g+1)$ **h** $-6t(3t+2s)$ **i** $-11e(2e+3)$

6 a $(a+2)(a+3)$ **b** $(x-1)(4+x)$ **c** $(x+4)(5-x)$
 d $(x-9)(x+1)$ **e** $(c+d)(a-b)$ **f** $(x-3)^2$
 g $(r+s)(t-1)$ **h** $(y-6)(y+1)$

7 a $(x+4)(x+8)$ **b** $(x+1)(x-1)$
 c $(x+2)(x-8)$ **d** $(x-10)(x-6)$
 e $(d+7)(2d+13)$ **f** $(y-1)(x-y-5)$
 g $5(a+1)(3a+1)$ **h** $(n-3)(24-7n)$
 i $-(n-3)(n+2)$ **j** $-2(x-14)(x+19)$

8 a $2x^2$ and $-3xy$ have a common factor of x to be taken out.
 b $6x(2x-3y)$

9 a $2x^2$ units^2 **b** $2x(x-2)$ units^2 **c** $4x(x-1)$ units^2

EXERCISE 13C

1 a $(d+e)(d-e)$ **b** $(p+q)(p-q)$ **c** $(q+p)(q-p)$
 d $(x+y)(x-y)$ **e** $(x+3)(x-3)$ **f** $(x+10)(x-10)$
 g $(y+7)(y-7)$ **h** $(4x+3)(4x-3)$
 i $(2b+1)(2b-1)$ **j** $(6y+7)(6y-7)$
 k $(9+d)(9-d)$ **l** $(11+3y)(11-3y)$

2 a $3(x+5)(x-5)$ **b** $8(x+2)(x-2)$
 c $5(a+2)(a-2)$ **d** $5(x+1)(x-1)$
 e $9(b+10)(b-10)$ **f** $5(4+t)(4-t)$
 g $6(k+2)(k-2)$ **h** $15(1+y)(1-y)$
 i $r(r+7)(r-7)$ **j** $z(z+1)(z-1)$
 k $x^3(x+1)(x-1)$ **l** $xy(x+y)(x-y)$

3 a $(5a+b)(5a-b)$ **b** $(x+9y)(x-9y)$
 c $(4x+7y)(4x-7y)$ **d** $(5p+6q)(5p-6q)$
 e $(9j+k)(9j-k)$ **f** $(st+3)(st-3)$
 g $(7x+yz)(7x-yz)$ **h** $(2d+5e)(2d-5e)$

4 a $x(x+4)$ **b** $(x+3)(x-9)$ **c** $(x-1)(21-x)$
 d $(x+1)(17-x)$ **e** $(x+8)(x+6)$ **f** $-(x+7)(x+1)$

5 a $391 = 20^2 - 3^2$ **b** $391 = (20+3)(20-3) = 23 \times 17$
∴ 17 is a factor of 391

6 a Area of path
= (Area of path and area of pool) − (area of pool)
= $(L^2 - l^2)$ m^2

b **i** $L^2 - l^2 = (L+l)(L-l)$
ii $L = l + 2x$
∴ $(L+l)(L-l) = (l+2x+l)(l+2x-l)$
$= 2x(2l+2x)$
$= 4x(x+l)$

EXERCISE 13D

1 a $A_1 = a^2$, $A_2 = ab$, $A_3 = ab$, $A_4 = b^2$
b Total area $= (a+b)^2$ and
Total area $= A_1 + A_2 + A_3 + A_4 = a^2 + 2ab + b^2$

2 a perfect square, $(x+8)^2$ **b** not perfect square
c perfect square, $(a-3)^2$ **d** not perfect square
e perfect square, $(t-1)^2$ **f** not perfect square
g perfect square, $(1-x)^2$ **h** not perfect square
i perfect square, $(3-2x)^2$

3 a $(x-1)^2$ **b** $(x+5)^2$ **c** $(x+4)^2$ **d** $(x+8)^2$
e $(x-7)^2$ **f** $(x+11)^2$ **g** $(x-10)^2$ **h** $(x-9)^2$
i $(x-4)^2$ **j** $(x+6)^2$ **k** $(x-3)^2$ **l** $(x+12)^2$
m $(1+x)^2$ **n** $(2-x)^2$ **o** $(3+x)^2$

4 a $(2x-1)^2$ **b** $(5x+2)^2$ **c** $(8x-3)^2$ **d** $(3x+7)^2$
e $(6x-5)^2$ **f** $(4x-5)^2$

5 a $(x+1)^2$ **b** $(x+3)^2$ **c** $(x-2)^2$ **d** $(x-4)^2$
e $(2x+1)^2$ **f** $(2y-3)^2$

6 a perfect square, $(x+9)^2$ **b** not perfect square
c not perfect square **d** perfect square, $(3x-4)^2$
e perfect square, $(7x+1)^2$ **f** perfect square, $(x-2y)^2$

7 a $2(x+3)^2$ **b** $2(x-1)^2$ **c** $3(x+2)^2$
d $4(x-4)^2$ **e** $5(x+5)^2$ **f** $-(x-6)^2$
g $-(x+8)^2$ **h** $-2(x-10)^2$ **i** $-3(x-5)^2$

8 961, $\sqrt{961} = 31$

EXERCISE 13E

1 a 1, 7 **b** 5, 6 **c** 5, 12 **d** −2, 4
e −5, 4 **f** −8, −7 **g** −24, 2 **h** −10, 1

2 a $(x+2)(x+5)$ **b** $(x+3)(x+11)$ **c** $(x+2)(x+17)$
d $(x+1)(x+23)$ **e** $(x+4)(x+12)$ **f** $(x+4)(x+5)$
g $(x+3)(x+18)$ **h** $(x+4)(x+25)$ **i** $(x+7)(x+9)$

3 a $(x-4)(x-5)$ **b** $(x-4)(x-14)$ **c** $(x-6)(x-7)$
d $(x-4)(x-7)$ **e** $(x-1)(x-2)$ **f** $(x-7)(x-9)$
g $(x-8)(x-12)$ **h** $(x-2)(x-15)$ **i** $(x-3)(x-5)$

4 a No two integers have a sum of 3 and a product of −2.
b No two integers have a sum of 1 and a product of 2.

5 a $(x-6)(x+1)$ **b** $(x-9)(x+2)$ **c** $(x+16)(x-5)$
d $(x+9)(x-8)$ **e** $(x+11)(x-3)$ **f** $(x-11)(x+7)$
g $(x+19)(x-3)$ **h** $(x-10)(x+9)$ **i** $(x-21)(x+4)$
j $(x-5)(x+2)$ **k** $(x+9)(x-5)$ **l** $(x+12)(x-6)$

6 a $2(x+1)(x+2)$ **b** $4(x+2)(x+5)$
c $3(x+1)(x+3)$ **d** $4(x-3)(x+1)$
e $9(x+4)(x-1)$ **f** $5(x-1)(x-3)$

g $7(x-5)(x+1)$ **h** $5(x-11)(x+2)$
i $6(x-2)(x-10)$ **j** $3(x-6)(x+1)$
k $2(x-9)(x+2)$ **l** $10(x-5)(x+4)$

EXERCISE 13F

1 a $4a(a+2)$ **b** $3(b^2+5)$ **c** $6(x-6y)$
d $(p+5)(p+7)$ **e** $(x-18)(x+1)$ **f** $(x-4)(x+1)$
g $-x(x+49)$ **h** $h^2(1+2h)$ **i** $st(t-2)$
j $-2(x^2+9)$ **k** $(r+7)(r-6)$ **l** $5x(2+x)(2-x)$
m $3z(1-7y)$ **n** $a(a+16b)$ **o** $(y-3)(y-4)$
p $9(x-2)(x+1)$ **q** $6(y-3)^2$ **r** $x(x+2)^2$

2 a $(x-5)^2$ **b** $(x+13)(x-13)$ **c** $4(c+2)(c-2)$
d $10(1+y)(1-y)$ **e** $5(x+5)(x-5)$ **f** $12(f+1)^2$
g $(9y+7x)(9y-7x)$ **h** $(d-3)(d-8)$ **i** $2(x-8)^2$

3 a $pq(q-p-1)$ **b** $4b^2(b-1)$ **c** $cd(cd-8)$
d $y(4+3y)(4-3y)$ **e** $(x+1)(x-1)$ **f** $f(e+g)(e-g)$
g $(x-5)(2-x)$ **h** $(x+y)(9-y)$ **i** $(s+t)(s-t)$
j $k(k-100)$ **k** $6(x+7)(x-2)$ **l** $m(7n+m)(7n-m)$

REVIEW SET 13A

1 a $3ab$ **b** $3(x+1)$
2 a $x(x-3)$ **b** $3n(m+2n)$ **c** $ax^2(x+2)$
3 a $-2x(x+16)$ **b** $(t+2)(d-4)$ **c** $(x-1)(x-2)$
d $(x+3)(2x-5)$ **e** $3(g+1)(g-2)$ **f** $(b-c)^2$

4 a not perfect square **b** perfect square, $(4x+3)^2$

5 a $(x+5)(x-5)$ **b** $(10+k)(10-k)$
c $(3+4x)(3-4x)$ **d** $(y+x)(y-x)$
e $(3a-2b)(3a+2b)$ **f** $6(x+2)(x-2)$

6 a $(x+2)^2$ **b** $(x-5)^2$ **c** $(x+7)^2$
7 a $(2x+5)^2$ **b** $(3x-1)^2$ **c** $5(x-2)^2$

8 a $(x+7)(x+3)$ **b** $(x+7)(x-3)$ **c** $(x+3)(x-7)$
d $(x-2)(x-3)$ **e** $4(x+3)(x-1)$
f $(x+4)(x+9)$ **g** $(x+4)(x+5)$
h $2(x+5)(x-6)$ **i** $3(x-2)(x-8)$

REVIEW SET 13B

1 a $2y$ **b** $2(x-2)$
2 a $2x(x+3)$ **b** $-2x(y+2)$ **c** $(x+3)(x-1)$
3 a $xy(y+4)(y-4)$ **b** $3(x-10)^2$ **c** $(a+2)(p-q)$
d $2cd(2d-3c)$ **e** $(k-3)(k-2)$ **f** $x(x+2)$

4 No two integers have a sum of −2 and a product of 4.

5 a $(5+x)(5-x)$ **b** $(2a+3b)(2a-3b)$
c $(7+3z)(7-3z)$

6 a $(x-6)^2$ **b** $(n-3)^2$ **c** $(x+4)^2$ **d** $(x-3)^2$
e $(x-5)^2$ **f** $(x-5)(x-7)$

7 a $2(x+1)^2$ **b** $x(x+4)(x-4)$ **c** $2(3x-1)^2$

8 a $(x+5)(x+7)$ **b** $(x+7)(x-5)$ **c** $(x-5)(x-7)$
d $2(x-7)(x+5)$ **e** $(x-6)(x-5)$ **f** $(x-10)(x+2)$
g $(x-11)(x-3)$ **h** $4(x-11)(x+2)$ **i** $2(x+9)(x-4)$

EXERCISE 14A

1 a 20:05 **b** 9116 **c** Tue, Wed, Thur
d 3 hours 1 minute **e** 9 minutes **f** 09:26

2 a no **b** yes **c** 5 days
d **i** Dec 24 **ii** Dec 25, Jan 1

Answers **673**

3 a £35   b i 105 kg   ii £90   c i £135   ii £30
4 a i 69.00 inhabitants   ii 79.03 inhabitants
   b 2010, 2011, 2012, 2013, 2014, 2015, 2016
   c i South Korea   ii Japan   d Russia
5 a 35%   b 59%   c 'Poor'   d 500 guests
6 a There is no distance between a town and itself. We shade the boxes along the diagonal to show this.
   b i 23 miles   ii 25 miles   iii 55 miles   iv 13 miles
   c From closest to Flint to furthest away.
   d

	Colwyn Bay	Conwy	Flint	Llanfairfechan	Mostyn	Pensarn	Rhyl	Talacre
Colwyn Bay		15	32	23	23	6	11	19
Conwy	15		47	8	38	21	25	33
Flint	32	47		55	9	26	22	13
Llanfairfechan	23	8	55		46	29	34	42
Mostyn	23	38	9	46		17	13	5
Pensarn	6	21	26	29	17		5	13
Rhyl	11	25	22	34	13	5		8
Talacre	19	33	13	42	5	13	8	

7 a 19 585 female students   b Computer Science
   c i Business & administrative studies
      ii Veterinary science
   d ≈ 62.2%   e 16 subject areas   f ≈ 60 400 students

EXERCISE 14B

1 a concert   b ≈ 1000 people   c ≈ 29.7%
2 a i £150 000   ii £254 000   b £88 000
   c ≈ 81.1%
3 a 2700 students   b ≈ 26%   c 2015
4 a i ≈ 38%   ii ≈ 19%
   b i ≈ 29 children   ii ≈ 10 children
5 a 15 flights   b Friday   c 31 flights
6 a i 5 trucks   ii 1 bicycle
   b No, as there were 18 cars and 20 other vehicles.
7 a ≈ 1500 performances   b ≈ 350 performances
   c 2012
8 a i 7.6%   ii 15.9%   b Sauvignon
   c i $2.16 billion   ii $2.67 billion
9 a £483 billion   b 16%
   c i £115.92 billion   ii £67.62 billion
10 a i 10°C   ii 20°C   b i 25°C   ii 2 pm
   c decreasing; the line is sloping downwards at 4 pm.
   d 20°C
11 a The volume of water in a lake from March to September.
   b i ≈ 22 300 Ml   ii ≈ 24 800 Ml
   c i May   ii August   d ≈ 25 200 Ml
   e ≈ 18.2% increase

EXERCISE 14C

1 a

Level of achievement	Tally	Frequency
A	|||	3
B	|||||	5
C	||||| ||||| |||||		16
D	|||	3
E			1
	Total	28

   b 16 students   c $\frac{5}{28}$

   d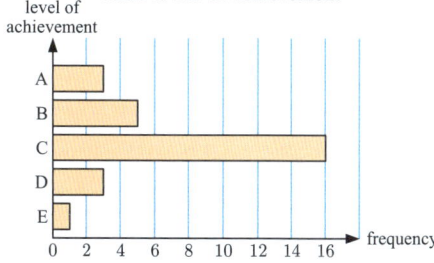
   Class levels of achievement

2 a

Attraction	Tally	Frequency
Side shows	||||| ||||	9
Farm animals	|||	3
Ring events	||||| ||	7
Dogs and cats	||	2
Wood chopping	||||	4
	Total	25

   b 16%

   c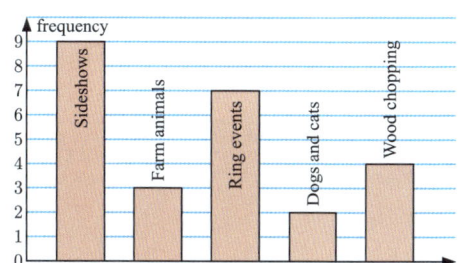
   Preferred show event

3 a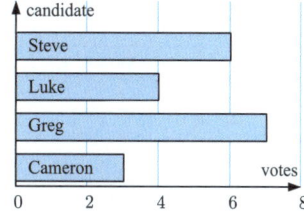
   Votes for team captain

   b Greg   c i 20%   ii 65%

4 a

Ice cream flavour	Tally	Frequency
Chocolate	||||| ||||| ||||| |||	18
Strawberry	||||| |||||	10
Vanilla	||||| ||||| |||	13
Lime	||||| ||||	9
	Total	50

   b 13 students
   c 18%
   d chocolate
   e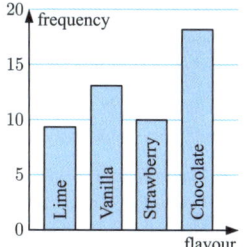
   Favourite ice cream flavours

5 a Eye colour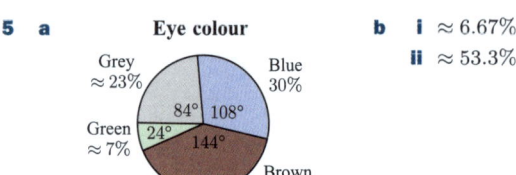

b i ≈ 6.67%   **ii** ≈ 53.3%

6 a 60 fines   **b** Driving offences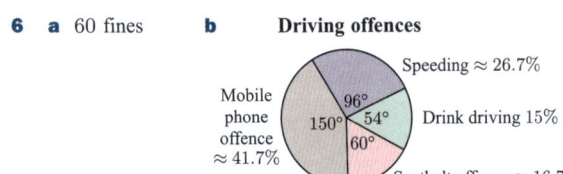

c i false   **ii** true   **iii** true

7 a

Response	Frequency
Excellent	8
Good	30
Fair	25
Poor	27

b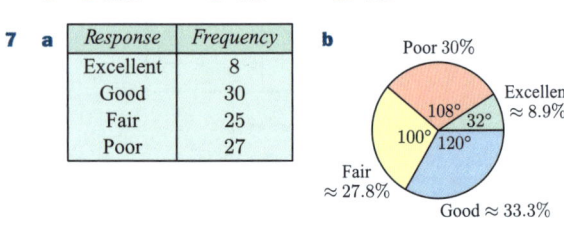

8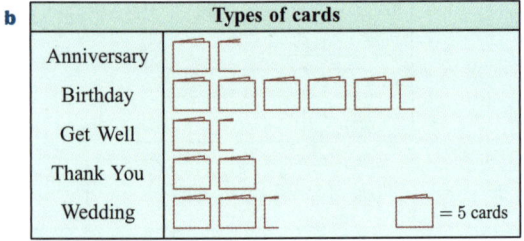

9 a 7 get well cards sold

b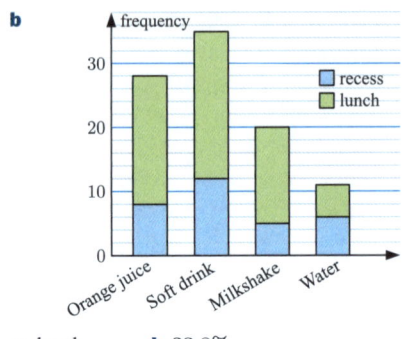

c 18.75%

EXERCISE 14D

1 a 4 students   **b** 6 students   **c i** east   **ii** south
  d class A   **e** 40%

2 a i 25 members   **ii** 25 members
  b adult members   **c** squash   **d** £2200

3 a 10 students   **b** Redstone
  c i Redstone   **ii** Hillsvale

4 a Tuesday   **b** 87.5%   **c** small
  d Monday; this is the only day that large outsold the other sizes.

5 a • Graph for soft drink (lunch) is incorrect.
  • Graph for water (lunch) too large, and graph for water (recess) too small. (The numbers were reversed.)

b

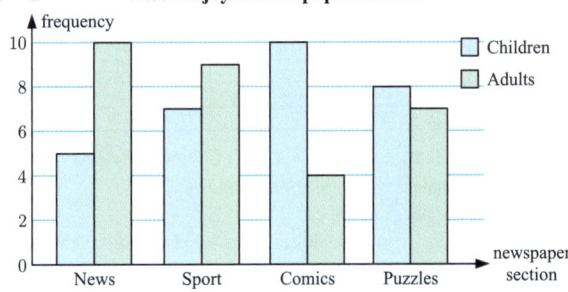

c lunch   **d** 23.8%

6 a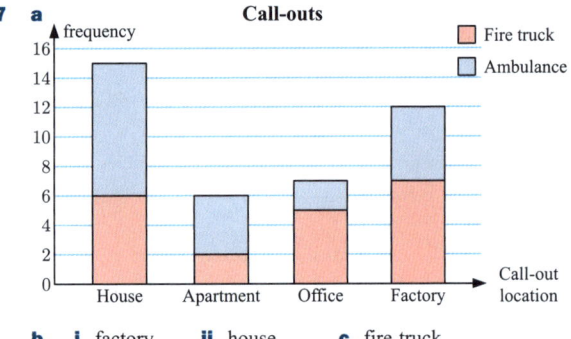

b i comics   **ii** news

c News and comics, as adults are more interested in the news whilst children are more interested in comics.

7 a Call-outs

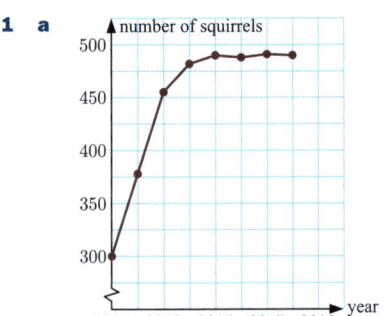

b i factory   **ii** house   **c** fire truck

EXERCISE 14E

1 a

b The squirrel population increased steadily from 2011 to 2015, at an ever decreasing rate. It then seemed to stabilise from 2015 onwards with only minimal change.

2 a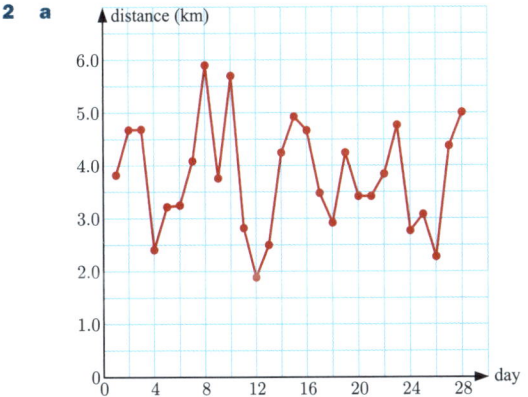

b 12 days   **c** 8th February

3 a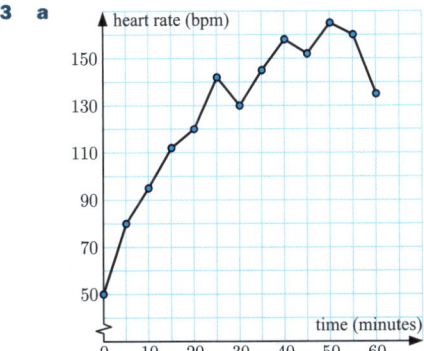

b 50 minutes
c There is an upward trend for the first 50 minutes, after which Jana's heart rate begins to decrease.

4 a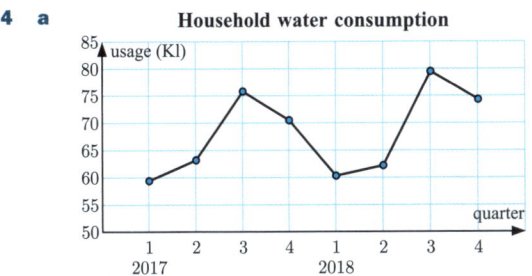

b i 3rd quarter   **ii** 1st quarter
People tend to use more water in summer (July to September) than in winter (January to March).

5 a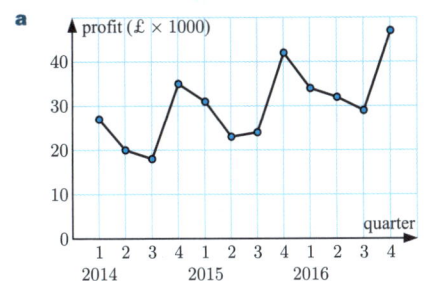

b The shop generally makes the most profit in the 4th quarter. This could result from increased sales in the holiday period.
c There is an increasing trend in profit.

REVIEW SET 14A

1 a 24%   **b** 30%   **c** Public transport, Schools, Parks
2 a £17   **b** £28   **c** 1.5 kg
3 a 5 litres   **b i** Friday   **ii** Wednesday
  c i 30 litres   **ii** 55 litres
4 a 200 males   **b** $\approx 55\%$
  c 65 - 74 years and 75 years and over
5 a 150 litres   **b** 50 litres   **c** 600 litres
6 a

b brown   **c** 47.5%
7 a £40   **b** April   **c** £15
8 a 360 students   **b**

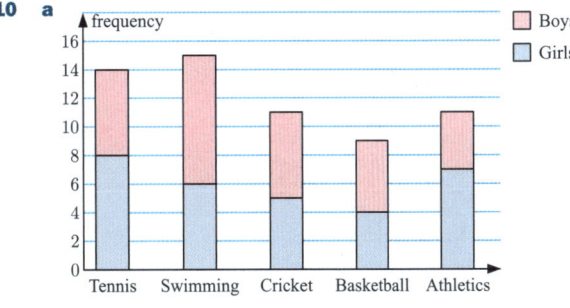

c As we do not know how many students are from School B, we cannot say whether this statement is valid. It would be appropriate to say that "School B had a higher proportion of students who scored a distinction than School A.".

9 a i 3 calf injuries   **ii** 5 calf injuries   **b** Panthers
  c i hamstring injuries   **ii** knee injuries

10 a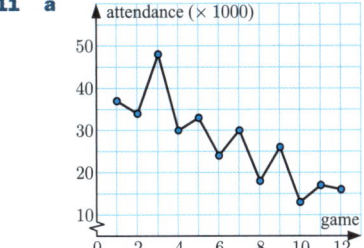

b girls   **c** 36.4%   **d** swimming

11 a

attendance graph

b The data shows a decrease in attendance rates as the season progresses.

REVIEW SET 14B

1 a i 865 km   **ii** 129 km   **b** Aberdeen   **c** 1379 km

2 a 31.0 g   **b** 45 mg   **c i** dietary fibre   **ii** 25 g
3 a i 28%   **ii** 24%   **b** 80%
   c i 500 million tonnes   **ii** 70 million tonnes
4 a 955 billion barrels   **b** 120 billion barrels   **c** $\approx 15.7\%$
5 a i 8 fish   **ii** 9 fish
   b

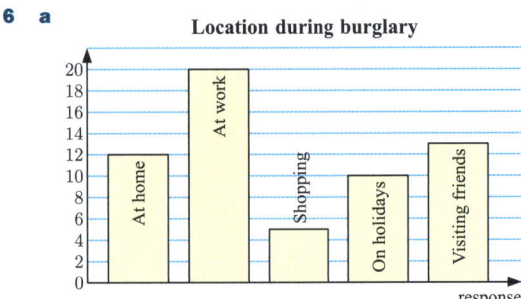

Fish catches

6 a

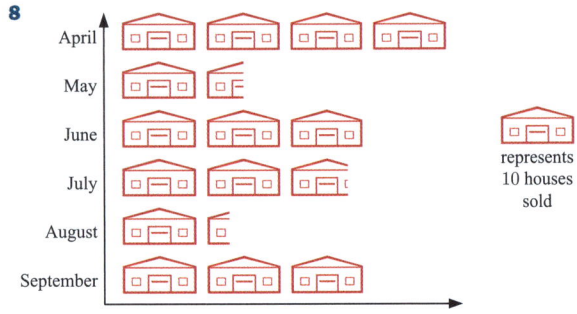

Location during burglary

b $8.\dot{3}\%$

7 a There is a slowly increasing trend.
   b International sales have increased and are now much larger than the UK sales.
   c 2017; international exposure could be a reason for the increase in international sales.

8

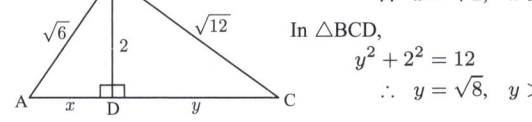

represents 10 houses sold

9 a 7 hours   **b** 1 hour   **c** REM sleep   **d** Wednesday
10 a • missed day 7 from the line
   • did not connect day 13 to day 14
   • did not title the graph
   b Liverpool's daily rainfall

c The trend is generally increasing throughout the fortnight but has decreased from day 13 to day 14.

EXERCISE 15A

1 a $x = \pm 3$ **b** $x = \pm 7$ **c** $x = \pm 6$ **d** $x = 0$
   e $x = \pm 1$ **f** $x = \pm\sqrt{17}$ **g** $x = \pm\sqrt{23}$
   h $x = \pm 10$ **i** no real solutions   **j** no real solutions
   k $x = \pm\sqrt{27}$ **l** no real solutions
2 a $x = \pm 2$ **b** $x = \pm 3$ **c** $x = \pm 5$
   d $x = \pm 4$ **e** $x = \pm 6$ **f** $x = \pm 5$
   g $x = \pm\sqrt{50}$ **h** $x = \pm\sqrt{44}$ **i** $x = \pm\sqrt{12}$
3 a $x = \pm 3$ **b** $x = \pm 4$ **c** $x = \pm 2$
   d $x = \pm 4$ **e** $x = \pm\sqrt{30}$ **f** $x = \pm\sqrt{20}$
   g $x = \pm 3$ **h** $x = \pm\sqrt{7}$ **i** $x = \pm\frac{1}{\sqrt{5}}$

EXERCISE 15B

1 a 5 cm   **b** ≈ 8.60 m   **c** ≈ 8.54 cm
   d ≈ 6.32 m   **e** ≈ 4.24 m   **f** ≈ 10.0 cm
2 a 6 cm   **b** ≈ 5.66 cm   **c** ≈ 6.71 m
   d ≈ 8.49 km   **e** ≈ 4.49 cm   **f** ≈ 7.07 m
3 a ≈ 6.40 cm   **b** ≈ 5.74 m   **c** ≈ 4.36 cm
   d ≈ 5.29 m   **e** ≈ 8.60 mm   **f** ≈ 11.6 m
4 a $y = \sqrt{3}$ **b** $y = \sqrt{2}$ **c** $y = \sqrt{6}$
5 a $x = \sqrt{5},\ y = \sqrt{6}$ **b** $x = 2,\ y = \sqrt{13}$
   c $x = 5,\ y = \sqrt{26}$ **d** $x = 4,\ y = \sqrt{33}$
   e $x = 2\sqrt{51}$ **f** $x = \sqrt{37}$
6 a $x = \sqrt{27}$ **b** $x = \sqrt{52}$ **c** $x = 2$
7 $AC = \sqrt{39}$ m ≈ 6.24 m
8 Hint: In $\triangle ABD$, $x^2 = 6 - 2^2$
$\therefore\ x = \sqrt{2},\ x > 0$
In $\triangle BCD$,
$y^2 + 2^2 = 12$
$\therefore\ y = \sqrt{8},\ y > 0$
In $\triangle ABC$, $AC^2 = (\sqrt{6})^2 + (\sqrt{12})^2 = 18$ $\therefore\ AC = \sqrt{18}$
9 a $\sqrt{17}$ cm   **b** $\sqrt{29}$ m   **c** $\sqrt{41}$ m

EXERCISE 15C

1 b and **f** are right angled.
2 a right angled at C   **b** not right angled   **c** right angled at P
3 yes
4 a The ground is flat and the street lamp is straight.   **b** no

EXERCISE 15D

1 $\sqrt{89} \approx 9.43$ cm   **2** $\sqrt{92.48} \approx 9.6$ cm
3 ≈ 4.71 m   **4** $\sqrt{34} \approx 5.83$ cm   **5** ≈ 3.48 km
6 ≈ 8.38 m   **7** 15 km   **8** ≈ 38.2 m
9 $8^2 + 11.55^2 = 197.4025$ and $14.05^2 = 197.4025$
$\therefore$ is right angled.
10 a Train A: 135 km, Train B: 210 km   **b** ≈ 250 km
11 ≈ 58.0 m   **12** ≈ 8.66 cm   **13** 8 cm   **14** ≈ 7.52 m
15 ≈ 2.72 m   **16** ≈ 10.6 m   **17** ≈ 42.4 cm $\times$ 42.4 cm
18 ≈ 6.96 cm $\times$ 20.9 cm   **19** ≈ 18.5 cm

REVIEW SET 15A

1 a $x = \pm 9$ **b** $x = \pm 5$ **c** no real solutions
2 a $x = 3\sqrt{5}$ **b** $x = 12$ **c** $x = \sqrt{32}$ **3** ≈ 9.2 cm

Answers 677

4 **a** no, $8^2 + 9^2 \neq 11^2$ **b** yes, $(\sqrt{2})^2 + 3^2 = (\sqrt{11})^2$
5 **a** $y = \sqrt{10}$ **b** $x = 4$, $y = \sqrt{65}$ 6 ≈ 7.60 m
7 **a** Alex: 5 km, Boris: 2 km **b** ≈ 5.39 km
8 **a** ≈ 2.68 m **b** ≈ 9.88 m
9 **Hint:** Find the two lengths that make up the base of the triangle, then use Pythagoras in the largest triangle.
10 1.6 m 11 7.54 m

REVIEW SET 15B
1 **a** $x = \pm\sqrt{5}$ **b** $x = \pm\sqrt{12}$ **c** $x = \pm\sqrt{13}$
2 **a** $x = 6$ **b** $x \approx 7.81$ **c** $x \approx 10.2$
3 ≈ 6.4 m 4 $x = \sqrt{7}$, $y = 4$
5 right angled at B as $(8)^2 + (\sqrt{17})^2 = 8 + 17 = 25 = 5^2$
6 **a** $x = 40$ **b** **i** ≈ 2.26 m **ii** ≈ 9.05 m
  c ≈ 22.6 m
7 **a** AB = $\sqrt{85}$ m ≈ 9.22 m
  b AB = $\sqrt{33}$ cm ≈ 5.74 cm
8 ≈ 1.42 m 9 ≈ 21.5 m
10 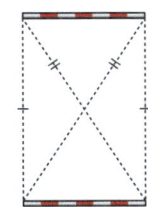 This means that the shape made by the poles and the lines connecting the ends of the poles is a rectangle.
∴ the poles are parallel.

EXERCISE 16A
1 **a** 38.8 cm **b** 17.8 cm **c** 11.9 km **d** 29 m
  e 11.6 cm **f** 30 m **g** 17.6 m **h** 99.6 km
  i 19.1 m
2 **a** ≈ 22.0 cm **b** ≈ 25.1 m **c** ≈ 25.7 cm
  d ≈ 10.7 m **e** ≈ 441 m **f** ≈ 14.3 cm
  g ≈ 85.7 cm **h** ≈ 22.8 cm **i** ≈ 75.4 cm
3 **a** ≈ 8.09 cm **b** ≈ 19.6 mm **c** ≈ 8.09 cm
4 **a** 24.4 m **b** 97.6 m **c** £966.24 5 31.1 m
6 **a** **i** ≈ 565.5 m **ii** ≈ 502.7 m **b** ≈ 62.8 m
7 **a** 69.38 m **b** 146.08 m 8 133 times
9 £1531.53 10 155 cm
11 **a** ≈ 39.0 m **b** 14 posts **c** £936.16

EXERCISE 16B.1
1 **a** 49 cm^2 **b** 7.5 mm^2 **c** 13 cm^2
  d 186.34 mm^2 **e** 5.46 cm^2 **f** 1.02 m^2
2 **a** 30 cm^2 **b** ≈ 31.2 cm^2 **c** ≈ 26.9 cm^2
  d 12 cm^2 **e** ≈ 27.7 cm^2 **f** ≈ 4.33 cm^2

EXERCISE 16B.2
1 **a** ≈ 28.3 m^2 **b** ≈ 38.5 cm^2
2 **a** ≈ 12.6 cm^2 **b** ≈ 56.5 m^2 **c** ≈ 26.2 m^2
  d ≈ 30.2 cm^2

EXERCISE 16B.3
1 **a** 112 m^2 **b** 39 m^2 **c** 74 m^2
  d 84 cm^2 **e** 31.5 cm^2 **f** 189 cm^2
2 **a** 36 cm^2 **b** 65 cm^2 **c** 387 cm^2
3 **a** ≈ 6.85 cm^2 **b** ≈ 30.6 cm^2 **c** ≈ 30.9 cm^2
  d ≈ 6430 m^2 **e** ≈ 113 cm^2 **f** ≈ 36.9 cm^2

EXERCISE 16B.4
1 2475 kg 2 **a** 160 tiles **b** £1504
3 **a** ≈ 9.42 m^2 **b** £74.46
4 ≈ 2.85 m^2 5 $\approx 46.2\%$ 6 $\approx 21.5\%$
7 The square has the larger area, by 6.25 cm^2. A rectangle of fixed perimeter will have maximum area when its side lengths are equal.
8 **a** **i** ≈ 245.74 cm^2 **ii** ≈ 214.47 cm^2
  iii ≈ 163.08 cm^2 **iv** ≈ 180.96 cm^2
  b ≈ 804.25 cm^2, area of whole pizza ≈ 804.25 cm^2
9 $\approx 25.5\%$
10 **a** 4 m **b** £73.60 **c** ≈ 5.85 m^2 **d** $\approx 8.58\%$

EXERCISE 16C.1
1 **a** 54 cm^2 **b** 121.5 cm^2 **c** 576.24 mm^2
2 **a** 276 cm^2 **b** 6880 mm^2 **c** 8802 m^2
3 **a** ≈ 198 m^2 **b** ≈ 496 cm^2 **c** ≈ 148 cm^2
4 **a** 360 cm^2 **b** 340 m^2 **c** ≈ 9840 cm^2
5 **a** ≈ 1.73 m^2 **b** 13 056 cm^2 **c** $\approx 63 500$ cm^2
6 **a** ≈ 1010 m^2 **b** $\approx £25 300$
7 £3717 8 26.52 m^2

EXERCISE 16C.2
1 **a** ≈ 207 cm^2 **b** ≈ 339 cm^2 **c** ≈ 196 cm^2
  d ≈ 56.7 m^2 **e** ≈ 124 m^2 **f** ≈ 79.5 cm^2
2 **a** ≈ 204 cm^2 **b** ≈ 56.5 m^2 **c** ≈ 298 cm^2
  d ≈ 49.5 mm^2 **e** ≈ 302 cm^2 **f** ≈ 95.1 m^2
3 **a** ≈ 50.3 cm^2 **b** ≈ 707 m^2 **c** ≈ 201 m^2
  d ≈ 145 km^2 **e** ≈ 84.8 cm^2 **f** ≈ 115 mm^2
4 ≈ 18.1 cm^2 5 **a** $l \approx 5.39$ **b** ≈ 33.8 cm^2
6 **a** ≈ 179.1 cm^2 **b** ≈ 238.8 cm^2 **c** ≈ 216.8 m^2
7 **a** ≈ 251 m^2 **b** 11 cans **c** £577.50
8 **a** $\approx 5.15 \times 10^8$ km^2 **b** $\approx 3.65 \times 10^8$ km^2
  c **i** $\approx 1.89\%$ **ii** $\approx 6.50\%$
9 £1100 10 ≈ 213 cm^2

EXERCISE 16D.1
1 **a** 96 m^3 **b** 343 mm^3 **c** 12 cm^3
2 **a** ≈ 352 cm^3 **b** ≈ 35.3 m^3 **c** ≈ 1.54 m^3
3 **a** 45 cm^3 **b** 320 cm^3 **c** 704 cm^3
  d 432 cm^3 **e** 288 cm^3 **f** ≈ 2320 cm^3
4 **a** 2000 cm^3 **b** ≈ 0.982 cm^3 **c** ≈ 183 cm^3
  d 2196 cm^3 **e** ≈ 2380 cm^3
5 15 915 handles
6 **a** **i** ≈ 3.93 m^3 **ii** ≈ 9.05 m^3 **b** ≈ 13.0 m^3
  c £1842.42
7 **a** 32.5 m^2 **b** 195 m^3
8 **a** ≈ 5.94 m^2 **b** ≈ 3.56 m^3
9 **a** ≈ 2.54 cm^2 **b** $\approx 0.006\,36$ m^3 **c** ≈ 9.54 kg

EXERCISE 16D.2
1 **a** 32 m^3 **b** ≈ 66.0 cm^3 **c** 120 cm^3
  d ≈ 233 m^3
2 **a** ≈ 37.7 m^3 **b** ≈ 738 cm^3 **c** ≈ 42.7 m^3
  d ≈ 1.89 m^3
3 8820 m^3 4 ≈ 2.48 m^3 5 ≈ 258 cm^3
6 611 garden stakes

678 Answers

EXERCISE 16D.3
1 **a** ≈ 268 cm^3 **b** ≈ 102 cm^3 **c** ≈ 82.3 cm^3
  d ≈ 1.07 m^3
2 **a** ≈ 905 mm^3 **b** ≈ 718 cm^3
3 **a** ≈ 80.4 cm^3 **b** ≈ 101 m^3 **c** ≈ 22.4 m^3
4 ≈ 0.905 m^3 5 ≈ 0.116 m^3 6 ≈ 15.0 cm^3
7 **a** 30.912 cm^3 **b** ≈ 905 mm^3 **c** 34 spheres
  d $\approx 0.484\%$

EXERCISE 16E
1 **a** 22.05 kl **b** ≈ 23.6 ml **c** ≈ 9.11 ml **d** 300 ml
2 368 bottles 3 ≈ 24.4 ml
4 **a** 13 500 m^2 **b** $\approx 23\,000$ kl
5 **a** ≈ 53.8 litres **b** ≈ 618 km 6 ≈ 1410 kl
7 ≈ 888 ml 8 **a** 1000 cm^3 **b** 6.20 cm **c** ≈ 484 cm^2
9 ≈ 0.707 m 10 **a** ≈ 18.1 kl **b** ≈ 0.177 m

REVIEW SET 16A
1 **a** perimeter $= 42$ m, area $= 60$ m^2
  b perimeter ≈ 256 mm, area ≈ 3320 mm^2
  c perimeter $= 28$ cm, area $= 40$ cm^2
  d perimeter $= 9.2$ m, area ≈ 4.51 m^2
2 **a** 9.1 m **b** £305 3 ≈ 565 m
4 **a** ≈ 105 cm^2 **b** 30 cm^2 **c** ≈ 177 cm^2
5 **a** 182.88 cm^2 **b** ≈ 2630 mm^2 **c** ≈ 102 cm^2
6 **a** ≈ 729 cm^3 **b** ≈ 303 cm^3 **c** ≈ 65.4 cm^3
7 **a** ≈ 209 cm^3 **b** ≈ 227 cm^2
8 **a** 5196 cans **b** ≈ 154 m^2
9 **a** ≈ 212 m^2 **b** $\approx 46\,700$ m^3 **c** $\approx$ £11 900 000
10 **a** 375 kl **b** 350 kl
11 **a** $\approx 2\,460\,000$ m^3 **b** $\approx 6.56 \times 10^6$ tonnes **c** ≈ 327 kl

REVIEW SET 16B
1 **a** perimeter ≈ 20.8 cm, area $= 20$ cm^2
  b perimeter ≈ 17.9 m, area ≈ 19.6 m^2
  c perimeter ≈ 19.8 m, area $= 22.5$ m^2
  d perimeter ≈ 41.7 cm, area ≈ 37.7 cm^2
2 ≈ 31.9 cm 3 ≈ 1.21 m^2
4 **a** ≈ 138 mm^2 **b** ≈ 692 cm^2 **c** 3024 m^2
  d ≈ 37.7 m^2
5 **a** $\approx 10\,900$ km **b** $\approx 3.79 \times 10^7$ km^2 6 2842 m^3
7 **a** ≈ 339 cm^3 **b** ≈ 1270 cm^3 **c** $\approx 14\,100$ cm^3
8 ≈ 12.3 km 9 No
10 **a** $\approx 23\,979.2$ cm^3 **b** 5184.2 cm^2 11 $\approx 53\,100$ kl

EXERCISE 17A.1
1 **a**

Input number	Calculation	Output number
1	4×1	4
2	4×2	8
3	4×3	12
4	4×4	16

b

Input number	Calculation	Output number
4	$4 + 3$	7
6	$6 + 3$	9
12	$12 + 3$	15
26	$26 + 3$	29

c

Input number	Calculation	Output number
0	$(0 + 2) \times 3$	6
1	$(1 + 2) \times 3$	9
2	$(2 + 2) \times 3$	12
5	$(5 + 2) \times 3$	21

d

Input number	Calculation	Output number
3	$(3 - 3) \times 2$	0
4	$(4 - 3) \times 2$	2
10	$(10 - 3) \times 2$	14
15	$(15 - 3) \times 2$	24

2 Input $\to \times 2 \to$ Output $\to -2 \to$
  $1 \to 2 \to 0$
  $2 \to 4 \to 2$
  $3 \to 6 \to 4$
  $4 \to 8 \to 6$

3 **a** Input Output
  $0 \xrightarrow{+3} 3 \xrightarrow{\times 2} 6$
  $1 \xrightarrow{+3} 4 \xrightarrow{\times 2} 8$
  $3 \xrightarrow{+3} 6 \xrightarrow{\times 2} 12$
  $5 \xrightarrow{+3} 8 \xrightarrow{\times 2} 16$

b Input Output
  $2 \xrightarrow{\div 2} 1 \xrightarrow{+4} 5$
  $6 \xrightarrow{\div 2} 3 \xrightarrow{+4} 7$
  $10 \xrightarrow{\div 2} 5 \xrightarrow{+4} 9$
  $18 \xrightarrow{\div 2} 9 \xrightarrow{+4} 13$

c Input Output
  $2 \xrightarrow{+10} 12 \xrightarrow{\div 3} 4$
  $8 \xrightarrow{+10} 18 \xrightarrow{\div 3} 6$
  $14 \xrightarrow{+10} 24 \xrightarrow{\div 3} 8$
  $23 \xrightarrow{+10} 33 \xrightarrow{\div 3} 11$

4 **a** A $M = 2n - 1$ B $M = \dfrac{(n + 7)}{3}$ **b** 2

EXERCISE 17A.2
1 **a** Input Output
  $\square \xrightarrow{\times 2} \square \xrightarrow{+2} 10$
  $\square \xrightarrow{\times 2} 8 \xrightleftharpoons[-2]{+2} 10$
  $4 \xrightleftharpoons[\div 2]{\times 2} 8 \xrightleftharpoons[-2]{+2} 10$
  The input number was 4.

b i Input Output
  $\square \xrightarrow{\times 2} \square \xrightarrow{+2} 2$
  $\square \xrightarrow{\times 2} 0 \xrightleftharpoons[-2]{+2} 2$
  $0 \xrightleftharpoons[\div 2]{\times 2} 0 \xrightleftharpoons[-2]{+2} 2$
  The input number was 0.

Answers 679

 ii Input → ×2 → □ → +2 → 4
     □ → ×2 → 2 ⇌ +2/−2 → 4
     1 ⇌ ×2/÷2 → 2 ⇌ +2/−2 → 4
     The input number was 1.

2 **a** {1, 3, 11}   **b** {2, 3, 4}   **c** {0, 2, 5}
3 **a** If $x^2 = k$ then $x = \pm\sqrt{k}$.
   b **i** ±1   **ii** ±2   **iii** ±4

EXERCISE 17B

1 **a** $A = 5 \times 15$ **b** $A = 5p$ **c** $A = tp$
2 **a** $A = 2000 + 150 \times 8$ **b** $A = 2000 + 150w$
   c $A = 2000 + dw$ **d** $A = P + dw$
3 **a** $C = 40 + 60 \times 5$ **b** $C = 40 + 60t$
   c $C = 40 + xt$ **d** $C = F + xt$
4 **a** $P = 10 \times 3 - 1(15 - 10)$ **b** $P = 3c - 1(20 - c)$
   c $P = 3c - 1(a - c)$
5 **a** $D = 4 \times 6 + 2 \times (4 - 1)$ **b** $D = 5m + 3 \times (5 - 1)$
   c $D = 8m + b(8 - 1)$ **d** $D = mp + b(p - 1)$
6 **a** $G = 2 \times (3 - 1) + 3 \times (2 - 1)$
   b $G = 3 \times (5 - 1) + 5 \times (3 - 1)$
   c $G = 4 \times (4 - 1) + 4 \times (4 - 1)$
   d $G = m(n - 1) + n(m - 1)$
7 **a** $P = (2x + y)$ cm   **b** $P = (4x + 6)$ m
   c $P = (2\pi r + 2d)$ m
8 **a** $A = 2ab + \frac{\pi a^2}{2}$ **b** $A = ar + \frac{\pi r^2}{2}$
   c $A = aw + \frac{\pi w^2}{4}$ **d** $A = 2ar - \pi r^2$
   e $A = \frac{\pi(b^2 - a^2)}{8} + \frac{a\sqrt{b^2 - a^2}}{2}$
9 **a** $V = Al$ **b** $V = \frac{\pi d^2 h}{4}$ **c** $V = \frac{abc}{2}$
   d $V = \frac{2}{3}\pi r^3$ cm³   **e** $V = (\frac{2}{3}\pi r^3 + \pi r^2 h)$ m³
10 **a** $A = 2ab + 2bc + 2ac$ **b** $A = 6a^2 + 8ab$
   c $A = \pi r^2 + 2\pi rh + \pi rs$ **d** $A = 10rh + 6r^2 + 4\pi r^2$
11 **a** **i** The sector is $\frac{\theta}{360}$ of the circle with radius s.
   ∴ the arc is $\frac{\theta}{360} \times 2\pi s$.
12 Hint: Area of end section $= \pi R^2 - \pi r^2$
         $= \pi(R + r)(R - r)$

EXERCISE 17C

1 **a** ≈ 26.4 cm   **b** ≈ 17.8 cm   **c** ≈ 127 m
2 **a** 19.6 m   **b** ≈ 4.52 s
3 **a** ≈ 129 cm²   **b** ≈ 7.14 m
4 **a** ≈ 4260 cm³   **b** ≈ 1.06 cm   **c** ≈ 4.99 mm
5 **a** ≈ 707 cm²   **b** ≈ 39.9 cm
6 **a** $P = 4x$ cm, $x = 3$ **b** $P = 2\pi x$ cm, $x \approx 1.91$
7 **a** $A = \frac{\sqrt{3}}{4}x^2$ cm²   **b** 8 cm
8 **a** ≈ 1.34 s   **b** 81 cm

EXERCISE 17D

1 **a** $y = 7 - x$ **b** $y = x - 3$ **c** $y = \frac{1}{2} - \frac{1}{2}x$
   d $y = 2 - \frac{2}{5}x$ **e** $y = 5 - \frac{3}{4}x$ **f** $y = 2x - 8$
   g $y = 2 - \frac{2}{7}x$ **h** $y = 10 - \frac{5}{2}x$ **i** $y = \frac{2}{3}x + 4$
2 **a** $x = r - p$ **b** $x = \frac{z}{y}$ **c** $x = \frac{d - a}{3}$
   d $x = \frac{d - 2y}{5}$ **e** $x = \frac{p - by}{a}$ **f** $x = \frac{y - c}{m}$
   g $x = \frac{s - 2}{t}$ **h** $x = \frac{m - p}{q}$ **i** $x = \frac{6 - a}{b}$
3 **a** $y = \frac{t - z}{5}$ **b** $y = \frac{c - p}{2}$ **c** $y = \frac{a - t}{3}$
   d $y = \frac{n - 5}{k}$ **e** $y = \frac{a - n}{b}$ **f** $y = \frac{a - p}{n}$
   g $y = \frac{4 - c}{x}$ **h** $y = \frac{6 - w}{a}$ **i** $y = \frac{m + k}{t}$
4 **a** $z = \frac{b}{ac}$ **b** $z = \frac{q}{p}$ **c** $z = \frac{a}{d}$
   d $z = \frac{2d}{3}$ **e** $z = \frac{7n}{k}$ **f** $z = -\frac{pt}{q}$
   g $z = \pm\sqrt{2a}$ **h** $z = \pm\sqrt{bn}$ **i** $z = \pm\sqrt{m(a - b)}$
5 **a** $a = \frac{F}{m}$ **b** $r = \frac{C}{2\pi}$ **c** $d = \frac{V}{lh}$
   d $K = \frac{b}{A}$ **e** $h = \frac{2A}{b}$ **f** $T = \frac{100I}{PR}$
6 $h = \frac{A - 2\pi r^2}{2\pi r}$ or $h = \frac{A}{2\pi r} - r$
7 **a** $r = \sqrt{\frac{A}{\pi}}$ **b** $x = \pm\sqrt{aN}$ **c** $k = \pm\sqrt{\frac{M}{5}}$
   d $x = \sqrt[3]{\frac{n}{D}}$ **e** $x = -\sqrt{\frac{y + 7}{4}}$ **f** $Q = \pm\sqrt{P^2 - R^2}$
8 **a** $a = d^2 n^2$ **b** $l = 25T^2$ **c** $a = \pm\sqrt{b^2 + c^2}$
   d $d = \frac{25a^2}{k^2}$ **e** $l = \frac{gT^2}{4\pi^2}$ **f** $b = \frac{16a}{A^2}$
9 **a** $x = \frac{c - a}{3 - b}$ **b** $x = \frac{c}{a + b}$ **c** $x = \frac{a + 2}{n - m}$
   d $x = -\frac{a}{b + 8}$ **e** $x = \frac{a - b}{1 - c}$ **f** $x = \frac{e - d}{r + s}$
10 **a** $x = \frac{1 - 4y}{3}$ **b** $x = \frac{6 + z}{7}$ **c** $x = \frac{5a + b}{b - a}$
   d $x = \frac{k - 7}{2k - 1}$ **e** $x = \frac{6}{2 - 3y}$ **f** $x = \frac{2m}{4m + n}$
11 **a** $a = \frac{2 - bP}{P}$ **b** $r = \frac{8 - qT}{T}$ **c** $q = \frac{Ap - B}{A}$
   d $x = \frac{3 - Ay}{2A}$

EXERCISE 17E

1 **a** $\theta = \frac{360A}{\pi r^2}$
   b **i** ≈ 63.7°   **ii** ≈ 105°   **iii** ≈ 214°
2 **a** $a = \frac{d^2}{2bK}$ **b** **i** $a = \frac{9}{7}$ **ii** $a = \frac{81}{5}$
3 **a** $t = (H - 1)^2$
   b **i** 1 year   **ii** 4 years   **iii** $6\frac{1}{4}$ years

4 a $r = \sqrt[3]{\dfrac{3V}{4\pi}}$
  b i ≈ 2.12 cm **ii** ≈ 5.76 cm **iii** ≈ 62.0 cm
5 a $v = \sqrt{u^2 + 2as}$ **b i** ≈ 20.6 m/s **ii** ≈ 52.9 m/s
6 a $\approx 58.8\%$ **b** $w = \dfrac{Pl}{100-P}$ **c** 9 matches
  d 7 consecutive matches
7 a $\approx 2.01 \times 10^{20}$ Newtons **b** $d = \sqrt{\dfrac{Gm_1 m_2}{F}}$
  c i $\approx 1.50 \times 10^{11}$ m **ii** $\approx 1.43 \times 10^{14}$ m
8 b $p = \dfrac{3g}{g-1}$ **c** $0g$ $0p$, $4g$ $4p$

EXERCISE 17F
1 a 4 **b** 7 **c** 10 **d** 13 **e** 16 **f** $3n+1$
2 a 6 **b** 11 **c** 16 **d** 21 **e** 26 **f** $5n+1$
3 a 7 **b** 10 **c** 13 **d** 16 **e** 19 **f** $3n+4$
  g 34 **h** 304
4 a 4 **b** 10 **c** 16 **d** 22 **e** 28 **f** $6n-2$
  g 58 **h** 598
5 a i 4 **ii** 9 **iii** 16 **iv** 25 **b** $S_n = n^2$
6 a i 3 **ii** 7 **iii** 15 **iv** 31
  b As $3 = 2^2 - 1$, $7 = 2^3 - 1$, $15 = 2^4 - 1$, and so on, $S_n = 2^n - 1$

REVIEW SET 17A
1 a $M = \left(\dfrac{n}{2}\right)^2$
  b

Input number (n)	Calculation	Output number (M)
i 2	$\left(\dfrac{2}{2}\right)^2$	1
ii 4	$\left(\dfrac{4}{2}\right)^2$	4
iii 8	$\left(\dfrac{8}{2}\right)^2$	16
iv 12	$\left(\dfrac{12}{2}\right)^2$	36

2 a 4 **b** 7 **c** 3 **d** 1
3 a i $V = 6 \times 8$ litres **ii** $V = 8n$ litres
  iii $V = ln$ litres
  b $V = 25 + ln$ litres
4 a 90 km/h **b** 3900 km **5** $A = \dfrac{\sqrt{3}}{2}a^2 + 3ab$
6 a $x = \dfrac{3p-n}{m}$ **b** $x = \dfrac{5y}{7}$
7 a $k = T^2 + l^2$ **b** $k = -\sqrt{\dfrac{P+r}{2}}$
8 a $V = \tfrac{1}{12}\pi x^3$ m³ **b** ≈ 1.79 m
9 a 8 amperes **b** $r = \dfrac{E - IR}{I}$ **c** 0.15 ohms
10 a 4 **b** 7 **c** 10 **d** 13 **e** 16 **f** $3n+1$
11 a i ≈ 283.2 K **ii** ≈ 183.2 K **iii** ≈ 338.7 K
  b $F = \tfrac{9}{5}(K - 273.15) + 32$
  c i 104°F **ii** -459.67°F **iii** -99.67°F

REVIEW SET 17B
1 a 1 **b** 9 **c** -3
2 a $B = 15 + 25 \times 5$ **b** $B = c + 25p$ **c** $B = c + mp$

3 a $E = 2 \times (3-2) + 2 \times (5-2)$
  b $E = 2 \times (4-2) + 2 \times (8-2)$
  c $E = 2(m-2) + 2(n-2)$
4 a $M = 37$ **b** $r = 8$ **5** $V = 4x^2 y$ cm³
6 a $a = \dfrac{B+f}{d}$ **b** $a = \dfrac{9Q^2}{t^2}$ **c** $a = \dfrac{5 - G^2}{G^2}$
7 a $b = \dfrac{a}{a-1}$ **b** $b = \tfrac{3}{2}$; $3 \times \tfrac{3}{2} = 3 + \tfrac{3}{2} = 4\tfrac{1}{2}$ ✓
8 $(5n+3)$ matchsticks
9 a i $6 = 2 \times 3$ **ii** $12 = 3 \times 4$
  iii $20 = 4 \times 5$ **iv** $30 = 5 \times 6$
  b $S_n = n(n+1)$
10 a 1000 joules **b** $v = \sqrt{\dfrac{2E}{m}}$ **c** 8 m/s

EXERCISE 18A.1
1 a 1.3 **b** 0.9 **c** 1.15 **d** 0.65
  e 1.12 **f** 0.925
2 £4.20/kg **3** 80 people **4** £64 800 **5** 44 minutes
6 £9724 **7** 772.8 m

EXERCISE 18A.2
1 a 10% increase **b** 15% decrease **c** 40% decrease
  d 50% increase
2 a 15% increase **b** 28% decrease **c** 16.$\dot{6}$% increase
  d 7.5% increase **e** 12.5% decrease
3 a i $\approx 36.2\%$ **ii** $\approx 19.0\%$ **iii** $\approx 13.8\%$
  b $\approx 84.5\%$
4 a Germany: 0.364% increase, Italy: 4.30% increase, France: 5.19% increase, Spain: 4.25% increase, Greece: 2.73% decrease, Portugal: 1.90% decrease, United Kingdom: 7.89% increase
  b i United Kingdom **ii** Greece

EXERCISE 18A.3
1 a 20 cm **b** 60 kg **c** £125
  d 4000 litres **e** 50 000 people
2 80 nations **3** ≈ 2475 black rhinoceroses **4** £420

EXERCISE 18B.1
1 a £33.60 **b** £77.50 **c** £162 **d** £416 **e** £57.60
2 £1.17
3 a £3.68 per kg **b** £1.44 each **c** £2.64 each
4 £4.69 per kg **5 a** £27 **b** £12
6 33.$\dot{3}$% mark-up **7** 80% mark-up
8 a 462.5% mark-up **b** £37
9 £17.59 per kg **10** £1.05

EXERCISE 18B.2
1 a £56 **b** £120 **c** £156 **d** £262.50
2 £392 **3** £17 **4** £697
5 a £171 **b** £296 **c** £801 **6** 2 extra days
7 8% discount **8** 22.5% discount
9 £29 **10** £80 **11** £140

EXERCISE 18B.3
1 a £10 **b** £60 **2 a** £9 **b** £54
3 £306 **4** £76.40

EXERCISE 18C
1 £40.92 **2** ≈ 124.7 cm **3** £504 **4** 27 025 people

Answers

5 £42 **6 a** £336 000 **b** ≈ £310 000
7 £1.40 **8** 9.18% **9** 20% mark-up **10** ≈ 3.61 kg
11 No, $0.87 \times 1.13 \approx 0.9831$
So the yield after 2 years is only 98.31% of the original yield.
12 No, $1.2 \times 1.2 \times 1.2 = 1.728$
Julia's allowance will be 72.8% higher when she is 12.

EXERCISE 18D
1 ≈ £533.03 **2** £2426.76 **3** £3400
4 a ≈ £1842.38 **b** 5 years **5** £1655.36
6 ≈ 1.38% decrease **7** ≈ 1.70% depreciation
8 a ≈ £268 000 **b** ≈ 0.748% depreciation

EXERCISE 18E
1 a £600 **b** £400 **c** £22 000 **d** £37 500
2 a £3120 **b** £14 720 **c** £9000 **d** £24 150
3 a £31 000 **b** £645.84
4 a £8695 **b** £144.92 **5** 5 years **6** £3500

EXERCISE 18F
1 £4410 **2 a** £7498.58 **b** £498.58
3 a £7311.62 **b** £811.62
4 £1270.50 **5 a** £7827.36 **b** ≈ 26.2%
6 a £9478.84 **b** 5 years
7 a Option 1: £10 160 after 3 years
Option 2: £10 077.67 after 3 years
Option 1 should be chosen as you would have about £82 more than option 2.
b Yes. Option 1: ≈ £11 600, Option 2: ≈ £11 754.62
After 5 years, option 2 would be worth £154.62 more than option 1.
8 a £2407.26 **b** Yes, but only by about £23.75.
9 a After n years, the amount in the account would be
$A = 5000 \times \underbrace{1.04 \times 1.04 \times \ldots \times 1.04}_{n \text{ times}}$
$= 5000 \times 1.04^n$
b i £5624.32 **ii** £6579.66 **c** 12 years

REVIEW SET 18A
1 a 6.25% increase **b** 28.6% decrease
2 54 900 people **3** £53.20 **4** ≈ £532 000
5 4 hectares **6** £2883.60
7 a £23 506.89 **b** £21 156.20 **8** £840
9 a £6700 **b** £139.59
10 a £10 099.82 **b** £2099.82
11 a As the caravan is depreciating, the multiplier will be 0.84. After n years, the value of the caravan will be
$V = 15\,000 \times \underbrace{0.84 \times 0.84 \times \ldots \times 0.84}_{n \text{ times}}$
$= 15\,000 \times 0.84^n$
b £6273.18 **c** 7 years

REVIEW SET 18B
1

Offence	Old fine	New fine
Speeding	£200	£210
Drink driving	£840	£882
Not wearing seatbelt	£260	£273
Illegal parking	£50	£52.50

2 £41.65 **3** ≈ £53.30 **4** £80

5 a £100 **b** £20 **6** 6 cm
7 a ≈ 26.0% **b** 1950 customers
8 a £1824.98 **b** £2298.95
c Yes, she would have earnt an extra £477.45.
9 £20 100
10 Option A: £6400, Option B: £6312.38
Option A would make you £87.62 more than option B.
11 a £34 333.36 **b** 7 years

EXERCISE 19A
1 a 5 miles in 1 hour **b** 15 pounds in 1 hour
c 7 litres in 1 second **d** 99 pence for 1 litre
e 30 kg in 1 hour **f** 14 grams in 1 minute
g 96 euros in 1 day **h** 66 metres in 1 second
i 21 ml in 1 hour
2 a £/h **b** miles/h **c** pence per litre
d words/min **e** °C/min
3 a 16 km per litre **b** 52 miles per hour
c 3.5 litres per s **d** £1.09 per litre **e** £17 per hour
4 75 beats/min **5** 39 megajoules/day
6 a Annie: ≈ 0.56 miles per min, Victoria: 0.5 miles per min
b train
7 a Xinsong: £8 per hour, Jay: £9 per hour **b** Jay
8 a 400 litres per day **b** 8000 litres
9 a £7.40 per hour **b** £140.60 **10** 24 300 litres
11 a 15 trees **b** 16 hours
12 a £3 per metre **b** £81 **c** 150 m
13 a i 18.5 km/litre **ii** ≈ 5.41 litres/100 km
b 80 litres **c** £108
14 a i £1200 **ii** £2080 **b** 64 people

EXERCISE 19B.1
1 a 25 miles/h **b** 30 miles/h **c** 12 miles/h
d 900 miles/h
2 Jason's average speed is 72.5 miles/h which is over the speed limit.
∴ he has broken the law.
3 a 96 miles **b** ≈ 1.69 hours (1 h 41 min)
4 ≈ 1.6 m/s **5 a** 350 km **b** 2 hours
6 a Yiren: ≈ 2.67 m/s, Sean: 2.5 m/s **b** Yiren by 50 s

EXERCISE 19B.2
1 a Yes, as the travel graph is a straight line.
b 100 miles **c** 50 miles/h
2 a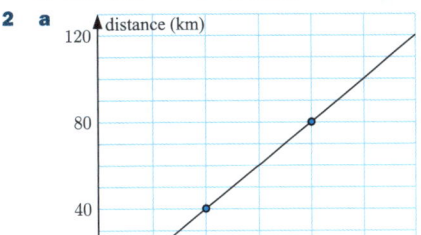
b 60 km
c 5 hours
d 20 km/h

3 a 100 km **b** 2 hours **c** 50 km/h **d** 100 km
e 1 hour **f** 100 km/h **g** 200 km **h** 3 hours
i ≈ 66.67 km/h

EXERCISE 19C

1 **a** 4 g per cm^3 **b** 9 g per cm^3 **c** 22.6 g per cm^3
2 ≈ 1.06 g per cm^3
3 die **B** (6.25 g/cm^3 compared with ≈ 5.79 g/cm^3)
4 Petrol has a lower density than water so it will float on water. So, the upper layer is petrol.
5 The density of the doorstop is ≈ 1.33 g/cm^3, which is higher than the density of water. So, the doorstop will sink in water.
6 **a** 428.4 g **b** ≈ 56.0 cm^3
7 **a** 540 g **b** 82.4 g
8 ≈ 5.58 g
9 **a** ≈ 5.31 cm^3 **b** ≈ 166.7 cm^3
10 **a** ≈ 13.7 cm^3 **b** ≈ 8.48 g/cm^3

EXERCISE 19D

1 50 N/m^2 **2** 20 N **3** 80 cm^2 **4** ≈ 66.7 N/cm^2
5 **a** ≈ 0.0566 N/cm^2 **b** ≈ 212 N
6 **a** 500 N/m^2 **b** ≈ 166.7 N/m^2

EXERCISE 19E

1 **a** £3.80 per ball **b** £1.69 per kg **c** 0.56 pence per g
   d 79.2 pence per litre **e** £2 per m **f** 124 pence per litre
2 **a** 400 g at $\approx$ £1.28 per 100 g
   b 200 ml at $\approx$ £0.90 per 100 ml
   c 4 boxes at $\approx$ £0.51 per box
   d 36 pack at $\approx$ £0.25 per tablet
   e 50 m at £0.73 per 10 m
   f 250 g at 8.6 cents per 10 g
3 **a** small toothpaste: £0.29 per 10 g
       large toothpaste: $\approx$ £0.25 per 10 g
   b 160 g **c** Yes, rate is $\approx$ £0.19 per 10 g.

EXERCISE 19F

1 **a** 180 Australian dollars **b** 450 Australian dollars
   c ≈ 30.56 pounds **d** ≈ 611.11 pounds
2 **a** 35.60 pounds **b** 240.30 pounds
   c ≈ 1011.24 euros **d** ≈ 1685.39 euros
3 ≈ 232.83 pounds
4 Yes, Estelle has 637 British pounds and the cost of accommodation for a 5 night trip is 600 British pounds.
5 **a** 3580 New Zealand dollars **b** 2200 New Zealand dollars
   c ≈ 1229.05 pounds

EXERCISE 19G.1

1 10 800 litres/hour
2 **a** 1 beat/s **b** 3600 beats/h **c** 86 400 beats/day
3 **a** 125 ml/s **b** 450 litres/h
4 **a** 0.3 m/day **b** 0.0125 m/h **c** 12.5 mm/h
5 **a** 1225 g/week **b** 1.225 kg/week
6 300 N/m^2 **7** 6800 kg/m^3

EXERCISE 19G.2

1 **a** 108 km/h **b** 252 km/h **c** 18 km/h
2 **a** 20 m/s **b** 40 m/s **c** 2.5 m/s
3 **a** 720 km/h **b** 162 km/h **c** 97.2 km/h
   d 2880 km/h

4 **a** ≈ 13.89 m/s **b** ≈ 30.56 m/s **c** ≈ 5.83 m/s
   d 150 m/s
5 **a** ≈ 10.42 m/s **b** ≈ 37.52 km/h
6 **a** ≈ 37.11 km/h **b** ≈ 62.07 km/h **c** ≈ 55.38 km/h
   d 6 km/h
7 **a** 300 m in 40 seconds = 7.5 m/s = 27 km/h
   b If Sophie could maintain a speed of 27 km per hour for one hour, she would have run 27 km. However, it is unlikely that she could do this.

REVIEW SET 19A

1 14 litres per minute **2** **a** 2.5 beats/s **b** 194.4 km/h
3 train (96.25 km/h versus ≈ 84.71 km/h)
4 **a** 150 km **b** 2 hours **c** 50 km/h **5** 10 minutes
6 **a** 13 km/h **b** ≈ 3.61 m/s **7** 2.8 g/cm^3
8 **a** 3560 Singapore dollars **b** ≈ 367.98 pounds
9 2.4 N/m^2 **10** ≈ 0.014 g/cm^3
11 80 g at $\approx$ £0.27 per 10 g **12** ≈ 833.3 cm^3

REVIEW SET 19B

1 £27.50 per hour **2** **a** ≈ 22.22 m/s **b** 12.5 mm/month
3 **a** £3.50 per kg **b** £8.75
4 **a** **i** 66 km/h **ii** 18 km/litre
   b It is cheaper for Trent to ride his motorcycle.
5 41 pence per bar
6 Density of object ≈ 0.71 g/cm^3, which is less than the density of water. So, the object would float.
7 75 N **8** **a** 1.3 kg at $\approx$ £4.23 per kg **b** 1.3 kg packet
9 **a** 20 miles **b** 30 min **c** 40 miles/h
   d 52 miles/h **e** at C and D
10 6550 Mexican pesos **11** ≈ 84.8 g
12 **a** 50 km/h **b** 6.5 hours **c** ≈ 13.9 m/s

EXERCISE 20A

1 P(4, 2), Q(−1, −3), R(3, −1), S(−2, 5), T(7, −3)
2 **a** 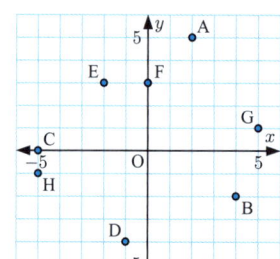   **b** A - first,
B - fourth,
C - on x-axis,
D - third,
E - second,
F - on y-axis,
G - first,
H - third

3 **a** quadrants 1 and 3 **b** quadrants 2 and 4
4 **a** 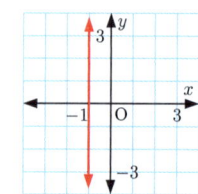 **b**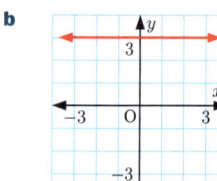

All points on a vertical line 1 unit left of the y-axis.

All points on a horizontal line 3 units above the x-axis.

Answers 683

c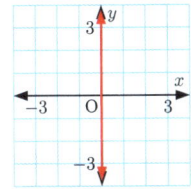
All points on the y-axis.

d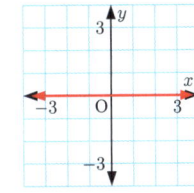
All points on the x-axis.

e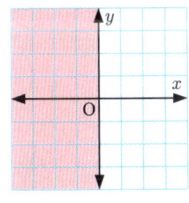
All points left of the y-axis.

f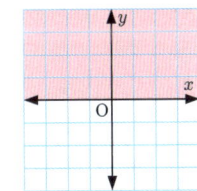
All points above the x-axis.

g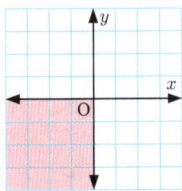
All points in quadrant 3.

h
All points in quadrant 2.

5 a A is $(-2, 3)$, B is $(4, 3)$, C is $(4, -1)$
   b D is $(-2, -1)$
6 a P is $(5, 1)$, Q is $(4, 3)$ **b** R is $(0, 1)$
   c S is $(4, -1)$

EXERCISE 20B

1 a m is the independent variable. I is the dependent variable.
b

m	0	1	2	3	4	5	6	7	8	9	10
I	300	320	340	360	380	400	420	440	460	480	500

c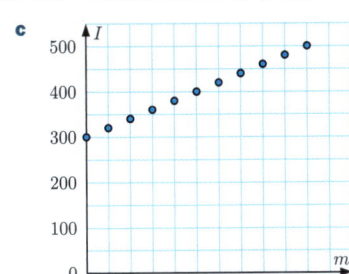

d yes
e No, as we do not have half members, etc.
f £20

2 a

n	0	1	2	3	4	5	6	7	8
S	3	5	7	9	11	13	15	17	19

b n is the independent variable. S is the dependent variable.
c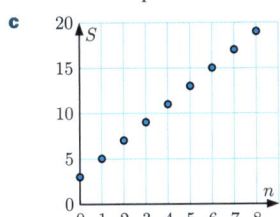

d yes
e Yes, we can pour $\frac{1}{2}$ bottle in the container.
f 2 litres
g 8 litres
h 6 bottles

3 a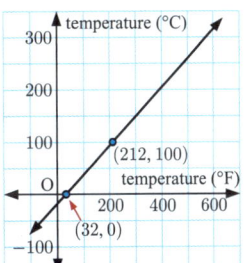

b $-40°$F is $-40°$C
c i $\approx 175°$C
   ii $\approx 63°$C

d

°F	0	180	302	392
°C	-18	82	150	200

EXERCISE 20C.1

1 a $\frac{2}{3}$ **b** -1 **c** $\frac{1}{3}$ **d** $-\frac{4}{3}$ **e** $\frac{1}{6}$
   f undefined   **g** -3 **h** $-\frac{1}{8}$ **i** 0

2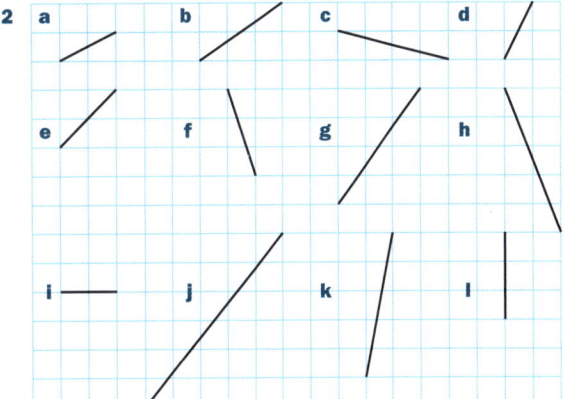

3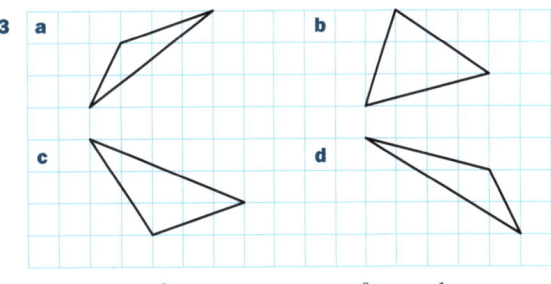

4 a $\frac{1}{6}$ **b** $\frac{2}{5}$ **c** 14   **d** $-\frac{3}{4}$ **e** $\frac{1}{5}$
5 a i 0   ii $\frac{1}{5}$ iii $\frac{3}{5}$ iv 1   v 2   vi 4
   vii undefined   viii -4 ix $-\frac{1}{2}$
   b i 0   ii undefined   iii increase in magnitude
6 a OP, PQ, RS, TU   **b** QR, ST, UV   **c** TU
   d ST   **e** VW   **f** PQ
7 a 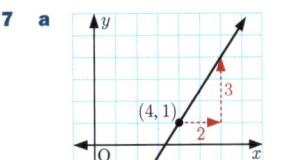   **b**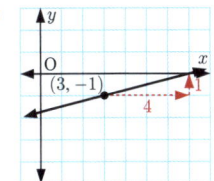

c 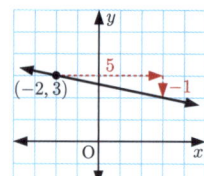 **d**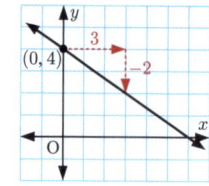

e 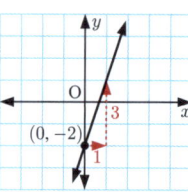 **f**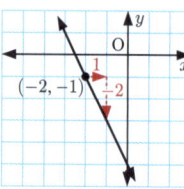

g 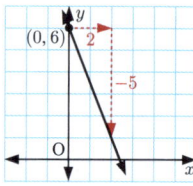 **h**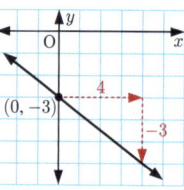

8 a 3 **b** $-\frac{1}{2}$ **c** 3 **d** 1 **e** $-\frac{1}{2}$ **f** -1

EXERCISE 20C.2

1 a $\frac{4}{3}$ **b** $-\frac{5}{6}$ **c** 0

2 a 3 **b** $-\frac{1}{2}$ **c** -6 **d** -5 **e** undefined **f** 0
  g 1 **h** $\frac{5}{2}$ **i** $-\frac{1}{3}$ **j** $\frac{3}{4}$ **k** $\frac{2}{3}$ **l** $\frac{9}{4}$

3 b $a = -14$ **4 b** $a = 12$

EXERCISE 20D

1 a i $\frac{2}{3}$ **ii** -1 **iii** $\frac{2}{3}$ **iv** $\frac{11}{9}$ **v** $\frac{1}{9}$ **vi** 1
  b AB and CD

2 a gradient of PS $= 3$, gradient of QR $= 3$
    ∴ PS is parallel to QR
  b gradient of PQ $= -\frac{1}{7}$, gradient of RS $= \frac{1}{4}$
    ∴ PQ is not parallel to RS.

3 a i gradient of AB $= 0$, gradient of CD $= 0$
    ∴ AB is parallel to CD
  ii gradient of AD $= \frac{3}{5}$, gradient of BC $= \frac{3}{5}$
    ∴ AD is parallel to BC
  b a parallelogram

4 a 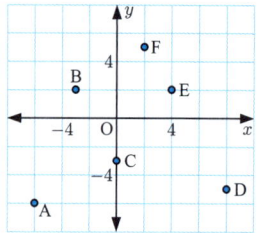 **b** BC and FD

5 a gradient of PQ $= \frac{1}{4}$, gradient of QR $= \frac{1}{4}$
    ∴ PQ is parallel to QR
  b As Q is a point on both PQ and QR, and PQ is parallel to QR, then the points P, Q, and R are collinear.

6 a $b = 5$ **b** $b = \frac{30}{7}$

EXERCISE 20E

1 a x-intercept is 1, y-intercept is -2
  b x-intercept is -3, y-intercept is -1
  c x-intercept is 4, y-intercept is 3
  d x-intercept is 3, y-intercept is 2
  e x-intercept is 2, y-intercept is -4
  f x-intercept is 0, y-intercept is 0

2 a x-intercept is -1, y-intercept is 1, gradient is 1
  b x-intercept is 3, y-intercept is 4, gradient is $-\frac{4}{3}$
  c x-intercept is -4, y-intercept is 3, gradient is $\frac{3}{4}$

3 a 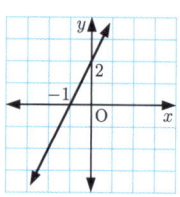 **b**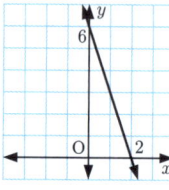
  gradient is 2   gradient is -3

  c 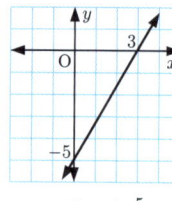 **d**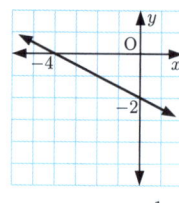
  gradient is $\frac{5}{3}$ gradient is $-\frac{1}{2}$

EXERCISE 20F

1 a $y = 4x$ **b** $y = x + 2$ **c** $y = -2x$
  d $x + y = 3$ **e** $y = 2x + 1$ **f** $y = 2x - 1$

2 a yes **b** no **c** yes **d** no **e** yes **f** no

3 a 5 **b** $-\frac{1}{2}$ **4 a** 9 **b** 4

5 a -6 **b** 10 **c** x-intercept is 10, y-intercept is -6

EXERCISE 20G.1

1 a i

x	-3	-2	-1	0	1	2	3
y	-3	-2	-1	0	1	2	3

  ii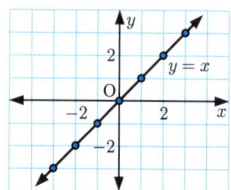

  b i

x	-3	-2	-1	0	1	2	3
y	3	2	1	0	-1	-2	-3

  ii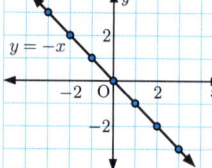

Answers 685

c i

x	−3	−2	−1	0	1	2	3
y	−6	−4	−2	0	2	4	6

ii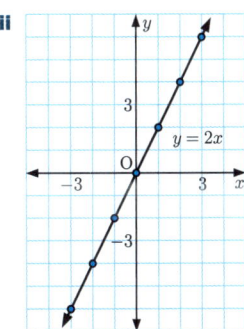

d i

x	−3	−2	−1	0	1	2	3
y	6	4	2	0	−2	−4	−6

ii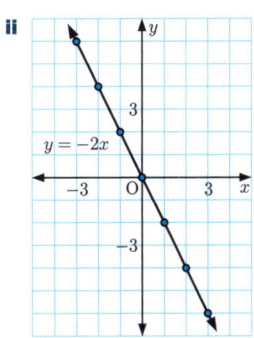

e i

x	−3	−2	−1	0	1	2	3
y	−7	−5	−3	−1	1	3	5

ii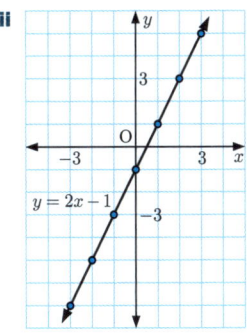

f i

x	−3	−2	−1	0	1	2	3
y	$1\frac{1}{4}$	$1\frac{1}{2}$	$1\frac{3}{4}$	2	$2\frac{1}{4}$	$2\frac{1}{2}$	$2\frac{3}{4}$

ii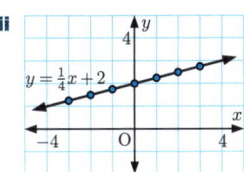

g i

x	−3	−2	−1	0	1	2	3
y	$2\frac{3}{4}$	$2\frac{1}{2}$	$2\frac{1}{4}$	2	$1\frac{3}{4}$	$1\frac{1}{2}$	$1\frac{1}{4}$

ii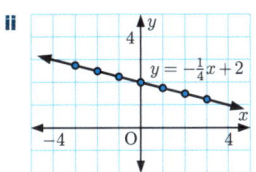

h i

x	−3	−2	−1	0	1	2	3
y	5	4	3	2	1	0	−1

ii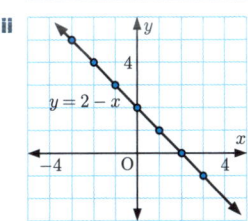

i i

x	−3	−2	−1	0	1	2	3
y	3.5	2	0.5	−1	−2.5	−4	−5.5

ii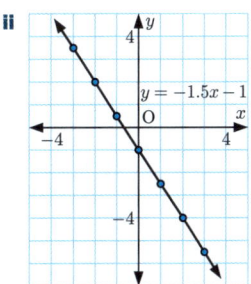

2 a i

x	−3	−2	−1	0	1	2	3
y	0	1	2	3	4	5	6

ii 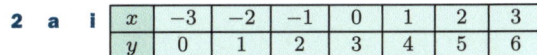   **iii** gradient = 1, y-intercept = 3

b i

x	−3	−2	−1	0	1	2	3
y	7	5	3	1	−1	−3	−5

ii 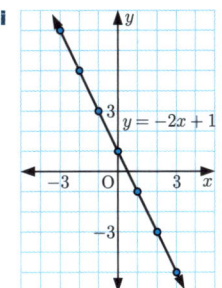   **iii** gradient = −2, y-intercept = 1

c i

x	−3	−2	−1	0	1	2	3
y	−2	$-1\frac{2}{3}$	$-1\frac{1}{3}$	−1	$-\frac{2}{3}$	$-\frac{1}{3}$	0

ii 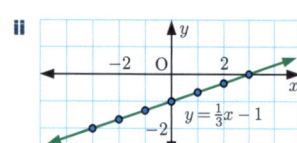 **iii** gradient $= \frac{1}{3}$, y-intercept $= -1$

EXERCISE 20G.2

1 **a** gradient is 4, y-int. is 8 **b** gradient is -3, y-int. is 2
  c gradient is -1, y-int. is 6 **d** gradient is -2, y-int. is 3
  e gradient is 0, y-int. is -2 **f** gradient is -3, y-int. is 11
  g gradient is $\frac{1}{2}$, y-int. is -5 **h** gradient is $-\frac{3}{2}$, y-int. is 3
  i gradient is $\frac{2}{5}$, y-int. is $\frac{4}{5}$ **j** gradient is $\frac{1}{2}$, y-int. is $\frac{1}{2}$
  k gradient is $\frac{2}{5}$, y-int. is -2 **l** gradient is $-\frac{3}{2}$, y-int. is $\frac{11}{2}$

2 **a** 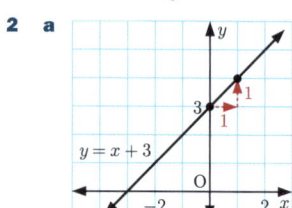 **b**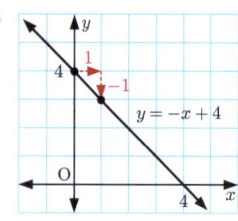
  c 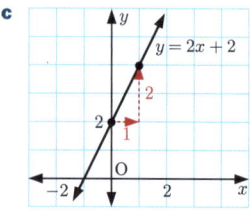 **d**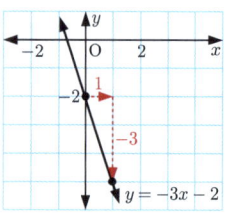
  e 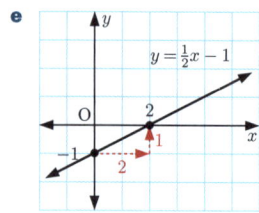 **f**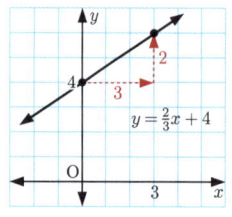
  g 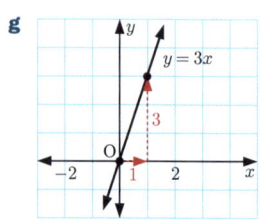 **h**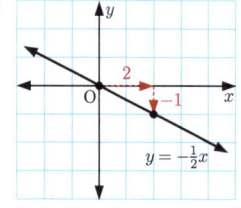
  i 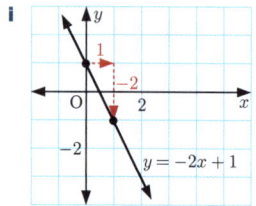 **j**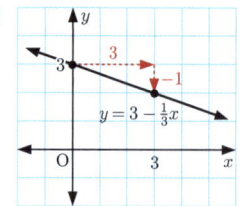

3 **a** **i** gradient is $-\frac{2}{3}$
    ii y-intercept is 2
    iii x-intercept is 3
  c yes

k 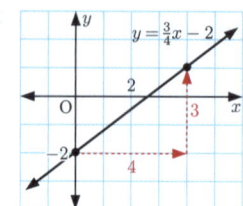 **l**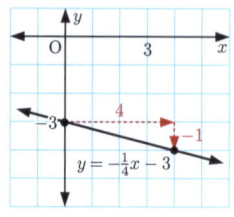

b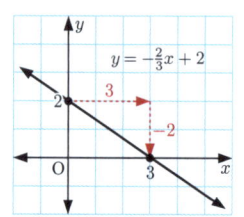

EXERCISE 20H

1 **a** 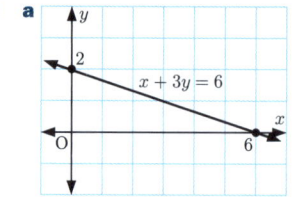 **b**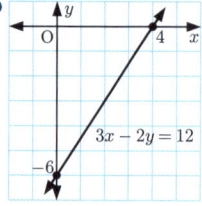
  c 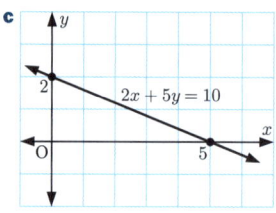 **d**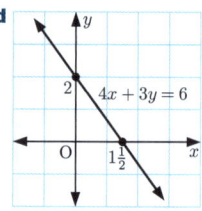
  e 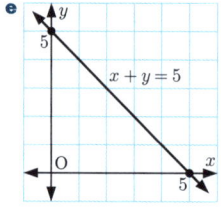 **f**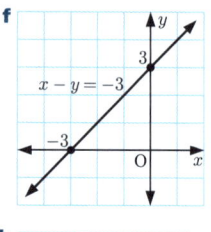
  g 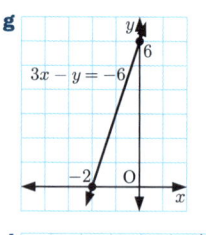 **h**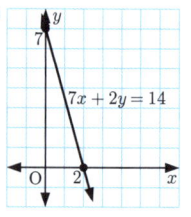
  i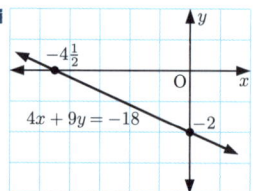

2 **a** **i** 5   **ii** -3 **b** **i** yes   **ii** no

Answers 687

c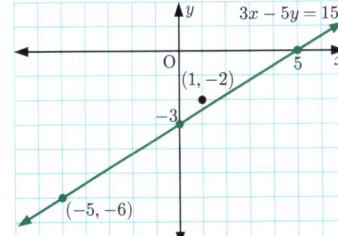

3 a i 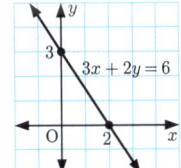   **ii**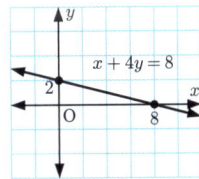

iii 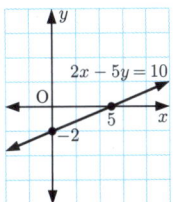   **iv**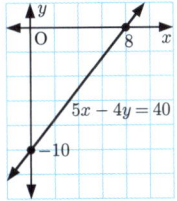

b i $-\frac{3}{2}$ **ii** $-\frac{1}{4}$ **iii** $\frac{2}{5}$ **iv** $\frac{5}{4}$

The gradient is $-\dfrac{\text{coefficient of } x}{\text{coefficient of } y}$.

c $y = -\dfrac{A}{B}x + \dfrac{C}{B}$ $\therefore$ gradient $= -\dfrac{A}{B}$

4 a 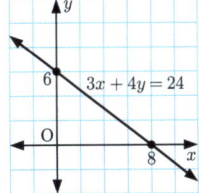   **b** D

EXERCISE 20I

1 a 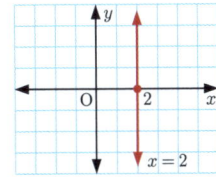   **b**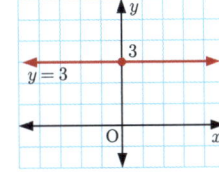

gradient is undefined   gradient is 0

c 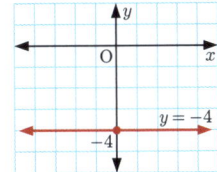   **d**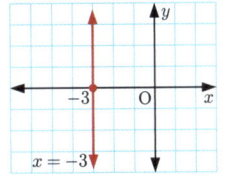

gradient is 0   gradient is undefined

e 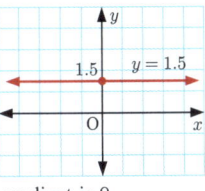   **f**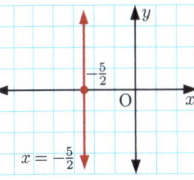

gradient is 0   gradient is undefined

g 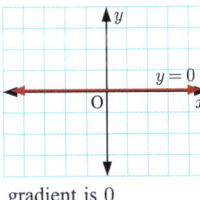   **h**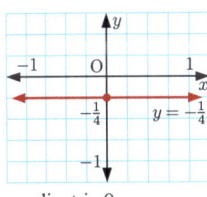

gradient is 0   gradient is 0

2 a 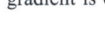   **b i** A(5, 0), B(0, −3)

ii $\frac{3}{5}$

3 a 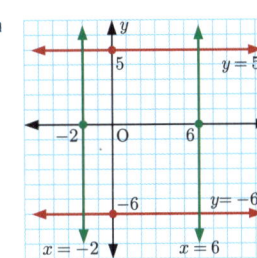   **b i** rectangle

ii 88 units^2

iii $\frac{11}{8}$ and $-\frac{11}{8}$

EXERCISE 20J

1 a $y = 3x + 4$ **b** $y = 7x - 1$ **c** $y = -x + \frac{1}{3}$
  d $y = \frac{1}{2}x$ **e** $y = 5$ **f** $y = -\frac{5}{2}x - \frac{2}{3}$

2 a $y = 2x + 6$ **b** $y = -3x - 4$ **c** $y = 5x + 14$
  d $y = \frac{3}{4}x + 6$ **e** $y = 4$ **f** $y = -\frac{1}{2}x - \frac{13}{2}$

3 a $y = 2x + 2$ **b** $y = \frac{2}{5}x + 3$ **c** $y = -x - 1$
  d $y = -2x + 4$ **e** $y = \frac{1}{2}x - \frac{1}{2}$ **f** $y = -\frac{2}{5}x$

4 a $y = 4x - 5$ **b** $y = -3x + 7$ **c** $y = \frac{3}{8}x - 1$

5 a $y = 2x + 3$ **b** $y = -2x + 12$ **c** $y = \frac{2}{3}x - 2$
  d $y = -3$ **e** $y = \frac{3}{2}x - 5$ **f** $y = -\frac{4}{3}x + \frac{1}{3}$

6 a $y = 3x - 2$ **b** $y = -\frac{1}{3}x + 4$ **c** $y = 7$
  d $y = -\frac{2}{5}x - 2$ **e** $y = \frac{5}{4}x + \frac{5}{2}$ **f** $x = -4$

7 a $V = \frac{1}{2}t + 1$ **b** $N = 3 - d$ **c** $C = \frac{1}{2}t + 2$

8 a $y = 3x - 2$ **b** $y = -\frac{1}{4}x - 1$ **c** $2x - 5y = -15$

9 a $y = -\frac{3}{5}x + 11$ **b** $y = -\frac{3}{5}x + \frac{27}{5}$

10 a $y = 4x + 13$ **b** $y = 5$ **c** $y = 4x - 23$ **d** $y = -3$

REVIEW SET 20A

1 A(2, 3), B(−2, 2), C(3, 0), D(−3, −4), E(1, −3)

2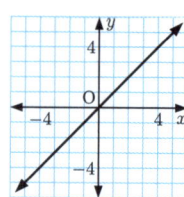

3 a

x	-3	-2	-1	0	1	2	3
y	13	10	7	4	1	-2	-5

b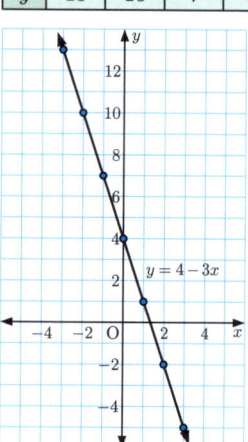

c gradient $= -3$, y-intercept $= 4$

4 a $\frac{2}{5}$ **b** $-\frac{4}{3}$

5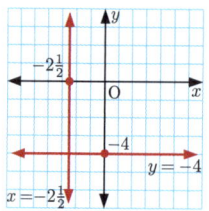

6 a Time t, is the independent variable.
Charge C, is the dependent variable.

b

t	0	1	2	3	4
C	70	190	310	430	550

c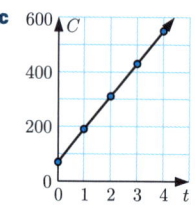

d yes

e Yes, you can have a 1.5 hour meeting, for example.

f £120

7 a gradient is 4, y-intercept is -3
b gradient is -1, y-intercept is 2
c gradient is $\frac{3}{2}$, y-intercept is $-\frac{1}{2}$

8 a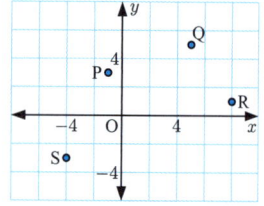

b i yes **ii** no
c trapezium

9 a x-intercept is -3, y-intercept is 6, gradient is 2
b x-intercept is 5, y-intercept is -3, gradient is $\frac{3}{5}$
c x-intercept is 6, y-intercept is 4, gradient is $-\frac{2}{3}$

10 a i gradient = 2, y-intercept = -3
ii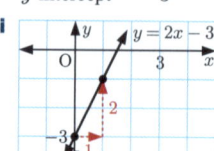

b i gradient = $\frac{1}{3}$, y-intercept = 1
ii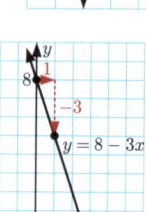

c i gradient = -3, y-intercept = 8
ii (graph of $y = 8 - 3x$)

11 a (graph of $y = \frac{3}{2}x - 2$)

b (graph of $2x - 3y = 30$)

c (graph of $y = -\frac{1}{4}x + 5$)

d (graph of $4x + 3y = -36$)

12 a $y = x - 3$ **b** $y = -\frac{2}{3}x + \frac{2}{3}$ or $2x + 3y = 2$

REVIEW SET 20B

1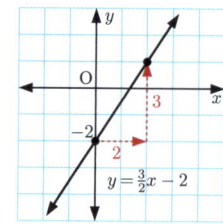

2 $x + y = 3$

3 a Time t, is the independent variable.
Volume V, is the dependent variable.

b

t	0	2	4	6	8	10
V	400	360	320	280	240	200

c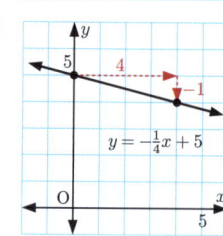

d yes

e Yes, we can measure the volume after $\frac{1}{2}$ minute, for example.

f A decrease of 20 litres.   **g** 350 litres   **h** $11\frac{1}{2}$ min

4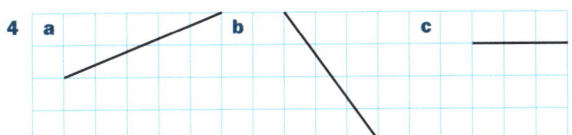

5 a 2 **b** $-\frac{7}{3}$ **c** $\frac{4}{5}$ **d** $-\frac{5}{3}$

6 a 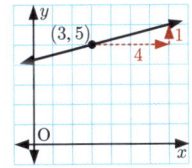 **b**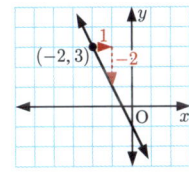

7 a yes **b** yes

8 a 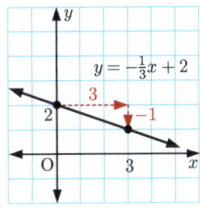 **b**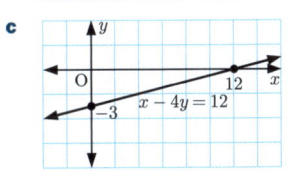

c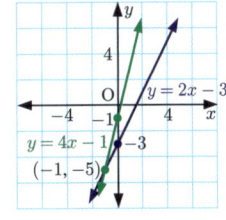

9 a $y = -\frac{3}{4}x + 3$ **b** $P = r + 4$ **10** B

11 a 2 **b** -3 **c** $y = \frac{3}{2}x - 3$

12 a $y = 2x - 3$ **b** $y = -\frac{1}{2}x + 2$ **c** $y = -x + 2$
  d $y = \frac{3}{2}x - 5$

EXERCISE 21A

1 a no **b** yes **c** yes **d** yes
2 a $x = 3,\ y = 1$ **b** $x = 7,\ y = 4$ **c** $x = 3,\ y = 5$
  d $x = 3,\ y = 9$ **e** $x = 3,\ y = 1$ **f** $x = 4,\ y = 2$
  g $a = 1,\ b = 0$ **h** $p = 5,\ q = 2$ **i** $x = -5,\ y = 16$

EXERCISE 21B

1 a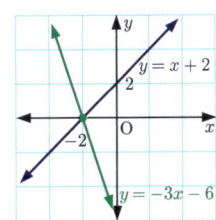

2 a 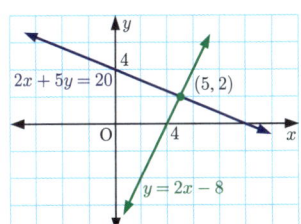 $x = 5,\ y = 2$

b 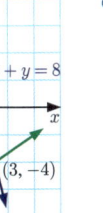 **c**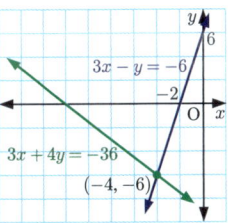

$x = 3,\ y = -4$ 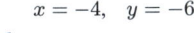 $x = -4,\ y = -6$

3 a 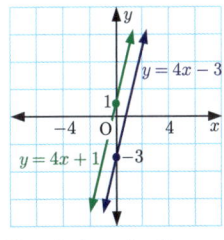 **b**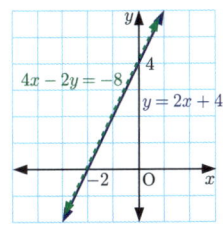

No solutions, the lines are parallel.   Infinitely many solutions, the lines are coincident.

4 a $l_1:\ y = \frac{1}{2}x + 1,\quad l_2:\ 2x + 3y = 12$
  b $x \approx 2.6,\ y \approx 2.3$

EXERCISE 21C

1 a $x = 4,\ y = 5$ **b** $x = -1,\ y = 3$ **c** $x = 1,\ y = 3$
2 a $x = 0,\ y = -4$ **b** $x = 4,\ y = 0$
  c $x = -5,\ y = 3$ **d** $x = -\frac{13}{4},\ y = -\frac{11}{4}$
  e $x = 8,\ y = 6$ **f** $x = \frac{5}{3},\ y = 0$
3 a no solutions **b** $x \approx 2.66,\ y \approx 2.51$
  In **a** the lines are parallel.

EXERCISE 21D

1 a $x = 3,\ y = 5$ **b** $x = -2,\ y = -3$
  c $x = 1,\ y = -6$ **d** $x = -7,\ y = -3$
  e $x = 0,\ y = 3$ **f** $x = -\frac{1}{2},\ y = \frac{1}{2}$
2 a $x = \frac{2}{5},\ y = -\frac{7}{5}$ **b** $x = \frac{5}{4},\ y = 3$
  c $x = \frac{25}{17},\ y = \frac{10}{17}$
3 a $(2, 5)$ **b** $(3, 1)$ **c** $(0, 5)$ **d** $(0, 4)$
4 a No solutions as the lines are parallel.
  b There are infinitely many solutions as the lines are coincident.

EXERCISE 21E.1

1 a $6x = 6$ **b** $-y = 8$ **c** $5x = 7$
  d $-6x = -30$ **e** $8y = 4$ **f** $-2y = -16$
2 a $x = 2,\ y = -1$ **b** $x = -2,\ y = 5$ **c** $x = 3,\ y = 2$
  d $x = -2,\ y = -1$ **e** $x = 5,\ y = -3$ **f** $x = 4,\ y = -3$

EXERCISE 21E.2

1 a $9x + 12y = 6$ **b** $-2x + 8y = -14$
  c $25x - 5y = -15$ **d** $-21x - 9y = 12$

e $8x + 20y = -4$ **f** $-3x + y = 1$

2 a $x = 3, \ y = 2$ **b** $x = 8, \ y = 7$ **c** $x = -2, \ y = 3$
  d $x = 5, \ y = -1$ **e** $x = -\frac{5}{2}, \ y = 0$ **f** $x = 4, \ y = -\frac{3}{2}$

3 a $x = -3, \ y = 1$ **b** $x = 2, \ y = 1$
  c $x = 3, \ y = 1$ **d** $x = 3, \ y = 2$ **e** $x = 5, \ y = 2$
  f $x = 4, \ y = 1$ **g** $x = -1, \ y = -3$
  h $x = 2, \ y = -5$ **i** $x = 19, \ y = -17$

4 a Yes, dividing an equation by a non-zero constant is equivalent to multiplying the equation by a fraction, which will not change its solutions.
  b Dividing $5x - 10y = 15$ by 5, we obtain $x - 2y = 3$.
  c $x = 5, \ y = 1$

5 a Infinitely many solutions.
  b No solutions as the lines are parallel.

EXERCISE 21F

1 97 and 181 **2** $81\frac{1}{2}$ and $118\frac{1}{2}$ **3** 9 and 17
4 hammer £14, screwdriver £6
5 adult's ticket £28, child's ticket £12
6 19 five pence and 14 twenty pence coins
7 14 rabbits and 21 pheasants
8 74 one litre and 23 two litre cartons
9 a $x = 3, \ y = 5$ **b** 8 cm²
10 a $x = \frac{3}{5}, \ y = 1$ **b** 36 cm

REVIEW SET 21A

1 $x = 2, \ y = 5$ **2** $x = -1, \ y = -3$ **3** $x = 8, \ y = -13$
4 a $x = -5, \ y = -7$ **b** no solutions
5 a $x = 4, \ y = -2$ **b** $x = 2, \ y = -1$
6 pencil £0.24, ruler £0.50
7 a $y = 29 - 4x$ **b** $x = 8, \ y = -3$
8 $x = 3, \ y = -4$ **9 a** $x = 5, \ y = 4$ **b** 13 cm²

REVIEW SET 21B

1 a 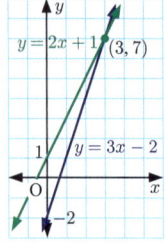 **b** $x = 3, \ y = 7$

2 $x \approx -0.3, \ y \approx 2.8$ **3** $x = 4, \ y = 4$ **4** $x = 2, \ y = 1$
5 a $x = -1, \ y = -7$ **b** $x = 2, \ y = -1$
6 37 and 48 **7** 13 ten pence and 8 fifty pence coins
8 a $x = -9, \ y = 23$ **b** no solutions
9 a $x = \frac{14}{11}, \ y = \frac{57}{11}$ **b** $\frac{125}{11}$ cm ≈ 11.4 cm

EXERCISE 22A

1 a 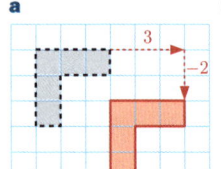 $\binom{3}{-2}$ **b** 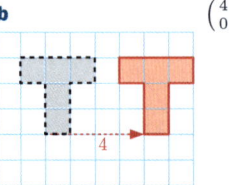 $\binom{4}{0}$

  c $\binom{0}{-3}$ **d** $\binom{-3}{1}$

  e $\binom{-1}{-4}$ **f** $\binom{-3}{3}$

2 a 5 units right, $\binom{5}{0}$ **b** 3 units down, $\binom{0}{-3}$
  c 2 units right, 2 units down, $\binom{2}{-2}$
  d 4 units left, 1 unit up, $\binom{-4}{1}$
  e 3 units right, 3 units down, $\binom{3}{-3}$
  f 3 units left, 3 units down, $\binom{-3}{-3}$

3 a $\binom{1}{4}$ means that the object moves 1 unit to the right and 4 units upwards.
  b $\binom{-2}{3}$ means that the object moves 2 units to the left and 3 units upwards.
  c $\binom{3}{0}$ means that the object moves 3 units to the right.
  d $\binom{-1}{-3}$ means that the object moves 1 unit to the left and 3 units downwards.
  e $\binom{0}{5}$ means that the object moves 5 units downwards.

4 a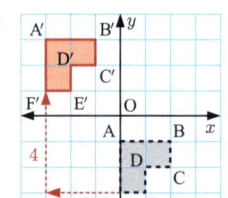
  $A'(-3, 1)$
  $B'(-2, -1)$
  $C'(-3, -2)$

  b
  $A'(0, 1)$
  $B'(1, 1)$
  $C'(2, 0)$
  $D'(-1, 0)$

  c
  $A'(-3, 3)$
  $B'(-1, 3)$
  $C'(-1, 2)$
  $D'(-2, 2)$
  $E'(-2, 1)$
  $F'(-3, 1)$

d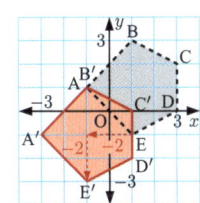

A′(−3, −1)
B′(−1, 1)
C′(1, 0)
D′(1, −2)
E′(−1, −3)

5 a $\begin{pmatrix} -4 \\ 2 \end{pmatrix}$ **b** $\begin{pmatrix} 6 \\ 1 \end{pmatrix}$ **c** $\begin{pmatrix} 2 \\ 3 \end{pmatrix}$

6 a no   **b** yes, $\begin{pmatrix} 2 \\ -3 \end{pmatrix}$ **c** no

7 a I, $\begin{pmatrix} -5 \\ -6 \end{pmatrix}$ **b** D, $\begin{pmatrix} 7 \\ 5 \end{pmatrix}$ **c** H

EXERCISE 22B.1

1 a 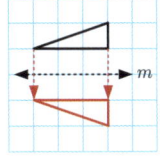   **b**

c 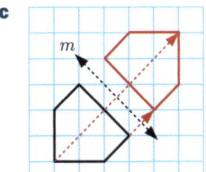   **d**

e 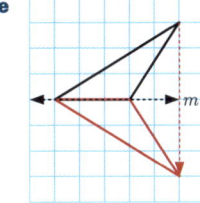   **f**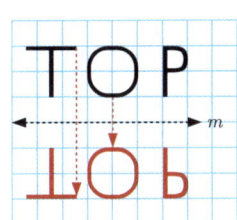

2 a and **c** represent reflections.

3 a 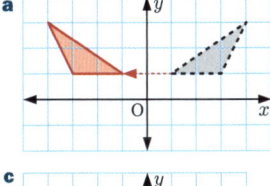   **b**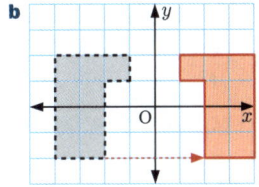

c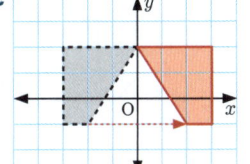

4 a 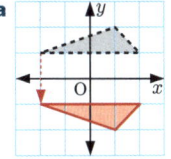   **b**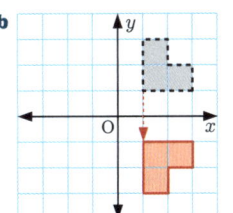

c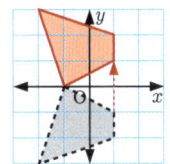

5 a reflection   **b** not a reflection   **c** not a reflection

6 a A and C   **b** x-axis   **c** A and D

7 a, b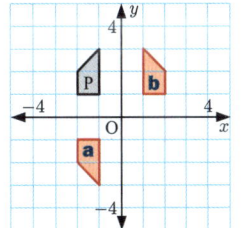

c, d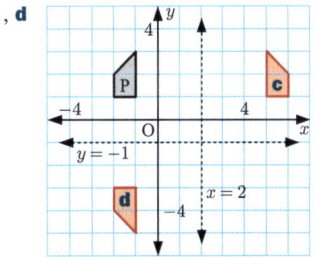

e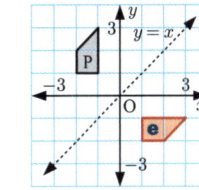

8 a $y = 0$ (x-axis)   **b** $x = -2$ **c** $y = 3$ **d** $y = -x$

EXERCISE 22B.2

1 a 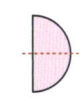   **b**   **c**

d 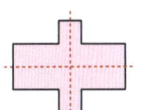   **e** 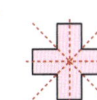   **f**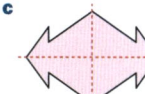

infinite

2 a 4 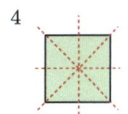   **b** 3   **c** 2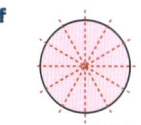

d 2 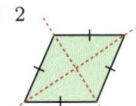   **e** 5 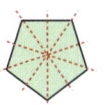   **f** 1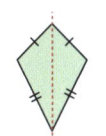

EXERCISE 22C.1

1 a 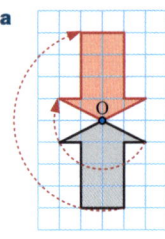 **b**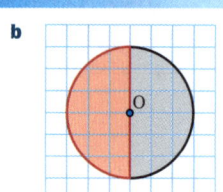

c 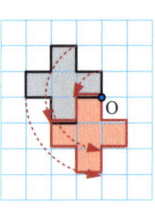 **d**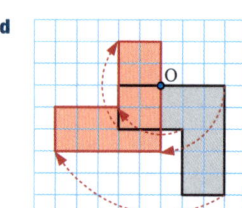

e 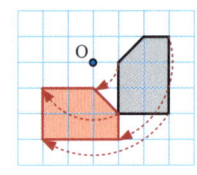 **f**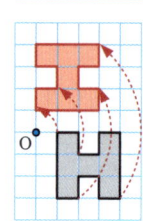

2 H, I, M, N, O, S, W, X, and Z either remain the same or become a different letter under a rotation less than 360°.

3 a D   **b** B   **c** C   **4** Rotate 75° anticlockwise about O.

5 a P(0, −2)
     Q(2, −1)
     R(2, −4)

b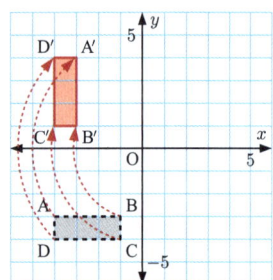

c P′(2, 0)
     Q′(1, 2)
     R′(4, 2)

6 a, b 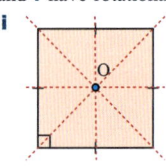   **c** A′(−3, 4)
                                      B′(−3, 1)
                                      C′(−4, 1)
                                      D′(−4, 4)

7 a C   **b** 180°

8 a centre O, clockwise through 90°
  b centre (2, 1), anticlockwise through 90°
  c centre (0, 3), through 180°

EXERCISE 22C.2

1 b, d, and f have rotational symmetry.

2 a i 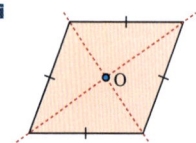   **b i** (image)

  ii 4   **ii** 2

c i 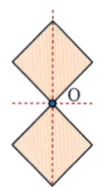   **d i**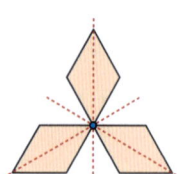

  ii 2   **ii** 3

e i 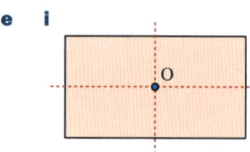   **f i**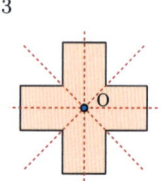

  ii 2   **ii** 4

g i 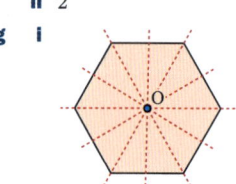   **h i**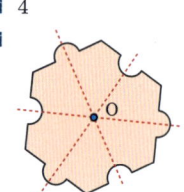

  ii 6   **ii** 3

3 a 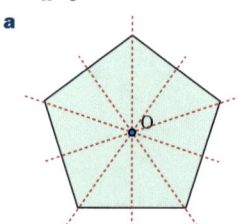   **b**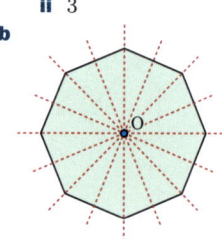

Note: There may be other answers.

EXERCISE 22D.1

1 a 2   **b** 3   **c** $\frac{3}{2}$ **2 a** $\frac{1}{2}$ **b** $\frac{1}{6}$ **c** $\frac{3}{4}$

3 a enlargement with scale factor 2
  b reduction with scale factor $\frac{1}{2}$
  c enlargement with scale factor 3
  d reduction with scale factor $\frac{1}{3}$
  e enlargement with scale factor $\frac{3}{2}$
  f reduction with scale factor $\frac{2}{3}$

EXERCISE 22D.2

1 a 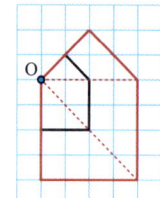   **b**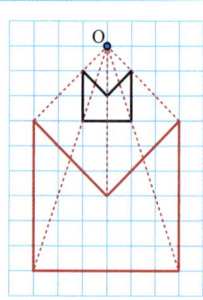

c 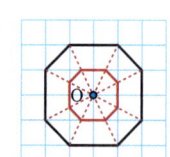   **d**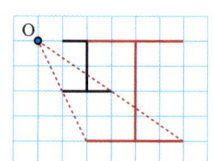

Answers 693

e / **f**

2 a / **b** / **c**

3 a

b Scale factor is $\frac{5}{2}$.

REVIEW SET 22A

1 a / **b**

2 a $\begin{pmatrix} -4 \\ -2 \end{pmatrix}$ **b** $\begin{pmatrix} 6 \\ -2 \end{pmatrix}$ **c** $\begin{pmatrix} 2 \\ -4 \end{pmatrix}$

3

4 $x = 1$

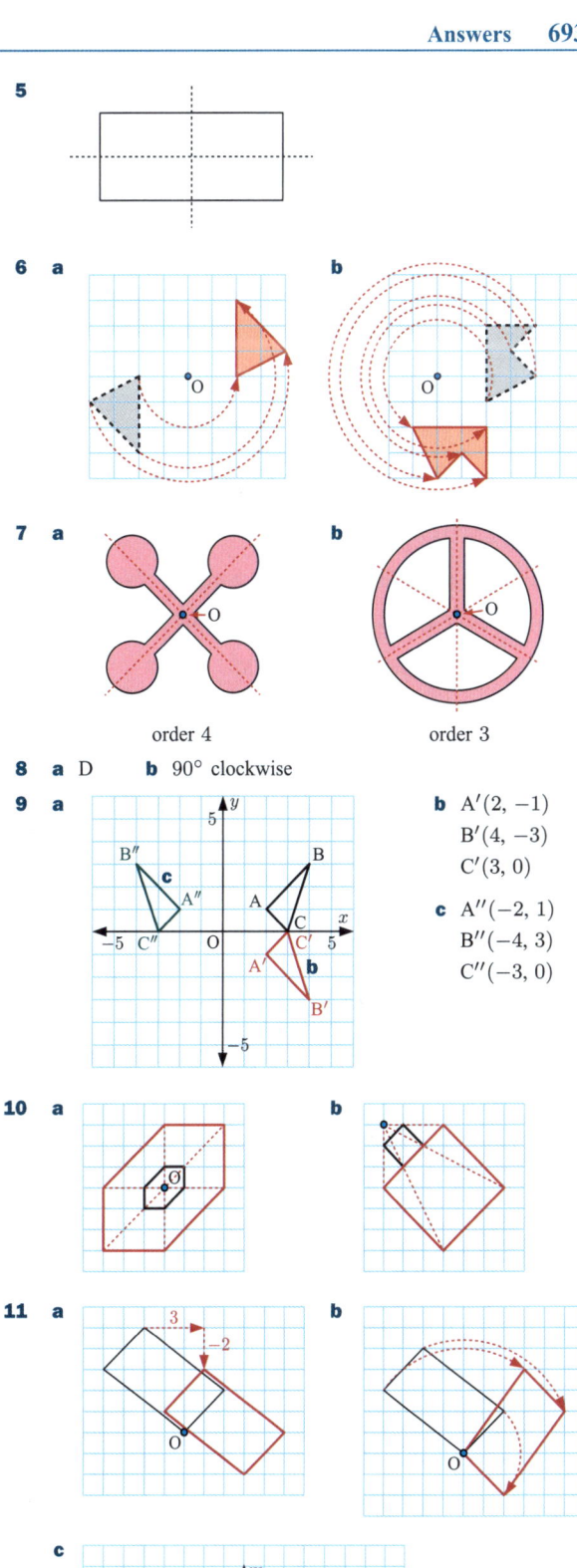

5

6 a / **b**

7 a order 4 **b** order 3

8 a D **b** 90° clockwise

9 a **b** $A'(2, -1)$, $B'(4, -3)$, $C'(3, 0)$ **c** $A''(-2, 1)$, $B''(-4, 3)$, $C''(-3, 0)$

10 a / **b**

11 a / **b** / **c**

d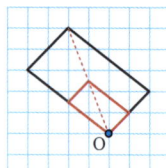

12 a 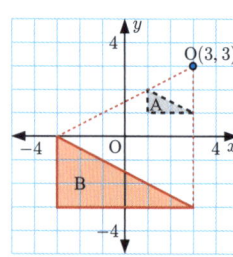   **b** Scale factor is 3.

REVIEW SET 22B

1 a $\binom{3}{-2}$ **b** a translation 3 units left and 2 units upwards

2 a, b 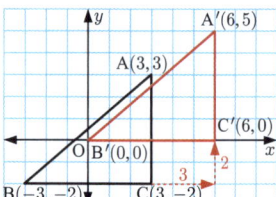   **c** $A'(6, 5)$, $B'(0, 0)$, $C'(6, 0)$

3 a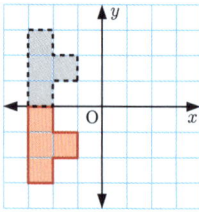

b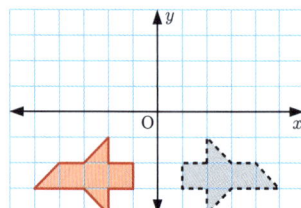

4 a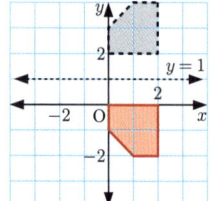

5 a 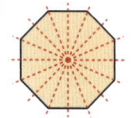   **b** 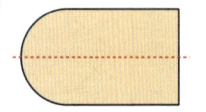   **c** no lines of symmetry exist

6 a $P(-4, -3)$
$Q(-2, 0)$
$R(-1, -2)$
$S(-1, -3)$

c $P'(3, -4)$
$Q'(0, -2)$
$R'(2, -1)$
$S'(3, -1)$

b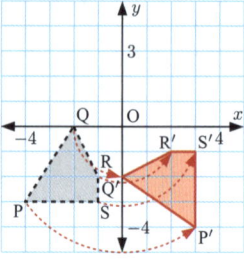

7 a 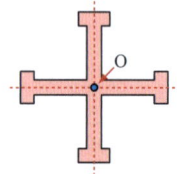   **b** 4

8 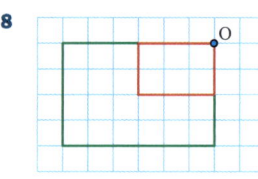   **9**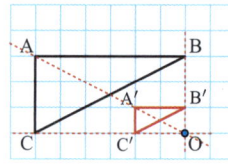

10 a 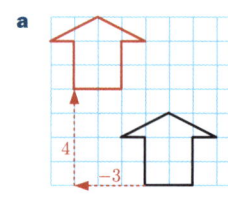   **b**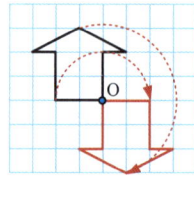

c 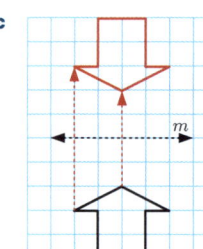   **d**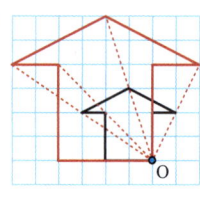

11 a i B   **ii** D   **iii** C   **b** A reflection in the y-axis.
12 centre $(1, 0)$, rotation is clockwise through $90°$

EXERCISE 23A

1 a similar   **b** $\dfrac{A'B'}{AB} = \dfrac{3}{4}$, $\dfrac{B'C'}{BC} = \dfrac{2}{3}$ ∴ not similar
  c not equiangular ∴ not similar   **d** similar
2 no, $\dfrac{140}{100} \neq \dfrac{100}{60}$
3 a true   **b** false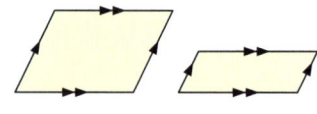
  c true   **d** false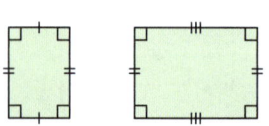

4 a $x = 8.75$ **b** $x = 4.8$ **c** $x \approx 6.86$ **d** $x = 7.5$
  e $x = 96$ **f** $x = 127$

5 **a** 6 cm **b** 60
6 **a** $x = 4.8$ **b** $x = 9$ **c** $x = 100$ **d** $x = 16$
7 **a**

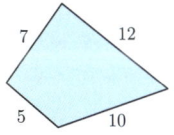

Note: Other answers are possible.

b

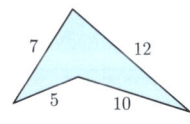

8 no

EXERCISE 23B.1
1 **a** $V\hat{U}W = X\hat{Y}W$, $V\hat{W}U = X\hat{W}Y$ {vertically opposite}
 b $F\hat{G}H = F\hat{D}E$, angle F is common.
 c $R\hat{S}T = R\hat{Q}P$, $S\hat{R}T = Q\hat{R}P$ {vertically opposite}
 d $A\hat{B}C = E\hat{D}C$ {equal alternate angles},
    $A\hat{C}B = D\hat{C}E$ {vertically opposite}
 e $L\hat{K}M = L\hat{N}J$, angle L is common
 f $VX = 15$ m, $XY = 34$ m {Pythagoras}
    $\therefore$ $YZ = 2WV$, $XY = 2XW$, $XZ = 2XV$
2 **Hint:** Find the unknown angles in terms of α.

EXERCISE 23B.2
1 **a** $x = 5$ **b** $x = 2.8$ **c** $x = \frac{36}{11} \approx 3.27$
 d $x = \frac{20}{3} \approx 6.67$ **e** $x = 7$ **f** $x = 7.2$
2 **a** $A\hat{B}C = E\hat{D}C$, $A\hat{C}B = E\hat{C}D$ {vertically opposite}
 b $x = 7$
 c area $\triangle ABC = 56$ cm^2, area $\triangle EDC = 14$ cm^2
3 7.5 m

EXERCISE 23C
1 **a** 7 m **b** 7.5 m **2** ≈ 9.22 m **3** 180 cm or 1.8 m
4 **a** ≈ 2.67 m **b** ≈ 5.67 m **5** 39 m **6** ≈ 117 m

EXERCISE 23D
1 **a** congruent **b** not congruent **c** not congruent
 d congruent
2 **B** and **E**
3 **B** and **C** **4** **a** 10 cm **b** 98° **c** 37.8 cm

EXERCISE 23E
1 **a** congruent {SSS} **b** congruent {RHS}
 c No, equal sides not in corresponding positions ($\alpha \neq \beta$).
 d congruent {AAcorS} **e** congruent {AAcorS}
 f Not congruent, no equal sides.
2 **a** No, insufficient information. **b** congruent {SAS}
 c congruent {SAS} **d** not congruent
 e congruent {AAcorS} **f** congruent {AAcorS}
3 **C** **4** **B** and **D**
5 **a** **i** $\triangle ABC \cong \triangle PQR$ {AAcorS}
   ii $P\hat{R}Q = A\hat{C}B$, $BC = QR$, and $AB = PQ$
 b **i** $\triangle JKL \cong \triangle XZY$ {SSS}
   ii $J\hat{K}L = X\hat{Z}Y$, $K\hat{J}L = Z\hat{X}Y$, and $K\hat{L}J = Z\hat{Y}X$
 c **i** not congruent

 d **i** $\triangle RST \cong \triangle YXZ$ {SAS}
   ii $R\hat{S}T = Y\hat{X}Z$, $T\hat{R}S = Z\hat{Y}X$, and $RS = YX$
 e **i** $\triangle ABC \cong \triangle EDC$ {AAcorS}
   ii $A\hat{B}C = E\hat{D}C$, $BC = DC$, and $AB = ED$
 f **i** $\triangle PQT \cong \triangle SQR$ {AAcorS}
   ii $P\hat{Q}T = S\hat{Q}R$, $PQ = SQ$, and $PT = SR$
 g **i** $\triangle ABC \cong \triangle DFE$ {AAcorS}
   ii $B\hat{A}C = F\hat{D}E$ and $BC = FE$
 h **i** not congruent

EXERCISE 23F
1 **a** In triangles ABD and CDB:
   • $A\hat{D}B = C\hat{B}D$ {equal alternate angles}
   • $A\hat{B}D = C\hat{D}B$ {equal alternate angles}
   • BD is common to both triangles
   $\therefore$ $\triangle ABD \cong \triangle CDB$ {AAcorS}
  Equating corresponding angles, $D\hat{A}B = B\hat{C}D$.
 b Opposite angles of a parallelogram are equal.
2 **c** In a kite, one diagonal bisects one pair of opposite angles.
3 **c** The diagonals of a square are equal in length.
4 **e** The opposite sides of a rhombus are parallel.

REVIEW SET 23A
1 similar
2 No, even if corresponding side lengths of rhombuses are always in the same ratio, the rhombuses are not always equiangular.
3 **a** $x \approx 6.47$ **b** $x \approx 1.71$ **4** ≈ 10.4 m
5 **a** $A\hat{B}E = A\hat{C}D$ {equal corresponding angles}, angle A is common.
 b 7.2 cm
6 **a** not congruent **b** congruent **c** congruent
7 **a** congruent {SAS}
 b Not congruent, equal sides are not in corresponding positions (and $\alpha \neq \beta$).
8 **c** In a kite, one pair of opposite angles are equal.
9 **a** $\triangle DEF \cong \triangle STU$ {AAcorS}
 b $D\hat{F}E = S\hat{U}T$, $FE = UT$, and $DE = ST$
10 **a** Hint: Use SAS. **b** Hint: Use SSS.
 c A and D are 50 cm^2. B and C are 150 cm^2.

REVIEW SET 23B
1 $x = 2.8$ **2** **B** and **D**
3 **a** $A\hat{X}B = D\hat{X}C$ {vertically opposite}
   $X\hat{A}B = X\hat{D}C$ {equal alternate angles}
   $X\hat{B}A = X\hat{C}D$ {equal alternate angles}
 b $B\hat{A}C = D\hat{E}C$, $A\hat{C}B = E\hat{C}D$ {vertically opposite}
 c For $\triangle QRS$ and $\triangle TRP$,
   $R\hat{Q}S = R\hat{T}P$ {given}, angle R is common
4 **a** $A\hat{B}C = E\hat{D}C$, $A\hat{C}B = E\hat{C}D$ {vertically opposite}
 b 3.2 cm
5 **a** No, no equal sides. **b** congruent {RHS}
 c congruent {SAS}
6 **c** The base angles of an isosceles triangle are equal.
7 $R\hat{S}Q = R\hat{P}T$, angle R is common, $\therefore$ $x = 12$
8 **a** **i** 3 cm **ii** 5 cm **b** 128 cm^2 **9** ≈ 66.7 m
10 **e** In a kite, one of its diagonals bisects the other.

EXERCISE 24A

1 a

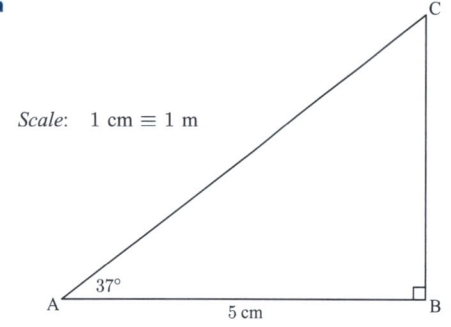

Scale: 1 cm ≡ 1 m

  b i ≈ 3.75 m   **ii** ≈ 6.25 m
2 ≈ 49 m high
3 a

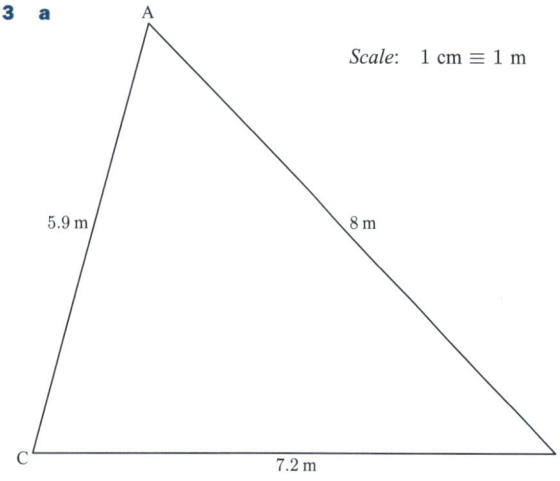

Scale: 1 cm ≡ 1 m

  b $\widehat{A} \approx 60°$, $\widehat{B} \approx 45°$, $\widehat{C} \approx 75°$
4 ≈ 8.1 m

EXERCISE 24B

1 a i AC **ii** BC **iii** AB
  b i RS **ii** RT **iii** ST
  c i LM **ii** MN **iii** LN
2 a b units **b** c units **c** c units **d** b units

EXERCISE 24C

1 a $\cos 18° \approx 0.95$ **b** $\sin 66° \approx 0.91$ **c** $\tan 23° \approx 0.42$
2 a AB ≈ 6.8 cm, BC ≈ 5.7 cm, AC ≈ 3.7 cm
  b i ≈ 0.838 **ii** ≈ 0.544 **iii** ≈ 1.54
  c i ≈ 0.839 **ii** ≈ 0.545 **iii** ≈ 1.54
3 a ≈ 7.86 cm
  b i ≈ 0.469 **ii** ≈ 0.883 **iii** ≈ 0.532
  c i ≈ 0.469 **ii** ≈ 0.883 **iii** ≈ 0.532
4 a

θ	$90° - \theta$	$\sin \theta$	$\cos \theta$	$\sin(90° - \theta)$	$\cos(90° - \theta)$
70°	20°	0.940	0.342	0.342	0.940
35°	55°	0.574	0.819	0.819	0.574

$\sin(90° - \theta) = \cos\theta$, $\cos(90° - \theta) = \sin\theta$
  b i $\sin(90° - \theta) = \dfrac{a}{c} = \cos\theta$

  ii $\cos(90° - \theta) = \dfrac{b}{c} = \sin\theta$

5 a i ≈ 0.788 **ii** ≈ 0.616 **iii** ≈ 1.28 **iv** ≈ 1.28

  b

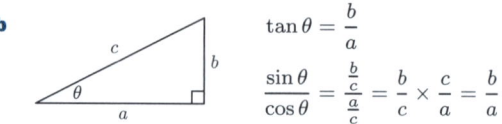

$\tan\theta = \dfrac{b}{a}$

$\dfrac{\sin\theta}{\cos\theta} = \dfrac{\frac{b}{c}}{\frac{a}{c}} = \dfrac{b}{c} \times \dfrac{c}{a} = \dfrac{b}{a}$

6 a ≈ 0.358 **b** ≈ 0.934 **c** ≈ 0.384

EXERCISE 24D

1 a $\cos 38° = \dfrac{x}{a}$ **b** $\sin 66° = \dfrac{x}{b}$ **c** $\sin 68° = \dfrac{x}{c}$
  d $\tan 71° = \dfrac{d}{x}$ **e** $\cos 42° = \dfrac{e}{x}$ **f** $\tan 51° = \dfrac{f}{x}$
2 a $x \approx 7.99$ **b** $x \approx 8.34$ **c** $x \approx 4.46$
  d $x \approx 6.40$ **e** $x \approx 6.18$ **f** $x \approx 5.92$
  g $x \approx 8.46$ **h** $x \approx 2.57$ **i** $x \approx 14.40$
  j $x \approx 9.77$ **k** $x \approx 11.71$ **l** $x \approx 8.75$
  m $x \approx 17.70$ **n** $x \approx 13.47$ **o** $x \approx 13.87$
3 a $\theta = 52°$, $x \approx 7.9$, $y \approx 6.2$
  b $\theta = 27°$, $x \approx 9.0$, $y \approx 4.1$
  c $\theta = 53°$, $x \approx 19.0$, $y \approx 11.5$

EXERCISE 24E

1 a $\sin\theta = \tfrac{4}{7}$ **b** $\theta \approx 34.8°$
2 a $\theta = 30°$ **b** $\theta \approx 58.0°$ **c** $\theta \approx 41.4°$ **d** $\theta \approx 37.6°$
  e $\theta \approx 39.5°$ **f** $\theta \approx 57.3°$ **g** $\theta \approx 45.1°$ **h** $\theta \approx 48.6°$
  i $\theta \approx 56.0°$ **j** $\theta \approx 52.9°$ **k** $\theta \approx 51.3°$ **l** $\theta \approx 30.5°$
3 a $\phi \approx 55.2°$, $\theta \approx 34.8°$, $x \approx 5.7$
  b $\alpha \approx 56.3°$, $\beta \approx 33.7°$, $x \approx 10.8$
  c $a \approx 39.5°$, $b \approx 50.5°$, $x \approx 7.5$

EXERCISE 24F.1

1 ≈ 17.5 m **2** ≈ 37.5 m **3** $\theta \approx 18.7°$ **4** ≈ 66.4°
5 ≈ 2.45 m each **6** ≈ 31.6 m above the water **7** ≈ 27.2 m
8 ≈ 55.2° **9** ≈ 19.9 cm **10** ≈ 96.4° and 83.6°
11 ≈ 0.982 m **12** ≈ 8.12 m

EXERCISE 24F.2

1 ≈ 036° **2** ≈ 236°
3 a ≈ 427 m **b** ≈ 111° **c** ≈ 291°
4 a ≈ 2.24 km, 063° **b** ≈ 3.16 km, 162°
  c ≈ 3.16 km, 072° **d** ≈ 4.47 km, 207°
  e ≈ 3.61 km, 304°

EXERCISE 24G.1

1 a $P(\cos 26°, \sin 26°)$ **b** $P(0.899, 0.438)$
2 a i 0.2588 **ii** 0.9659 **b** ≈ 3.73
3 a $NP = \sin 45° \approx 0.707$, $ON = \cos 45° \approx 0.707$
  Since $NP = ON \neq OP$, $\triangle ONP$ is isosceles.
  b By Pythagoras, $ON^2 + NP^2 = OP^2$
  $\therefore \; ON^2 + ON^2 = 1 \quad \{NP = ON\}$
  $\therefore \; 2\,ON^2 = 1$
  $\therefore \; ON = \sqrt{\tfrac{1}{2}}$
  $\therefore \; ON = NP = \sqrt{\tfrac{1}{2}}$
  c $\sqrt{\tfrac{1}{2}} = \tfrac{1}{\sqrt{2}} \approx 0.707$
  d $P(\tfrac{1}{\sqrt{2}}, \tfrac{1}{\sqrt{2}})$ **e i** $\tfrac{1}{\sqrt{2}}$ **ii** $\tfrac{1}{\sqrt{2}}$ **iii** 1

4 a A(1, 0), B(0, 1) **b i** 0 **ii** 1 **iii** undefined
c i 1 **ii** 0 **iii** 0
5 a OP = OA = 1 {equal radii}
∴ △OPA is isosceles.
But the included angle is 60°.
∴ △OPA is equilateral.
b ON = $\frac{1}{2}$
c By Pythagoras, ON² + PN² = OP²
∴ $(\frac{1}{2})^2$ + PN² = 1² {NP = ON}
∴ PN² = $\frac{3}{4}$
∴ PN = $\sqrt{\frac{3}{4}}$
d Using technology, $\sqrt{\frac{3}{4}} = \frac{\sqrt{3}}{2}$ **e** P($\frac{1}{2}, \frac{\sqrt{3}}{2}$)
f i $\frac{1}{2}$ **ii** $\frac{\sqrt{3}}{2}$ **iii** $\sqrt{3}$ **g** O$\widehat{P}$N = 30°
h i $\frac{\sqrt{3}}{2}$ **ii** $\frac{1}{2}$ **iii** $\frac{1}{\sqrt{3}}$

EXERCISE 24G.2

1 a $\frac{1}{2}$ **b** 0 **c** $\sqrt{3}$ **d** $\frac{1}{\sqrt{2}}$ **e** 0 **f** 1
2 a sin 30° + cos 60° = $\frac{1}{2} + \frac{1}{2} = 1$
b sin² 30° + cos² 30° = $(\frac{1}{2})^2 + (\frac{\sqrt{3}}{2})^2 = \frac{1}{4} + \frac{3}{4} = 1$
c cos² 45° + sin² 45° = $(\frac{1}{\sqrt{2}})^2 + (\frac{1}{\sqrt{2}})^2 = \frac{1}{2} + \frac{1}{2} = 1$
d sin 30° cos 60° + sin 60° cos 30° = $(\frac{1}{2})(\frac{1}{2}) + (\frac{\sqrt{3}}{2})(\frac{\sqrt{3}}{2})$
   = $\frac{1}{4} + \frac{3}{4} = 1$
e sin² 30° + sin² 45° + sin² 60° = $(\frac{1}{2})^2 + (\frac{1}{\sqrt{2}})^2 + (\frac{\sqrt{3}}{2})^2$
   = $\frac{1}{4} + \frac{1}{2} + \frac{3}{4} = \frac{3}{2}$
3 a $\frac{3}{4}$ **b** $\frac{1}{\sqrt{3}}$ **c** 3 **d** 2 **e** $\frac{3}{4}$ **f** 1
g $\sqrt{3}$ **h** $\frac{1}{2}$ **i** $2\frac{1}{2}$
4 a $x = 4$ **b** $a = 6\sqrt{3}$ **c** $k = 7$ **d** $c = 8\sqrt{3}$
e $x = \frac{6}{\sqrt{3}}$ **f** $y = 2$

REVIEW SET 24A

1 a

[Diagram of right triangle with R (right angle), Q (32°), 4 cm along RQ, P at top. Scale: 1 cm ≡ 1 m]

b ≈ 4.7 m
c ≈ 4.72 m
2 a BC **b** AC **c** AB
3 a PR ≈ 2.2 cm, QR ≈ 4.5 cm, PQ ≈ 5.0 cm
b i sin 26° ≈ 0.44 **ii** cos 26° ≈ 0.9
iii tan 26° ≈ 0.49
Using a calculator,
sin 26° ≈ 0.438, cos 26° ≈ 0.899, tan 26° ≈ 0.488
4 sin 49° = $\frac{x}{p}$
5 a $x ≈ 64.3$ **b** $x ≈ 11.0$ **c** $x ≈ 4.2$
6 a $\theta ≈ 34.7°$ **b** $\theta ≈ 38.7°$ **c** $\theta ≈ 53.1°$
7 $\theta ≈ 52.4°$, $\phi ≈ 37.6°$, $x ≈ 16.4$ **8** ≈ 13.7 m
9 $\theta ≈ 18.4°$ **10** ≈ 36.7 cm
11 a ≈ 107 km from her starting point **b** ≈ 306°
12 a $\frac{4}{3}$ **b** $\frac{3}{4}$

REVIEW SET 24B

1 a q **b** p **c** p **d** q
2 a ≈ 0.588 **b** ≈ 0.809 **c** ≈ 1.38
3 a $x ≈ 10.90$ **b** $x ≈ 6.66$ **c** $x ≈ 7.81$
4 a $\theta ≈ 52.1°$ **b** $\theta ≈ 27.5°$ **c** $\theta ≈ 45.7°$
5 a Ella should use sin 13° = $\frac{5}{x}$ instead of tan 13° = $\frac{5}{x}$.
b $x ≈ 22.2$
6 ≈ 2.84 m **7 a** $y ≈ 56.8$ **b** $y ≈ 18.9$ **c** $y ≈ 63.0$
8 $\theta = 59°$, $a ≈ 5.4$, $b ≈ 10.5$ **9** AC ≈ 25.7 cm
10 a 52.4 m
b short building ≈ 51 m, tall building ≈ 103 m
11 a

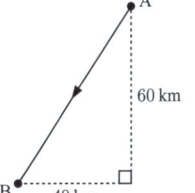

b ≈ 72.1 km
c 034°

12 a $x = 12$ **b** $x = \frac{5}{\sqrt{2}}$ **c** $x = 9\sqrt{3}$

EXERCISE 25A

1 a quadratic equation **b** not a quadratic equation
c quadratic equation **d** quadratic equation
e not a quadratic equation **f** quadratic equation
2 a 1 and 6 **b** -4 and 2 **c** 0 and 5 **d** 3 **e** -2 and $\frac{1}{2}$
3 If $x^2 + 5 = 0$, then $x^2 = -5$.
But x^2 is always $\geqslant 0$. ∴ there are no real solutions.

EXERCISE 25B

1 a $a = 0$ or $c = 0$ **b** $b = 0$ or $d = 0$
c At least one of a, b, and c is 0. **d** $x = 0$
e $x = 0$ or $x = 3$ **f** $x = 0$
g $x = 5$ or $y = 0$ **h** $x = 0$ or $y = 0$
2 a $x = 0$ or 1 **b** $x = 0$ or -5 **c** $x = 0$ or -2
d $x = 1$ **e** $x = 0$ or 4 **f** $x = 0$ or -3
g $x = 0$ or $-\frac{1}{2}$ **h** $x = 0$ or $\frac{3}{4}$ **i** $x = 0$ or $-\frac{5}{3}$
3 a $x = 1$ or 5 **b** $x = -2$ or 4 **c** $x = -3$ or -7
d $x = -7$ or 11 **e** $x = 0$ or 8 **f** $x = -12$ or 5
g $x = 0$ or -7 **h** $x = -\frac{1}{2}$ or 3 **i** $x = -6$ or $\frac{1}{3}$
j $x = -\frac{1}{2}$ or -6 **k** $x = 3$ **l** $x = 31$ or -11
m $x = -4$ or $\frac{1}{4}$ **n** $x = 0$ or $-\frac{3}{7}$ **o** $x = 2$ or $-\frac{4}{3}$

EXERCISE 25C.1

1 a $x = 0$ or 1 **b** $x = 0$ or 13 **c** $x = 0$ or -8
d $x = 0$ or -3 **e** $x = 0$ or -2 **f** $x = 0$ or 5
g $x = 0$ or 12 **h** $x = 0$ or -7 **i** $x = 0$ or 4
j $x = 0$ or $\frac{7}{2}$ **k** $x = 0$ or 5 **l** $x = 0$ or -4
2 a $x = 0$ or 3 **b** $x = 0$ or 10 **c** $a = 0$ or 1
d $x = 0$ or -6 **e** $x = 0$ or $-\frac{7}{2}$ **f** $x = 0$ or 5
g $x = 0$ or 8 **h** $y = 0$ or 6 **i** $x = 0$ or $\frac{6}{5}$
j $x = 0$ or -6 **k** $x = 0$ or 2 **l** $x = 0$ or 1

EXERCISE 25C.2

1 a $x = \pm 4$ **b** $x = \pm 3$ **c** $x = \pm 7$ **d** $x = \pm 8$
e $x = \pm 12$ **f** $x = \pm \sqrt{5}$

2 If $x^2 + 4 = 0$ then $x^2 = -4$.
But $x^2 \geqslant 0$ ∴ no real solutions.

3 **a** $x = \pm 2$ **b** $x = \pm 6$ **c** $x = \pm 3$ **d** $x = \pm 2$
e $x = \pm 2$ **f** $x = \pm 2$

4 **a** $x = \pm \frac{2}{3}$ **b** $x = \pm \frac{1}{3}$ **c** $x = \pm \frac{2}{5}$ **d** $x = \pm \frac{1}{2}$
e $x = \pm \frac{3}{2}$ **f** $x = \pm \frac{1}{2}$

EXERCISE 25C.3

1 **a** $x = 2$ or 5 **b** $x = -2$ or -4 **c** $x = -1$ or -10
d $x = 2$ or 6 **e** $x = 1$ or 4 **f** $x = 3$ or 8
g $x = -5$ **h** $x = -3$ or 6 **i** $x = 2$ or -9
j $x = 11$ **k** $x = 3$ **l** $x = -1$ or 6
m $x = 4$ or -15 **n** $x = 3$ or -21 **o** $x = -4$ or 16
p $x = 5$ or 14

2 **a** $x = 3$ or -5 **b** $x = 6$ **c** $x = -5$ or 4
d $x = 3$ or 4 **e** $x = 7$ or -6 **f** $x = -2$ or -6

3 **a** $x = -6$ or 2 **b** $x = 15$ or -1 **c** $x = 1$ or 2
d $d = 7$ or -4 **e** $x = 5$ or -4 **f** $x = 1$ or -8
g $x = 8$ or -3 **h** $x = 3$ or -4 **i** $k = 9$ or -5
j $x = 8$ or 2 **k** $x = 1$ **l** $y = 20$ or -1

4 **a** $x = 3$ or -5 **b** $x = 8$ or -5 **c** $x = 3$ or -4
d $x = -2$ or 3 **e** $x = 1$ **f** $x = 2$ or -4

EXERCISE 25D

1 The number is 6 or -7. **2** The number is 0 or 5.
3 The number is 8 or -7.
4 The integers are 7 and 8 or -8 and -7.
5 The numbers are 9 and 11 or -9 and -11.
6 The numbers are 11 and 7 or -7 and -11.
7 The width is 8 cm.
8 **a** $x = 5$ **b** The perimeter is 30 cm.
9 The base is 11 cm long. **10** The perimeter is 50 cm.
11 **b** $x = 3$
12 **b** **i** $x = 4$ **ii** The perimeter is 40 cm.
13 The outer radius is 7 cm.
14 **a** $0 < x < 5$
b Area of the icing is the total area minus the filling
$= 8 \times 5 - (8-x)(5-x)$
$= 40 - (40 - 13x + x^2)$
$= (-x^2 + 13x)$ cm^2

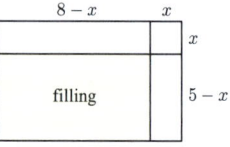

c 30% of $40 = 12$
∴ $-x^2 + 13x = 12$
∴ $x^2 - 13x + 12 = 0$
d two solutions, $x = 12$ or 1
e 1 cm (12 cm would be larger than the cake)

EXERCISE 25E.1

1 **b** and **c** are quadratic functions.
2 **a** $y = 19$ **b** $y = 82$ **c** $y = 34$
3 **a** $y = 0$ **b** $y = 5$ **c** $y = -15$ **d** $y = 12$

EXERCISE 25E.2

1 **a** $x = -1$ or 3 **b** $x = 1$ **c** $x = -2$ or 4
2 **a** $x = -3$ **b** $x = -2$ or -3 **c** $x = 1$ or 4
d no real solutions

EXERCISE 25F

1 **a**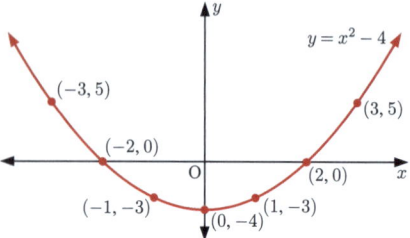

b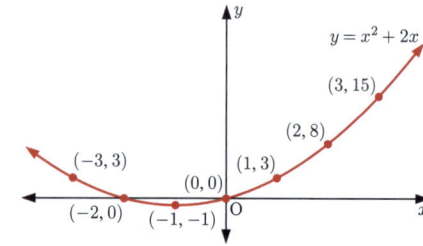

c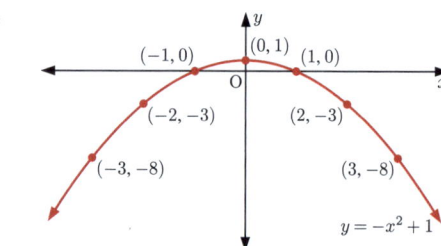

d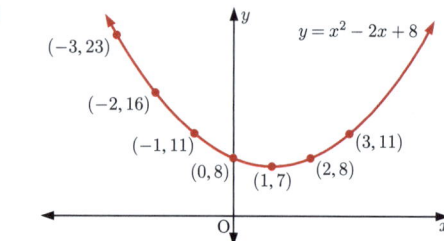

e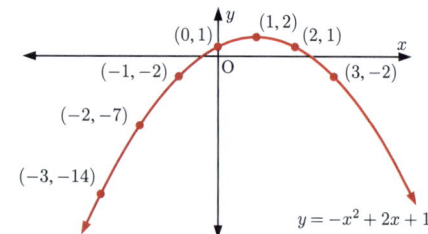

f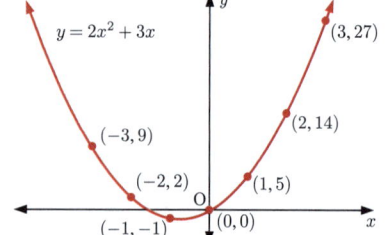

Answers 699

g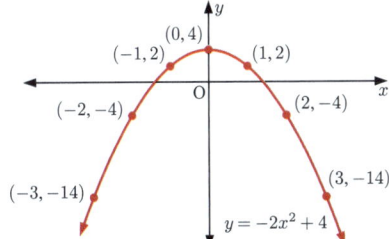

h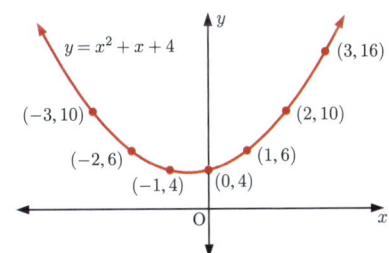

i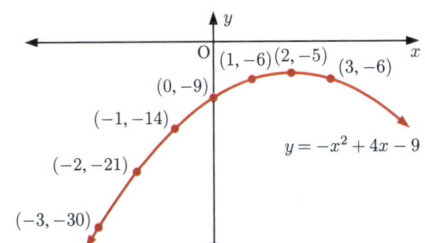

2 a i a, b, d, f, h **ii** c, e, g, i

   b "The graph of $y = ax^2 + bx + c$ has shape ∪ if $a > 0$, and shape ∩ if $a < 0$."

EXERCISE 25G.1

1 a x-intercepts -1 and 3, y-intercept -3
   b x-intercept 2, y-intercept -4
   c x-intercepts -4 and 1, y-intercept -2

2 a 6   **b** -4 **c** -7 **d** 12   **e** -15 **f** -32

3 a -2 and 3 **b** 0 and 4 **c** -3 and 7
   d 1 **e** 3 and -4 **f** -5

4 a 0 and 5 **b** -3 and 3 **c** -2
   d -2 and 7 **e** 6 **f** -1 and 2

5 a x-intercepts -5 and 4, y-intercept -20
   b x-intercept 4, y-intercept 16
   c x-intercepts -3 and 8, y-intercept -24

6 a $x = -3$ and $x = 2$ **b** $x^2 + x - 6 = (x+3)(x-2)$

7 a 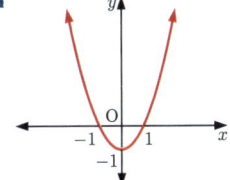   **b**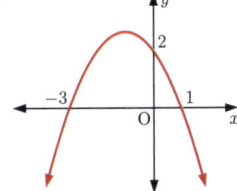

c 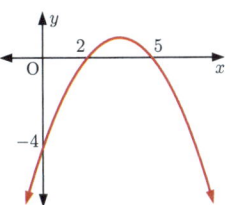   **d**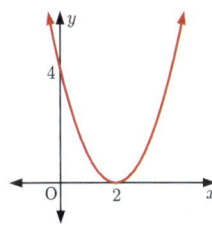

8 a i $a = 1$ **ii** -4
     iii -2 and 2

   b i $a = 1$ **ii** -3
     iii 1 and -3

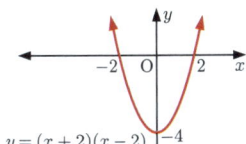   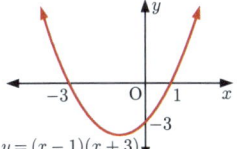

c i $a = 1$ **ii** 4
   iii -2 (touching)

d i $a = -1$ **ii** 2
   iii 2 and -1

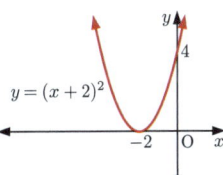   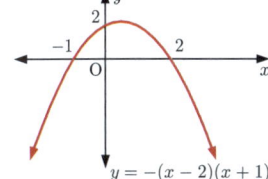

e i $a = 3$ **ii** 3
   iii -1 (touching)

f i $a = -3$ **ii** -12
   iii 4 and 1

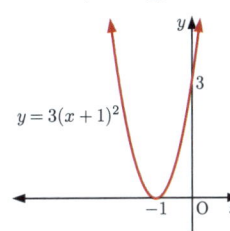   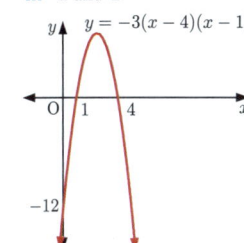

g i $a = 1$ **ii** -3
   iii 3 and -1

h i $a = 1$ **ii** -8
   iii -4 and 2

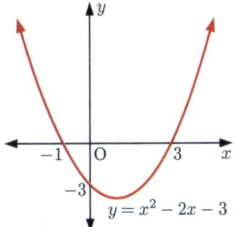   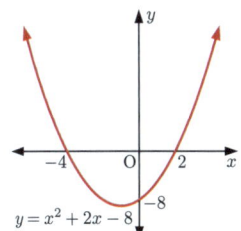

i i $a = -1$
   ii 6
   iii -6 and 1

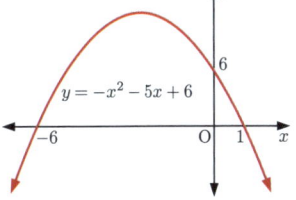

EXERCISE 25G.2
1 **a** $x \approx 0.6$ and 3.4 **b** $x \approx -2.3$ and 1.3
2 **a** $x \approx -4.4$ and -1.6 **b** $x \approx -4.7$ and -1.3
 c $x = -3$

EXERCISE 25H
1 **a** $x = 2$ **b** $x = 1$ **c** $x = -2$ **d** $x = \frac{3}{2}$
 e $x = -\frac{5}{2}$ **f** $x = 2$
2 **a** **i** 2 and 4 **ii** $x = 3$ **b** **i** -1 and 5 **ii** $x = 2$
 c **i** -3 and 3 **ii** $x = 0$
 d **i** 0 and -5 **ii** $x = -\frac{5}{2}$
 e **i** -4 **ii** $x = -4$ **f** **i** -6 and 9 **ii** $x = \frac{3}{2}$
3 3

EXERCISE 25I
1 **a** $(2, -3)$ **b** $(1, 4)$ **c** $(-3, 0)$
2 **a** $x = 2$ **b** $(2, -9)$
3 **a** -2 and 6 **b** $x = 2$ **c** $(2, -16)$
4 **a** **i** x-intercepts 0 and 2, y-intercept 0
  ii $x = 1$
  iii $(1, -1)$
  iv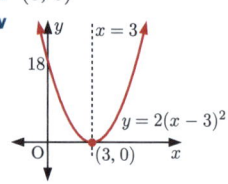
 b **i** x-intercepts -3 and 1, y-intercept 3
  ii $x = -1$
  iii $(-1, 4)$
  iv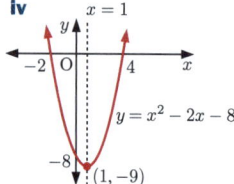
 c **i** x-intercept 3, y-intercept 18
  ii $x = 3$
  iii $(3, 0)$
  iv

  $y = 2(x - 3)^2$
 d **i** x-intercepts -2 and 4, y-intercept -8
  ii $x = 1$
  iii $(1, -9)$
  iv

  $y = x^2 - 2x - 8$
 e **i** x-intercepts 0 and 4, y-intercept 0
  ii $x = 2$
  iii $(2, 4)$
  iv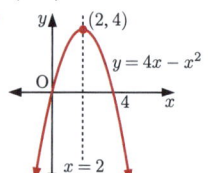
 f **i** x-intercepts -3 and 2, y-intercept 6
  ii $x = -\frac{1}{2}$
  iii $(-\frac{1}{2}, \frac{25}{4})$
  iv

  $y = -x^2 - x + 6$

REVIEW SET 25A
1 **a** quadratic equation **b** not a quadratic equation
 c not a quadratic equation
2 **a** $x = 0$ or 3 **b** $x = -5$ or 2 **c** $x = -7$

3 **a** $x = 0$ or -6 **b** $x = \pm 5$ **c** $x = -10$ or 3
 d $x = -3$ or 8 **e** $x = -1$ or 7 **f** $x = -7$ or 5
4 The number is -10 or 11.
5 square 16.8 cm by 16.8 cm, rectangle 11.2 cm by 12.6 cm
6 **a** $y = -11$ **b** $x = 6$ or -3
7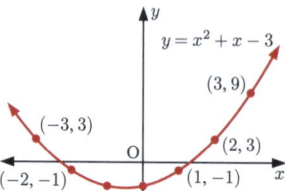

8 **a** 0 and -4 **b** -7 and 4 **9** $x \approx 0.2$ and $x \approx 4.8$
10 **a** $x = -\frac{3}{2}$ **b** $x = 2$
11 **a** x-intercepts -6 and 4, y-intercept -24
 b $x = -1$
 c $(-1, -25)$
 d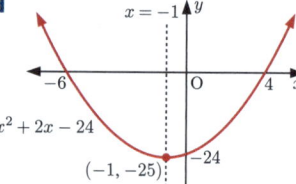

REVIEW SET 25B
1 If $x^2 + 9 = 0$, then $x^2 = -9$.
 But x^2 is always $\geqslant 0$. $\therefore$ there are no real solutions.
2 At least one of p, q, and r is 0.
3 **a** $x = 0$ or -4 **b** $x = \pm \frac{2}{3}$ **c** $x = \pm 4$
4 **a** $x = -3$ or 9 **b** $x = -2$ or 10 **c** $x = -1$ or -7
5 The rectangle is 7 m by 12 m.
6 The numbers are $8, 10, 12$ or $-12, -10, -8$.
7 $x = -7$ or 6
8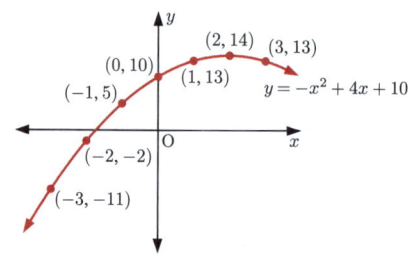

9 **a** x-intercepts -7 and 3, y-intercept -21
 b x-intercepts 3 and 10, y-intercept 30
10 **a** $a = 1$ **b** y-intercept -15 **c** -3 and 5

$y = x^2 - 2x - 15$

11 -8
12 **a** $A(-2, 0)$, $B(1, 0)$ **b** $x = -\frac{1}{2}$ **c** $(-\frac{1}{2}, -\frac{9}{4})$

Answers

EXERCISE 26A.1

1 B (a straight line through O)

2 a i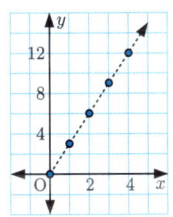
  ii directly proportional $k = 3$

b i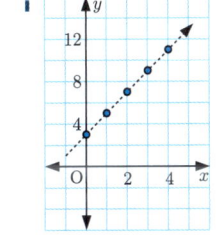
  ii not directly proportional

c i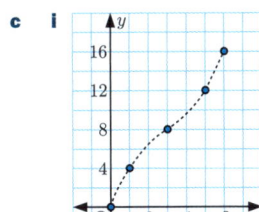
  ii not directly proportional

d i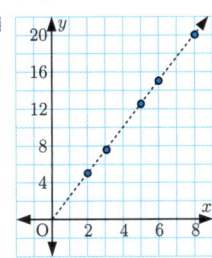
  ii directly proportional $k = 2.5$

3 a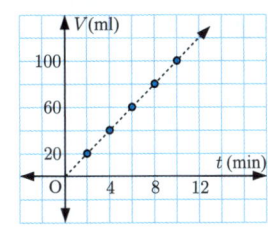
  b The graph of W against h is a straight line through the origin.
  c $W = 12.5h$

4 a

Time (t minutes)	0	2	4	6	8	10
Volume of water leaked (V ml)	0	20	40	60	80	100

  b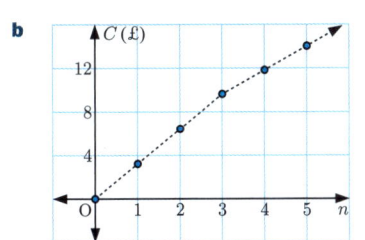
  c yes, $k = 10$

5 a

Number of items (n)	0	1	2	3	4	5
Cost (£C)	0	3.20	6.40	9.60	11.80	14

  b
  c no

6 a i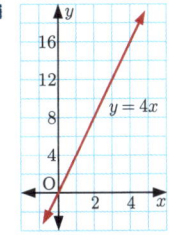
  ii directly proportional $k = 4$

b i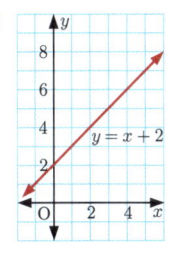
  ii not directly proportional

c i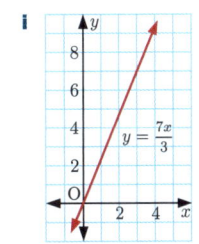
  ii directly proportional $k = \frac{7}{3}$

d i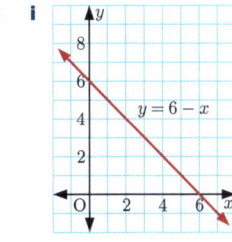
  ii not directly proportional

7 a y is doubled **b** y is trebled
  c x is doubled **d** x is halved
  e y is increased by 20% **f** x is decreased by 30%
  g cannot be determined **h** cannot be determined

8 a 2π is a constant so the graph of C against r is a straight line through the origin.
  b $k = 2\pi$
  c i C is doubled **ii** r is increased by 50%

EXERCISE 26A.2

1 a $y = 140$ **b** $x = 1$ **2 a** $y = 60$ **b** $x = 21$

3

x	5	15	30
y	20	60	120

4 a 4 litres **b** 121.5 km

5 a i $R = 0.006l$ **ii** 0.3 ohms **iii** 500 cm
  b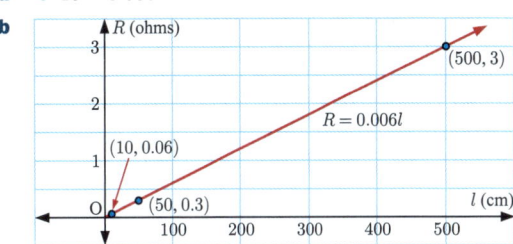

6 a 45 m² **b** 9 litres **7 a** $4l$ lollies **b** $\frac{150}{l}$ days

8 a 78.4 m/s **b** ≈ 10.2 s

9 a Yes, it is a straight line through the origin.
  b i €240 **ii** £150

10 a 10.5 cm **b** 1.2 m

EXERCISE 26B

1 a, d

x	1	2	4	8
$\frac{1}{x}$	1	$\frac{1}{2}$	$\frac{1}{4}$	$\frac{1}{8}$
y	40	20	10	5
xy	40	40	40	40

b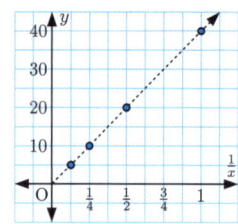

c The points of $y = \dfrac{1}{x}$ form a straight line through the origin,
∴ y is directly proportional to $\dfrac{1}{x}$
∴ y is inversely proportional to x.

2 a $xy = 12$ **c** $xy = 84$

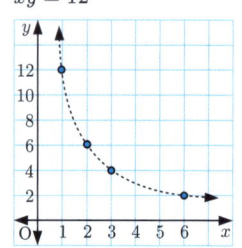 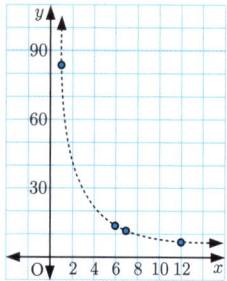

b not inversely proportional

3 **B** and **C**

4 a If d is fixed, s is **inversely** proportional to t.
b If t is fixed, s is **directly** proportional to d.
c If s is fixed, d is **directly** proportional to t.

5 a y is multiplied by $\tfrac{1}{4}$ **b** y is multiplied by $\tfrac{9}{7}$
c x is multiplied by $\tfrac{3}{8}$

6 a $y = 8$ **b** $x = \tfrac{2}{5}$ **7 a** $p = 18$ **b** $n = 12$

8

x	6	10	50
y	25	15	3

9 8 days   **10 a** 3750 chairs   **b** 16 hours
11 a The area of a rectangle is fixed.
∴ $xy = 60$, ∴ inversely proportional.

b

x	5	6	15
y	12	10	4

c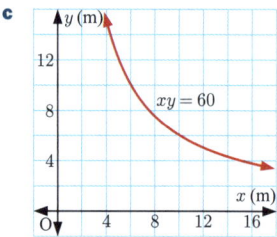

12 a 20 units   **b** 4 cm³
c

13 a 23 min 20 s   **b** 8 players   **c** 40 min

REVIEW SET 26A

1 B (a straight line through the origin)
2 a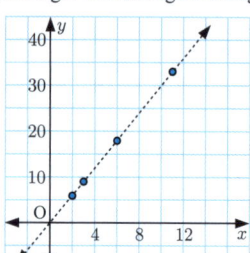

b The graph of y against x is a straight line through the origin. ∴ $y \propto x$

c $y = 3x$

3

x	12	60	80
y	3	15	20

4 a $xy = 650$ **b** $y \approx 54.2$ **c** $x \approx 4.33$
5 M is trebled   **6** 72 people
7 For $x = 4, 10, 100$ we have $xy = 200$ but for $x = 50$ we have $xy = 250$. ∴ $xy \neq$ constant.
∴ not inversely proportional.
8 $a = 0.4$, $b = 4$ **9** 36 days   **10** 1 filling

REVIEW SET 26B

1 $a = 14$, $b = 21$ **2** ≈ 229 km
3 a Each P value is 36 times each n value.   ∴ $P \propto n$
b $P = 36n$
c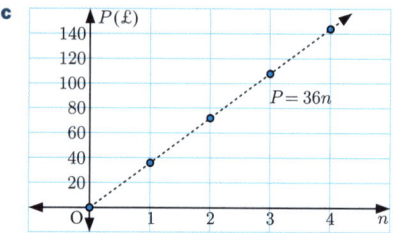

4 a D   **b** B   **5 a** $k = 19$ **b** $y = 119.7$
6 a The graph is a straight line through the origin. ∴ $y \propto x$
b 7.6 cm
7 a not inversely proportional   **b** $y = \dfrac{36}{x}$ **8** $x = 8$
9 If $y \propto \dfrac{1}{x}$ and we increase y by 25% ($\times \tfrac{5}{4}$) then

$\tfrac{5}{4} y = \tfrac{5}{4} \times \dfrac{k}{x} = \dfrac{k}{\tfrac{4}{5}x}$.

So, increasing y by 25% gives x a reduction of 20%.
10 27 minutes

EXERCISE 27A

1 a

x	-6	-4	-2	-1	1	2	4	6
y	-1	-1.5	-3	-6	6	3	1.5	1

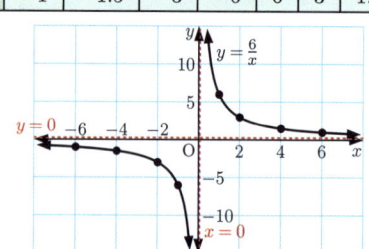

b

x	−6	−4	−2	−1	1	2	4	6
y	1	1.5	3	6	−6	−3	−1.5	−1

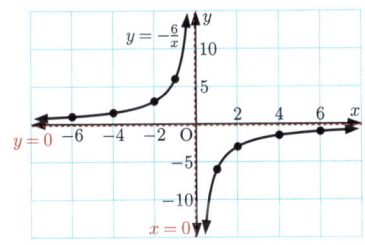

2 a $y = \dfrac{1}{x}$:

x	−4	−2	−1	1	2	4
y	−0.25	−0.5	−1	1	0.5	0.25

$y = \dfrac{2}{x}$:

x	−4	−2	−1	1	2	4
y	−0.5	−1	−2	2	1	0.5

$y = \dfrac{4}{x}$:

x	−4	−2	−1	1	2	4
y	−1	−2	−4	4	2	1

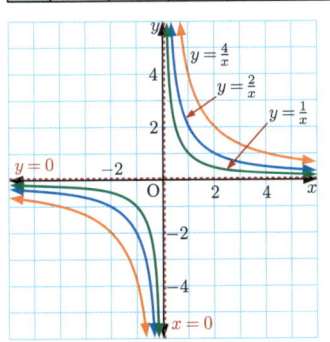

b For $k > 0$, the graphs are in quadrants 1 and 3 only. As k becomes larger the graphs move further from the origin.

3 a $y = -\dfrac{1}{x}$:

x	−4	−2	−1	1	2	4
y	0.25	0.5	1	−1	−0.5	−0.25

$y = -\dfrac{2}{x}$:

x	−4	−2	−1	1	2	4
y	0.5	1	2	−2	−1	−0.5

$y = -\dfrac{4}{x}$:

x	−4	−2	−1	1	2	4
y	1	2	4	−4	−2	−1

b For $k < 0$, the graphs are in the 2nd and 4th quadrants only. As $|k|$ becomes larger, the graphs move further from the origin.

4 a $y = \dfrac{6}{x}$ **b** $y = \dfrac{15}{x}$ **c** $y = -\dfrac{36}{x}$

EXERCISE 27B

1 a cubic   **b** not a cubic   **c** cubic
   d cubic   **e** not a cubic   **f** cubic

2 a

x	−2	−1	0	1	2
y	−6	1	2	3	10

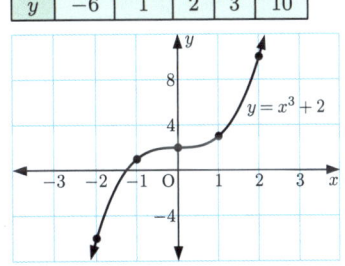

b

x	−2	−1	0	1	2
y	−2	2	0	−2	2

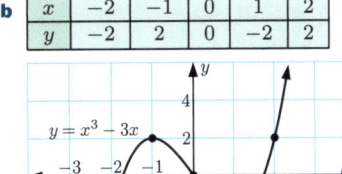

c

x	−2	−1	0	1	2
y	0	−3	0	3	0

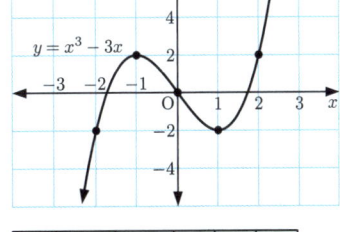

d

x	−2	−1	0	1	2
y	−12	−2	0	0	4

e

x	−2	−1	0	1	2
y	−8	−3	−4	−5	0

f

x	-2	-1	0	1	2
y	17	1	-3	-1	1

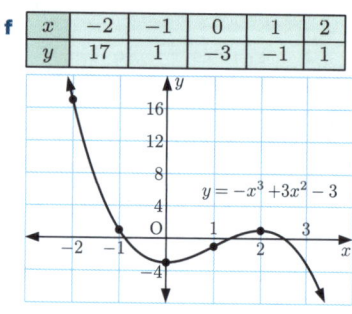

3 a 0, 2, and 6 **b** -5, -2, and 1 **c** 0 and -4
  d 0 and 2 **e** 1 and -3 **f** -5
4 a E **b** C **c** F **d** B **e** D **f** A
5 $x \approx -2.3$, 0.2, or 2.1

EXERCISE 27C
1 a 4.5 m/s **b** after ≈ 3.3s
2 a i 50 puffins **ii** 72 puffins **b** ≈ 3.8 years
3 b ≈ 2.3 m
4 a i 1.32 mg **ii** ≈ 0.86 mg
  b ≈ 1.6 mg after ≈ 2 hours

REVIEW SET 27A

1

x	-9	-3	-1	1	3	9
y	-1	-3	-9	9	3	1

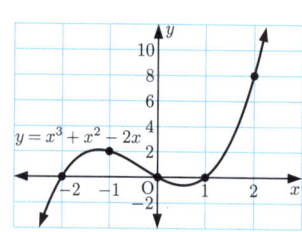

2 a $y = -\dfrac{32}{x}$ **b** When $x = 4$, $y = -8$, not 8.

3

x	-2	-1	0	1	2
y	0	2	0	1	8

4 a C **b** B **c** A
5 a $\approx 13.0°$C **b** $\approx 14.2°$C at 3 pm
  c $\approx -6.5°$C, 24 hours after midnight **d** ≈ 11.4 hours

REVIEW SET 27B

1

x	-12	-6	-3	-1	1	3	6	12
y	1	2	4	12	-12	-4	-2	-1

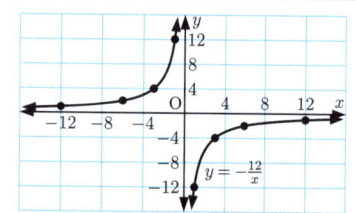

2 a $y = \dfrac{24}{x}$ **b** $y = -\dfrac{30}{x}$
3 a 0, -1, and 5 **b** 0 and 3 **c** -2
4 a $x \approx -1.2$, 0.7, and 2.5 **b** ≈ 2.3
5 a ≈ 18 m/s **b** 20 s

EXERCISE 28A
1 a 7, 13, 19, 25 **b** 40, 35, 30, 25 **c** 5, 10, 20, 40
  d 250, 50, 10, 2
2 a 4, 7, 10, 13 **b** 3, 8, 13, 18 **c** 18, 14, 10, 6
  d 2, 5, 10, 17 **e** 3, 4, 7, 12 **f** 2, 4, 8, 16
3 a 13, 21, 29, 37 **b** 85
  c start at 13, then add 8 each time
4 $c = 8$ **5 a** 14 **b** 136 **c** -14 (8th term)
6 21 **7** 25, 55, 115 **8 a** n^2 **b** 144
9 a 1, 4, 7, 10, 13 **b** start at 1, then add 3 each time
10 a 1, 3, 6, 10, 15
  b i $\frac{1}{2}(1)^2 + \frac{1}{2}(1) = \frac{1}{2} + \frac{1}{2} = 1$ ✓
    $\frac{1}{2}(2)^2 + \frac{1}{2}(2) = 2 + 1 = 3$ ✓
    $\frac{1}{2}(3)^2 + \frac{1}{2}(3) = \frac{9}{2} + \frac{3}{2} = 6$ ✓
    $\frac{1}{2}(4)^2 + \frac{1}{2}(4) = 8 + 2 = 10$ ✓
    $\frac{1}{2}(5)^2 + \frac{1}{2}(5) = \frac{25}{2} + \frac{5}{2} = 15$ ✓
   ii $\frac{1}{2}(10)^2 + \frac{1}{2}(10) = 50 + 5 = 55$
11 a 39 is the 7th term of the sequence.
   b 8 is *not* a member of the sequence.
   c -52 is the 20th term of the sequence.
12 a 21 **b** 41 is the 6th term of the sequence.
   c 80 is *not* a member of the sequence.
13 a 29, 34 **b** 20, 12 **c** $-\frac{1}{2}$, 0 **d** 64, 128
   e 25, 32 **f** 34, 42 **g** 48, 62 **h** 60, 75
14 a i Gemma's rule is "the nth term is the nth prime number".
   ii 13, 17
   b If Harry is thinking 2, 3, 5, 8, 12, 17, 23, ...
     the 6th and 7th terms are 17, 23.
     ($+1$ $+2$ $+3$ $+4$ $+5$ $+6$)

EXERCISE 28B.1
1 a The sequence is arithmetic with common difference 3.
   b The sequence is *not* arithmetic.
   c The sequence is *not* arithmetic.
   d The sequence is arithmetic with common difference -7.
2 a 4 **b** 8 **c** -5 **d** $2\frac{1}{2}$
3 a $\square = 16$ **b** $\square = 34$ **c** $\square = 15$, $\triangle = 3$
   d $\square = 13$, $\triangle = -5$
4 a The sequence is arithmetic with common difference 2.
   b The sequence is *not* arithmetic, as the number of dots added each time is not constant.
5 a 1 **b** 11 **c** -10 **d** -9

EXERCISE 28B.2
1 a 7 **b** $7n + 3$
2 a $3n + 5$ **b** $5n + 9$ **c** $41 - 6n$ **d** $8n - 3$
   e $4n - 11$ **f** $31 - 9n$
3 a $6n - 1$ **b** 119
4 a $34 - 3n$ **b** 4
   c -20 is the 18th term of the sequence.

Answers 705

5 a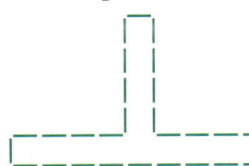
4 houses

b i 8 **ii** 14 **iii** 20 **iv** 26
c $6n + 2$
d 92

EXERCISE 28C

1 a The sequence is geometric with common ratio 2.
b The sequence is *not* geometric.
c The sequence is geometric with common ratio $\frac{1}{3}$.
d The sequence is *not* geometric.

2 D

3 a 5 **b** 7 **c** $\frac{1}{2}$ **d** 2 **e** $\frac{1}{10}$ **f** $\frac{3}{2}$

4 a $b = 16$, $c = 32$ **b** $b = \frac{1}{2}$, $c = \frac{1}{8}$ **c** $b = \frac{5}{3}$, $c = \frac{5}{9}$

5 a 3, 9, 27, 81
b 3, 9, 27, 81, ...
    ×3 ×3 ×3
∴ the sequence is geometric with common ratio 3.

6 a $\frac{1}{4}$, $\frac{1}{16}$, $\frac{1}{64}$, $\frac{1}{256}$
b $\frac{1}{4}$, $\frac{1}{16}$, $\frac{1}{64}$, $\frac{1}{256}$, ...
    ×$\frac{1}{4}$ ×$\frac{1}{4}$ ×$\frac{1}{4}$
∴ the sequence is geometric with common ratio $\frac{1}{4}$.

7 a 100 **b** 625

EXERCISE 28D

1 a 7, 11, 18 **b** 16, 26, 42 **c** 16, 25, 41
d 17, 28, 45 **e** 26, 41, 67 **f** 23, 36, 59

2 A, D

3 a □ = 9 **b** □ = 10 **c** □ = 2, ▽ = 14
d □ = 7, ▽ = 20 **e** □ = 6, ▽ = 21
f □ = 4, ▽ = 13

4 107 **5 a** 8, 12, 20 **b** 136 **6** 46

7 a $5 + b$ **b** 4th term: $b + (5 + b) = 5 + 2b$
∴ 5th term: $(5 + b) + (5 + 2b) = 10 + 3b$
c $b = 11$

8 a i $a + 9$ **ii** $a + 18$
b $a = 7$, sequence is 7, 9, 16, 25, 41

9 a i $x + y$ **ii** $3x + 5y$ **b** $x = 2$, $y = 10$

REVIEW SET 28A

1 a 8, 13, 18, 23 **b** 19, 12, 5, −2
2 a 3, 8, 15, 24 **b** 120 **3** 33, 42
4 a □ = 25 **b** □ = 21, △ = 9
5 a $3n + 10$ **b** $69 - 9n$
6 a $7n + 5$ **b** 61
c 100 is *not* a member of the sequence.
7 a 4 **b** $\frac{2}{3}$
8 a $b = 100$, $c = 500$ **b** $b = \frac{9}{16}$, $c = \frac{27}{64}$
9 a 18, 29, 47 **b** 22, 36, 58
10 a $a + b$ **b** 4th term: $b + (a + b) = a + 2b$
5th term: $(a + b) + (a + 2b) = 2a + 3b$
6th term: $(a + 2b) + (2a + 3b) = 3a + 5b$
7th term: $(2a + 3b) + (3a + 5b) = 5a + 8b$
c $a = 1$, $b = 6$
11 a 1, 4, 9, 16, 25

b The numbers of dots in the figures are 1, 4, 9, 16, ... which is 1^2, 2^2, 3^2, 4^2, ...
So the rule is: the nth figure is n^2.
c 225

REVIEW SET 28B

1 a 9, 11, 13, 15 **b** 4, 14, 30, 52
2 a $7n + 18$ **b** 88 **3** 4, 10, 28
4 a 12 **b** $-\frac{1}{2}$

5 a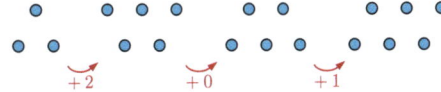
b i 10 **ii** 16 **iii** 22 **iv** 28
c $6n + 4$ **d** 124 matchsticks

6 a The sequence is geometric with common ratio 5.
b The sequence is *not* geometric.

7 a $\frac{3}{2}$, $\frac{9}{4}$, $\frac{27}{8}$, $\frac{81}{6}$
b $\frac{3}{2}$, $\frac{9}{4}$, $\frac{27}{8}$, $\frac{81}{16}$, ...
    ×$\frac{3}{2}$ ×$\frac{3}{2}$ ×$\frac{3}{2}$
∴ the sequence is geometric with common ratio $\frac{3}{2}$.

8 a □ = 7, ▽ = 19 **b** □ = 4, ▽ = 9

9 a
    +2 +0 +1
∴ the sequence is *not* arithmetic, as the number of dots added each time is not constant.

b
    +1 +1 +1
∴ the sequence is arithmetic with common difference 1.

10 a 15, 26, 41 **b** 108

11 a i Start with 3, then add 3 each time. **ii** 18, 21
b i Chelsea believes this is a Fibonacci-type sequence, so each term is the sum of the previous two terms.
ii The 6th and 7th terms are 39 and 63.

EXERCISE 29A

1 a not likely to happen
b slightly more than equal chance of happening
c very unlikely to happen

2 a Lions **b** Strikers **c** 41% = 0.41 **d** false

3 a equal chance of happening as not happening
b certain **c** impossible **d** very unlikely to happen

4 a Patricia **b** Rob
c 100% = 1, which means it is certain that one of the students will win the race.
d 35% = 0.35
e It is very unlikely that Rob will win the race.
f i $P(E) = 0.22$
ii E' is the event that Edward will *not* win the race.
iii $P(E') = 0.78$

5 a S' is the event that it will *not* snow tomorrow.

b $P(S') = 0.97$
c **i** very unlikely to happen   **ii** very likely to happen

EXERCISE 29B.1
1 a $\frac{47}{100} = 0.47$ **b** $\frac{53}{100} = 0.53$
2 a $\frac{17}{190} \approx 0.0895$ **b** $\frac{24}{190} \approx 0.126$
3 a $\frac{38}{78} \approx 0.487$ **b** $\frac{4}{78} \approx 0.0513$ **c** $\frac{57}{78} \approx 0.731$
4 a 46 nights
  b **i** $\frac{29}{121} \approx 0.240$ **ii** $\frac{85}{121} \approx 0.702$ **iii** $\frac{82}{121} \approx 0.678$
5 a $\frac{1}{5}$ of people surveyed do not have a pet, so the experimental probability that a randomly selected person *does* have a pet is $1 - \frac{1}{5} = \frac{4}{5} = 0.8$.
  b $x = 14$ **c** ≈ 0.15
6 a $x = 7$, $y = 8$ **b** ≈ 0.475

EXERCISE 29B.2
1 a ≈ 0.325
  b Connor could increase the number of trials by continuing to visit the same bakery.
2 a **i** ≈ 0.771 **ii** ≈ 0.779
  b Melia's estimate is more likely to be accurate since her experiment had a larger number of trials.
3 a **i** ≈ 0.0324 **ii** ≈ 0.291
  b **i** 597 people played the game. 412 people did not win a prize, 149 people won a minor prize, and 36 people won a major prize.
  ii ≈ 0.310
  c The estimate in **b ii** is likely to be more accurate since it has a larger number of trials.

EXERCISE 29B.3
1 a [tree diagram: students (125) → boys (68) → left-handed (8), right-handed (60); girls (57) → left-handed (5), right-handed (52)]
  b **i** ≈ 0.0877 **ii** ≈ 0.912
2 a [tree diagram: cars (277) → white (99) → small (57), large (42); not white (178) → small (109), large (69)]
  b ≈ 0.424
3 a [tree diagram: reservations (198) → Saturday (102) → lunch (35), dinner (67); Sunday (96) → lunch (33), dinner (63)]
  b ≈ 0.344

EXERCISE 29C
1 a $\mathcal{E} = \{\text{heads, tails}\}$ **b** $\mathcal{E} = \{\text{red, blue, yellow, green}\}$
2 a $\mathcal{E} = \{1, 2, 3, 4, 5, 6\}$ **b** $A = \{2, 3, 5\}$
3 a $\mathcal{E} = \{$A, B, C, D, E, F, G, H, I, J, K, L, M, N, O, P, Q, R, S, T, U, V, W, X, Y, Z$\}$
  b **i** $A = \{$B, C, D, F, G, H, J, K, L, M, N, P, Q, R, S, T, V, W, X, Y, Z$\}$
  ii $B = \{$E, N, T, V$\}$

EXERCISE 29D
1 a $\frac{1}{5}$ **b** $\frac{2}{5}$ **c** $\frac{3}{5}$
2 a $\frac{1}{8}$ **b** $\frac{1}{2}$ **c** 0
3 a $\frac{4}{7}$ **b** $\frac{3}{7}$ **c** 1
4 a $\frac{1}{13}$ **b** $\frac{1}{2}$ **c** $\frac{1}{4}$ **d** $\frac{16}{52} = \frac{4}{13}$
5 For a 4-year cycle: **a** $\frac{124}{1461}$ **b** $\frac{120}{1461}$ **c** $\frac{113}{1461}$
6 a E' is the event that the chosen muffin is not blueberry $\therefore$ it is raspberry.
  b $P(E) = \frac{1}{3}$ **c** $P(E') = \frac{2}{3}$
7 a **i** $\frac{3}{37}$ **ii** $\frac{8}{37}$ **iii** $\frac{26}{37}$
  b 1. This is the sum of all possible outcomes.   **c** $\frac{3}{11}$
8 a **i** $\frac{3}{14}$ **ii** $\frac{4}{7}$ **b** **i** $\frac{3}{13}$ **ii** $\frac{4}{13}$ **9** $\frac{1}{9}$

EXERCISE 29E.1
1 a [Venn diagram: $\mathcal{E}$ with set A containing 4, 6; outside: 1, 2, 3, 5]
  b $\frac{1}{3}$
2 a [Venn diagram: sets A and B; A only: 3, 7; intersection: 2, 6; B only: 4, 8; outside: 1, 5]
  b **i** $\frac{1}{4}$ **ii** $\frac{3}{4}$
3 a [Venn diagram: sets A and B; A only: 3, 6, 9, 15, 18, 21; intersection: 4, 12, 16, 24; B only: 8, 20, 23, 25, 22, 17; outside: 1, 2, 5, 7, 10, 11, 13, 14, 19]
  b **i** $\frac{8}{25}$ **ii** $\frac{2}{25}$ **iii** $\frac{4}{25}$ **iv** $\frac{2}{5}$
4 a $\frac{3}{26}$ **b** $\frac{1}{13}$ **c** $\frac{11}{26}$ **d** $\frac{15}{26}$

EXERCISE 29E.2
1 a $\frac{4}{7}$ **b** $\frac{3}{7}$
2 a 46 students   **b** **i** $\frac{3}{46}$ **ii** $\frac{10}{23}$ **iii** $\frac{21}{46}$
3 a [Venn diagram: sets C and P; C only: (34); intersection: (11); P only: (8); outside: (5)]
  b **i** $\frac{13}{58}$ **ii** $\frac{53}{58}$ **iii** $\frac{21}{29}$ **iv** $\frac{4}{29}$ **v** $\frac{17}{29}$
4 a [Venn diagram: sets F and R; F only: (150); intersection: (150); R only: (200); outside: (0)]
  b **i** $\frac{3}{10}$ **ii** $\frac{3}{10}$

EXERCISE 29F

1 A and B, A and D, A and E, A and F, B and D, B and F, C and D

2 **a** No, since P(A and B) $\neq 0$. **b** 0.75   3 P(Y) = 0.4

4 If C and D were mutually exclusive, then P(C or D) = 0.6 + 0.7 = 1.3, which is not possible.

EXERCISE 29G

1 **a**

1st child	2nd child
B	B
B	G
G	B
G	G

b

1st	2nd	3rd
A	B	C
A	C	B
B	A	C
B	C	A
C	A	B
C	B	A

c

1st coin	2nd coin	3rd coin	4th coin
H	H	H	H
H	H	H	T
H	H	T	H
H	H	T	T
H	T	H	H
H	T	H	T
H	T	T	H
H	T	T	T
T	H	H	H
T	H	H	T
T	H	T	H
T	H	T	T
T	T	H	H
T	T	H	T
T	T	T	H
T	T	T	T

d

1st	2nd	3rd	4th
A	B	C	D
A	B	D	C
A	C	B	D
A	C	D	B
A	D	B	C
A	D	C	B
B	A	C	D
B	A	D	C
B	C	A	D
B	C	D	A
B	D	A	C
B	D	C	A
C	A	B	D
C	A	D	B
C	B	A	D
C	B	D	A
C	D	A	B
C	D	B	A
D	A	B	C
D	A	C	B
D	B	A	C
D	B	C	A
D	C	A	B
D	C	B	A

2 **a**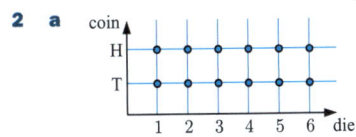

b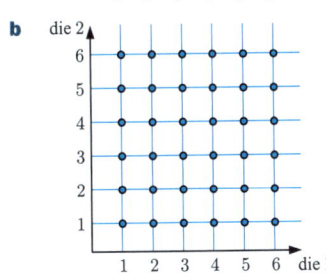

c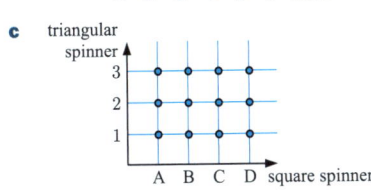

d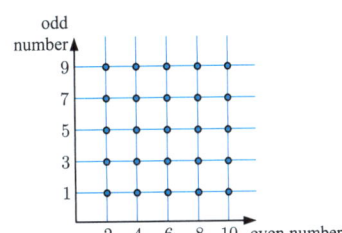

3 **a** 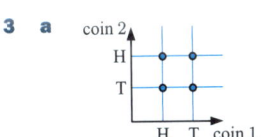   **b** **i** $\frac{1}{4}$ **ii** $\frac{1}{2}$ **iii** $\frac{3}{4}$

4 **a** 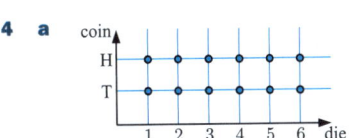   **b** **i** $\frac{1}{12}$ **ii** $\frac{7}{12}$ **iii** $\frac{2}{3}$ **iv** $\frac{5}{12}$ **v** $\frac{1}{4}$ **vi** $\frac{3}{4}$

5 **a** 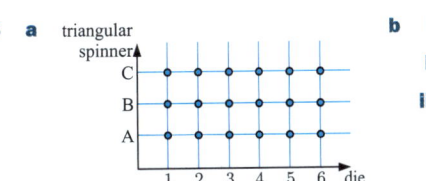   **b** **i** $\frac{1}{18}$ **ii** $\frac{1}{6}$ **iii** $\frac{2}{9}$

6 **a** $\frac{1}{24}$ **b** $\frac{1}{12}$ **c** $\frac{2}{3}$ **d** $\frac{3}{8}$ **e** $\frac{1}{3}$ **f** $\frac{5}{8}$
   g $\frac{1}{6}$ **h** $\frac{7}{12}$ **i** $\frac{1}{2}$ **j** $\frac{1}{2}$

7 **a** $\frac{2}{49}$ **b** $\frac{24}{49}$ **c** $\frac{11}{49}$ **d** $\frac{3}{7}$ 8 $\frac{21}{64}$

EXERCISE 29H

1 $\frac{1}{10}$ 2 $\frac{1}{3}$ 3 $\frac{1}{8}$

4 **a** 0.18   **b** 0.28

5 **a** 0.03   **b** 0.68   **c** 0.17

6 **a** 0.343   **b** 0.147   **c** 0.063

EXERCISE 29I

1 **a** $\frac{10}{21}$ **b** $\frac{5}{21}$

2 **a** **i** $\frac{2}{15}$ **ii** $\frac{1}{3}$ **iii** $\frac{4}{15}$ **iv** $\frac{4}{15}$
   b 1. This is the sum of all possible outcomes.

3 **a** **i** $\frac{2}{15}$ **ii** $\frac{1}{3}$ **b** $\frac{8}{15}$

4 **a** $\frac{1}{3725} \approx 0.000\,268$ **b** $\frac{147}{7450} \approx 0.0197$
   c $\frac{3577}{3725} \approx 0.960$

5 **a** 0.56   **b** 0.12   6 **a** $\frac{1}{22}$ **b** $\frac{7}{44}$

7 **a** $\frac{5}{28}$ **b** $\frac{1}{56}$ **c** $\frac{5}{56}$ 8 **a** $\frac{8}{65}$ **b** $\frac{1}{13}$ **c** $\frac{28}{195}$

EXERCISE 29J.1

1 **a**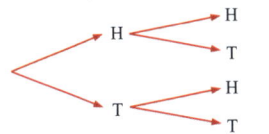

b die → coin tree: 1–6, each to H, T

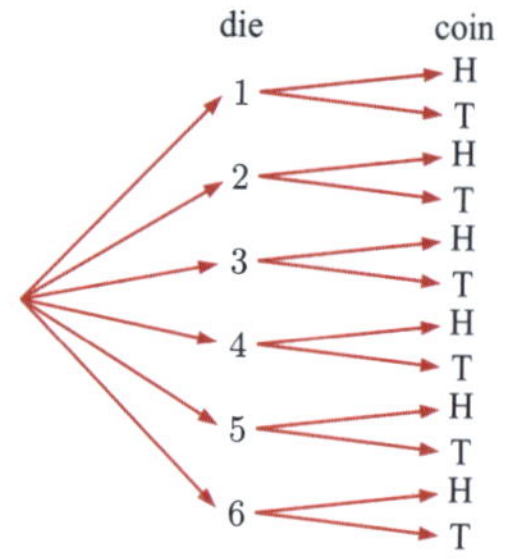

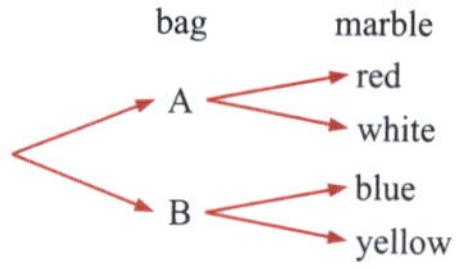

c first child, second child, third child tree: B/G branching to B/G to B/G

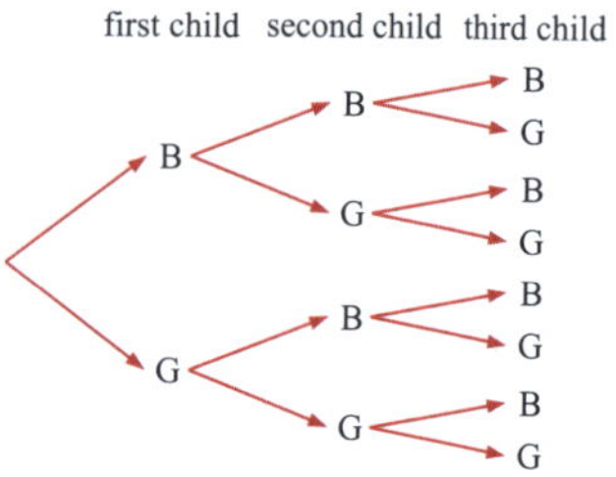

2 bag → marble: A → red, white; B → blue, yellow

3 hat → first ticket → second ticket: A, B, C each → pink/white → pink/white

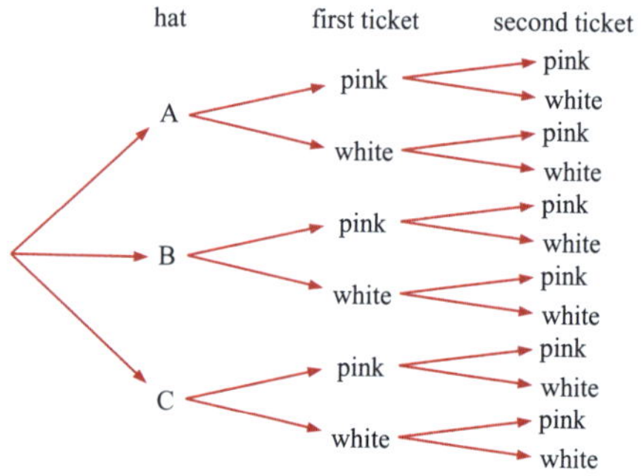

EXERCISE 29J.2

1 a first spin, second spin: B ($\frac{2}{5}$) → B ($\frac{2}{5}$), Y ($\frac{2}{5}$), G ($\frac{1}{5}$); Y ($\frac{2}{5}$) → B, Y, G; G ($\frac{1}{5}$) → B, Y, G

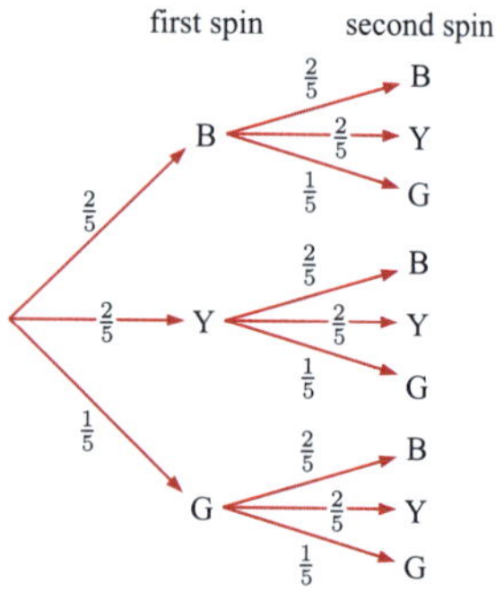

b i $\frac{4}{25}$ **ii** $\frac{1}{25}$ **iii** $\frac{16}{25}$ **iv** $\frac{16}{25}$

2 a track condition → result: muddy ($\frac{1}{4}$) → win ($\frac{2}{5}$), lose ($\frac{3}{5}$); not muddy ($\frac{3}{4}$) → win ($\frac{1}{20}$), lose ($\frac{19}{20}$)

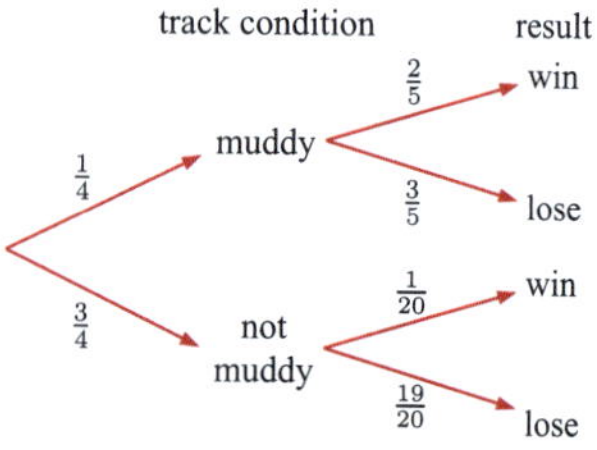

b $\frac{11}{80}$

3 0.752 **4** 0.034 **5** $\frac{23}{60}$ **6** $\frac{1}{5}$ **7** $\frac{23}{42}$

8 a 1st semi-final, 2nd semi-final, grand final tree:
A (0.4) → C (0.35) → A (0.55), C (0.45); D (0.65) → A (0.47), D (0.53)
B (0.6) → C (0.35) → B (0.42), C (0.58); D (0.65) → B (0.54), D (0.46)

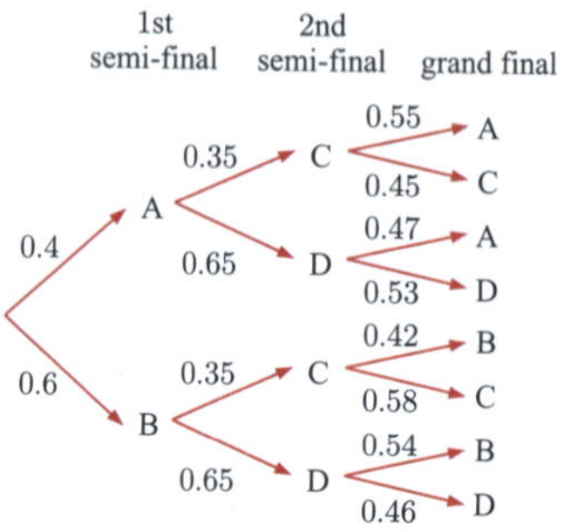

b 0.1992 **c** player D **d** player B

EXERCISE 29K

1 27 goals **2** 15 days **3** 200 times **4** 30 times
5 a 300 people **b i** 0.55 **ii** 0.29 **iii** 0.16
c i 4125 people **ii** 2175 people **iii** 1200 people
6 150 times

REVIEW SET 29A

1 a very likely to happen
b slightly less than equal chance of happening

2 a ≈ 0.422
b The accuracy of the estimate can be improved by using a larger sample size.

3 a $\frac{1}{13}$ **b** $\frac{7}{26}$ **c** $\frac{7}{26}$

4 a coin vs spinner grid (T and H rows, spinner 1–5)
b i $\frac{1}{10}$ **ii** $\frac{3}{5}$

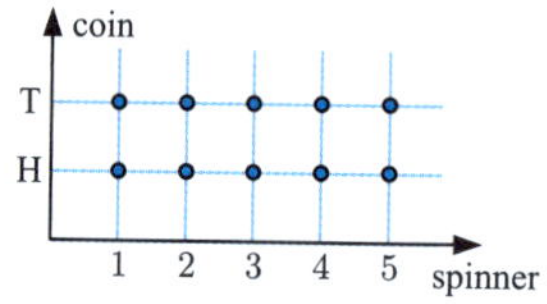

5 a 156 teenagers

b

Reason	Frequency	Relative frequency
Homework	29	≈ 0.186
Social networking	43	≈ 0.276
Playing games	15	≈ 0.0962
Streaming music	69	≈ 0.442
Total	156	≈ 1

c i ≈ 0.186 **ii** ≈ 0.558

6 a 0.2375 **b** 0.0375 **c** 0.0125 **d** 0.7125

7 a bag → lolly: A ($\frac{1}{2}$) → mint ($\frac{5}{8}$), chocolate caramel ($\frac{3}{8}$); B ($\frac{1}{2}$) → mint ($\frac{1}{2}$), chocolate caramel ($\frac{1}{2}$)

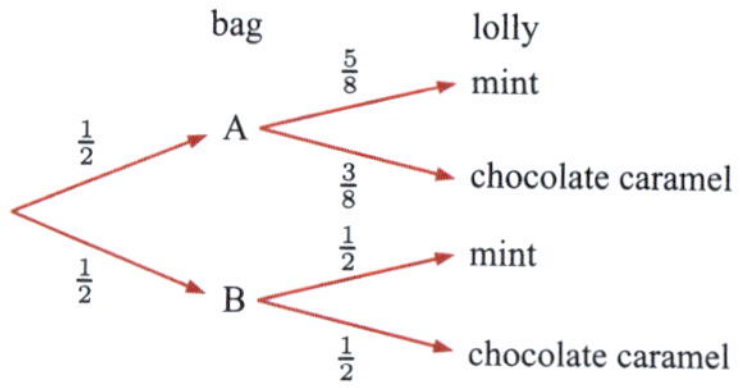

b i $\frac{7}{16}$ **ii** $\frac{9}{16}$

8 ≈ 62 waves **9** $\frac{5}{33}$
10 a 0.73 **b** ≈ 266 long showers
11 a foggy ($\frac{1}{20}$) → delayed ($\frac{9}{10}$), not delayed ($\frac{1}{10}$); not foggy ($\frac{19}{20}$) → delayed ($\frac{1}{12}$), not delayed ($\frac{11}{12}$)

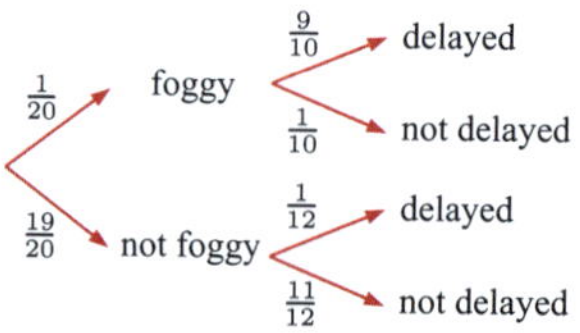

b i $\frac{9}{200}$ **ii** $\frac{149}{1200}$ **iii** $\frac{1051}{1200}$

Answers 709

    c A flight is either delayed or not delayed, so these cases cover all possibilities and the probabilites sum to 1.

12 a 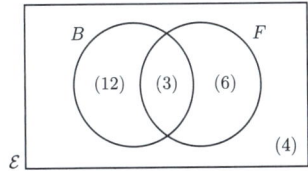   **b**   **i** $\frac{4}{25}$ **ii** $\frac{12}{25}$

13 a 0   **b** 0.14

REVIEW SET 29B

1 If two events are independent then the occurrence of one does not affect the occurrence of the other.

2 a ≈ 0.60 **b** ≈ 0.40

3 a $x = 3$ **b**   **i** ≈ 0.0263 **ii** ≈ 0.289 **iii** ≈ 0.5

4 a 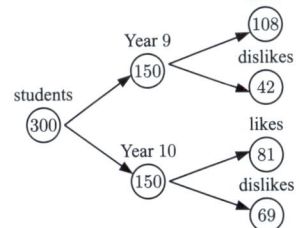   **b** ≈ 0.28

5 a 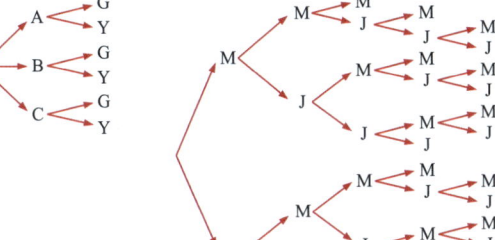   **b**

6 a

1st digit	2nd digit	3rd digit	
1	6	9	✓
1	9	6	✓
6	1	9	
6	9	1	
9	1	6	
9	6	1	✓

    b $\frac{1}{2}$

7 a 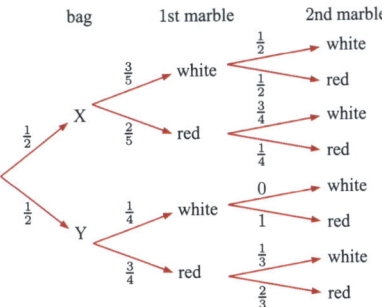   **b** $\frac{11}{20}$

8 a 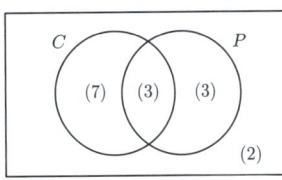   **b** $\frac{2}{15}$

9 6 goals

10 a $\frac{13}{55}$

    b   **i** Increases, as the number of tickets of the desired colour does not decrease with each draw.
       ii $\frac{37}{121}$

11 a 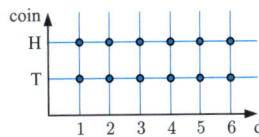   **b**   **i** $\frac{1}{4}$ **ii** $\frac{5}{12}$ **iii** $\frac{1}{3}$ **iv** $\frac{3}{4}$

12 a 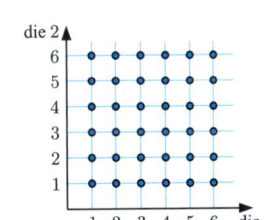   **b**   **i** $\frac{1}{36}$ **ii** $\frac{11}{36}$ **iii** $\frac{1}{6}$ **iv** $\frac{1}{4}$

    c 375 times

13 $P(X \cup Y) = 0.8$

EXERCISE 30A

1 a all boxes of cereal filled by the machine
   b It would not be practical to check the volume of cereal in every box. It would take too long and cost too much.

2 a census   **b** sample   **c** census   **d** sample   **e** sample

3 a 820 students   **b** 40 students

4 a It would be too time consuming to inspect all 375 rooms for Hugh to sensibly perform a census.
   b 375 rooms   **c** 15 rooms
   d Hugh is more likely to gain a better overall assessment of all of the rooms on each floor by randomly selecting one room from each floor. If he had selected all the rooms on one floor, these may have all been cleaned by the one cleaner, so this sample may not be representative of the whole population.

EXERCISE 30B

1 a 45 shoppers   **b** 18 shoppers   **c** $\approx 15.6\%$
   d positively skewed

2 a 10 employees   **b** 2 appointments   **c** $\approx 4.44\%$
   d positively skewed   **e** It is an outlier.

3 a 18
   b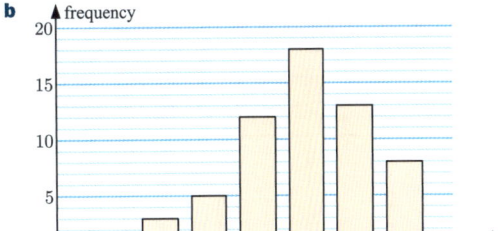

c slightly negatively skewed, no outliers

4 a

Number of TV sets	Frequency
0	6
1	7
2	4
3	2
4	1
Total	20

b Number of TV sets in students' households

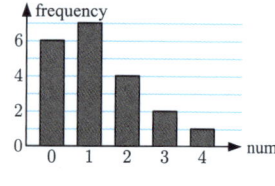

c positively skewed, no outliers

d 6 households

e 15%

5 a

No. of toothpicks	Frequency
47	1
48	5
49	10
50	23
51	10
52	9
53	2
Total	60

b Number of toothpicks in boxes

c approximately symmetrical **d** ≈ 38.3%

6 a Number of phone calls

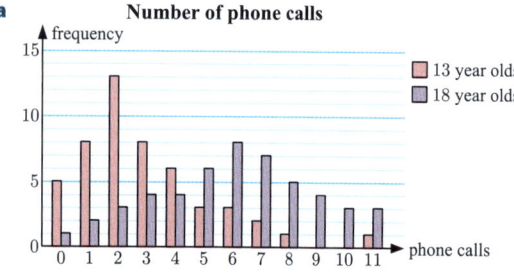

b The 13 year olds data set is positively skewed while the 18 year olds data set is more symmetrical. The 18 year olds generally made more phone calls.

7 a

Number of eggs	Frequency	
	Rhode Island Red	Buff Orpington
0	1	8
1	2	7
2	3	2
3	8	3
4	4	1
5	3	0

b Number of eggs collected

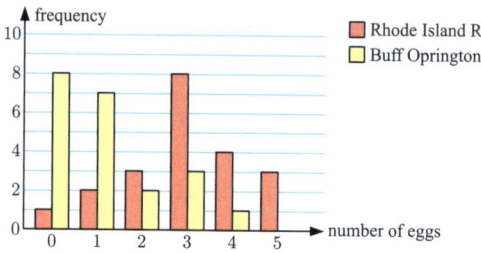

c Rhode Island Red: slightly negatively skewed
Buff Orpington: positively skewed

d The Buff Orpington: it has a lower modal value (0 eggs) and most of its data is lower than that of the Rhode Island Red.

8 a i building A   **ii** building B

b i

Number of occupants	Frequency
0	2
1	14
2	9
3	6
4	4
5	0
6	0

ii

Number of occupants	Frequency
0	0
1	3
2	4
3	11
4	5
5	3
6	2

c Number of people in each building

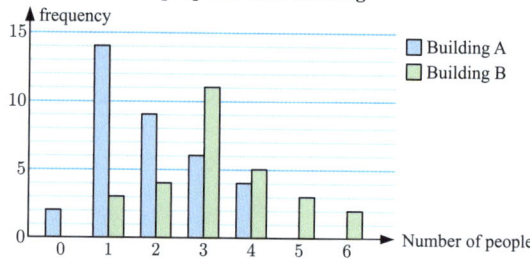

d Building A is positively skewed while building B is approximately symmetrical.

EXERCISE 30C

1 a approximately symmetrical **b** $14 \leqslant d < 16$ m **c** 35%

2 a *Weight* can take any value in the given range, and is measured, not counted.
b approximately symmetrical
c $85 \leqslant w < 90$ kg. More people in the volleyball squad have weights between 85 kg and 90 kg than in any other interval.
d 28 players   **e** ≈ 21.4% players

3 a positively skewed

b i, ii

Height (h mm)	Frequency
$20 \leqslant h < 40$	4
$40 \leqslant h < 60$	17
$60 \leqslant h < 80$	15
$80 \leqslant h < 100$	8
$100 \leqslant h < 120$	2
$120 \leqslant h < 140$	4

c i 6 seedlings   **ii** 30%
d i ≈ 754 seedlings   **ii** ≈ 686 seedlings

4 a i Girls: approximately symmetrical
Boys: negatively skewed
ii Girls: $15 \leqslant t < 15.5$ s, Boys: $15 \leqslant t < 15.5$ s
b boys **c** boys
d The boys data set, as the 1 runner in the $11.5 \leqslant t < 12$ s interval is separated from the main body of the data.

EXERCISE 30D.1

1 a i 25 **ii** 24 **iii** 30
b i ≈ 13.3 **ii** 11.5 **iii** 8
c i ≈ 10.3 **ii** 10 **iii** 11.2
d i ≈ 429 **ii** 428 **iii** 415, 427

2 a data set A: $\bar{x} \approx 7.73$ data set B: $\bar{x} \approx 8.45$
b data set A: 7 data set B: 7
c The data sets are the same except for the last value. Since the last value of A is less than the last value of B, the mean of A is less than the mean of B.
d The middle value of both data sets is the same, so the median is the same.

3 a mean: £295 900, median: £262 500, mode: £240 000
b The mode is the second lowest value, so does not take the higher values into account.
c The median; it is unaffected by large values.

4 a mean: ≈ 3.11 mm, median: 0 mm, mode: 0 mm
b i 15 and 27
ii The outliers should not be removed unless they are a result of a recording error.

5 a 44 points **b** 44 points **c** ≈ 40.6 points
d i increase, as 42 is greater than the mean calculated in **c**
ii 40.75 points

6 a 1 head **b** 1 head **c** 1.4 heads

7 a

Donation (£)	Frequency
1	7
2	9
5	2
10	4
20	8
Total	30

b 30 donations
c i $\approx £7.83$ **ii** £2 **iii** £2
d the mode

8 a i 4.25 ducklings **ii** 5 ducklings **iii** 5 ducklings
b Yes, it is negatively skewed.
c The mean is lower than the mode and median.

9 a i Leeds: 3.475 nights, Manchester: 4.075 nights
ii Leeds: 2 nights, Manchester: 3 nights and 4 nights
iii Leeds: 3 nights, Manchester: 4 nights
b the Manchester hotel

10 a Group A: mean $= 19.91$ cm, median $= 19.85$ cm
Group B: mean ≈ 17.86 cm, median $= 17.6$ cm
b The mean and median are higher for the plants which were fertilised than for those which were not. This suggests that the fertiliser is effective.

11 a Store A: mean $= £7120$, median $= £6230$
Store B: mean $\approx £10\,700$, median $= £10\,270$
b Store B
c The mean; all values should be accounted for in the prediction.

12 a Trattoria Casaro

b

Trattoria Casaro		Sergio's Italian	
Score	Frequency	Score	Frequency
0	1	0	0
1	0	1	0
2	0	2	2
3	2	3	8
4	5	4	16
5	8	5	22
6	14	6	25
7	12	7	26
8	18	8	16
9	23	9	10
10	20	10	4

c Trattoria Casaro: mean ≈ 7.62, median $= 8$, mode $= 9$
Sergio's Italian: mean ≈ 6.14, median $= 6$, mode $= 7$
d The numerical measures show that Trattoria Casaro generally had better review scores, which agrees with our initial observation.

13 26 goals **14** 7.875 **15** 48 trees

EXERCISE 30D.2

1 $\approx 32.1 °C$
2 a $165 \leqslant s < 170$ km h^{-1}
b i 76 serves **ii** not possible **iii** 73.5%
c ≈ 167 km h^{-1}
3 a 23 times **b** 16 times **c** ≈ 24.1 runs

EXERCISE 30E

1 a 7 **b** 14 **c** 7.7 **2** range $= 12$ seeds
3 a ≈ 6.3 tows **b** 7 tows **c** 6 tows
4 a 11 items **b** 11 items
c No, the range of the data with the outlier is much larger than the range of the main body of data.
5 a i Kylie: 11, Chris: 12.5 **ii** Kylie: 17, Chris: 9
b Chris **c** Kylie
6 a i Year 6: 4 visits, Year 10: 2 visits
ii Year 6: 4 visits, Year 10: 8 visits
b i the Year 6 class **ii** the Year 10 class

REVIEW SET 30A

1 a census **b** sample
2 positively skewed
3 a discrete
b no
c

Number of bedrooms	Frequency
1	1
2	10
3	6
4	3
Total	20

d Number of bedrooms in students' houses

4 a 14 **b** 13.8 **c** 14 **d** 17
5 18 males
6 a approximately symmetrical
b $50 \leqslant m < 51$ g; more eggs weigh between 50 g and 51 g than any other interval.

c

Mass (m g)	Frequency
$48 \leqslant m < 49$	1
$49 \leqslant m < 50$	1
$50 \leqslant m < 51$	16
$51 \leqslant m < 52$	4
$52 \leqslant m < 53$	3

d 25 eggs
e ≈ 50.8 g

7 $a = 3$, $b = 4$
8 a Davis: mean = £115.28, range = £30.10
Douglas: mean = £102.37, range = £61.00
b the Davis family **c** the Douglas family
9 a C, A, B **b** C, B, A

REVIEW SET 30B

1 a All the customers of the shop.
b The first 10 customers on Monday morning.
2 a 14.55 **b** 14.5 **c** 14, 15 **3** 13.56

4 a

Number of people	Frequency
7	2
8	4
9	7
10	8
11	5

b Number of people at judo class

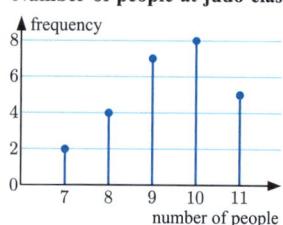

c slightly negatively skewed
5 a 32 students
b i ≈ 9.84 pieces **ii** 11 pieces **iii** 10 pieces
c Litter pieces picked up by students

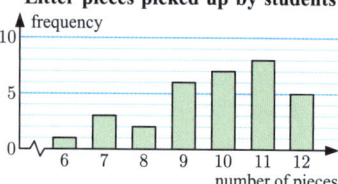

d negatively skewed

6 a Ez To Do lists

Number of stars	Frequency
1	2
2	8
3	17
4	10
5	11

Producktivity

Number of stars	Frequency
1	9
2	3
3	6
4	18
5	15

b Ez To Do lists: mean ≈ 3.42, median = 3, mode = 3
Producktivity: mean ≈ 3.53, median = 4, mode = 4
c Producktivity
d As both have the same star system, they both have the same range, so we gain no useful information.

7 $\bar{x} = 43.5$ people

8 a approximately symmetrical **b** 27 babies **c** 70%
d about 17 babies

9 a Nia's class

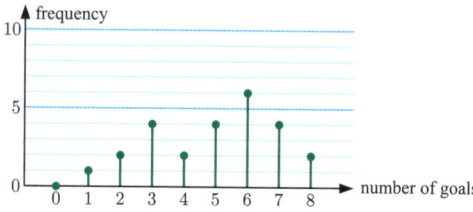

Rex's class

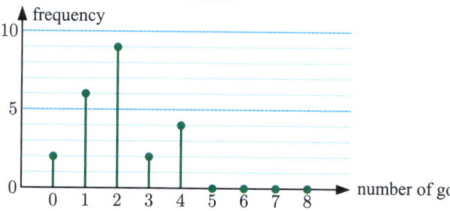

b Nia's: slightly negatively skewed
Rex: positively skewed
c i Nia's class **ii** Rex's class
d Nia: median = 5 goals, range = 7 goals
Rex: median = 2 goals, range = 4 goals

EXERCISE 31A

1 a D **b** C **c** E and H **d** E and A
2 a E **b** A **c** yes, B and C
d Yes, generally more hours of training results in higher productivity (from the graph).
3 a D **b** F **c** 2 **d** 3
4 a

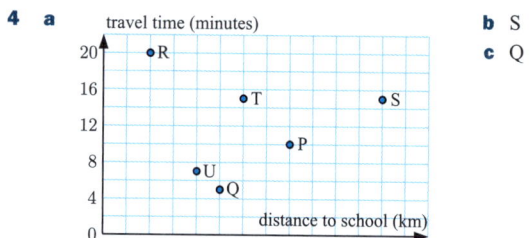

b S
c Q

d Student R. They live the shortest distance from school, yet take the most time to get there.

EXERCISE 31B.1

1 a i negative association **ii** linear
  iii strong **iv** no outliers
b i no association **ii** not linear
  iii zero **iv** no outliers
c i positive association **ii** linear
  iii moderate **iv** one outlier
d i positive association **ii** linear
  iii weak **iv** no outliers
e i positive association **ii** not linear
  iii moderate **iv** one outlier
f i negative association **ii** not linear
  iii strong **iv** no outliers

2 a as x increases, y increases   **b** as T increases, d decreases
  c If there is no association between two variables then the points on the scatter graph are randomly scattered.

3 a   **b** A moderate, positive, linear correlation.

4 a i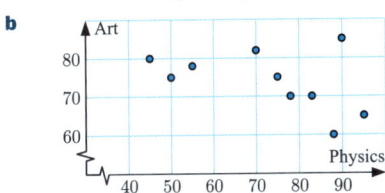
  ii A moderate, positive, linear correlation.
  b
  A very weak, negative, linear correlation (virtually zero correlation).

5 a
  b There is a weak, negative, linear correlation between sales and temperature.

EXERCISE 31B.2

1 a Causal, applying more fertiliser to the plant causes it (and its fruit) to grow more.
  b Not causal, likely dependent on temperature.
  c Causal, if a person drives to work then they will use more petrol each week the further they have to travel.
  d Not causal, dependent on the size or population of the city.

2 a i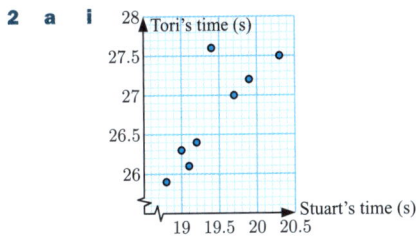

  ii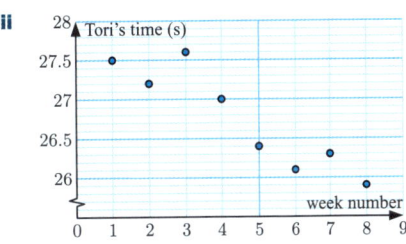

  iii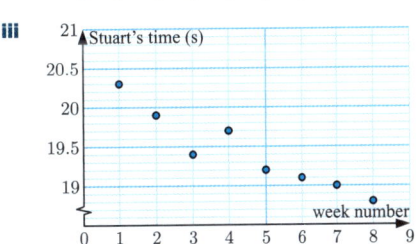

  b i Strong, positive, linear correlation. There is one outlier that lies above the general trend of the data.
  ii Moderate, negative, linear correlation.
  iii Strong, negative, linear correlation.
  c No; both Tori's times and Stuart's times are strongly related to a third variable: *week number*.

EXERCISE 31C

1 a $\overline{x} = 7.2$, $\overline{y} = 10$ **b, d**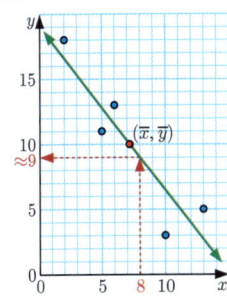
  c negatively correlated
  e when $x = 8$, $y \approx 9$

2 a $\overline{x} = 7$, $\overline{y} = 85$
  b, c, d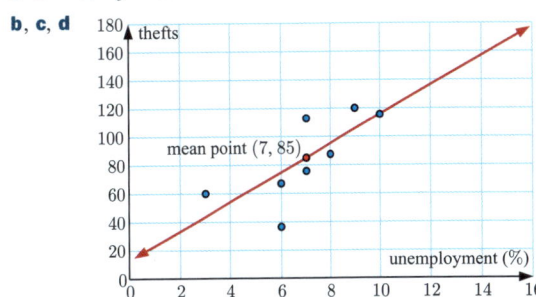
  e i ≈ 170 thefts
  ii Unreliable as it is an extrapolation well beyond the upper pole.

3 a $\overline{x} = 50$, $\overline{y} = 38$
  b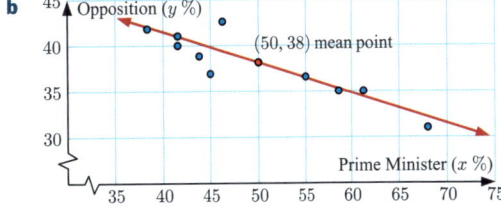

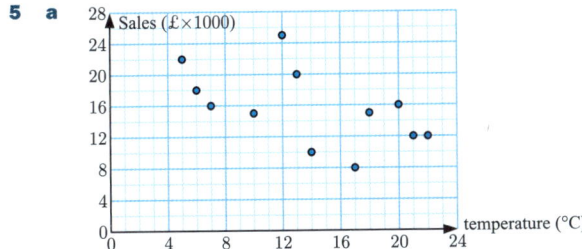

c 39%

4 a (13.3, 78.8)

b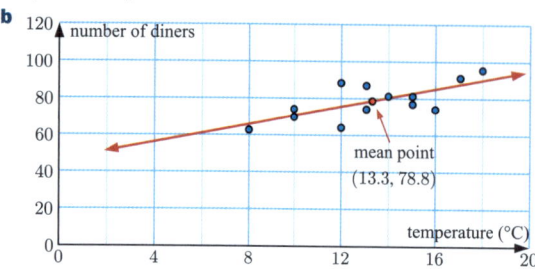

c i 73 diners **ii** 56 diners

d We expect the first estimate to be reliable as it is an interpolation, but the second may not be reliable as it is an extrapolation.

5 a, b, c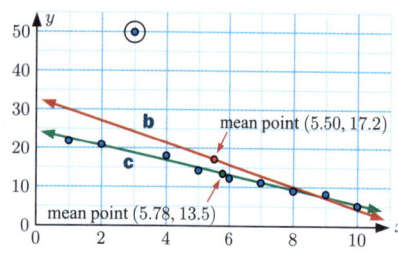

d The line of best fit in **b** is skewed towards the outlier and thus is not very close to the points to the left of the mean point.

The line of best fit in **c** however, is very close to all of the points except the outlier. This line of best fit therefore better describes the trend seen in the main body of data.

REVIEW SET 31A

1 There is a weak, negative, linear correlation between the variables.

2 a E **b** A

c The greater the time spent on player fitness, the smaller the chance of player injury.

3 a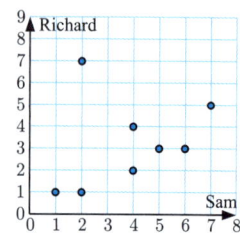

b There is a weak, positive, linear correlation between the variables.
There is one outlier that lies above the general trend.

c No; both variables are related to the number of moves made in the game.

4 a, b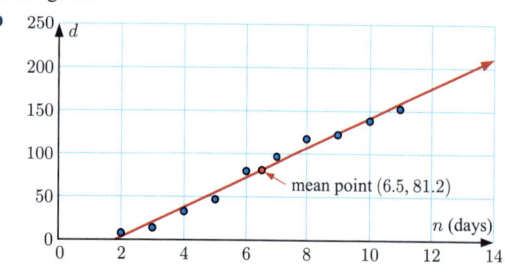

c We estimate that there will be about 210 diagnosed cases on day 14. This estimate is very unreliable as 14 is outside the poles. The medical team have probably isolated those infected at this stage and there could be a downturn which may be very significant.

5 a, b, c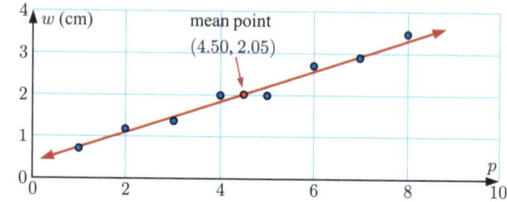

d i ≈ 5.7 cm
  ii As $p = 14$ is outside the poles, this prediction could be unreliable.

6 a Negatively correlated. The more games won, the higher the team's position on the ladder (and so the value for *Position* is smaller).

b 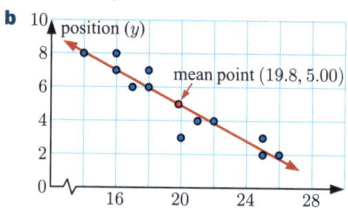 **c** $\overline{x} \approx 19.8$, $\overline{y} = 5.00$
  d 4th

REVIEW SET 31B

1 a J **b** E **c** D and G **d** C

2 a i D **ii** H

b As the number of pages increases, the number of chapters generally increases.

3 a $\overline{x} = 9$, $\overline{y} = 15$

b, d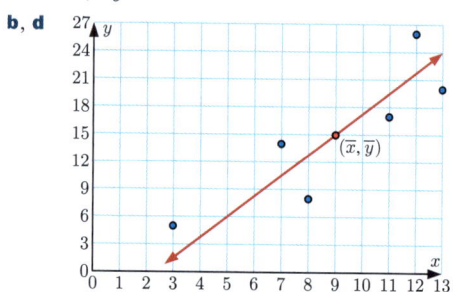

c positively correlated

e $y \approx 6$. Reasonably accurate by interpolation.

4 a No; we expect a periodic relationship between the month of the year and temperature. We expect temperatures at the beginning and end of the year to be similar. This is not a linear pattern.

b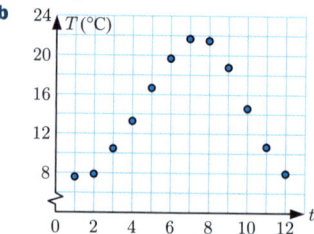

Answers

5 a

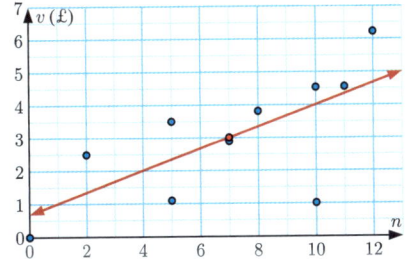

b There is a weak, positive, linear correlation between the variables. A causal relationship between the variables exists.

c $\bar{x} \approx 7.00$, $\bar{y} \approx 3.02$

d **i** £7.43
ii This prediction is unlikely to be reliable as it is an extrapolation and the correlation between the variables is weak.

6 a **i** Positive, as numbers get larger they have more possible factors.
ii Weak, even amongst large numbers there exist primes, squares, cubes, and so on which have a small number of factors.

b

Number	1	2	3	4	5	6	7	8	9	10
Number of factors	1	2	2	3	2	4	2	4	3	4

Number	11	12	13	14	15	16	17	18	19	20
Number of factors	2	6	2	4	4	5	2	6	2	6

c

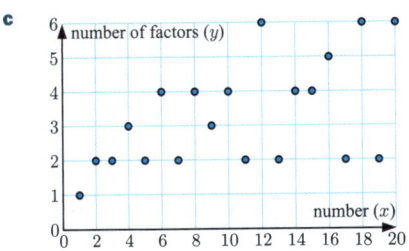

d weak, positive, linear correlation

7 a, c

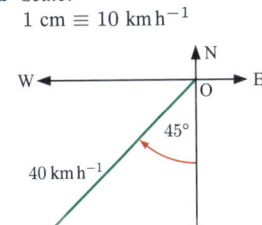

b The outlier is the point $(35, 9.1)$.

d when $I = 55$, $R \approx 5.0$

e £120 000 gives a rate of -1.09 which is meaningless. So, £120 000 is outside the data range of this model.

EXERCISE 32A

1 a *Scale*: $1\text{ cm} \equiv 10\text{ km h}^{-1}$

b *Scale*: $1\text{ cm} \equiv 10\text{ m s}^{-1}$

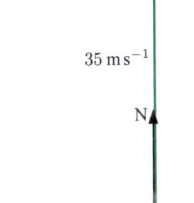

c *Scale*: $1\text{ cm} \equiv 10\text{ m}$

d *Scale*: $1\text{ cm} \equiv 10\text{ m s}^{-1}$

2 a *Scale*: $1\text{ cm} \equiv 15\text{ Newtons}$

b *Scale*: $1\text{ cm} \equiv 15\text{ Newtons}$

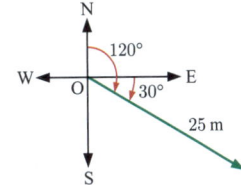

3 a *Scale*: $1\text{ cm} \equiv 20\text{ km h}^{-1}$

b *Scale*: $1\text{ cm} \equiv 10\text{ km}$

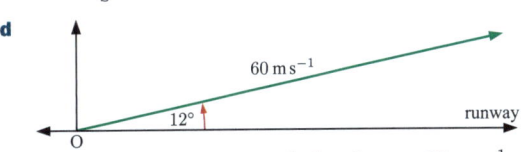

c *Scale*: $1\text{ cm} \equiv 30\text{ km h}^{-1}$

EXERCISE 32B

1 a **a**, **c**, and **e**; **b** and **d**   **b** **a**, **b**, **c**, and **d**
c **a** and **b**; **c** and **d**   **d** none are equal   **e** **a** and **c**; **b** and **d**

2 a b **b** a **c** −b **d** −a
3 a true **b** false **c** false **d** true **e** false **f** true

EXERCISE 32C

1 a 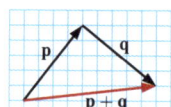 **b** (diagram)

c (diagram) **d** (diagram)

e (diagram) **f** (diagram)

2 a $\overrightarrow{QS}$ **b** $\overrightarrow{PR}$ **c** $\overrightarrow{PQ}$ **d** $\overrightarrow{PS}$

3 a i 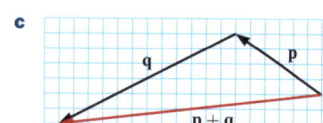 **ii**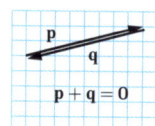

b For any two vectors **p** and **q**, we can form a parallelogram.

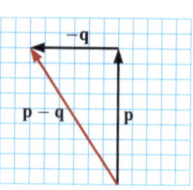

Now, $\overrightarrow{OA} + \overrightarrow{AC} = \overrightarrow{OC}$ and $\overrightarrow{OB} + \overrightarrow{BC} = \overrightarrow{OC}$
$\therefore \overrightarrow{OA} + \overrightarrow{AC} = \overrightarrow{OB} + \overrightarrow{BC}$
$\therefore \mathbf{p} + \mathbf{q} = \mathbf{q} + \mathbf{p}$

4 a $\overrightarrow{PS} = \mathbf{a} + \mathbf{b}$ **b** $\overrightarrow{PR} = \mathbf{a} + \mathbf{c}$ **c** $\overrightarrow{QT} = \mathbf{c} + \mathbf{d}$
d $\overrightarrow{PT} = \mathbf{a} + \mathbf{c} + \mathbf{d}$

5 a c **b** f **c** g **d** g **e** g

EXERCISE 32D

1 a 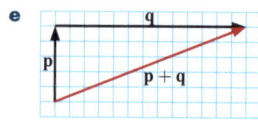 **b**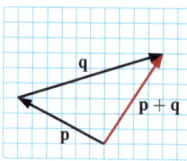

c 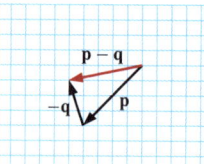 **d**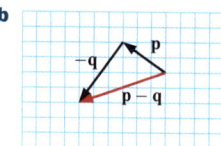

e 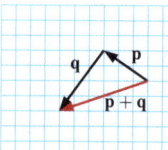 **f**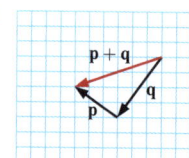

2 a $\overrightarrow{QS}$ **b** $\overrightarrow{PR}$ **c** 0 (zero vector) **d** $\overrightarrow{RQ}$ **e** $\overrightarrow{QS}$
f $\overrightarrow{RQ}$

3 a $\overrightarrow{AB} = \overrightarrow{AO} + \overrightarrow{OB}$
$= (-\mathbf{a}) + \mathbf{b}$
$= \mathbf{b} - \mathbf{a}$

b i $\overrightarrow{BC} = \mathbf{c} - \mathbf{b}$
ii $\overrightarrow{CA} = \mathbf{a} - \mathbf{c}$

4 a d **b** g **c** d **d** a **e** −c **f** −g

EXERCISE 32E.1

1 a  **b** 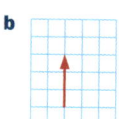 **c** 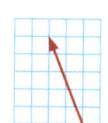 **d** (diagram)

2 a $\begin{pmatrix} 4 \\ 2 \end{pmatrix}$ **b** $\begin{pmatrix} 0 \\ -3 \end{pmatrix}$ **c** $\begin{pmatrix} -3 \\ -4 \end{pmatrix}$ **d** $\begin{pmatrix} -6 \\ 0 \end{pmatrix}$
e $\begin{pmatrix} -6 \\ 4 \end{pmatrix}$ **f** $\begin{pmatrix} 2 \\ -4 \end{pmatrix}$

3 a $\begin{pmatrix} 3 \\ 4 \end{pmatrix}$ **b** $\begin{pmatrix} -1 \\ 2 \end{pmatrix}$ **c** $\begin{pmatrix} 2 \\ -1 \end{pmatrix}$ **d** $\begin{pmatrix} 0 \\ -7 \end{pmatrix}$

4 a $k = -4$ **b** $k = -1$ **c** $k = 4$

EXERCISE 32E.2

1 a $\begin{pmatrix} 3 \\ 8 \end{pmatrix}$ **b** $\begin{pmatrix} 4 \\ 10 \end{pmatrix}$ **c** $\begin{pmatrix} 7 \\ -1 \end{pmatrix}$ **d** $\begin{pmatrix} -2 \\ 5 \end{pmatrix}$
e $\begin{pmatrix} -4 \\ -6 \end{pmatrix}$ **f** $\begin{pmatrix} -8 \\ -2 \end{pmatrix}$

2 a $\begin{pmatrix} 5 \\ -4 \end{pmatrix}$ **b** $\begin{pmatrix} 5 \\ -4 \end{pmatrix}$ **c** $\begin{pmatrix} 1 \\ -4 \end{pmatrix}$ **d** $\begin{pmatrix} 1 \\ -4 \end{pmatrix}$
e $\begin{pmatrix} 0 \\ -6 \end{pmatrix}$ **f** $\begin{pmatrix} 0 \\ -6 \end{pmatrix}$ **g** $\begin{pmatrix} 4 \\ -6 \end{pmatrix}$ **h** $\begin{pmatrix} 3 \\ -7 \end{pmatrix}$

3 a $\begin{pmatrix} 4 \\ 2 \end{pmatrix}$ **b** $\begin{pmatrix} -3 \\ -4 \end{pmatrix}$ **c** $\begin{pmatrix} -1 \\ 2 \end{pmatrix}$

4 a $\begin{pmatrix} 6 \\ 2 \end{pmatrix}$ **b** $\begin{pmatrix} 5 \\ 2 \end{pmatrix}$ **c** $\begin{pmatrix} -4 \\ 5 \end{pmatrix}$ **d** $\begin{pmatrix} -6 \\ -4 \end{pmatrix}$
e $\begin{pmatrix} 8 \\ -3 \end{pmatrix}$ **f** $\begin{pmatrix} 11 \\ -3 \end{pmatrix}$

5 a $\begin{pmatrix} 1 \\ 6 \end{pmatrix}$ **b** $\begin{pmatrix} -5 \\ 1 \end{pmatrix}$ **c** $\begin{pmatrix} -6 \\ 4 \end{pmatrix}$ **d** $\begin{pmatrix} -2 \\ 10 \end{pmatrix}$
e $\begin{pmatrix} -4 \\ -2 \end{pmatrix}$ **f** $\begin{pmatrix} 2 \\ -10 \end{pmatrix}$

6 a $\sqrt{17}$ units **b** 6 units **c** $\sqrt{13}$ units
d $\sqrt{26}$ units **e** $\sqrt{20}$ units **f** $\sqrt{37}$ units

EXERCISE 32F

1 a $\begin{pmatrix} 3 \\ -3 \end{pmatrix}$ **b** $\begin{pmatrix} -8 \\ -1 \end{pmatrix}$ **c** $\begin{pmatrix} -4 \\ -6 \end{pmatrix}$ **d** $\begin{pmatrix} -7 \\ -3 \end{pmatrix}$
e $\begin{pmatrix} 11 \\ -2 \end{pmatrix}$ **f** $\begin{pmatrix} 8 \\ 1 \end{pmatrix}$

Answers 717

2 a i $\begin{pmatrix} -2 \\ -3 \end{pmatrix}$ **ii** $\sqrt{13}$ units

b i $\begin{pmatrix} 5 \\ -2 \end{pmatrix}$ **ii** $\sqrt{29}$ units

c i $\begin{pmatrix} -3 \\ -4 \end{pmatrix}$ **ii** 5 units

d i $\begin{pmatrix} -12 \\ 5 \end{pmatrix}$ **ii** 13 units

3 a In $\overrightarrow{AB} + \overrightarrow{BC}$ we move from A to B then from B to C. This is a movement from A to C.
$\therefore \overrightarrow{AC} = \overrightarrow{AB} + \overrightarrow{BC}$

b $\begin{pmatrix} 1 \\ 8 \end{pmatrix}$

4 $\begin{pmatrix} -3 \\ -3 \end{pmatrix}$ **5** $\begin{pmatrix} -4 \\ 9 \end{pmatrix}$

EXERCISE 32G

1 a $\begin{pmatrix} 6 \\ 15 \end{pmatrix}$ **b** $\begin{pmatrix} -5 \\ 20 \end{pmatrix}$ **c** $\begin{pmatrix} -12 \\ 0 \end{pmatrix}$ **d** $\begin{pmatrix} 2 \\ 7 \end{pmatrix}$

e $\begin{pmatrix} 8 \\ -28 \end{pmatrix}$ **f** $\begin{pmatrix} -3 \\ \frac{2}{3} \end{pmatrix}$ **g** $\begin{pmatrix} 7 \\ 11 \end{pmatrix}$ **h** $\begin{pmatrix} 10 \\ -13 \end{pmatrix}$

2 a i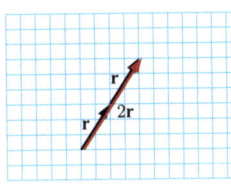

ii $2\mathbf{r} = \begin{pmatrix} 4 \\ 6 \end{pmatrix}$

b i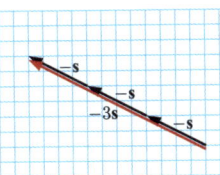

ii $-3\mathbf{s} = \begin{pmatrix} -12 \\ 6 \end{pmatrix}$

c i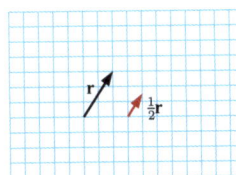

ii $\frac{1}{2}\mathbf{r} = \begin{pmatrix} 1 \\ \frac{3}{2} \end{pmatrix}$

d i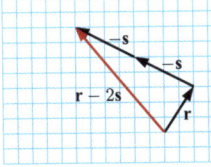

ii $\mathbf{r} - 2\mathbf{s} = \begin{pmatrix} -6 \\ 7 \end{pmatrix}$

e i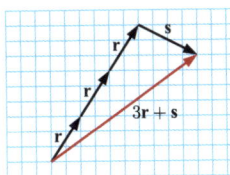

ii $3\mathbf{r} + \mathbf{s} = \begin{pmatrix} 10 \\ 7 \end{pmatrix}$

f i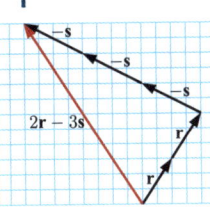

ii $2\mathbf{r} - 3\mathbf{s} = \begin{pmatrix} -8 \\ 12 \end{pmatrix}$

g i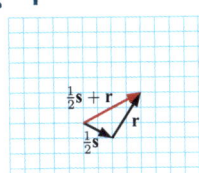

ii $\frac{1}{2}\mathbf{s} + \mathbf{r} = \begin{pmatrix} 4 \\ 2 \end{pmatrix}$

h i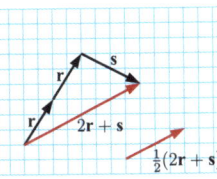

ii $\frac{1}{2}(2\mathbf{r} + \mathbf{s}) = \begin{pmatrix} 4 \\ 2 \end{pmatrix}$

3 a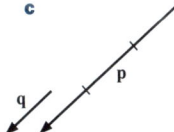

b   **c**

d   **e**

4 $\mathbf{c} = 3\mathbf{a}$, $\mathbf{d} = -2\mathbf{b}$, $\mathbf{e} = \mathbf{a} + 2\mathbf{b}$, $\mathbf{f} = -\mathbf{a} - 3\mathbf{b}$

5 a $2\mathbf{a}$ **b** $-2\mathbf{b}$ **c** $\mathbf{a} - \mathbf{b}$

REVIEW SET 32A

1 a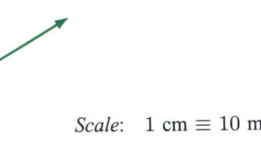

Scale: $1 \text{ cm} \equiv 10 \text{ m s}^{-1}$

b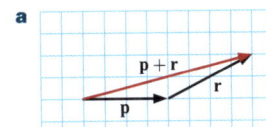

Scale: $1 \text{ cm} \equiv 10 \text{ m}$

2 a   **b**

3 a $\overrightarrow{CA} = -\mathbf{c}$ **b** $\overrightarrow{AB} = -\mathbf{a} + \mathbf{b}$

c $\overrightarrow{OC} = \mathbf{a} + \mathbf{c}$ **d** $\overrightarrow{BC} = -\mathbf{b} + \mathbf{a} + \mathbf{c}$

4 a $\overrightarrow{AD}$ **b** $\overrightarrow{BD}$ **c** $\mathbf{0}$ (the zero vector)

5 a   **b**

c  **d**

6 a $\begin{pmatrix} 3 \\ 7 \end{pmatrix}$ b $\begin{pmatrix} 2 \\ 6 \end{pmatrix}$ c $\begin{pmatrix} -12 \\ 54 \end{pmatrix}$ 7 $\overrightarrow{PR} = \begin{pmatrix} 2 \\ 9 \end{pmatrix}$

REVIEW SET 32B

1 a **p** and **q** have the same length.
  b $|\mathbf{p}| = 2|\mathbf{q}|$, **p** and **q** are parallel, and are in the same direction.

2 $\overrightarrow{BA} = -\overrightarrow{AB}$, $\overrightarrow{BA}$ has the same length as $\overrightarrow{AB}$, but has opposite direction.

3 a $\begin{pmatrix} 3 \\ 2 \end{pmatrix}$ b $\begin{pmatrix} 0 \\ 4 \end{pmatrix}$ c $\begin{pmatrix} -5 \\ -2 \end{pmatrix}$

4 a $\begin{pmatrix} 6 \\ -6 \end{pmatrix}$ b $\begin{pmatrix} 7 \\ -16 \end{pmatrix}$ c $\begin{pmatrix} 0 \\ 2 \end{pmatrix}$

5 a $\begin{pmatrix} -1 \\ -9 \end{pmatrix}$ b $\sqrt{34}$ units

6 a $\overrightarrow{CD} = -\mathbf{p}$ b $\overrightarrow{BM} = \tfrac{1}{2}\mathbf{q}$ c $\overrightarrow{MD} = \tfrac{1}{2}\mathbf{q} - \mathbf{p}$
  d $\overrightarrow{AD} = \mathbf{q}$

7 a $\begin{pmatrix} 11 \\ 1 \end{pmatrix}$ b $\begin{pmatrix} 1 \\ 12 \end{pmatrix}$ c $\sqrt{34}$ units

8 a $\overrightarrow{AB} = \begin{pmatrix} 3 \\ -4 \end{pmatrix}$ b 5 units

INDEX

24-hour time	71
acre	64
acute angle	180
addition law of probability	576
adjacent	487
algebra	88
algebraic equation	163
algebraic flowchart	152, 339
algebraic inequality	165
allied angles	186
alternate angles	186
angle	179
angle of rotation	450
angles at a point	182
apex	213, 233
appreciation	367
approximation	76
arc	224
arc length	309
area	312
area of sector	315
arithmetic sequence	551
average	94
average speed	379
axis of symmetry	520
bar chart	264
base	20
base angle	213
bearing	192
BEDMAS	24
bimodal	603
bisect	195
bivariate statistics	616
block solid	239
capacity	65, 331
cardinal direction	192
Cartesian plane	397
categorical data	273
causality	621
census	593
centre	223
centre of data	602
centre of enlargement	456, 457
centre of rotation	450
chord	224
circle	223
circumference	308
class interval	600
closed figure	206
co-interior angles	186
coefficient	91
collinear	177
common difference	551
common factor	29
common multiple	30
common ratio	554
complementary angles	182
complementary events	562
component form	637
composite bar chart	278
composite number	28
composite shape	315
compound interest	370
concurrent lines	177
cone	233, 322
congruent	473, 476
constant term	91
continuous variable	600
converse of Pythagoras' theorem	299
coordinates	397
correlation	618
corresponding angles	186
cubic function	540
cylinder	233, 321
decimal number	44
denominator	36
density	383
dependent events	581
dependent variable	616
depreciation	367
derived units	59
diagonal	207
diameter	224
difference	94
difference between two squares	144
directly proportional	526
discount	363
discrete variable	594
disjoint	576
distance-time graph	381
distributive law	136
elevation	238
elimination	431
enlargement	441, 455
equal ratio	103
equation	90
equation of a line	410
equidistant	195
equilateral triangle	209
equivalent fractions	37
error interval	80
estimate	76
event	568
exchange rate	389
expanded form	44
expansion laws	121
expectation	586
experimental probability	563
exponent	20
expression	90
extrapolation	624
face	232
factor	27
factor tree	28
factorisation	248, 250
Fibonacci sequence	555
Fibonacci-type sequence	555
force	386
formula	338
four digit notation	71
fraction	36
frequency	563
frequency histogram	600
frequency table	594
frequency tree	567
fundamental theorem of arithmetic	28
geometric construction	194
geometric sequence	554
gradient	402
gradient formula	405
hectare	63
highest common factor	29
horizontal asymptote	539
hypotenuse	294
image	440
imperial system	60, 62
improper fraction	36
independent events	579
independent variable	616
index laws	119
index notation	20
inequality	148
inflation	367
instantaneous speed	379
integer	12
interest rate	368
International System of Units	59
interpolation	624
inverse operation	19, 151
inverse trigonometric functions	493
inversely proportional	531
isometric projection	237
isosceles triangle	209, 213
kilogram	67
kite	218
like term	91, 92
line	177
line graph	265
line of best fit	622
line of symmetry	449
linear equation	148
linear relationship	399, 400
litre	65
lowest common denominator	40
lowest common multiple	30
lowest terms	38
magnitude	640
major arc	224
mark-up	362
mass	67
mean	94, 602
mean point	623
median	603
member	548
metric system	60
minor arc	224

mirror line	445	proportionality constant	527	standard form	127
mixed number	36	pyramid	233	statistics	592
modal class	600	Pythagoras' theorem	295	straight angle	180
mode	603	quadrant	398	subject	338, 347
multiple	27	quadratic equation	506	substitution	430
multiple bar chart	277	quadratic function	514	sum	94
multiplier	358, 359	quadratic trinomial	257	supplementary angles	182
mutually exclusive	576	quadrilateral	216	surface	232
natural number	12	quotient	94	surface area	319
negative index law	125	radius	224	survey	593
negatively skewed	595	range	610	symmetrical distribution	595
net	234	rate	376	table of values	412
newton	386	ratio	100	tally and frequency table	273
Null Factor law	507	rational equation	160	tangent	224
number line	12	real number	36	tapered solid	328
number sequence	548	reciprocal	42, 125	term	91, 548
numerator	36	reciprocal function	538	term-to-term rule	548
object	440	rectangle	218, 312	terminating decimal	53
oblique projection	237	rectangular hyperbola	539	theoretical probability	569
obtuse angle	180	recurring decimal	53	three point notation	179
one figure approximation	82	reduction	441, 455	time series data	281
opposite side	487	reflection	441	transformation	440
order of operations	24	reflex angle	180	translation	441
order of rotational symmetry	454	regular polygon	207	translation vector	442
ordered pair	397	relative frequency	563	transversal	185
ordinal direction	192	repeated division	28, 29	trapezium	218, 312
origin	397	revolution	180	tree diagram	583
outcome	563	rhombus	218	trial and error	427
outlier	596	right angle	180	triangle	208, 312
parabola	515	right angled triangle	294, 487	trigonometric ratios	488
parallel lines	177	rotation	441, 450	trigonometry	486
parallelogram	218, 312	rotational symmetry	453	true bearing	193
percentage	49	rounding off	76, 77	turning point	521
percentage change	358	sample	593	unit circle	498
perfect square	23, 142	sample space	568	unit cost	387
perimeter	308	scalar	630	unitary method	52
perpendicular	182	scale diagram	108	value added tax (VAT)	365
perpendicular bisector	195	scale factor	109, 455	variable	90
pictogram	264	scalene triangle	209	variation	610
pie chart	264	scatter graph	616	vector	630
plan	238	scientific notation	127	velocity	630
plane	177	sector	224	Venn diagram	571
plane figure	206	segment	224	vertex	206, 232, 521
point	176	semi-circle	224	vertical asymptote	539
polygon	206	set notation	568, 569	vertical line chart	595
population	593	significant figure	76	vertically opposite	185
position vector	631, 641	similar	464	volume	65, 324
position-to-term rule	548	simple interest	368	x-intercept	516
positively skewed	595	simplest form	103	y-intercept	516
power	20	simplifying fractions	38	zero vector	632
pressure	386	simultaneous equations	426		
prime factorisation	28	solid	232		
prime number	28	solution	149		
prism	233	speed	379		
probability	561	sphere	322, 330		
product	94	spread	610		
pronumeral	88	square	218		
proper fraction	36	square number	23		
proportion	105	square root	23		